31st International Colloquium
Plastics Technology

7–8 September 2022

INSTITUTE FOR PLASTICS PROCESSING
in Industry and Craft at
RWTH Aachen University

Download proceedings
At www.ikv-aachen.de/login you can also download the proceedings until 31 January 2023.
Please note that the proceedings are only available in English languange.

Presentations in English can be downloaded as well if authorised by the authors.

Username: ikv-colloquium
Password: BKAZ#2022

IKV - Institute for Plastics Processing in
Industry and Craft at RWTH Aachen University

Chair of Plastics Processing
Prof. Dr.-Ing. Christian Hopmann

Seffenter Weg 201 | 52074 Aachen | Germany
Phone: +49 241 80-938 06 | Fax: +49 241 80-922 62
zentrale@ikv.rwth-aachen.de | www.ikv-aachen.de

www.ikv-colloquium.com

Bibliographic information published by the Deutsche Nationalbibliothek

The Deutsche Nationalbibliothek lists this publication in the Deutsche Nationalbibliografie;
detailed bibliographic data are available in the Internet at http://dnb.d-nb.de.

© Shaker Verlag 2022
All rights reserved. No part of this publication may be reproduced, stored in a retrieval system,
or transmitted, in any form or by any means, electronic, mechanical, photocopying, recording
or otherwise, without the prior permission of the publishers.

Printed in Germany.

ISBN: 978-3-8440-8662-1

SHAKER VERLAG Shaker Verlag GmbH | Am Langen Graben 15a | 52353 Düren
Phone: +49 2421/99 011-0 | Telefax: +49 2421/99 011-9
Internet: www.shaker.de | E-mail: info@shaker.de

Preface

Dear guests of the 31st IKV colloquium,

we are living in times of crisis. The COVID-pandemic is not yet over, climate change is affecting our landscapes, the long period of peace in Europe has come to an end and an economic recession is expected. Under these circumstances, the transition to a circular economy for plastics is an enormous challenge, but one that is essential for both our environment and our economy. Clearly, the pollution of oceans, air and landscape caused by plastics and other materials must be stopped. For far too long, this issue has been ignored and neglected, leading to the current pressure on the plastics sector.

One may complain about some exaggerations and untruths being spread about plastics in this context, but if we are honest with ourselves, we have to admit that for too long the plastics sector has neither thought nor acted sustainably, provoking this contradiction and the current public indignation. Instead of complaining, it is now our task to develop solutions and counter the criticism – whether justified or not – with sustainable processes and products. From a scientific point of view, a sustainable and, in particular, CO_2-neutral plastics economy is possible. We 'just' have to implement it.

On the way there, numerous technical challenges and problems need to be solved, which requires significant efforts in research and development. Well-established topics such as lightweight construction and sustainable product design are just as important as the digitalisation of value chains and materials and new and improved recycling and recovery processes. At the colloquium, we present our ideas and results on all these topics and many more. We invite you to check their applicability and feasibility and to join us in our discussions.

Sincere welcome to the 31st IKV colloquium!

Christian Hopmann

Contents

		page
PI	Plastics and hydrogen technology	1
S1	Towards autonomous injection moulding production	7
S2	Recyclable barrier systems for sustainable products	45
S3	Enhanced process and material knowledge in foam extrusion	81
S4	Added value potential through digital infrastructure	125
S5	Control of material flows in packaging recycling	161
S6	Mixing and flow processes in rubber processing	205
S7	Production and operation of hydrogen pressure vessels	253
S8	Production of high-precision optical lenses	303
S9	Dimensioning methods for failure mechanisms of technical thermoplastics	345
S10	Resilient injection moulding process control	383
S11	Long-fibre-reinforced plastic battery housing	421
S12	Joining of innovative and sustainable material combinations	467
S13	Development of resilient injection moulded products	507
S14	Additive manufacturing of large components	547
S15	Modelling of time-scaled properties of FRP	595

Plastics and hydrogen technology

Dr.-Ing. Kai Fischer

Institute for Plastics Processing (IKV) at RWTH Aachen University

Dr.-Ing. Kai Fischer

Dr. Fischer, born in 1977, studied mechanical engineering with a focus on aerospace engineering at RWTH Aachen University. He worked as a research assistant at the Institute of Plastics Processing at RWTH Aachen University (IKV) from 2005 to 2010, where he obtained his doctorate with honours in the field of high-volume production of components made of fibre-reinforced plastics with thermoset matrix. From 2010 to 2015, he headed the department "Fibre Reinforced Plastics/Polyurethanes".

Since 2015, he is responsible for the strategic planning and cross-departmental coordination of all technologies related to lightweight engineering as scientific director of the IKV. Since 2012, he is also a member of the management board of the Aachen Center for Integrative Lightweight Production (AZL) of RWTH Aachen University and since 2013 managing director of AZL Aachen GmbH, a provider of services for business and technology development in the field of lightweight engineering and strategic consulting. Dr. Fischer has published approximately 200 publications as an author. Since 2013, he is lecturer at RWTH Aachen University for the lecture "Fibre Reinforced Composites".

1 Plastics and hydrogen technology

Kai Fischer[1]
[1]*Institute for Plastics Processing (IKV) at RWTH Aachen University*

Abstract

Since its foundation the IKV aims for providing sustainable solutions to the value chain of plastics products. This comprises a more efficient design and production of plastics products as well as improving the use of sustainable technologies, in which plastics have the role of an enabler. One societal key-aspect driving technological developments in terms of sustainability is the usage of climate-neutral energy over all areas of daily life. Hydrogen technology plays a major role to reach these ambitious targets. The IKV has been active in developing technologies related to hydrogen systems since a long time, in the younger past especially in the field of storage and conversion of hydrogen for mobile applications. The plenary talk will present various topics and activities of the IKV embedded in its network of science and industry to help further expand the application fields for plastics and make hydrogen more competitive and sustainable.

1.1 Introduction

Hardly any other topic is exerting such an influence on current social discourse as the environmentally aware use of resources. In this context, hydrogen represents an alternative to fossil fuels and raw materials, and is being used in various industries and applications. Hydrogen could be produced in different ways from different resources with various energy origins. In terms of sustainability, the aim must be to produce and use so-called "green" hydrogen, which comes from renewable sources, such as solar or wind energy. This is the only way to make hydrogen production climate neutral.

Hydrogen is used as feed stock for the chemical industry, as process gas in industrial plants and as an energy carrier for the generation of thermal and electrical energy. The use as feed stock and as process gas exhibits the vast majority of the hydrogen consumption today. As feed stock, hydrogen is e.g. used in the ammonia and fertiliser industry. The use as process gas has a high potential for reducing carbon emissions, e.g. the reduction of iron in steel production is realised by substituting CO_2-emitting coal by water-emitting hydrogen. Thermal energy could be generated by burning hydrogen like other fossil fuels or bio-fuels. Electrical energy can be generated by converting hydrogen in fuel cells.

Estimations for the future development are mostly expecting a growth in the segments related to the conversion of hydrogen to heat and electricity. The generation of hydrogen will switch from technologies using non-renewable and/or non-CO_2-neutral resources and processes to "green" technologies, e.g. electrolysis using renewable electrical energy. The hydrogen storage and distribution will change significantly, as a decentralised distribution network as well as decentralised stationary and mobile storage systems must be developed. Significant improvements of efficiency in generating, transporting, storing and converting hydrogen will only be stimulated by innovations enabled by materials and correlated processing technologies. More than ever a smart combination of different materials enables efficiency improvements in both production of complex components and their performance in a critical environment. As it has been valid for nearly all industries, plastics with their capability for producing highly-integrated components will belong to the enablers for increasing the economical and ecological efficiency of hydrogen systems in the future. IKV´s activities are aiming to expand the know-how for identifying, developing, manufacturing and operating plastic products to realise a more sustainable and cost-efficient hydrogen economy.

1.2 IKV´s activities

The IKV follows different activities for the development of solutions as well as for networking within industry and academia. In addition to publicly funded research projects, currently mainly in the area of cost reduction in the production and operation of Type-IV pressure vessels and plastic-based fuel cells, bilateral development and consulting projects are carried out. In the recent past, IKV has already been involved in various research projects on the development of components for fuel cells and the storage of hydrogen in CFRP pressure vessels. However, due to the dimension of the overall topic, current research at IKV is specifically seeking synergies and is being deeply embedded in activities of RWTH Aachen University and other partners – for example, as a member of the RWTH Centre for Sustainable Hydrogen Systems, as a partner of the Cluster for Future "Hydrogen" sponsored by the Federal Ministry of Education and Research (BMBF), or as the initiator of a technology study conducted with 20 industrial partners to identify new potential for plastics in the hydrogen economy. In the following, the various activities are outlined briefly.

1.3 Centre for sustainable hydrogen systems: Establishment of a uniform hydrogen research infrastructure at RWTH Campus

The Centre for Sustainable Hydrogen Systems was founded in spring 2021 to bundle research in the Aachen region on the topic of hydrogen. More than 50 professors from RWTH Aachen University and the Jülich Research Centre met at the kick-off workshop and presented their research and development projects on the topic. The consortium, already building on more than 100 recent research projects, agreed on the joint vision of a unified hydrogen research infrastructure with the aim of making the Aachen region a leading hydrogen region. The Centre serves in particular to promote networking between scientists working on hydrogen technologies in Aachen and Jülich. IKV is member of the Centre.

1.4 Hydrogen cluster for future: "Innovation valley of hydrogen technologies"

The Hydrogen Cluster for Future is one of seven regional innovation networks that won the "Clusters4Future" ideas competition organised by the Federal Ministry of Education and Research (BMBF). The project, coordinated by RWTH Aachen University, is intended to pool the already existing high level of expertise in the field of hydrogen technologies in the Aachen and Jülich regions, and to promote regional research and development activities along the entire value-added chain of hydrogen as a resource: from production, storage and distribution to utilisation. The joint research pursues an interdisciplinary approach, e.g. the objectives and effectiveness of concrete projects on technological, economic and social networking. The Hydrogen Cluster for Future is funded by the BMBF with up to 45 million euros. As a project partner, IKV is actively involved the network: 24 research institutes of RWTH Aachen University and Jülich Research Centre, together with an advisory board of 16 industrial partners and 47 other research partners from all over Germany, are pooling their know-how in the cluster to research hydrogen technologies. In cooperation with partners, the IKV is developing a methodology for monitoring the operational safety of mobile CFRP pressure vessels for hydrogen storage at low cost and with low effort.

1.5 Market and technology study hydrogen: Potentials of a future technology for the plastics industry

To identify R&D-topics in a structured approach, IKV started a joint technology analysis with industrial partners in October 2021. The results of the project, which will be finalised in October 2022, will provide important impetus for the development, qualification and production of more efficient systems for a competitive and sustainable hydrogen economy. Part of the consortium are amongst others the companies 3M Deutschland GmbH, AGC Chemicals Europe Ltd., Air Liquide S.A., ALLOD Werkstoff GmbH Co. KG, BASF

Polyurethanes GmbH, Brabender GmbH Co. KG, Covestro Deutschland AG, EBG group GmbH, Freudenberg SE, ENGEL AUSTRIA GmbH, Evonik Industries AG, Freudenberg SE, Georg Fischer AG, Getzner Werkstoffe GmbH, HUEHOCO GROUP Holding GmbH Co. KG, Klöckner DESMA Elastomertechnik GmbH, LANXESS Deutschland GmbH, MOCOM Compounds GmbH Co. KG, Robert Bosch GmbH, WILO SE and YIZUMI Germany GmbH. The project is divided in three phases. First the different segments of the hydrogen economy have been analysed. The information (280 result slides in phase 1) on market sizes, growth expectations and specific technological boundary conditions has been the basis for setting a focus for the second phase. This was conducted by a voting by the industry consortium. In phase 2, which was finalised just a week before writing these lines here, the selected segments have been analysed for the specific technologies being used for generating, transporting, storing and converting hydrogen. The systems have been analysed and a system-breakdown was conducted. The result of phase 2 is a very extensive description of the function and the requirements of key-components used in the systems (480 result slides in phase 2). Based on this well-grounded knowledge, short-, mid- and long-term opportunities and challenges for plastics in the hydrogen economy are identified and translated into a road-map for future developments and research projects. The consortium of the study should act as a nucleus, for establishing a cross-industry network contributing to both: grow the opportunities for plastics technologies in the hydrogen economy and accelerate the penetration of this sustainable technology into the broad societal everyday use to save our climate.

Abbreviations

Notation	Description
BMBF	Federal Ministry of Education and Research
CFRP	Carbon-fibre-reinforced plastics

Towards autonomous injection moulding production

Moderator: Dr.-Ing. Micha Scharf, Phoenix Contact GmbH & Co. KG

Content
---

1 Automation of the setup and start-up process for injection moulding 9

M. Scharf[1]
[1] *Phoenix Contact GmbH & Co. KG*

2 Effective data-selection strategy for efficient model-based injection moulding setup with transfer learning 19

Ch. Hopmann[1], M. Schmitz[1], Y. Lockner[1]
[1] *Institute for Plastics Processing (IKV) at RWTH Aachen University*

3 Enabling shopfloor optimisation in injection moulding through digital shadows 34

Ch. Hopmann[1], M. Schmitz[1], P. Sapel[1]
[1] *Institute for Plastics Processing (IKV) at RWTH Aachen University*

Dr.-Ing. Micha Scharf

After graduating from RWTH Aachen University in 2005 with a degree in mechanical engineering specialising in plastics and textile technology, Dr.-Ing. Micha Scharf worked and received a doctor's degree at the Institute of Plastics Processing at RWTH Aachen University in the field of extrusion die design and flat film extrusion. From 2010 to 2012, he was Technology Manager at Phoenix Feinbau GmbH Co. KG in Lüdenscheid, where he built up extrusion manufacturing among other things. In 2012, he moved to the parent company Phoenix Contact GmbH Co. KG in Blomberg. As a Master Specialist Plastics Processing, he deals with all topics related to the qualification of injection moulds as well as troubleshooting and process optimisation in the production of plastic parts.

1 Automation of the setup and start-up process for injection moulding

M. Scharf[1]
[1] Phoenix Contact GmbH & Co. KG

Abstract

The autonomy level of production processes has been constantly evolving over time. For injection moulding, four levels of control are considered regarding their potential for automation. While the production planning could be automised completely, the provision of operating resources, either for maintenance or production, requires manual labour to some extent depending on the production equipment and conditions in the particular company. The consecutive flow of material needs careful supervision to ensure high and constant material quality. Finally, the production process needs to be started with an adequate set of machine parameters, whose acquisition is still depending on human judgement and guidance. Significant process parameters such as the maximum injection pressure or the melt cushion may hint at stable process conditions for production. Finally, a complete automation of the injection moulding process is still unavailable. An automated start- and set-up procedure offers the highest potential for further development.

1.1 Introduction

Autonomous industrial production, which means production without human intervention, is the declared goal of the changes in industrial production that have become known as Industry 4.0. The four stages of industrialisation describe the automation of different aspects of industrial production.

Industry 1.0 means the sufficient provision of energy, whether by wind turbines or water wheels, steam engines, combustion engines or by electric motors. This has long been implemented in injection moulding machines, which are usually powered by electricity.

Industry 2.0 means the specifically organised flow of materials and the defined sequence of operations. Thanks to automatic granulate feeding, the injection moulding cycle, that is run over and over again, and the automatic ejection of the individual plastic parts this stage is implemented as well.

Industry 3.0 means the use of machine controls that independently execute defined process steps and feedback controls that make predefined adjustments in the correcting variables, when deviations from target specifications are detected. Such controls have also been installed in injection moulding machines for several decades and enable re-adjustment within an injection moulding cycle.

Industry 4.0 means automating the configuration of a production and the decisions to be made to gain further benefit. In relation to the injection moulding process, for example, setting parameter sets have to be found without human input and the sufficient quality of plastic parts have to be determined in order to adjust the setting data and thus the injection moulding process in case of quality deviations. This is not currently state of the art.

In terms of injection moulding production, there are many other configuration processes and decision-making processes. Injection moulding production comprises four levels, as shown in (Figure 1.1):

- Level 1 is the organisational level of manufacturing control.
- Level 2 is the physical level of provision and maintenance of operating equipment (injection mould, injection moulding machine and peripheral equipment).
- Level 3 is the physical level of material flow for plastic parts production.
- Level 4 is the procedural level of the injection moulding process.

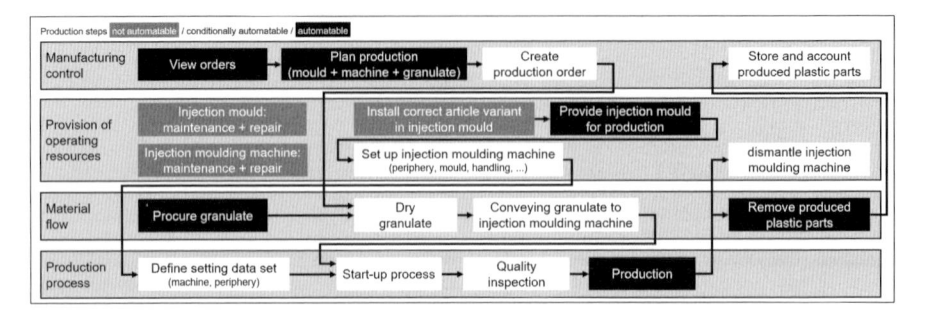

Figure 1.1: Levels of injection moulding production

1.2 Level 1: Manufacturing control

From the technological point of view, the manufacturing control level can be completely automated [GR20]. However, the detailed planning of production orders, short-term rescheduling due to mould failures or prioritisation is still carried out manually by the schedulers in most plastics production facilities.

Similarly, the transport of individual plastic parts, storage and accounting in the ERP system can be automated in principle, but are still often carried out manually.

1.3 Level 2: Provision of operating resources

Two areas have to be distinguished for the provision of operating resources:

- General, regular maintenance of injection moulds, injection moulding machines and peripheral equipment.
- Specific provision of operating resources for a production order.

Due to the large number of tests and activities to be performed, maintenance cannot be automated in principle. However, as this maintenance can generally be planned, it does not fundamentally prevent autonomous injection moulding production.

The specific provision of operating resources for a production order can be automated to some extent, provided that suitable surrounding conditions are set in the company. The first important point is the design of the injection moulds. If a separate injection mould is available for each plastic article, no manual work is required prior to provision. If, on the other hand, a conversion mould concept is used in which different articles are to be produced with the same injection mould by replacing certain cavity components, manual conversion by a mould mechanic is usually required. There are individual mould concepts in which rotatable sprue bushes, cavity parts or entire cavities can be adjusted automatically by means of an adjustment device. For most injection moulds in use, this is not state of the art. Figure 1.2 shows an injection mould from Phoenix Contact GmbH Co. KG, Blomberg, Germany, with cassettes, each containing 14 different sets of 4 cavities on the ejector side and the nozzle side. The cassettes are re-positioned by the control system of the injection moulding machine using step motors and then clamped pneumatically in the target position, so that a fully automatic mould changeover takes place on the injection moulding machine.

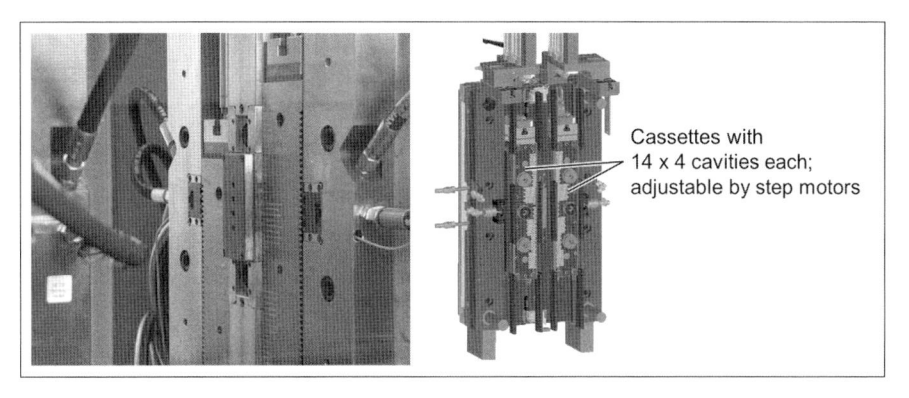

Figure 1.2: Injection mould with automatic positioning of the cassettes

The next step is to set up the injection moulding machine. As a demonstrator from Stäubli, Bayreuth, Germany, already showed at K2016, transport of the mould to the injection moulding machine, insertion of the mould into the injection moulding machine, fixing of the mould on the clamping platens and connection of temperature control hoses, pneumatic hoses and hot runner plugs can be completely automated [Löh16]. However, this requires extensive standardisation in the injection mould dimensions, the positioning and number of connections of the cooling circuits, the use of quick coupling systems, and the installation of various positioning aids and actuators in each injection moulding machine. This is associated with high investments that probably only pay off in large-scale production.

To remove the produced plastic parts or cold runner manifolds from the injection mould without damage, a suitable gripper hand may have to be mounted and connected to the handling system and aligned to the mould. Automation has already been technically implemented in some injection moulding production facilities via magazines for the various gripper hands, but, similar to the automation of the setup process for injection moulds, this is associated with high costs for standardisation and high investments for the magazine and article-specific gripper hands. In addition, a fully automatic gripper hand change and a subsequent start-up not controlled by a process mechanic risk a non-gripping of the plastic parts in the event of minor positioning errors or even a collision between the gripper hand and the injection mould.

1.4 Level 3: Material flow

Once the injection mould has been installed in the injection moulding machine, further manual setup operations are required. For supply with hygroscopic granulate, this starts a few hours before the start of production with selecting an appropriate granulate dryer, filling, setting and switching on the dryer. If an automatic granulate conveying system is used, this continues with the correct connection of the feed line to the coupling station and checking that there are no granulate residues from the previous production in the pipelines and in the pellet hopper on the injection moulding machine. Here, automation of the mechanical work is conceivable, but not of the necessary control tasks.

1.5 Level 4: Production process

Once all mechanical preparations for the setup have been completed, the correct setting data have to be entered into the control system of the injection moulding machine, the handling system and, if necessary, other peripheral equipment. The first obstacle to autonomous injection moulding production is that the correct setting data set has to be established or experimentally determined. There is currently no established fully automated procedure, as an evaluation of the part quality and process stability is always required to assess the suitability of a setting data set. If a suitable setting data set is available, it can be automatically loaded into the control system of the injection moulding machine, but only to a limited extent into the control systems of all peripheral equipment. For example, hot runner control units often communicate neither with the injection moulding machine's control system nor with a higher-level master computer system such as the Arburg master computer system (ALS), in which setting data records can be stored, managed and made available centrally.

Once both the mechanical setup and the data-related setup have been completed, the injection moulding process is started up. Here, the plasticising unit and, if necessary, also the hot runner system must first be purged with fresh melt to remove material residues from the previous production. Modern machine control systems provide automated processes, e. g. for purging the plasticising unit. However, the entire purge process cannot be carried out autonomously by the injection moulding machine, because it has to be started manually by a process mechanic and, at least subsequently, a process mechanic must remove the ejected melt in front of the machine nozzle and remove spray residues from the cavities and, if necessary, collection plates from between the mould halves.

If the injection moulding machine is now ready for production, start-up production begins, during which reject parts are initially produced until the plasticising unit, the hot runner system and the injection mould have sufficiently approached their thermal equilibrium state and a sufficiently constant residence time of the plastic melt has been achieved. Ensuring that this state is reached is currently usually done by rejecting a minimum number of shots, which is determined on the basis of experience. Alternatively, analysis and control systems in the control system of some injection moulding machines can be used, which show deviations of the actual process data from known target process data defined as "good" by the process mechanic or which minimise the process data deviations and thus hopefully also the component quality deviations by intervening in the setting values (e. g. the changeover volume) [URLa, URLb]. Figure 1.3 shows for an exemplary injection moulding process the curves of process and quality data during repeated start-up of the process.

If the end of the start-up process and thus the achievement of a quasi-stationary state of the injection moulding process is to be detected or even predicted based on an automated analysis of characteristic process data values recorded by the injection moulding machine in each cycle (e. g. maximum injection pressure or melt cushion), it is generally not sufficient to compare the current characteristic process data values with specified target values. Due to external disturbing influences, the repeated start-up of a process always results in slightly different final values despite identical setting values of the injection moulding machine, as shown in Figure 1.4.

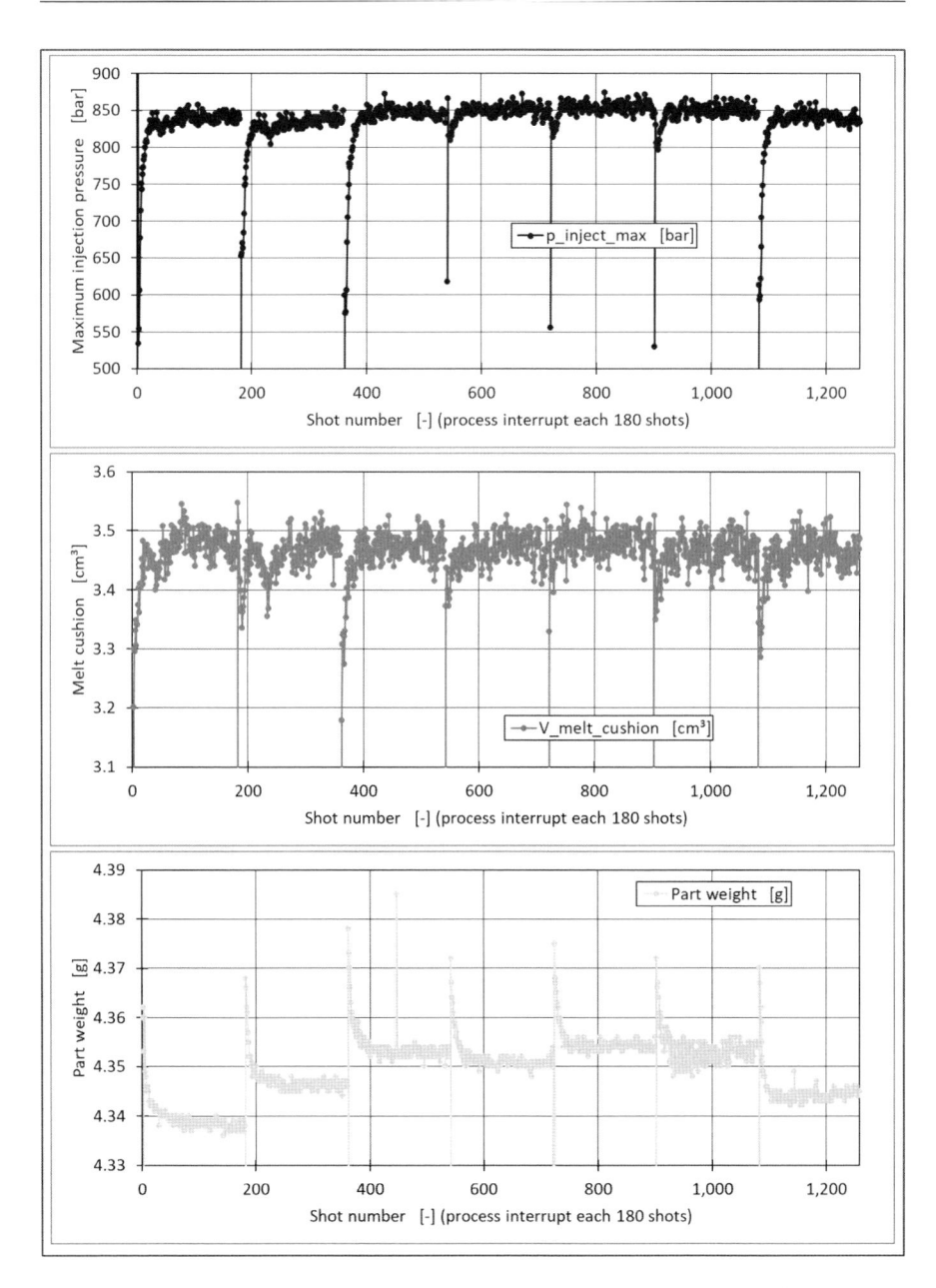

Figure 1.3: Process data and quality data curves during repeated start-up of an injection moulding process

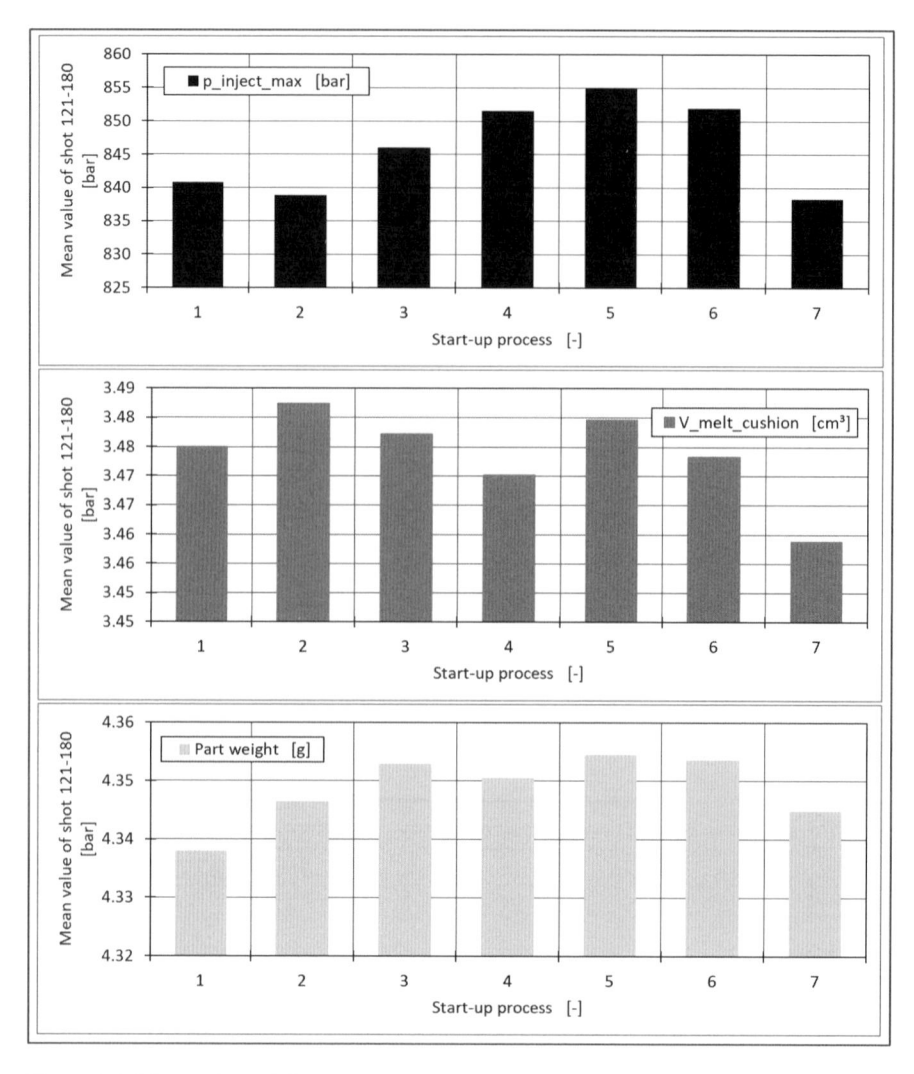

Figure 1.4: Mean values of the process and quality data at the end of the start-up processes (for each start-up process mean values of shots 121-180)

It is more target-oriented to evaluate the course of the process data. A simple possibility is the evaluation of local extrema of a propagated average value (Figure 1.5).

An alternative is the approximation of the course of the process data since the beginning of the start-up process by a function (e. g. exponential function with negative exponent), the prediction of the final value the function presumably will reach after a very long running time of the process and the final evaluation whether a sufficient approximation to the predicted final value has already been reached with the current shot (Figure 1.6).

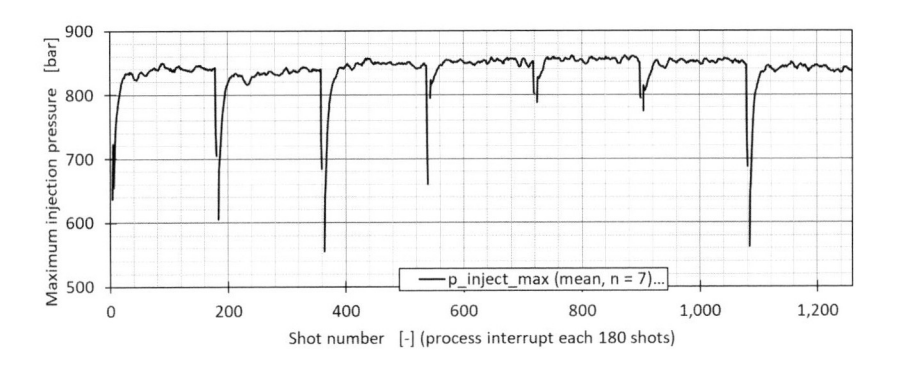

Figure 1.5: Propagated average of process and quality data (n = 7)

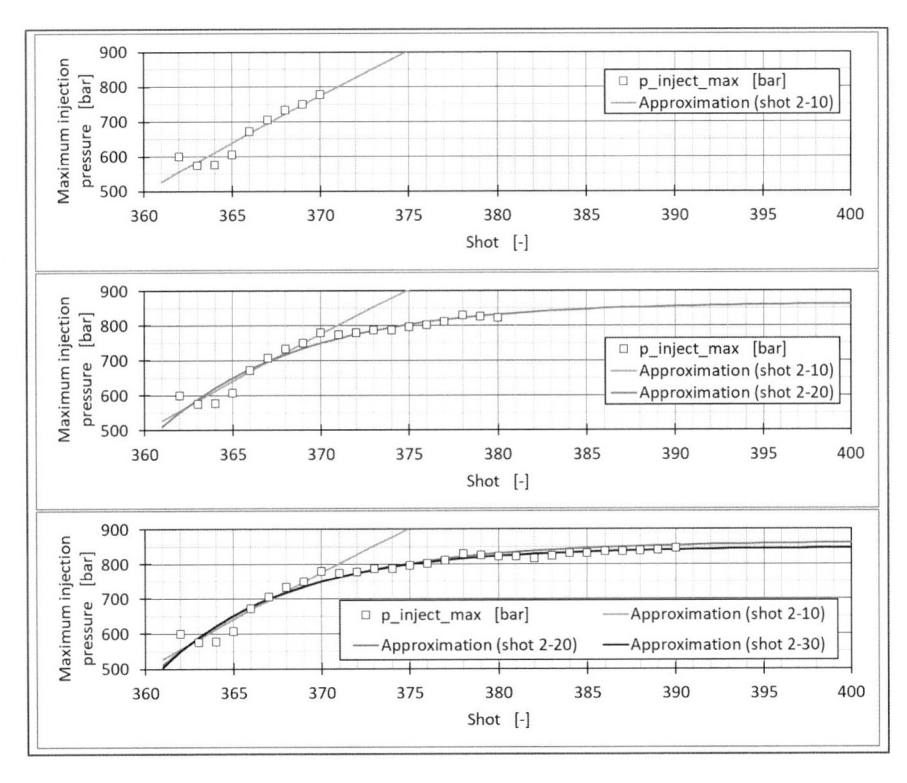

Figure 1.6: Predicted course of the maximum injection pressure
(based on the first 10 / 20 / 30 shots after start-up)

In principle, it should be noted that to assess whether the start-up process is complete or not, the process data curves can only be meaningfully evaluated, if the process data of the injection moulding process and the quality data of the plastic parts correlate strongly.

1.6 Conclusions and outlook

Fully autonomous injection moulding production is not possible with the current technical state of the art. Automation of many aspects of injection moulding production, on the other hand, is technically possible, but only makes limited economic sense. This includes, above all, the mechanical setup process for installing the injection mould in the injection moulding machine. The most significant steps that cannot currently be automated are the determination of the setting data set and the start-up of the injection moulding process. In both cases, a decision is required as to whether the part quality achieved is adequate or not. Due to the lack of automated quality control of the components in injection moulding production, this decision can only be made by a process mechanic. However, by automatically analysing the process data of the injection moulding machine, well-founded assistance can be provided.

References

[GR20] Gerlach, C.; Ressmann, A.: Der Kunde bestimmt den Takt. *Kunststoffe* 110 (2020) 10, p. 20–24

[Löh16] Löh, D.: Trends in der Automatisierung auf der K2016 – Einsatzspektrum von Robotern wächst. *Plastverarbeiter* 67 (2016) 12, p. 14–16

[URLa] N.N. *APC plus – Stabilität, Präzision und Kosteneffizienz in der Fertigung. Firmenschrift KraussMaffei.* URL: https://www.kraussmaffei.com/media/download/cms/media/digitaleprodukte/apc-plus/imm-fl-apcplus-de.pdf, 03.11.2021

[URLb] N.N. *iQ weight control – Prozessschwankungen intelligent ausgleichen. Firmenschrift Engel.* URL: https://www.engelglobal.com/fileadmin/master/Downloads/Broschueren/iQWeightControl_de.pdf, 03.11.2021

Symbols

Symbol	Unit	Description
p	bar	Melt pressure
V	cm^3	Melt volume

Abbreviations

Notation	Description
ALS	Arburg Leitrechnersystem

2 Effective data-selection strategy for efficient model-based injection moulding setup with transfer learning

Ch. Hopmann[1], M. Schmitz[1], Y. Lockner[1]
[1] Institute for Plastics Processing (IKV) at RWTH Aachen University

Abstract

The setup of an injection moulding process can be very complex and sometimes needs iterations to find a suitable process point. Machine learning models like artificial neural networks can help to optimise the setup procedure but require a significant amount of training data to be fitted to the process. Currently, transfer learning is being researched as a strategy to reuse already collected data in a model finetuning approach to significantly lower the mandatory data generation effort.

A practitioner commonly only has a rough idea in advance which data or model from an unknown process shall be reused which, however, is crucial to streamline the process setup procedure even before the experimental sampling starts. To achieve a better understanding regarding the transferability of injection moulding data from one process to another, this work investigates whether solely based on a-priori known process characteristics of old and new processes an estimation regarding the transfer learning success can be given. Previously generated transfer learning results based on simulation data are used together with 25 metadata parameters in a modelling approach. A variety of machine learning models are probed. Two model classes undergo cross-validation with holdout datasets and are analysed regarding their feature importance.

Lasso regression and *AdaBoost* with support vector regression as base estimators are chosen after model selection as suitable learning strategies. In a "Leave-One Out Cross-Validation", *AdaBoost* achieves a R^2 score of 0.805 and a mean absolute error of 0.025, in relation to R^2 values in the used dataset. The *AdaBoost* model qualifies as rough estimator for the resulting model quality for a finetuning transfer learning approach and could help practitioners to efficiently select datasets for transfer learning before modelling. In the analysis of the models' feature importances, theoretical cooling time, maximal wall thickness and part width were determined as most influential on the predictions. The developed methodology needs to be further investigated and validated with experimental transfer learning results currently being prepared at Institute for Plastics Processing (IKV).

2.1 Introduction

Increasingly higher customer demands and smaller failure tolerances on produced injection moulding parts challenge manufacturers in high-wage countries and call for innovations to remain competitive on international markets [BJS+11]. The injection moulding process as the most important process for manufacturing complex plastic parts is regularly subject to evaluation and optimisation. An efficient, reproducible and reliable procedure to find suitable machine setting parameters during process setup could remarkably lower process costs.

Manufacturers commonly rely on expert knowledge by employees to define a process point [MVS13]. However, intelligent approaches already exist which make use of the benefits of machine learning models to learn the complex relations between machine, process and quality parameters in injection moulding and ultimately optimise the setup process [IAB13, CFG+20]. Especially artificial neural networks have been implemented and validated for quality prediction, error diagnosis and scheduling or process optimisation [ABP17, WLS+19, KKW+18, FSSC18]. After a training phase with collected process data, the fitted model can be used in a second step for the calculation of a setting parameter set, given a certain quality parameter value of the part. Usually, evolutionary algorithms are used to suggest suitable parameter values [CFG+20, BMBJ19].

As the data generation for the model training usually is a time-consuming and costly process, current research is looking into the possibility to use simulation data together with experimental data to reduced the required experimental effort [HHT18]. Furthermore, even the transfer between different processes could be interesting to raise attractiveness in industrial applications [LH21]. The procedure of data reuse in machine learning called *transfer learning*, however, remains a challenge as influences on the transfer success are data domain specific and often still need to be discovered. Ideally, users and manufacturers should have a rough idea about the aptitude of a given dataset for transfer learning before the mould is actually sampled in production.

In this work, first steps towards explainable knowledge transfer between injection moulding processes with machine learning methods are attempted. Chapter 2.2 introduces transfer learning as modelling approach for injection moulding processes and illustrates the used dataset. Chapter 2.3 defines the approach how model quality for transfer learned processes shall be predicted solely based on process characteristics already known before the process setup. Chapter 2.4 describes the machine learning models evaluated for the modelling task. In chapter 2.5, model selection is conducted to choose the best model class(es) for the given dataset. To achieve an unbiased estimate regarding the possible generalisation qualities of the chosen model class(es), a cross-validation is performed in chapter 2.6. Important influencing factors on the prediction quality are analysed and interpreted. Finally, chapter 2.7 concludes with a brief summary of the work and gives an outlook for this research.

2.2 Transfer learning for different part geometries

Previous to this work, a high number of experiments have been conducted to benchmark the finetuning transfer learning approach between two injection moulding processes. *Lockner and Hopmann* presented work that compared the transferability of artificial neural networks to an unknown process manufacturing a different part [LH21]. The models trained with 59 simulation datasets were transferred to the 60th target domain, which was chosen to be a 4x2 original toy building block. The target part is illustrated in Figure 2.1 together with the other variants. The injection moulding simulation software Cadmould from Simcon kunststofftechnische Software GmbH, Würselen, Germany, was used for simulation. In total, 60 different part geometries were available as data domains, all variants of toy building blocks. Six machine setting parameters, namely injection volume flow, cooling time, packing pressure, packing pressure time, melt temperature and coolant temperature were used as input parameters. The weight of the produced part served as quality parameter for the regression task.

The experiments were performed for 20 discrete, artificially reduced amounts of target domain data for finetuning of the target task estimator which models the new injection moulding process. Furthermore, the displayed data series for the conventional approach and the complete model transfer are averaged from a multitude of model trainings with varying orders of training data and initial neuron weights [LH21]. Generally, transfer learning is beneficial when a small amount of data samples is available from a target domain (see Figure 2.2, left). This is the case for an injection moulding process setup: In this setting, a new process shall be characterised and a suitable set of machine setting parameters shall be found for mass production. Therefore, Transfer learning shows potential to reduced the timely effort for an injection moulding process setup.

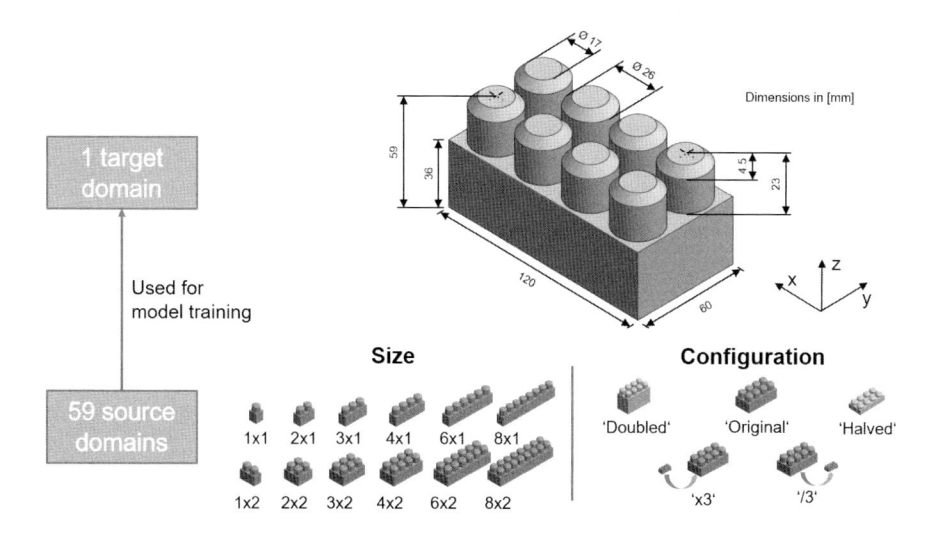

Figure 2.1: Source domains and target domain of the transfer learning results used for the analysis

While this methodology still needs to be confirmed for experimentally generated real datasets, the transfer success from the used 59 source domains was compared and analysed. Figure 2.2, right illustrates the model qualities for four different source configurations of the 4x2 toy building block. Depending on the configuration of the 4x2 toy building block in the source domain, the resulting model quality after finetuning with some target domain samples varied significantly. This was expected as transfer learning is assumed to work well, if source and target domain show a certain amount of similarity [PY10].

However, as in all other use cases for transfer learning, similarity is a task-specific concept. The more similar the data domains, the better the transfer learning results. Therefore, in order to save as much time and effort in advance to the setup of an injection moulding process, the most suitable data domains or models, respectively, from a database for transfer learning should be a-priori determinable. In the following, a method is presented and evaluated that predicts the model quality only based on the known process characteristics after the design phase without the need to perform real experiments.

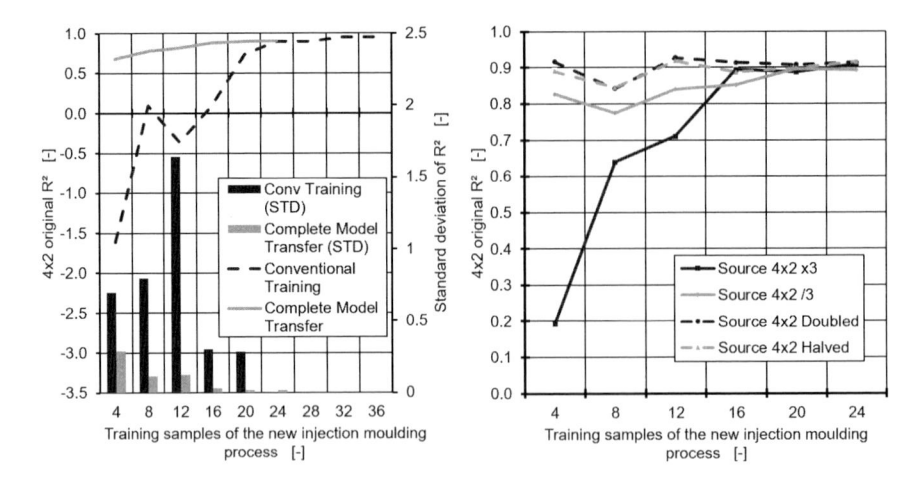

Figure 2.2: Transfer Learning results for geometry variation (left) and source domain specific results (right)

2.3 Modelling approach to predict the transfer learning result

Most information and characteristics of an injection moulding process are already known before the mould is produced. These information which are usually not part of a process modelling approach are considered *metadata* in the following. Various parameters can be considered metadata, e. g. all features that can be classified as (peripheral) influence on the injection moulding process.

As influence factors on the injection moulding process and the resulting part quality are generally considered the injection moulding machine, the material, the mould, the processing conditions or parameters and the production environment [Sch09]. Sometimes, automation and the human operator are considered as well. These influence factors can be further broken down to concrete parameters. For the used datasets, besides the machine setting parameters, the part geometries were varied. Therefore, a selection of different parameters describing the parts and subsequent processing conditions was chosen to represent the metadata of each process respectively (see Table 2.1).

Nr.	Description	Unit	Min	Max
1	Length	[mm]	10.00	720.00
2	Width	[mm]	10.00	180.00
3	Height	[mm]	19.67	177.00
4	Volume	[mm^3]	462.84	3899162.07
5	Injection volume	[mm^3]	462.84	3,899,162.07
6	Average wall thickness	[mm]	0.65	5.94
7	Maximal wall thickness	[mm]	0.67	12.00
8	Minimal wall thickness	[mm]	0.50	6.00

9	Flow path length	[mm]	20.10	583.20
10	Surface	[mm^2]	1,414.80	1,321,238.17
11	Surface-volume-quotient	[1/mm]	0.34	3.06
12	Flow-path-length-wall-thickness-quotient	[-]	19.15	112.83
13	Minimal-maximal-quotient (wall thickness)	[-]	0.50	1.00
14	Theoretical cooling time	[s]	0.71	230.01
15	Plasticising performance	[mm^3/s]	47.04	60,450.06
16	PACYNA 1 (thin walledness)	[log(Lw/mwd)]	0.99	1.49
17	PACYNA 2 (elongation)	[log(rd/rdw)]	0.20	0.60
18	PACYNA 3 (volume bulkiness)	[log(l*b*h/Vw)]	0.60	0.80
19	Main moment of intertia 1	[$kg*mm^2$]	0.02	20,771.85
20	Main moment of intertia 2	[$kg*mm^2$]	0.02	167,574.63
21	Main moment of intertia 3	[$kg*mm^2$]	0.02	170,906.26
22	Studs per row	[-]	1.00	8.00
23	Rows	[-]	1.00	2.00
24	Shoulder height	[mm]	7.67	69.00
25	Stud height	[mm]	12.00	108.00

Table 2.1: Metadata for the source and target domains

Parameters like length, width and height of the part geometries are used to include the general dimensions of the part. The wall thickness is an important parameter for the packing and cooling phase of the process while the flow path length co-determines the fillability of the cavity during injection. Some quotients are used to calculate specific values for the mould. The PACYNA identification numbers, regularly used by manufacturers to classify moulds, contain information about the thin walledness (PACYNA 1), elongation (PACYNA 2) and volume bulkiness (PACYNA 3) of the part. Furthermore, part specific parameters like the studs per row or complex parameters like the theoretical cooling time (t_c) are used. t_c is calculated as follows [JM04]:

$$t_c := \frac{d^2}{\pi \cdot a} \cdot \ln(\frac{4}{\pi} \cdot \frac{\theta_M - \theta_W}{\theta_E - \theta_W}) \tag{2.1}$$

The thermal diffusivity a was set to 0.08877 mm^2/s for the calculation. A melt temperature θ_M of 240 °C and a cavity wall temperature θ_W in the mould of 40 °C was assumed. The average part temperature at the time of ejection θ_E as estimated at 80 °C.

These input parameters were extracted for all part geometries as indicated in Figure 2.3. Volume and injection volume have been set to the same values in order to account for randomness in the model fit in the analysis later on. Before the parameters could be used, the values needed to be processed together with the respective parameter values of the transfer learning target data domain. This needed to be done as the used transfer learning results in the modelling approach refer to a combination of source and target domain. The processing should yield the differences between source and target domain per influence parameter and therefore describe the (unweighted) similarity between the two injection moulding processes. For the first approach, the absolute difference (ABS) between two parameter values was chosen to represent the respective offset.

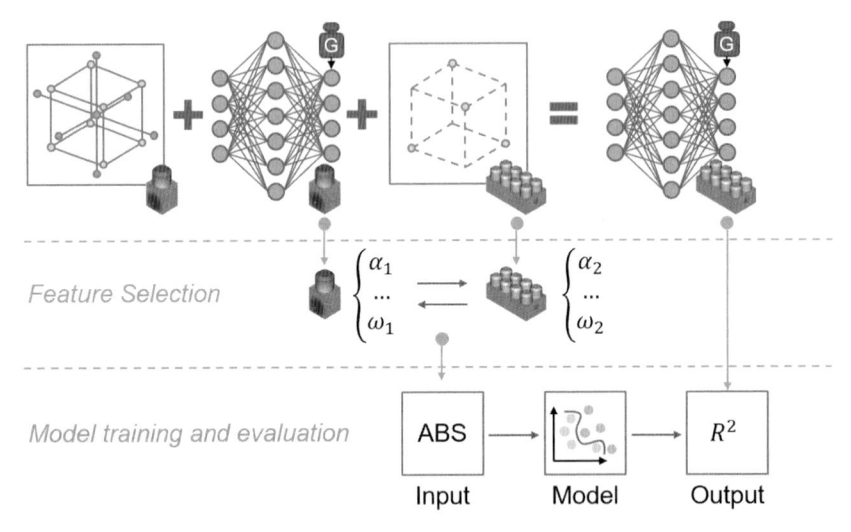

Figure 2.3: Transfer Learning results for geometry variation (left) and source domain specific results (right)

In the last step of the modelling approach, the calculated similarity between the processes (column-vectors) and the transfer learning results, representing input and output data, were utilised for model training. The respective transfer learning results for the provision of 12 data samples was used in the following (see Figure 2.2). The database was chosen regarding highest attractiveness in case of successful modelling: for 12 data samples, many approaches yielded good absolute results for the model quality, while the difference to the conventional model training was quite large.

2.4 Probed models for the correlation of metadata and transfer learning results

Machine learning models have specific characteristics that make them more or less suitable for certain modelling tasks. It is not possible to determine without information about the respective dataset, whether a model class will perform better than another. Therefore, six different machine learning models were evaluated to learn the correlations between metadata and transfer learning results in this setting. All models were implemented in Python, version 3.7.5, and initialised with the open-source library scikit-learn, version 0.24.1. The results were evaluated with the scikit-learn metrics *r2_score*, *mean_squared_error* and *mean_absolute_error*. The degree of determination and mean squared error are common quality functions in this area [TL14, TGH+18, IAB13, ZEAE14]. The learners were chosen regarding easy interpretability of the input feature importance for the following analysis. The implementations of the models can be reviewed in the source code of the open-source framework [URL18].

Lasso Regression is a variant of linear regression that aims to improve interpretability of the fitted model by performing feature selection inherently, therefore reducing the influencing variables to a minimal number. *Lasso Regression* can effectively reduce the risk of model overfitting to the given training dataset by applying regularisation to the regression function. *Polynomial Regression* does not only use input feature linearly for regression, but also transform the input feature space into a higher dimensional dataset. *Support Vector Regression* is the for regression tasks adapted implementation of support vector machines which were initially designed for classification tasks [Shm10]. By a learned set of hyperplanes in a high-dimensional feature space, data is being separated into classes for prediction. *Random Forest Regression*, *AdaBoost Regression* as well as *GradientBoost Regression* are so called *ensemble* techniques [Bre01, FS97, PS10, HTF08]. That means, all of the named modelling strategies make use of base learners that are either learned at the same time in an uncorrelated manner (Random Forest) or sequentially, intelligently adapting either the data or model parameters between each model training (AdaBoost and GradientBoost).

2.5 Model selection for optimised results

Previously mentioned models presumably differ in their suitability to correlate metadata and transfer learning results. In order to evaluate the modelling approach, firstly a model selection needs to be conducted. For this purpose, a nested hyperparameter search was designed (see Figure 2.4).

Based on a six folds cross-validation (CV), a nested hyperparameter tuning is marked by its consideration of the hyperparameter search itself as a variable. The inner cross-validation (inner CV) is conducted with the framework *optuna* using the *study* class [ASY+19]. As indicated in Figure 2.4, each cross-validation iteration for one model type results in a best model, which is evaluated with a holdout dataset. In this way, an unbiased estimate regarding the generalisation capability of the machine learning model can be achieved. This is done six times for a varying composition of the training and holdout datasets, to avoid sampling effects which are usually apparent in small datasets. At the end, the nested hyperparameter tuning will result in six unbiased model quality results estimated per model type, meaning 36 results in total. All experiments are conducted with a pseudo-randomised data order for repeatability.

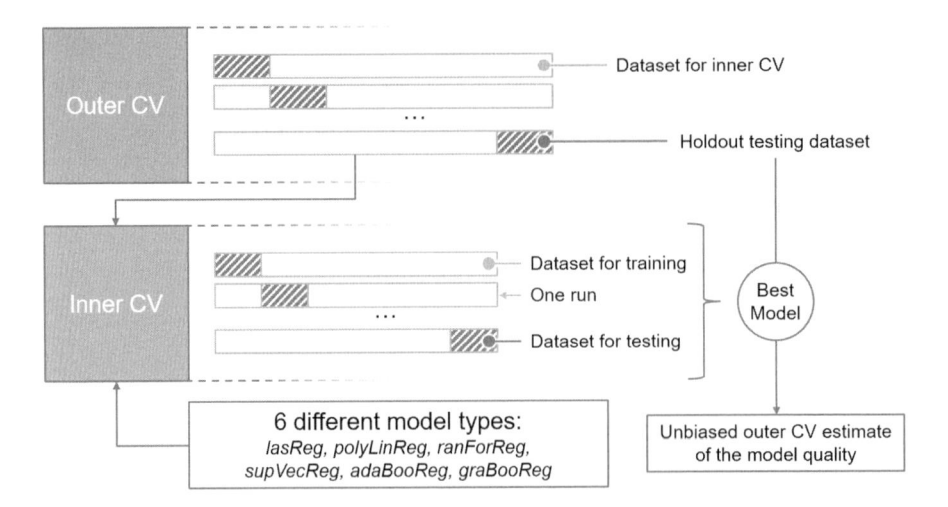

Figure 2.4: Nested hyperparameter tuning for model (pipeline) selection

Figure 2.5 illustrates the modelling results. Displayed are the averaged model qualites over six holdout test datasets (outer CV) with the standard deviations. The upper bar chart shows the mean-squared-error results (MSE) and below the degrees of determination (R^2) are denoted. The results for polynomial regression have been omitted due to bad quality.

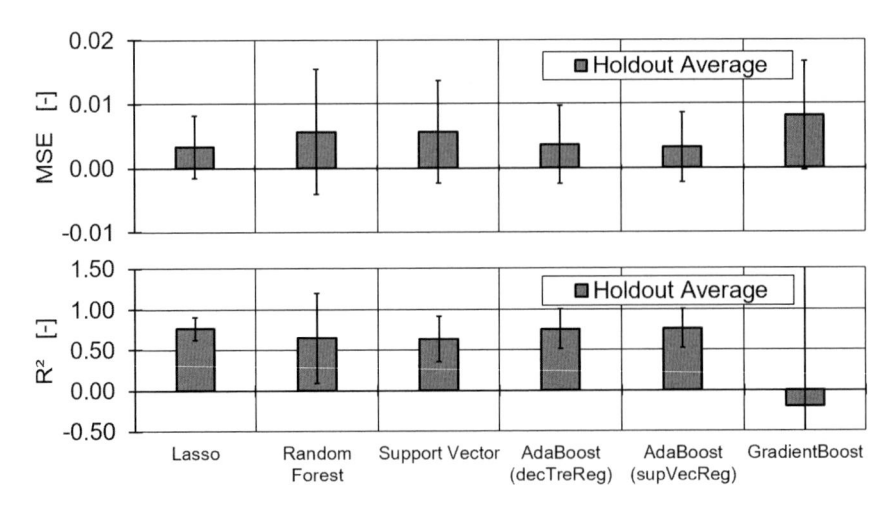

Figure 2.5: Holdout testing set prediction results for the six-Fold outer cross-validation

Both evaluation functions for the model quality determine *AdaBoost Regression* with Support Vector Regression as base estimators as best model pipeline. On average, the pipeline predicted a residual MSE of 0.0032 with the second smallest standard deviation of 0.0055 of the probed models. Only *Lasso Regression* performed slightly better with 0.0049 as standard deviation, yet it did produce a worse result for the average generalisation capability in case of MSE with 0.0033. However, *Lasso-Regression* did perform better in terms of degree of determination with a result of 0.7667 on average and a standard deviation of 0.1418. Especially the standard deviation is significantly smaller than for the rest of the models with *AdaBoost* in combination with Support Vector Regression in second place with a result of 0.2413 as standard deviation.

As both *Lasso Regression* and *AdaBoost Regression* showed good results, both model pipelines will be chosen for further examination. In the following, the best parameters found for both models will be used to train the models with the complete dataset.

2.6 Model quality estimation and feature importance

In the previous chapter, *Lasso Regression* as well as *AdaBoost* with Support Vector Machines as base estimators had been chosen for further investigation. The best hyperparameters found in the prior cross-validation of both models are utilised now in a so-called "Leave-One-Out cross-validation" (LOOCV). From the presented dataset of size 59, 58 data samples will be used to fit both models with the chosen hyperparameters (see Table 2.2). The remaining 59th sample will be used for evaluation. This is aligned with the use case for injection moulding: One new process shall be optimised, therefore all known processes can be used for the model training. As the previous model selection step was designed to choose the modelling strategy for the correlation of metadata and transfer learning success, an unbiased approximation of the model's generalisation quality is required. To achieve this for a fixed hyperparameter set, each of the 59 samples will once be used for prediction, hence a LOOCV is conducted.

Hyperparameter	Value
Lasso Regression	
Alpha	0.114845
Maximal iterations	2759
AdaBoost (Support Vector Regression)	
Kernel	Linear
C	1.0
Degree	3
Epsilon	0.1
Number of estimators	5
Learning rate	0.024438
Loss	Exponential

Table 2.2: Hyperparameter sets for Lasso and AdaBoost Regression (support vector regression)

Figure 2.6 illustrates the results for *Lasso Regression* and *AdaBoost Regression*. Listed on the x-axis are the single test samples for each iteration of the LOOCV. The y-axis represents the R^2 values that were determined for the finetuning transfer learning approach, when the test domain was used as source domain together with the 4x2 original configuration toy building block.

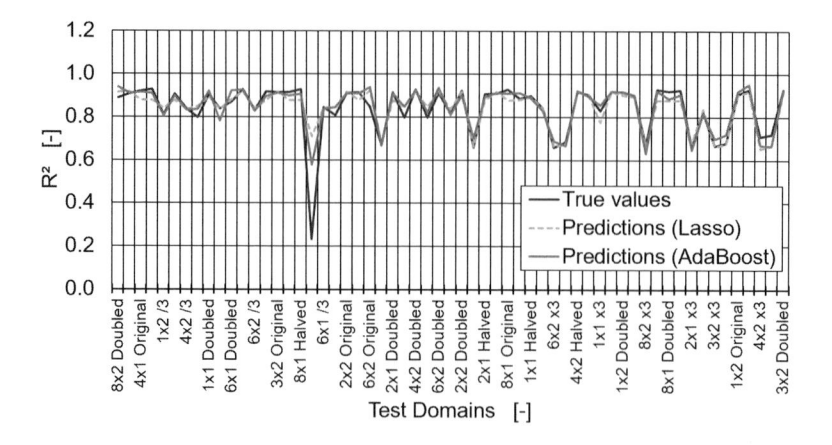

Figure 2.6: LOOCV test sample results for Lasso and AdaBoost Regression

It can be observed, that the predictions for the transfer learning results with the test domains appear to be more accurate for the *AdaBoost* model than for *Lasso Regression*. The *Lasso Regression* seems to be less able to adapt to the variances in the test domains. This can be confirmed by the calculation of the degree of determination for the comparison of true values and predictions: *Lasso Regression* solely yielded a score of 0.667 for the degree of determination, which is lower than its score for the nested hyperparameter tuning procedure, where even less data was used for the model training (compare Figure 2.5). A possible explanation for this could be the regularisation inherent in the model, which prevents the model from finding a good representation of the correlations in the data when the training set grows and becomes more diverse.

On the other hand, *AdaBoost* scores an R^2 of 0.805, which is a significantly better result compared to the nested hyperparameter tuning evaluation (compare Figure 2.5). *AdaBoost*, in comparison to *Lasso Regression*, is not a single model but a ensemble technique: A discrete number of estimators, here five (comp. Table 2.1), is trained sequentially to account for possible deficiencies in a single estimator. Each model in the ensemble is contributing to the final prediction by a calculated percentage, based on its single prediction performance. Therefore, *AdaBoost* could be more suited to learn the complex relationship between metadata and model quality by transfer learning. Considering all test domains equally important, the mean absolute error for the predictions by *AdaBoost* is 0.025. The *AdaBoost* regressor is therefore well suited for a first, rough estimation for the suitability of a source domain for a finetuning transfer learning approach.

However, both models can be easily evaluated regarding feature performance for the predictions. *Lasso Regression*, on the one hand, is a variant of linear regression. Therefore, the parameter weights can simply be extracted from the regression equation. *AdaBoost*, on the other hand, uses support vector machines for regression as base estimators. For linear kernels, the scikit-learn allows the direct extraction of summarised parameter weights per model in the ensemble. To extract an overall value for each parameter, a weighted sum was calculated with the models' prediction contribution percentages as factors. In this way, the parameters are weighted in relation to the prediction performance of *AdaBoost*. Figure 2.7 exhibits respective feature importances (y-axes) for each metadata parameters, indicated by its number in Table 2.1. The data series' affiliation to primary or secondary y-axis is specified by directed arrows.

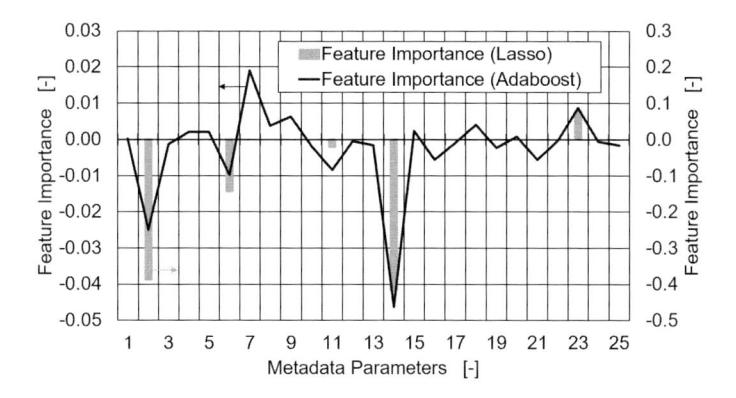

Figure 2.7: Feature Importances for *Lasso Regression* and *AdaBoost Regression*

Parameter numbers 4 and 5 represent volume and injection volume. Both importance values are either set to 0 (*Lasso*) or 0.001997 (*AdaBoost*). Randomness in the model fit can therefore be ruled out. Dominating parameters for both *Lasso* and *AdaBoost* are the theoretical cooling time (nr. 14) and the part width (nr. 2). Furthermore, average wall thickness, surface-volume-quotient and the amount of rows are recognised by both modelling approaches as influences on the prediction with the given model hyperparameters. It is not possible directly translate the absolute values of the feature importance analysis to the importance of a parameter. The parameters should rather be compared and its feature importance evaluated against each other.

Negative feature importance values confirm the assumption that similar metadata parameters in source and target domain lead to better transfer learning results. The bigger the differences in the parameters, the worse the results. E. g. part width, in this case, correlates with the parts' configuration and therefore the overall size of the part. Equally to the average wall thickness, these parameters influence the injected melt volume into the mould's cavity, and therefore the necessary cooling time or effectiveness of shrinkage compensation in the packing phase. This is especially important for the part weight as quality parameter in the herein used dataset. Even though the theoretical cooling time was used as complex parameter, the assumptions for melt, cavity and ejection temperature were constant for all simulations. Therefore, variance in the values for theoretical cooling time was only composed of the variance in the maximal wall thickness d, which is parameter nr. 7.

This is probably the explanation for a positive value for the maximal wall thickness within the *AdaBoost* model. As the importance of the theoretical cooling time, the square of the maximal wall thickness respectively, is overestimated, it is reduced by the non-squared value. It could be expected that the feature importance for the maximal wall thickness would be negative for an input feature space without the theoretical cooling time as parameter. The flow path length and number of rows exhibit the same peculiar behaviour as they result in a positive value. Balancing behaviour could be suspected, as higher width of the parts correlates with longer flow paths and a different number of rows in the part.

2.7 Conclusions and outlook

In this work, initial experiments have been conducted to approximate model quality of injection moulding process models with finetuning as transfer learning approach just by metadata of the processes, which is available before the experiments. It should be determined, if an a-priori data selection for optimised transfer learning success with this approach can be performed too. For the experiments, a previously generated, extensive dataset with transfer learning results for injection moulding simulations of different part geometries was used as dataset.

25 metadata parameters have been declared and used to correlate the transfer learning results. Several machine learning models have been probed as suitable learners for this task. *Lasso Regression* and *AdaBoost* with support vector regression as base estimator have been both chosen for further investigation. While *Lasso Regression* performed worse in a LOOCV performance validation with an R^2 of 0.667, *AdaBoost* showed a good performance with an R^2 of 0.805. Latter model could therefore be used for a rough dataset selection before transfer learning. The theoretical cooling time, maximal wall thickness respectively, as well as the part width have been determined as main influencing factors on the prediction quality.

Further experiments with varying datasets and input parameter constellations have to be performed to confirm these initial results. While the methodological development has been performed with a simulation dataset, the approach needs to be validated with experimental transfer learning results. Necessary preparations are ongoing at the IKV. Eventually, an increase of the prediction quality could also be achieved by reduction of the dataset variance, which needs to be verified in an adapted experiment. Also, if no interpretation of the influence factors is required, black box models like artificial neural networks could be an option to further raise the prediction quality.

Acknowledgements

This research was funded by the Deutsche Forschungsgemeinschaft (DFG, German Research Foundation) under Germany's Excellence Strategy - EXC-2023 Internet of Production - 390621612. We would like to cordially extend our thanks to the DFG.

References

[ABP17] ADEMUJIMI, T. T.; BRUNDAGE, M. P.; PRABHU, V. V.:. *A Review of Current Machine Learning Techniques Used in Manufacturing Diagnosis* p. 407–415. Springer International Publishing. 2017

[ASY+19] AKIBA, T.; SANO, S.; YANASE, T.; OHTA, T.; KOYAMA, M.: Optuna: A Next-generation Hyperparameter Optimization Framework. *Proceedings of the 25th ACM SIGKDD International Conference on Knowledge Discovery amp; Data Mining*Anchorage, AK, USA. 2019

[BJS+11] BRECHER, C.; JESCHKE, S.; SCHUH, G.; AGHASSI, S.; ARNOSCHT, J.; BAUHOFF, F.; FUCHS, S.; JOOSS, C.; KARMANN, O.; KOZIELSKI, S.; ORILSKI, S.; RICHERT, A.; RODERBURG, A.; SCHIFFER, M.; SCHUBERT, J.; STILLER, S.; TÖNISSEN, S.; WELTER, F.:. *Integrative Produktionstechnik für Hochlohnländer*. Springer Verlag. Berlin, Heidelberg. 2011

[BMBJ19] BENSINGH, R. J.; MACHAVARAM, R.; BOOPATHY, S. R.; JEBARAJ, C.: Injection molding process optimization of a bi-aspheric lens using hybrid artificial neural networks (ANNs) and particle swarm optimization (PSO). *Measurement* 134 (2019) Unknown issue, p. 359–374

[Bre01] BREIMAN, L.: Random Forests. *Machine Learning* 45 (2001), p. 5–32

[CFG+20] CAO, Y.; FAN, X.; GUO, Y.; LI, S.; HUANG, H.: Multi-objective optimization of injection-molded plastic parts using entropy weight, random forest, and genetic algorithm methods. *Journal of Polymer Engineering* 40 (2020) 4, p. 360–371

[FS97] FREUND, Y.; SCHAPIRE, R. E.: A Decision-Theoretic Generalization of On-Line Learning and an Application to Boosting. *Journal of Computer and System Sciences* 55 (1997), p. 119–139

[FSSC18] FAZEL, M. H. Z.; SADAT, A. A. A.; SOTUDIAN, S.; CASTILLO, O.: A state of the art review of intelligent scheduling. *Artificial Intelligence Review* 53 (2018), p. 501–593

[HHT18] HOPMANN, C.; HEINISCH, J.; TERCAN, H.: Injection Moulding Setup by Means of Machine Learning Based on Simulation and Experimental Data. *ANTEC 2018 - The Plastics Technolog Conference*, Orlando, Florida, USA. 2018

[HTF08] HASTIE, T.; TIBSHIRANI, R.; FRIEDMAN, J.: Boosting and Additive Trees. In: *The Elements of Statistical Learning*. New York, NY, USA: Springer, 2008

[IAB13] INIESTA, A. A.; ALCARAZ, J. L. G.; BORBÓN, M. I. R.: Optimization of injection molding process parameters by a hybrid of artificial neural network and artificial bee colony algorithm. *Revista Facultad de Ingenería Universidad de Antioquia* 67 (2013), p. 43–51

[JM04] JOHANNABER, F.; MICHAELI, W.: *Handbuch Spritzgießen*. München: Carl Hanser Verlag, 2nd edition, 2004

[KKW+18] KIM, D.-H.; KIM, T. J. Y.; WANG, X.; KIM, M.; QUAN, Y.-J.; OH, J. W.; MIN, S.-H.; KIM, H.; BHANDARI, B.; YANG, I.; AHN, S.-H.: Smart Machining Process Using Machine Learning: A Review and Perspective on Machining Industry. *International Journal of Precision Engineering and Manufacturing-Green Technology* 5 (2018), p. 555–568

[LH21] LOCKNER, Y.; HOPMANN, C.: Induced network-based transfer learning in injection molding for process modelling and optimization with artificial neural networks. *Int J Adv Manuf Technol* 112 (2021) 11-12, p. 3501–3513

[MVS13] Meiabadi, M. S.; Vafaeesefat, A.; Sharifi, F.: Optimization of Plastic Injection Molding Process by Combination of Artificial Neural Network and Genetic Algorithm. *Journal of Optimization in Industrial Engineering* 6 (2013) 13, p. 49–54

[PS10] Pardoe, D.; Stone, P.: Boosting for Regression Transfer. *27th International Conference on Machine Learning.* Haifa, Israel, 2010

[PY10] Pan, S. J.; Yang, Q.: A Survey on Transfer Learning. *IEEE Transactions on Knowledge and Data Engineering* 22 (2010), p. 1345–1359

[Sch09] Schiffers, R.: *Verbesserung der Prozessfähigkeit beim Spritzgießen durch Nutzung von Prozessdaten und eine neuartige Schneckenhubführung.* Dissertation, 2009

[Shm10] Shmilovici, A.: Support Vector Machines. In: Maimon, O.; Rokach, L. (Editor): *Data Mining and Knowledge Discovery Handbook.* Boston, Massachusetts, USA: Springer Verlag, 2010

[TGH+18] Tercan, H.; Guajardo, A.; Heinisch, J.; Thiele, T.; Hopmann, C.; Meisen, T.: Transfer-Learning: Bridging the Gap between Real and Simulation Data for Machine Learning in Injection Molding. *51st CIRP Conference on Manufacturing Systems.* Stockholm, Sweden, 2018

[TL14] Tsai, K.-M.; Luo, H.-J.: An inverse model for injection molding of optical lens using artificial neural network coupled with genetic algorithm. *Journal of Intelligent Manufacturing* 28 (2014) 2, p. 473–487

[URL18] N.N. *Keras FAQ: Frequently Asked Keras Questions.* URL: https://keras.io/getting-started/faq/, 19.07.2018

[WLS+19] Weichert, D.; Link, P.; Stoll, A.; Rüping, S.; Ihlenfeldt, S.; Wrobel, S.: A review of machine learning for the optimization of production processes. *The International Journal of Advanced Manufacturing Technology* 104 (2019), p. 1889–1902

[ZEAE14] Zahra, M. M.; Essai, M. H.; Abd Ellah, A. R.: Performance Functions Alternatives of Mse for Neural Networks Learning. *International Journal of Engineering Research Technology* 3 (2014), p. 967–970

Symbols

Symbol	Unit	Description
d	mm	Critical wall thickness
a	mm^2/s	Thermal diffusivity
t_c	s	Theoretical cooling time
θ_W	$°C$	Cavity wall temperature
θ_E	$°C$	Ejection temperature
θ_M	$°C$	Melt temperature

Abbreviations

Notation	Description
ABS	Absolute difference between two metric values
CV	Cross validation
LOOCV	Leave-one-out cross validation
MSE	Mean squared error
MAE	Mean absolute error
R^2	Degree of determination

3 Enabling shopfloor optimisation in injection moulding through digital shadows

Ch. Hopmann[1], M. Schmitz[1], P. Sapel[1]
[1] Institute for Plastics Processing (IKV) at RWTH Aachen University

Abstract

In Industry 4.0 data-based decision support is necessary to achieve certain benefits. The benefits of a data-based decision-support lie in human-independent and evidence-based decision-making, handling complex tasks, and increased process efficiency for fulfilling a specific task. The challenges for realising such a decision-support are the availability of valid data that often is stored in a distributed manner and suitable models. Gathering all the relevant data efficiently can be exhausting, especially in complex domains with multiple relations and constraints. Typically, the production planning and control domain is complex for scheduling purposes, as already a few assets can lead to an enormous number of possible solutions. This contribution demonstrates how Industry 4.0 can support decision-makers to fulfil their tasks in complex domains by using Digital Twins and Digital Shadows.

3.1 Introduction

With the advancement of production machines and peripherals to sophisticated data driven devices, new potentials arise. On the one hand, those potentials include imminent engagement with the process regarding its setup and control. On the other hand, they provide precise knowledge about the produced parts and the corresponding production process. Regarding the latter, the challenge is not to collect the data but also to select the desired data and put it into context for immediate and future use. The prerequisite for any data proceeding is the representation of assets in the digital world with so-called Digital Twins. While Digital Twins provide all necessary data of an asset, a purposive data acquisition and data processing can be executed through Digital Shadows. They provide a conceptual framework towards a common understanding of economical but context-comprehensive data accumulation. Although strictly complying with structural requirements when addressing very context-specific tasks might seem undue, an agreement on minimal structural requirements of such data tasks is contributing to the vision of an extensive "Internet of Production". This vision promotes cross-domain collaboration concerning data usage, knowledge exchange and process optimisation as well as making historic data reusable for future tasks. This contribution addresses the topic of data-based decision support within the injection moulding production with the aim of an optimal set up order regarding setup costs. The first section gives an introduction of Digital Twins and their implementation as Asset Administration Shells. Following, the state of research on Digital Shadows within the research project "Internet of Production" (IoP) at RWTH Aachen University and their differentiation to Digital Twins is presented in section 3.3. The next section 3.4 introduces the use case "Minimisation of setup costs" and illustrates the information a decision-maker requires to fulfil this task. Subsequently, a demonstration of an implemented Digital Twin and Digital Shadow for this use case and the implementation of a blueprint for establishing a general data flow is shown. This contribution closes with a conclusion and outlook towards future research opportunities.

3.2 Digital twins as virtual representation of the physical production system

3.2.1 Towards Industry 4.0 with digital twins

Looking back to the historical industrial evolution let identify three milestones that significantly changed production and production management. At the end of the 18th century, the first water- and steam-powered machinery increased manufacturing speed. Following, an electrically powered mass production built the second milestone in industrial evolution in 1870. The rise of information technology in the 1970s gradually led to computer integrated manufacturing as the third milestone that enables automated production [HVHB17]. In 2011, the fourth industrial revolution (Industry 4.0) was introduced as a strategy for facing Germany's challenges on a globalised market, which primarily arise in high labour costs compared to other countries. Within Industry 4.0, the economical manufacturing of products with lot size is one general goal (so-called mass customisation) to provide the customer with a wide range of individual products. Moreover, production systems should evolve from automated to autonomous environments so employees can focus on tasks, where computers provide unsatisfactory results [NN13]. As injection moulding is not suitable for manufacturing in lot size one, the main benefits for Industry 4.0 in injection moulding are robust production planning and control decisions (e.g., among other things, reducing the production orders lot sizes), a faster and improved process setup, and a robust production process, ideally executed autonomously by the involved assets [HS21]. An asset can be physical (e.g. injection moulding machine, mould, periphery), but also virtual like a bill of material, a production order or a technical drawing. Realising an autonomous production strictly requires a coordinated system with high efforts in process transparency and thus communication between the single assets [Hof19]. Because the given quantitative relationships inside the production are usually too complex for a human understanding, systems are needed to manage this complexity and support the humans. Therefore, all relevant information and subsequently data have to be gathered and provided at the right time to the right place. Thus, a transfer of all relevant production assets into the digital world is necessary [Lu17]. The resulting co-existence of real production assets and their virtual counterpart leads to Cyber-Physical Production Systems (CPPS), where Digital Twins represent relevant assets and their current status in the virtual world. The Digital Twin is a holistic representation of an asset that comprises its properties [Hur20]. Hence, a Digital Twin provides a universal, non use case related representation of an asset in the digital world. Furthermore, the Digital Twin includes models that can be used, e.g. by Digital Shadows. Whereas the Digital Twin only denotes a concept for the virtual representation of assets, the Asset Administration Shell is a specific and standardised method for implementing Digital Twins [URL18].

3.2.2 Asset Administration Shells as implementation of Digital Twins

An Asset Administration Shell (AAS) acts as Digital Twin and represents an asset in the virtual world by providing relevant characteristics and endpoints for data exchange.

In this context, "an asset is an item, thing or entity that has potential or actual value to an organization" [DIN17] regarding its information content.

The general purpose of an AAS is a standardised administration of the asset's characteristics, parameters and status over its entire life cycle, so the asset becomes discoverable, explorable and accessible for other assets. The AAS comprises two main components: the manifest and the component manager (Figure 1) [TA17].

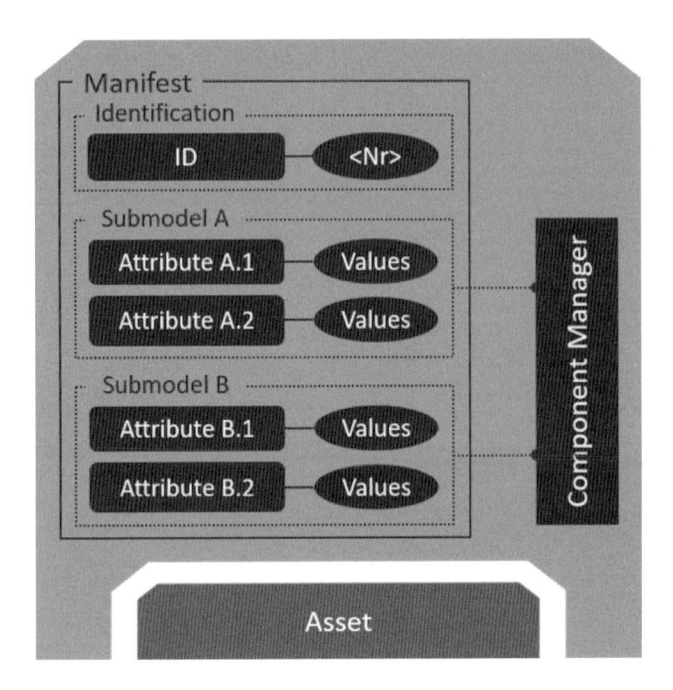

Figure 3.1: Depiction of a general AAS, based on [TA17]

The manifest of the AAS covers all relevant attributes of an asset. In the vision of Industry 4.0, those attributes are formulated in a standardised manner and include semantics. For this reason, globally accessible dictionaries like ECLASS[1] or the electro-technical vocabulary of the IEC 60050[2] exists. In these dictionaries, the attributes are extended with a global unique identifier (UID). The unique identification ensures unequivocal access to the attribute and its meaning and acts as one main building block for realising autonomous communication. The advantage of UID is that the addressing of several attributes is independent of their name, which differs in different languages, domains or companies (e. g. the terms "production order", "manufacturing order", and "work order" can be used synonymously). Within the AAS, an encapsulation of specific attributes into sub-modules is possible. A sub-module logically groups attributes for different use cases or domains, e.g. the injection moulding process [WGE+17]. The AAS is not a closed system, but expandable with additional attributes and sub-models if necessary for describing the asset in the digital world. Moreover, the AAS is independent of any external data format or storage as the shell only administrates the regarding asset [WGE+17]. The second building block of an AAS is the component manager. The component manager tethers the sub-models to the network so services can operate with the relating data and information. With the establishment of the AAS, a tangible implementation for Digital Twins that fully represents assets in a virtual environment is given. However, for fulfilling particular task, only a subset of all accessible data is required (e. g. for scheduling production orders to reach minimal costs). The gathering and (pre-)processing of the required data were executed by Digital Shadows that will be introduced in the next section.

---

[1] https://www.eclass.eu/
[2] https://www.electropedia.org/

3.3 Conceptual foundations of digital shadows

When discussing the conceptual foundations of Digital Shadows (DS), the differentiation of the term from the related Digital Twin (DT) is essential. With increasing real-world data acquisition and modelling applications, the term Digital Twin has become general linguistic use for anything digitally representing a physical object. In research, the definition has to be more precise and selective. One common approach that this contribution follows distinguishes between a simple model, a DS and a DT based on the information flow between the digital and the physical object, as shown in Figure 3.2.

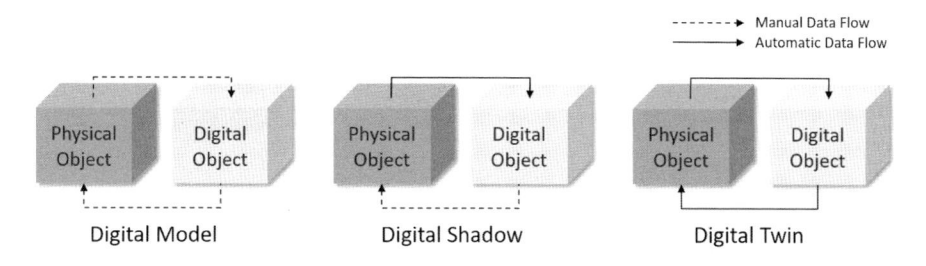

Figure 3.2: Difference between a model, a Digital Shadow, and a Digital Twin, based on [KKT+18]

A simple model has no automated interaction with the physical object and is only updated manually during model revisions. A DS can, as the name implies, follow certain behaviour of the physical object. It shadows the object through an automated data flow from the object to the model. A DT amplifies this definition by the ability to influence the physical object automatically. Therefore, the model and the object become twins and adapt to each other regardless of the two changes [KKT+18].

3.3.1 The digital shadow as contextual abstraction

The contextual abstraction of a DS can be shown best by following the metaphor of a shadow. A shadow depicts an abstraction of the real object depending on the viewpoint and the lighting. Therefore, the context and viewpoint in which an object is dealt with can define which of the object's properties are reproduced in the DS and which are not. The result is a DS composed for one specific object in one specific context (Figure 3.3). In order to comply with the vision of an Internet of Production of cross-domain collaboration and extensive usage of collected data, it is then crucial to concur in conceptual foundations creating those DS. This agreement on the basic data architecture ensures the DS' correct understanding and interoperability when accessed in other applications. Only then, DS of similar or equal objects and contexts are effectively usable for shopfloor, cross-domain, and cross-industry applications.

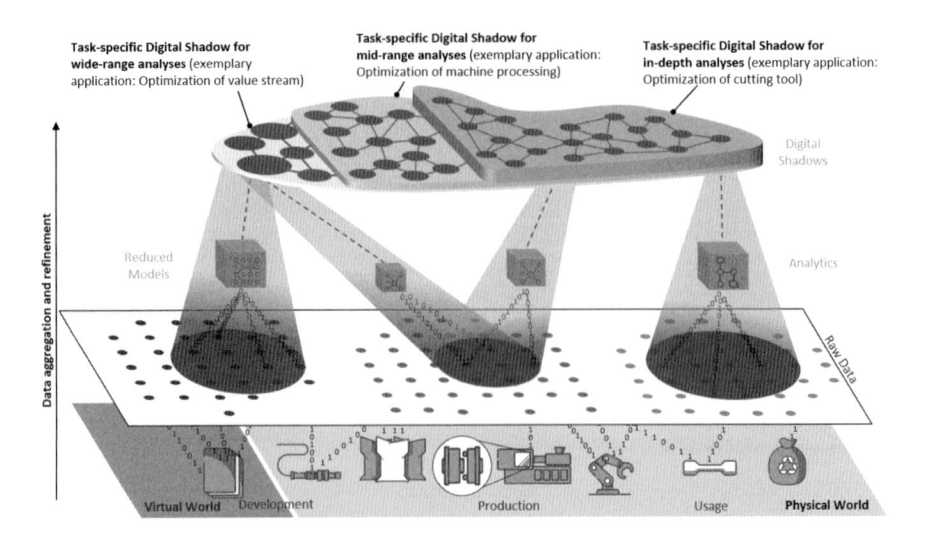

Figure 3.3: Digital Shadows as a contextualised abstraction of real world objects, based on [NN18]

The discussed foundations of data flow and contextual abstraction result in certain minimal structural and behavioural requirements regarding DS. Those requirements have been consolidated and phrased by the research group at RWTH previously as the following [SHH+20]:

Digital Shadows...
- must contain domain-specific knowledge
- are always context-specific
- enable cross-domain analyses
- require an adapted IT infrastructure within a production environment
- do not manipulate the real system

3.3.2 Meta-model of a digital shadow

In order to ensure that the conceptual foundations presented before are considered broadly and correctly, a comprehensive and precise framework has to be provided. Such a framework must contain specific boundaries regarding all minimum requirements, while leaving freedom for further research and conceptual extension. A first approach to such a framework has been defined through a series of interdisciplinary workshops in the Cluster of Excellence "Internet of Production". These workshops lead to consent in a meta-model of a Digital Shadow following the formal modelling guidelines of an UML class diagram (Figure 3.4). Through the class diagram, all constraints regarding minimal components of the DS as well as corresponding cardinalities can be realised in a distinct and normed manner. This meta-model can thus be seen as an approach to a DS's minimum viable product (MVP).

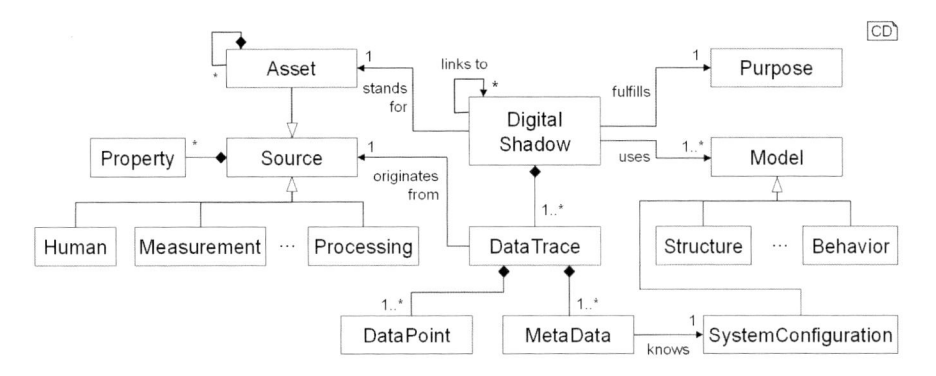

Figure 3.4: Metamodel of a Digital Shadow displayed as a class diagram [BBD+21]

A Digital Shadow only exists to fulfil one specific purpose by gathering data from one or many assets and generating insights through the usage of models. An asset can be either physical or virtual. As an asset can provide data, it acts as a data source. The specification of the data of every source as property characterises the source. The entire properties of an asset are located in the manifest of the AAS. Moreover, data sources can be of different origin like data from humans, measurements of sensors and processed data from models. A Digital Shadow always uses one or multiple models with a sub-division into different categories, e. g. structural or behavioural models. A structural model classifies the composition of the required data (like tables and their column headers), while behaviour models represent the expected actions of a system, e. g. a simulation model or mathematical functions. Furthermore, the configuration of the observed system is also a model as it defines the framework of this system. Data comprising the configuration of a system are specified as metadata. Combined with generated, single data points from the sources and appropriate models, contextualised data traces were built to meet the mentioned purpose. The linkage of Digital Shadows to a higher-level DS, e. g. for comparing different results for a scheduling task regarding an optimisation criterion (costs or due date), is possible. The following section introduces the building of Digital Shadows and Asset Administration Shells by transferring the single elements of the meta-model to a real-world application.

3.4 Enabling potentials for shop-floor optimisation through digital shadows

3.4.1 Digital Shadows as Decision Support for an optimal Setup of Moulds

In Industry 4.0 new benefits arise by providing service modules for a specific domain or purpose. Ideally, an implementation of service modules via "drag and drop" in a running enterprise is feasible without any configuration effort. The service modules establish the data flow autonomously and are utilisable directly. Therefore, the basis for this are standardised Asset Administration Shells comprising standardised attributes. In this case, the service is decoupled from the company-specific naming of the single attributes of an asset.

In an enterprise, the shopfloor is the place, where the demands from the customers are transferred into several production orders for fulfilment. Within production planning and control, scheduling of those production orders takes place. Ideally, the manufacturing of all production orders is on time, with the right quality and at the lowest costs. Unfortunately, meeting those objectives simultaneously in an optimal manner is rare through their competing nature and leads to the so-called "magic triangle of production" [KD21].

For example, increasing the quality of a product (e. g. by using better raw materials) commonly leads to higher costs, while faster processing time often negatively affects the product's quality. In this context, the usage of Digital Shadows as decision support for optimal scheduling regarding the different objectives can be beneficial by comparing different schedules that consider different targets. Hereinafter, a demonstration of data-based decision support for scheduling production orders via Digital Shadows is shown. Because in injection moulding it is not compulsory that changeover times of one mould to another are identical with the reverse setup direction, a Digital Shadow estimates the optimal schedule with relation to minimising the mould setup time. Therefore, the Digital Shadow computes a setup matrix containing all recorded setup processes and presents an optimal schedule under consideration of all open jobs. Figure 3.5 depicts the instantiated meta-model for this application.

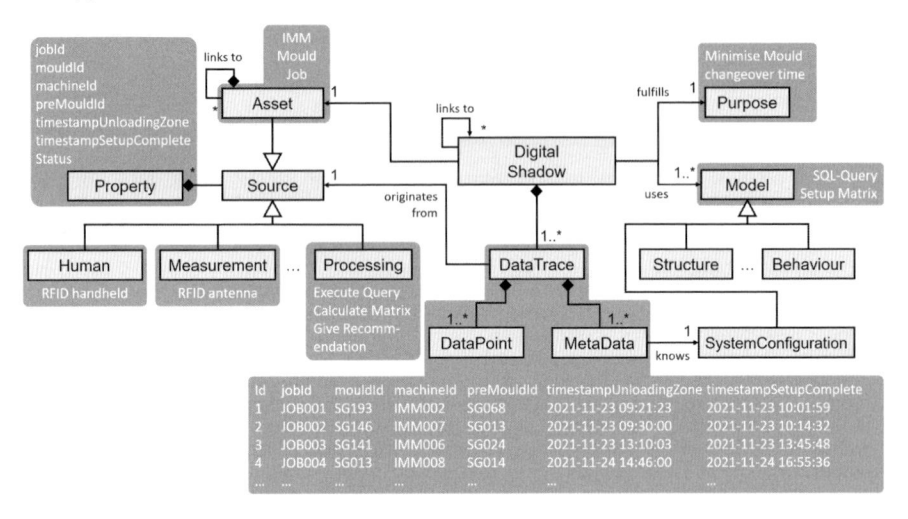

Figure 3.5: Use case transferred to the DS metamodel

The considered assets in this use case are the injection moulding machines, the moulds, and open jobs. In this case, the injection moulding machines and moulds relevant property is their identifier (*machineId* and *mouldId*). The properties for the jobs are, besides the identifier (*jobId*), the timestamp of the beginning (*timestampUnloadingZone*) and the end of the setup process (*timestampSetupComplete*), the job status (*status*), and the identifier of the mould that was equipped before the current job (*preMouldId*). The moulds are equipped with an RFID tag, whereas the machines are fitted with an RFID antenna. This antenna covers the area in front of every machine, where the unloading station for the moulds is located. As soon as a mould arrives at the unloading station, the RFID antenna scans the moulds tag automatically and forwards the timestamp of the moulds arrival and the corresponding injection moulding machine to a MES. After mounting the mould, the worker scans the mounted mould with a RFID handheld to receipt the setup. Both the measured and manually entered values represent single data points. For providing decision support regarding an optimal schedule, multiple steps have to be conducted. First, all relevant machines concerning the technical and scheduling capabilities have to be checked as not every machine has to be considered in the calculation. Second, a structural model builds the setup matrix for every combination on a single machine. After that, a behaviour model calculates the average setup duration by using the recorded data-points. Finally, by an ascend sorting of the single average setup duration, a recommendation for an optimal setup with relation to the minimal expected setup times has been pointed out.

3.4.2 Implementation of a blueprint via Node-RED

One core challenge to providing data to a service module lies in gathering the right data, which generally are not stored in one single database or system, but a distributed manner. To solve this issue, the component manager of the Asset Administration Shell comprises the endpoints for the data-points. The endpoints and the connection of the data-points to a resulting data-trace are mandatory to establish a flow of data and decision support. A graphical, web-based tool that supports users to build the component manager and to establish such a flow of data is called Node-RED, an open source community project. Node-RED encompasses a comprehensive collection of standard nodes that represent a feature (e. g. input, switches, functions, parser and output). For specific use cases, Node-RED is easily extendable by special nodes, e. g. OPC UA server and clients or SQL databases interfaces. For the Asset Administration Shell, Node-RED is can act as the component manager because the specification of endpoints (e. g. IP address) for some nodes is essential. Furthermore, function nodes are used where the user can insert code that operates with the received input data and forward the results to an output node. Figure 3.6 shows the general data flow in Node-Red in the field of Digital Shadows and the Asset Asset Administration Shell. Whereas the *Ontology*, the *AAS_Submodel*, and *Model* contain the domain specific knowledge, *Query* and *Datatrace* belong to the Digital Shadow.

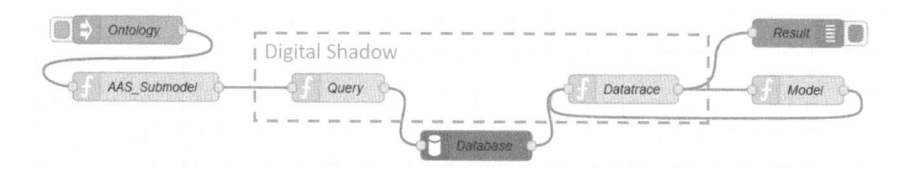

Figure 3.6: General Data Flow showing a Digital Shadow in Node-Red

A domain ontology acts as a foundation by providing the relevant classes, their related attributes and their relations. An Asset Administration Shell represents one or multiple classes of this ontology and commonly are divided into several logical sub-models. These two nodes represent the actual system and are equal to the Digital Twin. A Digital Shadow is introduced using actual settings and values for a specific use case. Because the single data-points of an Asset Administration Shell's sub-model are often stored in different databases, the Digital Shadow executes queries to gather the desired values. The component manager of the Asset Administration Shells provides the endpoint of each data-point so the Digital Shadow can extract the values and build a data-trace. If data processing is necessary, the Digital Shadow can use suitable models, e. g. for pre-processing and calculation of data, so a new data trace is built. After the pre-processing and the calculation steps, the Digital Shadow provides the result to tackle the specific purpose. With this result, a decision maker is able to fulfil the task with the support of data.

Transferring to a real use case of production scheduling in an injection moulding plant, the setup in Node-Red is as follows (Figure 3.7):

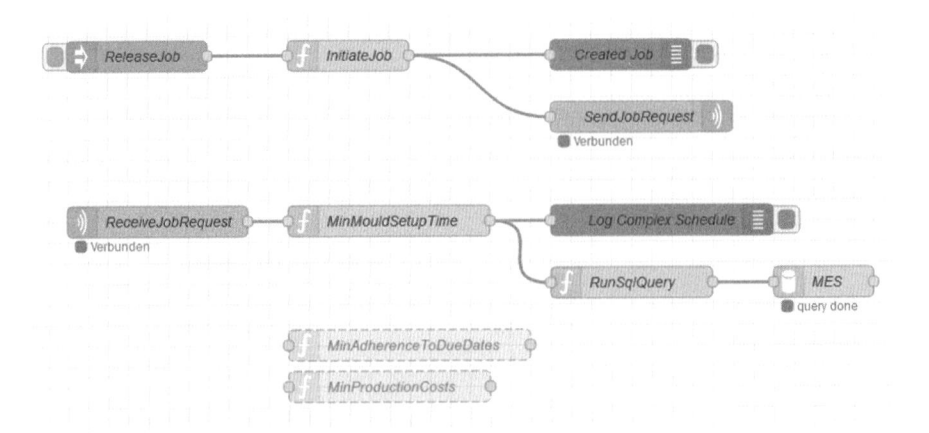

Figure 3.7: Enabling reduction of mould changeover costs using use case specific models for injection moulding in Node-Red

The top flow instantiates a production job and transmits the job request via MQTT (SendJobRequest). By listening to the related MQTT topic, the ReceiveJobRequest node on the bottom flow takes the input and forwards it to the node called MinMouldSetupTime. Besides the jobID, the transferred inputs are the job's planned processing time and the corresponding mould. Based on historical data, the MinMouldSetupTime node comprises the setup matrix, which includes the average time for the setup process of one mould to another. The setup matrix is a condensed data-trace, as it is updated frequently with the latest accrued setup time. The received job request is compared with the actual schedule and the setup matrix. By studying the actual schedule, the current moulds equipped on the machines were pointed out. Considering the requested mould in the open production job, the MinMouldSetupTime node presents the machine, where the estimated setup time is minimal. If no historical data is available, because a from-to combination does not exist, the algorithm inside the MinMouldSetupTime node chooses the next free injection moulding machine. By executing the RunSqlQuery function, the result is forwarded to the MES database. Due to the high complexity caused by many constraints in production (e. g. planning restrictions, due dates, priorities of customers), this example only deals with gaining the optimal mould setup time and does not take other constraints into account.

Beneficial for users is the implementation of a blueprint that is applicable to many tasks. With the modular usage of nodes, users can cover different use cases through switching nodes. For example, suppose the desired optimisation criterion minimises production costs or adherence to delivery dates. In that case, the nodes MinAdherenceToDueDates or MinProductionCosts could be connected to the flow instead of the node responsible for minimising the mould setup times. Hence, changing the nodes lead to another content of the Digital Shadow.

3.5 Conclusions and outlook

Enabling Shopfloor optimisation through Digital Shadow needs data, information, and suitable models that provide decision support for typical tasks in the production environment. Therefore, all relevant assets are mapped to the digital world. In this world, the holistic representation of an asset is conducted by Digital Twins. One implementation of the Digital Twin is the Asset Administration Shell that comprises the properties of an asset in the manifest as well as the endpoints for the data access in the component manager. To solve a specific problem, usually a fraction of accessible data is needed. For that reason, Digital Shadows were introduced. Digital Shadows exists only for a specific purpose. They gather the relevant data from different sources, either directly from the Digital Twin or by using the endpoints provided by Asset Administration Shells and use models to process the data to information and knowledge. Setting up the flow of data and information to models and decision-makers is important and could be a challenging task. An flexible testing environment introduced in this work is Node-RED, a web-based tool that enables the data flow by connecting different nodes on a graphical user interface. Further work will address service oriented composition of node networks that enables a fast and modular setup of the desired data flow.

Acknowledgements

Funded by the Deutsche Forschungsgemeinschaft (DFG, German Research Foundation) under Germany's Excellence Strategy – EXC-2023 Internet of Production – 390621612.

References

[BBD+21] Becker, F.; Bibow, P.; Dalibor, M.; Gannouni, A.; Hahn, V.; Hopmann, C.; Jarke, M.; Koren, I.; Kröger, M.; Lipp, J.; Maibaum, J.; Michael, J.; Rumpe, B.; Sapel, P.; Schäfer, N.; Schmitz, G. J.; Schuh, G.; Wortmann, A.: A Conceptual Model for Digital Shadows in Industry and Its Application. In: Ghose, A.; Horkoff, J.; Silva Souza, V. E.; Parsons, J.; Evermann, J. (Editor): *Conceptual Modeling*. Cham: Springer International Publishing and Imprint Springer, 2021

[DIN17] N.N. *Asset Management - Übersicht, Leitlinien und Begriffe*. Standard, Berlin, 2017

[Hof19] Hoffmann, M.: *Smart Agents for the Industry 4. 0: Enabling Machine Learning in Industrial Production*. Wiesbaden: Springer Fachmedien Wiesbaden GmbH, 2019

[HS21] Hopmann, C.; Schmitz, M.: *Plastics Industry 4.0: Potentials and applications in plastics technology*. Munich and Cincinnati, OH: Carl Hanser Verlag and Hanser Publications, 2021

[Hur20] Hurson, A. R.: *The Digital Twin Paradigm for Smarter Systems and Environments: The Industry Use Cases*Issn Ser. San Diego: Elsevier Science & Technology, 2020

[HVHB17] ten Hompel, M.; Vogel-Heuser, B.; Bauernhansl, T. (Editor): *Handbuch Industrie 4.0 Bd.4: Allgemeine Grundlagen*. SpringerLink Bücher. Berlin, Heidelberg: Springer Vieweg, 2nd edition, 2017

[KD21] Kletti, J.; Deisenroth, R.: *Lehrbuch für digitales Fertigungsmanagement*. Berlin, Heidelberg: Springer Berlin Heidelberg, 2021

[KKT+18] Kritzinger, W.; Karner, M.; Traar, G.; Henjes, J.; Sihn, W.: Digital Twin in manufacturing: A categorical literature review and classification. *IFAC-PapersOnLine* 51 (2018) 11, p. 1016–1022

[Lu17] Lu, Y.: Industry 4.0: A survey on technologies, applications and open research issues. *Journal of Industrial Information Integration* 6 (2017), p. 1–10

[NN13] N.N.:. Umsetzungsempfehlungen für das Zukunftsprojekt Industrie 4.0: Abschlussbericht des Arbeitskreises Industrie 4.0, 2013

[NN18] N.N.: Internet of Production: Cluster of Excellence Funding Line (2018),

[SHH+20] Schuh, G.; Häfner, C.; Hopmann, C.; Rumpe, B.; Brockmann, M.; Wortmann, A.; Maibaum, J.; Dalibor, M.; Bibow, P.; Sapel, P.; Kröger, M.: Effizientere Produktion mit Digitalen Schatten. *Zeitschrift für wirtschaftlichen Fabrikbetrieb* 115 (2020) s1, p. 105–107

[TA17] Tantik, E.; Anderl, R.: Integrated Data Model and Structure for the Asset Administration Shell in Industrie 4.0. *Procedia CIRP* 60 (2017), p. 86–91

[URL18] Boss, B.; Malakuti, S.; Lin, S.; Usländer, T.; Clauer, E.; Hoffmeister, M.; Stojanovic, L.: *Digital Twin and Asset Administration Shell Concepts and Application in the Industrial Internet and Industrie 4.0: An Industrial Internet Consortium and Plattform Industrie 4.0 Joint Whitepaper*. Plattform Industrie 4.0 and Industrial Internet Consortium, 2018

[WGE+17] Wagner, C.; Grothoff, J.; Epple, U.; Drath, R.; Malakuti, S.; Gruner, S.; Hoffmeister, M.; Zimermann, P.: The role of the Industry 4.0 asset administration shell and the digital twin during the life cycle of a plant. *2017 22nd IEEE International Conference on Emerging Technologies and Factory Automation*. Piscataway, NJ, 2017

Recyclable barrier systems for sustainable products

Moderator: Dr.-Ing. Montgomery Jaritz, Ionkraft GmbH

Content
---

1 Plasma technology as enabler for more recycling in the plastic packaging industry 47

M. Jaritz[1]
[1] Ionkraft GmbH

2 In-Plasma Air2Air: Sustainable PECVD barrier coatings on plastic films 55

R. Dahlmann[1], P. Alizadeh[1]
[1] Institute for Plastics Processing (IKV) at RWTH Aachen University

3 Recyclates for food packaging - Preventing migration with PECVD coatings 67

L. Kleines[1], R. Dahlmann[1]
[1] Institute for Plastics Processing (IKV) at RWTH Aachen University

Dr.-Ing. Montgomery Jaritz

Dr.-Ing. Montgomery Jaritz studied industrial engineering at the RWTH Aachen University. He started working at the Institute for Plastics Processing in 2010 as a student research assistant. Dr. Jaritz gathered international work experience in Chuzhou, China, where he completed an internship at Bosch and Siemens household appliances GmbH. After completing his doctorate in plasma technology at IKV, he co-founded IonKraft in 2021. The company specialises in the development of plasma processes and reactors. The vision is to produce fully recyclable packaging for demanding products from the chemical industry.

1 Plasma technology as enabler for more recycling in the plastic packaging industry

M. Jaritz[1]

[1] Ionkraft GmbH

Abstract

Plastic packaging must harmonise a high level of product protection with complex supply chains. Packaging solutions made of multi-material composites are therefore particularly widespread. The individual layers oftentimes consist of different plastics.Cheaper structural plastics such as PP or PE are combined with more expensive barrier materials such as EVOH or PA. After use, these individual layers are difficult to separate and thus often unusable for material recycling. The application of plasma-polymer coatings is one way of functionalising plastic packaging in a variety of ways without impairing their recyclability. In the scenario of packing chemicals even higher demands must be applied. In addition to the requirements for recyclability, a barrier effect against oxygen and volatile substances as well as chemical resistance to the filling material must be guaranteed. The spin-off IonKraft from the IKV is currently transferring insights from fundamental research into practical application to challenge this demanding task.

1.1 Introduction

Plastics are produced in a conventional polymerisation process by gradually joining monomers together to form long chains, the polymers. Plasma polymerisation, on the other hand, is a non-specific polymerisation of fragments formed in a plasma, resulting in a polymer structure consisting of partially preserved and new functional groups from a gaseous monomer. Molecular fragments formed in the plasma are adsorbed on the substrate surface and form a coating. With this so-called plasma-enhanced chemical vapour deposition (PECVD), thin functional layers can be produced on surfaces. For temperature-sensitive substrates such as plastics, only low-temperature plasmas can be used. One way to technically generate a low-temperature plasma is to ignite the plasma under low-pressure conditions. The excitation of gases into a plasma by means of microwave radiation has nowadays become particularly important for the purposes of industrial application, as high layer deposition rates can be achieved in this way.

Organosilicon compounds, such as hexamethyldisiloxane (HMDSO), are used as basic building blocks (monomers) for the layer deposition of transparent plasma-polymerised SiOx coatings as gas barriers, as scratch- and wear-resistant coatings and organosilicon coatings (SiOCH) as adhesion-promoting coatings.

Despite many advantages, plasma technology for barrier finishing of plastic containers is nowadays used almost exclusively in the beverage industry. One of the reasons for this is that beverage bottles are containers with a small volume and simple geometry. This makes the process design simple. An efficient PECVD barrier coating of large-area substrates such as packaging films, canisters or barrels, on the other hand, is significantly more difficult and is therefore not yet widely used.

1.2 Barrier solutions for demanding filling goods

Chemicals, many of which are hazardous substances, have high demands on their packaging, some of which are laid down in EU law in the "Accord européen relatif au transport international des marchandises Dangereuses par Route" (ADR, European Agreement concerning the International Carriage of Dangerous Goods by Road). In addition to the mechanical stability of the containers, these include a high barrier performance against gases and aromatic substances, but above all oxygen and water vapour, as well as a high resistance to acidic and alkaline media.

Since metal is too heavy, glass too fragile and both are also expensive, often only containers made of plastic come into question for this application. However, there is no inexpensive plastic that can meet all the above requirements at the same time. Therefore, containers made by co-extrusion blow moulding are used, whose walls consist of several layers of different plastics. In this case, a mixture of hard polyethylene (HD-PE), polyamide (PA) and ethylene-vinyl alcohol co-polymer (EVOH) is used. HD-PE is very resistant to non-oxidising acids, alkalies, oils and greases, but has a high permeability to certain gases and solvents, which is why EVOH and PA are used as a barrier layer. This packaging is not environmentally friendly because only monomaterial solutions can be material recycled. In the EU, there are already collection systems for empty containers for products of, for example, agrochemicals after use by the farmer, which achieve high collection rates. However, the multi-material containers described above can only end up in landfills or incineration. Furthermore, PA and EVOH cost more than three times as much per tonne as HD-PE. Thus, this multi-material packaging is significantly more expensive than packaging made only of HD-PE.

Besides the use of non-recyclable plastic multi-material solutions, the only other option for barrier finishing of plastics is fluorination or coating using plasma technology. In fluorination, the containers are placed in a vacuum chamber and exposed to a fluorine gas mixture under exclusion of atmosphere. Due to its high reactivity, fluorine partially replaces hydrogen atoms on the material surface. However, this only creates an adsorption barrier, which can only prevent the migration of media such as solvents. Fluorine is also toxic and highly corrosive, which makes this technology unattractive and expensive. Furthermore, there are more and more studies that attribute negative environmental and health impacts to fluorinated plastics.

Plasma polymer coatings, on the other hand, are not only inexpensive, they can also be produced using harmless chemicals and require only a small amount of energy and materials. Using PECVD, fluorine-free silicon-based barrier coatings can be produced, which can act as a passive barrier to prevent the migration of any substances. Thus, the coatings are simultaneously suitable, for example, as a solvent barrier, a barrier against the migration of contaminants from a plastic recyclate into a filling material or as a gas barrier, for example, against oxygen, hydrogen or carbon dioxide, which opens up a wide range of possible applications. However, the silicon-based coatings available on the market proved to be unsuitable for the application of storing sensitive and aggressive filling goods due to some technical challenges that have not been solved so far.

The packaging of certain chemicals requires a particularly high barrier performance against the migration of the filling material from the plastic packaging. According to ADR, for example, a maximum permissible permeation of the substance through the packaging of 0.008 g/(l·h) at 23 °C applies to flammable substances (flash point < 61 °C). This means that particularly high-performance barrier coating systems are required. In addition, silicon oxide barrier coatings known to date are susceptible to hydrolysis and thus not resistant to products with an elevated pH value, which limits their possible applications in the field of packaging.

A homogeneous and pore-free layer deposition on the entire surface is a necessary condition to achieve a high barrier performance. This can only be achieved by a homogeneous energy density of the plasma in the entire container. Microwaves are absorbed by plasma, whereby the intensity decreases quadratically with the distance to the energy source. This can result in large differences in the densities of reactive particles and their energies in front of the substrate to be coated if the design is poor. This is less critical in small containers. The larger and geometrically more complex a container becomes, for example in the form of a canister, the more difficult it is to implement a homogeneous and closed layer. Furthermore, due to the process, plasma processes cannot simply be transferred from one substrate to the next. If the shape of the substrate or the material properties change, the process parameters have to be adapted. In industry, this process development has so far been carried out purely empirically, which is often not effective and at the same time time time time-consuming and expensive.

1.3 Development approach from fundamental research

The great potential of plasma technology for the packaging industry and many other applications was also recognised by the German Research Foundation (DFG) in 2010, so that it approved an application from RWTH Aachen University, Ruhr University Bochum and the University of Paderborn to set up and fund a collaborative research centre (SFB) entitled "Pulsed high-performance plasmas for the synthesis of nanostructured functional layers". Collaborative Research Centres are long-term research institutions at universities, designed to last up to twelve years, in which scientists work together within the framework of an interdisciplinary research programme. They make it possible to work on particularly innovative, demanding, elaborate and long-term research projects by coordinating the resources at the participating universities.

Since the beginning, the Institute for Plastics Processing at RWTH Aachen University (IKV) has been conducting research within this SFB alongside fourteen other institutes. The focus of the IKV's research work is on solving the aforementioned, until recently unsolved, challenges with regard to the scalability of plasma processes, barrier performance and the transfer of the previously predominant empirical layer development to a plasma analytics and simulation-based development concept.In many years of research, among others, in this research community, we have been able to build up extensive process knowledge that has enabled us to overcome the technical challenges cited.

1.4 Process parameters and reactor setup

Microwaves are the only type of excitation that offer an industrially relevant layer deposition rate. However, a homogeneous gas and energy distribution on the surface to be coated with regard to large containers with a volume of 5 l, 20 l or even 100 l is a challenge for which no solution for the deposition of silicon-containing layers has been available so far.

In addition to the design of the gas distribution, the energy coupling also plays a decisive role with regard to the plasma homogeneity. At the IKV, a method was researched to make the plasma homogeneity measurable and thus adjustable by means of optical emission spectroscopy (OES) [Fra12]. This uses the fact that plasmas emit light whose intensity correlates directly with the energy density of the plasma. Since the wavelength of the emitted light is also characteristic of the atom from which it originates, the particle density distribution of different components of the plasma can also be evaluated individually using OES, which allows spatially resolved conclusions to be drawn about the mixing ratio of the process gases.

This methodology enables us to design the reactor and the process in such a way that it is also possible to deposit high-quality barrier layers in large-volume containers with complex geometries.

The desired functionality can only be achieved through good coordination of the influencing and control variables in a plasma process. As soon as the geometry or chemical composition of a substrate changes, adjustments to the process parameters are usually necessary. Furthermore, process parameters can at most be transferred between completely identical plasma reactor systems.

1.5 Coating development based on intrinsic plasma parameters

The approach to process and layer development prevalent among all players in the plasma technology market is purely empirical. Together with the RUB Bochum, an alternative approach was developed at the IKV based on the setting of intrinsic plasma properties. As a criterion for the transfer of plasma processes, for example, the so-called ion fluence was researched as a characteristic value for the plasma intensity [JHB+17, Beh16]. The ion fluence is defined as the number of ions that impinge on the surface to be treated during pretreatment. This can be calculated from the electron density, which in turn can be measured using so-called Langmuir probes. Our investigations have shown that intrinsic plasma parameters are suitable for process transfer, so that we can transfer our proven coating systems to new systems and substrates by means of plasma-agnostic processes.

In plasma-polymer barrier layers, residual stresses form during the layer growth process with increasing layer thickness, which lead to cracks in the layer above a certain thickness [Bah17, JHB+17]. It is therefore not possible to increase the barrier performance by increasing the layer thickness. Furthermore, depending on the deposition conditions, open nanopores are formed in the layers, which can negatively influence the barrier [JHW+20]. Through extensive studies of the layer growth mechanisms [BBB+13, BBM+14, GMB+18], the elongation capacity [KDH+14], the permeation mechanisms [MBFB10, WWJ+17, KWJ+17, KJM+17, Beh16, DHJK16], the deposition with different monomers [MSH+18] for layer deposition as well as different multilayer system architectures [BBM+14], we succeeded in developing coating systems with chemically stable high-barrier properties that meet the demands of the chemical industry.

With the spin-off IonKraft from the IKV, this technology is now being transferred from science to application. IonKraft was founded in April 2021, and since then the team has further developed the coatings and built a reactor in the IKV's technical centre that can equip hollow plastic packaging with a volume of up to 20 L with a chemically resistant barrier function. Figure 1.1 shows an exemplary comparison of the O2 barrier effect of the IonKraft barrier systems with known competitor solutions. PE-HD bottles for use in the field of agricultural chemistry were coated here. It can be seen that the coatings can improve the barrier of the bottles by a factor of approx. 1760 and can thus offer a higher performance than the established competitor solutions.

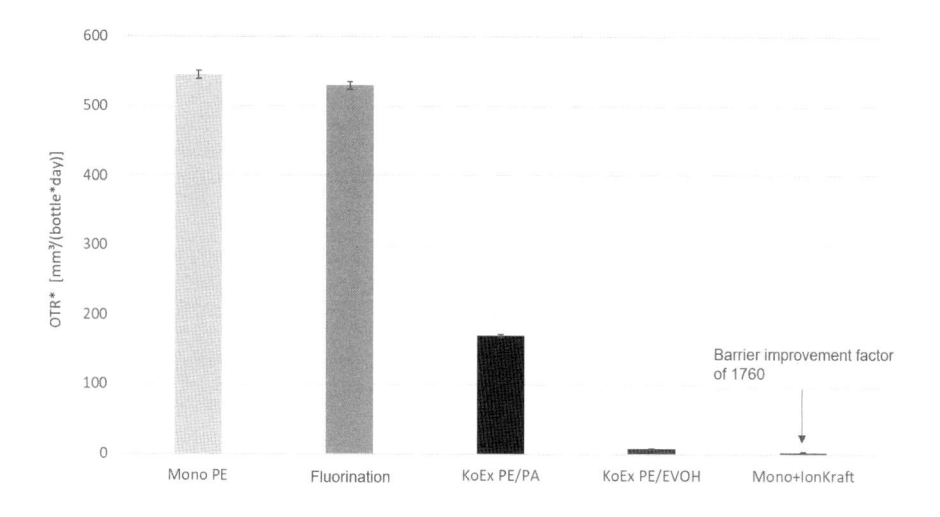

Figure 1.1: Barrier performance of the IonKraft coating against O2 and benchmarking against competitor solutions for 1 L PE-HD bottles

All IonKraft coatings are deposited in a single process and within a few seconds. The total coating thickness always remains in the range of a maximum of 200 nm, so that with a container wall thickness of 2 mm, the wafer-thin layers make up just 0.001 % of the total volume. This means that these layers are not noticeable in the recycling process.

References

[Bah17] BAHROUN, K.: *Prozessentwicklung zur Abscheidung dehnbarer Barriereschichten auf PET mittels Plasmapolymerisation*. RWTH Aachen University, Dissertation, 2017

[BBB+13] BAHRE, H.; BAHROUN, K.; BEHM, H.; STEVES, S.; AWAKOWICZ, P.; BÖKE, M.; HOPMANN, C.; WINTER, J.: Surface pre-treatment for barrier coatings on polyethylene terephthalate. *Journal of Physics D: Applied Physics* 46 (2013) 8, p. 084012

[BBM+14] BAHROUN, K.; BEHM, H.; MITSCHKER, F.; AWAKOWICZ, P.; DAHLMANN, R.; HOPMANN, C.: Influence of layer type and order on barrier properties of multilayer PECVD barrier coatings. *Journal of Physics D: Applied Physics* 47 (2014) 1, p. 015201

[Beh16] BEHM, H. W.: *Untersuchung von Plasmaprozessen und deren Einfluss auf die Verbundeigenschaften von mittels Plasmapolymerisation beschichtetem Polypropylen*. RWTH Aachen University, Dissertation, 2016

[DHJK16] DAHLMANN, R.; HOPMANN, C.; JARITZ, M.; KIRCHHEIM, D.: Wirkmechanismen bei der Barriereausrüstung von Kunststoffen mit Hilfe von PECVD. *Vakuum in Forschung und Praxis* (2016) 28,

[Fra12] FRAGSTEIN UND NIEMSDORFF, F.: *Wechselwirkungen zwischen Plasmaprozessen und thermoplastischen Oberflächen: Interactions between plasma processes and thermoplastic surfaces*: Zugl.: Aachen, Techn. Hochsch., Diss., 2011. Aachen: Mainz1. aufl. edition, 2012

[GMB+18] GEBHARD, M.; MAI, L.; BANKO, L.; MITSCHKER, F.; HOPPE, C.; JARITZ, M.; KIRCHHEIM, D.; ZEKORN, C.; DE LOS ARCOS, T.; GROCHLA, D.; DAHLMANN, R.; GRUNDMEIER, G.; AWAKOWICZ, P.; LUDWIG, A.; DEVI, A.: PEALD of SiO2 and Al2O3 Thin Films on Polypropylene: Investigations of the Film Growth at the Interface, Stress, and Gas Barrier Properties of Dyads. *ACS applied materials & interfaces* 10 (2018) 8, p. 7422–7434

[JHB+17] JARITZ, M.; HOPMANN, C.; BEHM, H.; KIRCHHEIM, D.; WILSKI, S.; GROCHLA, D.; BANKO, L.; LUDWIG, A.; BÖKE, M.; WINTER, J.; BAHRE, H.; DAHLMANN, R.: Influence of residual stress on the adhesion and surface morphology of PECVD-coated polypropylene. *Journal of Physics D: Applied Physics* 50 (2017) 44, p. 445301

[JHW+20] JARITZ, M.; HOPMANN, C.; WILSKI, S. S.; KLEINES, L. I.; BANKO, L.; GROCHLA, D.; LUDWIG, A.; DAHLMANN, R.: Comparative study of the residual stress development in HMDSN-based organosilicon and silicon oxide coatings. *Journal of Physics D: Applied Physics* (2020),

[KDH+14] KIRCHHEIM, D.; DAHLMANN, R.; HOPMANN, C.; BAHROUN, K.; BEHM, H.: Elongation properties of multilayer PECVD barrier coatings on PET films. *Journal of Plastics Technology* (2014) 10,

[KJM+17] KIRCHHEIM, D.; JARITZ, M.; MITSCHKER, F.; GEBHARD, M.; BROCHHAGEN, M.; HOPMANN, C.; BÖKE, M.; DEVI, A.; AWAKOWICZ, P.; DAHLMANN, R.: Transport mechanisms through PE-CVD coatings: Influence of temperature, coating properties and defects on permeation of water vapour. *Journal of Physics D: Applied Physics* 50 (2017) 8, p. 085203

[KWJ+17] KIRCHHEIM, D.; WILSKI, S.; JARITZ, M.; MITSCHKER, F.; GEBHARD, M.; BROCHHAGEN, M.; BÖKE, M.; BENEDIKT, J.; AWAKOWICZ, P.; DEVI, A.; HOPMANN, C.; DAHLMANN, R.: Temperature-dependent transport mechanisms through PE-CVD coatings: Comparison of oxygen and water vapour. *Journal of Physics D: Applied Physics* 50 (2017) 39, p. 395302

[MBFB10] MICHAELI, W.; BAHROUN, K.; VON FRAGSTEIN, F.; BEHM, H.: Plasma-Assisted Barrier Coating. *Bioplastics Magazine* 5 (2010) 4, p. 24–25

[MSH+18] MITSCHKER, F.; SCHÜCKE, L.; HOPPE, C.; JARITZ, M.; DAHLMANN, R.; DE LOS ARCOS, T.; HOPMANN, C.; GRUNDMEIER, G.; AWAKOWICZ, P.: Comparative study on the deposition of silicon oxide permeation barrier coatings for polymers using hexamethyldisilazane (HMDSN) and hexamethyldisiloxane (HMDSO). *Journal of Physics D: Applied Physics* 51 (2018) 23, p. 235201

[WWJ+17] WILSKI, S.; WIPPERFÜRTH, J.; JARITZ, M.; KIRCHHEIM, D.; MITSCHKER, F.; AWAKOWICZ, P.; DAHLMANN, R.; HOPMANN, C.: Mechanisms of oxygen permeation through plastic films and barrier coatings. *Journal of Physics D: Applied Physics* 50 (2017) 42, p. 425301

Abbreviations

Notation	Description
ADR	European Agreement concerning the International Carriage of Dangerous Goods by Road
DFG	German Research Foundation
EVOH	Ethylene-vinyl alcohol co-polymer
HMDSO	Hexamethyldisiloxane
IKV	Institute for Plastics Processing at RWTH Aachen University
MFC	Mass flow controller
OES	Optical emission spectroscopy
PE-HD	High density polyethylene
PECVD	Plasma enhanced chemical vapour deposition
PET	Polyethylene terephthalate
PA	Polyamide
PTFE	Polytetrafluorethylen
SFB	Collaborative research centre
SiOCH	Organo silicon
SiO_x	Silicon oxide

2 In-Plasma Air2Air: Sustainable PECVD barrier coatings on plastic films

R. Dahlmann[1], P. Alizadeh[1]
[1] Institute for Plastics Processing (IKV) at RWTH Aachen University

Abstract

Commercially used plastics, such as polyethylene terephthalate (PET), do not have a sufficient barrier function against gases and flavourings for certain applications, so they have to be equipped with an additional barrier. In film production, multi-layer films made of different plastics, each of which fulfils a specific function (e.g. barrier against water, barrier against gases, etc.), are therefore particularly widespread. Since these composites can not be separated from each other economically, the post-consumer waste is not suitable for mechanical recycling.

One way of equipping plastics with a barrier while maintaining their recyclability is plasma enhanced chemical vapour deposition (PECVD). This is a vacuum-based coating process in which a few tens of nanometres thick optically transparent coatings with high barrier efficiency (often SiO_x layers) are polymerised on a substrate with the help of a plasma. However, the technical implementation of this technology in the field of flexibles has not yet succeeded in an economically attractive solution. The aim of this study is therefore to investigate technical concepts that can be applied in order to improve previous attempts to making the technology accessible to the packaging market.

Approaches to the efficient process regulation from PECVD processes implemented on a large industrial scale are described and transferred to the present plant concept.Analytical as well as experimental approaches are described that give guidance in designing the complex system layout.

2.1 Introduction

Flexible packaging must balance high product protection with complex supply chains. Requirements exist with regard to their mechanical integrity and flexibility, as well as with regard to their barrier effect against gases and economical production. In the area of food packaging for example, plastics ensure that food lasts longer and thus contribute to less food waste. Packaging solutions made of multi-material composites are therefore particularly widespread. The individual layers can consist of different plastics, metallised layers or paper. After use, these individual layers are difficult to separate and thus often not compatible with material recycling. Hardly any other material has such a high potential for mechanical recycling as thermoplastics and the circular economy in Germany already offers a suitable platform for making such plastics available for the production of new parts. However, the prerequisite for this is that plastics can be separated by the type of material, which makes mono-material solutions necessary.

Plasma enhanced chemical vapour deposition (PECVD) is one way of functionalising plastic films in a variety of ways without impairing their recyclability [YM05]. In technical low-pressure plasmas, the process temperature rises only slightly above ambient temperature due to the increased mean free path length through the vacuum, largely independent of the mixing ratio of the process gases [Bin08]. In this way, thermally sensitive materials can also be coated. In the case of film coating, however, the closed vacuum systems required for this operate as a batch process on a roll-to-roll basis, which significantly reduces the attractiveness and acceptance of this technology. Due to the high process costs compared to technologies like co-extrusion, the technology has not yet become established in the packaging market for plastics films.

Several inventors from research and industry have elaborated concepts to make the vacuum coating technology more attractive economically by attempting an inline process layout [HKY90, SWYK87, IUN+85]. The required working pressure in the reactor chamber is achieved in all concepts by pre-chambers and gate systems that lead to sequential pressure reduction. However, the systems described in these patents have apparently never been able to establish themselves for the coating of plastic films. Besides the more difficult control of the process due to leakages, particle formation and displacement are likely to be causes for this.

The project partner Bühler AG (former Leybold Heraeus GmbH) was also already involved in the continuous coating of thermoplastic web material in the 1970s and can therefore contribute first-hand experience to the project. The reactor described in the patent DE2747061A1 of Leybold Hereaus GmbH from 1979 (Figure 2.1) was built and put into operation at that time.

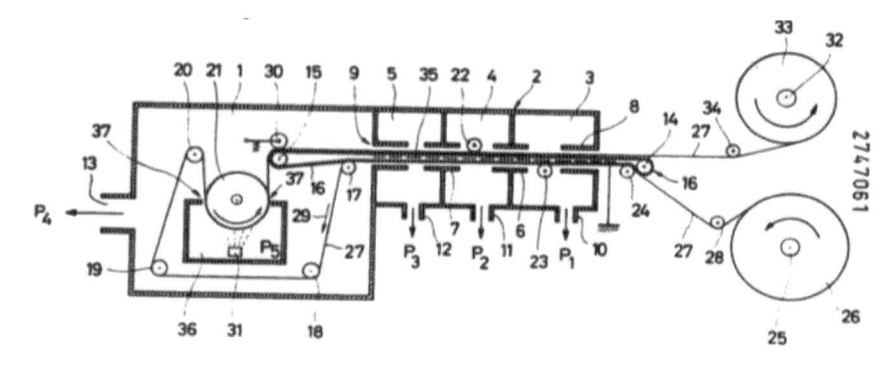

Figure 2.1: Air2Air concept described in the patent DE2747061A1 [Wal79]

In this concept, the film is fed in and out of the reactor through the same pre-chambers. this results in advantages with regard to the vacuum periphery. The film is electrostatically applied to a guide belt that runs along during operation. At each airlock there are gaps between the film and the airlock wall. Films were coated with AlOx. In order to further enhance the economic efficiency compared to earlier system concepts the transfer project "In-Plasma-Air2Air" from SFB TR 87 was launched. Its aim is to develop an experimental reactor that significantly reduces the downtime due to cleaning and evacuation of the process chamber by means described in the following section.

2.2 Experimental

In order to evaluate the theoretical assumptions a simplified two-chamber vacuum measurement system has been set up (Figure 2.2). All parts are standardised vacuum equipment with a nominal diameter of 160 mm. The pumping station used provides a nominal suction capacity of 65 m^3/h.

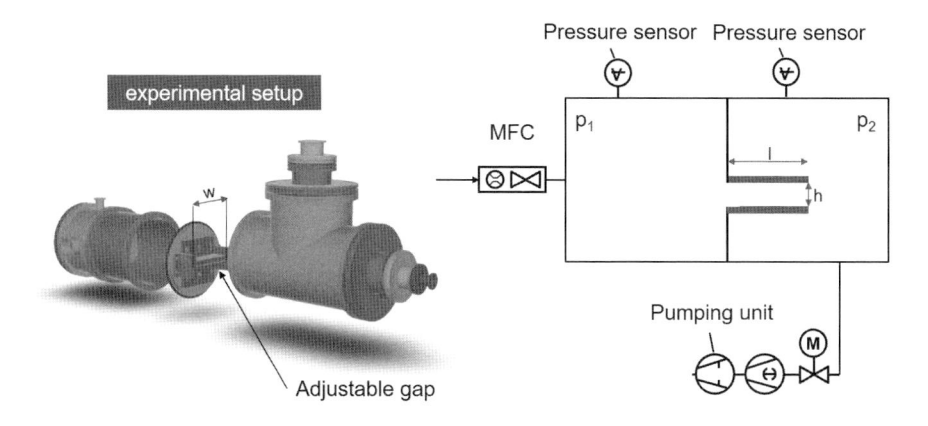

Figure 2.2: Experimental two-chamber vacuum system

The two chambers are connected via a flange with an adjustable cross-sectional opening (width: $w = 90$ mm, length: $l = 45$ mm) and an adjustable height gap (0 mm $\leq h \leq$ 20 mm). Each chamber is equipped with a pressure sensor. A mass flow controller (MFC) is connected to the outer chamber. The pumping unit is connected to the inner chamber. The pressure state therefore can be described precisely, since the gas load must be constant as a volume flow at every point in the system (law of continuity) [O'H03]. The gas exchange between the chambers is described exclusively by the height, width and length of the gap. The height of the gap can be adjusted on the flange. The direct influence of the gap height was determined by repeatedly measuring the pressures p_1 and p_2 in the chambers and the volumetric flow rate $\dot{Q}$ between them.

Resulting pressure values have been measured for gap heights of 0.1 mm, 0.25 mm, and 0.5 mm. Additionally, flow rates of 0 sccm, 2,500 sccm, 5,000 sccm and 6,000 sccm have been considered (sccm - standard cubic centimeters per minute). All measured pressure values were determined after 10 minutes and in triplicate.

2.3 The in-plasma concept

In order to be able to develop a plant concept that fulfils the 6 success criteria mentioned, a look was first taken at a successful example of a low-pressure plasma process implemented on a large industrial scale and suitable for mass production. One such example is the PECVD barrier finishing of PET bottles. The principle was developed at IKV in 1986 and a patent application was filed [Ple89]. The reactor design can be seen in Figure 2.3.

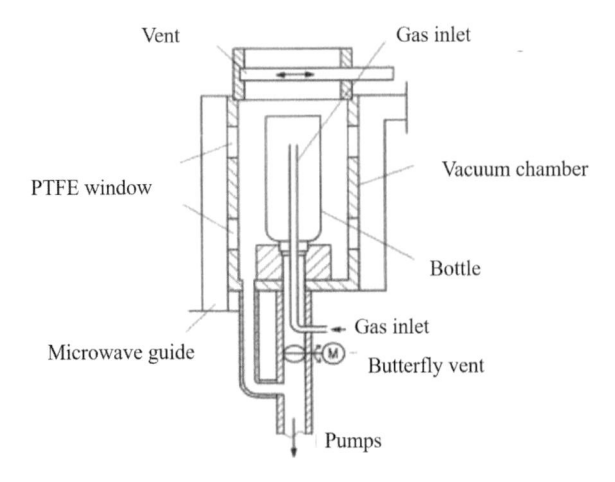

Figure 2.3: Reactor design for the inner coating of hollow bodies

The whole reactor chamber including the bottle's inside volume are evacuated simultaneously. Subsequently process gases are introduced into the bottle via a gas lance. When a stationary state between evacuation and process gas inlet has been achieved at a stable pressure, microwave power is introduced via the slotted wave guides. The plasma is excited exclusively on the inner surface of the bottle. This concept has two particularly noteworthy advantages, which have probably made a major contribution to its commercial success. Firstly, it is a so-called in-plasma process. The microwaves penetrate through the bottle wall into the bottle interior and ignite the plasma on the inner surface of the bottle. This distinguishes it from the usual arrangements with downstream or remote plasma excitation as Figure 2.4 illustrates. Microwaves are absorbed by plasma, whereby the intensity decreases quadratically with the distance to the energy source (indicated by the colour gradient in Figure 2.4). This results in differences in the densities of reactive particles and their kinetic energy on contact with the substrate.

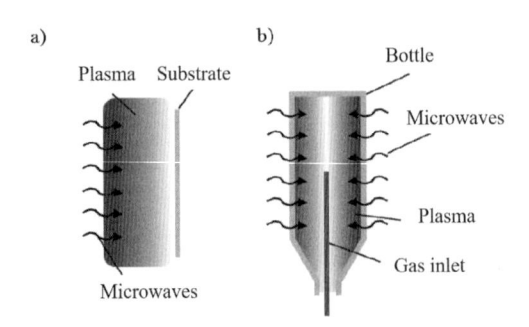

Figure 2.4: Remote (a) and In-Plasma (b) concepts

With the in-plasma concept, the plasma has the highest energy, the highest degree of ionisation and thus the maximum achievable layer deposition rate directly on the inside of the substrate to be coated, which makes this configuration extremely efficient. On industrial PET bottle coating systems up to 40.000 bottles can be coated per hour [URL21]. The second major advantage is that the bottle spatially limits the plasma reaction. Therefore, the coating of system parts (reactor walls, screens, etc.) is significantly reduced, so that there is little downtime due to cleaning processes. Only the gas inlet is also coated, but can be replaced quickly.

2.4 Developed Air2Air reactor concept

Derived from the requirements in section 2.1 and from the previously investigated patents, the reactor concept shown in Figure 2.5 has been elaborated.

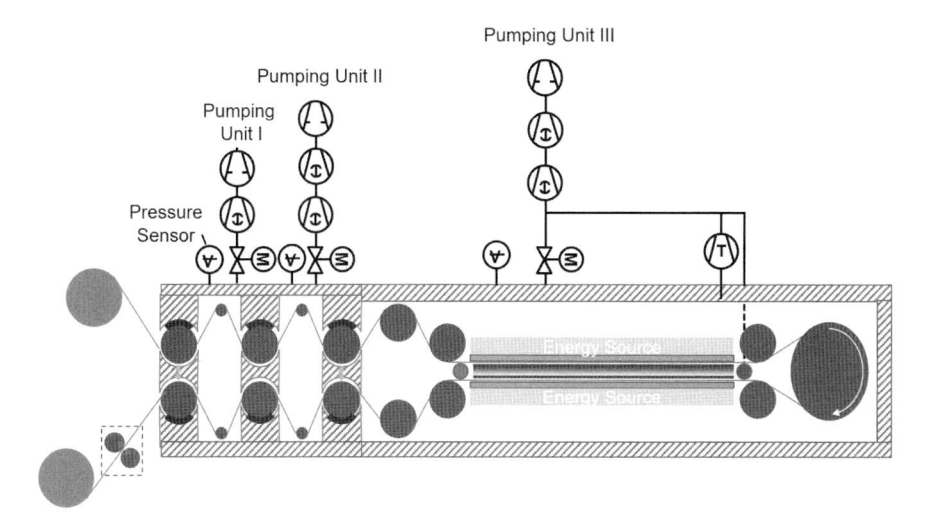

Figure 2.5: Multi-chamber PECVD plant concept [DJ21]

Using the in-plasma concept as an example and taking into account the requirements mentioned in section 2.1, the concept shown in Figure 2.5 was developed. In order to minimise the required acquisition investment, the film is fed in and out via the same pre-chambers, following the concept from patent DE2747061A1 [Wal79]. A narrow film guide is intended to counteract possible fluttering of the film in the process, since flow velocities in the range of the speed of sound must be expected [WAW88], especially at the inlet to pre-chamber 1. For this purpose, the contact surface of the film on the sluice rollers and the angle to the flow direction are maximised by means of deflection rollers. Furthermore, the film guidance in the reactor chamber is designed in such a way that a closed reaction chamber results: The plasma ignites exclusively in the area between the two films, so that neither the rollers nor the reactor chamber walls outside the plasma volume will be coated. Microwaves are guided into the chamber through a wave-guide and into the reaction chamber from outside through the film via slits or PTFE windows. This results in no coating removal on the energy sources (usually the most contamination is to be expected here) and maximises the coating deposition rate in the reactor (in-plasma concept). The fact that the film passes through the plasma zone twice increases the throughput.

By using sealing gas between the reaction chamber and the intermediate chamber, both the entry of foreign gas into the reaction zone and the diffusion of layer-forming radicals into the intermediate zone can be avoided. Here, oxygen should be used as a sealing gas, since it is already introduced into the reactor chamber as an additional reaction gas for the layer deposition. In this way, an influence on the layer formation can be minimised. To prevent particles from entering the reactor, a sealing gas barrier with filtered compressed air is provided at the outermost airlock of pre-chamber 1.

2.5 Analytical calculation of the pressure state in a multi-chamber system

In order to estimate the necessary suction capacity and pumping power, analytical calculations as well as numerical simulation have been conducted. Atmospheric pressure (1013.25 mbar) is to be reduced to fine vacuum (approx. 0.008 mbar). Vacuum generation is complicated by the fact that there are chamber leaks due to the transfer openings and the vacuum system cannot be completely sealed against the atmosphere. If a vacuum chamber has a defined leak, the leakage flow increases continuously with the pressure difference until the speed of sound is reached inside the leakage opening and the flow is blocked. In case of a flow blockage, the leakage flow remains constant at a given inlet pressure, even if the pressure difference is increased further. Flow blocking occurs at approximately half the inlet pressure [WAW88]. After flow blockage, additional pumping power immediately leads to further pressure reduction. Therefore, a minimum pumping power relative to the size of the transfer orifices is required to perform a power-efficient pressure reduction. The relationship between the pressure in a vacuum chamber and the applied suction power of the pumps S_2 can be approximated with the use of Equation 2.2 [KSW84].

$$p_2 = \frac{\dot{Q}_{1 \to 2}}{S_2} \qquad (2.1)$$

With gas flow rate $\dot{Q}$:

$$\dot{Q} = C \cdot (p_1 - p_2) \qquad (2.2)$$

In order to describe the flow conductance C it is necessary to define the pressure conditions distinctively in regards to the dominating type of flow and the occurrence of flow blockage. For the investigated experiments and the underlying application flow blockage is expected and favoured in the sense of maximum pressure reduction. As can be seen in Figure 2.6, the experiments are operated most probably in the area of laminar flow within the framework of the experiments carried out.

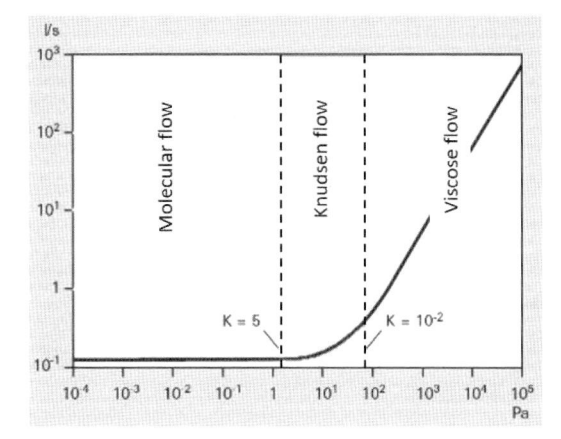

Figure 2.6: Type of flow in dependence of the pressure and conductance

The critical laminar flow conductance factor C^*, which determines the flow behavior in the event of blockage is defined by the condition of the gas and the apertures geometry (Equation 2.3). The describing variables are the narrowest cross-sectional area A_{min}, the respective pressures p_1 and p_2, the molar mass of the in-flowing gas M_{molar}, the general gas constant R, the Temperatur of the flowing gas T_0 as well as the ratio of the specific heat capacities of air $\kappa = C_p/C_v$.

$$C^* = A_{min} \cdot \frac{p_1}{p_1 - p_2} \cdot \frac{2}{\kappa + 1} \cdot \sqrt{\frac{2\kappa}{\kappa + 1} \cdot \frac{RT_0}{M_{molar}}} \tag{2.3}$$

An approximation for the conductance in the case of a molecular flow state is given with Equation 2.4. In accordance with Figure 2.6 the Conductance is not dependent on the pressure and represents a constant that is described solely by the geometry of the aperture [Laf98].

$$C_{molecular} = 11.6 \cdot A_{min} \tag{2.4}$$

In principle, the pressure could be reduced to the working pressure of the process chamber in a single stage in a pre-chamber with a constant leakage flow. In this case, however, doubling the suction power only halves the pressure. If, for example, the ratio between inlet pressure and pre-chamber pressure is $1/10$, doubling the pumping power causes a pressure difference of $1/20$. If, on the other hand, two pre-chambers are used, a pressure difference of $1/100$ can be achieved with the same pumping power, since the mass flow into the 2nd pre-chamber is already reduced to $1/10$ with the same transfer opening and the 2nd pre-chamber also causes a pressure difference of $1/10$.

2.6 Experimental evaluation

The applicability of the analytical calculation of the conductance is evaluated in the present section on the basis of experimental measurements. The ability to mathematically describe the conductance of complex apertures would be a great advantage in the design of multi-chamber vacuum systems. The experimental results obtained on the test system displayed in Figure 2.2 are visualised in Figures 2.7 to 2.9.

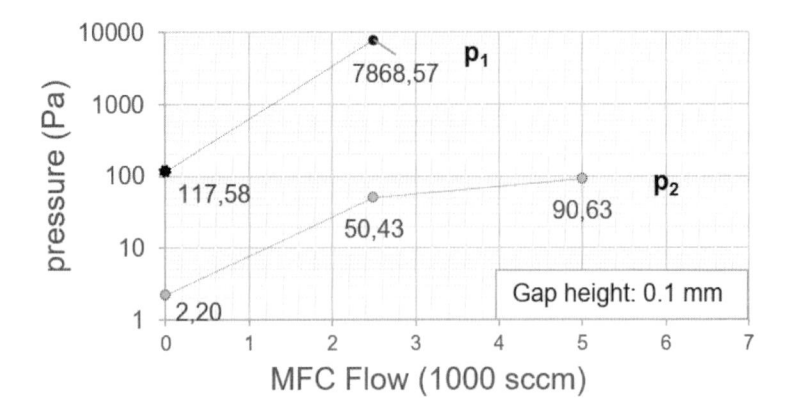

Figure 2.7: Pressure experiments using a 0.1 mm gap

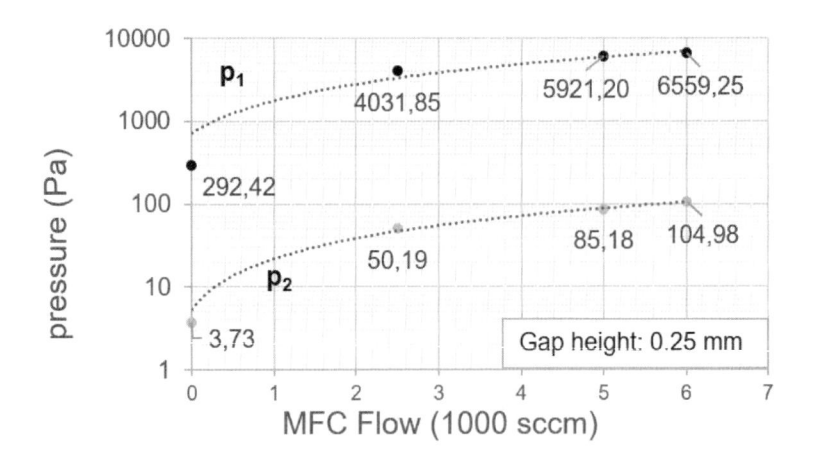

Figure 2.8: Pressure experiments using a 0.25 mm gap

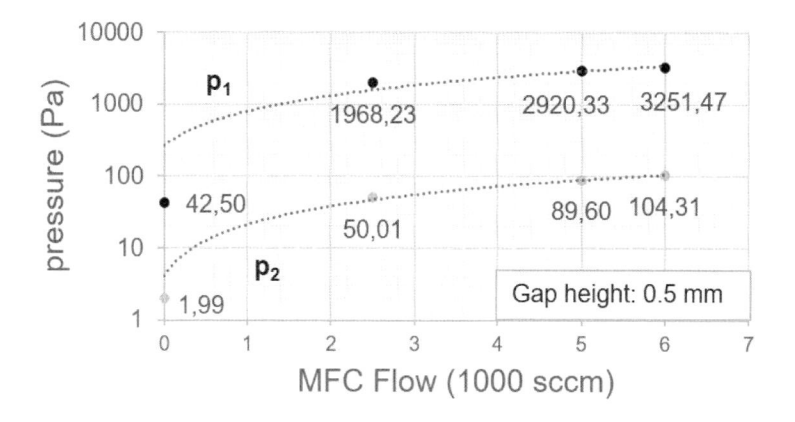

Figure 2.9: Pressure experiments using a 0.5 mm gap

It is noticeable that in all measurements the pressure at the pump-side chamber p_1 is distinctly lower than the pressure in the averted chamber p_2. Furthermore, p_1 is almost unchanged for all measuring points recorded. The pressure p_2 in the averted chamber on the other hand is influenced to a large extend by the gap height as well as by the set volume flow. This and the fact that pressure ratios between the chambers are well below 0.5 are clear indications that flow blockage occurs at all test points regardless of the volume flow set. Therefore, no further pressure reduction can be achieved for p_1. Pressure ratio increases for decreasing gap height and decreases for increasing gas flow.

With the given equations in section 2.5 it is possible to compare the measured conductance C (Equation 2.2), the theoretical minimum conductance for molecular flow $C_{molecular}$ (Equation 2.4) and the theoretical maximum conductance for viscose flow at blockage C^* (Equation 2.3). Table 2.1 shows the conductance values for the respective aperture gap heights. Each value has been averaged for all recorded pressure values in regard to the underlying mass flows.

Conductance	Height: 0.1 mm	Height 0.25 mm	Height 0.5 mm
$C_{molecular}$: theor. min. (eq. 2.4) [m³/h]	3.76	9.40	18.79
C: measured (eq. 2.2) [m³/h]	0.83	0.93	2.27
C^*: theor. max. (eq. 2.3) [m³/h]	8.60	21.46	43.79

Table 2.1: Comparison of the theoretical maximum flow conductance for blocked laminar flow, measured conductance values and the theoretical minimum conductance for molecular flow

Apparently, the complex flow conditions are not mapped accurately since the measured conductance values are lower than the theoretical minimum despite the fact, that the pressure interval investigated during the experiments is well above the molecular flow range. Even for the Knudsen a superposition of the two flow states should be observed [Dah02].

Possible reasons for this can be found, for example, in the mathematical description of the conductance in the case of a molecular flow state. Both conductance formulae do not take the length of the aperture into account, which is likely to lead to further reduction. Similar behaviour is observed for pipes with increasing length. Another important finding is, that for the achieved pressures a mean free path length in the range of the gaps height is established, which leads to a shift of the Knudsen flow state towards higher pressures. Knudsen flow is achieved for a limiting length of 0.1 mm between 100 Pa and 10 Pa. Therefore a superposition of both the molecular and the viscose flow state is necessary to describe the conductance.

2.7 Conclusions and outlook

It can be concluded that for a roll width of 400 mm, quite low effective gap heights have to be selected in order to be able to generate fine vacuum (p < 0.01 mbar) in the reaction area with realistic pumping effort.

Further investigations must be carried out to ensure that the plant design is appropriate. The simplifications in the analytical calculations do not correspond to the complexity of the pressure condition in the narrow apertures. With decreasing pressure, interactions between particles play a less important role. Therefore, numerical simulation approaches, that operate on a molecular level have to be taken into consideration as an complementary measure.

Analytical descriptions can still be taken into consideration in regard to the aperture geometry between the chambers. In particular, the gap length offers further room for describing the decreased measured conductivity values. For a full description of the conductance over the possible pressure range a further increase of the flow into the chamber is necessary. With data for a full scale experiment a refined description for three-dimensional narrow apertures is likely to being found. The mathematical prediction of the complex flow states will help making multi-chamber vacuum systems available for further applications.

Acknowledgements

This research was funded by the Deutsche Forschungsgemeinschaft (DFG, German Research Foundation) under the Sonderforschungsbereich Transregio 87 (SFB TR-87). We would like to cordially extend our thanks to the DFG. Furthermore, we would like to express our gratitude to AEPT (Lehrstuhl für Allgemeine Elektrotechnik und Plasmatechnik, RUB) and our project partners from the industry.

References

[Bin08]   BINKOWSKI, D.:. Plasmapolymere Barriereschichten für Kunststoffe: Verfahren, Materialien und Eigenschaften. 2008

[Dah02]   DAHLMANN, R.: *Permeation through Plasma Polymerized Layers and Plasma Coating of Plastics Pipes and Hollow Bodies*. RWTH Aachen University, Dissertation, 2002

[DJ21]   DAHLMANN, R.; JARITZ, M.: *EP3771747A1: Method and system for continuous vacuum-based processing of web material*. Patent. 2021

[HKY90]   HEISABURO HIROSHIMA SHIPYARD & ENG. W. FURUKAWA; KANJI HIROSHIMA SHIPYARD & ENG. W. WAKE; YOSHIO HIROSHIMA TECHNICAL INST. SHIMOZATO:. Patent. 1990

[IUN+85]   IMADA, K.; UEONO, S.; NOMURA, H.; TOHKAI, M.; HATA, Y.; KATO, K.: *US4551310A: Continuous vacuum treating apparatus Abstract*. Patent. 1985

[KSW84]   KIESER, J.; SCHWARZ, W.; WAGNER, W.: On the vacuum design of vacuum web coaters. *Thin Solid Films* 119 (1984) 2, p. 217–222

[Laf98]   LAFFERTY, J. M. (EDITOR): *Foundations of vacuum science and technology*A Wiley-Interscience publication. New York, NY: Wiley, 1998

[O'H03]   O'HANLON, J. F.: *A user's guide to vacuum technology*. Hoboken, NJ: Wiley-Interscience, 3rd edition, 2003

[Ple89]   PLEIN, P.: *DE 3632748 C2: Verfahren zur Beschichtung von Hohlkörpern*. Patent. 1989

[SWYK87]   SHIMOZATO, Y.; WADA, T.; YANAGI, K.; KATO, M.: *US4655168A: Continuous vacuum deposition apparatus with control panels for regulating width of vapor flow*. Patent. 1987

[URL21]   N N: *Streckblas-Beschichtungseinheit InnoPET FreshSafe Block*. URL: https://www.khs.com/produkte/maschinen-anlagen/detail/streckblas-beschichtungseinheit-innopet-freshsafe-block/, 13.09.2021

[Wal79]   WALTER, H.: *DE2747061A1: Vacuum chamber for coating strip - has air lock chambers to permit continuous coating*. Patent. 1979

[WAW88]   WUTZ, M.; ADAM, H.; WALCHER, W.: *Theorie und Praxis der Vakuumtechnik*. Wiesbaden: Vieweg+Teubner Verlag, 4., verbesserte edition, 1988

[YM05]   YASUDA, H.; MATSUZAWA, Y.: Economical Advantages of Low-Pressure Plasma Polymerization Coating. *Plasma Processes and Polymers* 2 (2005) 6, p. 507–512

Symbols

Symbol	Unit	Description
A_{min}	mm	Cross sectional area
C	l/s	Conductance factor
h	mm	Gap height
κ	$-$	Isentropic exponent
l	mm	Gap length
M_{molar}	g/mol	Molar mass
p_1	Pa	Pressure open side
p_2	Pa	Pressure pump side
$\dot{Q}$	$mbar * l/s$	Gas flow rate
S	l/s	Suction power
T_0	K	Temperature of the gas
w	mm	Gap width

Abbreviations

Notation	Description
MFC	Mass flow controller
PECVD	Plasma enhanced chemical vapour deposition
PET	Polyethylene terephthalate
PTFE	Polytetrafluorethylen
TMP	Turbo molecular pump

3 Recyclates for food packaging - Preventing migration with PECVD coatings

L. Kleines[1], R. Dahlmann[1]
[1] Institute for Plastics Processing (IKV) at RWTH Aachen University

Abstract

The European Commission's Plastics Strategy, published in January 2018, sets the goal to recycle at least half of the plastic waste by 2030 and thus to substantially increase the share of recyclates used in newly produced plastic products. Although the packaging sector in Germany processes the largest amount of plastic in a sector comparison, the success of the European initiative to increase the use of recyclates is currently hindered by, among other things, the very limited possibility of using recycled plastics in the food sector. One of the main reasons for this is the risk of migration of chemical substances from the used packaging into the food and a subsequent lack of safety of the food contact materials evaluated in an assessment of safety by the European Food Safety Authority (EFSA). Sufficient product protection in terms of toxicity, taste and odour changes as well as contamination of various kinds cannot be guaranteed for most materials especially when using post-consumer recyclates (PCR).

One solution to this problem is to explore and further develop the potential of functional PECVD (plasma enhanced chemical vapour deposition) coatings as migration barriers for contaminants from recyclates. In plasma and surface technology at the IKV, intensive research is currently being carried out in the field of developing highly functional coating systems for the application field of recycled packaging in the food sector. Together with an industrial partner from recyclate production, a processor of rigid food packaging and a research partner from interface analysis, close-to-application solutions are being elaborated.

The focus for this work is primarily on non-PET materials such as polypropylene (PP), as this is where the biggest hurdles lie for the use of recyclates in food contact. Basic research and at the same time application-oriented experiments and theoretical models are used to understand different effects and processes in the system consisting of recycled plastic, the PECVD coating and a foodstuff. Findings from this are subsequently transferred to and validated with a real packaging case. In this context, perspectives and experiences from material technology, processing technology and approval organisation are taken into account and a methodology for evaluating the performance of PECVD coatings in a real application is developed.

In the following, challenges in the use of recyclates in packaging applications are first discussed. Subsequently, plasma technology as a solution technology for the described application as well as regulatory contexts for functional barriers are being described. Finally, with describing the analysis of the performance of PECVD coatings as a migration barrier, the approach followed for the development of highly functional coating systems research is shown.

3.1 Use of recyclates for packaging applications

Plastic recyclates that originate both from the processing of waste from the use or consumption of products by an end consumer (post-consumer recyclates, PCR) and from direct recycling of production and processing waste from industry (post-industrial recyclates, PIR) are mainly reused in the construction and packaging sector. With a share of approx. 24 % (approx. 474 kt) of the total amount of recyclates reused in Germany, packaging is the second most important area of application. This quantity thus currently represents about 10.9 % of newly produced plastic packaging (see Figure 3.1). [Con20]

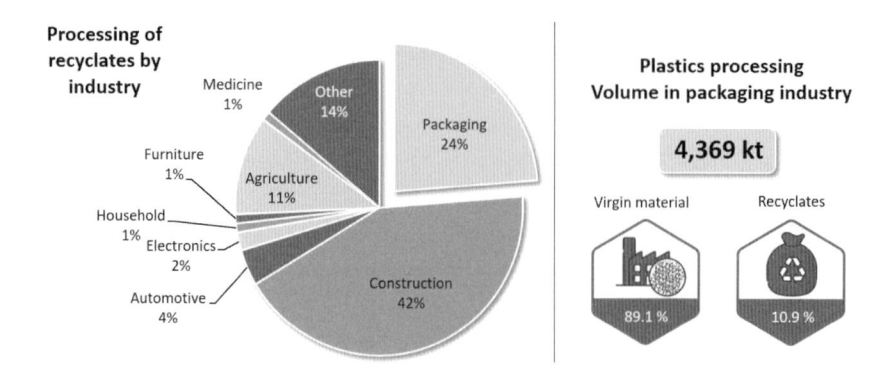

Figure 3.1: Processing share of recyclates by industry (left) and processing volume in packaging industry (right) in 2019 in Germany [Con20]

This share is expected to increase significantly, especially in line with the European goals as part of the Circular economy action plan adopted by the European Commission in 2020. Although the packaging sector processes the largest amount of plastics in Germany (approx. 4.4 Mo. tonnes, 31 % of the total volume processed), the reuse of plastics, especially from the post-consumer waste stream (about 85 % of the collected plastic waste in Germany), is currently severely limited by various barriers (e.g. concerning availability, costs, legal obstacles), especially for use in the area of food contact materials [Con20, Ges19]. Large food and beverage companies are now focusing on the strongly increased use of recycled materials in the future for reasons of sustainability and environmental protection, which significantly increases this conflict and the demand for products that are suitable for the food sector.

One of the predominant reasons for these restrictions is the risk of impurities (contaminants) remaining in the recycled material and migrating into the food with which the material comes into contact (Figure 3.2). The prevention of contaminant migration is therefore one of the most important aspects for the safe use of recycled plastics for food contact applications. Migration is also therefore the subject of testing of recycled materials by, for example, the European Food Safety Authority (EFSA) for approval for this application. For this reason, the detection, analysis and minimisation of contamination as well as the investigation and modelling of migration of substances in plastics are the subject of intensive research. [FS08, SW20] Recycling processes for the production of PCR polypropylene have only developed further in recent years, with the result that recyclates have not been on the market for long and especially are not yet of such a high quality for a safe use in cosmetics and food industries [URL21]. Since they are not yet suitable for food applications due to the insufficient quality after the recycling process, polypropylene recyclates are the subject of the developments in IKV.

Figure 3.2: Virgin material (center) compared to two different recyclate qualities

3.2 Plasma technology as a solution for the use of recycled materials in food contact

A promising approach to address the issue of potential migration of contaminants from food contact materials into the food is to decouple the PCR products from the food by means of a functional barrier. A suitable process in which nanoscale functional coatings can be applied quickly, non-destructively and in an environmentally friendly manner is through a plasma enhanced chemical vapor deposition (PECVD) process. By this, highly cross-linked silicon oxide gas barrier coatings (SiOx coatings) can be deposited on the PCR substrate with the potential of simultaneously acting as a barrier against substance migration (see Figure 3.3).

Other advantages of PECVD coatings for this application compared to other barrier technologies, such as the coextrusion of multilayer composites of different plastics, metallisation or wet chemical processes, are its high degree of environmental friendliness, low operating and material costs as well as the homogeneity of the coatings with optical transparency. Additionally, the nanoscale coatings produced in this process do not impair the recyclability of plastics, which was most recently officially confirmed in 2019 in the minimum standards for measuring the recyclability of packaging by the Stiftung Zentrale Stelle Verpackungsregister [Sti19].

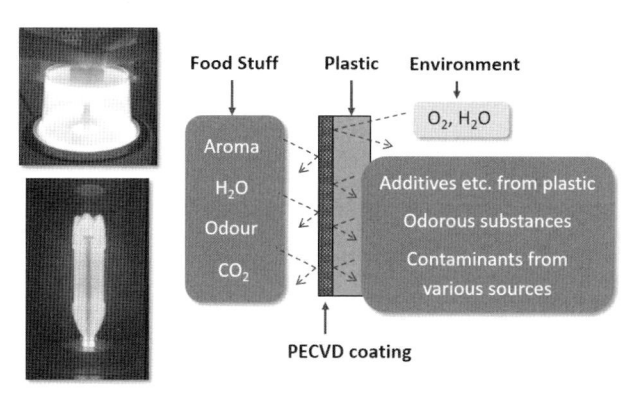

Figure 3.3: PECVD coatings as migration barriers

3.3 Functional barriers in the context of regulatory aspects

The use of post-consumer recyclates as plastic packaging material poses some significant challenges, which are mainly related to the fact that due to the unclear conditions during the use phase as well as in the recycling process, the properties and composition of the PCR material are in many ways unknown and inhomogeneous. On the one hand, this results in difficulties in the processing of plastics, but on the other hand also in considerable restrictions for the use of the recyclates, especially in contact with foodstuffs. Due to the unassessable complex properties of the original materials and conditions during initial use, storage, waste collection and the recycling process, no generally valid statement can currently be made regarding the type and concentration of contaminants or additives and fillers. In addition, it must be taken into account that the PCR materials do not consist of 100 % polypropylene due to non-ideal waste sorting, but are contaminated by other materials and therefore the chemical composition of the recyclates cannot be adequately defined and fluctuates.

As a result, the recyclability of PCR plastics for the food sector needs to be assessed. For the American market this assessment is regulated by the U.S. Food and Drug Administration (FDA) in a recommendation for industry (see [Foo06]) and for the European market in an EU regulation of the European Commission (see REGULATION (EC) No. 282/2008, [EU 08]). The authorisation procedures in Europe are carried out by the European Food Safety Authority (EFSA). At the moment it is not legally possible to use PCR PP for packaging applications with direct food contact, as the necessary documents for food safety cannot be issued due to the currently valid guidelines of the EU (EFSA). This means that the entry requirement for the legally required conformity work cannot be fulfilled.

According to REGULATION (EU) No. 10/2011, however, it is possible to use PCR material for food packaging under the condition of a "functional barrier" (see Figure 3.4, which is located between the food and the PCR material and can thus reduce or completely prevent a possible migration into the food [EU 11]. Such a "functional barrier" could be achieved by coating with PECVD. According to EU regulations, the use of recycled PCR materials behind functional barriers is not covered by the EU Recycling Regulation 282/2008, which regulates recycled plastic materials and articles intended to come into contact with food. Rather, this particular type of food contact material is assessed similarly to substances in multilayer packaging, where the barrier is a layer within food contact materials preventing the migration of substances from behind that barrier into the food. Behind a functional barrier, non-authorised substances may be used, provided they fulfil certain criteria and their migration remains below a given detection limit. [EU 11]

In this context, the aim of the developments at IKV is to fully exploit the potential of functional PECVD coatings for use as a migration barrier in order to overcome the regulatory challenges.

3.4 Approach to the development of functional PECVD coatings as a migration barrier

Particular challenges in the development of functional PECVD coatings as a migration barrier are especially the knowledge of

- the interaction between a PCR material of unknown and varying composition and a plasma-polymerised coating, as well as
- the migration processes of contaminants in the system consisting of recycled plastic, the PECVD coating and a foodstuff.

Therefore those are in the main focus for the successful research on the application potential of PECVD coatings as migration barriers.

Comparative material characterisation

To adress the first point, a comparative analysis of the PCR and virgin material have to be executed to highlight the challenges in coating development on the materials. Due to the recycling process and the previous use phase, PCR material differs from virgin material particularly in its chemical composition and degree of contamination. The PCR material is often contaminated with other plastics, contaminants and foreign particles. This results in heterogeneous properties and a defect-ridden surface. Since the substrate properties have a decisive influence on the functional coatings applied by the PECVD process, it can be assumed that barrier coatings develop fundamentally differently on PCR material than on virgin material.

In comparative studies of the material properties of PCR and virgin polypropylene from industrial production, the following differences and challenges were identified:

a Figure 3.4 shows representative curves of the second heating from DSC investigations (DSC: Differential Scanning Calorimetry) for the virgin and the PCR material. Two differences in particular can be highlighted: Firstly, the melting point of the virgin material is slightly higher than that of the recycled material, which indicates that the PCR material is a mixture of different PP types or that the recycled material is possibly already in a more degraded state than the virgin material. On the other hand, there is an additional endothermic peak at 125 °C for the PCR material. The peak can be attributed to an admixture of a foreign plastic present in the material due to the preparation of the recyclate. For high density polyethylene, the melt temperature is typically between 125 and 135 °C. The presence of small amounts of PE-HD within the PCR material is therefore likely and leads to inhomogeneity of the material.

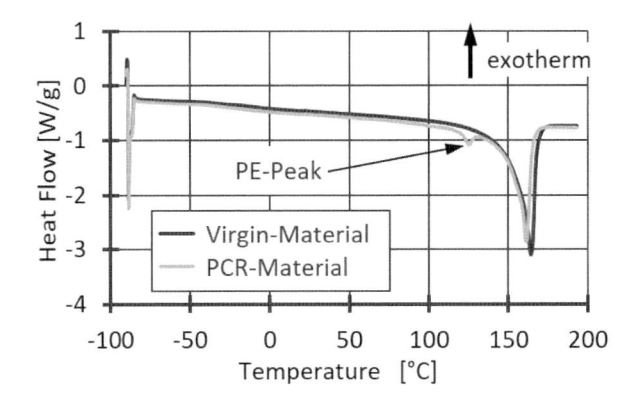

Figure 3.4: Result DSC measurements comparing PCR and Virgin material

b From the frequency sweep obtained from oscillation tests in the plate-plate rheometer, conclusions can be drawn about the molecular state of the materials. The results of the frequency sweep are shown in Figure 3.5. Here, the storage modulus G' and the loss modulus G" are plotted logarithmically for both materials as a function of frequency. In order to make statements about the molecular mass distribution (MMD) and the average molecular mass (AMM) of a polymer, the cross over point of G' and G" is of interest. If two polymers have the same average molar mass but different MMD, then the intercept G' = G" for the sample with a wider MMD shifts down along the y-axis. If, on the other hand, two polymers have the same MMD but different mean molar masses, the cross over point shiftes along the x-axis to a lower frequency range for a higher mean molar mass [Mez16]. Since the exact MMD or mean molar mass is not known for either the PCR or Virgin material, the interpretation of the G' = G" intercept should be considered only as a qualitative indicator of differences between the materials. Figure 3.5 shows that the cross over point for the PCR material is lower on the y-axis and at a higher frequency than the intercept for the Virgin material. This indicates a difference in the manufacturing process of the PCR material: Recycling processes are often accompanied by a shortening of the macromolecular chains, since the material is damaged, for example, by aging or by the preparation process itself (remelting, melt filtration, shredding of the packaging). In addition, the material has been processed at least once into a product. Aging and reprocessing cause chain scission within the material, which not only reduces the average molecular weight, but also affects the MMD.

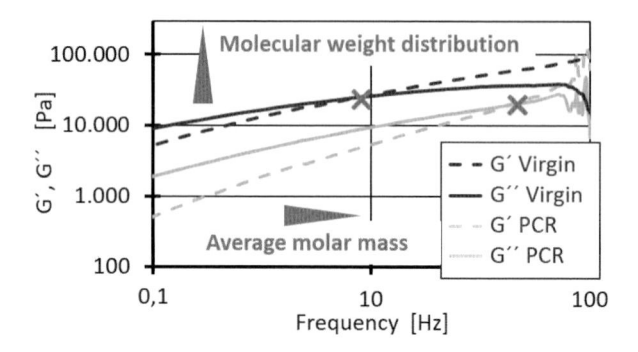

Figure 3.5: Frequency sweep from oscillation experiments comparing PCR and virgin material

c When comparing the surface structure of films extruded from PCR and Virgin material as well as of thermoformed packaging cups, strong differences in the average surface roughness can be observed (see Figure 3.6). This varies for the investigated materials between a few nanometers up to 50 nm, which exceeds the usual thickness of a PECVD barrier coating on smooth surfaces. This results in a challenge in the deposition of barrier coatings, as they may not be able to grow consistently at roughness peaks, which causes defects. In addition, the AFM (Atomic Force Microscope) images as well as the visual inspection show strong inhomogeneities in the materials (specks, color differences, flaws, etc.).

Virgin-Material	Extruded Films	PCR-Material
R_a = 12,65 nm		R_a = 8,5 nm
R_a = 40,57 nm	Thermoformed Cups	R_a = 51,43 nm

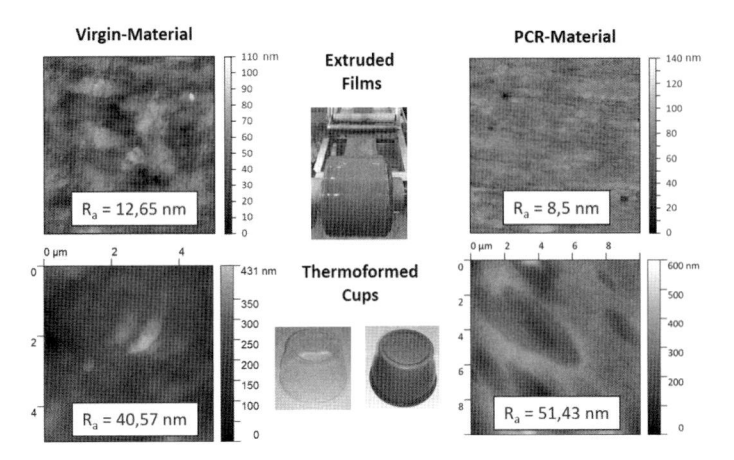

Figure 3.6: Comparison of surface structures through AFM (Atomic Force Microscope) pictures for PCR and Virgin Material

In summary, the comparative material study shows that the PCR material is characterized by fluctuating properties compared to the Virgin material, which are difficult to access. Contamination by foreign plastics (polyethylene) as well as changes in the molecular structure of the recyclates were detected. As a consequence, this leads to strongly fluctuating surface properties such as structure and chemical composition, which in turn influence the polarity and wettability of the surface. This, together with the presence of weak interfaces due to the material degradation already employed, potentially complicates coatability and coating adhesion in the PECVD process. It is expected that contamination, if any, may migrate to the surface and enter the vapor phase in the process. This results in very complex interactions between substrate and plasma, which require further investigation.

Coating development

To address the aforementioned challenges, the development of coating systems consisting of different layers that take on different functionalities seems to be suitable. The various properties and possible applications of PECVD coatings are discussed in many research articles. Especially silicon-based coatings are considered to have advantageous properties, so that they are already established in industrial processes for the functionalization of material surfaces. [DGT+08, DHJK16, LOS+13]

For the generation of gas barriers on plastics, silicon oxide barrier coatings (SiO_x coatings) have become established both industrially and on the research side, as they can improve the permeation properties of plastics by several orders of magnitude. In the literature, SiO_x layers are described as "hard[...], highly crosslinked[...] and brittle[...] gas barrier coatings with a low C content, good barrier properties, low ductility and smooth surfaces" [Bah17]. They are especially suitable for food packaging modification because the deposited layers have glass-like barrier properties on polymer surfaces, with significant cost advantages over "higher barrier" materials such as EVOH. [DGT+08] Those coatings are industrially used to reduce the permeation of permanent gases such as oxygen, carbon dioxide and water vapor and therefore have the potential to significantly reduce migration processes as demanded in using recyclates. Since the deposited SiO_x layers are usually only a few nanometers thick, the coated packaging can be introduced into existing recycling processes without further precautions. SiO_x coatings have been approved by the U.S. Food and Drug Administration (FDA) as an acceptable food contact material. [WF08] The SiO_x coatings can be influenced by different process parameters in the PECVD process. The amount of power input and the power pulsing have been have a significant influence on the permeation properties of the barrier coating. For the deposition of SiO_x barrier layers, according to literature data a high degree of fragmentation of the monomer is required. [Bah17] This can be achieved by a high power input into the PECVD process and with the help of a high oxygen content. [Beh16, DGT+08, Mit18]

As noted earlier, the surface characteristics of the recycled materials, such as defect areas, impurities, significant roughness, and molecular degradation, potentially affect coatability. A possible solution to this problem is the use of silicon organic intermediate coatings, which are intended to open up the possibility of decoupling the properties of the SiO_x barrier coatings from the properties of the underlying substrates. Organic SiOCH layers, which are described as "relatively soft, weakly cross-linked, elastic silicon organic layers with a high C content, a high surface roughness, and a high ductility and permeability" [Bah17], are suitable as an intermediate layer. Here, the SiOCH layers should be deposited on the substrate surface prior to the SiO_x layer, which can relieve film stresses in the barrier coating. Organic SiOCH layers are deposited using the same monomers that are used to synthesize SiO_x coatings. [JHB+17] However, in the deposition of SiOCH coatings, only small amounts of oxygen are added in the literature or no oxygen is introduced at all, so the process is carried out using only the precursor gas. The plasma process of SiOCH layers is characterized by low fragmentation of the monomer. [Bah17] Due to the lower energy input and oxygen admixture in the PECVD process, SiOCH layers have a higher organic content than SiO_x coatings. [Beh16] Besides the possibility to relieve layer stresses in the rigid SiO_x layers, further advantages open up when using layer systems with SiOCH interlayers: For example, it was found in [JHB+17] that the coating adhesion of a PP/SiO_x system can be improved with the help of a SiOCH interlayer. Similar advantages of a SiOCH/SiO_x coating system were also demonstrated in other studies. [Beh16] In the case of a direct SiO_x coating deposition, etching processes can occur on the substrate surface due to the high oxygen content in the PECVD process, so that the application of a SiOCH interlayer can protect the already molecularly degraded substrate surface. [Mit18] Moreover, by combining a SiOCH and a SiO_x layer deposited with the same precursor, the barrier effect can be enhanced. Lastly, SiOCH interlayers can smooth out impurities and defects in the substrate surface, decoupling the SiO_x barrier from the substrate material and allowing it to grow more uniformly.

Initial investigations have already shown that the growth behavior of SiO_x barrier coatings is strongly dependent on the structure of the organosilicon coatings, and thus also influences the barrier properties of the coating system (see Figure 3.7). It has proven to be advantageous to deposit the silicon organic interlayers under very low energy input in the process (low microwave input power as well as low pulse on time for the input of the pulsed microwaves). The investigation of the coating morphology by means of electron microscopic images clearly shows that largely closed and fine-grained SiOCH coatings can be generated at low energy input, which are capable of overgrowing surface defects of the substrate. A low pulse-on time and power coupling leads to a lower fragmentation of the precursor monomer. In contrast to the low energy deposited films, the SiOCH films under high energy input are much more granular and the surfaces are rougher. When a silicon-oxide barrier layer is deposited afterwards, the granular structure is also formed in it, resulting in defects in the layer between the grown islands. This phenomenon subsequently leads to a weak barrier effect of the SiOx layer: Since diffusion through the barrier coating is is defect-controlled, gas particles can reach the substrate unhindered at high defect density and the barrier effect is reduced.

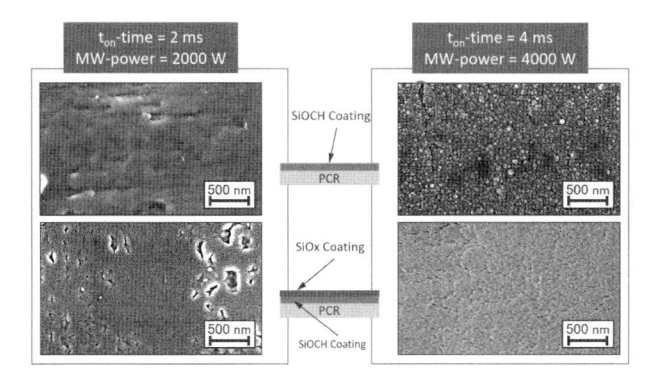

Figure 3.7: FESEM (Field Emission Scanning Electron Microscope) images comparing the influence of different SiOCH interlayers

The first investigations show that the approach of depositing multilayer systems, in which the different coating types fulfill different functions, is promising. However, there is still a need for further development, which requires on the one hand the refinement of the individual coating structures, and on the other hand the development of suitable coating thicknesses and coating architectures. For example, a system of several dyads consisting of intermediate and barrier layers is conceivable in order to compensate even better for existing surface defects and roughness.

Migration analysis

As mentioned above, further important steps (besides the coating development) in the development of migration barriers are the analysis and understanding of migration processes of contaminants in the system consisting of recycled plastic, the PECVD coating and a foodstuff. In order to be able to evaluate the performance of the coatings with regard to a migration barrier, a method for the model-based investigation of the migration of contaminants in the system must be developed. The test methodology represents an essential basis for further development and should be based on the procedure of the regulatory authorities in order to ensure the industrial relevance of the investigations.

The EFSA guidelines require an experimental determination of cleaning efficiency of recycling processes, where the ability to reduce potential contaminants in the process is tested. For this, plastic is highly contaminated with model chemicals as substitute contaminants. The contaminated plastic is then introduced into the recycling process and the residual concentration of the surrogate contaminants is determined after the process, resulting in the decontamination/cleaning efficiency of the recycling process. [Eur11]

In EU Regulation 10/2011, which generally regulates the use of food contact plastics and is applicable to the use of functional barriers, both the testing of an overall migration limit and specific migration limits of substances are described. The test includes the contact of the material to be tested with a food simulant and general rules are described such as aspects of sample preparation, the choice of food simulants and the contact conditions when using food simulants. Subsequently, the residual concentration of the substances in the food simulant is determined and thereby compliance with the regulated migration limits is checked.

Testing the suitability of plasma-polymerised coatings as a migration barrier when using recyclates can be well implemented by combining the two test methods (see Figure 3.8):

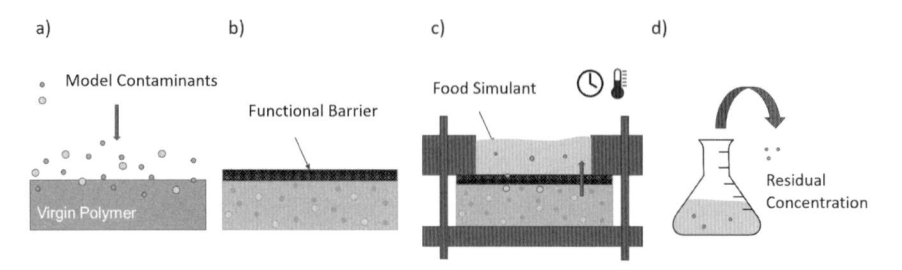

Figure 3.8: Migration testing of specifically contaminated plastics with functional barriers from a PECVD process

The following steps are to be carried out:

a Contamination of material with model chemicals/substitute contaminants
b Deposition of a functional barrier in the PECVD process
c Contact with food simulants under special storage conditions
d Evaluation of the residual concentration of the model contaminants in the food simulant with respect to compliance with migration limits.

Knowledge on migration to the respective interface in the system of recyclate and plasma-polymerised coating under variation of important coating properties is the basic prerequisite for a model-based description of the migration processes. Research on boundary layer migration is therefore carried out, which allows modelling of the migration and the transfer through the boundary layer into a medium. Finally, based on the previous steps, the performance of the PECVD coatings as a migration barrier must be studied. To ensure the quality of the processes for the application, concepts must then be developed in close cooperation with the industrial partners that can be transferred to the real application. Alongside this, the coating performance will be validated on a real packaging product made from PP recyclates to confirm the suitability of the technology for the application.

3.5 Conclusions and outlook

The reuse of high-quality recycled plastics in food contact packaging applications - apart from PET - is not yet established. A major reason for this is the approval regulations of the European Food Safety Authority (EFSA), which set high requirements for the acceptable migration of contaminants from the packaging into the food. Plasma technology presents itself as a problem solver for this issue: the use of functional PECVD coatings as migration barriers for contaminants from recyclates not only has a high level of potential from a process technology perspective, but also from a regulatory point of view. The comparative material analysis of PCR and virgin material has shown that for the development of suitable PECVD processes and coatings, there are various challenges that have to be considered in the development of migration barriers. They can only be solved by the development of multilayer coating systems, where the application of SiO_x barrier layers in combination with low energy applied SiOCH interlayers is very suitable. There is currently a need for further development in the design of suitable coating structures and architectures. Along with the development of PECVD processes, the evolution of suitable migration analyses for the present system should be pursued in order to be able to investigate and evaluate the performance of different coating systems.

Acknowledgements

This research was funded by the Deutsche Forschungsgemeinschaft (DFG, German Research Foundation) under the Sonderforschungsbereich Transregio 87 (SFB TR-87). We would like to cordially extend our thanks to the DFG. Additionally we would like to thank our partners GIZEH Verpackungen GmbH Co. KG, DSD – Duales System Holding GmbH Co. KG and Professor Grundmeier and his team from Universität Paderborn.

References

[Bah17] BAHROUN, K.: *Prozessentwicklung zur Abscheidung dehnbarer Barriereschichten auf PET mittels Plasmapolymerisation.* RWTH Aachen University, Dissertation, 2017

[Beh16] BEHM, H.: *Investigation of plasma processes and their influence on the composite properties of polypropylene coated by means of plasma polymerization.* RWTH Aachen University, Dissertation, 2016

[Con20] CONVERSIO MARKET & STRATEGY GMBH: *Stoffstrombild Kunststoffe in Deutschland 2019: Studie im Auftrag der BKV GmbH mit Unterstützung versch. Trägerverbände*, 2020

[DGT+08] DEILMANN, M.; GRABOWSKI, M.; THEISS, S.; BIBINO, N.; AWAKOWICZ, P.: Permeation mechanisms of pulsed microwave plasma deposited silicon oxide films for food packaging applications. *Journal of Physics D: Applied Physics* 41 (2008), p. 135–207

[DHJK16] DAHLMANN, R.; HOPMANN, C.; JARITZ, M.; KIRCHHEIM, D.: Barriereausrüstung von Kunststoffen mittels PECVD. Untersuchung zu Wirkmechanismen der Sauerstoff- und Wasserdampftransmission. *Vakuum in Forschung und Praxis* 28 (2016), p. 36–41

[EU 08] EU COMMISSION: *Commission Regulation (EC) No 282/2008 on recycled plastic materials and articles intended to come into contact with foods and amending Regulation (EC) No 2023/2006*: Official Journal of the European Union, 27. März 2008

[EU 11] EU COMMISSION: *Commission Regulation (EU) No 10/2011 on plastic materials and articles intended to come into contact with food*: Official Journal of the European Union, 14. Januar 2011

[Eur11] EUROPEAN FOOD SAFETY AUTHORITY, EFSA: *Scientific Opinion on the criteria to be used for safety evaluation of a mechanical recycling process to produce recycled PET intended to be used for manufacture of materials and articles in contact with food*: EFSA Journal, 2011

[Foo06] FOOD AND DRUG ADMINISTRATION: *Guidance for Industry: Use of Recycled Plastics in Food Packaging (Chemistry Considerations)*: Food and Drug Administration, 2006

[FS08] FRANZ, R.; SIMONEAU, C.: *Modelling migration from plastics into foodstuffs as a novel and cost efficient tool for estimation of consumer exposure from food contact materials*: Fraunhofer-Institut für Verfahrenstechnik und Verpackung IVV - Final Synthetic Projekt Report zum EU-Projekt QLK1-CT2002-2390 "Foodmigrosure", 2008

[Ges19] GESELLSCHAFT FÜR VERPACKUNGSMARKTFORSCHUNG MBH: *Hemmnisse für den Rezyklateinsatz in Kunststoffverpackungen. Studie im Auftrag der Klimaschutzoffensive des Handels Handelsverband Deutschland - HDE - e.V.*, 2019

[JHB+17] JARITZ, M.; HOPMANN, C.; BEHM, H.; KIRCHHEIM, D.; WILSKI, S.; GROCHLA, D.AND BANKO, L.; LUDWIG, A.; BÖKE, M.; WINTER, J.; BAHRE, H.; DAHLMANN, R.: Influence of residual stress on the adhesion and surface morphology of PECVD coated polypropylene. *Journal of Physics D: Applied Physics* 50 (2017), p. 111–134

[LOS+13] LIU, C.-N.; OZKAYA, B.; STEVES, S.; AWAKOWICZ, P.; GRUNDMEIER, G.: Combined in situ FTIR-spectroscopic and electrochemical analysis of nanopores in ultra-thin SiOx-like plasma polymer barrier films. *Journal of Physics D: Applied Physics* 46 (2013), p. 84

[Mez16] MEZGER, T.: *Das Rheologie Handbuch: Für Anwender von Rotations- und Oszillazions-Rheometern*: TVincentz Network, 2016

[Mit18] MITSCHKER, F.: *Influence of plasma parameters in pulsed microwave and radio frequency plasmas on the properties of gas barrier films on plastics*. Ruhr-Universität Bochum, Dissertation, 2018 – supervisor: P. Awakowicz

[Sti19] STIFTUNG ZENTRALE STELLE VERPACKUNGSREGISTER: *Mindeststandard für die Bemessung der Recyclingfähigkeit von systembeteiligungspflichtigen Verpackungen gemäß § 21 Abs. 3 VerpackG.*, 2019

[SW20] SCHMID, P.; WELLE, F.: Chemical Migration from Beverage Packaging Materials — A Review. *Beverages* 6(2) (2020), p. 111–134

[URL21] N N: *Verpackung und Nachhaltigkeit, Materialfraktion Polypropylen*. URL: https://www.agvu.de/de/polypropylen-pp-147/, 16.11.2021

[WF08] WELLE, F.; FRANZ, R.: SiOx layer as functional barrier in polyethylene terephthalate (PET) bottles against potential contaminants from post-consumer recycled PET. *Food Additives and Contaminants: Part A* 25 (2008), p. 788–794

Abbreviations

Notation	Description
EFSA	European Food Safety Authority
FDA	U.S. Food and Drug Administration
PE	Polyethylene
PCR	Post-Consumer Recyclates
PECVD	Plasma Enhanced Chemical Vapour Deposition
PET	Polyethylene Terephthalate
PIR	Post-Industrial Recyclates
PP	Polypropylene

Improved understanding of processes and materials in foaming

Moderator: Dr. Andreas Peine, W. Köpp GmbH & CO. KG

---- **Content** ----

1   Discontinuous foaming of polymers versus foam extrusion   83

A. Peine[1]
[1] *W. Köpp GmbH & CO. KG*

2   From the iterative-analytically designed flow channel to the die concept
    of an inline extensional rheometer   96

Ch. Hopmann[1], N. Reinhardt[1]
[1] *Institute for Plastics Processing (IKV) at RWTH Aachen University*

3   Influence of the die design on the foam structure of foamed multilayer
    blown films   109

Ch. Hopmann[1], M. Stieglitz[1]
[1] *Institute for Plastics Processing (IKV) at RWTH Aachen University*

Dr. Andreas Peine

Dr. Andreas Peine has studied chemistry at the University of Cologne and in 1992 received his Doctorate in natural science for identifying temperature induced structural changes using linear temperature programmed pyrolysis in combination with infrared spectrometry, especially in polyvinyl ketones. In 1993 he started his career at Odenwald Chemie in Neckarsteinach, moving up to Head of Sales and RD working on thermoformed polyethylene foams used in automotive sealing tasks.

In 2006 his career path lead him into South China as Head of Manufacturing Asia and later on RD joining Armacell (Guangzhou) Limited, focusing on production of elastomeric foam for application in building and automotive industry. Continuing his career in Asia he joined Alpla China as Head of China, overseeing 6 plants manufacturing extrusion blow moulded packaging solutions.

In 2011 he left China, returning to Germany to join W. Köpp GmbH CO. KG. in Aachen. As Business Development Representative he worked on converting Köpp from a trading and conversion facility into a full-scale bun foam manufacturing company for elastomeric and thermoplastic bun foams with manufacturing locations in Germany and Romania as well as a Joint Venture in India.

1 Discontinuous foaming of polymers versus foam extrusion

A. Peine[1]
[1] W. Köpp GmbH & CO. KG

Abstract

W. Köpp GmbH & Co. KG is a small to middle size German enterprise located with the headquarters in Aachen. We have 4 locations serving our customers (Figure 1.1).

Figure 1.1: Locations

Founded in 1938 it is now the 3rd generation of the family running the business that has undergone quite some change in the last 10 years. Over a long period, trading foam products was our core business. With the third generation stepping into the business this changed and Köpp has become a strong and reliable manufacturer of those products that have been traded over so many years. Starting off with building up a joint venture together with Roop Polymer in India, Delhi we have begun to manufacture rubber bun foam for the Indian market. After that a manufacturing plant in Romania was targeted for a technology upgrade.

With 4 plants worldwide we now can supply rubber sponge products, cellular elastomer bun foams, cellular thermoplastic bun foam besides several other foam related products to our customers (Figure 1.2).

Figure 1.2: Turnover

Foaming of polymers (Figure 1.3) has a manufacturing tradition going way back more than 150 years starting with products like rubber sponge targeting very often to substitute natural sponge being a rather expensive and not always available produce [URL21b].

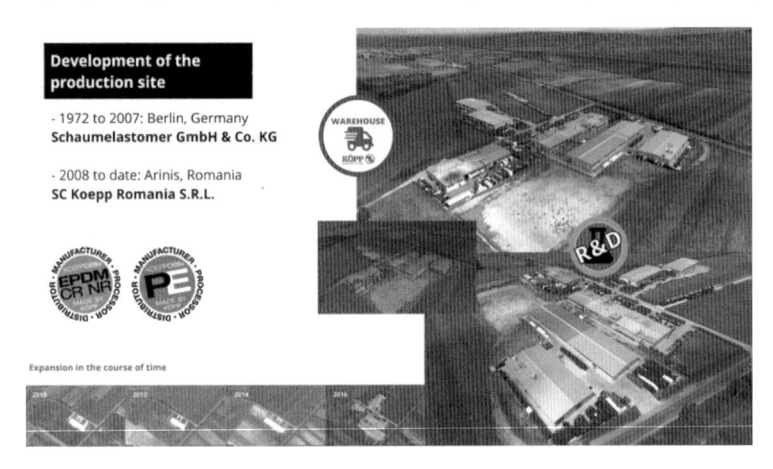

Figure 1.3: Plant View

Arriving in modern time several rather different technologies have evolved to foam polymers (Figure 1.4).

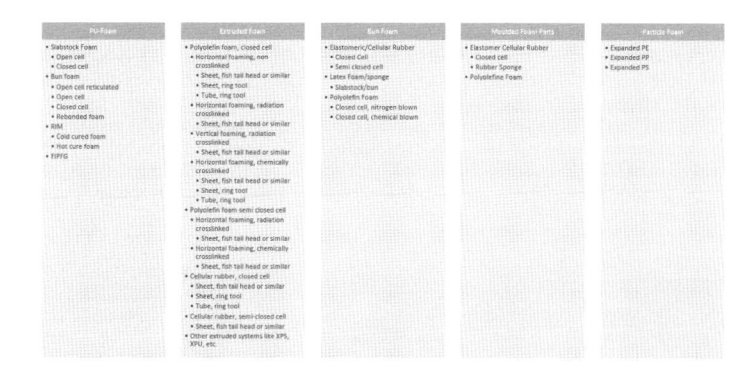

Figure 1.4: Foam

Several manufacturing technologies like PU foaming, extrusion foaming, bun foaming, moulded foam parts and particle foam parts are shown. This figure is rather more to show the diversity of the process then to claim being complete. This article focuses on the topics of extrusion foaming and bun foaming. The fundamental differences between the processes are highlighted.

In the production of thermoplastic foams by extrusion, the blowing agent is either added to the pellets in solid form (chemical blowing agent) or injected into the plastic melt in liquid or supercritical form during processing (physical blowing agent). Chemical blowing agents decompose during processing mainly into nitrogen (and in minor quantities also in CO, CO_2, H_2O, and NH_3, which is then dissolved in the melt under the pressure prevailing in the extruder in the same way as physical blowing agents. In both cases, a homogeneous, single-phase solution of polymer melt and blowing agent is present at the extruder outlet. When the melt is formed at the die, the pressure drop falls below the saturation pressure of the blowing agent, which causes the nucleation of cells.

The continuous extrusion foaming offers several possibilities to combine crosslinking (vulcanisation) or non-crosslinking with horizontal or vertical foaming technologies, different types of head geometries like fish tail, ring tool heads or flat head. This can be used to foam rubber compounds as well as thermoplastic compounds and can result in open to semiclosed cell or closed cell products (Figure 1.5).

Figure 1.5: PE Foaming Line (IKV)

Neglecting the latex sponge the discontinuous bun foaming is distinguishes in cellular rubber production and cellular thermoplastic production. For the first step a mould is filled based on volume 1:1 with a material matrix. During heating of the material, the decomposition of the chemical blowing agent is induced. The gas is captured under pressure. Based on pressure and temperature the gas becomes supercritical and is dissolved into the polymer. The general case always is a crosslinked or vulcanised product. A commonly found non-crosslinked PE bun, rather thick, especially used in the toy industry is not a discontinuous product step but a continuous extrusion production afterwards cut down to bun size (Figure 1.6).

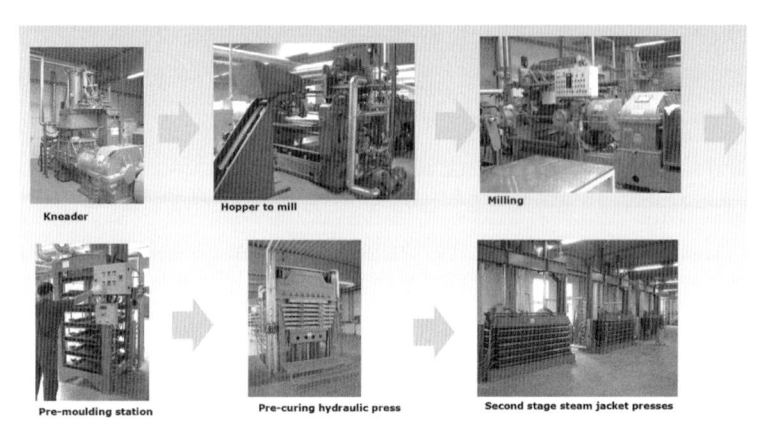

Figure 1.6: PE bun foam line [URL21c]

1.1 The shape of the cell created

Assumption:

    Two cells are started at the same time

    Are created in direct neighbourhood

    Are sharing one wall

Conclusion:

    Created at the same time

    With the same blowing agent

    Result in the same inside pressure of the cell

    Subsequently the pressure on both sides of the wall is the same

    The wall is a straight wall and can be described as a plain

Example drawings of such cells [Figure 1.7].

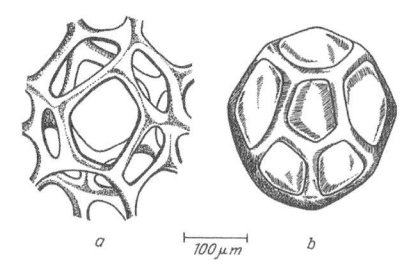

Figure 3.2 Schematic views of gas structural elements of rigid phenolic foam; (a) open gas structural element; (b) closed gas structural element

Figure 1.7: Cell Shapes [KS04]

Left side: rigid phenolic foam, also identical cell structure for reticulated PU foam

Right side: typical closed cell foam

The simplest approach is to consider the cell to be a sphere but in reality all types of x-hedrons can be found (Figure 1.8).

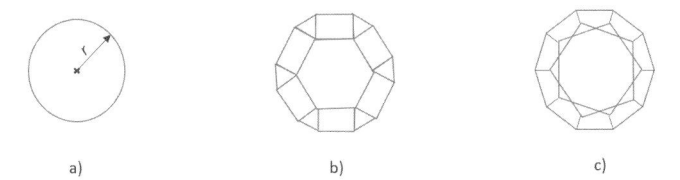

Ideal geometries of cells of various shapes [8]; (a) Sphere; (b) 14-hedron, with a surface composed of 6 squares and 8 hexagons; (c) 12-hedron, with a surface composed of regular pentagons only

Figure 1.8: Hedron [KS04]

Continuing with assumptions:

        The material has a unique temperature
        The foaming step is a fast-growing step.

Two cells in a neighbourhood, if at same time nucleated and created with the same blowing agent have the same size and the neighbouring wall is a plain, not a sphere. During growth of the cell the plain behaves like a biaxial stretched film. Now the big question is: what is happening during that period?

1.2 The influence of material properties during foaming

Continuing with assumptions:

>The material has a unique start temperature
>The foaming-step is a fast-growing step.

Now the big question is: what is happening during that period?

Case 1, the expansion of the material leads to dissipation of the inner energy over a larger volume.

Result:

>The temperature drops
>
>The viscosity increases
>
>The growing speed is slowing down.

Case 2, biaxial stretch can lead to order phenomena like crystallisation. Crystallisation energy can be set free.

Result:

>Partial temperature increases in regions of crystallisation
>
>The viscosity increases due to physical crosslinking
>
>The growing speed is influenced but how?

Case 3a, if using a chemical blowing agent that decomposes exothermic.

Result:

>The temperature will increase
>
>The viscosity will drop
>
>The growing speed will increase

Case 3b, if using a chemical blowing agent that decomposes endothermic or physical blowing agents

Result:

>The temperature will drop
>
>The viscosity will increase
>
>The growing speed will slow down

Case 4, due to crosslinking (vulcanisation) mechanical properties of the matrix will change.

Result:

The viscosity will increase

The growing speed will slow down

Even if you argue that in most cases only one blowing agent is used, the number of variables that influence the growing behaviour are huge. So far it was not possible to find an approach to calculate the behaviour based on used material properties.

It is therefore still common to develop based on experimental approach.

Mooney viscometers, moving die rheometers with pressure measurement included, Rheotens and other helpful instruments support this work but are mostly one dimensional access points of the foaming behaviour.

1.3 The continuous extrusion foaming case

The head, whatever design, has a pressure build up at one point and then a gradually dropping pressure towards the opening.

Then in many cases a temperature gradient with increasing temperature towards the head is set. This does mean that we do not have an adiabatic system. Due to a temperature profile set at the extruder the actual temperature of the matrix is depending on time and position.

In this scenario the matrix is arriving at the head. Here we have a pressure peak followed by a gradual drop of pressure the farther the material is exiting the head. This leads to a wider spread of the start point of foaming subsequently a wider cell size distribution (Figure 1.9).

Whether the foaming starts directly in the head or is later induced in a secondary oven step makes no difference. We have a non-adiabatic system.

A similar statement can also be given if you use additional nucleation agents. They can influence the total number of cells created but not the distribution of cell size.

It must be mentioned that particularly polyolefin matrixes, due to the lack of filler, have a strong observable effect by adding a nucleation agent.

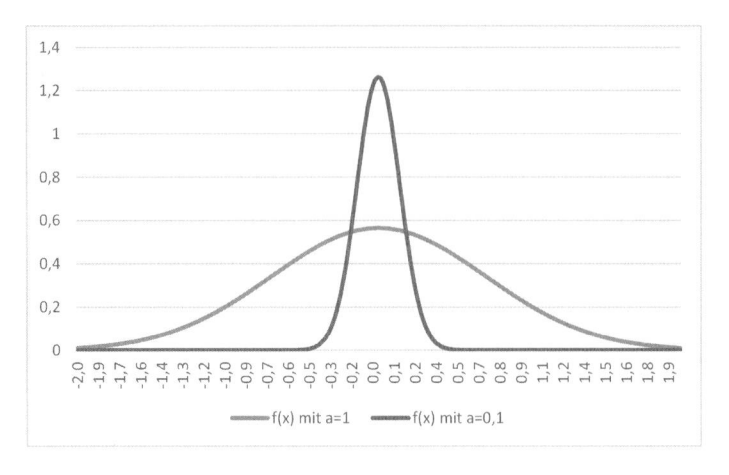

Figure 1.9: Distribution

Density of a centred normal distribution.

$$\delta_a(x) = \frac{1}{\sqrt{\pi a}} e^{-\frac{x^2}{a^2}} \qquad (1.1)$$

With a $\to 0$ the peak of the function will become higher but the area below the curve stays the same. [URL21a] This is a picture of a wide cell size distribution. The regional differences are clearly observable (Figure 1.10).

Figure 1.10: Rubber Sponge provided by Köpp GMBH und Co.KG

The presented case is a pressure less rubber sponge foaming.

1.4 The discontinuous foaming case

As already mentioned in a the first step a mould is filled with material. During heating of the material, the decomposition of the chemical blowing agent is induced. The gas is captured under pressure. Based on pressure and temperature the gas becomes supercritical and is dissolved into the polymer. There are some special cases where that does not happen, but they are not discussed today.

The opening of the mould happens very quickly. If process parameters are under control the temperature of the material is everywhere the same (adiabatic). Once the pressure is gone the gas will exit the polymer matrix and a cell nucleation will happen. (Coca Cola bottle effect).

The foaming is then based on an adiabatic nucleation step. In this way a narrow cell size distribution can be achieved. These recipes are very often with a larger filler content. An effect based on additional nucleation agents can be left aside as the filler is in large quantity and provides an excess amount of nucleation points.

Example of an adiabatic nucleation. The result is showing a nice unique cell size distribution. Differences observed in the diameters are depending on the plain in which the cells are cut, top middle or bottom (Figure 1.11).

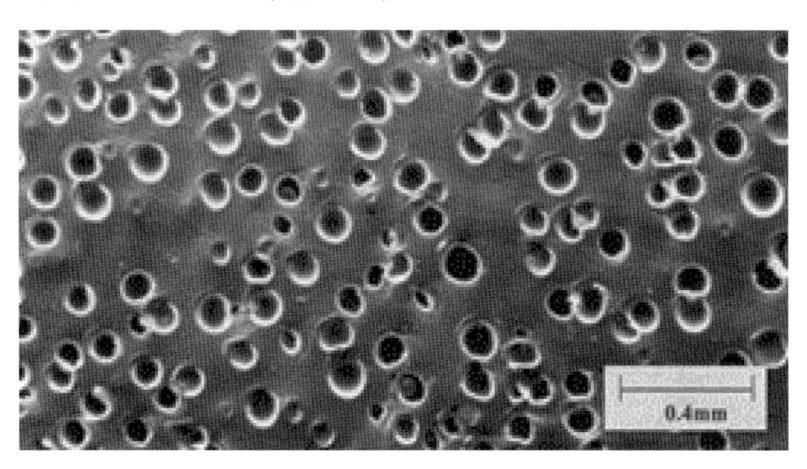

Figure 1.11: Adiabatic nucleation of polyethylene by high pressure autoclave process of the UK company Zotefoams PLC, Croyden

After the first step there is a second process step to fully grow the block to size (Figure 1.12) [URL21c].

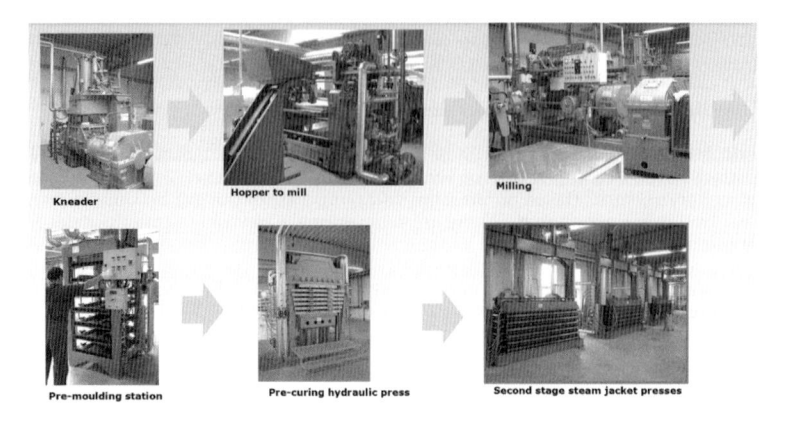

Figure 1.12: Second Step

In contrast to the two-stage process described so far continuous extrusion foaming is only a one step foaming process.

Comparing both cases (continuous extrusion foaming and discontinuous foaming) there is still the same statistics to be found. Both functions follow the principal of a normal distribution and the area beneath the graphs are the same. In the second case the standard deviation is just wider (Figure 1.9).

1.5 Types of cell size distribution

In some cases, it is difficult to see whether we have a wide cell size distribution due to overlaying effects. This can happen in case there is more than one gas participating in the foaming. For example, small quantities of water can create a bi-modal cell size distribution (Figure 1.13).

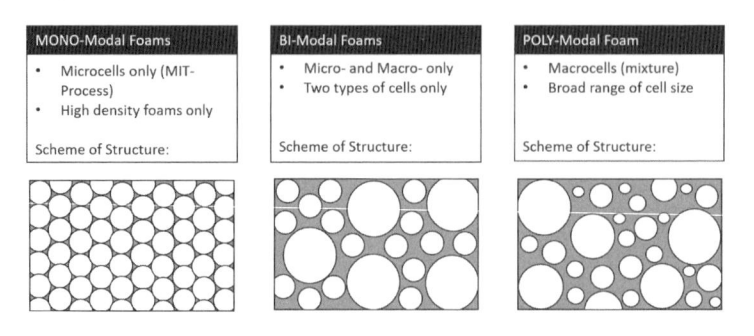

MONO-Modal Foams	BI-Modal Foams	POLY-Modal Foam
• Microcells only (MIT-Process) • High density foams only	• Micro- and Macro- only • Two types of cells only	• Macrocells (mixture) • Broad range of cell size
Scheme of Structure:	Scheme of Structure:	Scheme of Structure:

Figure 1.13: Modal [KS04]

Poly-modal means there is an extremely wide cell size distribution that is not really showing a maximum anymore. In such cases the standard deviation exceeds the range of measured values.

1.6 Conclusions

Current technology enables us to manufacture cellular rubber as well as thermoplastic products. It can be stated that the extrusion foaming technology gives us a fast continuous manufacturing method with a wider cell size distribution and in general a thickness limitation in contrast to discontinuous bun foam production that offers a more narrow cell size distribution and the possibility of a thicker final product. This statement valid only if same densities are compared.

1.7 Outlook

Extrusion process doning foaming with super critical gas injection are today already possible and provide much finer cell size and cell size distribution. These are running under different names like Mycell, Cellmould or ProFoam. Unfortunately, there are still thickness and density limitations.

In discontinuous foaming it would be of great advantage if physical blowing agents could be used.

But the biggest achievement would be if there would be a master theory to the physics of foaming that could be used on predict polymer matrix behaviour during the foaming.

References

[KS04]   KLEMPER, D.; SENDIJAREVIC, V.: *Handbook of polymeric foams and foam technology (2nd edition)*. Munich, Germany: Carl Hanser Verlag, 2004

[URL21a]   N.N.: *Normalverteilung - Wikipedia*. URL: https://de.wikipedia.org/wiki/Normalverteilung

[URL21b]   N.N.: *Prospekte - Koepp*. URL: https://www.koepp.de/de/service-center/prospekte/prospekte.php

[URL21c]   N.N.: *Über uns*. URL: https://ufm.hu/de/ueber-uns/

Symbols

Symbol	Unit	Description
δ	–	Density of normal distribution
α	–	Constant
r	mm	radius

2 From the iterative-analytically designed flow channel to the die concept of an inline extensional rheometer

Ch. Hopmann[1], N. Reinhardt[1]
[1]Institute for Plastics Processing (IKV) at RWTH Aachen University

Abstract

While the plastic melt in the extruder and in the die is primarily subjected to shear stress, in foam extrusion the material is subjected to biaxial elongation stress when the plastic foams. Consequently, the typical characteristic values used for material characterisation, such as shear viscosity or melt mass flow rate, do not allow any conclusions to be drawn about the foamability of the plastic material.

An iterative-analytical approach to determine the geometry of a rheometric flow channel for a biaxial extensional rheometer has been developed. The division of the channel into discret segments as well as the backward calculation of the circumference and the optimisation of the channel height lead to a flow channel with biaxial elongation of the melt. In order to be able to stretch the plastic melt equibiaxially within the flow channel of the inline extensional rheometer in a targeted manner, further optimisations and simulations are carried out. This opens up new possibilities for precise material characterisation under processing conditions not only in foam extrusion, but also in other process with biaxial stretching such as blown film extrusion or thermoforming. In order to transfer the iterative-analytical flow channel into an extrusion die that can be connected to an extruder different die concepts were developed and evaluated.

2.1 Introduction

Plastics processing technologies such as foam extrusion promise improved product properties in addition to cost and resource savings. Foamed plastics are characterised by a high thermal insulating effect as well as a high mechanical damping. Depending on the plastic and blowing agent used, the foam products can vary in their property profile (e.g. hardness, flexibility, density). It is particularly important to achieve a cell structure, that is as fine-pored and homogeneous as possible, as higher mechanical properties (e.g. compressive strength) of the foamed products are achieved through fine-cell foam morphologies. The cell structure depends on a number of different process and material parameters. To achieve small cell structures, for example, the deformation resistance of the plastic should be high enough to prevent a strong growth of the cells. The deformation resistance is influenced by the local deformation speed, the deformation history as well as the temperature [Kro99]. However, the extensional rheological properties and their characteristics are strongly dependent on the molecular polymer structure, its degree of branching and the molecular mass distribution. Thus, the extensional viscosity usually decreases at higher strain rates of the melt, but can even lead to strain hardening depending on the polymer structure [Bus10, Mü80, MSM05].

Although the extensional viscosity is one of the most important parameters for assessing the foamability of plastics, a comprehensive understanding of the interaction between the molecular structure of the plastic used and the resulting foam properties in foam extrusion is still lacking. In addition, many plastics (especially polyesters) have insufficient melt strength, so that the high melt deformations that occur during foaming often cause the cell walls to collapse. This undesirable effect can usually only be countered by modifying the plastics at the molecular level or by specially developed additives. However, when verifying this improvement, a foaming test only provides a qualitative rather than a quantitative statement. From an economic point of view, there is a need to constantly reduce the costs of manufacturing the foam products using existing plant technology and to efficiently drive forward product developments. For example, in order to be able to react to new customer requirements. Summarised, many plastics can only be foamed by taking into account narrow process limits or by using modified materials and additives. However, product development and the determination of a suitable process window is often an elaborate, iterative process.

The development of a practicable measurement methodology in the field of extensional rheology as well as the deepening of the process and material understanding of the influence of extensional rheology on foam production should make it possible in the future to adapt the process and material to the needs of the processor and to substitute additives or material modifications.

2.2 Iterative-analytical calculation of a rheometer geometry

An iterative-analytical design methodology was developed with the objective of determining a rheometer geometry that can reproduce the biaxial stretching process resulting from foaming. It is designed to be mounted to a foam extruder and enables a process-integrated strain rheological material characterisation. The approach is based on the idea of accelerating the melt in an annular gap-shaped flow channel in both the flow and circumferential directions (see Figure 2.1).

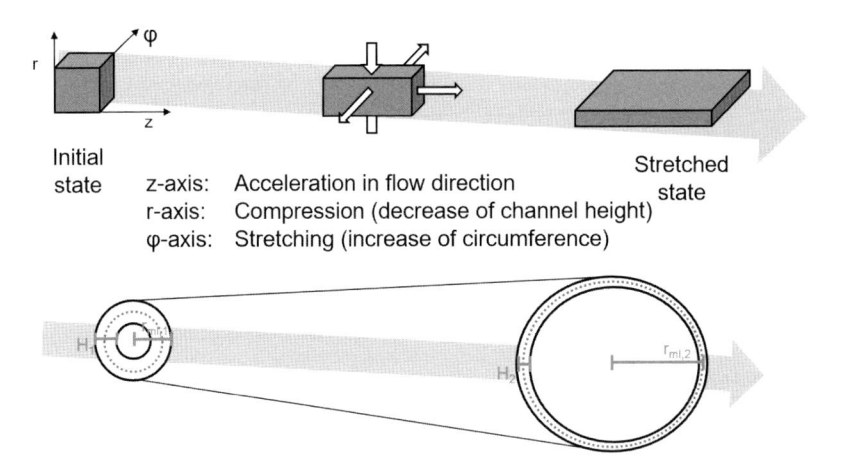

Figure 2.1: Biaxial deformation of a volume element and schematic representation of the inlet and outlet cross-section of the rheometer [HRS+21]

Acceleration in the direction of flow occurs due to a decreasing height of the flow channel. At the same time, the acceleration in the circumferential direction occurs through a widening of the flow channel. In order to achieve an approximately constant and equibiaxial strain rate, the geometry must be adapted to the local velocities. A calculation rule is used, which is based on the method of representative sizes and decomposes a complex flow channel into a number i of analytically solvable annular gap segments (AGS). Assuming an isothermal, laminar, stationary flow and a structural viscosity describable by means of the power law, the rheometer geometry is determined iteratively. Taking into account the stretching transverse to the direction of flow and specifying the desired transverse strain rate $\dot{\varepsilon}_{t,i}$ the radius of the centre line $r_{cl,i}$ of the flow channel is determined as a function of the rheometer length z. For $\dot{\varepsilon}_{t,i}$ it is assumed for simplification that Equation 2.1 applies:

$$\dot{\varepsilon}_{t,i} = \Delta_{circumference} \Delta_{time} = \frac{rel.\ change\ of\ circumference\ to\ AGS\ (i+1)}{average\ residence\ time\ in\ AGS\ i} \qquad (2.1)$$

With the value of $r_{cl,outlet}$ at the outlet as a design variable, $r_{cl,i}$ can be determined for all annular gap segments upstream. Consideration of the stretching in the flow direction is based on the height h_i of the flow channel at the respective discrete point of the flow path. Given the desired longitudinal strain rate $\dot{\varepsilon}_{l,i}$, the heights h_i of the flow channel are determined as a function of the rheometer length z. For $\dot{\varepsilon}_{l,i}$ it is assumed for simplification that the relationship in Equation 2.2 applies:

$$\dot{\varepsilon}_{l,i} = \frac{\Delta_{velocity}}{\Delta_z} = \frac{change\ in\ mean\ velocity\ to\ AGS\ (i+1)}{length\ of\ the\ AGS\ i} \qquad (2.2)$$

With the aid of the solver function of the Excel calculation programme (Microsoft Corporation, Redmond, USA) the squared error sum of nominal and actual values of $\dot{\varepsilon}_{l,i}$ is minimised by adjusting all h_i. Here, the residence times and radii $r_{cl,i}$ are automatically adjusted. The outer and inner walls of the flow channel with biaxial stretching effect are thus composed of individual points of the annular gap segments. Each point on the flow channel is defined by an r- and a z-coordinate (radius and rheometer length). The distance between two neighbouring points is again determined by the total length of the rheometer and the number of annular gap segments. The flow channel design is based on a resolution of 50 annular gap segments and a specified strain rate of 3 s^{-1}. The constant strain rate allows for a calculation of the extensional viscosity. The strain rate was chosen with regard to the total pressure loss and in accordance with the literature. Typical expansion rates in foam extrusion are estimated to be up to 9 s^{-1}, with the blowing agent-loaded melt expanding by a factor of 20 to 50 at the die exit [Gen05, Kro99]. Figures 2.2 and 2.3 show the changes in the flow channel contour of the first 20 mm of the extensional rheometer, which have been determined by the iterative-analytical design methodology. As the changes in the flow channel are most pronounced in the first section, the illustration is limited to this area. Based on the linear progression of the outer and inner wall of the initial flow channel contour (Figure 2.2, left), circumferential optimisation leads to an increase in the mean flow channel radius (Figure 2.2, right) at a constant flow channel height. The combination of circumferential and height optimisation leads to a decrease in the flow channel height with a simultaneous increase in the mean flow channel radius (Figure 2.3, left) and thus theoretically to a biaxial expansion of the melt. Due to the constant flow channel height, the first 7 mm of the flow channel are not taken into account in the transfer to a die concept. This results in the flow channel depicted in Figure 2.3 on the right side.

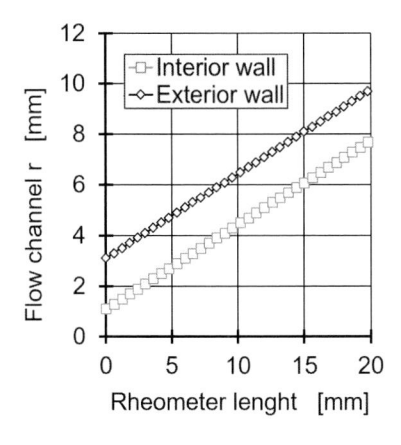

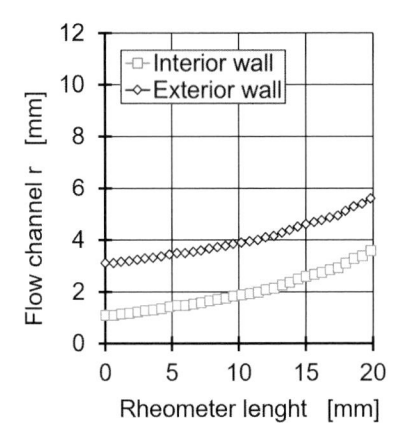

Figure 2.2: Development steps up to the flow channel contour with biaxial stretching effect: initial flow channel contour (left), flow channel after circumferential optimisation (right)

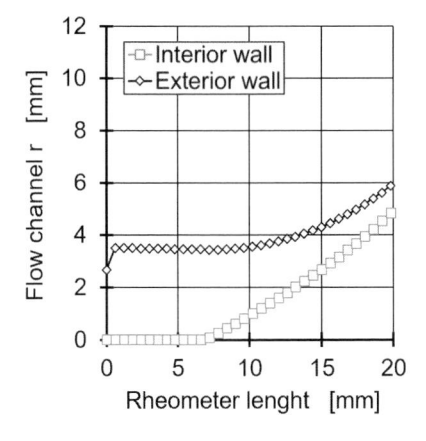

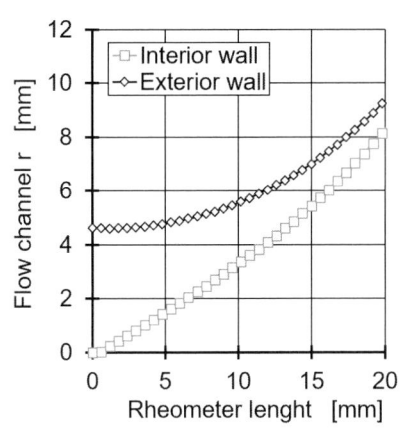

Figure 2.3: Development steps up to the flow channel contour with biaxial stretching effect: flow channel after circumferential and height optimisation (left), flow channel after eliminating the first 7 mm of the flow channel (right)

The contour executed as a rotated part around the z-axis provides the schematic representation of the complete measuring section with biaxial stretching effect (Figure 2.4). The pressure loss is also depicted.

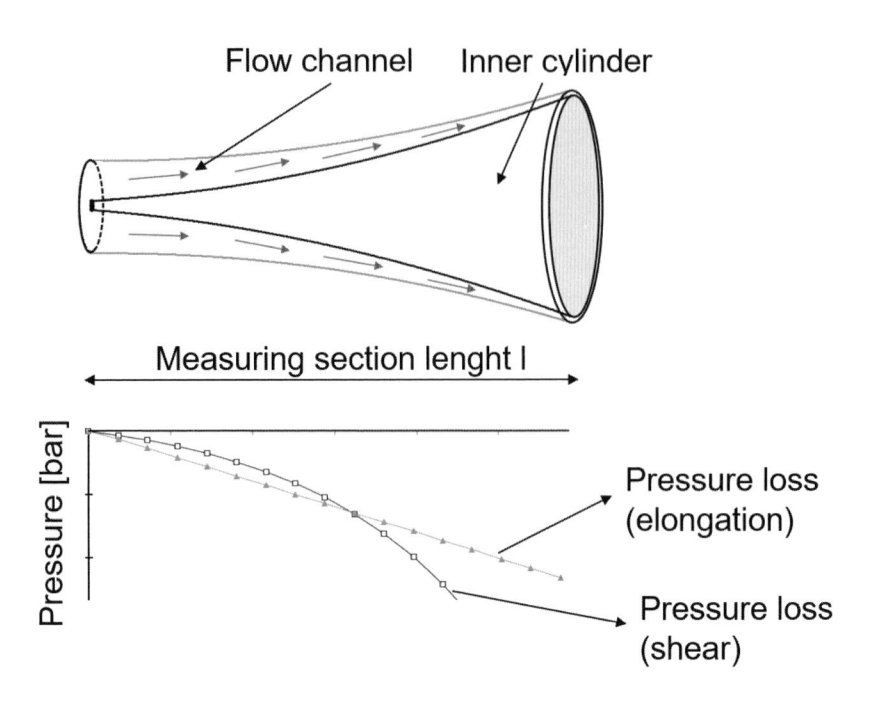

Figure 2.4: Schematic representation of the measuring section of the extensional rheometer (top); calculated pressure loss due to shear and strain based on the design methodology (bottom) [HRS+21]

To determine the biaxial extensional viscosity and thus to characterise the material close to the process, the total pressure loss is continuously recorded at three positions with pressure transducers along the measuring section during the extrusion process. This cumulative total pressure loss is composed of a shear and an elongation component (compare Figure 2.4). The pressure loss of the shear viscosity is strongly dependent on the flow channel geometry and can be determined in each annular gap segment using the formula for a slit die according to [Mic04] and subtracted from the total pressure loss. The resulting strain pressure drop (Δp_{elong}) can be determined according to Equation 2.3 and must be linear at constant strain rates.

$$\Delta p_{elong} = \Delta p_{measured} - \Delta p_{shear} \tag{2.3}$$

From the determined pressure loss of the strain, the extensional viscosity $\eta(\dot{\varepsilon})$ can thus be calculated according to Equation 2.4 with the strain rate ($\dot{\varepsilon}$) specified during the design:

$$\eta(\dot{\varepsilon}) = \frac{\Delta p_{elong}}{\dot{\varepsilon}} \tag{2.4}$$

2.2.1 Review of the design methodology using OpenFOAM

To verify the design methodology based on the simplified assumptions, the "real" behaviour of melt was simulated. The open-source flow simulation software OpenFOAM, OpenCFD Ltd./ ESI Group, Paris (France) was used for this purpose. From the comparison of simulation and iterative-analytical solution, correction factors of strain rates and pressure losses for the analytical calculation method can be determined. Figure 2.5 visualises the biaxial stretching effect of the iterative-analytical designed flow channel. It is sufficient to consider only one gap of the flow channel. A positive side effect is the significant reduction of the calculation time for the simulation. For simple illustration, the biaxial stretching caused by the flow channel contour is shown here on the basis of the deformation of the volume elements (particle cloud) added at the inlet along the flow channel. For the residence times 0 s, 0.25 s and 5 s the particles are shown.

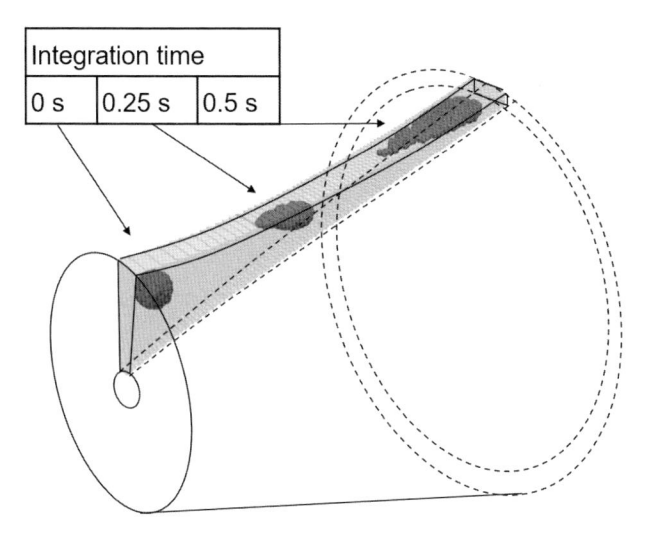

Figure 2.5: Illustration of the biaxial stretching effect emanating from the iterative-analytically designed flow channel with a length of 70 mm

At the flow channel inlet (residence time = 0 s), an ideally spherical particle cloud is present. Due to the stretching in the circumferential and extrusion direction, the particles are already arranged flatly in the middle of the flow channel (after approx. 0.25 s), whereby a greater expansion in the extrusion direction than in the circumferential direction is already visible at this point. At the end of the flow channel, the expansion of the particle cloud in the extrusion direction is almost twice as large as the expansion in the circumferential direction. It can be seen, that the melt is stretched both in the longitudinal and in the transverse direction by the determined flow channel geometry and thus a biaxial stretching is present in the flow channel contour considered. Even if equibiaxial stretching could not yet be achieved, the rheological design of a biaxial extensional rheometer could be carried out on the basis of the iterative-analytical design methodology. Before further efforts are made to optimise the biaxial stretching and to approach equibiaxial stretching ratios by means of different degrees of longitudinal stretching rates, the focus is first on the mechanical realisation of a rheometer die.

2.3 Die design for the extensional rheometer

In order to transfer the iterative-analytical flow channel into a die that can be connected to an extruder different concepts were developed and evaluated. The holistic presentation of the design process according to the VDI standard 2221 [NN93] is omitted here. Instead, selected aspects of the list of requirements are taken into account and discussed together with the concepts presented. The focus of the chapter is therefore the mechanical engineering solution for an inline elongation rheometer with biaxial stretching effect, which enables the characterisation of (gas-loaded) melts under process conditions. The requirements for the geometry of the die are derived from the calculated flow channel with biaxial elongation. The dimensions of the measuring section and thus the exact position and centring of the outer and inner wall in relation to each other are defined and must be ensured by the corresponding concepts. The connection of the rheometer die to the extruder of the foam extrusion unit is realised via appropriate spacer parts and flanges. Since the extensional rheometer can be used on any extrusion lines with only corresponding transition channels, the extruder connection is not shown in the illustrations and is not discussed further. Pressure sensor holes are integrated at three points along the flow path in order to determine the pressure losses due to elongation and shear. Due to the small length of the developed measuring section and the curved free-form surface of the flow channels outer wall, it is not possible to align and mount the pressure sensors flush with the flow channel. The pressure sensors are therefore placed in so-called pressure holes. A capillary with a diameter of 2 mm connects the pressure sensor chamber with the flow channel. This offers the advantage, that the pressure losses can be determined with pinpoint accuracy. When calculating the shear-viscous pressure losses, it is therefore not necessary to take the average over the membrane diameter of the sensor. In order to ensure that the temperature control is as close to the contour as possible, it is planed the research project to manufacture the extensional rheometer using the selective laser melting process (SLM). Since the precise production of the free-form surfaces of the flow channel is not possible with this process without further processing steps conventional manufacturing methods will be used for the production of the first prototype. In addition, the two rotationally symmetrical components of the flow channel (outer and inner cylinder) allow homogeneous tempering by means of conventional heating tapes or heating cartridges.

2.3.1 Concepts for centring the measuring section in a die design

The first die concept (Figure 2.6) provides for an attachment of the inner cylinder after the measuring section of the flow channel with a biaxial stretching effect. Coming from the extruder, the melt is conveyed through an inlet area of constant diameter, passes the tip of the inner cylinder head-on and is evenly guided into the annular measuring section. After passing through the metering section, the melt emerges from the annular outlet gap. The area where the melt exits is at the same time the area in which the mounting and centring of the inner cylinder is ensured. The cylindrical extension of the inner cone enables the fit in a plate, which in turn enables the fastening to the outer cylinder with screws and centring pins.

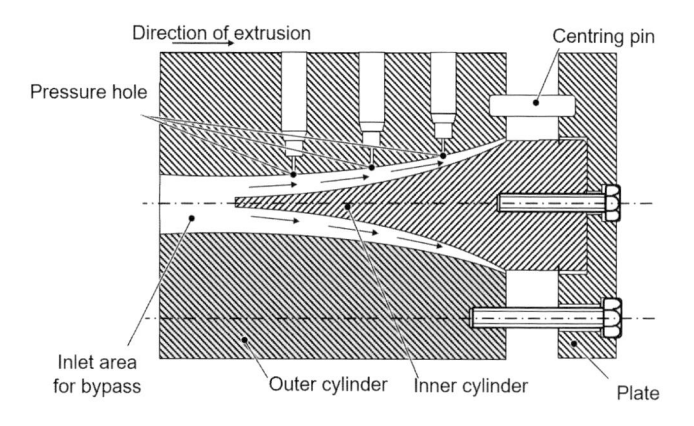

Figure 2.6: Concept 1 with an attachment after the measuring section

Die concept number 2 (Figure 2.7) shows a possibility of mounting the outer and inner cylinder in the area of the measuring section. Deviating from the rotational symmetry of the iterative-analytical design, the concept provides for a semicircular outer (housing) and inner cylinder, which are centred and mounted by using a plate. In order to reduce or exclude an influence of the biaxial strain by the base plate, the pressure sensors are integrated at the highest point orthogonal to the base plate in this concept. For the connection to the extruder and thus the transfer of the melt from the circular cross-section of the extruder cylinder to the semi-circular annular gap of the measuring section, a transition component with a complex inner contour is required. This makes the realisation of this concept even more complex in addition to the semicircle cylinders, that can only be produced on a simultaneous 5-axis milling machine.

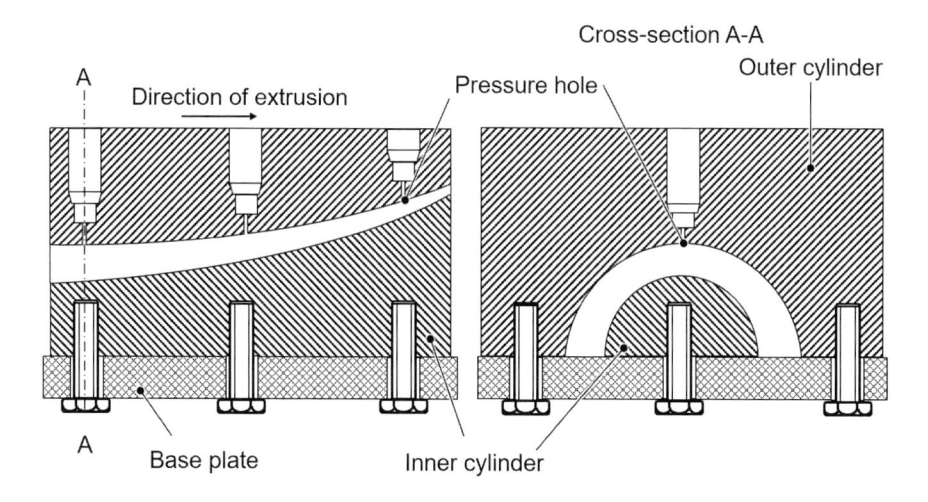

Figure 2.7: Concept 2 with an attachment in the middle of the measuring section

The mandrel dies, which are not only typically used in foam sheet production, are the basis for the die concept shown in Figure 2.8 The tip of the inner cylinder is attached to the outer cylinder by means of bars (so called spider legs). This concept thus represents a possibility of mounting directly in front of the measuring section. The difficult mountability (split outer housing), the additional design of the webs with regard to the forces acting on them and the flow redistributions in the measuring section caused by the webs are major disadvantages.

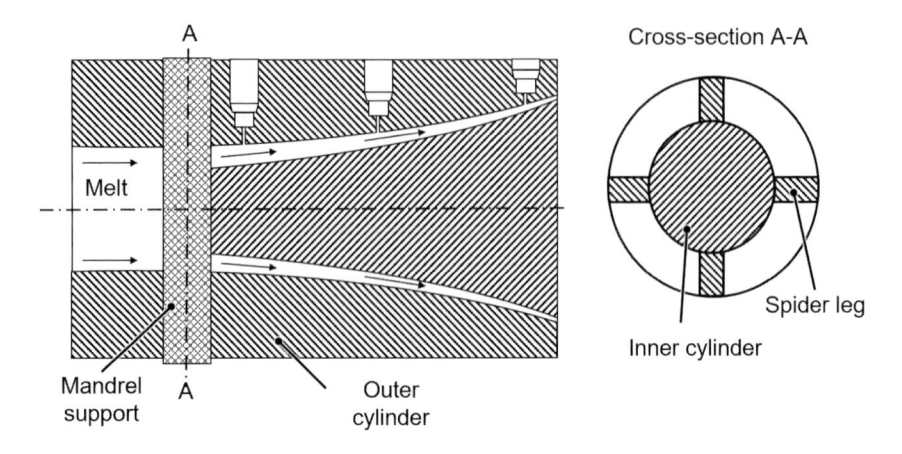

Figure 2.8: Concept 3 with an attachment before the measuring section

Three different approaches to a die concept have been presented. They all have in common that they contain the iterative-analytically designed flow channel, or at least a section of it. In addition, all these concepts allow a connection to an extruder for feeding with (gas-loaded) melt. In order to demonstrate the mechanical feasibility of a flange-mountable biaxial extensional rheometer for the first time without further increasing the degree of complexity, the aspect of inline suitability was excluded from the concept development for the time being. This will be taken into account in a new tool concept after the commissioning of the first prototype. Table 2.1 summarises the core aspects of the three concepts.

	Concept 1	Concept 2	Concept 3
Position of the mounting of the outer and inner cylinder in relation to the measuring section	After the measuring section	Parallel to the measuring section	Before the measuring section
Fixing and centring of the inner cylinder	Fixing with fitted plate at the end of the cylinder; centring via centring pins	Fixation and centring ensured via horizontal plate	Fixing and centring due to mandrel support
Influence and orientation of the melt by the mounting system	Influence only after leaving the measuring section	Influence of the melt flow in the boundary areas by the plate.	Influence due to the bars
Inline suitability	Only possible in bypass flow	Only possible in bypass flow	Possible by modifying the design

Table 2.1: Comparison of the three die concepts

2.4 Conclusions and outlook

A novel developed iterative-analytical design methodology was used to design an extensional rheometer, that can be used for the process-integrated determination of the biaxial extensional rheological material properties. In a first design step, a flow channel geometry for the implementation of the new elongation rheometer concept was determined for a biaxial strain rate of 3 s^{-1}. The iterative-analytical method utilises the subdivision of the flow channel into analytically solvable segments. The circumference of the segments was determined by backward calculation and the channel height of the flow channel was iteratively optimised using the method of least squares. A constant strain rate in extrusion direction and in the transverse direction can be realised, which results in a biaxial stretching effect. The visualisation conducted by the flow simulation OpenFOAM confirmed a biaxial stretching effect. However, the longitudinal strain rate is more pronounced than the transverse strain rate. In order to achieve equibiaxial stretching, the flow channel design will be iteratively further developed.

Based on the presented flow channel, the requirements for a die design were listed with rinciple solutions being determined and discussed in detail. Different concepts for the die design were carried out and evaluated with regard to its use as an inline rheometer.

A first prototyp according to the concept that provides for an attachment of the inner cylinder after the measuring section of the flow channel with a biaxial stretching effect has been manufactured and has yet to be commissioned. Further developments of the rheometer will be made and used to determine different biaxial strain rates in the foam extrusion process. Deepen knowledge of elongational values with and without blowing agent can thereby be obtained and thus used to investigate the correlation with the foaming ability.

Acknowledgements

The research project 21302 N has been sponsored as part of the "Industrielle Gemeinschaftsforschung und -entwicklung (IGF)" by the German Bundesministerium für Wirtschaft und Klimaschutz (BMWK) due to an enactment of the German Bundestag through the AiF. We would like to extend our thanks to all organizations mentioned.

The authors would also like to thank the companies Sabic Europe, Geleen (Netherlands) and Borealis AG, Vienna (Austria) for providing the test materials.

References

[Bus10]   BUSSMANN, M.: *Ein kalibrierbares integratives Modell zur Beschreibung des Schlauchbildungsprozesses in der Blasfolienextrusion.* Universität Duisburg-Essen, Dissertation, 2010

[Gen05]   GENDRON, R.: *Thermoplastic Foam Processing - Principles and Development.* Boca Raton, USA: CRC Press, 2005

[HRS+21] HOPMANN, C.; REINHARDT, N.; SCHÖN, M.; FRINGS; FACKLAM, M.: Verstreckung messen - Schaum verbessern. *Kunststoffe* 111 (2021),

[Kro99]   KROPP, D.: *Extrusion thermoplastischer Schäume mit alternativen Treibmitteln.* RWTH Aachen University, Dissertation, 1999

[Mic04]   MICHAELI, W.: *Beurteilung der Verschäumbarkeit von Polymeren anhand des biaxialen Spannungs-/Dehnungsverhaltens.* Institut für Kunststoffverarbeitung, RWTH Aachen, Schlussbericht zum IGF-Vorhaben Nr. 13188 N, 2004

[MSM05]   MÜNSTEDT, H.; STEFFL, T.; MALMBERG, A.: Correlation between rheological behaviour in uniaxial elongation and film blowing properties of various polyethylenes.. *Rheologica Acta* 45 (2005) 1, p. 14–22

[Mü80]   MÜNSTEDT, H.: Dependence of the Elongational Behaviour of Polystyrene Melts on Molecular Weight and Molecular Weight Distribution.. *Journal of Rheology* 24 (1980) 6, p. 847–867

[NN93]   N.N.: *VDI 2221: Methodik zum Entwickeln und Konstruieren technischer Systeme und Produkte.* Düsseldorf, Germany: Verein Deutscher Ingenieure, 1993

Symbols

Symbol	Unit	Description
$\dot{\varepsilon}_{l,i}$	$1/s$	longitudinal strain rate of the annular gap segment i
$\dot{\varepsilon}_{t,i}$	$1/s$	transverse strain rate of the annular gap segment i
h_i	mm	flow channel height of the annular gap segment i
l	mm	length of the measuring section of the rheometer
$\eta(\dot{\varepsilon})$	Pas	elongation viscosity
Δp_{elong}	bar	elongational pressure loss
$\Delta p_{measured}$	bar	measures pressure loss
Δp_{shear}	bar	shear pressure loss
r	mm	radius

Abbreviations

Notation	Description
AGS	annular gap segment
SLM	selective laser melting

3 Influence of the die design on the foam structure of foamed multilayer blown films

Ch. Hopmann[1], M. Stieglitz[1]
[1]*Institute for Plastics Processing (IKV) at RWTH Aachen University*

Abstract

The combination of multilayer blown film extrusion with the technology of foaming offers the possibility to use the advantages of both processes. For example, foamed films can be pro-duced for food and industrial packaging, which have a higher functionality (thermal insulation, mechanical damping etc.).

One of the most important parameters in the foaming of blown films is the resulting foam structure. The objective is to achieve a homogeneous foam with as fine a cell structure as pos-sible and a significant reduction in density. The cell structure of the foam produced during blown film extrusion depends on a multitude of influencing factors. In addition to material-side and process-side influences, the die used to form the melt tube also has a major influence on the foam structure.

The influencing factor in the extrusion die is the pressure loss in the melt at the die outlet. By adjusting the die outlet in different areas, the foam structure can be changed. Through an analytical calculation of the pressure loss, the residence time and other process variables such as the flow velocity, the effects of the different areas of the die design could be obtained. In general, an increase in pressure gradient at the die outlet leads to a high cell density and a decrease in cell size. Five novel die geometries were constructed, analytically calculated and manufactured, so that the foam structure could be investigated on produced blown film. The extrusion tests and subsequent analysis of the foam structure, in terms of density and cell size, provided results for the selection of a foaming optimised die design. With this die design, blown films with a fine and homogeneous cell structure could be produced in a stable process.

3.1 Introduction and motivation

The constantly increasing economic competition as well as the demand for a sustainable life cycle of packaging materials is confronting film producers with ever greater challenges [ESE20]. Due to the introduction of regulations in the packaging sector, material con-sumption is increasingly coming into focus. At the same time, the reduction in material consumption leads to a reduction in material costs, which account for up to 80% of film production costs [HHHH14].

The further development of technical processes offers the possibility to reduce the material input and thus has great potential to increase the economic efficiency of blown film extru-sion. The saving of plastics can be realised by foaming, which reduces the density. In this way, material costs and the amount of raw materials can be optimised [Sta06]. Foamed plas-tics are also characterised by improved properties in certain areas. For example, foaming increases the thermal insulating effect [Tol15].

The combination of plastic foaming with blown film extrusion offers a highly efficient pro-duction process for manufacturing novel film products (see Figure 3.1). Foaming in the middle layer of multilayer blown films offers several advantages. Due to the compact outer layer, all possibilities remain open about appearance, printability, haptics and possible lam-inating or sealing functions. Foamed packaging is used, for example, to protect sensitive food such as fruits, so that damage and thus premature spoilage is prevented. Because of the good thermal insulation of the foams, they are also used for packaging convenience food in order to keep the temperature change of the food as low as possible [URL21].

- Compact outer layer: mechanical strength, functionalisation (sealing, printing, barrier)
- Foamed middle layer: density reduction, functionalisation (thermal insulation, mechanical damping, high stiffness)
- Compact outer layer

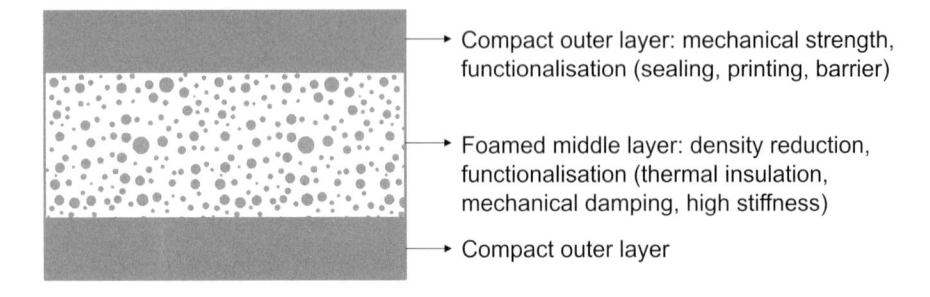

Figure 3.1: Multilayer film with foamed middle layer

The layer thickness ratio is largely responsible for the property profile of the foamed multilayer blown film. The foam structure should not be visible on the surface and considerable material savings must be aimed for. The resulting foam cells also influence the mechanical properties of the film, which depend on the foam structure. For this reason, the objective in foam extrusion is to achieve a structure, that is as fine-cell as possible, since this results in higher mechanical properties of the film compared to a large-cell structure. The foam structure depends on material parameters such as extensional viscosity and process parameters such as temperature or pressure, and also on the die technology. The process understanding of foam extrusion is not yet sufficiently, so that further investigations and developments are needed. In the further section, the optimisation of the die technology will be dis-cussed, which can be varied to achieve a fine-cell foam structure for blown films. In this con-text, the influence of the die design is of particular importance.

3.2 Foaming in the blown film extrusion

In plastics technology, foaming is achieved by adding a blowing agent. A distinction is made between chemical and physical foaming based on the mode of action. In physical foaming, a liquid or gaseous blowing agent (e.g. nitrogen or carbon dioxide) is injected directly into the plastic melt. However, this requires special extrusion technology and cannot be carried out with conventional extrusion lines. In previous research work at the IKV [MH10], a cost-effective retrofit option was developed to foam the middle layer of a multilayer blown film. The disadvantage of the additional extrusion technology in physical foaming is compensated by the significantly higher density reduction compared to chemical foaming.

In the chemical foaming process considered below, a chemical foaming agent is added in form of a masterbatch and following chemical reaction to achieve foaming of the plastic. The advantage of using chemical foaming agents is, that they can be used in conventional extrusion lines and allow a simplified process control compared to physical blowing agents [Dau08]. Many different chemical blowing agents are available on the market, but their use is limited by the low gas output, so that foamed films can only be produced with a film density of at least $350 kg/m^3$ [Weg05]. The theory of foam growth applies to both chemical and physical foaming, according to which it is divided into four stages (Figure 3.2).

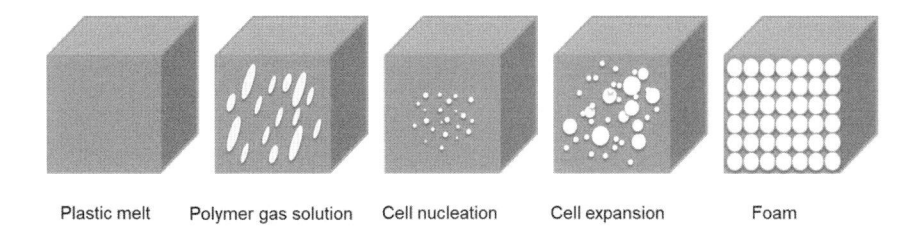

Figure 3.2: Schematic illustration of the sequence of foam extrusion [Gla17]

The process starts with the formation of a single-phase polymer-gas solution, which is achieved by good mixing of the blowing gas and the polymer melt. The solubility of the blowing gas in the polymer melt depends on the temperature, the pressure and the interactions be-tween polymer and blowing gas. Undissolved or excess gas leads to the formation of large cavities in the polymer matrix. In addition to solubility, diffusion is essential for the formation of a single-phase polymer-gas solution. Diffusion describes the distribution mechanism of the gas within the polymer melt and is driven by concentration differences in the mixture [Sta06, Zha10].

During subsequent cell nucleation, stable cell nuclei capable of growth are formed in the polymergas solution. Nucleation is initiated by a change in thermodynamic equilibrium in the form of a pressure loss at the die outlet or temperature increase [Zha10]. The change results in reduced solubility of the gas in the polymer melt. Parts of the gas dissolve from the melt and form a gas phase [Sta06].

Once a so-called critical cell radius is reached, this is called cell growth. This is largely diffusion-controlled and dependent on the gas exchange of the cell with the surrounding polymer matrix as well as the rheological behaviour of the polymer matrix [Kos96].

To ensure that the foam structure is as fine as required, the growing foam cells must be stabilised in time. This is done by increasing the viscosity of the polymer melt by cooling it down. A rapid melt cooling leads to a rapid viscosity increase and a short growth phase. Only parts of the gas diffuse into the cells during this time, and a low degree of foaming with small cells is achieved [Lan95, Sta06].

Besides the pressure loss mentioned before, the foam structure also depends on the process control. For blown film extrusion it was shown by IKV [Hop18], that the foam structure be-comes more fine-cellular with increasing film thickness. The production of thin films with small cells can be achieved conventionally by high blowing ratios and take-up ratios. But these ratios are very limited for blown films. For example, take up ratios greater than two lead to very strong stretching and thus enlargement of the foam cells. A technological approach to extrude thin films is to vary the gap width of the die used. This also increases the pressure loss at the die outlet. This is desirable because, as described in more detail above, the cell structure is improved. A significant reduction in the die outlet gap can result in the blown film extrusion process not running stable.

3.3 Research on the development of a new die design for improved foam structure

In order to avoid high degrees of stretching, which lead to thinning of the film and stretching of the foam cells and to further optimise the foam structure, it is necessary to investigate the influence of different die geometries in more detail. Achieving a fine-cell foam structure is important to achieve maximum density reduction while maintaining good mechanical film properties. The objective is therefore the development of a novel die design. The influence of the different die areas and the die gap is used to adapt the foam structure in a systematic way.

3.3.1 Design and analytical investigations of different blown film dies

In the first step, the standard die design, which was not optimised for foaming, was examined in detail and considerations were made, how the die design should be varied. The outlet of an extrusion die for blown film extrusion consists in most cases of a reduction of the diameter at a certain angle, a parallel zone as well as the outlet gap. These three design parameters, shown in Figure 3.3, influence the behaviour of the plastic melt in the extrusion die. The parallel zone length L_{St} designates the section in the die gap where there is no longer any reduction in the diameter. Thus, the die gap is constant over the parallel zone length. The die gap D_g corresponds to the thickness of the unfoamed film tube exiting the die. The angle determines how fast the diameter changes to the final die gap. After designing the reference die with an outlet gap of 0.7 mm and 1.5 mm, the effect of the change of the three parameters was investigated with help of analytical calculations. Furthermore, the basic shape of a conical die is also maintained. For the analytical calculations, the die is considered as an annular gap and further simplifications are made according to [Hop16]. The outer die diameter of the blow head is 80 mm and was kept constant as well as the mass throughput of 15 kg/h. The calculations are based on the material parameters of PE-LD.

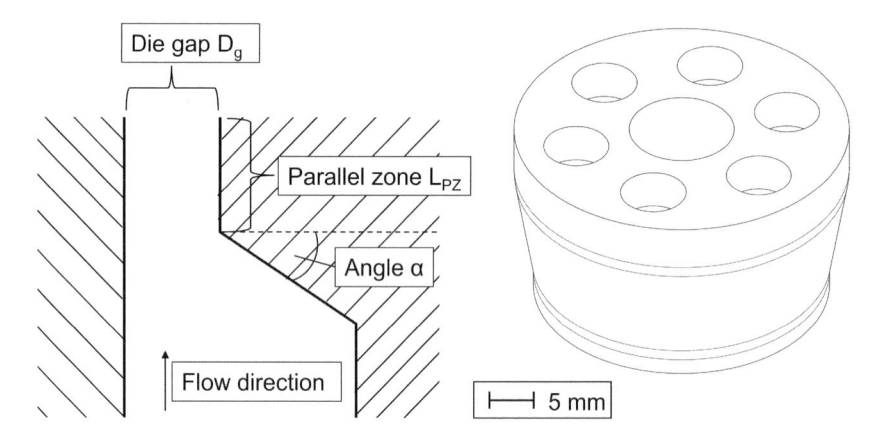

Figure 3.3: Schematic layout of the three design parameters of the die and CAD model of the die insert

Figure 3.4, Figure 3.5 and Figure 3.6 summarise the results of the analytical calculations of the pressure loss for the individual die areas. A larger pressure loss at the die outlet leads to a greater cell density and a simultaneous decrease in cell size [Hei02]. This is due to the nucleation rate, which increases exponentially with the pressure loss [Kos96]. The increase in cell density or decrease in cell size continues to decrease with increasing pressure loss. But when the pressure loss is low, only a few cell nuclei are nucleated. A pressure loss over a short distance as well as in a short time leads to a better foam structure. The two diagrams in Figure 3.4 are used to illustrate the effect of the varying parallel zone length (6.5, 11.5, 16.5 mm) on the pressure loss. The upper diagram shows the course of the die ge-ometry by means of the die radius. The influence of the parallel zone lengths on the pressure curve is illustrated below. The loss in pressure in the range of 20 - 30 mm die length is steeper with increasing parallel zone length. In the outlet area, the pressure curve of all three parallel zone lengths is about 10 bar. Furthermore, the loss in pressure starts later with the shorter par-allel zones, so that a longer parallel zone length leads to an increase in pressure loss.

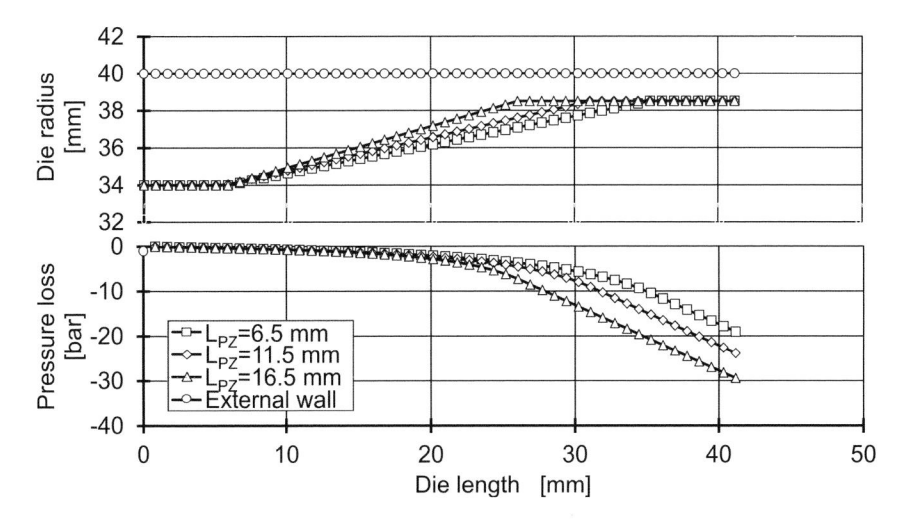

Figure 3.4: Effects of the parallel zone length on the pressure loss

Figure 3.5 shows the geometries of three dies with different gaps at the die outlet (1.5, 1, 0.5mm) in the upper image and the resulting pressure curve in the lower image. Due to the reduction of the die gap, the melt flow cross-sectional area decreases. This necessarily induces an increase in pressure. However, the pressure loss in the gap changes significantly more with the reduc-tion from 1 mm to 0.5 mm than with the reduction from 1.5 mm to 1 mm. This results in a pressure loss of 163 bar at the outlet with a die gap of 0.5 mm compared to 23 bar with a die gap of 1.5 mm.

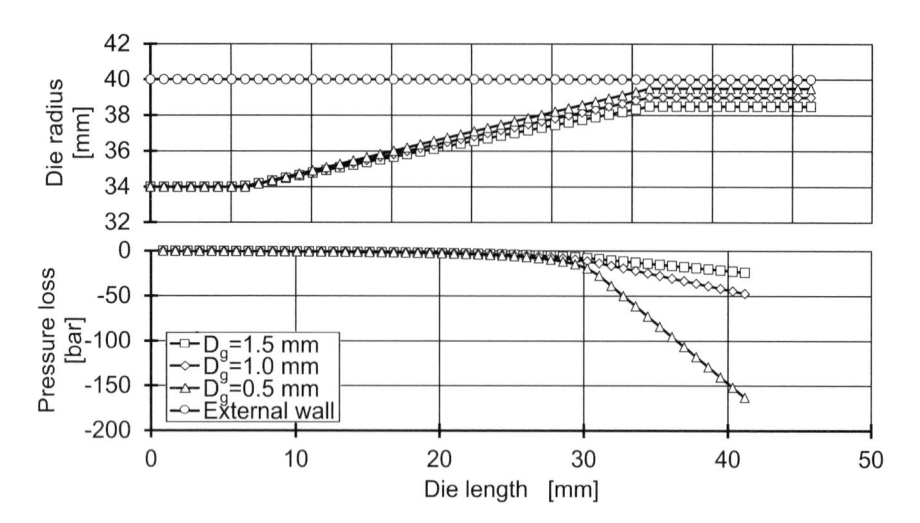

Figure 3.5: Effects of the die gap on the pressure loss

Finally, the images in Figure 3.6 show the influence of the angle, of the cross-sectional reduction, on the pressure curve. For this purpose, the pressures were calculated for four angles (80, 70, 60, 40°). The same effect results as with the variation of the parallel zone length. The loss increases with greater angels and starts earlier. However, the influence of the different angle is not as significant than with the variation of the parallel zone length or die gap.

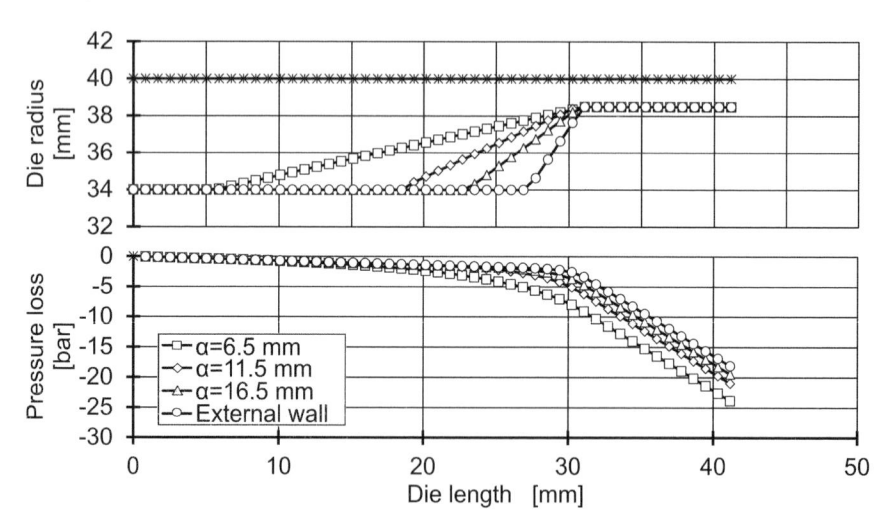

Figure 3.6: Effects of the angle on the pressure loss

Based on the effects of the three parameters angle, parallel zone length and die gap, a re-design of the die was made and then also analytically calculated. Figure 3.7 shows the die curves of all five dies. These differ from each other by the length of the parallel zone, the width of the outlet and the angle. In the process, care was taken to ensure that there was a large variance in the individual zones.

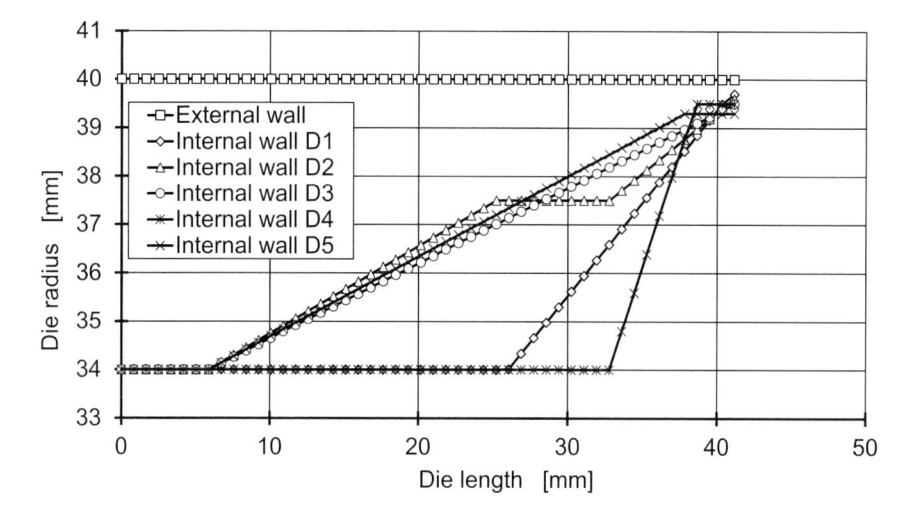

Figure 3.7: Gap shape of the re-designed dies

The differences between the re-designed dies in the three areas of outlet gap, parallel zone and angle are shown in Table 3.1.

Die Geometry				
Die Number	Die Gap: D_g [mm]	Parallel Zone Length: L_{pz} [mm]	Angle α [°]	Pressure Loss: δ_p [bar]
D1	Melt 0.3	0.0	70	34.4
D2	0.4	7.5	2 steps: 80 at z= 0-25 mm 76 at z =33-41 mm	36.4
D3	0.5	3.4	45	42.4
D4	0.5	0.0	80	38.3
D5	0.7	4.0	80	47.8

Table 3.1: Analytical calculation of pressure loss at $\dot{m}_{ges} = 15$ [kg/h]

In previous extrusion test with the reference die (D5) the extruder of the middle layer with a diameter of 35 mm (compared to the outer extruder D = 45 mm) reached its pressure limit of 300 bar at higher mass throughputs. To avoid this, the total pressure was reduced, especially for the small outlet gaps. Since the pressure gradient is also crucial for achieving a fine foam structure and a high nucleation rate, special attention was given to this parameter in the design, taking into it the knowledge gained previously. The total pressure loss of the individual dies is also shown in Table 3.1.

In the following, the results of the analytical calculations in Fig. 8 for the die with the 0.3 mm (D1) and in Fig. 9 for the die with the 0.7 mm (D5) outlet gaps are presented as examples. If the diagrams for the 0.3 mm gap are compared with those for the 0.7 mm gap, it can be seen, that with the die D1 the pressure loss occurs in a significantly shorter range and in a shorter time. The analytical calculations show, that with die D5 even a larger pressure loss results However, this is built up over a longer period of time. For the production of foamed blown films, it is expected that especially the novel dies (D1, D5) are characterised by a good foam structure, since the pressure loss with these dies occurs particularly quickly and over a short die length.

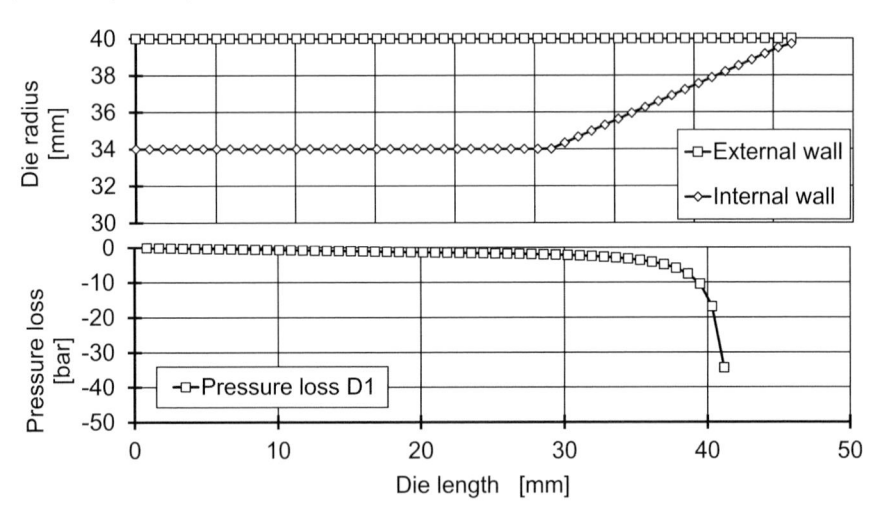

Figure 3.8: Influence of die design $D1(D_g = 0.3mm, L_{Pa} = 0mm, \alpha = 70)$ on pressure loss

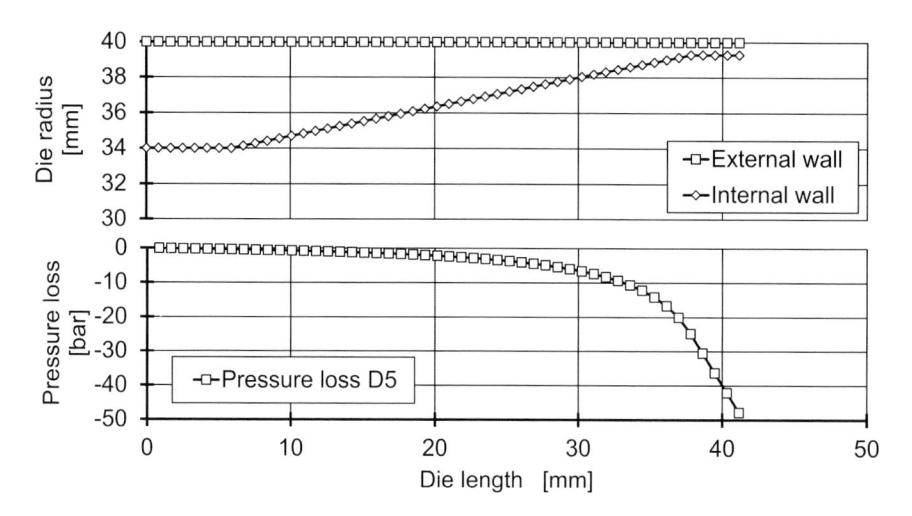

Figure 3.9: Influence of die design D5($D_g = 0.7mm$, $L_{Pa} = 4mm$, $\alpha = 80$) on pressure loss

3.3.2 Extrusion tests with different die designs

For the validation of the analytical investigation and to evaluate the process capability of the dies, extrusion tests were performed. The influence of the die design on the foam structure was carried out in a chemical foaming process. For this purpose, different proportions (3-15 %) of a chemical foaming agent (CFA) Hydrocerol from Avient, Avion Lake, Ohio, USA, were added to the plastic granulate in preliminary tests and the degree of foaming was investigated. Hydrocerol is a chemical foaming and nucleating agent, which is particularly suitable for the production of fine-cell foams. A chemical foaming agent improves the plasticisation of the plastic, which reduces the required heat input. This can save energy and improves the CO_2 balance of the process. The highest degree of foaming with a good foam structure was achieved with 7 wt.-% Hydrocerol and kept constant in the further tests.

For the production of a three-layer film, with compact outer layers and foamed middle layer, a three-layer blown film line from Kuhne Anlagenbau GmbH, St. Augustin, Germany, is used for these tests. The melt is distributed in an annular gap by a radial spiral distributor and exits through the re-designed and manufactured dies. To evaluate the influence of the die-dependent foaming behaviour during film production of the die design, the plastic material is kept constant for all film layers as well as for all tests. A low-density polyethylene 2102X0 from Sabic Europe, Geleen, Netherlands, was used. The MFI (190°C, 2.16 kg) of the material is $1.9g/10min$ [NN21]and the density is $921 kg/m^3$, which on the one hand reduces the foaming of the middle layer due to the high viscosity of the outer layers, as the added foaming agent reduces the viscosity of the middle layer. On the other hand, a long foam growth of the middle layer is avoided compared to a material with lower viscosity.

In order not to have to estimate the pressure loss at the die via the extruder pressure of the middle layer, a new die housing was designed and manufactured which enables the measurement of the melt pressure and the melt temperature approx. 60 mm before the melt exits.

Blown films were produced for the following analyses of the foam structures. The parameters of the test series are summarised in Table 2. For each die geometry, the mass flow rates are varied on two levels (8/11 kg/h), as they also have an influence on the pressure gradient and thus affect the cell size. For the individual process points, the parameters were kept constant, with the blow-up and take-up ratios playing a particularly important role. This ensured that films with a thickness of 125 µm were produced. The mass flow rate ratio of the two outer layers to the middle layer is two. The two outer layers were operated with the same mass flow rate of 1.

Material	Hydrocerol percentage [wt.-%]	Mass-throughput [kg/h]	Mass-throughput ratio [-]	Temperature All zones [°C]	Film Thickness [μm]	Blow-up-ratio [-]
PE-LD 2102X0	7	8/11	1:2:1	170	125	2

Table 3.2: Process parameters used in the extrusion trials

With the 10 test points, the influence of the die design on the process stability could be evaluated well. The process ran trouble-free with the reference geometry with an outlet gap of 0.7 mm (D5) as well as the two new dies with 0.5 mm (D3, D4). At the different process points, a three-layer structure with a foamed middle layer could be produced. The die D4 showed good production behaviour and better process stability compared to the die with the same outlet gap but different parallel zone and angle (D3). The dies with an outlet of 0.3 mm (D1) and 0.4 mm (D2) showed an increased number of film tears and it was not possible to set up a stationary process. However, film samples were still taken in order to be able to evaluate the foam structure afterwards. The very small gap meant that the expanding cells disrupted the very thin film in many places and thus no stability could be brought into the film tube.

3.3.3 Influence of the die design on the foam structure

The foam structure of foamed blown films is, next to the process stability during extrusion, the most important factor for the evaluation of the novel die designs. A fine-cell, homogeneous foam structure was aimed for. The two most important parameters for assessing foam quality are the density of the blown film and the cell size.

With the help of the density, the material savings can be evaluated. However, the density of the film does not allow direct conclusions to be drawn about the cell size or the volume of a single cell. To determine the total density of the three film layers, a circular sample of the film is taken. The thickness of the film is measured at 12 points equidistantly distributed on the radius. The film thickness is used to calculate the volume of the film. In addition, the weight of the sample is determined. With the presence of the two described quantities, the density of the total film can be determined (see Figure 3.10). Since the process points were kept constant for all the dies, the density of the total film can be compared directly with each other. It should only be noted that this is not the density of the foamed middle layer.

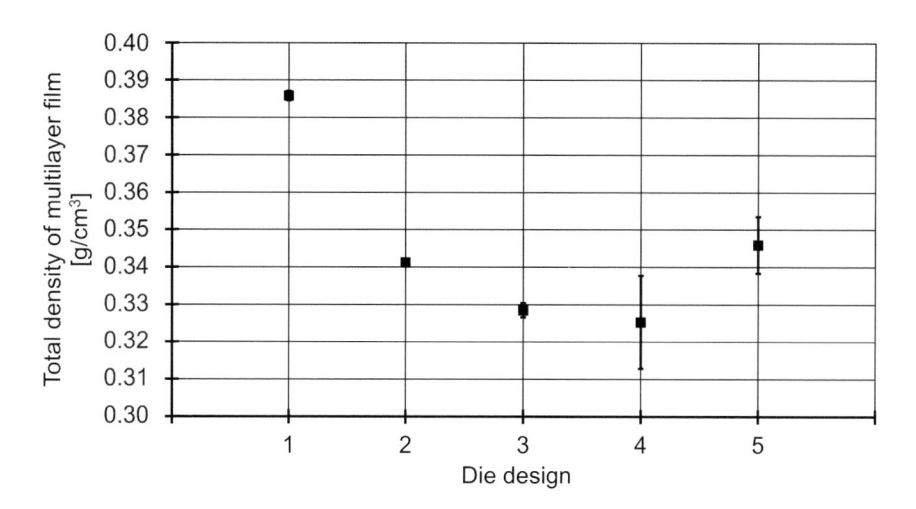

Figure 3.10: Influence of die design on the total density of the multilayer film

The total density of the multilayer film varies between 0.325 g/cm^3 and 0.386 g/cm^3 for all dies, so that the densities differ relatively little from each other. The highest value (0.386 g/cm^3) determined occurs with the die D1and the two lowest values of about 0,325 and 0.328 g/cm^3 were achieved by the dies with an outlet gap of 0.5mm (D3, D4). Since a low density can also be achieved with large cells, the next step is to examine the cell size. For the investigation of the cell size, a film sample was taken from each sample. For optical analysis images were taken with a camera COE-050-M-POE-050-IIR-C and the corresponding bitelecentric lens TC23036 from Opto Engineering, Mantova, Italy. The foam structures were then evaluated in two dimensions. For this purpose, lines were placed in the cells and the number of pixels along them was counted. The number of pixels is converted into a metric length by scaling. For the calculation of the cell area C_{area}, the cell is approximated as an ellipse (Equation 3.1). For each image, the cells were measured along the length and horizontally to the direction of the pull-off. C_a stands for the cell dimension along the pull-off direction and C_{trans} for the cell dimension transverse to the pull-off direction.

$$C_{area} = \frac{C_a * C_{trans} * \pi}{4} \qquad (3.1)$$

A comparison of the images of the foam structure (see Figure 3.11) with the cell sizes determined in Figure 3.12 illustrates, that the dies with the lowest outlet diameters (D1, D2) lead to the smallest cell size. The images are characterised by a very fine and homogeneous foam structure. This was to be expected, as the pressure gradient in the die gap increases with a smaller die gap. The dies D3 and D4, which produced the lowest film density achieve an average cell size. However, the D4 with no parallel zone is about 0.4 mm^2 below the other die. The reference die (D5) achieves the worst foam structure with the highest cell size on the images compared to the other structures. The relatively low standard deviation (see Figure 3.12) of all foam structures indicates a very homogeneous distribution of cell sizes.

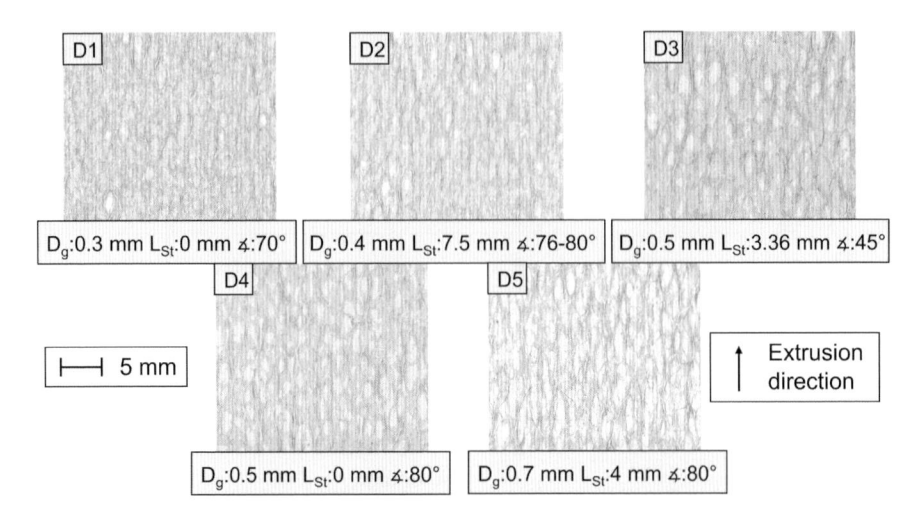

Figure 3.11: Images of the foam structure of the different die designs with a bitelecentric lens

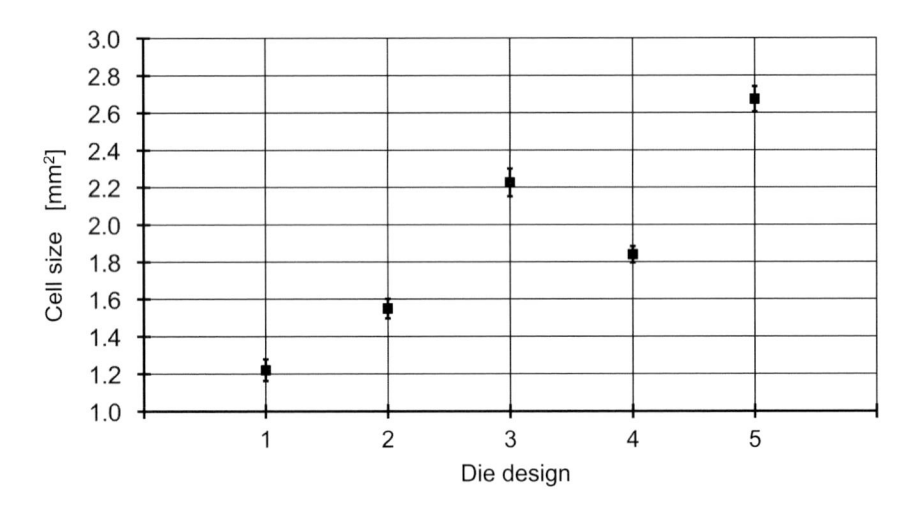

Figure 3.12: Influence of die design on the cell size

Due to the poor process stability of the dies (D1, D2), both flow channel designs are not suitable for further use in the production of blown film. Because the die (D4) is characterised by the smallest cell size ($1.84 mm^2$), suitable process stability and the lowest overall density ($0.33 g/cm^3$) compared to the other dies, it was selected as the final die.

3.4 Conclusions and outlook

For the development of a novel die design for foamed multilayer films, the influences of the three design parameters die gap, parallel zone length and angle were determined After the construction of five die designs, the influence on the pressure loss was shown by further analytical investigations. In the subsequent extrusion tests, problems occurred in the process stability, so that the dies with a small exit gap were not suitable here. In order to evaluate the foam structure, an optical measuring method was presented with which the cell size could be determined. With the finally selected extrusion die, blown films with a fine-cell foam structure could be produced and a stable process could be guaranteed. Furthermore, this die with a smaller outlet cross-section of about 0.5 mm is suitable for the production of thin films with a lower degree of stretching.

After the analysis and selection of the extrusion die, the further development of the extrusion technology should be followed by a focus on the influences on the material side. The molecular structure (molecular weight, molecular distribution, degree of branching) and the rheological properties (elongational viscosity) of the plastic have an influence on foam formation [Kro99, Sta06]. Therefore, a precise characterisation should be carried out to know especially the elongational viscosity of the plastics used. The determination of the elongation strength, which correlates with the elongation viscosity, can be done with the help of a membrane inflation rheometer.

Further developments in the field of foam extrusion are needed, as the application of foamed multi-layer packaging films will continue to expand thanks to the excellent thermal insulating effect in combination with the economic and environmentally friendly advantages of material reduction.

Acknowledgements

The research project (18977 N) of the Forschungsvereinigung Kunststoffverarbeitung is spon-sored as part of the "industrielle Gemeinschaftsforschung und -entwicklung (IGF)" by the German Bundesministerium für Wirtschaft und Energie (BMWi) due to an enactment of the German Bundestag through the AiF. We would like to extend our thanks to all organisations mentioned. In addition, the authors would also like to thank the companies Sabic Europe, Geleen, Netherlands and Clariant AG, Muttenz, Switzerland for providing the test materials.

References

[Dau08] DAUCH, M.: *IKV-Fachtagung „Kunststoffschäume - Neues aus Spritzgießen und Extrusion" Chemische Treibmittel - Eine Einführung in das Schäumen von Polymeren*. Aachen, 2008

[ESE20] EYERER, P.; SCHÜLE, H.; ELSNER, P.: *Polymer Engineering 3. Werkstoff- und Bauteilprüfung - Recycling - Entwicklung*. Berlin: Springer Verlag, 2020

[Gla17] GLADBACH, P.: *Herstellung von Schaumfolien aus Polystyrol mit Treibmittelgemischen*. Insti-tut für Kunststoffverarbeitung. RWTH Aachen University, unpublished Bachelor Thesis, 2017 – supervisor: R. Breuer

[Hei02] HEINZ, R.: *Prozessoptimierung bei der Extrusion thermoplastischer Schäume mit CO_2 als Treibmittel*. RWTH Aachen University, Dissertation, 2002

[HHHH14] HOPMANN, C.; HENNES, J.; HENNIGS, M.; HENDRIKS, S.: *Abschlussbericht zum IGF-Vorhaben Nr. 18977N*. Institut für Kunststoffverarbeitung, RWTH Aachen, Integrative Kunststofftechnik, 2014

[Hop16] HOPMANN, CH. AND MICHAELI, W.: *Extrusion Dies for Plastics and Rubber - Design and Engineering Computations (4th Edition)*. Munich: Carl Hanser Verlag, 2016

[Hop18] HOPMANN, C.: *Entwicklung eines Herstellverfahrens für physikalisch geschäumte Mehrschichtblasfolien*. Institut für Kunststoffverarbeitung, RWTH Aachen, 2018

[Kos96] KOSCHMIEDER, M.: *Untersuchung verschiedener Einflussgrößen bei der Schaumextrusion von Polyethylen mit CO2 als Treibmittel*. RWTH Aachen University, unpublished Diploma thesis, 1996 – supervisor: D. Kropp

[Kro99] KROPP, D.: *Extrusion thermoplastischer Schäumen mit alternativen Treibmitteln*. RWTH Aachen University, Dissertation, 1999

[Lan95] LANDROCK, A. H.: *Handbook of plastic foams: types, properties, manufature and applications*: Noyes Publications, 1995

[MH10] MICHAELI, W.; HILDBRAND, T.: Schaumextrusion mit OptiFoam – Potenziale physikalischer Treibmittel. *Blasformen Extrusionswerkzeuge* (2010) 4, p. 5–10

[NN21] N.N.: *SABIC® LDPE 2102XO, Datenblatt*. Sabic AG, Europa, 2021

[Sta06] STANGE, J.: *Einfluss rheologischer Eigenschaften auf das Schäumverhalten von Polypropyl-enen unterschiedlicher molekularer Struktur*. Universität Erlangen-Nürnberg, Dissertation, 2006

[Tol15] TOLINSKI, M.: *Additives for Polyolefins (2nd Edition)*. Amsterdam: Elsevier Science, 2015

[URL21] N.N.: *Was sind Schaumstoffe?* URL: https://verpackungen.de/info/was-sind-schaumstoffe, 18.09.2021

[Weg05] WEGNER, J.-E.: Additiv-Masterbatches für Schaumfolien. *Kunststoffe* 95 (2005) 1, p. 86–90

[Zha10] ZHANG, H.: *Scale-Up of Extrusion Foaming Process for Manufacture of Polystyrene Foams Using Carbon Dioxide*. Universität Toronto, Dissertation, 2010

Symbols

Symbol	Unit	Description
r	mm	Die radius
z	mm	Die length
Δp	bar	Pressure loss
$\dot{m}_{ges}$	kg/h	Total mass throughput
D_g	mm	Die gap
L_{Pa}	mm	Parallel zone length
α	°	Angle
C_a	mm	Cell dimension in pull-off direction
C_{trans}	mm	Cell dimension tansverse to the pull-off direction
C_{area}	mm	Cell surface area

Abbreviations

Notation	Description
CFA	Chemical Foaming Agent
MFI	Melt Flow Index
PE-LD	Low Density Polyethylen

… Session 4

Added value potentials through digital infrastructures

Moderator: Dipl.-Ing. Manfred Keuters, tetys GmbH & Co. KG

―――――――――― Content ――――――――――

1 Structure follows strategy: Why single-level planning solves more than the planning Problem.   127

A. Reßmann[1], M. Keuters[1]
[1] tetys GmbH & Co. KG

2 Effects of residual moisture on the processability of post-production PBT and implications on Process Monitoring and Control   137

Ch. Hopmann[1], J. Weber[1], P. Bibow[1]
[1] Institute for Plastics Processing (IKV) at RWTH Aachen University

3 Data-driven prediction the tensile strength of blown film due to digital shadows   148

Ch. Hopmann[1], D. Grüber[1]
[1] Institute for Plastics Processing (IKV) at RWTH Aachen University

Manfred Keuters

Manfred Keuters has been the general manager of GRP GmbH & Co. KG since 2010. GRP has been developing software for the shop-floor with a focus on plastics processing since 1979. Before 2010, Manfred Keuters worked in several leading and management positions. In 2016 he became Sales Director of FLS GmbH & Co. KG, which offers an extensive planning solution for medium-sized groups and companies. In 2021, GRP and FLS merged to tetys GmbH & Co. KG. Manfred Keuters is still authorised signatory of tetys GmbH & Co. KG. Furthermore, he founded tetys Plus GmbH as a sales organization for tetys GmbH & Co. KG.

1 Structure follows strategy: Why single-level planning solves more than the planning Problem.

A. Reßmann[1], M. Keuters[1]
[1] tetys GmbH & Co. KG

Abstract

The digital transformation of production companies has become a necessity due to mega trends such as mass customisation. A digitalisation strategy is key in order to implement digitised processes following the principle Structure follows Strategy. Planning philosophies, such as Just in Time (JIT), Load-Oriented Order Release (LOR), or Optimised Production Technology (OPT), are often too narrowly focused around one specific optimisation target and thus generalise badly to a variety of planning situations. Enterprise Resource Planning System (ERP) and Production Planning and Control System (PPS) solutions, on the other hand, often merely feature relatively coarse planning tools whose schedule has to be refined during detailed planning, which is typically executed within a separate system, the Manufacturing Execution Systems (MES). This two-tier planning in ERP and MES leads to problems such as over-inventory, inadequate time allocation, and an increased overhead and planning effort at the ERP level. Digitalisation providers are thus increasingly moving toward single-level planning approaches. Single-level planning offers the advantage that optimisation can be performed based on the original company goals, on-time delivery, and cost minimisation, rather than on proxy targets such as inventory or station lead time. Single-level planning can furthermore facilitate more complex planning tasks such as real time adjustments of the scheduled machine time based on forecasted or actual process speeds. It is furthermore able to consider complex interdependencies, for instance in the material supply. Thanks to direct feedback of the order status and the comprehensive planning within a single system, single-stage production planning can also respond more easily to disturbances. Mapping the actual process based on production data also enables artificial decision making and artificial troubleshooting. Furthermore, many MES systems allow searching for patterns in the production data and offer smart automation solutions for repetitive tasks.

1.1 Introduction

Today's digital value chains encompass a wide range of processes and business functions. Factories are currently in the process of a digital transformation toward a smart factory. This transformation is further accelerated by current mega topics and trends such as climate change (CO_2 savings, energy transition), a shortage of skilled workers, more individual products, and smaller batch sizes. Despite the listed challenges, the primary goal of each factory is to remain competitive. Thus, companies need to find the optimal balance between efficiency and flexibility. Developing a digitalisation strategy is thus essential to navigate through an almost unmanageable market of digitalisation solutions. Process patterns, such as engineering to order, configure to order, or select to order also have a major influence on the company's digital strategy.

Following the established principle of *Structure Follows Strategy*, companies can consider the aforementioned challenges as an opportunity to actively shape the digital transformation and remain competitive at the same time. Based on a defined company strategy, an flexible infrastructure is required. This digital infrastructure subsequently implements company-specific processes defined by the digital transformation strategy. This environment will be the start of our discussion in the following with a focus on production planning.

1.2 Market development and state of the art

Starting in the second half of the 1970s, digital data processing initially covered commercial areas such as payroll and financial accounting [URL21a]. Later, cost accounting, customer management, and other modules were added. With the increasing performance of computers, other more complex tasks became accessible for electronic data processing, such as the construction area, which was covered by CAD systems (Computer Aided Design) [URL21c].

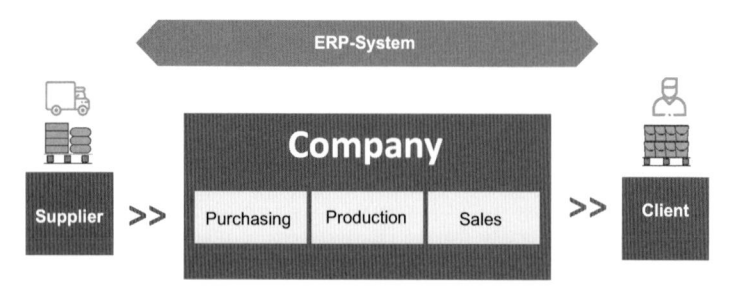

Figure 1.1: Basic Structure of ERP Systems

It was not until relatively late that it was realised that electronic data processing could also be used beneficially in the production sector. Thus, it was not until the 1980s that the so-called PPS came onto the market [Sch12]. Commercial management approaches were later continued under the term ERP. Although the term PPS/ERP is generally used today, there is no uniform definition of what it encompasses. If we assume a very broad definition, then almost all the concerns of a manufacturing company are subsumed under ERP, starting with the preparation of quotations, through sales order management, material planning, production planning, right up to purchasing. Depending on the structure and product range of a company, modules such as production planning may be added. ERP systems also include all commercial functions such as controlling, cost accounting, or balancing as well as customer interactions via social media or on-premise sales options.

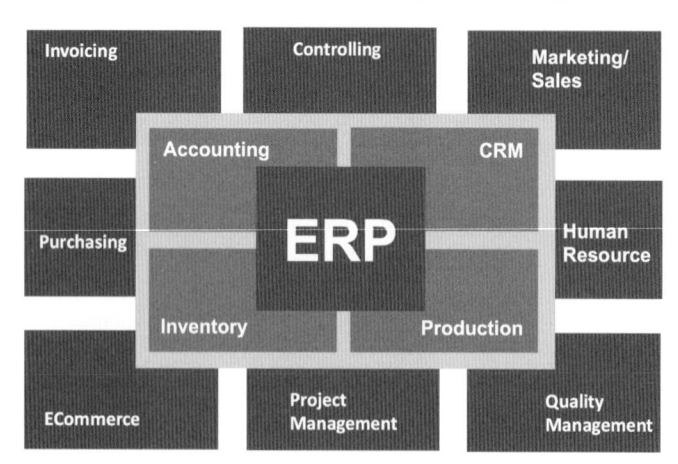

Figure 1.2: Typical modules of ERP systems (own representation based on [URL21d])

Today there are more than 600 providers of ERP systems on the German market [HG14, Gra11]. The market has thus become very confusing. With the industry focus on plastics processing, however, the portfolio can be significantly reduced. A central criterion of modern ERP systems is cloud capability. The key identified benefits are lower cost, scalability, fast and rapid implementation, improved accessibility, high availability, and easier update [URL18b, ANG16]. Therefore, decentralised working, virtual teams, and efficiency in the implementation of processes are a few reasons for this market development. ERP providers have promised users repeatedly that the use of these ERP systems would significantly increase productivity in companies because they would also be able to plan operations better than before. These promises have only been fulfilled to a very small extent, so that disappointment about the benefits of planning systems has generally spread [URL19, URL21d]. The essential processes in operations cannot be planned with the PPS systems commonly used today, or at least not sufficiently; moreover, most systems are too sluggish in their handling to provide the decision-making bases required at short notice in a timely manner [URL19, URL21d].

Parallel to the development of ERP systems and production planning software, a wealth of planning philosophies and methods have been developed and disseminated in recent years, some of which were used without the aid of computers and some with the aid of computers. KANBAN, JIT, OPT or LOOR are representatives of many others. A characteristic of all these philosophies and methods is that at their time they were propagated as the general solution to the internal planning problem, but came out of fashion and were replaced by the next philosophy in a relatively short time [URL21b]. The reason for this is obviously that each philosophy emphasizss one aspect while neglecting others. Most methodologies either aim for a short order throughput time or they focus on reducing inventory. For example, KANBAN reduces inventory between departments, but is unsuitable for complex product structures with a high vertical range of manufacture, especially for one-off and small batch production. KANBAN does not plan every resource in detail, because effectivity is gained by tailoring the work process, rather than the resources, to needs. Although this approach is more cost effective for the current situation, it does not support mid-term or long-term digital strategies. JIT focuses on reducing throughput times, but is not applicable without adequate warehousing, especially for bottleneck workplaces [SS13]. OPT attempts to build planning precisely from these bottleneck workplaces, but is not very suitable for managing changing bottlenecks, which are ultimately a sign of well-structured and utilised production. LOOR also aims to reduce the order throughput time: the order should only be started when it can be passed through the various workstations without long waiting times. Again, the more complex the product, the less suitable this method is. Moreover, the daily disruptions in the operational process prevent the theoretical benefits from materialising in practice [URL18a].

The conclusion to be drawn from this critical examination is that the widespread combination of PPS/ERP systems with an increasing number of production planning solutions is not yet a permanent solution in the planning area for manufacturing companies. Moreover, the relatively fast-changing planning philosophies indicate that there may not be a general solution for all planning problems in manufacturing companies. Essential planning functions, especially those required in day-to-day business, thus had to be solved with the help of other systems. These prerequisites initiated the development of production planning solutions in MES during the last decade. This in turn was favoured by the rapid increase in computer performance, which now made it possible to convey information in graphic rather than in tabulated form, which was additionally supported by colour processing. The first planning solutions in MES were initially only graphic display instruments for planning information from the PPS systems. As time went on and user requirements grew, more and more functionalities were shifted from the PPS systems to the production control stations. The situation today is perhaps best characterized as follows (see 1.1):

- The PPS/ERP systems, which are usually installed at considerable expense, have limited themselves to a rough planning level in internal planning.
- The task of detailed planning of work processes in production is increasingly being transferred to production planning solutions on MES level.
- Despite the rough planning level, the management and scheduling of materials remains a task for the PPS/ERP systems.

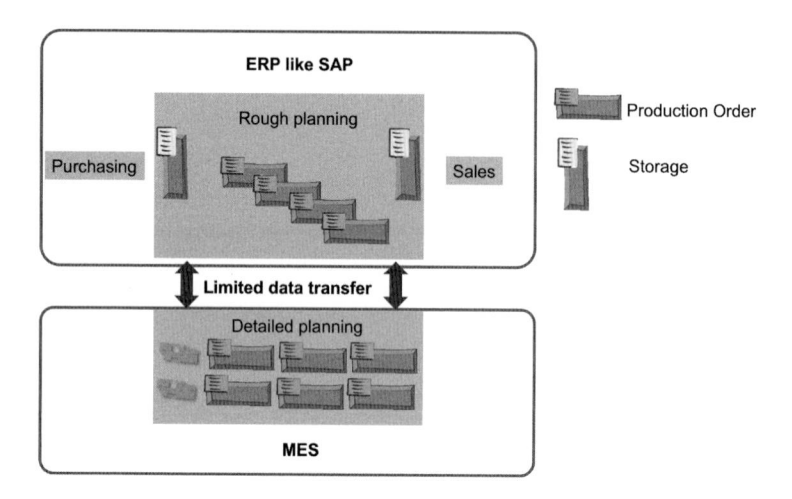

Figure 1.3: Current division of tasks between ERP and MES

An analysis of the current situation reveals two major weaknesses:

Firstly, neglecting important aspects of detailed planning leads to schedules in the rough planning in the PPS/ERP system that can no longer be realised in the detailed planning afterwards. If, for example, it still seems possible at the rough planning level to process a certain order quota within a workstation group on time, it may for example turn out during detailed planning that necessary peripherals are unavailable and the deadlines from rough planning are no longer tenable.

Secondly, the separation of material management in the PPS/ERP system on the one hand and production planning in the MES on the other hand leads to conflicts because the scheduling and provision of materials or assemblies are not coordinated with the short-term reactions in the production planning area. On the one hand, this leads to disruptions in the internal process because the required material cannot be provided in time. On the other hand, unnecessary storage space and additional capital is tied up by ordering material too early, which is often evident in the case of packaging material. This inadequate coordination between ERP systems and today's standard MES systems is frustrating, especially for the user, and brings with it all the disadvantages of a two-tier optimisation system. Following the market, ERP and MES providers today offer improved planning functions under the title Advanced Planning and Scheduling (APS), which can often be reduced to special cases of planning with derived objective functions [URL18a].

1.3 Single-level planning

One approach to solve the planning problem is single-level planning, which bundles all planning functionalities in one system. In contrast to the other planning systems described, single-level planning systems do not pursue derived objective functions, such as reducing lead times or minimising warehousing. For the area of production planning, single-level planning is dedicated to the original corporate goals, and these are:

- Meeting the promised delivery dates or, in the case of delivery from stock, maintaining the ability to deliver.
- Maximising the company's profit. However, since internal planning has no influence on sales revenues, profit maximisation in this case is equivalent to minimising costs. In general, adherence to delivery dates or readiness to deliver takes precedence over cost minimisation.

Let us, as an example, consider the proxy target *lead time of an order*. The lead time of an order can be shortened by additional shifts, e.g. on weekends. As a result, the goods can be delivered earlier, the invoice can be written earlier and the receipt of payment can be expected earlier. This reduces the interest burden on the capital tied up in the order. However, additional shifts are expensive. An optimiser solely based on lead time is not able to consider these additional costs and cannot balance lead time against cost. Single level planning, on the other hand, enables an holistic optimisation based on the primary company goals.

We observe that single-level planning changes the organisational structures. Using single-level planning will change the scope of the planning department. In the past, the planners had to master the complexity of the process and its interlinks by themselves. With single-level planning, the complex interlinks are reflected in the planning system, so that the planners can focus on defining the planning model rather than moving production orders around manually. The role of the planning department changes to an in-process controlling (e.g. planning and production) that acts based on data driven processes. On an abstract level, planning and scheduling become a business setting by management, and can be performed digitally.

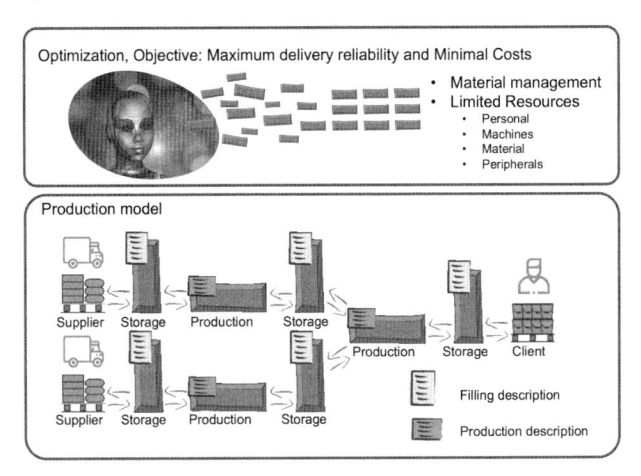

Figure 1.4: Optimisation with single-level planning

1.3.1 Saving potentials and concrete challenges in the market

The saving potentials of single-level automated planning are immense. Due to increasingly smaller batch sizes and at the same time steadily increasing product complexity, the challenges for production planning are growing. [NN21]. Above all, the automation of planning offers further advantages. After introducing a single-level planning system, the major optimisation potentials can be found in the following areas:

- Organisational errors: waiting for machine, mould, peripherals, and personnel.
- Reduction of inventory costs
- Reducing set-up time and costs
- Reducing rapid orders ("boss orders")
- Reducing extra shifts

Single-level production planning is flexible, transparent, and complex. It can be fully automated with today's (or the right) planning software. This offers a multitude of possibilities to increase the complexity of these planning tasks. Dependencies between several production orders can be considered more accurately [GR20].

Considering the post-production recycling process, an intelligent single-level planning approach could for instance plan the amount of virgin material to be dried, based on the expected amount of recycled material from sprues and scrap of a current production run (see chapter 4.2). Many processes are dependent on the input material in their processing speed (cycle time or take-off speed). After checking the input material and assigning it to a production order, the production time of the order can be adjusted based on a material dependent model of the achievable processing speed (see chapter 4.3).

1.3.2 Influences of direct feedback from the MES level on planning

Moving beyond the current state of the art, an increasing number of companies are implementing single-stage production planning with fast feedback from the shop floor. As modern production planning solutions combine material management as well as production planning at the MES level and their mutual dependencies, feedback from the shopfloor can be considered in the planning and becomes an important source for optimisation decisions.

An example: If an order is prioritised and deprives another order of material, a single-level planning solution can automatically postpone the "stolen" order until after the next material delivery. In this way, there is no need for a reconciliation of information with the ERP system, as it would be necessary with two-stage planning. The coupling of production planning and material management in a single-level planning also makes it possible to adapt the order proposals to the concretely planned production orders. This significantly reduces raw material inventory. Therefore, a direct integration of the MES level into production planning allows for significant saving potentials even in warehousing. A necessary basis for this fast feedback is the transparent and prompt feedback of the order status from production. Thus, order tracking in production is important to increase the potentials for single-level planning. The more accurately the actual production situation is mapped, the better it can be incorporated into production planning. Thus, the actual process situation is represented by all quantities that may be of importance to other processes. This means recording quantities, such as rejects, good parts, cycle times, processing speeds, and the degree of utilisation as well as downtimes, such as the maintenance and servicing process; and the quality process.

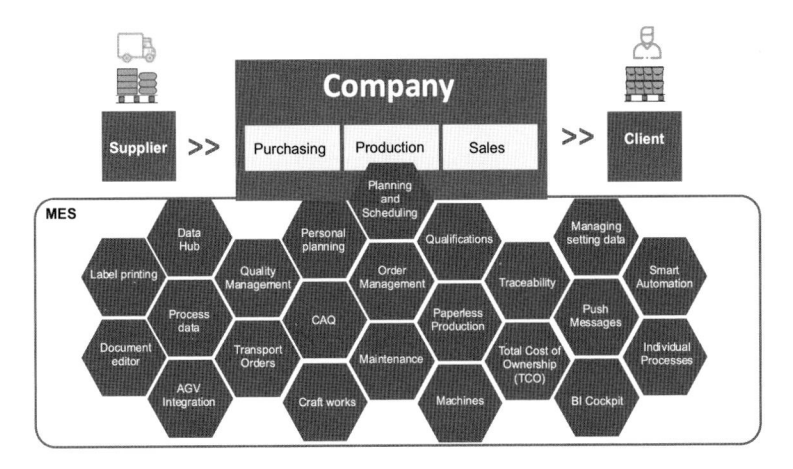

Figure 1.5: Typical MES tasks

1.4 Toward fact-based decision making and smart automation

The digital transformation of processes significantly increases the transparency of production and creates a culture of improvement based on figures, data, and facts. Production managers always have all the required key figures and thus the economic efficiency of their production in view. Weaknesses are revealed in concrete facts, e.g. which machine is producing at what speed and whether there are planned or unplanned downtimes. These facts that can be used to analyse the situation at any time and clearly plan future production targets. Data-driven processes allow more and more data to be considered in the decision space, and the smart automation of repetitive activities. The collection of process data and the management of setting data also allow fundamental optimisations.

But the next, deeper level of precisely scanned data like cavity pressure curves, or injection speed curves is also already being considered in many MES. These large data streams are searched highly efficiently for patterns and processes that could be automated. This smart automation goes far beyond the current automation level and supports setters, maintenance personnel, logisticians, and machine operators by reducing repetitive tasks in everyday work. The MES level now forms the digital infrastructure in production companies by digitising and orchestrating the business processes of production in these systems. Once the digital infrastructures have been created in production, a variety of optimisation methods can be applied digitally and thus effectively. The resulting process patterns are often industry- or process-specific. For this reason, it makes sense to look at the optimisation potential for company processes, especially in an institute specialising in manufacturing processes.

1.5 Digital transformation: Standard software vs. individual processes

As described in the introduction, the digitalisation of company-specific processes in a digital infrastructure is a logical consequence of individual corporate strategies for digital transformation.

When you start with your digitalisation strategy, the mapping of processes already begins with the software selection for the infrastructure. In principle, there are two extremes when it comes to software selection: using standard software with a *one size fits all*-approach for all customers, and implementing tailormade software solutions (e.g. in MS Access) based on the individual company processes. When selecting software for your production, these two worlds often seem to collide. Either your partly individual company processes meet standard software that forces companies into a standard organisation/workflow. Tailormade software, on the other hand, often sinks into interface issues due to numerous isolated solutions.

An example of how to address this problem is provided by the software ERNST from the company tetys GmbH Co. KG, displayed in 1.6. The goal of the ERNST process engine is to automate company-specific processes in a meaningful and individual way and to turn them into a powerful, smart automation via its platform. Especially in the consumer sector, there are already some platforms that allow processes and interfaces to be designed with little programming knowledge and the associated high level of abstraction. These process engines allow the implementation of processes with a low level of programming know how and address the requirements of the software user individually and transparently. This idea of data connections between different software modules and systems gives rise to the concept of platforms. These platforms offer prefabricated process patterns, standardised interfaces, and a clear reason for the user to use the various modules, which significantly exceeds the reasons for selling the individual software.

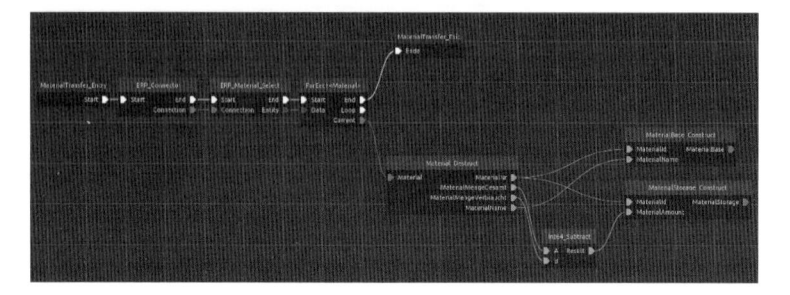

Figure 1.6: Example of implementing company-specific processes with ERNST

1.6 Summary and outlook

We must be aware that every step (KANBAN, JIT, LEAN, etc.) the production industry did was a necessary and useful step to achieve the next level of understanding and digitalisation. The concepts we now call Industrie 4.0 or Industrial Internet of Things (IIoT) are the next consequent step in a strategy to stay in touch with the markets. In production planning, many companies currently leave a lot of energy and optimisation potential in two-stage planning. This can be avoided by a holistic planning approach, such as the single-level planning.

Digital ecosystems play an increasingly important role in industry-driven digitalisation. These ecosystems allow for pre-considered paths to be taken, so that not every company has to reinvent the wheel for themselves. Another component of a modern digital infrastructure is the fast access to data from a variety of sources and the mapping of individual processes. Digital processes and infrastructures as described in this article offer enormous potential for optimisation in the manufacturing sector. Thus, digitalisation and digital infrastructures become more and more an enormous competitive advantage. Often the potentials in the individual areas, such as in planning or in the acquisition of production processes, are already significant. The interaction of the areas increases the optimisation potential even further. The rapid implementation and adaptation of digital processes also allows companies great flexibility. Digitalisation also paves the way for efficient production against the challenges of sustainability. We would like to encourage every company to develop a digital strategy that also includes sustainability requirements.

References

[ANG16] ABD ELMONEM, M. A.; NASR, E. S.; GEITH, M. H.: Benefits and challenges of cloud ERP systems – A systematic literature review. *Future Computing and Informatics Journal* 1 (2016) 1, p. 1–9

[GR20] GERLACH, C.; RESSMANN, A.: Der Kunde bestimmt den Takt. *Kunststoffe* 10 (2020), p. 20–24

[Gra11] GRAMMER, P. A.: *Der ERP-Kompass: ERP-Projekte zum Erfolg führen.* Heidelberg and Hamburg: mitp Verlag-Gruppe Hüthig Jehle Rehm, 1st edition, 2011

[HG14] HESSELER, M.; GÖRTZ, M.: *Basiswissen ERP-Systeme: Auswahl, Einführung & Einsatz betriebswirtschaftlicher Standardsoftware.* Dortmund: W3L-Verlag, 2014

[NN21] N.N.: *Manufacturing execution systems (MES): MES and Industrie 4.0*: VDI-Gesellschaft Kunststofftechnik, 2021

[Sch12] SCHUH, G.: *Produktionsplanung Und -Steuerung 1: Grundlagen der PPS.* Berlin, Heidelberg: Springer Verlag, 4th edition, 2012

[SS13] SINGH, R.; SOHANI, N.: Effect of Lean/JIT Practices and Supply Chain Integration on Lead Time Performance. *Journal of Supply Chain Management Systems* Volume 2 (2013),

[URL18a] HOFF, H.: *Welche Produktionsplanung soll es sein?* URL: https://www.it-production.com/produktionsmanagement/produktionsplanung-pps-erp-software/, 23.09.2021

[URL18b] KRAUSS, M.: *Enterprise-Resource-Planning (ERP) – Definition, Arten und Vorteile.* URL: https://www.maschinenmarkt.vogel.de/enterprise-resource-planning-erp-definition-arten-und-vorteile-a-779630/, 23.09.2021

[URL19] SCHRÖDER, M.: *Wenn das ERP an seine Grenzen stößt.* URL: https://www.com-magazin.de/praxis/business-it/erp-an-grenzen-stoesst-1683991.html, 23.09.2021

[URL21a] LEIMBACH, T.: *SAP: A 49-year history of success: The early years.* URL: https://www.sap.com/about/company/history/1972-1980.html, 23.09.2021

[URL21b] N.N.: *Das FEKOR Konzept: Whitepaper.* URL: https://www.fls.de/download/whitepaper-das-fekor-konzept/, 23.09.2021

[URL21c] N.N.: *Die Geschichte der CAD und CAM-Systeme: Von den Lochkarten über die parametrischen 3D-Modellierung bis heute - und darüber hinaus.* URL: https://www.west-gmbh.de/blog/die-geschichte-der-cad-und-cam-systeme, 23.09.2021

[URL21d] N.N.: *Wozu benötigt man ein ERP-System?* URL: https://www.erp-system.de/, 23.09.2021

Abbreviations

Notation	Description
APS	Advanced Planning and Scheduling
ERP	Enterprise Resource Planning System
IIoT	Industrial Internet of Things
JIT	Just in Time
LOOR	Load-Oriented Order Release
MES	Manufactureing Execution System
OPT	Optimised Production Technology
PPS	Production Planning and Control System

2 Effects of residual moisture on the processability of post-production PBT and implications on Process Monitoring and Control

Ch. Hopmann[1], J. Weber[1], P. Bibow[1]
[1]*Institute for Plastics Processing (IKV) at RWTH Aachen University*

Abstract

Plastic recycling has recently gained more attention both for economic and ecologic reasons. Against this background, manufacturers have been regrinding sprues and scrap parts and feeding them back into the injection moulding process. However, when working with hygroscopic materials, hydrolitic ageing due to residual moisture during processing can deteriorate the material processability and mechanical properties, rendering the material unfit for further reprocessing. Based on the example of polybutylene terephthalate (PBT) that was mechanically recycled five times, the influence of the drying temperature on residual moisture, and the effect of residual moisture on part weight and appearance was studied. After a sharp increase of part weight between virgin material and material recycled for the first time, moisture levels in the raw material decreased steadily, while part weight and injection pressure remained relatively stable. After exposing the material to ambient conditions for several weeks and recycling them twice again, processing properties would however change more significantly. This indicates that drying to manufacturer specifications restricts hydrolitic chain scissioning to a minimum. The results also indicate that post-production recycling of PBT is possible, provided moisture is in-control and the material handling ensures the material is not contaminated in the recycling process. Moisture is however not a parameter which is currently monitored, let alone actively controlled throughout the entire production and recycling process. Future systems could combine data from processing and ambient condition monitoring intelligently and thus help to reduce processing fluctuations due to moisture in reground material and at the same time prevent excess material ageing and energy consumption through unnecessary drying procedures.

2.1 Introduction and motivation

In injection moulding, it is customary to regrind post-production waste, such as sprues and scrap parts, and to directly recycle it into new parts [Mar06]. When the recycling stream is not handled properly, this may lead to process disturbances, especially when working with hygroscopic polymers that require drying prior to processing. The drying times and temperatures given in the material data sheets are usually based on the predried delivery condition of the virgin material [KB15]. When the material is exposed to ambient conditions, the initial humidity may be significantly higher [KB15]. To achieve the moisture content required for processing, the necessary drying times of material stored under uncontrolled conditions can thus multiply [KB15]. As residual moisture is often not measured before drying, there is thus a risk that reground material is not sufficiently dry. Residual moisture does not only adversely affect the processing behaviour of virgin plastic materials [HMW15]. Hydrolitic ageing due to residual moisture during processing can deteriorate the material processability and mechanical properties, rendering the material unfit for any further reprocessing cycles [Mar06].

In order to get a better understanding of how the reprocessing operation should be managed and what parameters should be monitored to ensure minimal material damage and process disturbance, we studied the influence of the drying temperature on the processing properties of PBT. To this end, PBT was dried following the upper and lower limit of the manufacturer specification and reprocessed several times by injection moulding. To judge the processability, the part weight, maximum injection pressure, and physical appearance were assessed.

Thus, this paper is structured as follows: Chapter 2.2 summarises relevant research concerning reprocessability and ageing of PBT. Chapter 2.3 describes the experimental approach and chapter 2.4 presents and discusses the main findings. The outlook presented in chapter 2.5 sets the results into context with the recycling demonstrator planned for the Plastics Innovation Center 4.0.

2.2 State of the art

The effects of moisture and mechanical recycling on technical polymers can be broadly differentiated into chemical and physical effects [HMW15]. Chemical effects are irreversible changes in molecular structure, such as chain scission or post-condensation, and may be caused by oxidative or hydrolytic reactions as well as elevated temperatures over a prolonged time [HMW15]. Physical effects are generally reversible effects, such as an elevated level of residual moisture [HMW15].

Chemical effects generally lead to a change in molecular weight. An increased molecular weight, in turn, leads to an increase in zero-shear viscosity as well as shear-dependent viscosity, as the transition time (i.e. the time the macromolecules need to disentangle under viscous flow conditions) increases [HMW15]. Moisture acts as a lubricant between the polymer chains and can thus lower the melt viscosity [HW15]. The changes in melt viscosity can cause a number of process and part quality issues in injection moulding, such as variations in shot weight, melt temperature, and product density [HMW15, HW15]. With regard to the product properties, chemical ageing can for instance cause discoloration, and a loss of desired mechanical properties [CV06]. Chain scission often causes embrittlement, as shorter chains can more easily arrange in semi-crystalline structures [CV06].

The actual ageing mechanism largely depends on the polymer and the environment conditions. At low oxygen concentrations, crosslinking is more likely, while high oxygen concentrations support chain scission reactions [CV06]. PBT has been reported to exhibit chain scission effects through transesterification when exposed to elevated temperatures [HMW15].

A small number of studies investigate the effect of material ageing on PBT blends. *Kuram et al.* (2014) reprocessed a blend of PBT/polycarbonate (PC)/acrylonitrile butadiene styrene (ABS) in equal proportions by weight five times by successive injection moulding and grinding. Fourier-transform infrared spectroscopy (FTIR) scans indicated that the chemical structure of the polymers was not altered significantly. Concerning the mechanical properties of the polymer, they reported an increase in impact strength and strain at break while tensile strength, yield strength, elastic modulus, flexural strength, flexural modulus, as well as the zero-shear-viscosity measured in terms of the Melt Flow Index (MFI) remained largely unaffected [KOY+14]. We can thus conclude that, given ideal processing conditions and in the absence of hydrolytic or thermo-oxidative stress, the PBT/PC/ABS blends can be recycled five times without any noteworthy degradation effects.

Sanchez (2007) investigated the effects of natural and accelerated ageing on PBT/PC blends. She reported that both ageing and reprocessing increased the MFI. Ageing alone significantly decreased elongation at break. However, reprocessing the aged material almost restored the initial elongation at break. *Sanchez* argues that ageing mainly caused material degradation at the surface and that during reprocessing, the degraded material was mixed with the intact material from the product core, alleviating the embrittlement effects [San07].

Lochem et al. (1996) investigated the recycling properties of glass fibre reinforced PBT with 30% fibre content. The study mainly focused on fibre-matrix adhesion and showed that it was severely impacted by humid ageing, even if the material was recompounded together with virgin PBT [LHL96].

Risch et al. (2000) analysed the influence of residual moisture on the viscosity of PBT samples after extrusion and compared it to samples that had undergone accelerated ageing. They could show that under processing conditions, PBT is sensitive to hydrolytic degradation. Processing a sample with 0.1% moisture caused a similar amount of chain scission and decrease in viscosity as ageing the material for 30 days at 85°C and 85% relative humidity. Processing material with a moisture content of 0.36% led to a viscosity reduction of up to 80%. They concluded that proper drying, as well as hydrolytic stabilisation, is an absolute necessity when processing PBT [RAA00].

Mielcki at al. introduced new factors into the Carreau/Arrhenius model for melt viscosity in order to incorporate thermally induced chemical ageing under inert conditions, and the viscosity reduction caused by residual moisture on polyamide (PA) and PBT [MWG+12, HMW15]. The complete equation is then given by:

$$\eta(t, T, c_{H_2O}, \dot{\gamma}) = \frac{\eta_0 \cdot a_T(T) \cdot a_{SC}(t,T) \cdot a_{PC}(c_{H_2O})}{(1 + (t_1 \cdot a_T(T) \cdot a_{SC}(t,T) \cdot a_{PC}(c_{H_2O}) \cdot \dot{\gamma})^2)^{(\frac{1-n}{2})}} \quad (2.1)$$

Where the *temperature shift factor* a_T describes the temperature dependence of the viscosity [HMW15]:

$$a_T(T, T_0) = e^{\left(\frac{E_0}{R} \cdot \left(\frac{1}{T} - \frac{1}{T_0}\right)\right)} \quad (2.2)$$

With the material parameter *flow activation energy* E_0, the reference temperature T_0 and the ideal gas constant R.

The *structural change factor* $a_{SC}(t,T)$ describes chemical ageing under inert conditions due to the partial temperatures T_i during the partial time spans t_i and is defined as [MWG+12]:

$$a_{SC}(t,T) = a_{SC,UL} - (a_{SC,UL} - 1) \cdot e^{-k_0 \sum \left[t_i \cdot e^{\left(-\frac{E_{0,SC}}{R \cdot T_i}\right)} \right]} \quad (2.3)$$

with the *upper limit of structural change* $a_{SC,UL}$ the *rate of structural change* k_0, the *flow activation energy* $E_{0,T}$ and the *Structural change activation energy* $E_{0,SC}$ being material dependent parameters [MWG+12]. The model was developed based on ageing trials of PA powder for additive manufacturing in solid state. The authors assume that the model would in principle still be valid for molten material, but that the model parameters would differ between melt and solid material [MWG+12].

Residual moisture forms hydrogen bonds with functional groups of the polymer chains, increasing the chain mobility and thus decreasing polymer viscosity [HMW15]. This physical effect of residual moisture is characterised by the *physical change factor* a_{PC}. The factor decreases linearly from 1 for a dry material with the material-dependent slope $a_{PC,IS}$ [HMW15]:

$$a_{PC}(c_{H_2O}) = 1 - a_{PC,IS} \cdot c_{H_2O} \qquad (2.4)$$

While this model is able to describe the material condition before entering the injection moulding process, it disregards thermal ageing and hydrolytic scissioning that may occur during the process. This is because during ageing trials, they did not observe a significant moisture effect on the thermally induced degradation [HMW15]. This may be true for typical drying temperatures of 80 - 120°C. However, based on the results of *Risch et al.* [RAA00], one would expect a significant change in material viscosity after processing, i.e. in the next recycling cycle, that cannot be explained by the model.

Viscosity variations are problematic for the injection moulding process, as higher material viscosities require higher pressures to fill the cavities [Kru15]. As a result, the melt is compressed more when injected and the part volume at ambient pressure changes. Furthermore, there is a reduced holding pressure effect at the end of the flow path [HFH17]. As a result, underfilling of the cavity occurs with unchanged process control [PWS20]. The opposite is true for a lower viscosity, which leads to a stronger compression of the melt in the holding pressure phase and consequently to the formation of burrs in the mould separation plane and an increased part weight [THH07]. The changed pressure distribution in the cavity also influences the local shrinkage behaviour in the holding pressure and cooling phase. This can lead to part warpage [HFH17].

2.3 Methods and materials

The material - Ultradur B4520-PBT by BASF, Ludwigshafen, Germany - was separated into two batches. The first batch was always dried at 120°C, the second batch was always dried at 80°C, both for 4h. The material was processed into tensile bars (dog-bone specimens) on an Arburg Allrounder 370A 600-170/170 from Arburg, Loßburg, Germany. The process parameters were kept the same during the entire test program. One to three days later, the material was reground in a cutting mill. The entire process was repeated four times within two weeks and another two times 8 weeks later (see Table 2.1).

In-between trials, moulded parts were stored in cardboard boxes under ambient conditions, allowing them to absorb moisture. Reground material was stored in the original material bag which was closed with tape. Dried material that was not processed within a few hours was vacuumed and sealed in airtight bags.

Batch	Processing Step	virgin	REC 1	REC 2	REC 3	REC 4	REC 5
80°C	pre-drying moisture measurement	1	7;8	9	11	63	65
	post-drying moisture measurement	2;4	8	11	14	63	65
	drying, injection moulding	4	8	10	14	63	65
	regrinding	7	9	11	60	64	
120°C	pre-drying moisture measurement	1	7;7	9	11	63	65
	post-drying moisture measurement	2;4	8	10	14	63	65
	drying, injection moulding	2	8	10	14	63	65
	regrinding	7	9	11	60	64	

Table 2.1: *Trial schedule (in days relative to start date)*

The material humidity was measured using an FMX HydroTracer moisture analyzer by Aboni, Schwielosee, Germany before and after drying. It should be noted that the moisture measurements were not taken optimally, though. The material was weighed into a freeze bag with zip closure in the technical center; HydroTracer measurements were taken in another room up to 18h later. The freeze bags are not completely airtight and, especially with the reground material, trace amounts of material would always remain in the bag, so that the actual material amount was probably less than the weight used for the moisture calculation. Moreover, samples for humidity before drying were taken before inserting the material into the dryer but the actual drying process would take place several hours later. It can thus be assumed that the reported moisture levels only possess indicative value.

To quantify the material degradation, we considered part weight and maximum injection pressure as easy-to-measure indications of material viscosity. Of each batch, 100 mouldings, each comprising two tensile bars and the cold runner system, were weighed directly after injection moulding. As the injection moulding machine had an open nozzle, there would sometimes be a smear or thread of melt attached to the sprue. It was usually clipped before weighting, but may nevertheless distort the weight readings toward higher weights, especially in the first two reprocessing steps.

2.4 Results and discussion

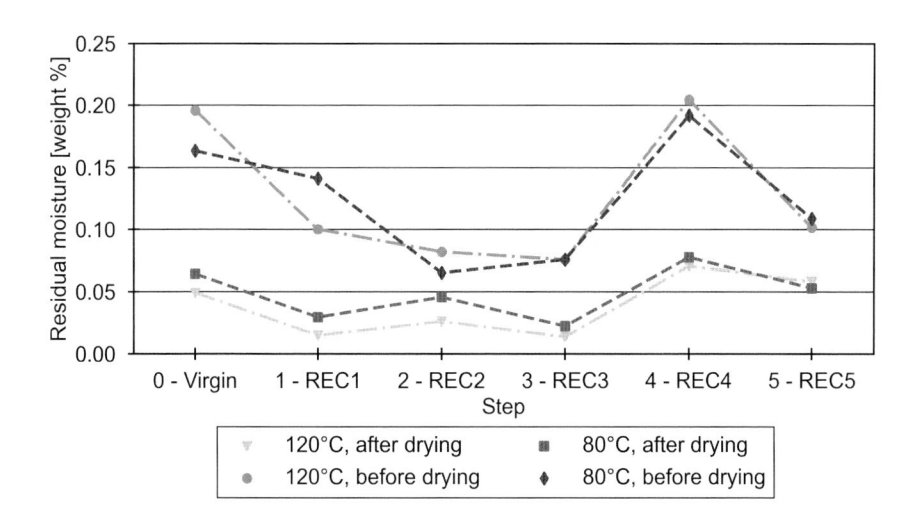

Figure 2.1: Evolution of material moisture content during the trials

Effect of the drying parameters on the moisture levels

Coming from a different material bag, the 80°C-batch exhibited a slightly lower moisture content than the 120°C-batch to begin with (see Figure 2.1. If two dates are provided in Table 2.1, the reported value is an average of both measurements). After drying, the moisture in the 120°C-batch was slightly lower throughout the first five processing cycles. It is however worth noting that both batches of the dried virgin material did not meet the moisture requirement of $< 0.04\%$ stated in the product data sheet. This may be because the material had already been stored for about 33 months before the trials. In the course of the trials, residual moisture within the material prior to drying decreased steadily, indicating that the material did not absorb as much moisture as lost during drying during the few days between the trials in the first three recycling runs despite the open storage. Residual moisture after drying, in turn, remained relatively stable. The material did however return to its initial moisture content during the 8-week storage period between the third and fourth recycling run. The moisture level was thus above specification during the last two recycling runs.

Development of part weight and injection pressure

Both part weight and injection pressure are considered good indicators for material viscosity. A lower viscosity causes a lower injection pressure. It also leads to a higher material compression during the back pressure phase and thus to heavier mouldings. As shorter chains and a higher moisture content lead to a lower viscosity, one would expect heavier mouldings for material with a higher moisture content and/or a higher number of reprocessing cycles. The weight distribution of the mouldings is displayed in Figure 2.2. Figure 2.3 displays the distribution of the maximum injection pressure[1].

As expected due to the overall lower moisture, the recycled 120°C mouldings were generally lighter and showed a higher injection pressure than the 80°C mouldings. This was however not true for the virgin material. This may either be because the 120°C-batch was still more humid and this was not reflected properly in the post-drying moisture measurement, or because the two virgin trials were not performed on the same day and some ambient conditions were thus different.

The part weight increased significantly from the virgin to the one time recycled material. Similarly, we can see a sharp decrease in maximum injection pressure. This behaviour was expected due to the relatively high moisture content in the virgin material and the accordingly likelihood of accelerated hydrolitic ageing during processing. However, part weight decreased slightly and injection pressure remained relatively stable over the successive two reprocessing cycles. A possible explanation is that the moisture content was so low that the physical moisture lubrication effect decreased or remained the same and the material did not significantly age during the recycling runs.

To validate that the weight increase between virgin and one-time recycled material can indeed be attributed to hydrolitic chain scissioning, the material was stored under ambient conditions for several weeks between the third and fourth recycling runs so that it could absorb a larger quantity of moisture. As seen in Figure 2.1, the moisture levels in recycling run four and five were similar to the moisture levels found in the virgin and one time recycled material, respectively. However, we did not observe another extreme change in part weight that could have been attributed to the elevated moisture content. A two sided t-test shows that there is no statistically significant difference between the part weights between the first and fifth recycling runs for both drying temperatures. However, the injection pressure decreased visibly both between the third and fourth as well as between the fourth and fifth reprocessing cycle. This indicates that there was a physical lubricating effect of moisture in the fourth reprocessing cycle, as well as hydrolitic ageing. The chain scissioning effects caused by processing the wet material in the fourth recycling run again caused a decrease in injection pressure in the fifth reprocessing cycle.

The difference between the fourth and fifth reprocessing cycle is not as pronounced as between the virgin and one-time recycled material. This may indicate that the material contains some chemical bonds that are more sensitive to ageing effects than others. After

---

[1] Due to problems in data acquisition via OPC UA, injection pressure curves were not recorded for the fourth recycling run of the 120°C-batch. The boxplots displayed in both figures contain approximately 100 mouldings per batch and recycling run, but do not necessarily represent the exact same production cycles.

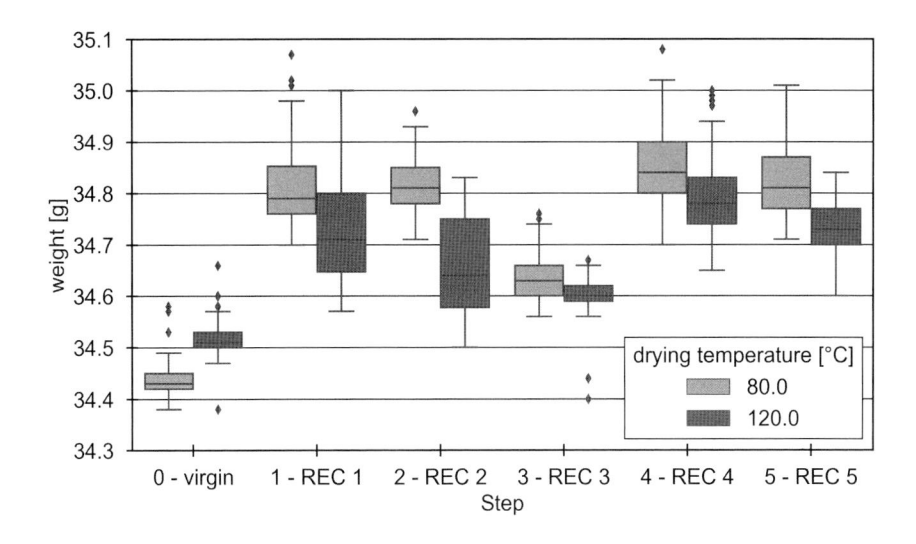

Figure 2.2: Evolution of part weight during trials

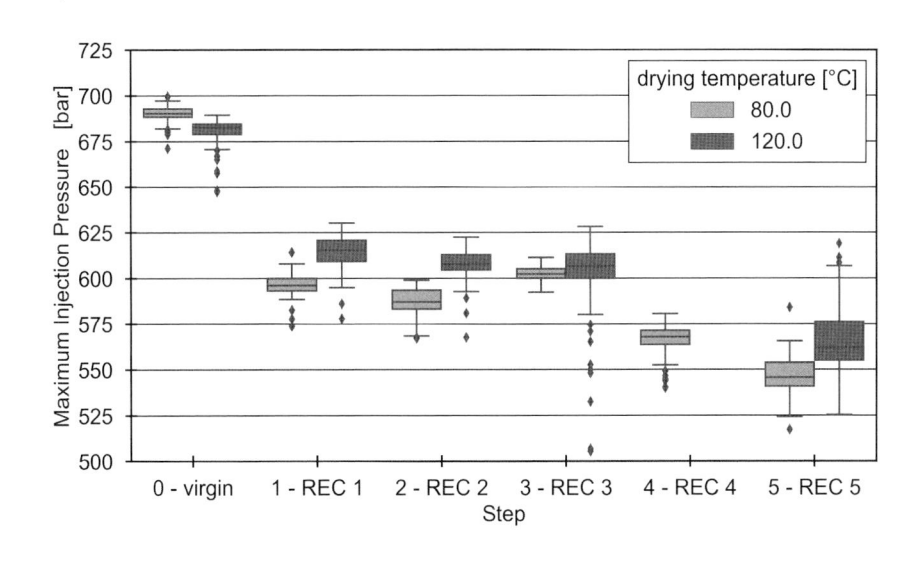

Figure 2.3: Evolution of maximum injection pressure during trials

severing the more sensitive bonds in the first reprocessing step, the material properties remain relatively stable in successive cycles, because the remaining chemical bonds are more robust against ageing.

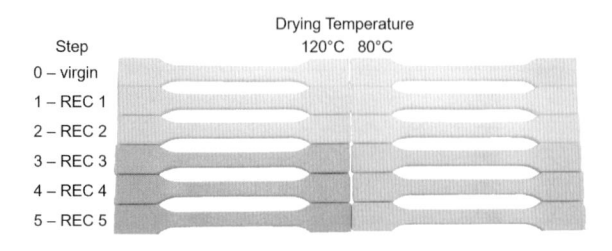

Figure 2.4: Physical appearance of the mouldings.

Physical appearance and further observations

Especially the 120°C material, which was always processed first, was contaminated with red Polyurethane foam during the first regrind, glass fibre reinforced poly(methyl methacrylate) (PMMA) (Arkema Elium 590) during the third regrind, and traces of low-density polyethylene (LDPE) foil during the fourth regrind. These materials had been processed in the mill before and could not be removed entirely by letting the mill run empty and vacuum cleaning the space below the sieve. This led to sporadic red dots throughout the trials and a notable greyish discoloration of the REC3, REC4 and REC5 specimens (see Figure 2.4). The 120°C-REC2 specimens appeared a bit darker and somewhat yellowish compared to the virgin material. All in all, discoloration was very subtle and presumably caused by contamination rather than material degradation.

2.5 Conclusions and outlook

The trials confirm the observations made by *Risch et al.* [RAA00]: PBT can be reprocessed several times without any major issues in processing, as long as residual moisture is within the recommended range. When reprocessing takes place within few days after the original moulding, it is sufficient to dry the material according to manufacturer specifications. After storing the material for several months, it is however advisable to check if the allowable processing moisture levels have been achieved. Hotter drying resulted in less weight gain of the product, but seemed to accelerate material discoloration (yellowing). The assumption that moisture and temperature can be considered independently when assessing the material viscosity made by *Mielcki et al.* [MWG+12, HMW15] does not seem appropriate when considering processing conditions in injection moulding.

The part geometry used for the trials presented here was simple, robust and easy to fill. Thin walled parts with long flow paths, as typically found in packaging applications, may be more sensitive to viscosity fluctuations and thus require tighter viscosity and ageing control and could be an interesting subject for further research.

Keeping moisture under control is not an easy aim to achieve without constant measurement. Laboratory moisture measurements are however laborious and time-consuming. A common alternative to frequent measuring and quality-checking would be to keep the entire process of regrinding, storing and drying under tight observation and control. This is not a trivial task, either: The material may be processed and reground at different places or even at different companies; companies may collect material to be reground for longer time spans; and the amount of reground material added to the stream may vary depending on the current rate of rejects. Digital systems such as MESs may become increasingly helpful in answering questions like: How long has the material been stored and should moisture be double checked due to the prolonged storage time? Do we expect the moisture to be so high that the material is unlikely to be sufficiently dry at the scheduled start of the next production run? Should we reschedule, or order fresh material? What material has been in the mill before and is it advisable to run some cleaning material through it?

Within its new technical centre, the Plastics Innovation Center 4.0, the Institute for Plastics Processing (IKV) plans to map large parts of the polyethylene terephtalate (PET) bottle production and recycling value chain to a laboratory scale demonstrator, comprising injection moulding and blow moulding operations as well as any necessary steps in material handling, such as storage condition monitoring, grinding, crystallising, and drying. Combined with an MES system for data acquisition and trial planning, the demonstrator will allow to pass on data on material properties and processing history along the value chain and to correlate them with the resulting process and quality data. Provided significant correlations exist, fluctuations in material quality could eventually be foreseen and compensated along the processing chain.

Acknowledgement

The presented research was conducted as part of the Plastics Innovation Center 4.0 project, which is funded by the State of North Rhine-Westphalia and by the European Regional Development Fund (EDRF). The authors would also like to thank Mr Pascal Bibow for the data acquisition and pre-processing; Ms Julia Jung, Ms Michelle Herbener, Mr Karem Hadla and Mr Ahmed Fakhr for their assistance in the trials; and Mr Jan Wolters for the moral support.

References

[CV06] COLIN, X.; VERDU, J.: Polymer degradation during processing. *Comptes Rendus Chimie* 9 (2006) 11-12, p. 1380–1395

[HFH17] HOPMANN, C.; FISCHER, T.; HEINISCH, J.: Online melt viscosity measurement during injection molding for new control strategies. *SPE ANTEC* (2017), p. 1560–1565

[HMW15] HEINZLER, F. A.; MIELICKI, C.; WORTBERG, J.: A viscosity model for technical polymers describing chemical and physical aging. *Polymer Engineering & Science* 55 (2015) 7, p. 1628–1633

[HW15] HEINZLER, F. A.; WORTBERG, J.: Qualitätsregelung beim Spritzgießen (Teil 1) : Rheologische Untersuchung prozessrelevanter Alterungseinflüsse. *Zeitschrift Kunststofftechnik* 11 (2015),

[KB15] KAST, O.; BONTEN, C.: Vorhersage des Trocknungsverhaltens hygroskopischer Kunststoffe. *Stuttgarter Kunststoffkolloquium* 24 (2015),

[KOY+14] KURAM, E.; OZCELIK, B.; YILMAZ, F.; TIMUR, G.; SAHIN, Z. M.: The effect of recycling number on the mechanical, chemical, thermal, and rheological properties of PBT/PC/ABS ternary blends: With and without glass-fiber. *Polymer Composites* 35 (2014) 10, p. 2074–2084

[Kru15] KRUPPA, S.: *Adaptive Prozessführung und alternative Einspritzkonzepte beim Spritzgießen von Thermoplasten.* Universität Duisburg-Essen, Dissertation, 2015

[LHL96] LOCHEM, J. H. V.; HENRIKSEN, C.; LUND, H. H.: Recycling Concepts for Thermoplastic Composites. *Journal of Reinforced Plastics and Composites* 15 (1996), p. 864–876

[Mar06] MARGOLIS, J. M.: *Engineering Plastics Handbook.* New York, USA: McGraw-Hill, 2006

[MWG+12] MIELICKI, C.; WEGNER, A.; GRONHOFF, B.; WORTBERG, J.; WITT, G.: Prediction of PA12 melt viscosity in Laser Sintering by a Time and Temperature dependent rheological model. *RTEjournal* 9 (2012),

[PWS20] PILLWEIN, G.; WILLNAUER, P.; STEINBICHLER, G.: Opening up a Broader Range of Applications for Recyclates: The iQ weight control Assistance System Ensures Higher Process Stability. *Kunststoffe International* 4 (2020), p. 26–30

[RAA00] RISCH, B. G.; AUVRAY, T.; AMMONS, D.: Effects of Moisture Content, CEC, and Processing Conditions on Mechanical Properties and Long-Term Reliability of PBT Fiber-Optic Buffer tubes. *58th Annual ANTEC Proceedings.* Orlando, FL, USA, 2000

[San07] SANCHEZ, E. M. S.: Ageing of PC/PBT blend: Mechanical properties and recycling possibility. *Polymer Testing* 26 (2007) 3, p. 378–387

[THH07] THEUNISSEN, M.; HOPMANN, C.; HEINISCH, J.: Compensating viscosity fluctuations in injection moulding. *PROCEEDINGS OF PPS-32: The 32nd International Conference of the Polymer Processing Society* volume 1914. Dresden, 2007

Symbols

Symbol	Unit	Description
a_{PC}	–	physical change factor
a_{sc}	–	structural change factor
$a_{PC,IS}$	–	material-dependent slope of a_{PC}
$a_{SC,UL}$	–	upper limit of structural change
a_T	–	temperature shift factor
c_{H_2O}	$weight\%$	residual moisture concentration
E_0	kJ/mol	flow activation energy
$E_{0,SC}$	J/mol	structural change activation energy
$E_{0,T}$	J/mol	flow activation energy
k_0	$1/s$	rate of structural change
R	$kJ/mol K$	ideal gas constant
t	s	time
t_i	s	partial time span
T	K	Temperature
T_0	K	reference temperature
T_i	K	partial temperature
$\dot{\gamma}$	$1/s$	shear rate
η	Pas	shear viscosity
η_0	Pas	zero-shear viscosity

Abbreviations

Notation	Description
ABS	acrylonitrile butadiene styrene
FTIR	fourier-transform infrared spectroscopy
IKV	Institute for Plastics Processing
LDPE	low-density polyethylene
MES	Manufacturing Execution System
MFI	Melt Flow Index
PA	polyamide
PBT	polybutylene terephthalate
PC	polycarbonate
PET	polyethylene terephtalate
PMMA	poly(methyl methacrylate)

3 Data-driven prediction the tensile strength of blown film due to digital shadows

Ch. Hopmann[1], D. Grüber[1]
[1]Institute for Plastics Processing (IKV) at RWTH Aachen University

Abstract

In addition to the optical and barrier properties, the mechanical properties are used as an important quality factor. The qualitative relationship between the influencing variables and the resulting film properties is largely known, but the findings are, depending on the model, highly machine-dependent. *Ohlendorf* [Ohl04] derives a property model through an empirical-statistical approach, which already allows a high prediction accuracy for the prediction of the Youngs's modulus and the film shrinkage. However, a prediction of the tensile strength is not yet possible with sufficient accuracy. An extended modeling approach therefore provides for modeling which, in addition to the known parameters, also takes into account the stretching and cooling behaviour in the tube formation zone.

3.1 Introduction and motivation

Around 40 % of the plastic processed throughout Europe is used in the production of packaging, of which plastic films make up a considerable amount [NN19c]. In Germany, too, the packaging sector represents an important branch of industry with an annual consumption of around 4.4 million tonnes. In addition to packaging, films are also used in medical products, in the automotive and electrical industries, and in the agricultural and construction sectors [Nen06]. Because plastic films can be further processed into a wide variety of products, machine producers are faced with increasing challenges: Advancing globalisation and the associated worldwide price competition. To remain competitive in the long term, it is often necessary to position oneself in the market by means of unique selling points such as increasing productivity with high product quality [Dis13].

Next to flat film extrusion, blown film extrusion is the most important production process for plastic films [Nen06]. Compared to flat film extrusion, the advantage of blown film extrusion is that the film width and thickness can be varied flexibly without having to modify the line or die technology. Biaxial properties of the film can also be adjusted in a targeted manner, which, however, requires a high level of process knowledge [Lim13]. Therefore, non-knowledge-based process control, only using machine parameters like take-up ratio, blow-up ratio or frost line ratio, usually does not lead to the desired film properties, as the shape of the foil geometry is not included in the modelling. In addition to the optical and barrier properties, the mechanical properties are used as an important quality criterion. The qualitative relationships between the influencing variables and the resulting film properties are largely known, but the findings are highly material-dependent and often machine-dependent, which means that a calculation of the mechanical properties without extensive, empirical investigation is only possible with reduced prediction quality [Ohl04].

Ohlendorf therefore developed a material-dependent calculation model which aims to predict the mechanical properties based on process parameters, thus enabling transferability to other machines [Ohl04]. However, a validation of the model shows that the transferability is only possible to a limited extent, which is due to an incomplete description of the process state, e.g. the thermal and stretching history of each segment of the film. Therefore, in the following an approach to increase the model quality is presented to increase the transferability of the empirical process model to other production machines.

For this purpose, it is first explained how the mechanical properties of blown films are influenced by processing. Subsequently, the process model according to Ohlendorf is explained and optimisation measures for increasing the model quality are shown.

3.2 Determination of the mechanical properties of blown films

The mechanical properties of blown films are set to a large extent in the tube formation zone between the die and the frost line. They can be qualitatively attributed to the biaxial orientation state and the crystallisation state of the material.

For the orientation of thermoplastics, they must be stretched above their softening temperature (retardation). In blown film extrusion, this occurs in the direction of extrusion if the take-off speed of the film above the frost line is greater than the exit speed of the melt from the die (take up ratio > 1). Similarly, orientation of the film in the circumferential direction occurs when the blow-up ratio is greater than one (blow-up ratio > 1), which is given by nozzle and bubble diameter. The faster the material is solidified, the lower the orientation retardation and therefore, the lower the decay of molecular orientation. It can thus be stated that the interplay of the cooling process and the course of longitudinal and transverse stretching determines the biaxial orientation state of the film. If the orientation behaviour of different materials is investigated, a qualitative correlation can be found: Materials with a lower molecular weight relax faster. In general, a higher degree of orientation of the macromolecules leads to an increase in the mechanical properties in the direction of orientation.

In addition to the formation of orientations, crystallisation can exert an influence on the mechanical properties. The more the film is stretched in the tube formation zone, the more spherulite growth is inhibited and existing superstructures are destroyed. The result is a hidden structure with a high degree of crystallisation. The faster the melt is cooled, the finer the resulting microstructure. To what extent thermoplastics crystallise at all and to what degree depends crucially on the structure of the macromolecule. Semi-crystalline thermoplastics with a linear chain structure, e.g. PE-HD, can reach a degree of crystallisation of up to 70 %, while branched chain structures (e.g. PE-LD) are not capable of a high packing density. However, a generally valid statement between the degree of crystallisation or the structure that forms and the resulting mechanical properties cannot be made [MHMS14].

3.3 Modelling for the prediction of mechanical properties in blown film extrusion

Analogous to other technical processes, there are different possibilities for choosing the input and output variables a model for predicting the mechanical film properties. A distinction must be made between machine, process and quality parameters. If the machine parameters are used as input parameters of the model, the model might not be transferable to other machines. Similarly, a process model reflects the relationship between process and quality parameters. Both physical and statistical modelling can be applied for both variants. Furthermore model building methods, e.g. "grey box models", allow a combination. Figure 3.1 shows the possibilities of model building for the calculation of mechanical properties.

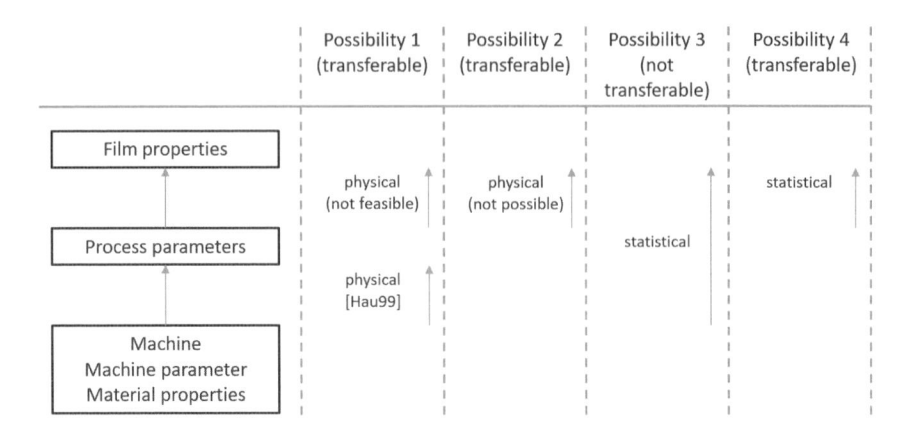

Figure 3.1: Possibilities of modelling for the prediction of film properties [Ohl04]

The first possibility consists in the purely physical modelling, which represents the ideal solution for the prediction of the mechanical properties with sufficient model quality. Hauk [Hau99] uses two-stage modelling to predict tensile strength. He succeeded in the first partial step, the physical calculation of the process state from the machine parameters, however, multiscale simulations are known which can derive the film properties physically conclusively from the process state, but currently they have unfeasibly high computational demands [Hau99]. As already explained, the mechanical properties depend significantly on the material- and process-dependent degree of orientation, crystallisation and their interaction, for which physically reasoned correlations, would have to be available. Due to the above-mentioned considerations, the second possibility, the physical modelling with selection of the process parameters as input variables, is not yet feasible today. A widely used method to predict film properties is the statistical modelling of the quality variables starting from the machine parameters. Since no process knowledge is developed about the behaviour of the melt in the tube formation zone, this model cannot be transferred to other machines. Furthermore, not all influences, such as material, machine or tool behaviour, are sufficiently captured and modelled. A statistical mapping from process to quality variables, which are valid for all processes and plants, allows the transfer of the model to other production plants if the modelling quality is sufficient. The disadvantage is also the experimental effort required to parameterise the model. However, since the modelling approach should be able to determine the mechanics based on the process state independently of the machine, the approach according to *Ohlendorf* is presented in the following and potential for improvement is shown.

3.3.1 The statistical process model according to Ohlendorf

The basis of the process model according to *Ohlendorf* is the statistical design of experiments, according to which machine parameters are systematically varied and lead to different process states and thus also to different mechanical properties of the film. Based on the resulting process state, the process parameters are collected and converted into dimensionless numbers in order to enable a machine-independent modelling. Subsequently, the film samples are examined for the required mechanical properties in longitudinal and transverse direction, e.g. tensile strength. Linear regression is used to determine the influence of the dimensionless numbers on the tensile strength. As a result, a linear model for the calculation of the tensile strength is established. The procedure for deriving the model can be taken from Figure 3.2.

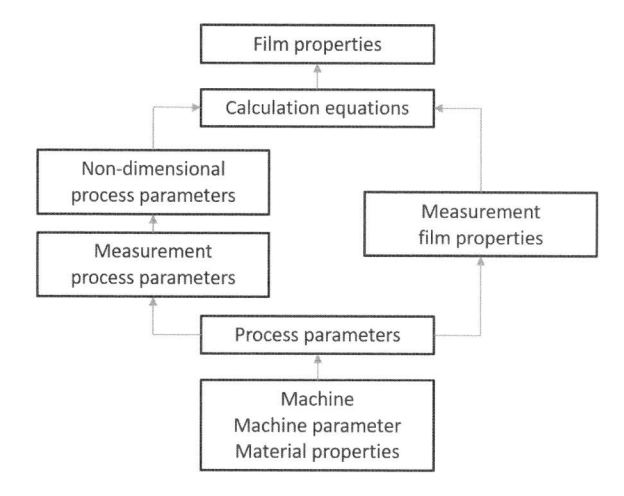

Figure 3.2: Procedure of the statistical process model [Ohl04]

First, the well-known, dimensionless numbers take-up ratio (TUR) and blow-up ratio (BUR) are introduced, which are regarded as a measure of the elongation in the extrusion direction and in the circumferential direction. The ratio of melt and frost line temperature describes the cooling of the tube forming zone in extrusion direction. In conjunction with the volume flow and frost line ratio, this ratio quantifies the effectiveness of the cooling ring. The influence of the film thickness on the mechanical properties is taken into account by the thickness ratio of the film thickness and the outlet gap width of the die.

$$\text{Take-up ratio:} \quad i_l = \frac{film\ line\ speed}{melt\ speed} \tag{3.1}$$

$$\text{Blow-up ratio:} \quad i_q = \frac{bubble\ diameter}{die\ outlet\ diameter} \tag{3.2}$$

$$\text{Temperature ratio:} \quad i_t = \frac{melt\ temperature}{frost\ line\ temperature} \tag{3.3}$$

$$\text{Volume flow ratio:} \quad i_v = \frac{melt\ volume\ flow}{cooling\ air\ flow} \quad (3.4)$$

$$\text{Frost line ratio:} \quad i_f = \frac{frost\ line\ height}{die\ outlet\ radius} \quad (3.5)$$

$$\text{Thickness ratio:} \quad i_d = \frac{film\ thickness}{gap\ width} \quad (3.6)$$

Ohlendorf establishes a linear relationship between the established dimensionless numbers and the mechanical characteristic values (Youngs's modulus, tensile strength, shrinkage). Applied to a specific material like LDPE LD 150 AC of ExxonMobil Corp., Inving, Texas, United States, the following linear calculation equation can be established for calculating the mechanics:

$$Mechanical\ parameter = l \cdot i_l + q \cdot i_q + t \cdot i_t + v \cdot i_v + f \cdot i_f + d \cdot i_d + k \quad (3.7)$$

Provided that the resulting process characteristics are linearly independent and the residuals are normally distributed, the regression coefficients l, q, t, v, f, d and k can be determined from the experimental space using linear regression.

To evaluate the quality of the model, the coefficient of determination r^2 is used, which can assume values between zero and one and assesses the quality the model. A coefficient of determination of one indicates that the linear relationship assumed by the model is matched perfectly by real world data. Furthermore, the average percentage deviation (a) between the forecast and the measured value is used as a further evaluation criterion.

Table 3.1 compares the coefficient of determination and the percentage deviation for predicting the modulus of elasticity, tensile strength and shrinkage in the longitudinal and transverse directions.

Direction	E-module		Tensile strength		Shrinkage	
	r^2 [%]	a [%]	r^2 [%]	a [%]	r^2 [%]	a [%]
Longitudinal	0.82	3.20	0.77	7.04	0.91	2.39
Cross	0.82	5.72	0.5	6.68	0.96	7.01

Table 3.1: Coefficient of determination and percentage deviation, LD 150 AC, IKV [Ohl04]

When analysing the model quality, it becomes clear that the model has the lowest accuracy when predicting tensile strength. In particular, the model does not provide sufficient accuracy transverse to the extrusion direction. It must therefore be stated that the statistical property model is suitable at most for rough estimation of the tensile strength.

Due to the modelling with process parameters, the model must be transferable to other machines. *Ohlendorf* therefore evaluates the transferability of the Young's modulus and the shrinkage behaviour to other production machines. Provided that the process condition is described sufficiently by the key figures introduced, no significant change in the model quality is to be expected. Table 3.2 shows an example of the coefficient of determination and the percentage deviation from the real data of the model, which were collected on a blown film line of Kuhne GmbH, Sankt Augustin.

Direction	E-module		Tensile strengh		Shrikage	
	r^2 [%]	a [%]	r^2 [%]	a [%]	r^2 [%]	a [%]
Longitudinal	0.12	14.61	-	-	0.55	19.73
Cross	0.27	26.81	-	-	0.85	18.73

Table 3.2: Coefficient of determination and percentage deviation, LD 150 AC, Kuhne GmbH [Ohl04]

Clearly, the model does not have sufficient quality for calculating the tensile strength. Furthermore, no sufficient transferability to other machines is guaranteed.

3.3.2 Derivation of development potentials

In the following chapter, possible deficits of *Ohlendorf's* approach will be identified. Subsequently, an approach to increase model quality is presented and evaluated. As explained earlier, the mechanical properties are decisively determined in the tube formation zone. To describe the geometric deformation of any given volume of melt, *Ohlendorf* uses the dimensionless numbers i_l, i_q, i_f, and i_d, each of which relates a geometric quantity at the outlet of the extrusion die and the state at the frost line, at which the mechanical properties are defined. Figure 3.3 illustrates that the listed key figures can have the same value with different characteristics of the film geometry.

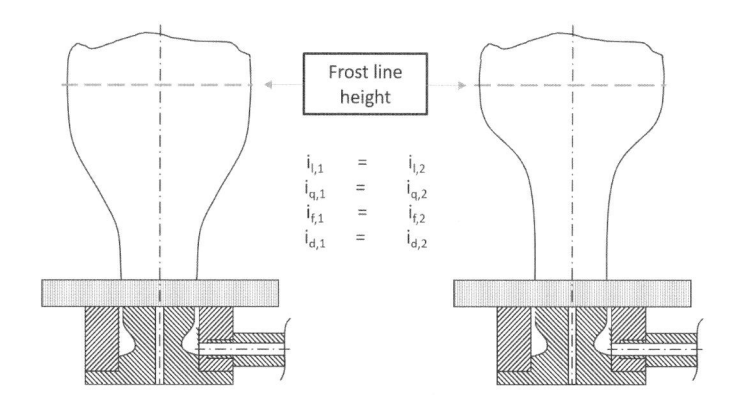

Figure 3.3: Same geometrical parameters with different bubble geometry

Reasons for different bubble geometries are different settings of the double-lip cooling ring or different lip designs. In addition changing process and environmental conditions, such as temperature or air flows in the production building, might change the film geometry under otherwise identical production conditions. This leads to the necessity to introduce further parameters which characterise the elongation across the bubble geometry.

A similar situation is found when considering cooling behaviour, which is described by the dimensionless numbers i_t and i_v, from which the exact cooling behaviour of the film in the tube formation zone cannot be inferred. Using only those two parameters in a linear regression model, at most a quadratic relationship between melt temperature and path length can be described. Figure 3.4 shows a representative cooling curve in the tube formation zone. The temperature does not decrease quadratically in the extrusion direction in accordance with the decreasing cooling capacity, but with a non-linear relationship (e.g. exponential decay). This circumstance is not considered in the work of *Ohlendorf*.

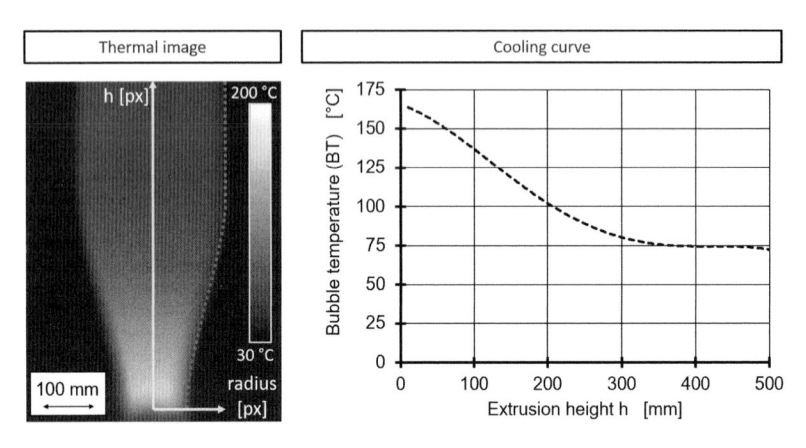

Figure 3.4: Non-linear, non-quadratic behaviour of bubble temperature

The process model must therefore be extended by further parameters so that an infinitesimal change in the film geometry can be assigned to an infinitesimal temperature gradient at any point in time. Therefore, new trials were conducted and significantly more data was collected on the behaviour of the film in the tube formation zone.

3.3.3 Investigations on the extension of the process model by digital process variables

The aim of these new investigations is to increase the model quality. This is made possible by continuously recording the film geometry and the cooling behaviour of the tube formation zone. The collected data should enable a data-driven modelling of the process to predict the mechanical properties, especially the tensile strength. This is based on the idea that if the elongation and the cooling behaviour in the tube formation zone are recorded, all the information for modelling is already available and the dimensionless numbers according to *Ohlendorf* no longer need to be used.

A centrally composed 2^k experimental design (see Table 3.3) is carried out to perform the experiments, which systematically varies the control variables of mass temperature, inflation ratio, film thickness and blower power. The material used is an LDPE (type 2102NOW, Sabic, Riyadh, Saudi Arabia). From the resulting film geometry and the cooling process, additional parameters are derived to supplement the process model.

Factor level	Mass temperature [°C]	BUR [-]	Film thickness [µm]	Blower power [%]
-1	200	2.4	100	15
0	215	2.8	150	25
1	230	3.2	200	35

Table 3.3: Factor levels

For the detection of the geometry of the tube formation zone, an optical detection or measuring system is necessary. In [Spi04] the use of a camera is proposed for this purpose. The geometry data is then derived from the generated photo material by (automatic) contour recognition. For this purpose, a camera of the model Basler ace 2 Basic of the company Basler AG, Ahrensburg, which offers the possibility of recording high-resolution images with 3840 x 2748 pixels (10.7 MP), is used. It is placed on a tripod at a defined height and defined distance from the tube formation zone. Furthermore, a suitable lightning of the tube formation zone is carried out by means of lamps, which are positioned behind a textile screen. During the tests, the film contour is recorded with one image per second. The bubble geometry is evaluated for each test point by a tool developed at IKV using MATLAB, MathWorks, Inc., Massachusetts, USA. The symmetry axis is assumed to be orthogonal to the annular gap die. Subsequently, the image coordinates are transformed into metric coordinates, since a later mathematical description of the film geometry should be physically meaningful in millimeters rather than pixels. Calibration is performed by recording a reference pattern of known geometry in the visible edge of the film bubble with subsequent data transformation. To model the film geometry as a function of the extrusion height, *Spirgatis* proposes a fifth-degree polynomial [Spi04], which also turns out to be an excellent fitting of the measured values in this application. Figure 3.5 shows the principal procedure for the survey of the film geometry using the example of the bubble geometry at low melt temperature and blower power as well as high blow-up ratio and film thickness.

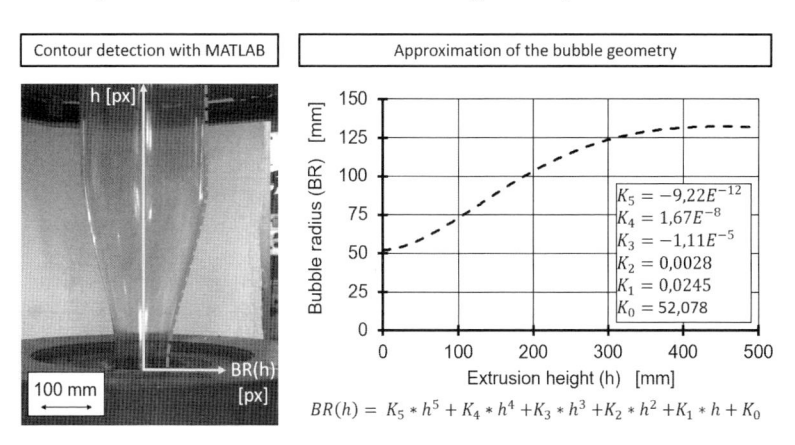

$$BR(h) = K_5 * h^5 + K_4 * h^4 + K_3 * h^3 + K_2 * h^2 + K_1 * h + K_0$$

$K_5 = -9{,}22E^{-12}$
$K_4 = 1{,}67E^{-8}$
$K_3 = -1{,}11E^{-5}$
$K_2 = 0{,}0028$
$K_1 = 0{,}0245$
$K_0 = 52{,}078$

Figure 3.5: Detection of the bubble contour and approximation by a 5th degree polynomial

In the the polynomial BR(h), there are six coefficients which describe the film geometry. Since the information of the die radius (h = 0 mm) and the bubble diameter at frost line height (h = 500 mm) is already contained in the coefficients i_q and i_f, the number of coefficients can be reduced to five (K_4' to K_0') by deriving the geometry function according to the extrusion height. The resulting function BR'(h) thus describes the expansion of the film bubble in the extrusion direction (h).

As already described, in addition to the description of the elongation, the cooling behaviour in the tube formation zone is relevant for the determination of the mechanical properties. To analyse the cooling behaviour, a thermal imaging camera of the type Flir SC 305 from Flir Inc., Wilsonville, Oregon, USA, with a resolution of 320 x 240 pixels and a maximum recording rate of three images per second is used. The temperatures in the image area are stored as a data matrix in a CSV file. To identify the relevant cooling process, a Python script is used which stores the image coordinates with the corresponding temperature value in an array. For the transformation of the image coordinates into metric coordinates, the geometry of the tube formation zone from the thermal and industrial camera is related to each other. Analogous to the acquisition of the geometry, the description of the cooling process is subsequently carried out by means of a fourth-degree polynomial as a function of the extrusion height.

Figure 3.6 shows the procedure for determining the cooling curve at the process point.

| Determination cooling curve | Approximation of the cooling curve |

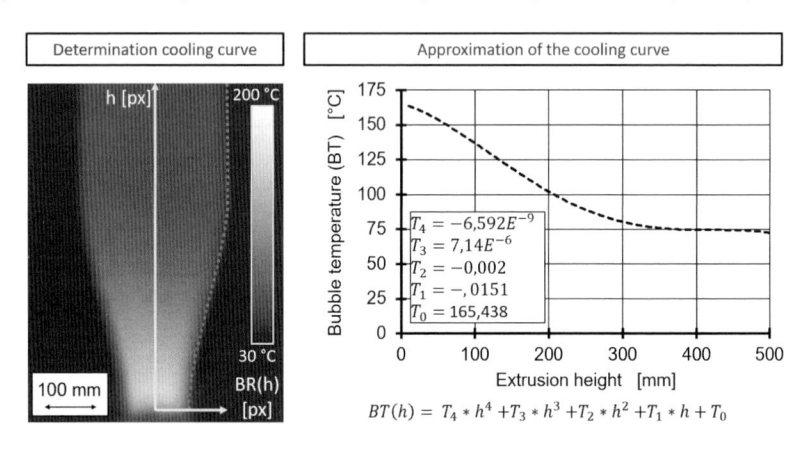

$T_4 = -6,592E^{-9}$
$T_3 = 7,14E^{-6}$
$T_2 = -0,002$
$T_1 = -,0151$
$T_0 = 165,438$

$$BT(h) = T_4 * h^4 + T_3 * h^3 + T_2 * h^2 + T_1 * h + T_0$$

Figure 3.6: Acquisition of the cooling process and approximation by a 4th degree polynomial

To determine the mechanical properties in the longitudinal and transverse directions, a Zwick Z10 universal tensile testing machine from Zwick GmbH Co. KG, Ulm, is used. For each test point 10 specimens are tested. Following DIN EN ISO 527-1 [NN19a], the specimen clamping length is 50 mm. The crosshead speed is 200 mm/min, reduced to 5 mm/min for the measurement of the modulus of elasticity. Due to the poor model quality for the determination of the tensile strength in *Ohlendorf's* modelling, an improved modelling for the tensile strength is considered.

After data analysis, the coefficients K_5 to K_0 are now available for each process point as a measure for the expansion of the tube formation zone and T_4 to T_0 as a measure for the cooling behaviour. There now are several possibilities to use these to improve the model quality.

When using so-called white box models, the system behaviour is clearly described by physical relationships, for example by differential equations. The advantage lies in the interpretability of the model and the high acceptance by the users. On the other hand, there is an enormous number of parameters needed to describe the physical relationships and a high numerical effort. In contrast, Blackbox models do not draw on any structured knowledge about the process. Instead, the modelling is completely data-based, which means that with a sufficient amount of data, even high complex differential equations can be represented. The disadvantage is that this model cannot be interpreted physically and no further understanding of the process can be built up. Also, extrapolation for a prediction outside of the captured data range is impossible in most cases. Grey-boy models combine the advantages of white- and black-box models. A well known example of grey box modelling is linear regression, which establishes the functional, statistical relationship between the input and output variables without using physical modelling. For this, an influence of a variable on the target variable must be assumed, e.g. in the context of a sensitivity analysis. If there is a basic understanding of the process, but the model quality is not sufficient, a black box model can be used. In this case, the main influences are first described mathematically (physically or statistically) and then optimized by data that is not part of the white or grey box model [NN19b]. Following *Ohlendorf*, the suitability of linear regression is first investigated to model the tensile strength.

For a linear regression there is the requirement that the coefficients must be linearly independent or not correlated. Table 3.4 shows the correlation matrix of the coefficients. Correlation coefficient of 1 or -1 means that there is a strictly linear relationship between the parameters. For a linear regression, a regression coefficient between -0.2 and 0.2 is desirable for (marked in grey).

	K_4	K_3	K_2	K_1	K_0	T_3	T_2	T_1	T_0
K_4	1	-0.988	0.941	-0.834	0.651	0.044	-0.021	-0.016	0.1
K_3		1	-0.981	0.904	-0.738	-0.17	0.147	-0.101	-0.004
K_2			1	-0.967	0.838	0.333	-0.315	0.267	-0.145
K_1				1	-0.942	-0.514	0.517	-0.484	0.363
K_0					1	0.617	-0.663	0.673	-0.597
T_3						1	-0.982	0.922	-0.788
T_2							1	-0.978	0.882
T_1								1	-0.956
T_0									1

Table 3.4: Correlation matrix of the determined coefficients

Due to the high correlation of the input parameters, they are not suitable for multiple regression, as sufficient accuracy cannot be achieved to predict the tensile strength. None of the parameters have a significant influence on the tensile strength in the longitudinal or transverse direction. The property model can therefore not be extended by adding further linear terms in the sense of linear regression. An alternative is the method of Partial Last Square Regression (PLS). The advantage of this method is that the input parameters of the model may be highly correlated or intercorrelated. If the PLS is carried out exclusively with the derived coefficients, without considering the "classical" parameters of the property model, a model quality of $r^2 = 0.602$ and $r^2 = 0.537$ results for the calculation of the tensile strength in the longitudinal and cross direction. Larger values of r^2 (close to 1) indicate a high model quality, low values (close to 0) indicate a low model quality. Considered on its own, the model is therefore not yet suitable for prediction.

However, this finding must be set in relation to the fact that, in addition to the added coefficients, no further elementary process parameters, such as the melt temperature or the take-off ratio, have yet been included in the PLS model. In this respect, an enormous potential of the collected parameters for the improvement of model quality can possibly be assumed. Another possibility is the use of the conventional model in combination with an artifactial neural network, which is likely able to represent also non-linear correlations of the derived coefficients with the tensile strength in a higher model quality (black box model as booster). A pure black box model including the parameters derived can also be considered.

3.4 Conclusions and outlook

The prediction of mechanical parameters during the blown film process is an important prerequisite for product development and quality optimization. Statistical or data-driven models are possibly suitable for modelling the influence of the process parameters on the mechanical properties. The problem of physically based models is the complex numerical solution with a high required number of input parameters. *Ohlendorf* derives a property model by an empirical-statistical approach, which already allows a high prediction accuracy for the prediction of the Young's modulus and the film shrinkage. However, a prediction of the tensile strength using this model is not yet possible with sufficient accuracy. This is due to an incomplete description of the tube formation zone, in which the mechanical properties are set. An extended model approach therefore should describe the stretching and cooling behaviour accurately with a high spatial resolution in the tube formation zone. In a new series of tests, the formation zone is therefore recorded by camera technology and the stretching of the film and the temperature gradient are described using polynomials. A correlation analysis of the polynomial coefficients obtained shows that they correlate strongly with each other. An extension of the property model by means of classical linear regression is therefore impossible. However, the method of Partial Last Square regression (PLS) shows that the coefficients determined in combination with the already established variables should be considered for further model building.

In further investigations, modern modelling methods, such as artificial neural networks, are to be examined. One possibility is the development of a hybrid model (black box model as booster), which uses the property model according to *Ohlendorf* as input in addition to the newly described polynomial coefficient. A purely data-driven modelling (black box model) should also be considered. Furthermore, the improved process model should be investigated to evaluate whether the transferability to other machines might be enabled by the extended description of the tube formation zone.

Acknowledgments

The construction of the Plastics Innovation Center 4.0 (PIC 4.0) is supported by funds from the state of North Rhine-Westphalia and the European Regional Development Fund (EFRE). We would like to express our gratitude to all institutions.

Furthermore, the authors would like to thank Sabic Europe B. V., Geelen, Netherlands, for the materials provided for experimental purposes.

References

[Dis13]   DISPAN, J.: *Kunststoffverarbeitung in Deutschland.* Company script, 2013

[Hau99]   HAUK, J.: *Entwicklung eines Simulationsprogrammes für den Schlauchfolienextrusionsprozess.* RWTH Aachen University, Dissertation, 1999

[Lim13]   LIMPER, A.: *Verfahrenstechnik zur Thermoplastextrusion.* München: Carl Hanser Verlag, 2013

[MHMS14]   MENGES, G.; HABERSTROH, E.; MICHAELI, W.; SCHMACHTENBERG, E.: *Menges Werkstoffkunde Kunststoffe.* München: Carl Hanser Verlag, 2014

[Nen06]   NENTWIG, J.: *Kunststofffolien: Herstellung – Eigenschaften – Anwendung.* München: Carl Hanser Verlag, 2006

[NN19a]   N.N.: *DIN EN ISO 527-1: Kunststoffe – Bestimmung der Zugeigenschaften – Teil 1: Allgemeine Grundsätze.* Berlin: Beuth Verlag, 2019

[NN19b]   N.N.: *Implementierung und Betrieb von Big-Data-Anwendung in der produzierenden Industrie – Durchführung von Big-Data-Projekten, VDI 3714 Blatt 1 – 7 (Entwurf).* Düsseldorf: VDI-Gesellschaft Kunststofftechnik, 2019

[NN19c]   N.N.: *Plastics – the facts 2019.* Report, Düsseldorf, 2019

[Ohl04]   OHLENDORF, F.: *Vorhersage der mechanischen Folieneigenschaften bei der Schlauchfolienextrusion.* RWTH Aachen University, Dissertation, 2004

[Spi04]   SPIRGATIS, J.: *Untersuchung des Einflusses des instationären konvektiven Wärmeübergangs bei der Folienherstellung auf die Folienqualität.* Universität Duisburg-Essen, Dissertation, 2004

Symbols

Symbol	Unit	Description
a	—	Percentage deviation
i_l	—	Take-up ratio
i_q	—	Blow-up ratio
i_t	—	Temperature ratio
i_v	—	Volume flow ratio
i_f	—	Frost line ratio
i_d	—	Thickness ratio
l, q, t, v, f, d, k	—	Regression coefficients
r^2	—	Coefficient of determination

Abbreviations

Notation	Description
BR(h)	Bubble radius
BUR	Blow-up ratio
PE-HD	High density polyethlylene
PE-LD	Low density polyethlylene
PLS	Partial Last Square Regression
TUR	Take-up ratio

Control of material flows in packaging recycling

Moderator: Dr. rer. nat. Ulrich Kückelmann, ALBA Recycling GmbH

---- Content ----

1 Actual market dynamics, technical and regulatory requirements   **163**

U. Kückelmann[1]
[1] *ALBA Recycling GmbH*

2 Functionalisation of polyolefin melts by means of atmospheric pressure plasma   **170**

R. Dahlmann[1], Ch. Hopmann[1], C. Wiesel[1], C. Wang[1], M. Schön[1]
[1] *Institute for Plastics Processing (IKV) at RWTH Aachen University*

3 Usage potentials for different qualities of recyclates in plastics packaging   **190**

R. Dahlmann[1], E. Berg[1], M. Schön[1]
[1] *Institute for Plastics Processing (IKV) at RWTH Aachen University*

Dr. rer. nat. Ulrich Kückelmann

Ulrich Kückelmann studied chemistry at the Ruhr-University Bochum, his focus had been analytical chemistry. He completed his doctorate there in 1998. Subsequently, he worked for 1 year as post-doc at the Institute for Spectrochemistry and Applied Spectroscopy, Dortmund.

In 1999, Dr Kückelmann became analytical laboratory manager at the German affiliate of the Swiss Flavor Fragrance producer Firmenich. At Firmenich, he moved from technical into account management.

In 2006, Dr Kückelmann left the personal care industry and started at C.H. Erbslöh's rubber and plastics division. After 2 years, he got responsible of C.H. Erbslöh's rubber plastics business unit, including the principal supplier management. In 2011, he moved to DuPont which Packaging Industrial Polymers business unit was a principal supplier. At DuPont PIP, Dr Kückelmann was sales market development consultant, in charge of several industrial market segments incl. polymer modification.

After 2 years as business product development manager at Everlam, which is a spin-off of DuPont's European Glass Laminating Solutions business, Dr Kückelmann changed end of 2018 to Interseroh Recycled-Resource (member of Alba Group) as business developer. Currently, he is responsible of sales business development at Alba Recycling's recythen® and procyclen® PE and HDPE PCR business.

1 Actual market dynamics, technical and regulatory requirements

U. Kückelmann[1]
[1] ALBA Recycling GmbH

Abstract

The PCR-PP market is structurally tight. Nevertheless, new applications are entering the market and existing applications of PCR-PP are increasing. It is shown why it is worthwhile to review the regulatory framework, existing norms and standards, and the current dynamics of the market.

1.1 Introduction

Use of plastic and plastic waste is the subject of public debate at present. Public awareness and interest in issues such as ocean plastics are very high, reinforced by intense media coverage. In comparison to materials such as paper, cardboard, glass and metal, plastic covers a wide range of different polymer types with different applications and properties. Due to the diversity of applications, plastics recycling may cause some confusion about what is mechanically recyclable. This is accompanied by confusing of non-existent labelling, various end markets and variations in what polymer types are collected for recycling. To differentiate between existing processes and future perspectives, it is helpful to focus on a single type of plastics and the recycling raw material stream thereof. Here, post-consumer polypropylene (PCR PP) is taken as an example, more specifically PP from the household collection of light weight packaging in Germany. Mechanical recycling of polypropylene from the collection of light weight packaging is well established in Germany. Regulatory boundaries are clear, and there are norms and standards in place to describe product qualities. This paper presents this framework and furthermore shows the actual dynamics of this market.

1.2 Norms and standards

The European Standard DIN EN 15343 is covering recycling processes and waste traceability and is the foundation of EUCertPlast quality certification scheme. Figure 1.1 shows examples of data to be recorded according to the end application. EUCertPlast was created by Plastics Recyclers Europe (PRE), European Association of Plastics Recycling und Recovery Organisations (EPRO), European Plastics Recyclers Association (EuPR) and Recovinyl. Furthermore, there is DIN EN 15347 which provides a scheme for the characterisation of plastics wastes, laying out those properties for which the supplier of the waste shall make information available to the purchaser, and identifying test methods where applicable. For the recyclates / re-granulated recycled plastics as such, there are individual standards available. E.g. DIN EN 13545 summarises characteristics and test methods for assessing and keeping record of the properties of recycled polypropylene. Similar standards also apply to other types of plastic such as polyethylene (PE), polystyrene (PS), polyethylene terephthalate (PET) and polyvinyl chloride (PVC). CEN/TR 15353 is the guideline for the development of standards for recycled plastics, it considers the general environmental aspects which are specific to the recycling process.

Origins	Material type/form
	Product type
	Type of waste e.g. pre-user, post user, demolition waste
	Where it came from (supplier identification)
	Date
	history of waste (e.g. known contact with hazardous substances)
Logistics	Collection (transporter/type of transport)
	Sorting
	Batch size, identification and marking
	Pre treatment (e.g. washing, grinding)
	Storage (e.g. outside)
Tests carried out before processing	EN 15347 Plastics recyclate characterisation of waste plastics
	Or as appropriate for the end use application
Process parameters	Details of the process used as appropriate
Tests carried out after processing	EN 15342
	EN 15344
	EN 15345
	EN 15346
	EN 15348
	Or any other standards as appropriate for the end use application
Intended [suitable] application	Details of appropriate or inappropriate applications
Other optional information as agreed between buyer and seller. NOTE This list is not exhaustive.	

Figure 1.1: DIN EN 15343 practical consideration for controlled waste and recycle streams [NN08]

Recently, a further guideline for technical specification of recycled plastics being traded digitally was released. It is called DIN SPEC 91446. This guideline may facilitate data quality standards which were partly missing for recycled plastics. Additionally, new parameters there like characterisation of odours may increase transparency along the value chain.

1.3 Regulatory

PP packaging applications are manouevring within boundaries, e.g. restrictions for food con-tact use. With the revision of the Packaging and Packaging Waste Directive (PPWD) in 2018, European legislation set a target of 50 % for recycling of plastic packaging by 2025, rising to 55 % by 2030. The revised calculation method for reporting on these targets applies stricter and more accurate approaches to the measurement of recycling rates of municipal and packaging waste. The Waste Framework Directive (WFD) requires (Article 8a) that producer responsibility schemes cover the full net costs of the separate collection of packaging (including for the clean-up of litter). In addition, the fees charged to producers are differentiated according to a range of criteria, including recyclability. EU member states and Extended Producer Responsibility (EPR) schemes need to further improve the quantity of plastics collected and sorted for recycling. EPR schemes should transfer the end-of life costs of products to producers, thus incentivising eco-design. The ERP system for packaging in the Netherlands is shown schematically in Figure 1.2.

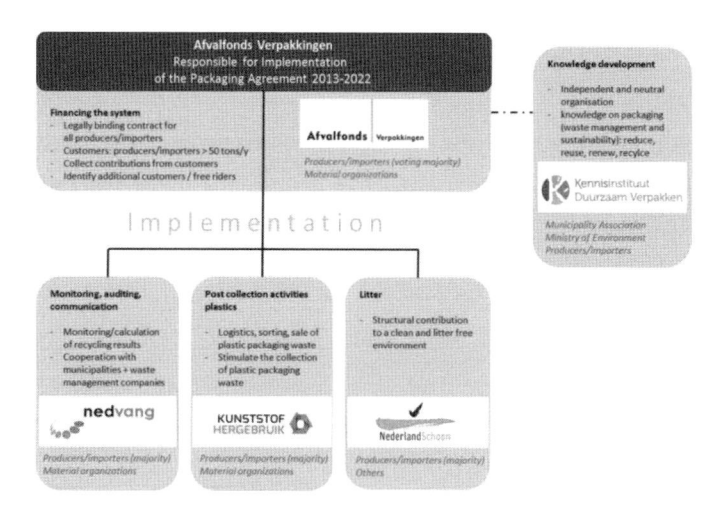

Figure 1.2: The Dutch EPR system for packaging (Source Nedvang)

The legal framework for EPR in Germany is the Packaging Act (VerpackG), which became effective beginning of 2019, replacing the former Packaging Ordinance. Now, a single national authority (Zentrale Stelle) is responsible for increasing the transparency of the overall system. The objective is to create an overall system for the national packaging waste disposal that eliminates competitive distortion. There are several obligations included, for instance regarding recyclability and recycling targets. For plastics material recycling, the minimum annual average rates are 58,5 % starting in 2019, and 63 % starting in 2022. In order to increase recyclability, the "Zentrale Stelle" defines requirements for recyclability along the value chain and thus measures for a more sustainable design of packaging (cf. Figure 1.3).

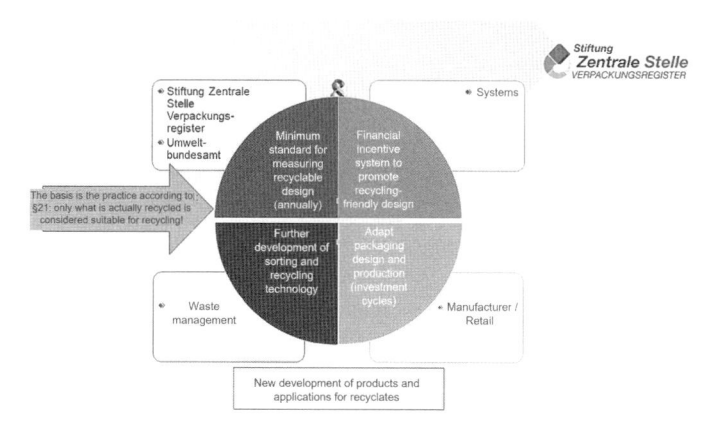

Figure 1.3: The „Zentrale Stelle" defines requirements for recyclability along the value chain [URL22a]

1.4 European / German PCR PP market dynamics

Plastics Recyclers Europe states for the period 2018-2020 that 1.2 million tonnes of recycled rigid polyolefins (PE-HD and PP) were produced annually in the EU28 + 2 (N, CH) [NN20], of which 0.7 million tonnes of PE-HD and 0.5 million tonnes of PP. In 2015, 93 % of EU member states had some kind of producer responsibility for packaging waste, as recommended by directive 94/62/EC, which should lead to some form of collection and recycling system. According to Eurostat, the recycling rate for PP and PE-HD packaging is 41,9 % [URL22b]. These figures are confirmed by Deloitte for rigid PP household packaging with a recycling rate of 42 % [NN17]. Nevertheless, it is a structurally short market, which is reflected in the price development of recycled PP (Figure 1.4).

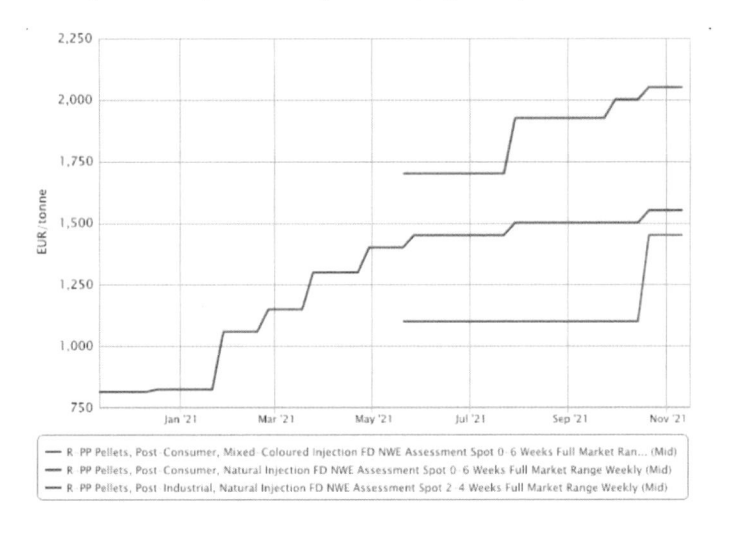

Figure 1.4: Price development of recycled PP in Europe in 2021 (Source ICIS)

In Germany, rigid PCR PP from household collection of light-weight packaging within the dual systems is a well established product. However, available volumes are limited [Bot12, CB20, DC12]. Figure 1.5 shows a detailed breakdown of the composition of light packaging waste. Ca. 2.5 Mio t of light-weight packaging are collected and sorted in the system. Estimations of rigid polyolefines content vary between 8 and 12 %, whereas about 2 % are PE-HD. That means, there are about 250.000 t of rigid PP packaging generated as sorting fraction. Considering that the sorting fraction still contains reasonable amount of impurities which needs to be eliminated or at least minimised, the remaining volume is low.

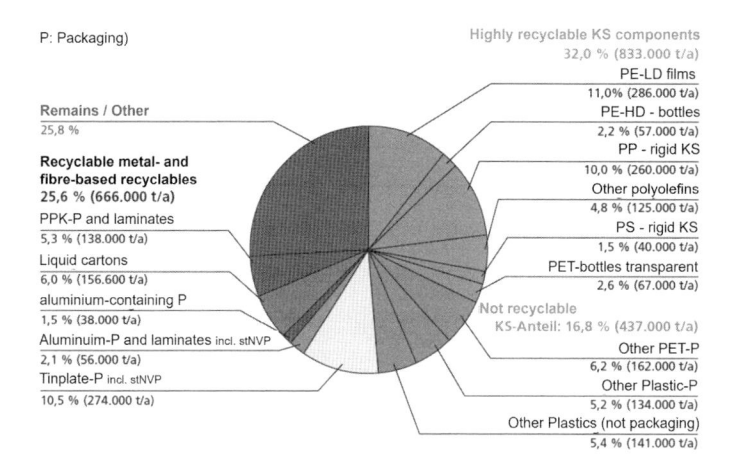

Figure 1.5: Composition of German light weight packaging waste [CB20]

Non-food packaging applications contain increasing amounts of PCR PP, there are double-digit growth rates for several years already. Figure 1.6 shows exemplary applications. Further-more, high quality of PCR PP allows the usage in technically demanding applications where manufacturers also have ambitous targets of recycled content. Accordingly, the increasing market demand meets a rather dry raw material source.

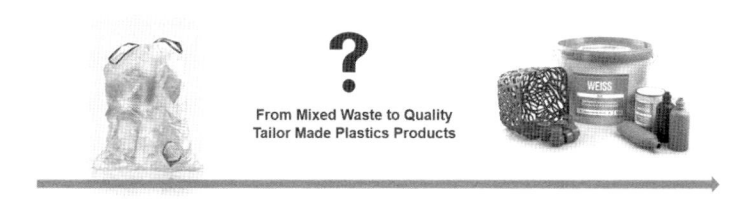

Figure 1.6: Typical applications of PCR PP (and PE-HD) (Source: Interseroh)

1.5 Conclusions

The characterisation of recycled plastics, in particular recycled PP, can be conducted based on comprehensive norms and standards which have been available for some time already, and are complemented by new guidelines like DIN SPEC 91446. EPR schemes determine the regulatory framework for packaging recycling in Europe, the German Packaging Act even describes minimum recycling rates for plastics, PP is included here. The demand of PCR PP has risen constantly increasing over the last years, the production of PCR PP has not kept pace, the market is structurally short for the time being.

References

[Bot12] BOTHE, D.: Mengenstromnachweise der dualen Systeme. *Müll und Abfall* (2012) 12, p. 664–678

[CB20] CHRISTIANI, J.; BECKAMP, S.: *Was können die mechanische Aufbereitung von Kunststoffen und das werkstoffliche Recycling leisten?*. Neuruppin: Thomé-Kozmiensky Verlag GmbH, 2020

[DC12] DEHOUST, G.; CHRISTIANI, J.: *Analyse und Fortentwicklung der Verwertungsquoten für Wertstoffe. Study for the setting of quotas for the separate household collection of packaging waste*. Dessau-Roßlau: Umweltbundesamt (UBA), 2012

[NN08] N.N.: *DIN EN 15343: Kunststoffe - Kunststoff-Rezyklate - Rückverfolgbarkeit bei der Kunststoffverwertung und Bewertung der Konformität und des Rezyklatgehalts*. Berlin: Beuth Verlag, 2008

[NN17] N.N.: *Deloitte Sustainability - Blueprint for plastics packaging waste: Quality sorting recycling*. Deloitte GmbH Wirtschaftsprüfungsgesellschaft, 2017

[NN20] N.N.: *Report on Plastics Recycling Statistics 2020*. Plastics Recyclers Europe, 2020

[URL22a] N.N. *Mindeststandard recyclinggerechtes Design*. URL: https://www.verpackungsregister.org/stiftung-behoerde/mindeststandard-21/grundlegende-informationen, 08.05.2022

[URL22b] N.N. *Packaging waste by waste management operations and waste flow (env_waspac)*. URL: https://ec.europa.eu/eurostat/databrowser/view/env_waspac/default/table?lang=en, 08.05.2022

Abbreviations

Notation	Description
PE	Polyethylene
PET	Polyethylene terephthalate
PCR PP	Post-consumer polypropylene
PP	Polypropylene
PS	Polystyrene
PVC	Polyvinyl chloride

2 Functionalisation of polyolefin melts by means of atmospheric pressure plasma

R. Dahlmann[1], Ch. Hopmann[1], C. Wiesel[1], C. Wang[1], M. Schön[1]
[1] Institute for Plastics Processing (IKV) at RWTH Aachen University

Abstract

Due to a difficult separation process of polyehtylene (PE) and polypropylene (PP), post consumer recyclates (PCR) of these materials contain several percent of the opposite material type. Compatibility in PE and PP blends such as post consumer recyclates is increased with the use of compatibilisers. An alternative method could be the plasma-treatment of polyolefins in order to generate polar functional groups for increased compatibility. It is therefore investigated whether the treatment of polyolefin melts is feasible using atmospheric pressure plasma. Measurements using FTIR-ATR show the formation of functional groups in polyolefins that were treated in both solid and molten state. The treatment in molten state also leads to a higher Carbonyl Index indicating higher amounts of functional groups. The high energy introduced into the polymer during plasma treatment leads to degradation of polypropylene chains.

2.1 Introduction

Waste polyolefins, especially polyethylene (PE) and polypropylene (PP), are the main components of waste plastics. The recycling of waste polyolefins is particularly significant in waste plastics recycling [GJL17]. However, due to similar chemical and physical properties, it is difficult to separate PE and PP in industry using conventional methods such as sink-float method and selective flotation technique [SNS08]. Furthermore, it is also hard to improve interfacial compatibility between PE and PP waste stream directly without adding compatibiliser as a result of its low surface energy. Using the expensive compatibiliser will increase the cost of recycling waste polyolefins in industry [LKJ+18]. At present, an alternative method to improve the interfacial compatibility between PE and PP waste stream using atmosphere pressure plasma is proposed, which is capable of introducing polar functional groups in PE and PP to increase its polarity. Most researchers have mainly focused on plasma modification of solid rather than molten polyolefin surface [JAW+21]. In addition, an ageing effect (i.e.: a degradation of the number of functional groups available over time) might be a challenge for application of plasma treated polyolefin [KKCP01]. Therefore, it is necessary to find a feasible way to rapidly functionalise PE and PP and reduce the ageing effect.

2.2 Objective

In this study, atmosphere pressure plasma is used to treat the surface of molten PE and PP with dry air as a gas source. The influence of the plasma treatment on chemical, physical, rheological and crystallisation properties of PE and PP surface is investigated systematically, which provides the scientific basis for the further study on improving the interfacial compatibility of PE and PP blends. Furthermore, the ageing effect of plasma-treated PE and PP samples in molten state is also evaluated.

2.3 Preparation and Characterisation

2.3.1 Materials

High density polyethylene (M453SE) was recommended and supplied by Saudi Basic Industries Corporation (SABIC) due to a low content of additives. The material has a density of 0.953 g/cm^3 (according to ISO 1183), melting point of 132 °C (according to ISO 11357-3) and a melt flow rate of 4.0 g/10 min (190 °C, 2.16 kg according to ISO 1133). Polypropylene (571P) was also supplied by SABIC, with a density of 0.905 g/cm^3 (according to ASTM D 1505), a melting temperature range of 160 °C to 170 °C and a melt flow rate of 5.7 g/10 min (230 °C, 2.16 kg according to ISO 1133) and a broad molecular weight distribution. The PP was also chosen due to a low content of additives.

2.3.2 Plasma generator

Plasmabeam (Plasmabeam standard, Diener electronic GmbH Co. KG, Germany) is an atmosphere pressure plasma system with a frequency of 20 kHz and a power of 300 W. The gas source is dry air with a pressure of 5 bar.

2.3.3 Sample preparation

The sample preparation contained two steps: melting stage and functionalisation stage. At first, the polytetrafluoroethylene (PTFE) container with 5.5 g PE raw material was heated in an oven (VT 5042 EK, Amikon GmbH, Borken, Germany) at 180 °C for 20 min. According to an earlier study, the change of chemical and molecular structure for Plasmabeam treated PE or PP was obvious at the range of treatment distance from 15 mm to 20 mm and the treatment time from 15 s to 20 s.

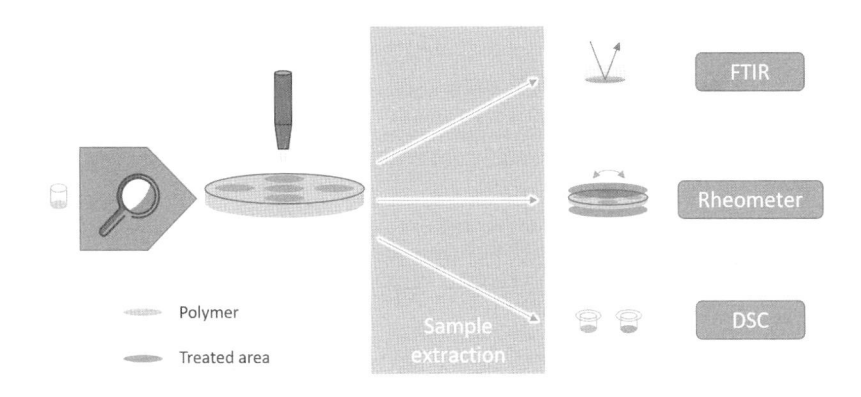

Figure 2.1: Sample preparation from treated polymer

Therefore, the molten PE was functionalised using Plasmabeam at a distance of 18 mm (between plasma nozzle and PE surface) for 20 s. The PP samples were first molten at 220 °C for 35 min in the oven, then an identical functionalisation process was carried out. The gas temperature was measured using temperature gauges (Z 251/1, HASCO Hasenclever GmbH & Co. KG, Lüdenscheid, Germany) with a resolution of 0.1 °C. The surface temperatures of PE and PP samples during functionalisation were detected using a Fluke 63 infrared mini thermometer. Before conducting all tests, the samples of plasma treated PE or PP in the molten state were cooled down to the room temperature. All samples for DSC were taken by scraping the needed amount of polymer from the plasma treated surface area according to Figure 2.1 and tested within several days. The samples for FTIR were taken by cutting the treated polymer while a larger sample with the treated surface was used for rheometry. Both FTIR and rheometry were conducted on the same day as the plasma treatment. PE that was treated in the molten state is defined as p-PE (molten) while plasma treated in the solid state was defined as p-PE (solid). This also applies for plasma treated PP as well. Material that has been molten and cooled down before the actual testing is referred to as "molten".

2.3.4 Optical emission spectroscopy (OES)

Optical emission spectroscopy (OES) was utilised to investigate and determine the active species in the plasma 'beam'. The measurement was done by EMICON high-resolution system (EMICON HR system, Plasus GmbH, Mering, Germany). The emission spectrum of the plasma 'beam' was inspected in the range of 200 nm – 1000 nm with a resolution of 1.5 nm. The recording interval and integration time amounted to 1000 ms and 100 ms. The software of PLASUS EMICON MC and PLASUS SpecLine was used to collect and analyse the emission spectrum of plasma 'beam'. The gas temperature of plasma was measured using temperature gauges (Z 251/1, HASCO Hasenclever GmbH & Co KG, Germany) with a resolution of 0.1 °C.

2.3.5 Fourier-transform infrared spectroscopy attenuated total reflectance (FTIR-ATR)

Fourier-transform infrared spectroscopy (Nexus 870, Thermo Fisher Scientific GmbH, USA) spectra were recorded at room temperature from 680 cm^{-1} to 4000 cm^{-1} with 64 scans for each sample in attenuated total reflectance (ATR) mode using a diamond crystal. Each sample was tested at least three times where each sample was taken manually. An inhomogenous surface due to plasma process complicated reproducibility. The Carbonyl Index (CI) describes the ratio of the peak intensities of carbonyl and olefin groups and is therefore used as a degree of functionalisation. For PE samples and PP samples it is adapted as shown in Equation 2.1 and 2.2, respectively [BSA+13, HZK+15]:

$$CI_{PE} = \frac{I_{1718}}{I_{720}} \qquad (2.1)$$

$$CI_{PP} = \frac{I_{1718}}{I_{1456}} \qquad (2.2)$$

where I_{1718} represented the peak intensity of C=O group at 1718 cm^{-1}. I_{720} represented the peak intensity of CH_2 group for PE at 720 cm^{-1} and I_{1456} represented the peak intensity of CH_3 for PP at 1456 cm^{-1}. The intensity of the peak was obtained according to the height of the absorbance using a constant baseline and a Gauss-Lorenz fitting-model.

2.3.6 Rheological test

The rheological properties were investigated with a plate-plate rheometer (MARS II, Thermo Haake GmbH, Karlsruhe, Germany). The PE samples were tested at 180 °C with frequencies from 0.628 rad/s to 628 rad/s at a strain of 10 %, while PP samples were tested at 200 °C with equal frequencies from 0.628 rad/s to 628 rad/s at a strain of 10 %. Each sample was treated using Plasmabeam at five different positions under 18 mm for 20 s. A plate with a thickness of around 2 mm cut from the surface was used to be tested.

2.3.7 Differential scanning calorimetry (DSC)

Differential scanning calorimetry (DSC Q1000, New Castle, USA) was used to represent the crystallisation behaviour of PE and PP samples. Every measurement was carried out three times for each material. The samples were treated using Plasmabeam at three different positions at a distance of 18 mm for 20 s. Thin films were scraped from the treated area. The mass of a testing sample was around 5.5 mg. Initially, the temperature was increased from 30 °C to 200 °C with a rate of 10 °C/min to eliminate the heating history and then kept isothermal at 200 °C for 0.5 min. Subsequently, the temperature was lowered to 30 °C with a rate of 10 °C/min and kept isothermal at 30 °C for 3.0 min. In the second heating process (2nd run) temperature was again increased from 30 °C to 200 °C with a rate of 10 °C/min. Half crystallisation temperature difference ($\Delta T_{1/2}$) represents crystallisation rate. The shorter the value of $\Delta T_{1/2}$, the faster the crystallisation.

2.4 Results and discussion

2.4.1 OES analysis

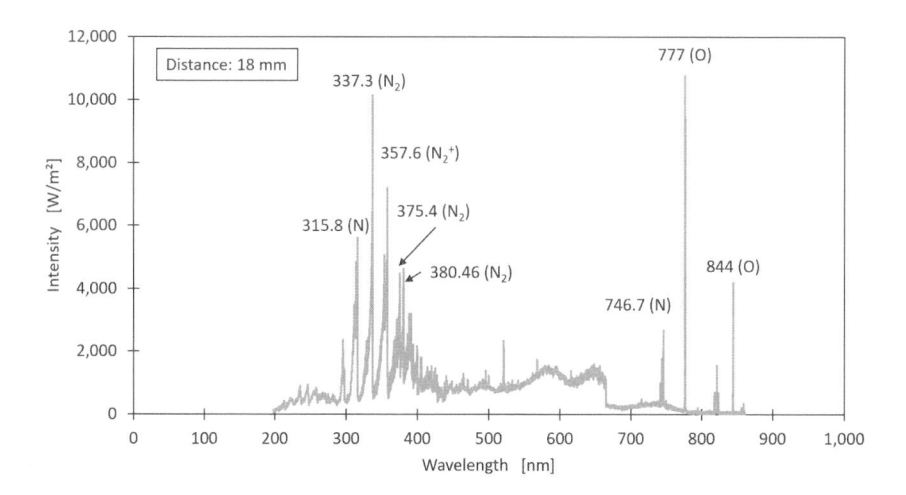

Figure 2.2: The OES spectrum of plasma at a distance of 18 mm

Before analysing the functionalisation of plasma-treated PE and PP samples, the type and intensity of the active species produced by the plasma device are determined. The distance between the tip of the plasma nozzle and the sample surface was 18 mm. The OES spectrum shown in Figure 2.2 illustrates that different active species are produced in the plasma torch, including nitrogen and oxygen species, which can provide polarity enhancement of PE and PP surfaces [DSK+20]. In the UV region ranging from 200 to 400 nm [JBH+16], N_2 transition and N^{2+} at 357.6 nm are the primary species. In addition, N_2 is detected at wavelengths of 315.8 nm, 337.3 nm, 357.6 nm, 375.4 nm and 380.46 nm, respectively [GCDT17]. At 309 nm, OH species are generated in atmospheric pressure plasma devices according to Zhu et al. [ZEG+17]. Due to the low moisture content in the dry air, almost no OH species are formed in this plasma torch [BBB+14]. The active atomic oxygen (O) from plasma was found at 777 nm and 844 nm. The degree of functionalisation is not only influenced by the different types of active species, but also by the gas temperature of the plasma. According to the thermometer measurement, the gas temperature of the plasma at a distance of 18 mm was 174 ± 2 °C, which is close to the melting point of PP and higher than that of PE. The surface temperature was around 153 ± 2 °C for p-PE (solid) and 168 ± 3 °C for p-PE (molten) while 161 ± 1 °C for p-PP (solid) and 171 ± 1 °C for p-PP (molten) when samples were treated at 18 mm for 20 s. Those temperatures are lower than the gas temperature of plasma.

2.4.2 Analysis of the influence of the melting process on PE and PP properties

In order to be able to quantify changes in properties caused by the melting process, a comparison is first carried out between the virgin materials and the plastics after a one-time melting and subsequent cooling down process. This means that subsequent changes in properties can only be attributed to the functionalisation process.

FTIR-ATR analysis

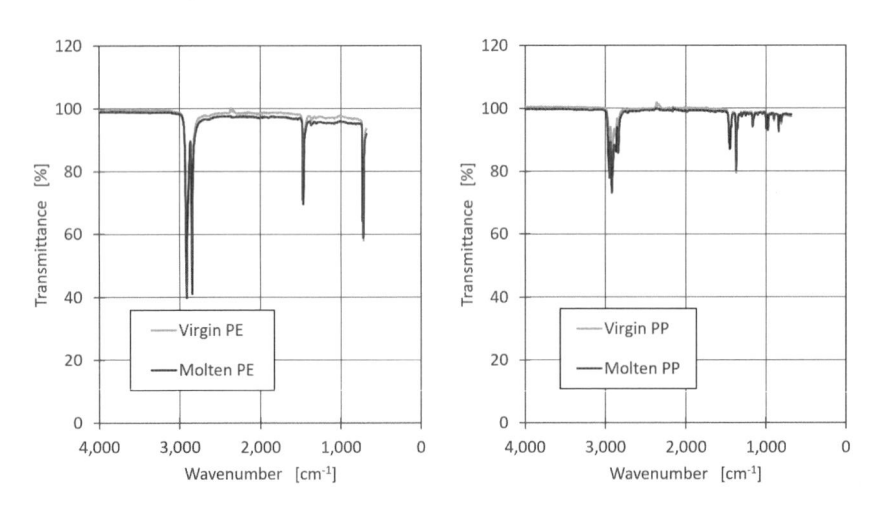

Figure 2.3: The ATR-FTIR spectra of virgin PE and once-molten PE samples (left), the spectra of virgin PP and once-molten PP samples (right)

Figure 2.3 displays the FTIR-ATR spectra of PE and PP samples, comparing a virgin sample and a sample collected after melting, respectively. It is obvious that no functional groups are generated during the melting process, which means the melting process has no effect on PE or PP the "portfolio" of functional group within the limit of detection.

Rheology analysis

The absolute value of the complex viscosity $|\eta^*|$ of once-molten PE shows no significant differences from virgin PE at all frequencies, shown in Figure 2.4. Furthermore, the storage modulus (G') and loss modulus (G'') curves of once-molten PE also overlap with those of virgin PE. Both results demonstrate that melting process has little effect on molecular chains of PE. Figure 2.5 shows that the absolute value of complex viscosity $|\eta^*|$ of once-molten PP strongly decreases compared to virgin PP at all frequencies. Thus, the melting process is shown to degrade polymer chains of PP. In general, the crossover point (G_x) between storage modulus (G') and loss modulus (G'') can illustrate the changes of molecular mass and molecular weight distribution. The ordinate position of G_x relates to the molecular mass distribution while the horizontal position of G_x relates the weight average molecular weight (M_w) [SM95]. The horizontal position of G_x of once-molten PP shifts to higher frequencies, which indicates that the M_w of PP decreases during the melting. Meanwhile, the ordinate position of G_x of once-molten PP almost stays the same with virgin PP, suggesting no changes of molecular mass distribution. Therefore, it is reasonable to consider once-molten PE or once-molten PP as a reference when analysing the effects of the functionalisation. By doing so, the influence of the melting process is eliminated.

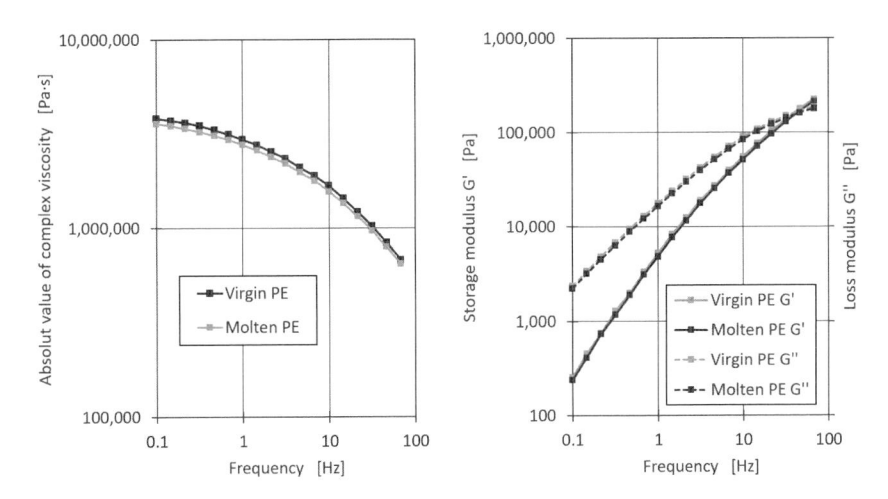

Figure 2.4: Absolute value of complex viscosity $|\eta^*|$ of (left), storage modulus (G') and loss modulus (G'') of PE samples (right)

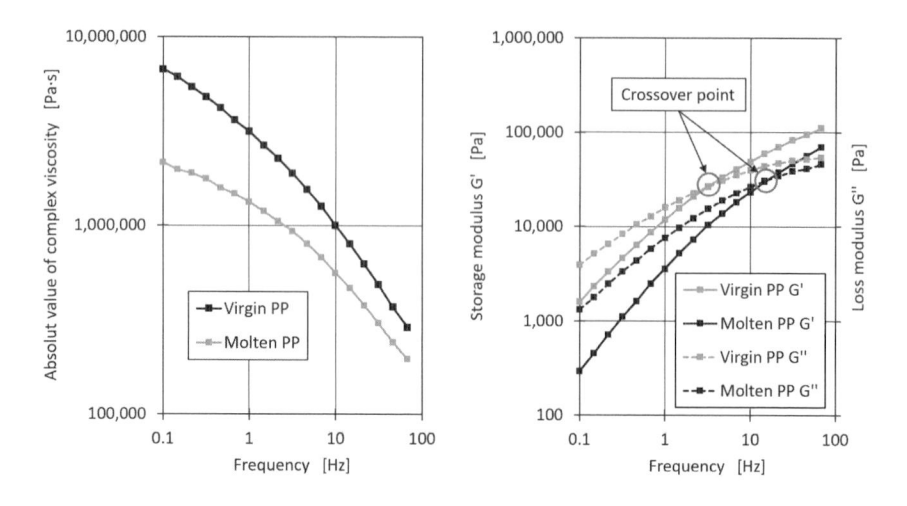

Figure 2.5: Absolute value of complex viscosity $|\eta^*|$ (left) and G' and G'' of PP samples (right)

Crystallisation analysis

In order to determine the variation of thermal properties of PE and PP samples caused by the melting process, the crystallisation behaviour is evaluated. If no significant deviation occurred within the three repetitions, the figures show a representative curve.

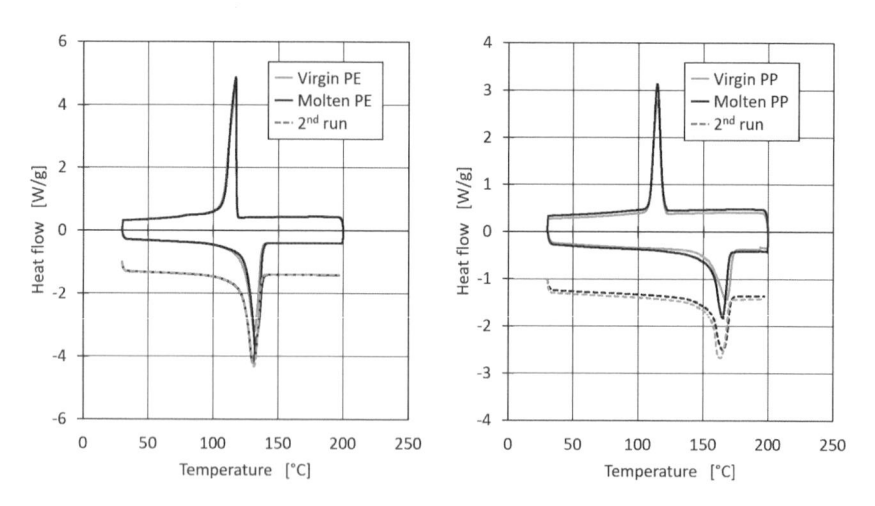

Figure 2.6: DSC curves of virgin and once-molten PE samples (left) and of virgin and once-molten PP samples (right)

Figure 2.6 shows that the shape of exothermic and endothermic peaks for virgin PE and once-molten PE almost keeps the same. Moreover, the value of some parameters between virgin PE and once-molten PE, such as the crystallisation peak temperature (T_c), crystallisation onset temperature ($T_{c,onset}$), enthalpy change of crystallisation (ΔH_c) and peak melting temperature (T_m), are almost identical. Both results, observed from Figure 2.7 (a)(b) and Table 2.1, demonstrate that the crystallisation behaviour of once-molten PE is not significantly changed during melting process (180 °C for 20 min). The value of $\Delta T_{1/2}$ for once-molten PE is close to that of virgin PE indicating the same crystallisation rate. It is therefore concluded that the melting process has a little or no effect on crystallisation behaviour of once-molten PE on the basis of same shape of exothermic and endothermic peaks between virgin PE and once-molten PE. The same conclusion can be drawn for PP samples based on the same peak shape and the proximity of some parameters.

Samples	T_c [°C]	$T_{c,onset}$ [°C]	ΔH_c [J/g]	T_m [°C]	$\Delta T_{1/2}$ [°C]
Virgin PE	117.4±0.05	117.8±0.02	193.1±0.65	130.7±0.1	6.97±0.45
Once-molten PE	117.3±0.3	117.9±0.1	190.5±3.6	130.9±0.5	7.03±0.4
Virgin PP	114.4±0.24	118.9±0.2	103.0±0.4	164.9±0.2	9.73±0.2
Once-molten PP	114.5±0.02	118.9±0.04	105.4±1.7	162.1±0.6	9.56±0.05

Table 2.1: Thermal parameters for PE and PP samples (2nd heating run)

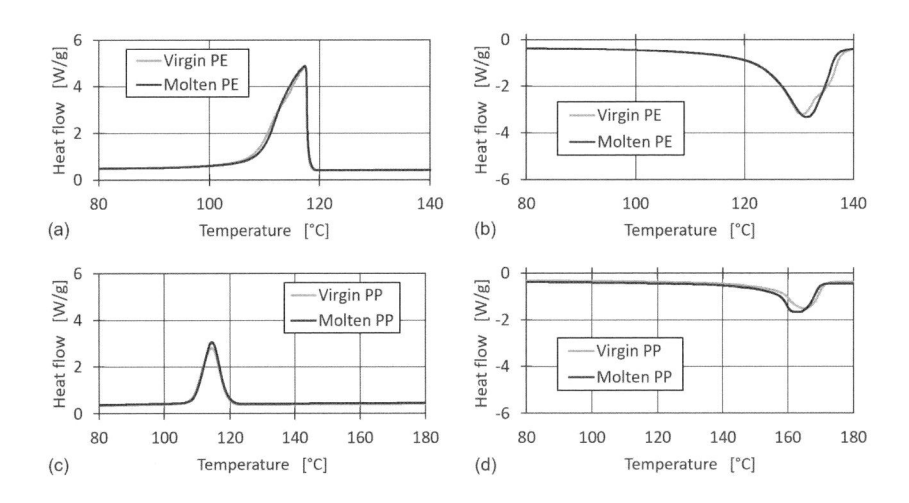

Figure 2.7: Detailed DSC curves of virgin and once-molten PE (top) and PP (bottom) samples (2nd heating run)

Interim conclusion

After the melting process, the portfolio of functional groups of once-molten PE (180 °C for 20 min) and once-molten PP (220 °C for 35 min) surface shows no change. Furthermore, the crystallisation behaviour between virgin and once-molten samples remains almost identical. Thermal degradation occurs for PP as seen in the decrease of complex viscosity and the shift of intersection point as well as T_m. It is reasonable to take once-molten PE and once-molten PP as the reference for the functionalisation stage to eliminate the effect of the melting process. Therefore, the variation of structure and properties of plasma treated PE and PP can be reliably attributed to the functionalisation stage.

2.4.3 Analysis of the functionalisation of plastic melts with an atmospheric pressure plasma

In this section, the PE and PP samples both in a solid and a molten state are treated using atmosphere pressure plasma at a distance of 18 mm for 20 s. The chemical structure, rheological properties and crystallisation behaviour of plasma treated PE and PP samples are investigated by FTIR-ATR, rheometer and DSC. The difference between plasma treated PE and PP samples in molten state and solid state is analysed afterwards. In addition, the ageing effect for plasma treated PE and PP samples was also studied.

FTIR-ATR analysis

It is apparent from Figure 2.8 (left) that the functionalisation groups, including carbonyl group (C=O) at 1718 cm^{-1} [KRH+20], ether group (C-O-C) at 1176 cm^{-1} and alcohol group (C-OH) at 1070 cm^{-1} [RHZ+11, LCFG91] and hydroxyl groups (-OH) at the range of 3200 cm^{-1} to 3600 cm^{-1} [AOC13] have been generated on the PE surface at the reaction distance of 18 mm. The functional groups are generated in both solid and molten state. This confirms that PE can be easily and quickly functionalised by surface oxidation processes in the molten state with atmospheric pressure plasma at a reaction time of 20 s [WFB+11].

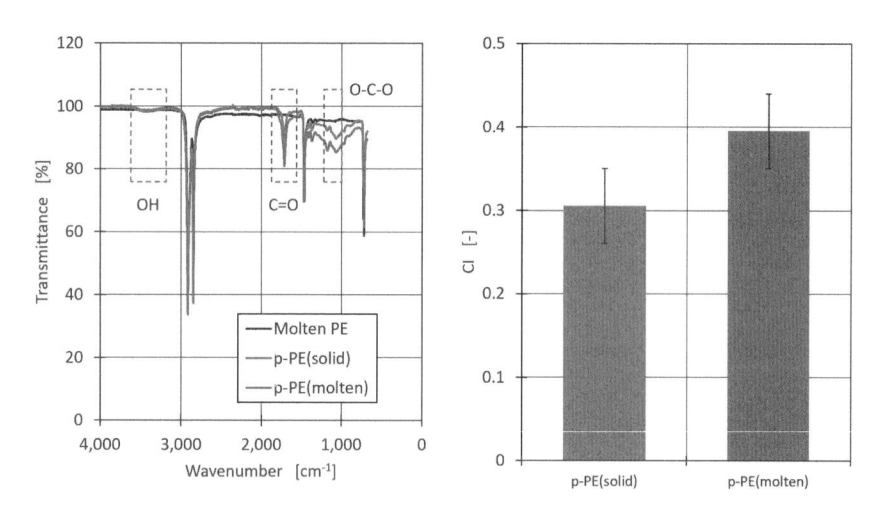

Figure 2.8: FTIR-ATR spectra of PE samples treated at solid state (p-PE(solid)) and at molten state (p-PE(molten)) (left), carbonyl index (CI) of p-PE(solid) and p-PE(molten) (right)

Furthermore, the molten state of the PE sample has a positive influence on the degree of functionalisation. The PE sample in the molten state compared to the solid state is more susceptible to reactions and therefore also to the action of the active species. This is because energetic species of the plasma are prone to attack amorphous regions rather than crystalline regions [OTEF99]. Therefore, the CI of p-PE (molten) (0.40 ± 0.03) is higher than that of p-PE (solid) (0.30 ± 0.05) at the same reaction conditions, as displayed in Figure 2.8 (right). As for the PP samples, the surface is also modified with plasma and a minor amount of C=O groups at 1718 cm^{-1}, seen in Figure 2.9 (left). Furthermore, ether, alcoholic and hydroxide groups were not detected. Theoretically, PP is more readily functionalised in the presence of plasma due to the existence of tertiary carbon atom. However, the degree of functionalisation of PP is not so pronounced due to the presence of antioxidants. In general, commercial PP compounds contain more additives than PE to prevent degradation. When the PP sample is treated, the molecular chain is bombarded by the active free radical generated from plasma. However, the antioxidants can capture the radicals and react with them to terminate the reaction between plasma and the PP molecular structure. Therefore, the polar groups of C=O are not clearly evident in the plasma treated PP sample surface. As already observed in the tests with PE, the PP samples show higher degrees of functionalisation in the molten state than in the solid state. The reason for this could be the mobility of the chains and the crystallinity. The CI of p-PP (molten) is 0.11 ± 0.005, about 30 % higher than that of p-PP (solid) with 0.08 ± 0.009.

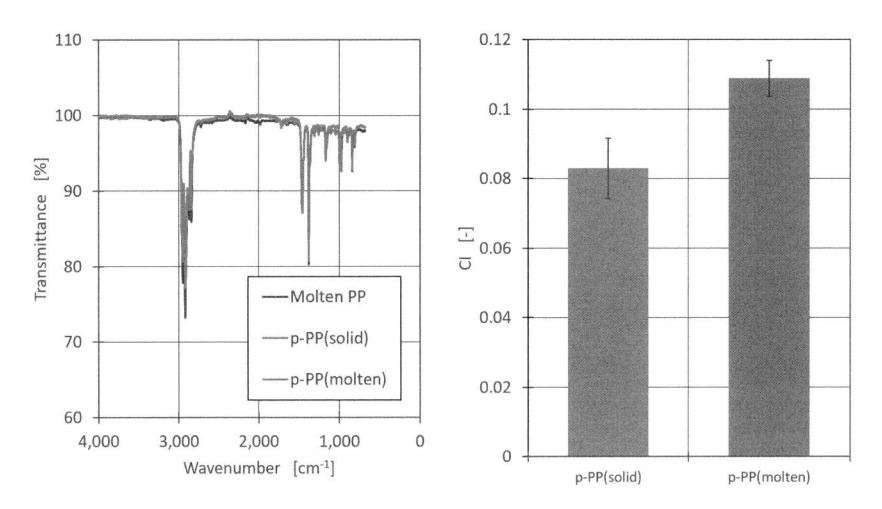

Figure 2.9: ATR-FTIR spectra of PP samples treated at solid state (p-PP(solid)) and at molten state (p-PP(molten)) (left), CI of p-PP(solid) and p-PP(molten) (right)

Rheology analysis

The rheology tests require a larger sample size than the previously discussed measurements. To ensure a significant and reproduceable amount of treated material, the samples of 2 mm height are treated in five different locations that cover most of the surface area. After the plasma treatment, the $|\eta^*|$ of p-PE (solid) and p-PE (molten) are higher than that of molten PE at the low frequency, as observed in Figure 2.10. Presumably, a cross-linked structure has formed in the plasma-treated PE samples, which increases the viscosity. Furthermore, at the same plasma treatment condition (18 mm-20 s), the cross-linked structure in p-PE (molten) is more pronounced than p-PE (solid) based on the higher $|\eta^*|$ of p-PE (molten).

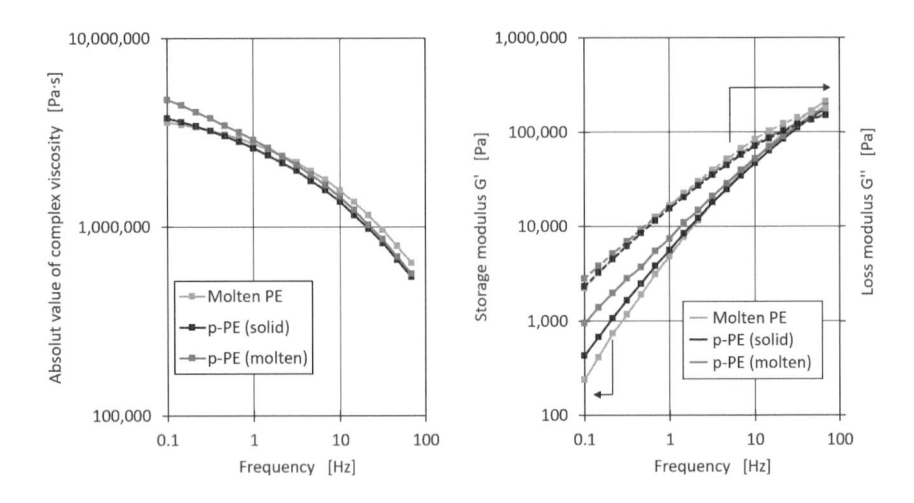

Figure 2.10: $|\eta^*|$ curve of p-PE (solid) and p-PE (molten) samples (left), G' and G'' curves of p-PE (solid) and p-PE (molten) samples (right)

According to Figure 2.8, it is noted that more polar groups, such as C=O, C-O-C and -OH, were generated in p-PE (molten) compared to p-PE (solid). Wang, C.Y. et al [WHHY21] reported that the three-dimensional structure (crosslinked structure) can be formed between active oxygen atom and PE to generate C-O-C groups. Therefore, it is reasonable to confirm that the formation of cross-linked structure based on the presence of such polar groups in the PE structure, and the extend of which is more pronounced in p-PE (molten).

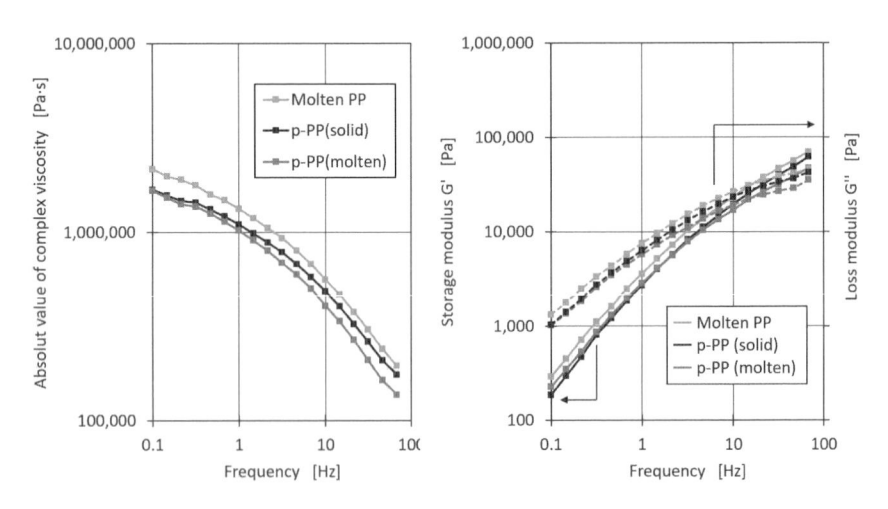

Figure 2.11: $|\eta^*|$ curve of p-PE (solid) and p-PE (molten) samples (left), G' and G'' curves of p-PE (solid) and p-PE (molten) samples (right)

Figure 2.11 shows that the $|\eta^*|$ of p-PP (solid) and p-PP (molten) are lower than that of once-molten PP, which proves that the degradation behaviour occurs in PP samples after plasma treatment [WFB+11]. Boyd, R.D. et al reported that chain scission and low molecular weight oxidized material can be created on PP surface after silent discharge plasma treatment, also called atmospheric dielectric barrier discharge [BKB97]. The degradation of p-PP (molten) is stronger than that of p-PP (solid), which causes the wide molecular weight distribution, as displayed in Figure 2.11.

Crystallisation analysis

Figure 2.12 shows exothermic and endothermic peaks of plasma treated PE samples. It can be observed from Figure 2.12 that the exothermic peaks of plasma treated PE samples become thinner and the positions of peaks move to low temperature direction compared to that of once-molten PE, indicating narrower crystallite dispersity and short molecular chains [HZK+15, KS70]. Therefore, degradation behaviour occurs in PE samples after plasma treatment. However, according to the value of half crystallisation temperature difference ($\Delta T_{1/2}$), crystallisation rate of plasma treated PE is faster than that of once-molten PE, which is a conflict result grounded in degradation behaviour. The main reason is the crosslinked structure formed in PE samples during plasma treatment under the force of oxygen-containing groups.

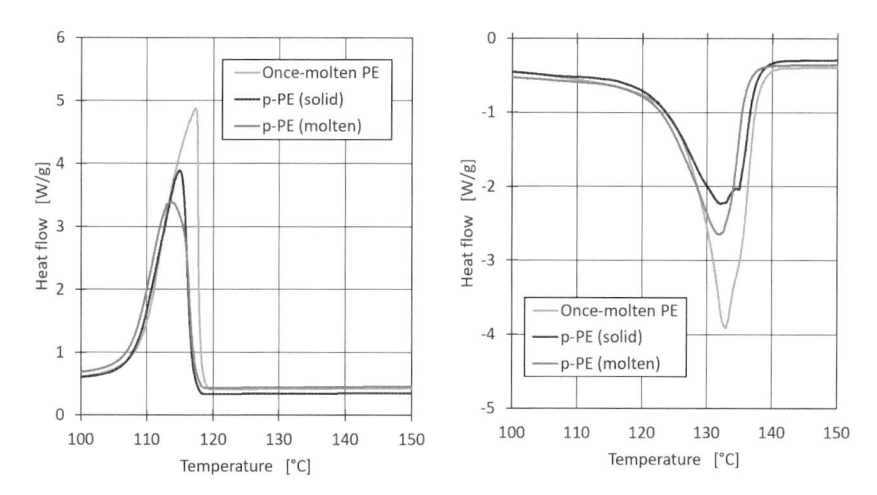

Figure 2.12: The DSC curves of p-PE (solid) and p-PE (molten) samples

There are two competition results happened in PE during plasma treatment. However, crosslinked structure is dominant formed than chain scission based on the increase of complex viscosity and storage modulus of plasma treated PE (observed from Figure 2.10). There is no significant crystallisation behaviour difference of plasma treated PE in the solid state and in the molten state. Although theoretically the amorphous region is more likely to be changed in molten state. The width at half-height of endothermic peak can be used to describe melting range. The increase of melting range means less perfect crystallites. The melting peak temperature (T_m) indicates crystallite size. A lower value of the T_m indicates a smaller size of the crystallites [RW97]. Figure 2.12 shows the reduction of T_m

and broader endothermic peaks of plasma treated PE, indicating smaller crystallites and less perfect crystallites.

Samples	T_c [°C]	$T_{c,onset}$ [°C]	ΔH_c [J/g]	T_m [°C]	$\Delta T_{1/2}$ [°C]
Once-molten PE	117.3±0.3	117.9±0.1	190.5±3.6	130.9±0.5	7.03±0.4
p-PE (solid)	114.8±0.15	116.9±0.01	146.0±3.5	128.5±0.2	8.4±0.37
p-PE (molten)	114.2±0.5	117.4±0.2	143.2±0.7	127.9±0.2	8.65±0.35

Table 2.2: Thermal parameters for plasma-treated PE samples at 18 mm for 20 s

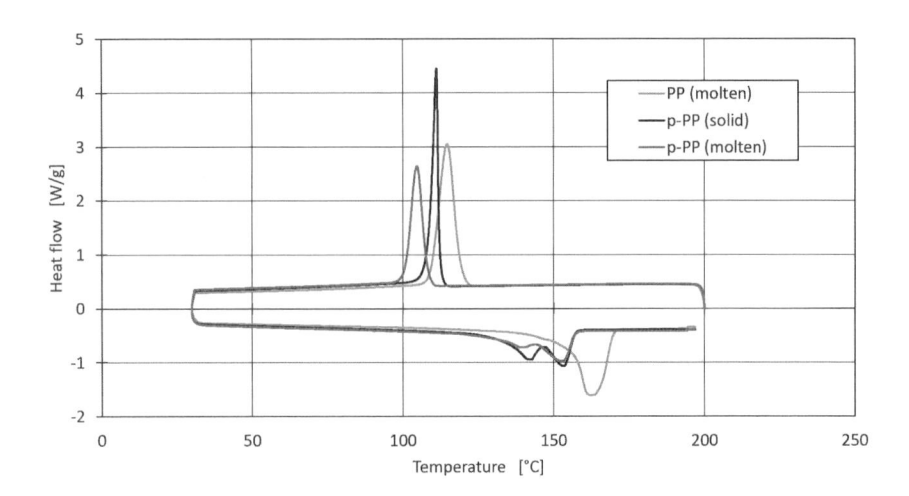

Figure 2.13: DSC curves of molten PP, p-PP(solid) and p-PP(molten) samples with second heating run

Figure 2.13 shows that the position of exothermic crystallisation peak of plasma treated PP shifts to low temperature, demonstrating the chain scission occurrence in plasma treated PP due to oxidative reactions [BBB+14]. The decrease of T_c, listed in Table 2.3, further verifies the above results. The reason is that short chains can be arrayed in lower temperature to form crystallite. The reduction of ΔH_c of plasma treated PP, summarised in Table 2.3, means the increase of irregularity of molecular chains caused by the introduction of oxygen-containing groups. Figure 2.13 displays that single endothermic peak of once-molten PP becomes two endothermic peaks of plasma treated PP due to reorganisation during heating. However, the segregation of high defective molecules is the main reason for peak doubling [RW97]. The decrease of T_m of plasma treated PP is due to the oxidative reactions on crystal surface that increase the surface free energy of the crystals [HL95]. A subsidiary peak appears in low temperature direction of main endothermic peak. This peak is probably related to the presence of -phase at this position [RW97]. Furthermore, the intensity of subsidiary endothermic peak of plasma treated PP in molten state decreases compared to that in solid state, meaning lower content of -phase. Because the -phase in PP sample is thermodynamically unstable, it can be recrystallised into more stable monoclinic or more perfect -modification. During heating process, the melting of -phase and its recrystallisation occurs simultaneously [SZCH92]. The exact mechanism should be researched further. The smaller value of $\Delta T_{1/2}$ for plasma treated PP in solid state is caused by the chain scission during plasma treatment compared with that of once-molten PP [RW97]. However, the increase value of $\Delta T_{1/2}$ for plasma treated PP in molten state is due to the disrupt of regularity of molecular chains on as a result of a relatively large number of oxygen-containing groups. While the first heating run shows different temperature history for the p-PP (molten) samples as seen in Figure 2.14, the second heating run shows no significant differences in the crystallisation behaviour. Melting temperatures have been calculated based on the second heating process.

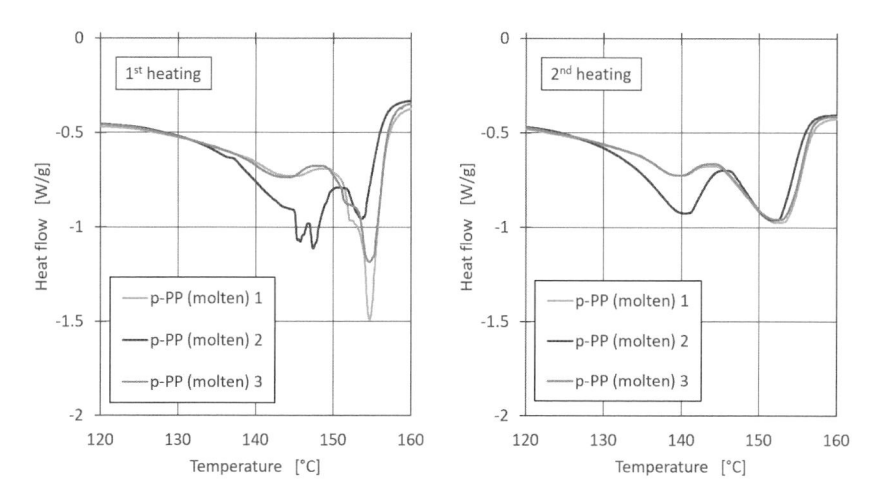

Figure 2.14: DSC curves of first and second heating process of p-PP(molten) samples

Samples	T_c [°C]	$T_{c,onset}$ [°C]	ΔH_c [J/g]	T_{m1} [°C] T_{m2} [°C]	$\Delta T_{1/2}$ [°C]
Once-molten PP	114.5±0.02	118.9±0.04	105.4±1.7	162.1±0.6 -	9.57±0.05
p-PP (solid)	110.1±0.3	111.3±0.02	78.3±0.9	140.6±0.4 152.9±0.3	7.3±0.2
p-PP (molten)	104.6±0.06	108.0±0.07	68.4±0.6	138.6±0.4 152.9±0.2	9.43±0.05

Table 2.3: Thermal parameters for plasma-treated PP samples at 18 mm for 20 s (Note: T_{m1} corresponds to the main melting peak, T_{m2} corresponds to the subsidiary melting peak)

2.4.4 Investigation of the temporal changes after functionalisation

Generally, a temporal effect or ageing effect is an unavoidable problem for the application of plasma-treated polymer surface, as the oxygen containing groups can migrate from the surface to the bulk of the material [PSDG09]. As time passes, the surface polarity decreases. Foerch, R. et al. reported that the oxygen/carbon ratio of linear low-density polyethylene

(PE-LLD) specimens exposed to static air-plasma (reaction time of 3 s) was found to decrease by 40 % when left in air for 45 days [FKW93]. Moreover, the oxygen of PE-LLD treated using remote plasma (reaction time of 5 s) has lost around 30 % of the initial amount on the surface within 5 days [FH92]. Arpagaus, C. et al stated that the initial ageing phase of PE-HD powder after short-time plasma treatment (0.13 s) occurred within 7 days and is characterised by a slower ageing rate [ARR05]. It should be noted that the external environment for storing plasma treated PE samples will also affect the ageing rate. The elevated temperature will accelerate the aging rate while polar liquid environment will hinder the ageing rate by controlling the surface rearrangement and diffusion of polar functional groups [KKCP01]. At present, there are three different assumptions of mechanisms for PE samples ageing process: a) the migration of oxygen-containing groups to subsurface or bulk of material; b) the association of polar groups at the surface; c) the desorption from the surface of low molecular weight oxidised molecules [RKH+11, WBD+93]. In this study, the CI in p-PE (molten) surface almost retains the same level even after 9 days, as observed in Figure 2.15, which means that the plasma treated surface at molten state effectively retards the ageing effect. There are studies suggesting that the crosslinked structure prevents the migration of oxygen-containing groups into the polymer matrix [GZ93].

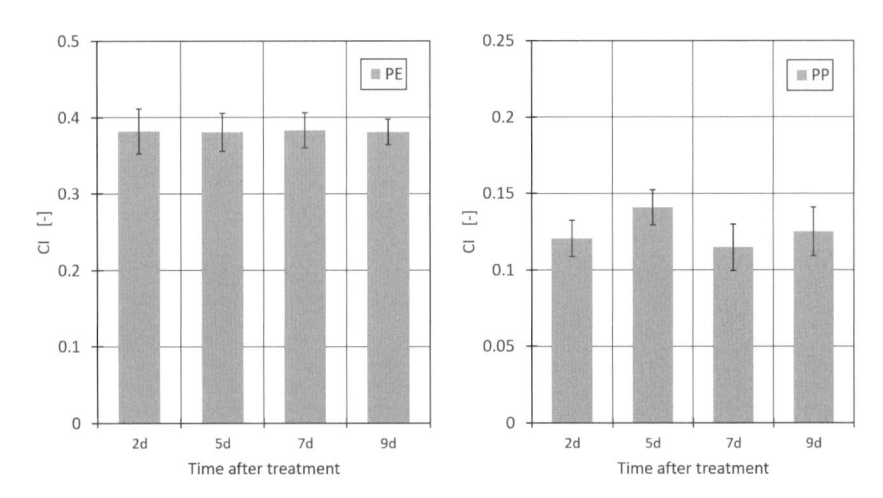

Figure 2.15: The CI of p-PE (molten) and p-PP (molten) after storage

The results for the investigation of the ageing phenomenon of p-PP (molten) show an unusual course (see Figure 2.15). The CI of p-PP (molten) first increases and then decreases. The exact mechanism remains unclear and must be considered in further series of investigations.

2.5 Conclusions and outlook

The transferability of functionalisation from solid PE and PP surfaces to the molten state was investigated. It could be shown that an atmospheric pressure plasma treatment of PE and PP in the molten state also results in a functionalisation. Polar groups, including C=O, C-O-C and -OH, are formed on the PE surface, while only C=O groups can be observed on the PP surface. The degree of functionalisation is increased when applied in the molten state compared to an application to the solid state. Moreover, the CI of plasma-treated PE in a molten state (p-PE(molten)) remains almost the same level after 9 days, indicating only a minor ageing effect. However, the functional groups are likely to be subject to changes in an unstable state right after plasma treatment. After plasma treatment, the $|\eta^*|$ of p-PE(molten) is higher than that of once-molten PE that has been subjected to an identical thermal load, presumably to the formation of a crosslinked structure. The $|\eta^*|$ of p-PP(molten) is lower than that of PP subjected an identical thermal load due to degradation behaviour. The decrease of ΔH_c from 190 J/g of once-molten PE to 143 J/g of p-PE(molten) is mainly caused by the formation of crosslinked structure after plasma treatment. The ΔH_c of once-molten PP decreases from 105 J/g to 68 J/g of p-PP(molten) primarily attributed to the degradation behaviour. This effect is well known and has to be taken into consideration during plasma treatment. This study investigates an approach for rapid functionalisation of PE and PP in the molten state. It is subject to further research if the functionalisation provides a positive effect on the homogeneity of PE/PP blends and can be used to improve the properties of post consumer recycled material.

Acknowledgement

The research project 21519 N of the Forschungsvereinigung Kunststoffverarbeitung has been sponsored as part of the "Industrielle Gemeinschaftsforschung und -entwicklung (IGF)" by the German Bundesministerium für Wirtschaft und Klimaschutz (BMWK) due to an enactment of the German Bundestag through the AiF. We would like to extend our thanks to all organisations mentioned. Furthermore, we would like to thank all the enterprises that support this research project with discussions, practical investigations and donations of materials.

C.Y. Wang acknowledges the financial support from the China Scholarship Council for funding her stay at RWTH Aachen University (Grant No. 202006150088).

References

[AOC13] AGUIAR, P.; OLIVIERA, ; CRUZ, S.: Modification of clarified polypropylene by oxygen plasma to improve the adhesion of thin amorphous hydrogenated carbon films deposited by plasma enhanced chemical vapor deposition. *Polymer Engineering Science* 53 (2013) 5, p. 1065–1072

[ARR05] ARPAGAUS, C.; ROSSI, A.; RUDOLF VON ROHR, P.: Short-time plasma surface modification of HDPE powder in a Plasma Downer Reactor – process, wettability improvement and ageing effects. *Applied Surface Science* 252 (2005), p. 1581–1595

[BBB+14] BEHM, H.; BAHROUN, K.; BAHRE, H.; KIRCHHEIM, D.; MITSCHKER, F.; BIBINOV, N.; BÖKE, M.; DAHLMANN, R.; AWAKOWICZ, P.; HOPMANN, C.; WINTER, J.: Adhesion of Thin CVD Films on Pulsed Plasma Pre-Treated Polypropylene. *Plasma Processes and Polymers* 11 (2014), p. 418–425

[BKB97] BOYD, R. D.; KENWRIGHT, A. M.; BADYAL, J. P. S.: Atmospheric nonequilibrium plasma treatment of biaxially oriented polypropylene. *Macromolecules* 30 (1997), p. 5429–5436

[BSA+13] BENÍTEZ, A.; SÁNCHEZ, J. J.; ARNAL, M. L.; MÜLLER, A. J.; RODRÍGUEZ, O.; MORALES, G.: Abiotic degradation of ldpe and lldpe formulated with a pro-oxidant additive. *Polymer Degradation and Stability* 98 (2013) 2, p. 490–501

[DSK+20] DARVISH, F.; SARKARI, N. M.; KHANI, M.; ESLAMI, E.; SHOKRI, B.; MOHSENI, M.; EBRAHIMI, M.; ALIZADEH, M.; DEE, C. F.: Direct plasma treatment approach based on non-thermal gliding arc for surface modification of biaxially-oriented polypropylene with post-exposure hydrophilicity improvement and minus aging effects. *Applied Surface Science* 509 (2020), p. 144815

[FH92] FOERCH, R.; HUNTER, D. H.: Remote nitrogen plasma treatment of polymers: polyethylene, nylon 6,6, poly (ethylene vinyl alcohol), and poly (ethylene terephthalate). *Journal of Polymer Science Part A: Polymer Chemistry* 30 (1992) 2, p. 279–286

[FKW93] FOERCH, R.; KILL, G.; WALZAK, M. J.: Plasma surface modification of polyethylene: short-term vs. long-term plasma treatment. *Journal of Adhesion Science and Technology* 7 (1993) 10, p. 1077–1089

[GCDT17] GAZAL, Y.; CHAZELAS, C.; DUBLANCHE-TIXIER, C.; TRISTANT, P.: Contribution of optical emission spectroscopy measurements to the understanding of TiO2 growth by chemical vapor deposition using an atmospheric-pressure plasma torch. *Journal of applied physics* 121 (2017), p. 123301

[GJL17] GEYER, R.; JAMBECK, J. R.; LAW, K. L.: Production, use, and fate of all plastics ever made. *Science Advances* 3 (2017) 7,

[GZ93] GAO, S.; ZENG, Y.: Surface Modification of ultrahigh molecular weight polyethylene fibers by plasma treatment. I. improving surface adhesion. *Journal of Applied Polymer Science* 47 (1993), p. 2065–2071

[HL95] HONG, S. G.; LIAO, C. M.: The surface oxidation of a thermoplastic olefin elastomer under ozone exposure: ATR analysis. *Polymer degradation and stability* 47 (1995), p. 437–447

[HZK+15] HE, G.; ZHENG, T.; KE, D.; CAO, X.; YIN, X.; XU, B.: Impact of rapid ozone degradation on the structure and properties of polypropylene using a reactive extrusion process. *RSC Advances* 5 (2015) 55, p. 44115–44120

[JAW+21] JARITZ, M.; ALIZADEH, P.; WILSKI, S.; KLEINES, L.; DAHLMANN, R.: Comparison of HMDSO and HMDSN as precursors for high-barrier plasma-polymerized multilayer coating systems on polyethylene terephthalate films. *Plasma Processes and Polymers* 18 (2021),

[JBH+16] JARITZ, M.; BEHM, H.; HOPMANN, C.; KIRCHHEIM, D.; MITSCHKER, F.; AWAKOWICZ, P.; DAHLMANN, R.: The effect of UV radiation from oxygen and argon plasma on the adhesion of organosilicon coatings on polypropylene. *Journal of Physics D: Applied Physics* 50 (2016),

[KKCP01] KIM, B. K.; KIM, K. S.; CHO, K.; PARK, C. E.: Retardation of the surface rearrangement of O2 plasma-treated LDPE by a two-step temperature control. *Journal of Adhesion Science and Technology* 15 (2001) 14, p. 1805–1816

[KRH+20] KEHRER, M.; ROTTENSTEINER, A.; HARTL, W.; DUCHOSLAV, J.; STIFTER, D.: Cold atmospheric pressure plasma treatment for adhesion improvement on polypropylene surfaces. *Surface and Coatings Technology* 403 (2020), p. 126389

[KS70] KAVESH, S.; SCHULTZ, J. M.: Lamellar and Interlamellar Structure in Melt-Crystallized Polyethylene. I. Degree of Crystallinity, Atomic Positions, Particle Size, and Lattice Disorder of the First and Second Kinds. *Journal of Polymer Science* 8 (1970), p. 243–276

[LCFG91] LIN-VIEN, D.; COLTHUP, N. B.; FATELEY, W. G.; GRASSELLI, J. G.: *The handbook of infrared and Raman characteristic frequencies of organic molecules*. Elsevier, 1991

[LKJ+18] LEE, B. J.; KLIMOVICA, K.; JIN, K.; LIN, T. W.; HOWARD, M. J.; ELLISON, C. J.; LAPOINTE, A. M.; COATES, G. W.; BATES, F. S.: Compatibilization of isotactic polypropylene (iPP) and high-density polyethylene (HDPE) with Ipp-PE multiblock copolymers. *Macromolecules* 51 (2018), p. 8585–8596

[OTEF99] OLDE RIEKERINK, M. B.; TERLINGEN, J. G. A.; ENGBERS, G. H. M.; FEIJEN, J.: Selective Etching of Semicrystalline Polymers: CF4 Gas Plasma Treatment of Poly(ethylene). *Langmuir* 15 (1999), p. 4847–4856

[PSDG09] PANDIYARAJ, K. N.; SELVARAJAN, V.; DESHMUKH, R. R.; GAO, C.: Modification of surface properties of polypropylene (PP) film using dc glow discharge air plasma. *Applied Surface Science* 255 (2009) 7, p. 3965–3971

[RHZ+11] REN, J.; HUA, X.; ZHANG, T.; ZHANG, Z.; JI, Z.; GU, N.: Grafting of telechelic poly(lactic-co-glycolic acid) onto O2 plasma-treated polypropylene flakes. *Journal of Applied Polymer Science* 121 (2011) 1, p. 210–216

[RKH+11] REZNICKOVA, A.; KOLSKA, Z.; HNATOWICZ, V.; STOPKA, P.; SVORCIKA, V.: Comparison of glow argon plasma-induced surface changes of thermoplastic polymers. *Nuclear Instruments and Methods in Physics Research B: Beam* 269 (2011) 2, p. 83–88

[RW97] RABELLO, M. S.; WHITE, J. R.: Crystallization and melting behaviour of photodegraded polypropylene– II. Re-crystallization of degraded molecules. *Polymer* 38 (1997) 1, p. 6389–6399

[SM95] SHROFF, R.; MAVRIDIS, H.: New measures of polydispersity from rheological data on polymer melts. *Journal of Applied Polymer Science* 57 (1995) 13, p. 1605–1626

[SNS08] SANGOBTIP, P.; NAPATR, K.; SOMSAK, D.: Combination of three-stage sink–float method and selective flotation technique for separation of mixed post-consumer plastic waste. *Waste Management* 28 (2008), p. 475–483

[SZCH92] SHI, G. Y.; ZHANG, X. D.; CAO, Y. H.; HONG, J.: Melting behavior and crystalline order of -crystalline phase Poly (propylene). *Macromolecular Chemistry and Physics* 194 (1992), p. 269–277

[WBD+93] WELLS, R. K.; BADYAL, J. P. S.; DRUMMOND, I. W.; ROBINSON, K. S.; STREET, F. J.: Plasma oxidation of polystyrene vs. polyethylene. *Macromolecular Chemistry and Physics* 7 (1993) 10, p. 1129–1137

[WFB+11] WANKE, C. H.; FEIJÓ, J. L.; BARBOSA, L. G.; CAMPO, L. F.; OLIVEIRA, R. V. B.; HOROWITZ, F.: Tuning of polypropylene wettability by plasma and polyhedral oligomeric silsesquioxane modifications. *Polymer* 52 (2011) 8, p. 1797–1802

[WHHY21] WANG, C. Y.; HE, G. J.; HUANG, W. T.; YIN, X. C.: Functionalization of molten polyethylene by ozonization and its application in wood fiber reinforced polyethylene composites. *Polymer Composites* (2021),

[ZEG+17] ZHU, J.; EHN, A.; GAO, J.; KONG, C.; ALEDEN, M.; SALEWSKI, M.; LEIPOLAD, F.; KUSANO, T.; LI, Z.: Translational, rotational, vibrational and electron temperatures of a gliding arc discharge. *Optics Express* 25 (2017) 17, p. 20243

Symbols

Symbol	Unit	Description
$|\eta^*|$	$Pa*s$	Absolute value of complex viscosity
G'	Pa	Storage Modulus
G''	Pa	Loss Modulus
G_x	Pa	Crossover Point
ΔH_c	J/g	Enthalpy change of crystallisation
M_w	g/mol	Weight average molecular weight
T_c	$°C$	Crystallisation Temperature
$T_{c,onset}$	$°C$	Crystallisation onset Temperature
$\Delta T_{1/2}$	K	Half crystallisation temperature difference
T_m	$°C$	Melting Temperature

Abbreviations

Notation	Description
CI	Carbonyl Index
DSC	Differential Scanning Calorimetry
FTIR-ATR	Fourier-Transform Infrared Spectroscopy - Attenuated Total Reflectance
OES	Optical Emission Spectroscopy
PCR	Post-consumer recyclate
PE	Polyethylene
PE-LLD	Polyethylene linear low density
PP	Polypropylene
PTFE	Polytetrafluoroethylene

3 Usage potentials for different qualities of recyclates in plastics packaging

R. Dahlmann[1], E. Berg[1], M. Schön[1]
[1] Institute for Plastics Processing (IKV) at RWTH Aachen University

Abstract

The reuse of PE-LD is currently limited to low-value applications, although polyolefins make up the majority of plastic production overall and life cycle assessments show considerable advantages in terms of resource conservation. The material quality is influenced, among other things, by the presence of non-volatile impurities. It is not clear to what extent these are contained in the recyclate and which of the respective impurities are predominantly present.

Two differently prepared recyclates were processed into blown films and observed under the light microscope. Energy dispersive X-ray spectroscopy (EDX), Fourier transform infrared spectroscopy (FT-IR) and differential scanning calorimetry (DSC) were used to analyse the impurities. The results show a variety of particles of different shapes and sizes, which can be attributed to both organic and inorganic components. A high number of impurities with dimensions far exceeding the filter pore width are suspected to be gel-like particles consisting of either cross-linked or high molecular weight material. Significant differences between areas with black spots and without black spots could not be detected with the methods used. PE-LD, PE-HD and PP were identified as polymeric components in the material, which already show first signs of degradation.

3.1 Introduction and motivation

The largest demand for plastics, at 26.5 % by mass of all plastic produced in Germany in 2019, lies in the packaging sector [NN20a]. Packaging often consists of a variety of polymers, additives, fillers, coatings and adhesives in order to achieve different functional properties in terms of packaged good protection. An overview of packaging production by segment and by plastic type used is provided in Figure 3.1. It is shown that low density polyethylene (PE-LD or PE-LLD) is the most commonly used polymer for packaging, especially as films [NN20a].

Packaging production by segments

- packaging films; 38%
- other; 5,40%
- barrels, canister, buckets; 5,40%
- closures; 7,80%
- cups, cans, boxes, pallets; 16,40%
- bottles and other plastics; 4,10%
- bottles PET; 11,20%
- bags; 11,70%

Packaging production by type of plastic used

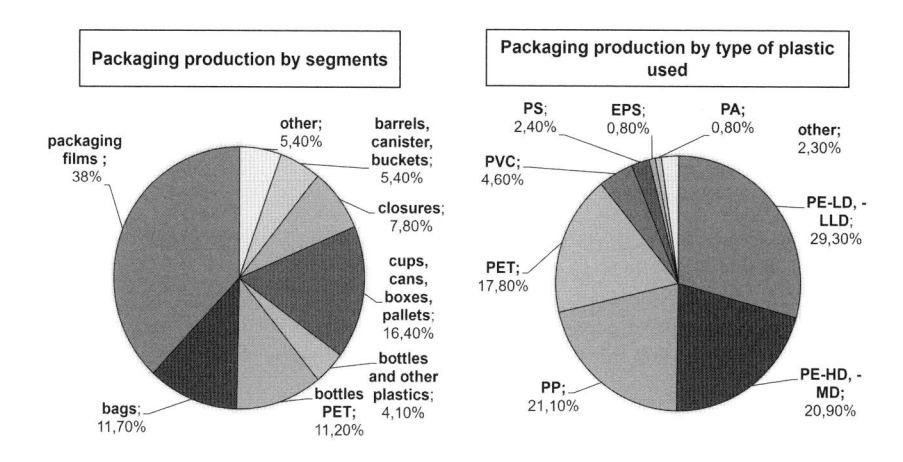

- PS; 2,40%
- EPS; 0,80%
- PA; 0,80%
- other; 2,30%
- PVC; 4,60%
- PET; 17,80%
- PP; 21,10%
- PE-HD, -MD; 20,90%
- PE-LD, -LLD; 29,30%

Figure 3.1: Overview of packaging production in Germany in 2018 [NN20a]

However, the proportion of recyclates used instead of primary raw materials in packaging applications is only 10.9 % by mass of processed plastic materials, with the majority coming from post-industrial waste [NN20b]. The main source for such recyclates is the recycling of films from the transport and industrial sectors, PET bottles and rigid packaging within Dual System activities [NN20c]. Film waste from the collection of light packaging (Ger. Leichtverpackungen, LVP) is only reprocessed to a share of 6 % in the recycling process to granulates [URL21]. Europe-wide political and social efforts to move away from linear recycling to-wards a circular economy (such as The Circular Economy for Flexible Packaging, CEFLEX) therefore require strategies for post-consumer polyolefin packaging, as light packaging made from polyolefins makes up the majority of plastics production.

By substituting new flexible plastics with post-consumer recyclates, both energy and CO_2 emissions can be saved compared to virgin material [NN08]. In particular, the recycling of PE-LD has the potential to strengthen the circular economy in the polymer sector, as existing life cycle assessments indicate significant benefits in terms of resource conservation [NN08]. Thus, the recycling of film packaging waste, which mostly consists of PE-LD, contributes to climate protection by reducing greenhouse gas emissions.

However, the recycling of many film packagings is hardly profitable so far, as the reuse of the secondary raw material is limited to low-value applications. Limitations exist with regard to the optical, mechanical and thermal properties of the recyclates as well as the odour [Mar16].

3.2 Mechanical recycling of film waste

At the end of their usage cycle, a considerable amount of plastic films are bundled by German recycling plants into a film fraction (DSD fraction 310) [NN18]. Films in fraction 310 are specified as used plastic items with a surface area larger than DIN A4. The low reusage rate of film waste is due to the fact that any fraction inevitably contains impurities (up to approx. 8 %) [NN18]. These have a different influence on the treatment process and thus the recyclate quality depending on their size and quantity. For example, most barrier layers as well as bonding agents and partly organic contaminants cannot be removed economically using established recycling methods [NN19]. Furthermore, film products contain labels, adhesives and colour pigments (e.g. inks) as well as additives (e.g. processing aids, antistatics) and fillers (e.g. minerals) in addition to the base materials [Nen06, NN19]. Overall, the various impurities change the recyclate properties depending on their type and concentration.

In addition to the direct influence of the impurities, the degradation of polyolefins also is affected by them. Increased degradation leads to worse material properties [CDI+07, EP07, ILKT76, MGWE06]. For example, metallic impurities or pigments used to color plastics can accelerate the chemical degradation reactions [EP07, GJ03, SS20]. The degradation of neat polyolefins during processing has already been studied by some authors. Studies on the reprocessing of PE-LD show the parallel occurrence of the degradation processes chain scission and chain branching, with the polymer showing a higher tendency to crosslink [Zwe16].

In summary, sorting challenges, plastic contamination and material degradation are the major barriers to material re-usage of polyolefin films. Recent innovations in recycling include improved sorting techniques, better washing processes, enhanced degassing systems and different filtration techniques to improve the quality of the polymer melt. Although these techniques reduce impurities in the granules, it is assumed that they cannot be completely removed. Particularly in the case of heavily contaminated waste fractions, there is a risk that effort of the recycling process exceeds the benefit, both ecologically and economically.

The influence of impurities in combination with the degradation of plastics is not yet sufficiently understood and studies with neat virgin materials cannot be directly transferred to recyclates. The aim of the investigations of the IKV is therefore to analyse the non-volatile impurities contained in commercially available film recyclates. The results obtained should form a basis for research into the degradation of recycled polyolefin blends and serve to optimise PE-LD processing conditions in film production.

3.3 Materials and methods for the investigation of impurities

For the purpose of the investigation, two different marketable film recyclates are extruded into blown films. The difference between the materials lies in the filter pore size during the recyclate preparation process. Material R1 was extruded through a filter with a mesh size of 100 µm, material R2 through a filter with a mesh size of 200 µm. A blown film line of the type KFB 45/600 from Kuhne, St. Augustin, is used to produce the test films. The process parameters of the test series are listed in Table 4.2. The use of an additional melt filter was deliberately omitted.

Mass-throughput [kg\h]	Die temperature [°C]	Blow-up ratio [-]	Film width [mm]	Film thickness [µm]
10	200	2,4	302	100

Table 3.1: Nominal process parameters used for film production

Different analytical methods can be used to identify the impurities contained in the films. Figure 3.2 shows the selected methodology for identifying the impurities in the granulate. The analytical methods are described in more detail below.

```
┌─────────────────────────────────┐       Two recyclates from recyclate processes
│   Production of blown films     │ ────▶ with mesh sizes of 100 μm and 200 μm
└─────────────────────────────────┘
                │
                ▼
┌─────────────────────────────────┐
│    Preparation of the Samples   │
└─────────────────────────────────┘
                │
                ▼
┌─────────────────────────────────┐
│        Light Microscopy         │ ────▶ Size and shape analysis
└─────────────────────────────────┘

┌─────────────────────────────────┐
│ Energy Dispersive X-ray         │ ────▶ Determination of the inorganic
│ Analysis (EDX)                  │       components
└─────────────────────────────────┘

┌─────────────────────────────────┐
│ Differential Scanning           │ ────▶ Determination of melting temperatures
│ Calorimetry (DSC)               │       and enthalpies
└─────────────────────────────────┘

┌─────────────────────────────────┐
│ Fourier Transmission Infrared   │ ────▶ Determination of chemical functionality of
│ Spectroscopy (FTIR)             │       organic components
└─────────────────────────────────┘
```

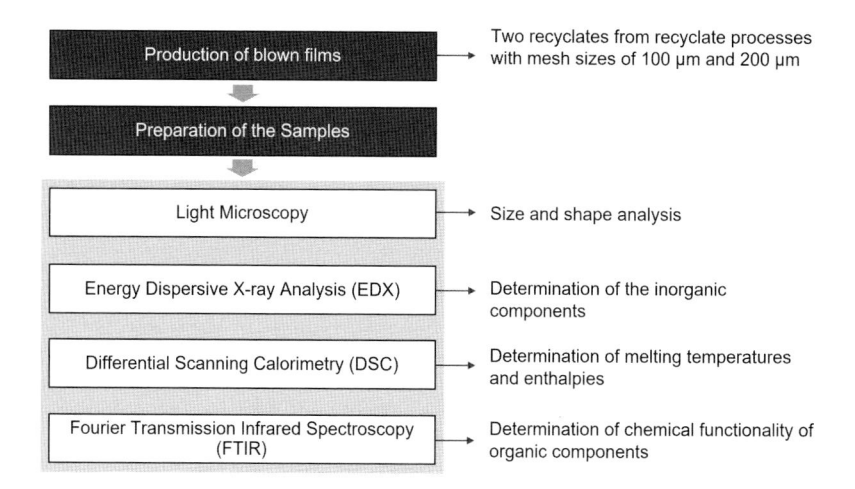

Figure 3.2: Analytical methods for testing impurities in film products

Samples are taken from the films at different locations and the impurities are first classified by means of light microscopy on the basis of their size, colour and shape. Five independent film sections are used for each material. Further preparation of the films is not necessary due to their sufficient transparency for transmitted light microscopy. Five areas per sample are imaged with a magnification of 10. Another 10 areas per sample are imaged with a magnification of 300. The open-source software Fiji Image J is used to analyze the microscopy images. Individual particles are measured and area distributions are created over all sample areas.

Energy dispersive X-ray spectroscopy (EDX) is used to obtain information about the inorganic components in the recyclates. Five different samples of each of the two materials are analysed. The samples are cut into strips along the extrusion direction and then about 30 film strips are joined into stacks. In order to be able to guarantee a clear surface for the analysis, one non-joined edge of each film stack is cut using a microtome. To stabilize the individual films during cutting, they are bonded in advance with a UV-curing epoxy resin and cooled during the cut. The experiments are carried out on a system consisting of a Sigma VP field emission scanning electron microscope, Carl Zeiss AG, Oberkochen, and further EDX equipment, Oxford Instruments plc, Abingdon.

In addition, local structural changes of the sample surface are investigated spectroscopically using Fourier transform infrared spectroscopy (FT-IR) by means of the ATR method (attenuated total reflection on a diamond crystal). When analyzing the impurities, they are isolated to minimize influences from the surrounding matrix as much as possible. All measurements are done with the FT-IR instrument Nexus 870 from Thermo Fisher Scientific GmbH, USA from 680 cm^{-1} to 4000 cm^{-1}. The results will provide further insight into the chemical functionality of organic components.

Differential scanning calorimetry (DSC) is used to obtain information about the polyolefins contained in the recyclate. All measurements are performed with the DSC Q1000 from TA Instruments, New Castle, USA. The mass of the film samples is approximately 2 mg and the mass of the granular samples is 4.1 mg. The measurements are carried out in the temperature range of 0 °C and 200 °C at heating and cooling rates are 20 °C/min. Average values of ten samples are used. The melt temperature T_m and the heat quantities H_m are determined in the second heating process.

The goal of the research depicted is to determine the number, size and nature of the contaminants. From this data, conclusions concerning the usage potentials for different qualities of recyclates in plastics packaging will be drawn.

3.4 Results and discussion

After the basic principles and characterisation methods have been considered in detail, the results of the investigations are discussed below.

3.4.1 Analysis of size, colour and shape

Initial visual examinations of the films without magnification, Figure 3.3, suggest a generally high particle density. Microscopy images with a magnification of 300 (cf. Figure 3.4) provide a qualitative overview of the types of these contaminants, which differ greatly in their color, size and shape. Sharp-edged particles, textile fibres and a variety of very small particles can be seen, which are difficult to specify in more detail at this point. Sharp-edged particles can possibly be attributed to brittle incompatible other polymers which broke into flakes during comminution and were able to pass through the filter due to their small size. Since the edges are still clearly visible after preparation and processing into blown films, the melting temperature of these polymers must be significantly higher than the processing temperature of T = 200 °C. Coloured defined areas can also indicate foreign poylmers into which colour masterbatches have previously been incorporated into the plastic melt. Other visible organic components are textile fibers. Depending on their length, these are either looped or straight. Such particles pass through the filter mesh due to their narrow shape and high flexibility. A detailed analysis of the particle sizes is not carried out due to the highly variable shape of the particles.

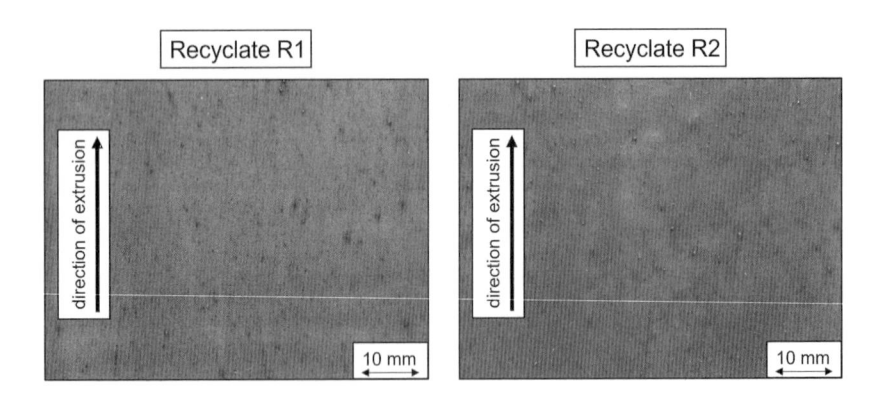

Figure 3.3: Analytical methods for testing impurities in film products

Regardless of the diversity of the particles, however, dark spots or larger accumulations of material dominate, which have significantly larger dimensions than the specified pore size of the melt filters of 100 µm or 200 µm. Such a particle in the film from the recyclate R1 (cf. Figure 3.4 a) has, for example, a minimum diameter of 280 µm and a maximum diameter of 480 µm. It is assumed that these particles have a gel-like consistency at extrusion temperature and can therefore pass through the filter deformed. These are possibly high-molecular or cross-linked structures consisting of PE. Such particles are referred to in the following as "black spots".

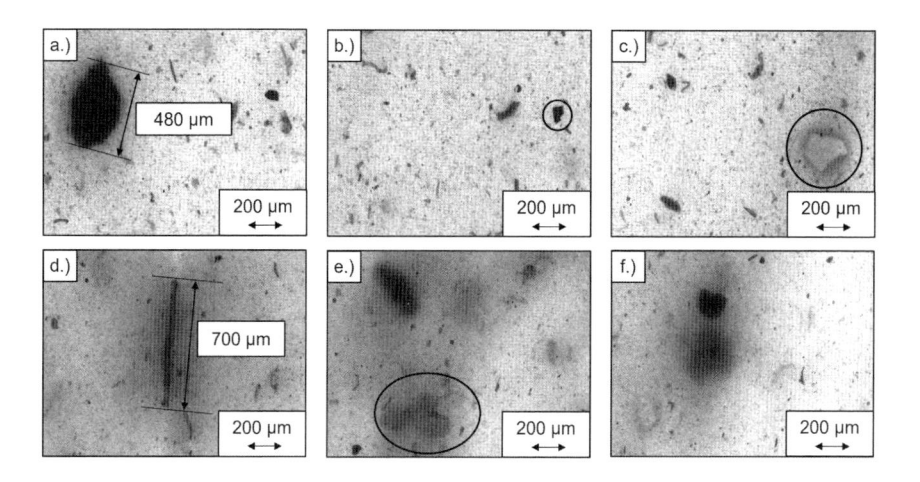

Figure 3.4: Images of the films from R1 at a magnification of 300

With the Fiji Image J software, binary images are generated from the microscopy images via global thresholding. The particle analysis therefore only identifies impurities whose size and colour is above the selected threshold value. To a certain extent, results will invariably reflect the choice of threshold and the lighting conditions and are therefore only suitable for comparative statements. Figure 3.5 shows an example of the selected threshold value, with the aim of dis-playing particles above the mesh size. With this, statements can be made about area distributions and the number of impurities of the two recyclates. In relation to the microscopy image with a size of about 32,300 x 24,300 µm^2, the results show an average of about 12.5 particles for R1 and 20 particles for R 2. These particles cover a total area of 11,495 µm^2 and 58,824 µm^2 on average. Accordingly, both the particle density in R2 is higher and the area fraction of such impurities is higher. It remains unclear at this point whether the soft melt components in R1 could be separated by means of the filter or whether these have split up into many smaller particles under higher shear.

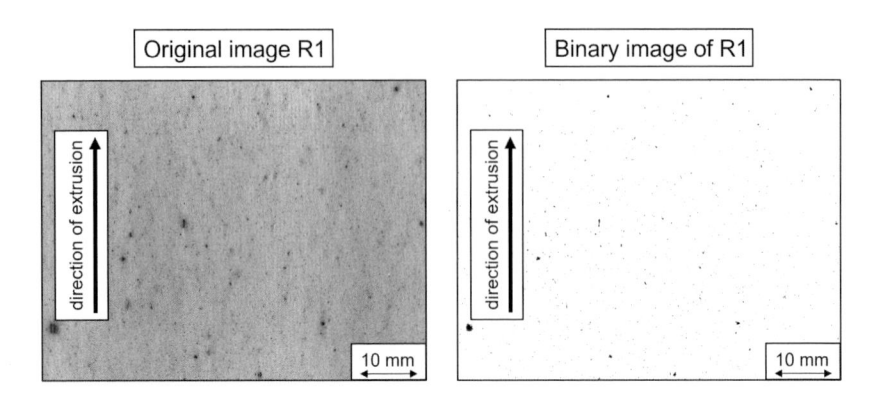

Figure 3.5: Images of the films without magnification

3.4.2 Chemical analysis

Scanning electron microscope (SEM) images are created for each material for four film samples in each case. In each SEM image with three to four film cross-sections, three spectra are examined with regard to the element distribution. In addition, a distribution image is created for each material. A SEM image of a film stack from R2 is shown as an example (cf. Figure 3.6). Investigations using SEM show that all visible particles are significantly smaller than 100 µm. This suggests that the black spots described above are of organic nature, as such contaminations usually appear soft and liquid and show less contrast. Furthermore, heating and cooling processes using a temperature-controlled microscopy stage with temperatures between $T = 105$ °C and $T = 240$ °C show unchanged dimensions and shapes of the black spots. This supports the assumption that the larger impurities are high molecular weight or cross-linked PE.

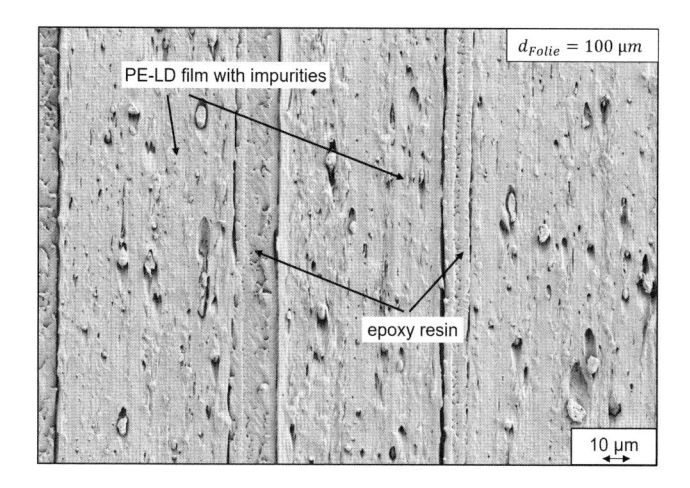

Figure 3.6: SEM image of a film stack from R1

The chemical analysis of the elements for each material shows similar intensities for all spectra and the two recyclates studied, as exemplified in Figure 3.7.

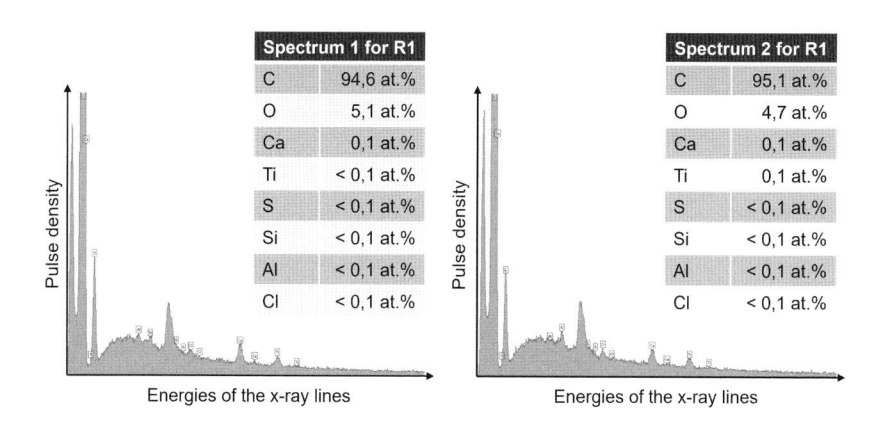

Spectrum 1 for R1	
C	94,6 at.%
O	5,1 at.%
Ca	0,1 at.%
Ti	< 0,1 at.%
S	< 0,1 at.%
Si	< 0,1 at.%
Al	< 0,1 at.%
Cl	< 0,1 at.%

Spectrum 2 for R1	
C	95,1 at.%
O	4,7 at.%
Ca	0,1 at.%
Ti	0,1 at.%
S	< 0,1 at.%
Si	< 0,1 at.%
Al	< 0,1 at.%
Cl	< 0,1 at.%

Figure 3.7: EDX element spectra 1 and 2 for the electron image of material R1

The element carbon (C) is dominant in all spectra as part of the polymer chain of PE. The presence of aluminium (Al) is due to the fact that chip bags, for example, are vapour-deposited with aluminium as a barrier. The same applies to the effect of silicon (Si) as a barrier layer. In addition, layered silicates such as talc are often added to film products, which also contain silicon. The layered silicate kaolin is also formed from silicon and aluminium. The addition of kaolin leads to an improvement in chemical resistance and electrical properties as well as to reduced water absorption. It also has a high degree of whiteness and is therefore widely used as a filler in paints and printing inks. In plastics processing, the fillers calcium carbonate and titanium dioxide are also used as white pigments or to improve the material properties. This can explain the occurrence of calcium (Ca) and titanium (Ti) in combination with oxygen (O) in the spectra. The detection of the element chlorine (Cl) can be attributed to the use of PVDC as a barrier layer in packaging films or polyvinyl chloride. In addition, chlorine may also be present in the recyclate in the form of salt as a component of the filling goods of packaging [Nen06]. To illustrate the distribution of the elements and the respective particle sizes, these are each shown individually as a distribution image in Figure 3.8. The elements silicon, chlorine, aluminium, calcium and titanium are clearly visible in the element images. Sulphur (S) can be excluded in large proportions in the films, as the element is located between the film layers and can therefore be assigned to the epoxy resin adhesive used.

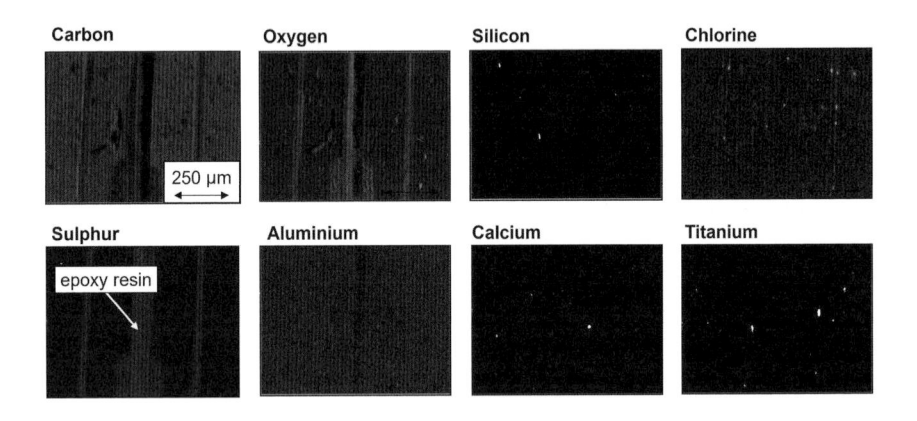

Figure 3.8: EDX distribution images of the elements from Recyclate 2

To determine the organic compounds of the recyclates and to investigate the influence of black spots, FT-IR measurements were carried out on extracted particles and on film areas without particles. Figure 3.9 shows an example of the IR spectra of two measurements each. The evaluation of the spectra show a high agreement with PE-LD. Nevertheless, a shoulder or a weak peak around 2950 cm^{-1} (antisymmetric vibration of the CH3 group) and around 1375 cm^{-1} (deformation vibration of the CH3 group) is recognizable, which indicate polypropylene.

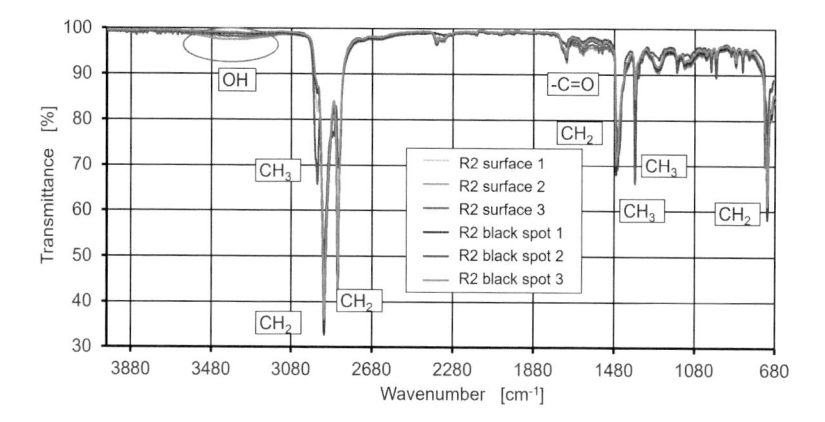

Figure 3.9: FTIR results using the example of R2

In addition, the samples show OH groups around 3300 cm^{-1}, which is an indication of aging by oxidation. Oxidation causes the formation of oxygen-containing groups. The peak around 1715 cm^{-1} indicates -C=O bonding. However, no clear differences are evident between the samples with and without black spot in terms of chemical composition.

3.4.3 Thermal analysis

The influence of the black spots on the thermal properties is analysed by means of DSC measurements. Only the cooling process and the second heating process are examined in more detail. A change in the molecular weight of the samples might become visible in different melting enthalpies. Measurements of ten samples with and without black spots are performed. As can be seen in Figure 3.10, three endothermic peaks are determined by DSC, which can be assigned to PE-LD. PE-HD and PP respectively. The PE-LD fraction in the recyclate melts at around 107.4 °C, the PE-HD fraction at 121.5 °C and the PP fraction at 160 °C. The results with and without black spots can be compared by DSC. However, no significant differences are discernible between the results with and without black spots. This applies to the melt temperatures as well as to the melt enthalpies. Smaller deviations with respect to the enthalpy can be attributed to the choice of the respective integration limits. A calculation of the degree of crystallization is not possible at this point due to the unknown composition of the samples.

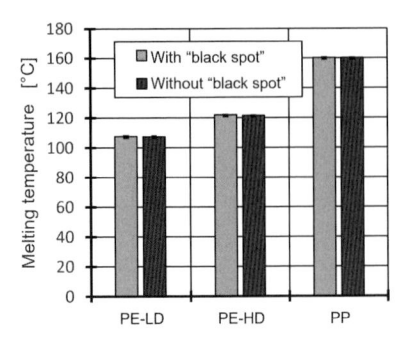

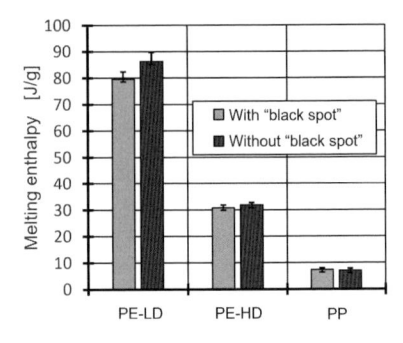

Figure 3.10: Influence of black spots on melt temperatures and enthalpies

3.5 Conclusions and outlook

The non-volatile impurities in commercially available film recyclates were analysed. It was shown that a big number of larger and smaller impurities cannot be filtered out using conventional filtration methods in the recycling process and cause localized defects in films. A large number of contaminants with dimensions far above the filter pore width were identified as black spots of an organic nature. It is assumed that these are gel-like particles consisting either of cross-linked or high molecular weight material. Unintuitively, a mesh size of 100 µm seems to reduce the relative area fraction of back spots compared to a mesh size of 200 µm. PE-LD, PE-HD and PP were identified as polymeric components in the material by means of DSC and FT-IR. Significant differences between areas with black spots and without black spots could not be detected by DSC and FT-IR. The results of the FT-IR measurements suggest first degradation phenomena in both types of samples due to the appearance of OH groups and -C=O groups. In addition, the elements silicon, chlorine, aluminum, calcium and titanium, which appear in packaging as fillers, barrier layers and packaged goods, were detected by means of EDX. An exact proportional composition of the organic and inorganic components could not be determined. In further investigations, ashing tests can be carried out to quantify the inorganic components. Statements about the degradation as well as missing information about the amount of PP, PE-HD and PE-LD in the recyclate can be provided by high-temperature gel permeation chromatography (HT-GPC) in combination with IR-detection. These investigations should be accompanied by investigations of well defined polyolefin blends. Furthermore, such specifically prepared blends can also be used for renewed analyses by means of FT-IR and DSC in order to establish calibration curves for typical recyclate compositions.

In summary, it remains to be weighed up whether a few large (visible) black spots or many small (less visible, but possibly more mechanically damaging) black spots fulfil the product specification for the respective application. A finer filter mesh size is therefore not necessarily equivalent to a higher purity of the recyclate and improved material properties. Unfortunately, beyond filter fineness, information about the type of filtration used by recyclers is not usually given out. However, the type of filtration system used (possibly also cascade of several filter units) may have a significant influence on the material quality. For the removal of flexible particles or black spots, a large-area filter with a depth filter medium could be used. In addition, further methods for the removal of foreign substances could be intensified or examined more closely in the future in order to improve the processing of recyclates. One possibility for improved purification of the materials is, for example, to adapt the washing process during material processing to remove non-volatile impurities.

Acknowledgements

The investigations received financial support from the German Bundesministerium für Bildung und Forschung (No. 02J20E540), to whom we extend our thanks.

References

[CDI+07] CAPONE, C.; DI LANDRO, L.; INZOLI, F.; PENCO, M.; SARTORE, L.: Micromechanical analysis of the in situ effect in polymer composite laminates. *Polymer Engineering and Science* 47 (2007) 11, p. 1813–1819

[EP07] EHRENSTEIN, G.; PONGRATZ, S.: *Beständigkeit von Kunststoffen*. Munich: Carl Hanser Verlag, 2007

[GJ03] GORGHIU, L.; JIPA, S. AND ZAHARESCU, T. S. R. M.: The effect of metals on thermal degradation of polyethylenes. *Polymer Degradation and Stability* 84 (2003) 1, p. 7–11

[ILKT76] IRING, M.; LÁSZLO-HEDVIG, S.; KELEN, T.; TÜDÓS, F.: Study of thermal oxidation of polyolefins. VI. Change of molecular weight distribution in the thermal oxidation of polyethylene and polypropylene. *Journal of Polymer Science* 57 (1976) 6, p. 55–63

[Mar16] MARTENS, H.: *Recyclingtechnik: Fachbuch für Lehre und Praxis*. Wiesbaden: Springer Vieweg, 2016

[MGWE06] MARSH, R.; GRIFFITHS, A. J.; WILLIAMS, K. P.; EVANS, S. L.: Degradation of recycled polyethylene film materials due to contamination encountered in the products' life cycle. *Journal of Mechanical Engineering Science* 47 (2006) 5, p. 593–602

[Nen06] NENTWIG, J.: *Kunststoff-Folien*. Munich, Vienna: Carl Hanser Verlag, 2006

[NN08] N.N.: *Recycling for Climate Protection - Results of the Study by Fraunhofer UMSICHT and INTERSEROH on CO_2 Savings through the Use of Secondary Raw Materials*. Fraunhofer-Institut für Umwelt-, Sicherheits- und Energietechnik UMSICHT, 2008

[NN18] N.N.: *Produktspezifikation 03/2018 Fraktions-Nr. 310-1*. DSD – Duales System Holding GmbH Co. KG, 2018

[NN19] N.N.: *Examination and testing of recyclability: Requirements and evaluation catalogue of the Institut cyclos-HTP for EU-wide certification*. Institut cyclos-HTP GmbH, 2019

[NN20a] N.N.: *Annual Report 2018/2019*. IK Industrievereinigung Kunststoffverpackungen e.V., 2020

[NN20b] N.N.: *Material flow diagram plastics in Germany 2019*. Conversio Market Strategy GmbH, 2020

[NN20c] N.N.: *Plastics-relevant waste streams in Germany 2019*. Conversio Market Strategy GmbH, 2020

[SS20] SCHYNS, Z.; SHAVER, M.: Mechnical Recycling of Packaging Plastics: A Review. *Macromolecular Rapid Communications* 42 (2020) 2, p. 1–27

[URL21] N.N.: *How to recycle more plastics in lightweight packaging?* URL: https://eu-recycling.com/Archive/20131, 14.10.2021

[Zwe16] ZWEIFEL, H.: *Stabilization of Polymeric Materials*. Berlin: Springer Vieweg, 2016

Abbreviations

Notation	Description
A	Aluminium
C	Carbon
Ca	Calcium
Cl	Chlorine
DIN	Deutsches Institut für Normung
DSC	Differential Scanning Calorimetry
DSD	Duales System Deutschland
EDX	Energy Dispersive X-ray Spectroscopy
FTIR	Fourier-Transform Infrared Spectroscopy
HT-GPC	High-Temperature Gel Permeation Chromatography
IR	Infrared
LVP	light packaging (Ger. Leichtverpackungen)
PE-HD	Polyethylene high density
PE-LD	Polyethylene low density
PP	Polypropylene
PVDC	Polyvinylidene chloride
R1	recyclate 1
R2	recyclate 2
S	Sulphur
Si	Silicon
T	Temperature
Ti	Titanium
O	Oxygen
UV	Ultraviolet

Mixing and flow processes in rubber processing

Moderator: Dr.-Ing. Markus Hesse, HF Mixing Group

———— Content ————

1 Investigation and optimisation of the material intake for internal mixers 207

M. Hesse[1]
[1] HF Mixing Group

2 Investigations on the influence of viscoelasticity on the swell behaviour and flow instabilities of carbon black-filled rubber compounds 220

Ch. Hopmann[1], F. Fey[1], V. Schöppner[2], D. Kleinschmidt[2], A. Limper[1], M. Schön[1]
[1] Institute for Plastics Processing (IKV) at RWTH Aachen University
[2] Kunststofftechnik Paderborn (KTP) of Paderborn University

3 Influence of process parameters on mixing time and properties of rubber compounds 238

Ch. Hopmann[1], M. Kostka[1], A. Limper[1], M. Schön[1]
[1] Institute for Plastics Processing (IKV) at RWTH Aachen University

Dr.-Ing. Markus Hesse

Dr.-Ing. Markus Hesse studied mechanical engineering at the University of Paderborn with a focus on plastics and process engineering. After his studies, Mr. Hesse completed his doctorate at the Institute for Plastics Technology KTP with a thesis on flow processes in tangential internal mixers.

In 2007, he joined the HF-Mixing Group, the world leader in rubber compounding technology, as Project Manager Sales. In 2013, Mr. Hesse took over the responsibility for the product development of mills and discharge extruders. In 2017, he became head of the technical centre in Freudenberg and since August 2021, Mr. Hesse is the Head of RD of the HF Mixing Group.

In addition to his professional activities, Mr. Hesse is a lecturer at the University of Ilmenau and a member of the board of trustees of IKV Aachen.

1 Investigation and optimisation of the material intake for internal mixers

M. Hesse[1]
[1] HF Mixing Group

Abstract

In order to maintain competitiveness, rubber companies are forced to constantly optimise their manufacturing processes. Especially in European countries with high wage and incidental costs, efficient manufacturing processes are of great importance. Of course, research and development is not carried out only to reduce costs. Aspects such as sustainability and environmental protection are becoming more and more important as drivers for new innovations. With a view to the mixing room, different development goals can be derived from this:

- High throughput
- High quality of the compound
- Energy efficiency / sustainability
- Low specific mixing costs

To achieve these goals, research and development work must be carried out on the machine and process side. This means state-of-the-art plant technology must be combined with efficient, intelligent process control.

The following paper is an example of such an approach. It describes an investigation of the material intake phase in internal mixers. The initial phase of the mixing process has a major influence on the throughput and quality of the compound. For this reason, an investigation and optimisation of this phase is very important to achieve the goals described above.

1.1 Introduction

Basically, there are three ways in which recipe ingredients can be added to the internal mixer. While the powdery ingredients are usually added automatically through pipe systems, the polymers bales are fed into the mixer through the feed door attached to the side of the mixer chute (see Figure 1.4). Liquid recipe ingredients such as plasticisers or liquid polymers are injected directly into the mixing chamber. In most cases, the ram is in its lower end position during this process.

If the entire mixing process is divided into mixing phases, the material intake is at the beginning.

Mixing Cycle						
Intake	Mastication	Incorporation	Dispersion	Distribution	Discharge

Figure 1.1: Mixing process - divided into mixing phases

For the following investigations, the intake phase begins with the addition of the polymer bales through the feed door and ends when the ram has reached its lower end position. For an effective mixing process, the intake phase should be as short as possible.

1.2 Intake behaviour of internal mixers

For rubber compounding, usually internal mixers are used. In these discontinuously operating machines, fillers, plasticisers, vulcanising agents and many other chemicals are incorporated into the raw polymer by counter-rotating rotors [Lim12, RS13].

In principle, internal mixers are divided into tangential and intermeshing systems.

Tangential rotors are freely rotatable, and comparable to a mixing mill, the rotor tips just do not overlap. For this reason, tangential mixer systems can be driven with different rotor speeds (friction). A particular advantage of tangential mixers is the good material intake. During rotation, large gaps are created between the rotors into which the material can be drawn (see Figure 1.2).

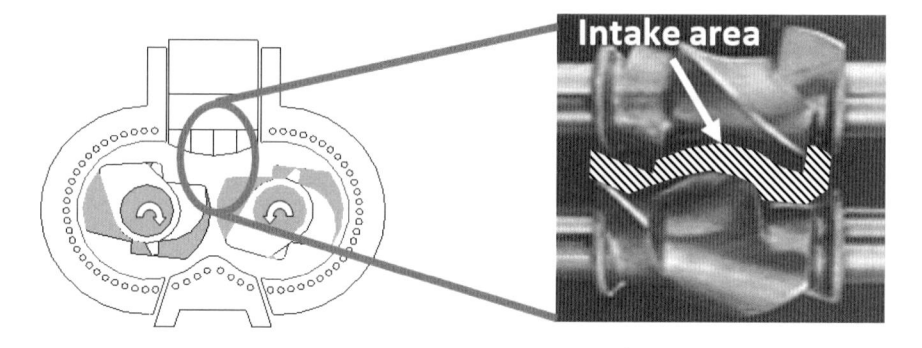

Figure 1.2: Intake area of tangential mixers

In the case of the intermeshing mixer, the rotor flights are overlapping each other. The high shear and elongational stress in the narrow gap area between the rotors lead to an increased distributive and dispersive mixing effect but worsen the intake behaviour of the machine (see Figure 1.3).

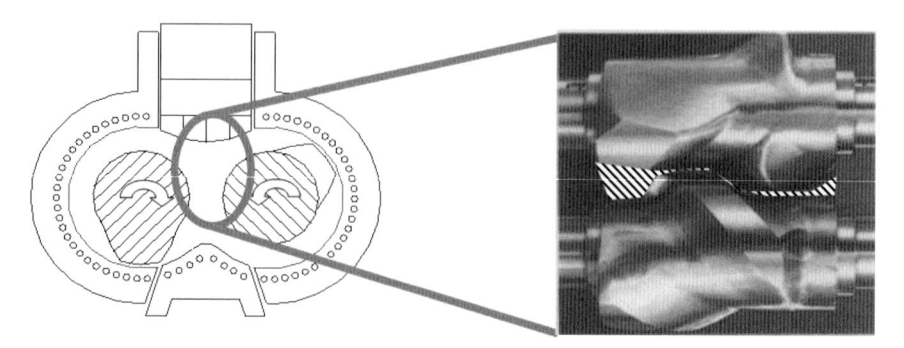

Figure 1.3: Intake area of intermeshing mixers

Since tangential systems usually show good intake behaviour, the focus of this investigation is on the intermeshing machines, where room for improvement has been identified.

1.3 Influence on the mixing quality

The compound quality is of decisive importance for rubber processing companies, as only the homogeneous distribution and dispersion of the recipe components can result in high-quality products [Lip19, MBR80]. For good dispersive mixing quality, it is important that the entire volume of material is exposed to sufficient shear and elongational stress. To achieve a homogeneous distribution of the recipe ingredients, it is important that all volume elements have carried out sufficient place changes. Inhomogeneity due to dead zones in the machine must be prevented at all costs.

After the material has been added to the mixer via the feeding door, the ram presses the compound into the mixing chamber. As long as the ram has not reached its lower end position, the feed chute of the mixer can be seen as a dead zone, as the material is only mixed within the chamber (see Figure 1.4).

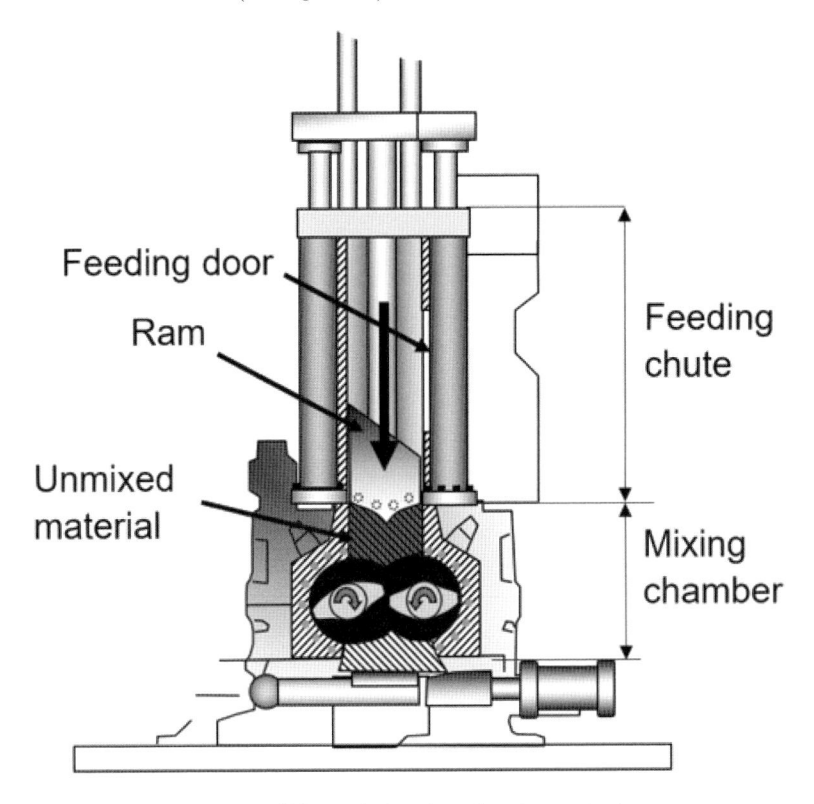

Figure 1.4: Schematic drawing of an internal mixer

If the ram only reaches its end position very slowly, this can already have a negative effect on the quality, as the active mixing time of the material in the chute is different to the material that was immediately drawn in. Figure 1.5 shows mixing curves with different ram setting times. For the comparison, two different rotor types were used at the same batch weight. In case of the mixer with the PES5 rotors, the ram needs approx. 40 s longer to reach its end position. This could already be an indication of a machine filled to highly and should at least be reflected in the mixing time.

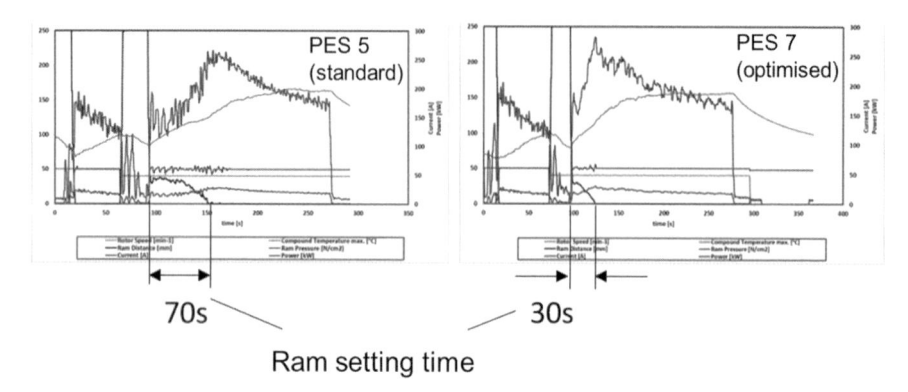

Figure 1.5: Comparison of the ram setting time

1.4 Influence of the intake on the cycle time

In addition to the positive effects on the material quality, an improved intake behaviour also results in economic advantages. If a mixing time reduction of 5 s due to an optimised intake is assumed, the following case study can be used to carry out an economic analysis. Example calculation: IM320E

Machine data: $IM320E$
Mixing chamber volume: $V = 332\ l$
Plant availability/year: $h = 6,000 \text{h}/year$
Cycle time: $t_Z = 300\ s$
Fill factor: $f = 0.65$
Mixing time reduction due to better intake: $t_R = 5\ s$
Material data:
Density: $\rho = 1.2\ \frac{kg}{dm^3}$

For the mixer described above, the better intake results in an increase in throughput of:

$$M = \frac{V \cdot f \cdot \rho \cdot h \cdot 3600}{(t_Z - t_R)} - \frac{V \cdot f \cdot \rho \cdot h \cdot 3600}{(t_Z)} \approx 316\ to/year \tag{1.1}$$

1.5 Influence of the intake on the fill factor

Not only the cycle time is influenced by the first mixing phase, also the optimal fill factor of the mixer depends on the intake behaviour. In order to understand the dependency between fill factor and intake behaviour, the main volume flows within the mixing chamber must be taken into account.

Basically, two main volume flows can be distinguished in an internal mixer. The volume flow A between the rotors and the two volume flows B along the mixing chamber (see Figure 1.6). It is obvious that A has a direct influence on the intake behaviour and the ram setting time of the machine.

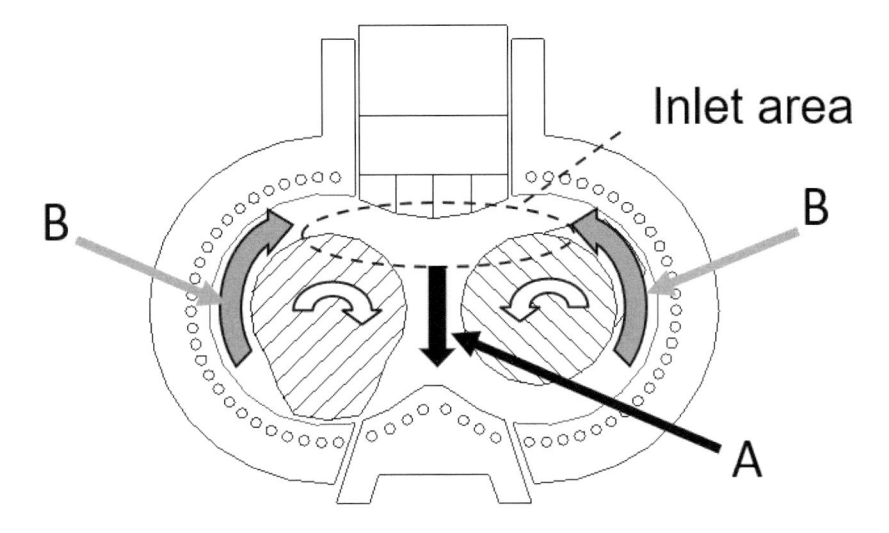

Figure 1.6: Volume flows

However, the volume flow A also has an influence on the achievable fill level of the mixer. When the ram is in its lower end position and the mixing chamber is closed, the volume flows A and B are in equilibrium B+B=A [LBG89]. Due to the narrow gap between the rotors, the conveying capacity A is limited. For this reason, the potential volume flow of B is always bigger than of A. The additional drag flow of the rotors along the mixing chamber is converted into pressure driven backflow in the inlet area of A (see Figure 1.6). If the pressure on the ram surface exceeds the force of the hydraulic power system, the ram is pushed up or does not reach its end position at all, which is a clear indication of an overloaded mixer. This means that with an increasing volume flow between the rotors, the possible fill factor of the mixer also increases.

1.6 Possibilities for influencing the intake behaviour of internal mixers

The explanations above have underlined the importance of an effective intake phase. Now the question arises how the intake can be influenced and optimised.

There are basically two ways. By adjusting the process parameters and by changing the rotor geometry. Both procedures are described in the following.

1.6.1 Influence of the process parameters on the intake behaviour

Through the tests described in the following chapter, it was possible to prove that the temperature and speed of the rotors have an influence on the intake behaviour of internal mixers.

Furthermore, the ram pressure influences the feeding. As a rule, higher ram pressure leads to shorter ram setting times. Since the pressure of the ram is limited for mechanical reasons and the system is usually always operated at maximum pressure during the intake phase, the optimisation potential for this parameter is limited. For this reason, the ram pressure will not be discussed further.

1.6.2 Experimental setup

To analyse the influence of the rotor temperature and the rotor speed, a laboratory mixer with a mixing chamber volume of 5 l was used. For the experiments, the feeding chute was separated from the mixing chamber by means of a metal plate (see Figure 1.7).

In the first step, the feeding chute is filled with the polymer. Three polymer types were used for the investigations: Natural Rubber (NR), Styrene-Butadiene Rubber (SBR), Polybutadiene Rubber (BR)

In the second step, the necessary settings are made.

- Speed of the rotors
- Temperature of the rotors

Since only the volume flow A is to be analysed, the discharge door of the mixer is open during the tests. The ram as a feed support was not used for the trials - similar experiments have been carried out at IKV [HKLB19].

In the last step, the metal plate is pulled out and the time is measured until the material has been completely drawn in.

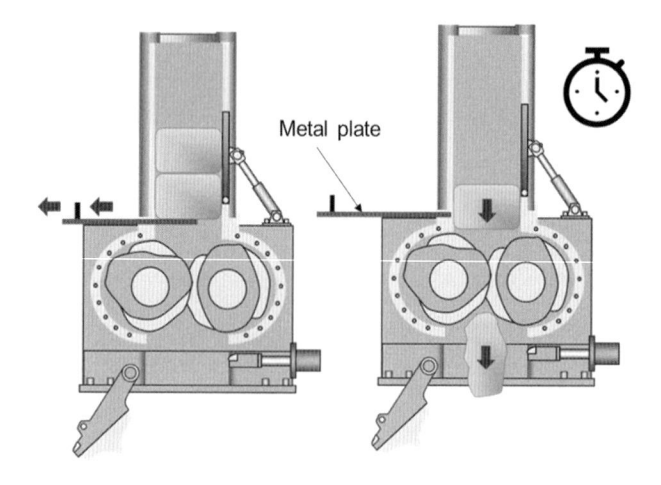

Figure 1.7: Experimental setup

During all tests the power consumption is recorded by the mixer control system. Also, the time for the polymer intake is measured. With the intake time and the weight of the polymer the throughput is calculated.

Influence of the rotor temperature

For the trials the rotor speed was constant at 30 rpm. In total three different rotor temperatures were tested: 25 °C, 40 °C, 55 °C.

In Figure 1.8 the analysis of the throughput rate of NR, BR and SBR at different rotor temperatures is shown. For NR and SBR a temperature of 40 °C shows the highest throughput. For both polymers the throughput can be increased by around 40 % compared to 55 °C rotor temperature. In general, the high rotor temperature of 55 °C seems to be not optimal for an effective intake. An increased tendency of wall sliding was observed at high surface temperatures, which reduces the volume flow A between the rotors (see Figure 1.6).

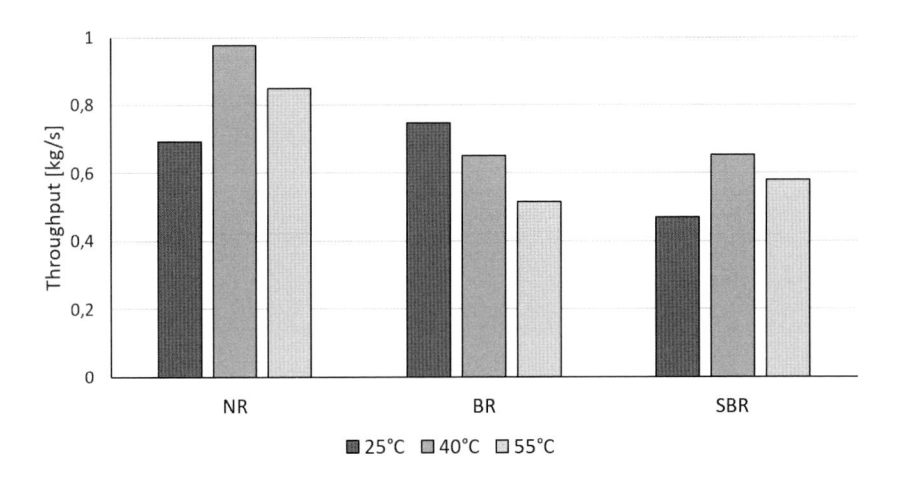

Figure 1.8: Influence of the rotor temperature on the polymer intake (without ram)

Influence of the rotor speed

For these trials the rotor temperature was constant at 40 °C. In total three different rotor speeds were tested: 10 rpm, 30 rpm, 60 rpm.

Figure 1.9 shows the results of the trials for NR, BR and SBR. As expected, the highest rotor speed shows the highest throughput rate for every polymer. But it can already be seen that the increase of the throughput for every polymer is different. For example, the increase in throughput for BR when doubling the rotor speed from 30 rpm to 60 rpm is at 135 % while the increase for NR is only 37 %.

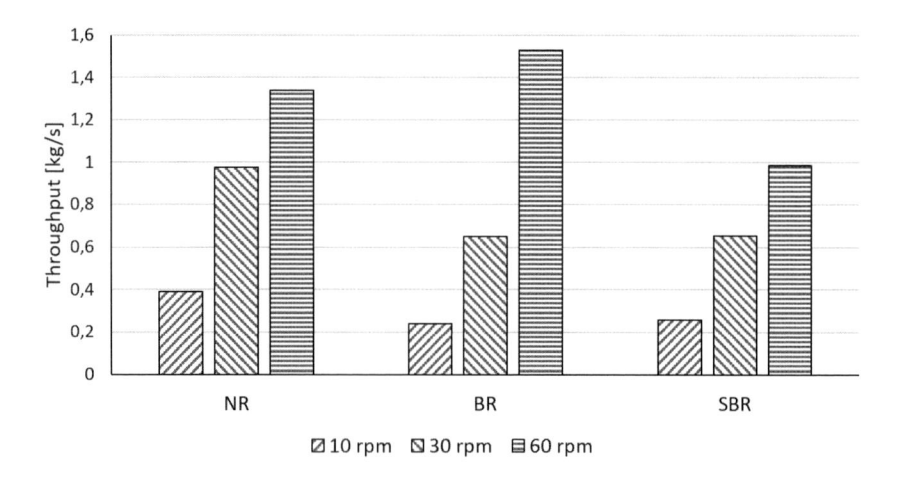

Figure 1.9: Influence of the rotor speed on the intake behaviour of the rotors (without ram)

To illustrate this more clearly, the throughput was converted from kg per second to kg per rotor revolution (see Figure 1.10).). In the case of SBR and NR, the throughput rate decreases with increasing rotor speed. This indicates that NR and SBR start to slide more on the rotor surface at high rotor speeds. The adhesive properties of BR do not change with rotor speed and the throughput remains constant. The low viscosity and elasticity of BR compared to SBR and NR reduces resistance in the intake phase. High rotor speeds do not cause the material to slip on the wall.

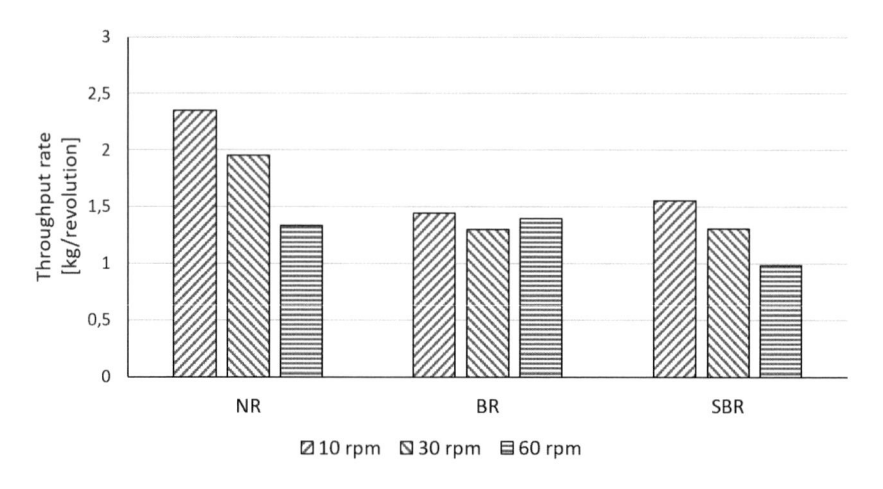

Figure 1.10: Influence of the rotor speed on the intake behaviour related to the rotor speed

In general, it can be said that a high rotor speed at the beginning of the intake phase brings an improvement for polymers that can be fed in without a ram.

1.7 Influence of the rotor geometry on the intake behaviour

In addition to the machine settings, the rotor design has a significant influence on the intake behaviour of the mixer. To illustrate this, two intermeshing rotor types are compared in terms of their intake behaviour.

- Standard geometry: PES5
- Intake-optimised geometry: PES7

Figure 1.11 shows the current consumption, the ram setting, the temperature development and the cumulative specific energy consumption as a function of time and rotor geometries PES5 and PES7.

The different intake behaviour of both geometries can already be derived from the current curve in the first 25 s of the mixing process. In this phase, SBR is fed into the machine. The ram is in its upper end position at this point, so not in contact with the polymer. In case of the PES7 geometry, the curve shows twice as high current peaks compared to PES5, which is a reliable indication for a better intake behaviour in this phase of the cycle. In the subsequent warm-up phase, the ram moves to its lower end position. After the polymer has reached a temperature of 100 °C, the hydraulic ram is raised and carbon black (N220) is added to the mixer. Once the filler has been fully dosed, the ram is moved downwards to incorporate the carbon black. Similar to the first 25 seconds of the mixing cycle, the power consumption increases again faster in case of the PES7. The better intake behaviour of the PES7 is particularly reflected by a 25 s shorter ram setting time.

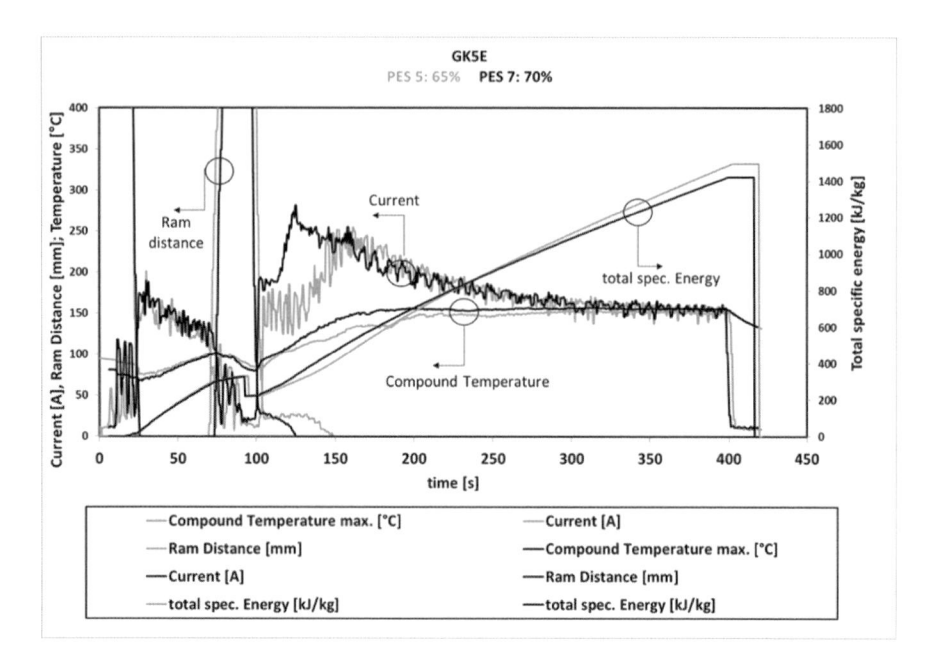

Figure 1.11: Current consumption, ram setting, temperature development and cumulative specific energy consumption as a function of time of the PES5 and PES7 rotor systems

In order to investigate the effect described in chapter 5 of the better intake behaviour on the optimum fill factor, a fill factor study was carried out. For this purpose, filling levels of 65 %, 70 % and 75 % were investigated in relation to ram settling times. Figure 1.12 shows that the PES7 has a 60 % shorter ram setting time for all investigated fill levels. From this result it can be concluded that the feed behaviour has been significantly improved. This should have a positive effect on the quality of the compound and will improve the throughput of the mixer.

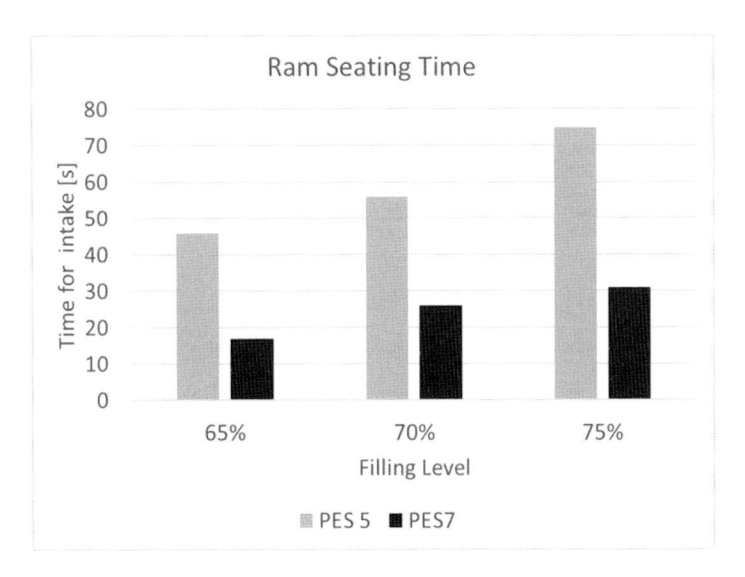

Figure 1.12: Ram setting times for the filling levels 65 %, 70 % and 75 % as a function of the rotor geometries PES5 and PES7

1.8 Conlusions and outlook

The investigations have underlined the importance of the feeding behaviour for intermeshing rotors. The volume flow through the gap area not only influences the ram setting time in the beginning of the mixing cycle, but it also has an impact on all subsequent mixing phases.

An example of this is shown in Figure 1.11. Despite the 5 % higher filling level in case of the PES7 rotors, the incorporation time of the carbon black is reduced by 25 seconds due to the improved intake behaviour. For this reason, more time is available for the dispersion of carbon black. A subsequent topographical analysis of the rubber surface confirmed a 20 % better carbon black dispersion due to the reduced black incorporation time [Weh98]. If an improvement in dispersion quality is not required, the cycle time can also be reduced. This either increases the throughput of the mixer line or leads to energy savings if the additional volume is not needed. In addition, the investigations have shown that the material intake can be optimised by changing the process parameters like rotor speed or rotor temperature. The first process phase can be significantly shortened by an optimal choice of these settings. However, there are no generally valid recommendations, rather the choice of parameters depends on the polymer type.

In the future, it would certainly be useful to investigate more polymers with regard to the feed behaviour in order to give clear recommendations for the adjustment of the mixer.

In addition, the structured detailed examination of the intake phases has proven to be very useful and effective. Further analysis of the incorporation, dispersion or distribution phase and their interactions would be very important for a holistic understanding of the mixing process.

References

[HKLB19] HOPMANN, C.; KOSTKA, M.; LIPSKI, A.; BAKIR, C.: Systematische Untersuchung des Einzugsverhaltens und der Mastikation im Innenmischer zur Ermittlung einer Abhängigkeit zwischen Prozessparametern und resultierenden Kautschukeigenschaften. *Gummi Fasern Kunststoffe* 72 (2019) 6, p. 206–211

[LBG89] LIMPER, A.; BARTH, P.; GRAJEWSKI, F.: *Technologie der Kautschukverarbeitung*. München and Wien: Carl Hanser Verlag, 1989

[Lim12] LIMPER, A.: *Mixing of Rubber Compounds*. München: Carl Hanser Verlag, 2012

[Lip19] LIPSKI, A.: *Fließvorgänge zwischen den Rotoren von tangierenden Kautschukinnenmischern*. RWTH Aachen University, Dissertation, 2019

[MBR80] MEISSNER, K.; BERGMANN, J.; REHER, E.: Über den Mischprozess im Innenmischer unter Beachtung viskoplastischer Werkstoffeigenschaften. *Plaste und Kautschuk* 27 (1980) 3, p. 147

[RS13] RÖTHEMEYER, F.; SOMMER, F.: *Kautschuktechnologie*. München and Wien: Carl Hanser Verlag, 2013

[Weh98] WEHMEIER, A.: *Entwicklung eines Verfahrens zur Charakterisierung der Füllstoffdispersion in Gummimischungen mittels einer Oberflächentopografie*. FH Münster, Diploma Thesis, 1998

Symbols

Symbol	Unit	Description
f	–	Fill factor
h	$h/year$	Plant availability/year
M	$to/year$	Increase in throughput
ρ	$\frac{kg}{dm^3}$	Density
t_R	s	Mixing time reduction
t_Z	s	Cycle time
V	l	Mixing chamber volume

Abbreviations

Notation	Description
BR	Polybutadien Rubber
NR	Natural Rubber
SBR	Styrene-Butadiene Rubber

2 Investigations on the influence of viscoelasticity on the swell behaviour and flow instabilities of carbon black-filled rubber compounds

Ch. Hopmann[1], F. Fey[1], V. Schöppner[2], D. Kleinschmidt[2], A. Limper[1], M. Schön[1]
[1] Institute for Plastics Processing (IKV) at RWTH Aachen University
[2] Kunststofftechnik Paderborn (KTP) of Paderborn University

Abstract

In rubber extrusion, the distinct viscoelasticity of highly filled compounds leads to various challenges. Depending on the formulation, die geometry and process parameters, flow anomalies as well as swell can negatively influence the quality and dimensional accuracy of the extrudate. Due to the extremely complex rheological material behaviour of rubber compounds, the design of dies and processes in industrial practice is not based on simulation and engineering calculations, but rather on empiricism and iteration. The results of established rheological measuring devices do not yet allow an exact prediction of the processing properties in the extrusion process (such as: die swell, presence of flow anomalies, etc.). In order to better understand the cause-effect relationships, three rubber compounds were examined with a rotation rheometer and a capillary rheometer. In addition, the swell of the extrudates of the capillary rheometer was determined. The amount of carbon black (CB) (used as an active filler) and the melt temperature were varied. The influence of the CB content and the melt temperature on the viscoelastic properties depend on the amplitude and frequency and cannot be generalised. While at low frequencies an increase in CB content or temperature leads to increasingly elastic material behaviour, this relationship is reversed at higher frequencies. In addition, increasing amounts of CB in the compound increase the relaxation time, but also reduce the swell.

2.1 Introduction

The quality characteristics of extrudates, such as dimensional accuracy and surface quality, depend, among other properties, on the process parameters, the extrusion die and the rubber compound [LBG89]. Die design is usually based on empiricism and iteration. A simulative design requires a great deal of effort, as extensive and process-related material characterisation is necessary [Mic09]. In addition, models do not take flow anomalies into account sufficiently. That is why rubber compounds tend to exhibit flow anomalies and swell effects depending on their formulation and the load history during processing. These phenomena are based on the distinct viscoelasticity of the rubber compound and the resulting complex flow behaviour [OR15].

With the aim of improving the predictability of extrusion instabilities and swell effects, the extent to which it is possible to transfer the measured results of established rheometric methods to the material behaviour in the extrusion process is analysed. In the results shown in this paper, the influence of carbon black (CB) content and melt temperature on various rheological parameters, swell behaviour and melt fracture was analysed.

2.2 Rheological peculiarities of filled rubber compounds

Compared to thermoplastics, rubber compounds generally have a more pronounced shear thinning behaviour. The shear-thinning of rubber compounds can be well described with the power law according to Ostwald and de Waele [Ost25, Wae23]. For a variety of rubber compounds, Leblanc et al. showed for a wide shear rate range over 8 decades that no viscosity plateau could be measured with different rheometer designs [Leb02, LB00]. With increasing temperature and decreasing pressure, the viscosity usually decreases [PGL95].

Rubber also reacts to shear with tensile stresses in the direction of flow and compressive stresses perpendicular to it. The force to stretch and accelerate the material F_{normal} is expressed with the first normal stress difference N_1 (Equation 2.1).

$$N_1 = \tau_{11} - \tau_{12} = \frac{F_{normal}}{A} \qquad (2.1)$$

The focus of this work is the entropy-elastic degradation of molecular orientations, which are caused in particular by deformations in cross-sectional constrictions. This degradation is described in terms of relaxation times. During relaxation, the shear modulus decreases with time and the time required to do so decreases with rising temperatures. With increasing temperature, the relaxation times become shorter because the molecules have a higher mobility [PGL95]. With dynamic loading, viscoelastic material also shows a phase shift by the angle δ. In this case, the deformation precedes the resulting stress by the phase angle. $tan(\delta)$ is called the loss factor. With δ the shear modulus can be divided into a real part G', the storage modulus, and an imaginary part G'', the loss modulus. $tan(\delta)$ is the ratio of loss modulus to storage modulus [RS13].

Fillers, especially CB, form three-dimensional filler-filler and filler-polymer networks which increases the shear modulus. The Payne effect describes the degradation of the networks as a result of the applied shear stress during processing [Pay62]. For NR it was shown that the temperature influence on the storage modulus decreases with increasing CB content, since the CB, in contrast to the polymer, does not experience a large temperature influence [OTP+08]. Under dynamic loading, the decrease in storage modulus with increasing strain rate is more pronounced in a material with high CB loading compared to a material with low CB content. This is due to the fact that the CB networks which are degraded in the process are responsible for a larger proportion of the storage modulus in materials with higher CB content. The correlation has also been demonstrated in SBR compounds [KMG11, RH21]. In HNBR it was shown that with increasing CB content not only the storage modulus increases, but also the loss factor decreases. On the one hand, the storage modulus is increased significantly more than the loss modulus by adding CB. On the other hand, the CB itself has a loss factor close to zero. The loss factor generally increases at higher deformations with the increasing degradation of the filler networks. With higher CB contents, the linear viscoelastic range becomes smaller or can no longer be observed [WSSO10]. In the case of SBR compounds, the material exhibits more elastic behaviour (low loss factor) in the linear-elastic range with increasing CB content and more viscous behaviour (high loss factor) in the non-linear range. A possible explanation is that with less polymer in the compound, there is less material capable of flow. At higher strains, the additional CB leads to increased dissipation by breaking down the CB network and slipping or disentangling the molecules. In addition, the relaxation time increases with rising CB content, because there are altogether more connection points between CB particles with each other and with the polymer, which requires more time for degradation [KMG11].

The viscoelastic behaviour leads to various challenges in rubber extrusion. When entering a nozzle, the molecules are oriented by the deformation. Orientations that do not relax within the die recover on exit (entropy elasticity) and lead to swell of the extrudate. The swell increases with increasing deformation rate, decreasing residence time respectively die length, decreasing melt temperature, increasing first normal stress difference and increasing relaxation time [Hor11, PGL95, Vla81]. A higher CB content results in a lower swell rate for NR, since proportionally less polymer is present that reacts elastically to deformation [OTP+08]. In addition to swell, flow instabilities increasingly occur when a critical shear stress is exceeded at the exit, leading to surface defects [Gie78, Gle82, Mer05]. The influence of the stress level [Bar64, CDV03, Gle82], the viscoelastic properties of polymers [Leb81, Vla74] and the fillers [Hor11, Leb81] on the flow instabilities have frequently been the subject of research. The effects occurring at nozzle exit cannot be explained by shear thinning behaviour [ML79, SG01], but by normal stress elasticity [Hor11, Leb81, Mei75] and relaxation behaviour [Hor11, Mei75]. It has been shown for various polymers that melt fracture occurs as soon as the first normal stress difference is more than 4.5 times the shear stress (stability criterion) [Gle82]. The ratio of normal stress difference and shear stress is also called Weissenberg number (Wi) [Poo12]. The extent to which the criterion can be applied to highly filled mixtures has not been conclusively determined. In high-pressure capillary rheometer measurements, the following correlation for the normal stress difference and the entrance pressure loss p_E was found empirically (Equation 2.2) [PGL95].

$$N_1(\dot{\gamma}) = \frac{1}{5} \cdot p_E(\dot{\gamma}) \qquad (2.2)$$

Due to the elaborate measurements and complex calculations, the quantitative relationships between the occurrence of the flow phenomena and the complex stress state of highly reinforced rubbers have been insufficiently investigated so far. Therefore, swell and the occurrence of flow instabilities during rubber extrusion are often compensated for by an experimental-iterative design of the extrusion dies and adjustment of the processing temperature or throughput [OR15].

2.3 Experiments

2.3.1 Material

Three sulphur-crosslinking ethylene-propylene-diene rubber (EPDM) compounds from WAGU Gummitechnik GmbH, Warstein, were used in the rheological investigations. The compounds were selected on the basis of standardised formulations of ISO 4097 [NN20] and have a high practical relevance for profile extrusion. The base polymer Keltan 6950 C has a Mooney viscosity of ML (1+4) (100 °C) 65 MU. The compounds differed only in the amount of CB N550 (60, 80, 100 phr).

2.3.2 Rheological material characterisation

Rubber process analyser (RPA)

The following amplitude and frequency sweeps as well as relaxation tests were carried out with a D-RPA 3000 from MonTech Werkstoffprüfmaschinen GmbH, Buchen, Germany. According to the specifications of the RPA, the strain can be set in the range of ±0.14 % to ±5000 % (±0.001 ° to ±360 °). The frequency can be continuously adjusted in a range between 0.001 Hz and 100 Hz, while the test temperature can be varied between room temperature and 232 °C. The test chamber consists of two heated chamber halves, which are biconical in design and have engaging flanks for force transmission. The rheological measurements were carried out at 80, 100 and 120 °C on test specimens with a weight of 5 g ± 0.01 g punched out of the rubber sheet.

Amplitude Sweep

To characterise the Payne effect as well as to detect the linear viscoelastic range (LVE range), amplitude sweeps were performed at a constant frequency of f = 1 Hz while varying the shear deformation. The shear deformation was carried out during the measurement in 40 steps in a range between 0.01 and 500 % at the defined temperatures.

Frequency sweep

Frequency sweeps were performed to determine the elasticity of the rubber compounds as a function of CB content and temperature. Due to the intrinsic non-linearity of filled rubber compounds, the frequency sweep was not performed in the LVE range. The measurements were performed at a strain of 10 % without pre-shear. The frequency sweep was performed in 40 steps in a frequency range from 0.05 to 50 Hz.

Stress relaxation

Based on stress relaxation tests, the relaxation behaviour as a function of the CB content and the test temperature is analysed varying the applied shear strain. For this purpose, the final shear strain was varied in a range of 10 - 350 % in 10 amplitude steps. A new specimen was used for each measurement. The measurement was carried out according to ISO 13145 [NN12]. Before the actual measurement, preconditioning took place at the set temperature for six cycles at the strain to be tested, which corresponds to the later measurement. In the subsequent relaxation step, the strain was directly applied by the RPA and kept constant for 36 seconds (no oscillation). During the relaxation step, the shear stress was recorded as a function of time.

High pressure capillary rheometer (HPCR)

The viscosity measurements were carried out on the RG 2002 high-pressure capillary rheometer from Göttfert Werkstoff-Prüfmaschinen GmbH, Buchen, Germany, on round hole nozzles with a nozzle diameter of 2 mm and different length to diameter (L/D) ratios (5, 10, 15) at the specified temperatures (80, 100, 120 °C). The shear rates were varied in a range between 5 and 500 1/s in 9 steps and applied from low to high shear rate. An evaluation of the extruded strands provides information about the swell behaviour of the EPDM compounds as a function of temperature and shear rate.

2.4 Results

2.4.1 Strain sweep

Amplitude sweeps were carried out for all three compounds to describe the Payne effect. Figure 2.1 shows the curve of the storage modulus $G^{'}$ as a function of strain for the different CB contents at a temperature T = 100 °C. The linear viscoelastic range (LVE range) is recognisable for all three compounds below a strain of 0.3 %. The LVE range shifts towards smaller deformations with increasing CB content at every test temperature. The position of the LVE area is not affected by the temperature.

The modelling of the deformation-dependent Payne effect is done by means of a logarithmic regression function of the following form (Equation 2.3).

$$G^{'}(\gamma) = a \cdot \ln(\gamma) + b \qquad (2.3)$$

In this function, γ represents the applied deformation, while a defines the slope and b the height of the storage modulus $G^{'}$. The coefficient of determination of the model is above 94 % for all test points, which indicates suitability.

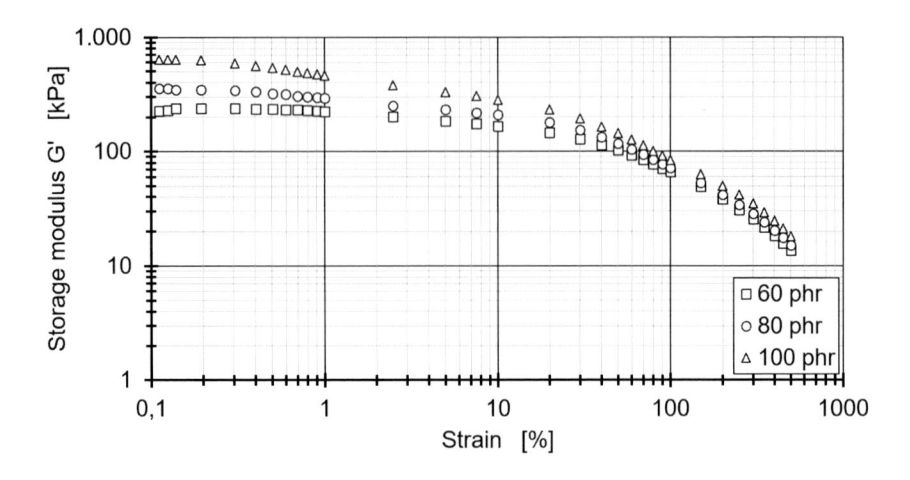

Figure 2.1: Curve of the storage modulus G' as a function of the strain depending on the CB content for 100 °C

A higher CB content leads to a stronger decrease of the slope (a) and to higher G' values (b) at all temperatures investigated (Figure 2.2). The Payne effect is more distinct with increasing CB content. Higher temperatures cause an opposite effect, since the mobility of the chain segments of the rubber polymer is increased with increasing material temperature. The lower the material temperature, the greater the influence of the CB content on the height of the G'-values (b) as well as on the course of the Payne effect (a). With increasing CB content, the temperature effect is more distinct. This is unexpected, as there is less polymer present that is more sensitive to temperature changes than CB.

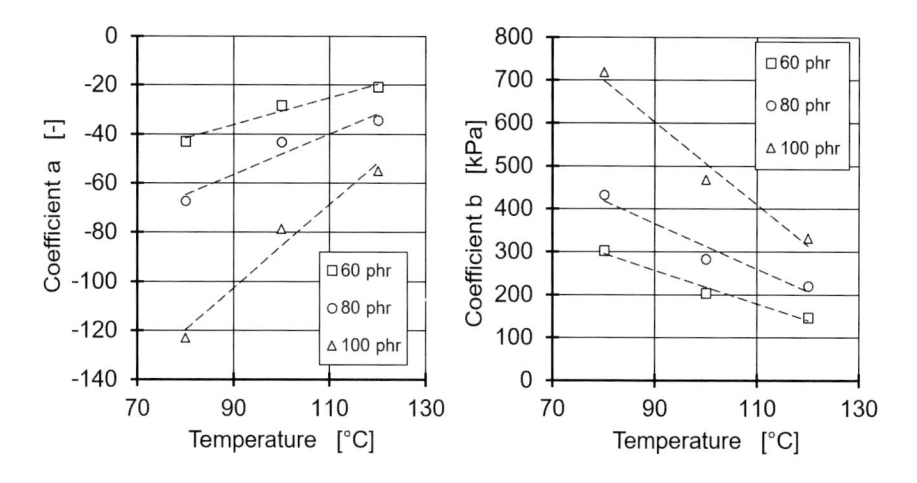

Figure 2.2: Coefficients a (left) and b (right) for modelling the Payne effect as a function of CB content and temperature

2.4.2 Frequency sweep

Figure 2.3 shows an example of the curve of the storage modulus G', the loss modulus G'' and the loss factor $tan(\delta)$ for 100 °C versus frequency for the different CB contents. At 80 and 120 °C, qualitatively almost identical curves were measured. Due to the reinforcing effect, G' increases at higher CB levels. G'' tends to increase with higher CB content, as more energy is dissipated by breaking down the CB networks. In contrast to G', G'' reaches a plateau as testing frequency increases. It is possible that the majority of the CB network has already been broken down during the measurement at the lower frequencies, so that the carbon black network contribution to dissipation is increasingly reduced. Due to the strong increase of G', the $tan(\delta)$ decreases with increasing frequency. While $tan(\delta)$ clearly decreases from 60 to 80 phr of CB, the comparison between 80 and 100 phr is ambiguous. Up to approx. 0.1 Hz, the $tan(\delta)$ of 100 phr is smaller and at higher frequencies it is larger than that of 80 phr. This correlation can also be seen at 80 and 120 °C. As the amount of CB in the mixture increases, the influence of the polymer and thus the proportion of viscous material behaviour decreases, so that a decrease in $tan(\delta)$ is expected. The increasing interactions of CB and polymer make it more difficult for the molecules to disentangle and also ensure less viscous behaviour. On the other hand, a higher CB content should lead to increasing dissipation, which is expected to increase G'' and $tan(\delta)$. Possibly, the influence of increasing dissipation in the polymeric region of the compound dominates at higher frequencies when the CB network is already broken down.

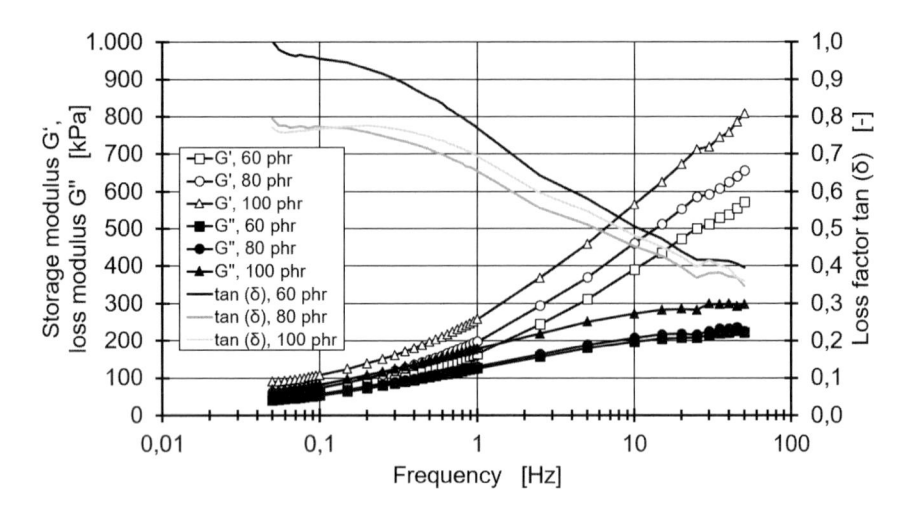

Figure 2.3: Frequency sweep of the 60, 80, 100 phr CB compound at 100 °C

Figure 2.4 shows an example of the frequency sweeps of the compound with 100 phr CB for the three temperatures 80, 100 and 120 °C. For the other mixtures, an analogous curve is obtained. As expected, a softening of the material is observed, i.e., $G^{'}$ drops significantly with increasing temperature. As with CB, the influence of temperature is also dependent on frequency. At low frequencies $tan(\delta)$ decreases with increasing temperature and from 0.3 Hz the opposite relationship exists. In principle, higher temperatures should favour flow and thus a viscous material response. $G^{''}$ decreases at higher temperatures. Due to a larger free volume, there is less friction at the same external load.

Overall, the material behaviour tends to become more elastic with increasing frequency, decreasing temperature and increasing CB content.

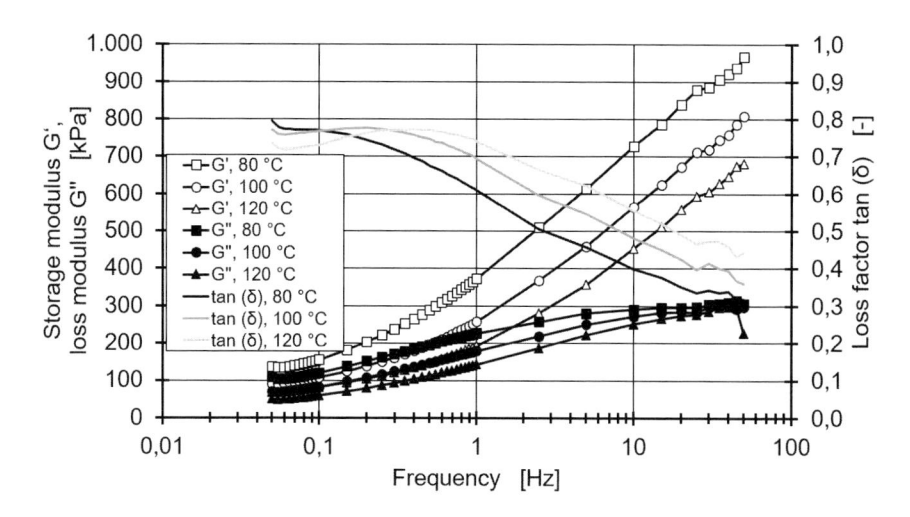

Figure 2.4: Frequency sweep of the 100 phr CB compound at 80, 100, 120 °C

2.4.3 Stress relaxation

The time-dependent shear modulus at 80 °C of the 80 phr compound is plotted against time in Figure 2.5. For the other compounds and measuring temperatures, the same qualitative correlations are obtained. For all measurement series, the curves can be well described with the following power function (Equation 2.4).

$$G(t) = G_1 \cdot t^{-r} \qquad (2.4)$$

In the double logarithmic representation, the function is a straight line. G_1 ($G(t = 1s)$) defines the height and $-r$ the slope. Only the measurements at 250 and 350 % strain deviate at very small times, as the RPA has to approach very high strains as quickly as possible.

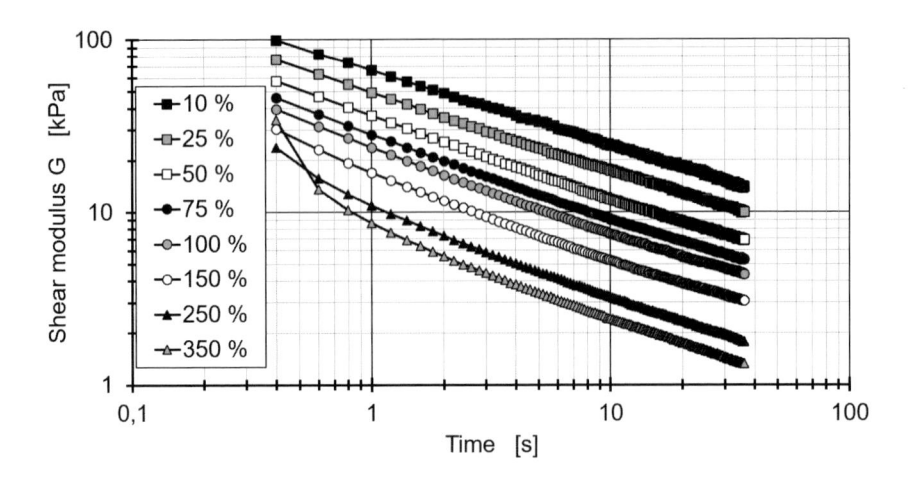

Figure 2.5: Time and shear deformation dependent module for 80 phr at 80 °C

The two model parameters G_1 and r are shown exemplarily for 25 and 75 % strain in Figure 2.5. In all measurement series, G_1 decreases with increasing temperature, increasing strain and decreasing CB content. This behaviour is explained by the fact that the material softens at higher temperatures, the CB networks are more strongly degraded during conditioning at high deformations and the added CB has a reinforcing effect.

The exponent r is a measure of the relaxation time. A larger r stands for a shorter relaxation time. In all measurement series, the relaxation time increases with increasing CB content (see Figure 2.6). The temperature influence is not clear in the measured data. There is a tendency the relaxation time to increase with increasing temperature. However, the example of 75 % strain (Figure 2.6, right) shows that isolated measurements deviate from this rule. At 250 and 350 %, there is no clear influence of temperature or CB content, especially on the relaxation time. The relaxation time also tends to be lower at higher strains.

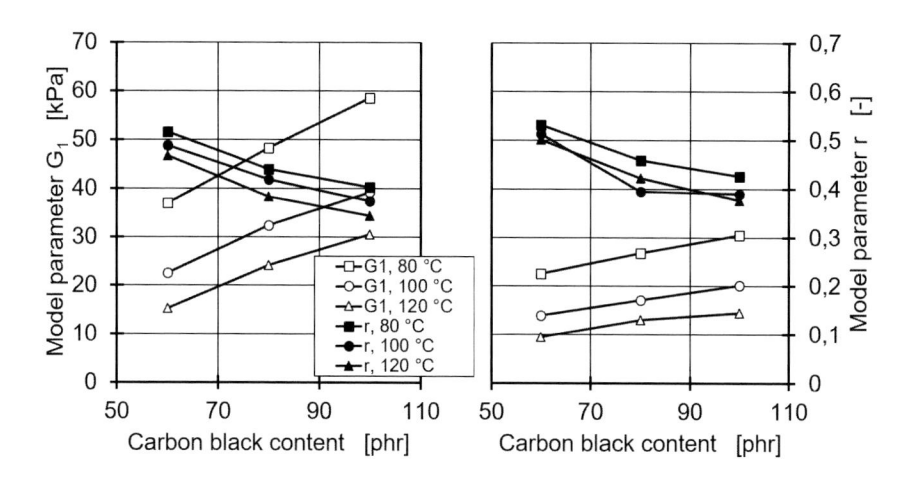

Figure 2.6: Model parameters of relaxation at 25 % strain (left) and 75 % strain (right)

2.4.4 High pressure capillary rheometer (HPCR)

When determining the viscosity using round hole nozzles, it is first necessary to correct the measured pressure for the entrance pressure loss. Figure 2.7 shows an example of the Ryder-Bagley plot [Sch18] for the EPDM compound with a CB content of 80 phr and a temperature of 100 °C. Since the slope of the regression line changes with increasing volume flow, wall sliding can be ruled out according to [Pel01]. With increasing nozzle length as well as shear rate the amount of pressure loss in the capillary is shifted to higher values.

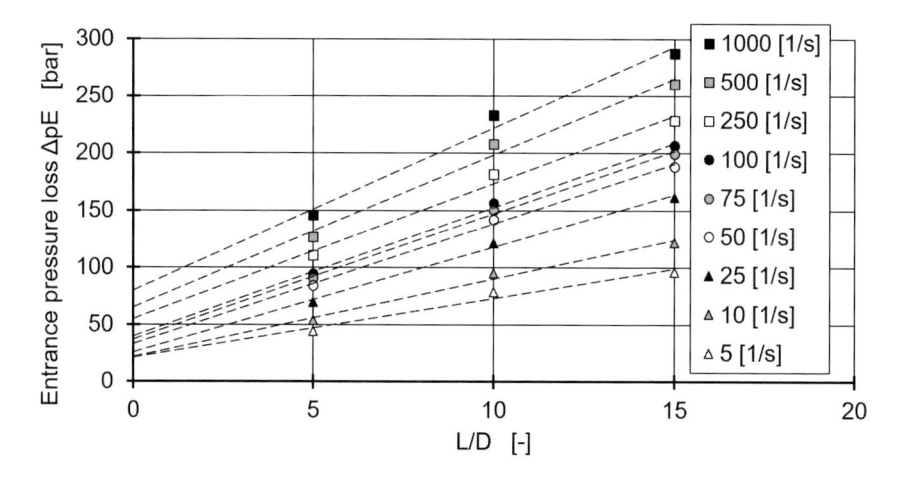

Figure 2.7: Ryder-Bagley plot for EPDM compound (80 phr CB) at 100 °C

The entrance pressure losses determined on the basis of the Ryder-Bagley plot form the basis for further evaluation with regard to the viscoelasticity of the rubber compounds. An increase in the CB content correlates with an increase in the inlet pressure loss, whereby from a CB content of 80 phr the entrance pressure loss is approximately constant at a given shear rate.

The correction method according to Weissenberg-Rabinowitsch [Sch18] must also be applied to convert the apparent shear rates assuming Newtonian material behaviour into true shear rates assuming shear thinning behaviour.

Using inlet pressure losses, the first normal stress difference N_1 is calculated (Equation 2.2). Figure 2.8 shows the flow curve after the corrections and N_1.

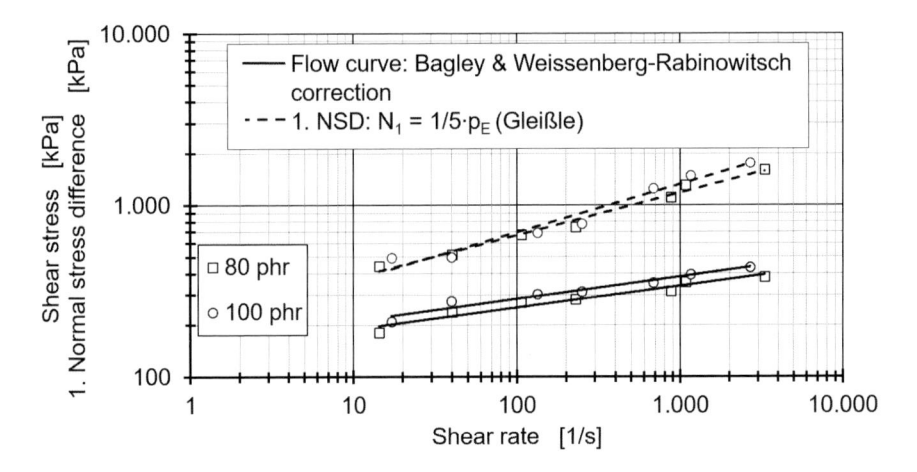

Figure 2.8: Curve of the shear rate-dependent flow curve (shear stress) and the first normal stress difference for 100 °C for a CB content of 80 and 100 phr

The compound with a CB content of 60 phr has not been considered in Figure 2.8, as the values are in a similar range to 80 phr. At constant CB content, the inlet pressure loss increases with decreasing temperature, since the mobility of the chain segments of the rubber polymer is reduced. This results in a higher absolute pressure and correspondingly a higher inlet pressure loss, which is why the shear stress as well as N_1 are shifted to higher values.

The frequency sweeps show that higher moduli are achieved with increasing CB content. This is also shown by the shear stresses in capillary flow and normal stress differences, as these are shifted to higher values with increasing CB content.

Swell behaviour

Following the HPCR measurements, the extrudate swell values were determined for all test points. Figure 2.9 shows the swell as a function of the residence time in the die for a temperature of 100 °C depending on the CB content. The residence time of the material in the die results from the geometry and the piston speed.

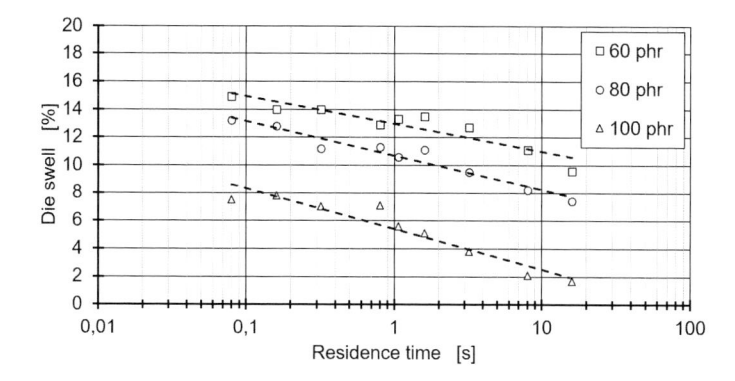

Figure 2.9: Influence of the CB content on the swell behaviour of the extrudates at a temperature T = 100 °C

The swell value increases with decreasing CB content. With decreasing CB content, there is correspondingly more polymer that has a marked tendency to swell. Furthermore, the tendency to swell increases with increasing shear rate (and thus with decreasing residence time). Since a higher shear rate results in a higher stress state in the material and at the same time the dwell time in the nozzle is shorter, the stresses in the nozzle can relax to a lesser extent than at a low shear rate.

The modelling of the swell value χ as a function of the residence time t_v is carried out by means of a logarithmic regression function of the following form.

$$\chi = c \cdot \ln(t_v) + d \tag{2.5}$$

In this function, c represents the slope and d the height of the swell.

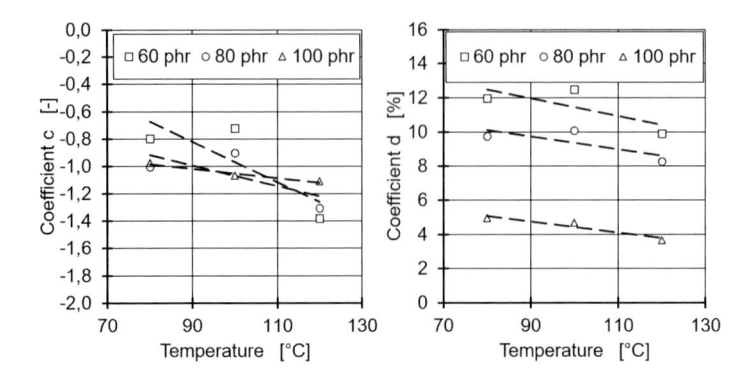

Figure 2.10: Coefficients c (left) and d (right) for modelling the swell behaviour as a function of the CB content as well as the temperature for the round hole nozzle $L/D = 20/2$

The extrudates with 60 phr CB were so soft that there are larger errors when measuring the swell values. Figure 2.10 shows that the temperature influence on the slope c of the regression function cannot be explained easily. The coefficient d and thus the swell tendency decrease with increasing CB content as well as increasing temperature. While the relaxation times increase with increasing CB content, the swell tendency decreases with higher CB content. This is contrary to expectation, as material that takes more time to relieve introduced stresses should show more swell at the same load. If there is less polymer in a given material, swell should decrease. The basic increase in normal stress differences that cause swell did not show up in the data when increasing the CB content.

Weissenberg number

Figure 2.11 shows the Weissenberg number versus the corrected shear rate as a function of the CB content for a temperature of 100 °C. In the shear rate range considered, the Weissenberg number is always below the critical limit of 4.5 for polymers (dashed line). Melt fracture did not occur in any of the measurements. Within these measurement data, the stability criterion can therefore be confirmed. However, the data present allow no improvement of the criterion.

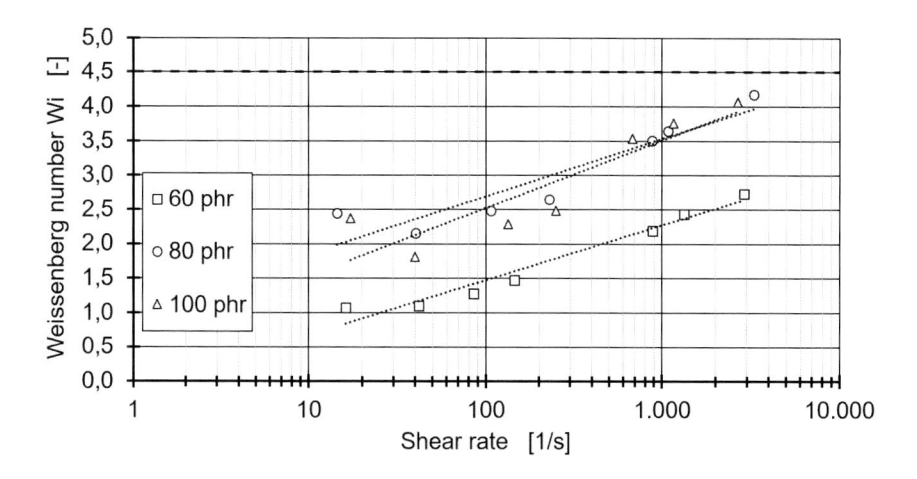

Figure 2.11: Influence of the CB content on the Weissenberg number as a function of the corrected shear rate at 100 °C

Furthermore, the Weissenberg number increases with increasing shear rate as well as CB content. The Weissenberg numbers for the rubber compounds with 80 and 100 phr show similar results, as the underlying entrance pressure losses and thus also the first normal stress difference between the two compounds do not differ significantly. With a constant CB content, the Weissenberg number is shifted to lower values with increasing temperature. Since an increase in temperature leads to a decrease in viscosity, the pressure in the capillary decreases at a constant shear rate. Accordingly, the prevailing shear stress and also the entrance pressure loss decrease and thus also the first normal stress difference or the Weissenberg number.

2.5 Conclusions and outlook

Laboratory analytical investigations were carried out by means of RPA and HPCR on EPDM rubber compounds with varying CB content from 60 to 100 phr to characterise the viscoelastic material behaviour. Table 2.1 summarises all the correlations found in this work. As expected, increasing the temperature has a comparable effect to lowering the CB content. Since swell decreases with increasing temperature and CB content, it is not possible to predict swell based on the parameters considered in this paper, such as the entrance pressure loss.

	a	b	G'	G''	$tan(\delta)$	G_1	r	relaxation time	p_E	swell
$CB \uparrow$	↓	↑	↑	↑	↑	↑	↓	↑	↑	↓
$T \uparrow$	↑	↓	↓	↓	O	↓	O	O	↓	↓

Table 2.1: Overview of all influencing variables

Frequency sweeps showed that only at very low frequencies the material becomes more elastic when more CB is present. Otherwise, the relationship is ambiguous and depends on the load. While at frequencies up to 0.3 Hz an increase in temperature makes the material more elastic, the opposite is true at higher frequencies. The relaxation times increase with increasing CB content. In addition, the relaxation time tends to be lower at higher strains and higher at higher temperatures. But only the influence of CB was unambiguous in this respect in the measurement data.

The increasing CB content reduces the swell, which seems to contradict the expectations from the relaxation tests, as a higher swell value is expected with the same residence time and longer relaxation time. A direct conclusion from relaxation tests to the swell behaviour is therefore not possible. Since no melt fracture occurred in any of the measurement series, the stability criterion could not be confirmed or disproved.

Based on the laboratory measurements, the HPCR measurements will be reproduced in the simulation environment openFOAM, whereby the underlying models for shear and extensional viscosity will be calibrated. On the basis of the flow velocity distribution and the associated residence time, it is to be shown how the swell tendency of rubber compounds can be determined using empirical methods. Further investigations are to be carried out to determine the conditions under which melt fracture occurs in CB-filled rubber compounds.

Acknowledgements

The research project (21531 N) of the Forschungsvereinigung Kunststoffverarbeitung has been sponsored as part of the "Industrielle Gemeinschaftsforschung und -entwicklung (IGF)" by the German Bundesministerium für Wirtschaft und Klimaschutz (BMWi) due to an enactment of the German Bundestag through the AiF. The realisation of these investigations was made possible by the supply of test material by WAGU Gummitechnik GmbH Warstein, Germany. We would like to extend our thanks to all organisations mentioned.

References

[Bar64]   BARTOS, D.: Fracture of polymer melts at high Shear stress. *Journal of applied physics* 35 (1964) 9, p. 2767

[CDV03]   COMBEAUD, D.; DEMAY, Y.; VERGNES, B.: Etude experimentale du defaut helicoïdal d'un polystyrène en ecoulement. *Rheologie* 4 (2003), p. 50–57

[Gie78]   GIESEKUS, H.: Sekundärströmungen und Strömungsinstabilitäten. In: *Praktische Rheologie der Kunststoffe*. Düsseldorf: VDI Verlag, 1978

[Gle82]   GLEISSLE, W.: Stresses in polymer melts at the beginning of flow instabilities (melt fracture) in cylindrical capillaries. *Rheologica Acta* 21 (1982), p. 484–487

[Hor11]   HORNIG, R.: Normalspannungselastizität bei FKM-Mischungen – Grundlegende rheologische und mischprozesstechnische Betrachtungen. Teil 1: Rheologische Untersuchungen an Elastomermischungen mit unterschiedlichen Rußkonzentration und Rußaktivität. *GAK Gummi Fasern Kunststoffe* 64 (2011) 12, p. 732–743

[KMG11]   KARRABI, M.; MOHAMMADIAN-GEZAZ, S.: The effects of carbon black-based interactions on the linear and non-linear viscoelasticity of uncured and cured SBR compounds. *Iranian Polymer Journal* 20 (2011) 1, p. 15–27

[LB00]   LEBLANC, J.; BARRES, C.: Recent developments in shear rheometry of uncured rubber compounds. *Polymer Testing* 19 (2000), p. 177–191

[Leb81]   LEBLANC, J.: Factors affecting the extrudate swell and melt fracture phenomena of rubber compounds. *Rubber chemistry and technology* 54 (1981) 5, p. 905–929

[Leb02]   LEBLANC, J.: Rubber-filler interactions and rheological properties in filled compounds. *Progress om Polymer Science* 27 (2002), p. 627–687

[Mei75]   MEISSNER, J.: Neue Messmöglichkeiten mit einem zur Untersuchung von Kunststoff-Schmelzen geeigneten modifizierten Weissenberg-Rheogoniometer. *Rheologica Acta* 14 (1975), p. 201–218

[Mer05]   MERTEN, A.: *Untersuchungen zu Fließinstabilitäten bei der Extrusion von Polymeren mit der Laser-Doppler-Anemometrie*. Universität Nürnberg-Erlangen, dissertation, 2005 – supervisor: H. Münstedt

[Mic09]   MICHAELI, W.: *Extrusionswerkzeuge für Kunststoffe und Kautschuk - Bauarten, Gestaltung und Berechnungsmöglichkeiten*. München: Carl Hanser Verlag, 2009

[ML79]   MÜNSTEDT, H.; LAUN, H.: Elongational behaviour of a low density polyethylene melt. Part II: transient behaviour in constant stretching rate and tensile creep experiments. comparison with shear data. temperature dependence of the elongational properties. *Rheologica Acta* 18 (1979), p. 492–504

[NN12]   N.N.: *DIN ISO 13145: Rubber – Determination of viscosity and stress relaxation using a rotorless sealed shear rheometer*. Berlin: Beuth Verlag, 2012

[NN20]   N.N.: *DIN ISO 4097: Rubber, ethylene-propylene.diene (EPDM) – Evaluation procedure*. Berlin: Beuth Verlag, 2020

[OR15]   OSSWALD, T.; RUDOLPH, N.: *Polymer Rheology: Fundamentals and Applications*. München and Wien: Carl Hanser Verlag, 2015

[Ost25]   OSTWALD, W.: Ueber die Geschwindigkeitsfunktion der Viskosität disperser Systeme. *Kolloid Zeitschrift* 36 (1925), p. 99–117

[OTP+08] OMNÈS, B.; THUILLER, S.; PILVIN, P.; GROHENS, Y.; GILLETS, S.: Effective properties of carbon black filled natural rubber: Experiments and modeling. Composites Part A: Applied Science and Manufacturing 39 (2008) 7, p. 1141–1149

[Pay62] PAYNE, A.: The dynamic properties of carbon black loaded natural rubber vulcanizates. part II. Journal of Applied Polymer Science 6 (1962), p. 368–372

[Pel01] PELZ, P.: Rheologie der Kautschukmischungen. In: Hempel, J. (Editor): Elastomere Werkstoffe. Weinheim: Freudenberg Forschungsdienste, 2001

[PGL95] PAHL, M.; GLEISSLE, W.; LAUN, H.: Praktische Rheologie der Kunststoffe und Elastomere. Düsseldorf: VDI-Verlag, 1995

[Poo12] POOLE, R.: The Deborah and Weissenberg numbers. The British Society of Rheology, Rheology Bulletin 53 (2012) 2, p. 32–39

[RH21] ROBERTSON, C.; HARDMAN, N.: Nature of carbon black reinforcement of rubber: Perspective on the original polymer nanocomposite. Polymers 13 (2021) 4, p. 538

[RS13] RÖTHEMEYER, F.; SOMMER, F.: Kautschuk Technologie. München and Wien: Carl Hanser Verlag, 2013

[Sch18] SCHRÖDER, F.: Rheologie der Kunststoffe. Theorie und Praxis. München: Carl Hanser Verlag, 2018

[SG01] SARKAR, D.; GUPTA, M.: Further investigation of the effect of elongational viscosity on entrance flow. Journal of reinforced plastics and composites 20 (2001) 17, p. 1473–1484

[Vla74] VLACHOPOULOS, J.: Die swell and melt fracture: effects of molecular weight distribution. Rheologica Acta 13 (1974), p. 223–227

[Vla81] VLACHOPOULOS, J.: Extrudate Swell in Polymers. Reviews on the Deformation Behavior of Materials 3 (1981) 4, p. 219–248

[Wae23] WAELE, A.: Viscometry and plastometry. Journal of the Oil Colour Chemists' Association 6 (1923), p. 33–88

[WSSO10] WONGWITTHAYAKOOL, P.; SIRISINHA, C.; SAE-OUI, P.: Cure and Viscoelastic Properties of HNBR Effects of Carbon Black Loading and Characteristics. KGK rubberpoint 63 (2010) 11, p. 506–512

Symbols

Symbol	Unit	Description
A	mm^2	cross-sectional area
a, b, c, d, r	−	parameter of the regression functions
δ	°	loss angle
F_{normal}	N	normal force
G	MPa	shear modulus
G'	MPa	storage modulus
G''	MPa	loss modulus
NSD	N/mm^2	normal stress difference
N_1	N/mm^2	first normal stress difference
p_E	bar	entrance pressure loss
τ_{ij}	N/mm^2	shear stresses
$tan(\delta)$	−	loss factor
T	°C	temperature
t	s	time
t_v	s	residence time
χ	%	swell value
γ	%	shear
$\dot{\gamma}$	$1/s$	shear rate
W_i	−	Weissenberg number

Abbreviations

Notation	Description
CB	carbon black
EPDM	ethylene-propylene-diene rubber
HPCR	high pressure capillary rheometer
L/D	length to diameter ratio
LVE	linear viscoelastic range
HNBR	hydrogenated acrylonitrile butadiene rubber
MU	Mooney unit
NR	natural rubber
phr	parts per hundred of rubber
RPA	Rubber Process Analyzer
SBR	styrene-butadiene rubber

3 Influence of process parameters on mixing time and properties of rubber compounds

Ch. Hopmann[1], M. Kostka[1], A. Limper[1], M. Schön[1]
[1]Institute for Plastics Processing (IKV) at RWTH Aachen University

Abstract

In the internal mixer, ingredients such as rubber, fillers, plasticisers and chemicals are processed to a compound. A consistent quality of the compound is elementary for the production of high-quality rubber products.

In order to analyse the processes that occur during compound production in the internal mixer, the mixing process can be subdivided into different mixing phases with varying mixing tasks. The time required until the completion of the mixing task is generally unknown. As an example, the carbon black incorporation and dispersion during the upside-down compound production (feeding of the carbon black and then the polymer) of an ethylene-propylene compound will be investigated.

The carbon black incorporation and dispersion are analysed as a function of the mixing chamber and rotor temperature, the rotor speed, the filling level and the type of carbon black used. The experimental investigations show that both, the process variables and recipe changes, have an effect on the carbon black incorporation and dispersion. The following findings can be noted:

- With increasing filling level, the time required for a complete carbon black incorporation rises.
- The higher the rotor speed, the faster the carbon black can be incorporated.
- If the mixer temperature is higher, the time required for carbon black incorporation decreases.
- The standard carbon black types N330 and N550 show only a marginal difference in terms of time required for carbon black incorporation.
- If higher temperatures are selected for the mixing chamber and rotor temperature, the dispersion quality decreases.
- If the rotor speed is elevated, the dispersion quality increases.

3.1 Introduction

Rubber compounds are produced discontinuously on roll mills or in internal mixers. Nowadays, compound production is mostly carried out in internal mixers due to the lower dust exposure, better reproducibility and automatability [RS13]. Basically, the rubber industry differentiates between tyres and technical elastomer products [Abt19]. The latter are used in numerous different sectors such as the automotive industry or medical technology.

Only by mixing rubber, fillers, plasticisers and chemicals the multifaceted properties of rubber products can be achieved. Up to 10,000 different compound formulations are used in the rubber industry, whose compositions are based on more than 30 types of rubber [Dic14, LBG89]. Rubber polymers composed of the same monomers have a large number of different types available on the market with their own specifications in terms of molecular weight and molecular weight distribution. In addition to the different polymers, 42 different standard types of carbon black exist [Dic14]. The ingredients are fed in different forms (e.g. bales, powder, granulate, liquid). Besides, the process control in the internal mixer influences the production of rubber compounds (e. g. [Hab16, HKB19, HKLB19, HKLF21a, HKLF21b, LBG89]). The mixing process is limited by boundary conditions such as the power limit of the internal mixer and a maximum allowed compound temperature to prevent premature cross-linking. Each recipe has its own mixing specification, which is largely developed empirically. In addition to the necessary expertise in compound production, the time required for empirical process development is high [Hab16]. To optimise the existing mixing processes, no extensive modifications of the existing plant technologies or large investments in new plant technology are required. Instead, large savings are possible through the systematic development or optimisation of the mixing processes.

In order to analyse the mixing process, it can be divided into different phases. Basically, a differentiation can be made between masterbatch and final mixing [RS13]. A further division of the mixing process into different mixing phases depends on the polymer to be processed. If, for example, natural rubber is being processed in a conventional mixing process, the mixing phases feeding of the polymer, mastication, addition of the plasticiser and fillers, incorporation of the recipe ingredients and compound homogenisation can be recognised. Subsequently cross-linking chemicals are added and distributed in the compound [HKLF21a]. When synthetic polymers are processed, mastication is often not necessary [Abt19]. The mixing process can also be carried out "upside down". In this case, the fillers (e.g. carbon black) are fed into the mixing chamber at the beginning of the mixing process together with plasticisers. Subsequently, the rubber follows [DW01]. This is followed by the same mixing phases as in a conventional mixing process. A possible advantage of "upside down" compounding is the reduction in mixing time. When considering the mixing phases separately from each other, it is essential to note that the final mixing state of one mixing phase is the input mixing state of the subsequent mixing phase.

Relevant quality criteria can be developed to evaluate the performance of the mixing phases. The feeding should be as short as possible and the polymer and possibly other components ("upside down") should be sufficiently well distributed in the mixing chamber. During mastication, the lowest possible energy input should reduce the polymer viscosity so that further processing is facilitated. Furthermore, a homogeneous mastication quality with regard to molecular weight and temperature should be achieved. During incorporation, the fillers (e.g. carbon black) must be completely incorporated in order to achieve a sufficiently good dispersion and distribution [HKLF21a].

The goal of the investigations was to find out to which degree mixing time can be reduced if the filling level as well as the rotor and mixing chamber temperature are varied.

3.2 Carbon black incorporation and dispersion

During carbon black incorporation, the carbon black is added into the rubber matrix. Subsequently, the dispersion and distribution of the carbon black in the polymer begins. Cotton further divides the carbon black incorporation into two phases. First, the carbon black agglomerates are enclosed by the polymer. The spaces between the agglomerates are filled with air. Then the rubber is pressed into the air-filled interstices and the contained air escapes [Cot85]. In addition to the space change processes, wall slip effects affect macroscopic flow processes. The cause of this (apart from the mixing chamber and rotor temperature) is the free carbon black which acts as a lubricant [LBG89]. The end of carbon black incorporation is described in the literature as being signified by the power reaching a maximum after carbon black addition [LBG89, Lim12]. Investigations carried out with the natural rubber grade SMR CV 60 from Resinex Deutschland GmbH, Zwingenberg, Germany, and the standard carbon black grade N550 from Cabot GmbH, Rheinfelden, Germany, showed that carbon black has been completely incorporated when the ram reaches its final position after the addition of carbon black [HKB19].

The dispersion phase ensures that carbon black agglomerates are broken up [HSM84]. Shear and elongation deformations during the phase cause the carbon black agglomerates to be destroyed [Sun92]. The applied shear stress is proportional to the compound viscosity and shear rate [HSM84]. In the case of carbon black dispersion, the compound is already heated up and a reduction in the viscosities due to rising temperatures should be avoided. Otherwise, only low stresses can be introduced, which are insufficient for filler dispersion. The homogenisation of the mixture takes place in the subsequent carbon black distribution phase [HSM84]. The carbon black dispersion and distribution is decisive for the properties of the rubber products [Lee85, Per84]. The properties are largely determined by the distribution and size of the carbon black agglomerates and the type and quantity of carbon black used.

3.3 Experimental analysis of the carbon black incorporation

An experimental analysis of carbon black incorporation is carried out in an intermeshing internal mixer of the type GK 1.5 E from Harburg-Freudenberger Maschinenbau GmbH, Freudenberg, which is equipped with PES 5 rotors and a pressure-controlled ram. The polymer used is Keltan 7752C from Arlanxeo Netherlands B.V., Geelen, Netherlands. In order to investigate the influence of the used carbon black on the carbon black incorporation, the ASTM standardised carbon blacks N550 and N330 from Cabot GmbH, Rheinfelden, were used. A carbon black content of 50 phr was used. The process of carbon black incorporation also depends significantly on the filling level used. Investigations were carried out with the following degrees of filling: 65 %, 70 %, 75 %, 80 %, 85 %.

The mixing process was carried out "upside down". At the beginning of the mixing process, the polymer and the carbon black were added together. The mixing time of all mixtures was 350 s. The batch temperature was determined with a thermometer after the end of the mixing process. Previous investigations have shown that the batch temperature rises sharply at the beginning of the mixing process. After reaching the temperature equilibrium between the batch temperature and the rotor and mixing chamber temperatures, the batch temperature no longer changes. The aim of the experimental investigation was to examine the influence of filling level, the rotor speed and the temperature control of the mixing chamber and rotors on the carbon black incorporation. The process parameters used can be found in Table 3.1.

Parameter	Unit	Value	
Rotor temperature (TR)	[°C]	30	70
Mixing chamber temperature (TM)	[°C]	30	70
Rotor Speed	[rpm]	40	80

Table 3.1: Process parameters used

3.4 Analysis of the influence of process control on the carbon black incorporation

In the experimental investigations, carbon black incorporation was considered to be completed when the ram reached its end position. The investigations have shown that the time required for the ram to reach the end position depends on the mixing chamber and rotor temperature, the filling level and the rotor speed.

If the mixing chamber and rotor temperature is 30 °C, there is a strong dependence on the filling level when using carbon black type N330 at a rotor speed of 40 rpm (Figure 3.1).

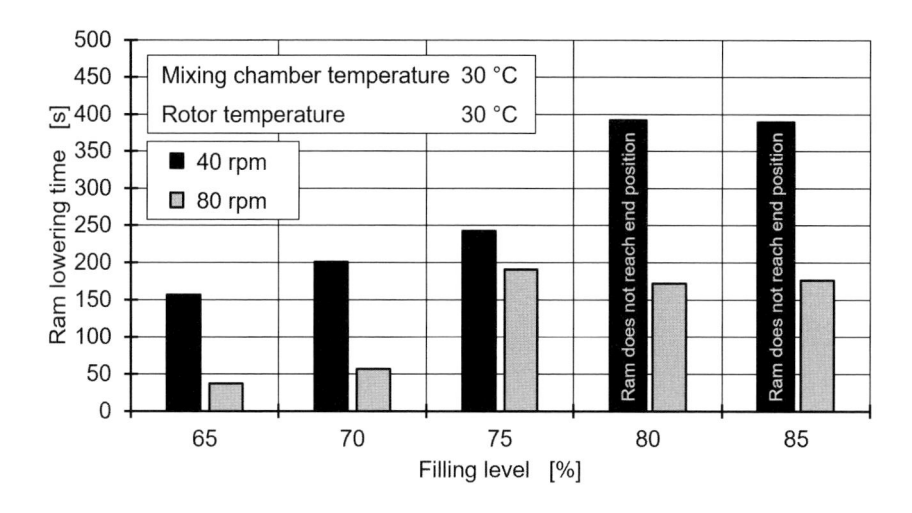

Figure 3.1: Influence of filling level and rotor speed on carbon black incorporation at a mixing chamber and rotor temperature of 30 °C

At a filling level of 65 %, the mixing time until the ram has reached its end position is 157 s. The remaining mixing time is by this reduced to an unacceptable minimum. As a benchmark, 30-45 s are given in the literature for the ram to reach its final position after the addition of filler. Longer mixing times until the ram reaches its end position can only be accepted if the compounds are highly filled [Lim12]. If the process is carried out at a rotor and mixing chamber temperature of 30 °C at a rotor speed of 40 rpm, the internal mixer is already overfilled at a filling level of 65 %. If the filling level is raised further, the time required to reach the final position of the ram increases. If the filling level is 70 %, the required mixing time increases by 27.8 % to 200.8 s. If the filling level is increased by another 5 %, the ram no longer reaches its final position and remains in the chute, 2 mm above the mixing chamber. Consequently, there are still components in the chute that do not participate in the incorporaton process, so that loose carbon black leaves the mixing chamber. If the filling level is raised again, the ram does not reach its final position even after a mixing time of 330 s. The mixing chamber is then filled with free carbon black. At a filling level of 85 %, the compound temperature determined after the mixture left the mixing chamber was approximately 103 °C. This means that the compound temperature is relatively low. If the rotor speed is doubled, the power introduced into the compound also increases by a factor of two. In addition, the rotor blades reach the end of the feeding chute more quickly after one revolution. This means that, compared to a rotor speed of 40 rpm, more carbon black per second can be transported into the mixing chamber. If the filling level is 65% at a rotor speed of 80 rpm, carbon black incorporation is completed after 37 s. Once carbon black incorporation is complete, the compound temperature is 127 °C. If the filling level is increased by 5 %, the time required for carbon black incorporation increases by 19 s to 56 s. If investigations are carried out at filling levels of 75 %, 80 % or 85 %, it becomes clear that the time required for carbon black incorporation is more than three times as long as at a filling level of 70 %, but does not increase further with increasing filling level. One reason for this could be that the compound temperature, despite a similar mixing time, increases significantly with increasing filler content. At a filling level of 75 %, the compound temperature is 136 °C. This rises to 148 °C at a filling level of 80 %. If the filling level is 85 %, a compound temperature of 164 °C could be determined. In addition to the relatively long mixing time due to the overfilling of the internal mixer, the compound temperature also shows that the mixing process with a filling level above 70 % is not practicable at a rotor and mixing chamber temperature of 30 °C each, as pre-crosslinking would occur in the internal mixer after the addition of the crosslinking chemicals.

If the rotor temperature is increased to 70 °C and the temperature of the mixing chamber is left at 30 °C, this has a substantial effect on the carbon black incorporation (Figure 3.2).

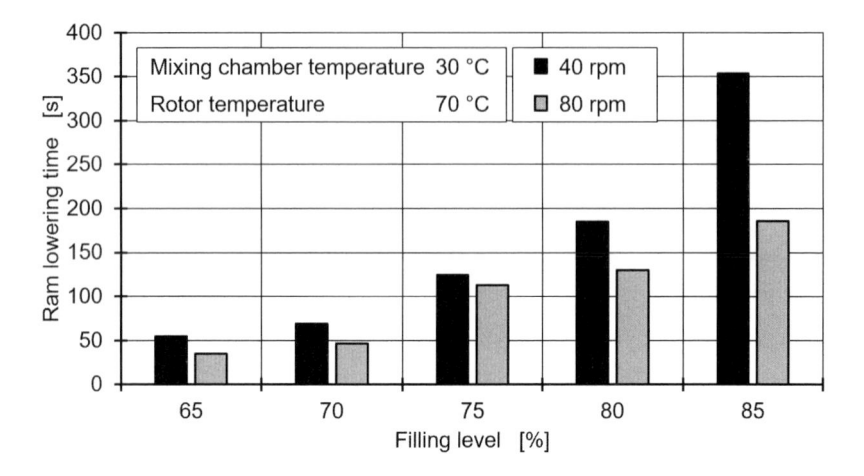

Figure 3.2: Influence of filling level and rotor speed on carbon black incorporation at a mixing chamber temperature of 30 °C and a rotor temperature of 70 °C

At a filling level of 65 %, a mixing time of 55 s is necessary to incorporate the carbon black. An increase in the filling level at a rotor speed of 40 rpm leads to an exponential increase in the time required for incorporation. When using a rotor speed of 80 rpm a linear relationship can be observed. At a filling level of 65 %, the ram needs 35 s to reach its end position. If the filling level is increased to 85 %, the carbon black is incorporated after 186 seconds. Consequently, it can be concluded that the filling level at a lower rotor speed has a greater influence on the incorporation of carbon black. A possible reason for this is that a higher rotor speed leads to a higher number of revolutions. This means that more novel surfaces of the compound are formed and the carbon black can be incorporated quicker. In addition, more place changing processes take place.

If the mixing chamber temperature is raised to 70 °C and the rotor is tempered to 30 °C, differences in the incorporation time are visible compared to the previous investigations (Figure 3.3).

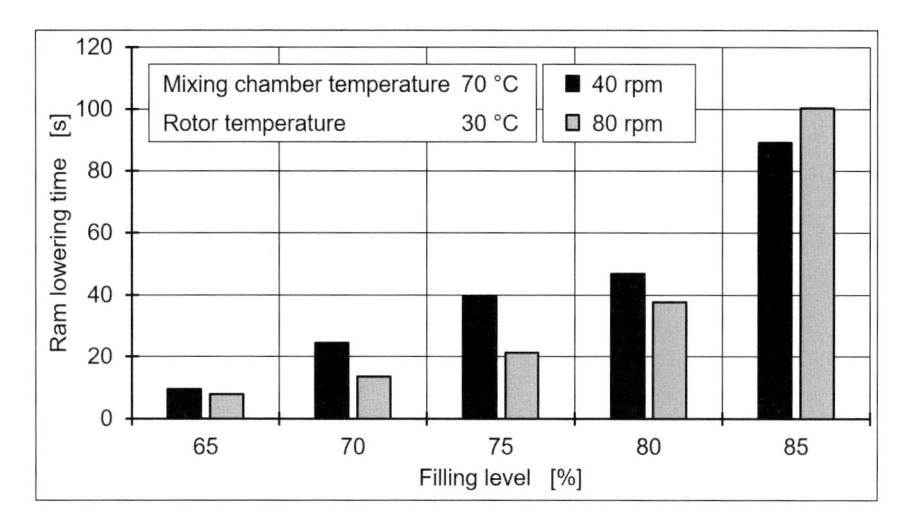

Figure 3.3: Influence of filling level and rotor speed on carbon black incorporation at a mixing chamber temperature of 70 °C and a rotor temperature of 30 °C

If, in contrast to the rotor, the mixing chamber is heated up, the compound temperature is also higher and thus the compound viscosity is lower. This can be explained by the larger heat-transferring surface of the mixing chamber compared to the rotors. In the case of the used mixer, the surface of the mixing chamber is approximately 4 times larger than that of the rotors. If the compound viscosity drops due to the higher compound temperatures, the carbon black can be incorporated quicker. At a rotor speed of 40 rpm, a mixing time of 40 s is necessary for the ram to reach its end position at a filling level of 75 %. In comparison, the incorporation time at a mixing chamber and rotor temperature of 30 °C each at a filling degree of 75 % and a rotor speed of 40 rpm is 243 s. Compared to the temperature control vice versa, the time required for incorporation increases exponentially with the filling level for both a rotor speed of 40 rpm and a rotor speed of 80 rpm.

Carrying out the investigations at a rotor and mixing chamber temperature of 70 °C each, it becomes clear that the time required for the ram to reach the final position is significantly shorter. If the filling level is 80 %, the time required for carbon black incorporation amounts to 32 s (Figure 3.4). Increasing the filling level to 85 % leads to a carbon black incorporation time of 41 s. By doubling the rotor speed to 80 rpm, the time required for the ram to reach its end position can be reduced. At a filling level of 80 % this is 20 s and increases to 25 s at a filling level of 85 %.

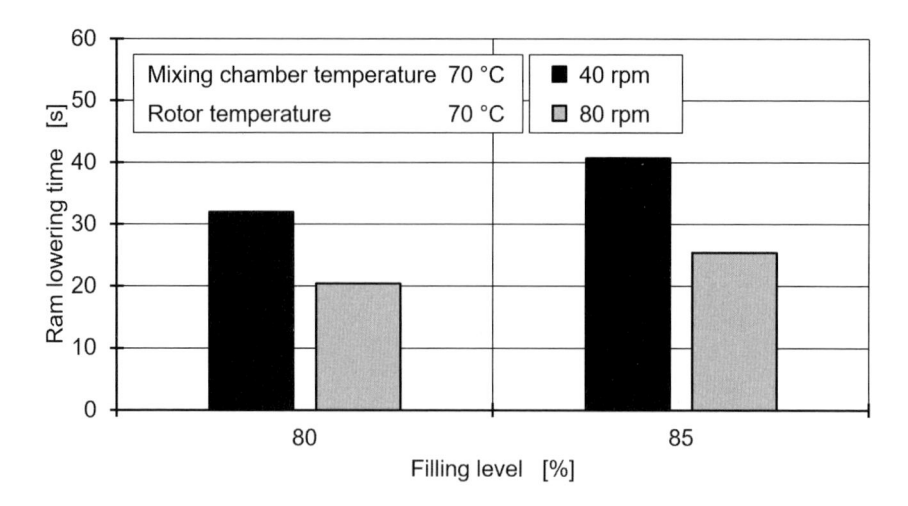

```
        60
             ┌─────────────────────────────────┬──────────────┐
        50   │ Mixing chamber temperature 70 °C│ ■ 40 rpm     │
  [s]        │ Rotor temperature          70 °C│ □ 80 rpm     │
             └─────────────────────────────────┴──────────────┘
        40
 time
        30
 lowering
        20
 Ram
        10

         0
              80                              85
                     Filling level  [%]
```

Figure 3.4: Influence of filling level and rotor speed on carbon black incorporation at a mixing chamber and rotor temperature of 70 °C

In general, it seems to be suitable to set the mixer temperature as high as possible, as the mixing time can be shortened due to the reduced incorporation time and a high filling level is possible. However, other boundary conditions such as a sufficiently good carbon black dispersion (Figure 3.8) must also be taken into account. The batch ejection temperatures using a rotor and mixing chamber temperature of 70 °C each, differ after the same mixing time depending on the used filling level and rotor speed (Table 3.2).

Rotor speed [rpm]	Filling level [%]	Batch drop temperature [°C]
40	80	131
	85	139
80	80	171
	85	173

Table 3.2: Batch drop temperatures using a rotor and mixing chamber temperature of 70 °C while processing

It can be seen that the compound temperatures at a filling level of 80 % and 85 % are quite high. Consequently, it must be assumed that the dispersion quality is insufficient and that pre-crosslinking of the compound in the mixing chamber will occur, if the crosslinking-chemicals will be added.

3.5 Analysis of the influence of the carbon black grade on the carbon black incorporation

If carbon black type N550 is used instead of carbon black type N330, it can be seen that the influence of the carbon black type on the carbon black incorporation time is almost negligible if the rotor and mixing chamber temperature are both 70 °C (Figure 3.5).

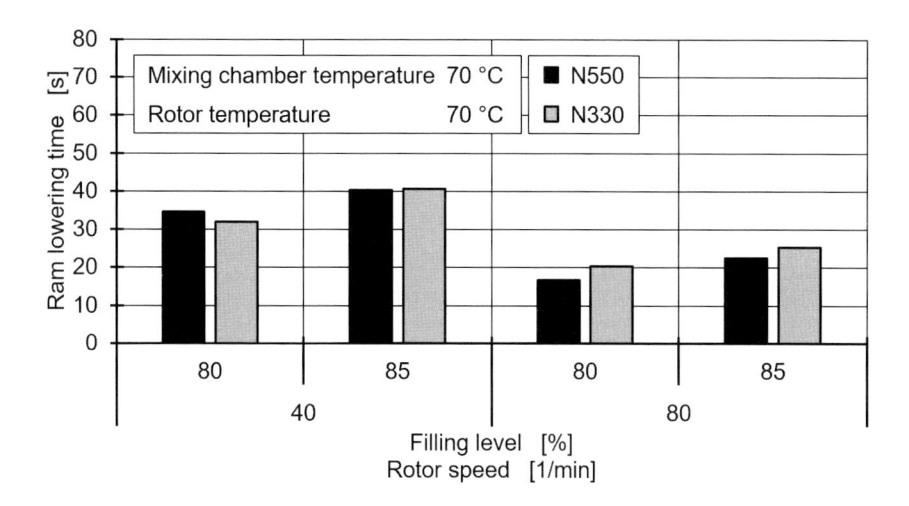

Figure 3.5: Influence of carbon black grade, filling level and rotor speed on carbon black incorporation

The standard carbon black N330 is a highly active carbon black. In contrast, carbon black type N550 is a medium activity carbon black that is usually easier to process because different carbon black types have varying influences on the compound viscosity [RS13]. Furthermore, it becomes clear that the correlations made with regard to the rotor speed and the filling level are independent of the carbon black grade. One possible explanation could be that the compound temperatures after complete carbon black incorporation are independent of the type of carbon black used (Table 3.3). It can be concluded that a rotor and mixing chamber temperature of 70 °C each at a rotor speed of 80 rpm are too high, as the high compound temperature would cause the compound to crosslink in the internal mixer after addition of the crosslinking system.

Rotor speed [rpm]	Filling level [%]	Batch drop temperature [°C]	
		N550	N330
40	80	130	130
	85	136	139
80	80	174	171
	85	174	173

Table 3.3: Batch drop temperatures using a rotor and mixing chamber temperature of 70 °C and different carbon black types

3.6 Analysis of the influence of process control in the internal mixer on the carbon black dispersion

The carbon black dispersion of batch containing the carbon black type N330 was investigated. The filling level during the mixing process was 80 %. After reaching the particular carbon black incorporation time, the samples were mixed for another 50 s before the crosslinking system was added and then mixed in for another 60 s. To investigate the carbon black dispersion, the mechanical properties (e.g. breaking stress and elongation, tear resistance, etc.) were determined. The measurement was carried out according to DIN 53504 at room temperature and a testing speed of 200 mm/min. Test specimens of type S2 were used, which were punched out of a 2 mm thick sample [NN17].

It can be seen that the process parameters during the production of the rubber compound in the internal mixer have an influence on the breaking stress. If the rubber compounds are produced with a cold mixing chamber and a cold rotor, the breaking stress is highest (Figure 3.6).

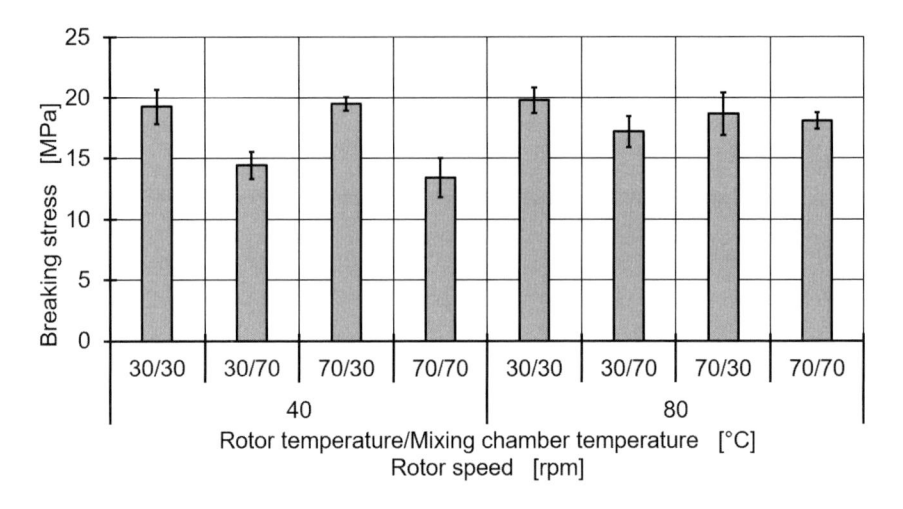

Figure 3.6: Influence of tempering and rotor speed of the tensile testing samples

In contrast, the compounds produced with a rotor temperature of 30 °C and a mixing chamber temperature of 70 °C as well as the compounds produced with a rotor and mixing chamber temperature of 70 °C each have a lower breaking stress. A possible explanation for this are different adhesion and sliding effects that occur depending on the rotor and mixing chamber temperatures used. In order to analyse this, the average power introduced was evaluated (Figure 3.7). The power introduced is compared for the cause that either the temperature of the rotors or the mixing chamber is increased. The average power input at a rotor and mixing chamber temperature of 30 °C each is 5 kW at a rotor speed of 40 rpm. If the mixing chamber temperature is increased to 70 °C, the average power input increases by 21 %. If, on the other hand, the rotor temperature is increased to 70 °C, the average power input drops by 14 % compared to the cold rotor. If the rotor is cold, the compound adheres to it and is transported through the gap between the rotors and the gap between the mixing chamber and the rotors with each rotor revolution. If, on the other hand, the rotor temperature is set to 70 °C, the mixture partially slides off the rotor surface. If the rotor speed is doubled, the same effects are visible. However, the size of the effects decreases. Both an increase in the rotor temperature and an increase in the mixing chamber temperature result in a detectable change in the average power.

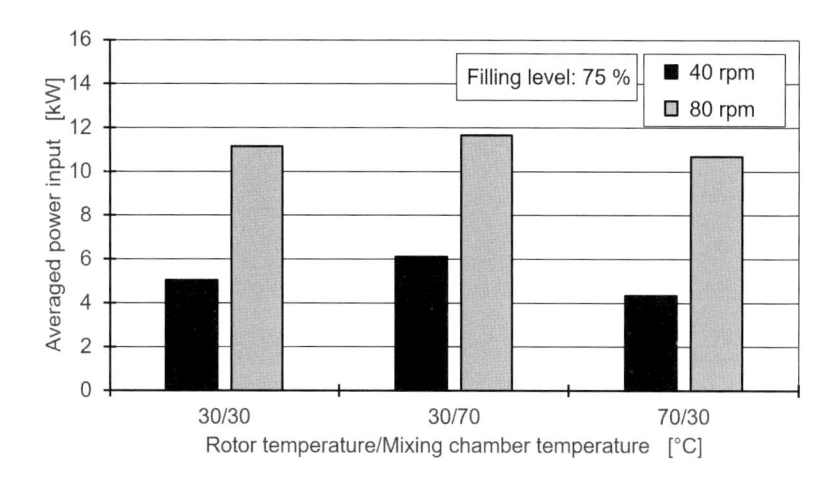

Figure 3.7: Influence on tempering on wall slip and adhesion as inferred by power input

In addition, topological investigations of the sample were carried out. The aim was to determine the number of carbon black defects and their size. Four different size ranges were detected: 2 5 μm, 5-10 μm, 10-15 μm and defects with a size above 15 μm. To see the largest different effects, samples were examined that were produced at a mixing chamber and rotor temperature of 30 °C and a mixing chamber and rotor temperature of 70 °C. Furthermore, the influence of the rotor speed on the quality of the carbon black dispersion was investigated. Almost no defects with a size of 15 μm can be observed, regardless of the mixer temperature and rotor speed used. It can also be seen that the largest number of carbon black defects has a size of 2-5 μm.

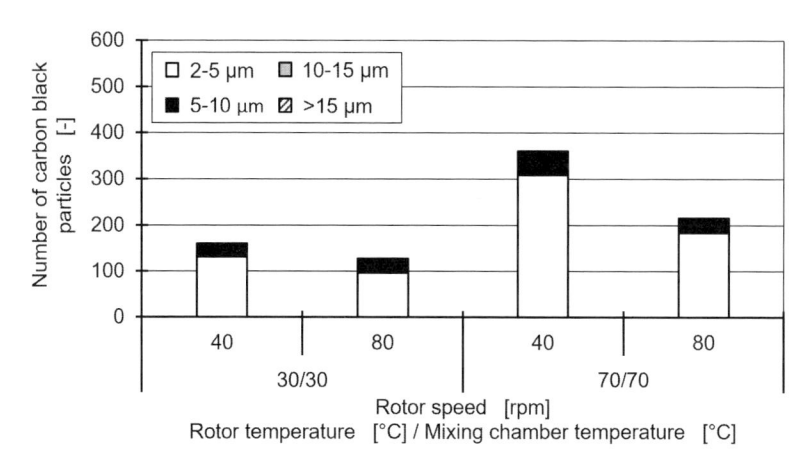

Figure 3.8: Influence of tempering and rotor speed on dispersion quality

Samples produced at a temperature of 30 °C for both the mixing chamber and the rotors show fewer defects, as adhesion between the batch and the mixing chamber wall plus the rotor surface occurs. If the mixtures are produced at a rotor speed of 40 rpm, the number of defects is marginally higher than at the rotor speed of 80 rpm. This is surprising because the batch temperature at a rotor speed of 40 rpm is lower than at a rotor speed of 80 rpm. If the viscosity decreases with increasing compound temperatures, less shear and elongation stress can be introduced in the filler dispersion phase and agglomerates remain in the compound. In order to investigate to what extent the compound viscosity differs at the different temperatures, further analyses are necessary. One likely reason why the number of defects is lower at a higher rotor speed is the number of revolutions. If the rotor speed is higher, more phantom transpositions may occur. These, along with the other process parameters, are expected to have an influence on the carbon black dispersion.

3.7 Conclusions and outlook

The investigations have shown that it is possible to analyse the carbon black incorporation and dispersion of ethylene-propylene-diene-rubber carbon black blends in the "upside-down" process systematically. Furthermore, it becomes clear that the parameters analysed (rotor and mixing chamber temperature, rotor speed, filling level, carbon black grade) have an influence on the mixing time until the carbon black is completely incorporated, under the condition that the time of complete carbon black incorporation is defined as reaching the final position of the ram.

The experimental trials have shown that the incorporation time increases with filling level. Whether this is a linear or exponential relationship depends on the rotor and mixing chamber temperature control and the rotor speed. In addition to the filling level, the incorporation time depends on the compound temperature. As the compound temperature increases, the compound viscosity decreases and the carbon black can be enclosed quicker by the rubber matrix. If the rotor speed is increased by factor two, the mixing time required for incorporation decreases. One reason for this is that the rotor blades can pick up carbon black that is still in the feeding chute twice as fast and transport it into the mixing chamber with each revolution.

Additionally, the influence of the type of carbon black used on the incorporation time was determined. The standard carbon black grades N330 and N550 were used for this purpose. The process parameters filling level and rotor speed were varied. The investigations showed that the carbon black type at a rotor and mixing chamber temperature of 70 °C each have only a minor influence on the mixing time required for incorporation.

In addition to the incorporation time, which is the main factor concerning the time efficiency, the carbon black dispersion was analysed. The investigations showed that the dispersion quality decreases with higher rotor and mixing chamber temperatures. It was also shown that the dispersion quality can be increased if the rotor speed is doubled.

In order to describe the process of carbon black incorporation and dispersion more general, further polymers and carbon black types should be selected and investigated. Subsequently, an optimisation of the individual mixing phases could be carried out with regard to the shortest possible duration of the respective mixing phase. Associated with this is a more energy-efficient mixing process design. In this way, plant capacities can be used in a targeted manner and resources can be saved at the same time. The carbon black dispersion and the mechanical properties should also be used as criteria for a suitable mixture production. If, for example, the filler incorporation phase is drastically shortened, an effect on the tensile properties due to insufficient filler incorporation is likely. In addition, the fingerprints of the optimised mixing phases will be evaluated. For example, there may be a temperature increase due to a higher viscous dissipation in the mixture caused by a higher power input. This higher power input can be caused by an increase in the rotor speed or a reduction in the compound temperature and thus a higher material viscosity.

Acknowledgements

The research project 20804 N of the Forschungsvereinigung Kunststoffverarbeitung has been sponsored as a part of the „industrielle Gemeinschaftsforschung und -entwicklung (IGF)" by the German Bundesministerium für Wirtschaft und Klimaschutz (BMWK) due to an enactment of the German Bundestag through the AiF. We would like to extend our thanks to all organisations mentioned. Furthermore, we would like to thank all the enterprises that support this research project with discussions, practical investigations and donations of materials.

References

[Abt19]   ABTS, G.: *Technologie der Kautschukverarbeitung*. München: Carl Hanser Verlag, 2019

[Cot85]   COTTON, G.: Mixing of Carbon Black with Rubber. II. Mechanism of Carbon Black Incorporation. *Rubber and Chemistry Technology* 58 (1985) 4, p. 774–784

[Dic14]   DICK, J.: *How to Improve Rubber Compounds*. Munich: Carl Hanser Verlag, 2014

[DW01]   DE, S.; WHITE, J.: *Rubber Technologist's Handbook*. Shawbury: Rapra Technology Limited, 2001

[Hab16]   HABERSTROH, E.: *Neuartige Methode zur Optimierung des Mischprozesses im Kautschuk-Innenmischer - Final Report*. Deutsches Institut für Kautschuktechnologie e.V., Final Report to IGF-research project 18244N, 2016

[HKB19]   HOPMANN, C.; KOSTKA, M.; BAKIR, C.: *Compounding - Influence of process parameters on the processing and resulting properties*. Aachen: IKV-Fachtagung "Rubber meets Science", 2019

[HKLB19]   HOPMANN, C.; KOSTKA, M.; LIPSKI, A.; BAKIR, C.: Systematische Untersuchung des Einzugsverhaltens und der Mastikation im Innenmischer zur Ermittlung einer Abhängigkeit zwischen Prozessparametern und resultierenden Kautschukeigenschaften. *Gummi Fasern Kunststoffe* 72 (2019) 6, p. 206–211

[HKLF21a]   HOPMANN, C.; KOSTKA, M.; LIMPER, A.; FACKLAM, M.: Analysis of the effects of various process parameters on the rubber compound production in the internal mixer. *Proceedings to the DKG Elastomer Synopsium*. Online-Conference, 2021

[HKLF21b]   HOPMANN, C.; KOSTKA, M.; LIMPER, A.; FACKLAM, M.: Systematische Untersuchung und Analyse des Mischprozesses - Auswirkungen unterschiedlicher Prozessparameter auf die Mischungsherstellung einer Naturkautschukmischung. *Gummi Fasern Kunststoffe* 74 (2021) 12, p. 30–36

[HSM84]   HESS, W.; SWOR, R.; MICEK, E.: The Influence of Carbon Black, Mixing, and Compounding Variables in Dispersion. *Rubber and Chemistry Technology* 57 (1984) 5, p. 959–1000

[LBG89]   LIMPER, A.; BARTH, P.; GRAJEWSKI, F.: *Technologie der Kautschukverarbeitung*. München and Wien: Carl Hanser Verlag, 1989

[Lee85]   LEE, B.-L.: Progress in Multiphase Rubber Processing: Controlled-Ingredient-Distribution Mixing. *Polymer Engineering and Science* 12 (1985) 25, p. 729–740

[Lim12]   LIMPER, A.: *Mixing of Rubber Compounds*. Munich: Carl Hanser Verlag, 2012

[NN17]   N.N.: *DIN 53504: Prüfung von Kautschuk und Elastomeren - Bestimmung von Reißfestigkeit, Zugfestigkeit, Reißdehnung und Spannungswerten im Zugversuch*. Berlin: Beuth Verlag, 2017

[Per84]   PERSSON, S.: Dispersion of Carbon Black. *Polymer Testing* 4 (1984) 1, p. 45–59

[RS13]   RÖTHEMEYER, F.; SOMMER, F.: *Kautschuk Technologie*. München and Wien: Carl Hanser Verlag, 2013

[Sun92]   SUNDER, J.: *Regelung und Optimierung des Mischprozesses von Elastomercompounds im Innenmischer*. RWTH Aachen University, Dissertation, 1992

Abbreviations

Notation	Description
TR	Rotor temperature
TM	Mixing chamber temperature

Production and operation of hydrogen pressure vessels

Moderator: Dr.-Ing. Jörg Strohhäcker, Hexagon Purus GmbH

--- Content ---

1 Challenges in the development of type-IV high-pressure vessels for high H_2 storage densities   255

J. Strohhäcker[1], C. Kaufhold[1], F. Richter[1], P. Judt[1], V. Strubel[2], P. Jatzlau[3], C. Beckmann[4], R. Schäuble[5]
[1] Hexagon Purus GmbH
[2] InnovationGreen
[3] RayScan Technologies GmbH
[4] Fraunhofer-Institut für Werkstoffmechanik IWM
[5] Fraunhofer-Institut für Mikrostruktur von Werkstoffen und Systemen IMWS

2 Detection of the laminate thickness increase during the production of type-IV hydrogen pressure vessels in the wet filament winding process 271

Ch. Hopmann[1], R. Müller[1], J. Fuchs[1], D. Schneider[1], K. Fischer[1]
[1] Institute for Plastics Processing (IKV) at RWTH Aachen University

3 Analysis of the thermal and mechanical stress of type-IV pressure vessels during hydrogen cycle tests   287

Ch. Hopmann[1], T. Gebhart[1], H. Çelik[1], K. Fischer[1]
[1] Institute for Plastics Processing (IKV) at RWTH Aachen University

Dr.-Ing. Jörg Strohhäcker

Dr.-Ing. Jörg Strohhäcker studied mechanical engineering with a specialisation in plastics technology at RWTH Aachen University. As a research assistant at the Institute for Plastics Processing, he worked on forming technology for continuous fibre-reinforced thermoplastics. This included vacuum forming, punch forming and double-diaphragm forming technology, which was eventually further developed for hollow bodies. This was then also the subject of his dissertation entitled "Manufacturing method for thermoplastic fibre-reinforced hollow structures".

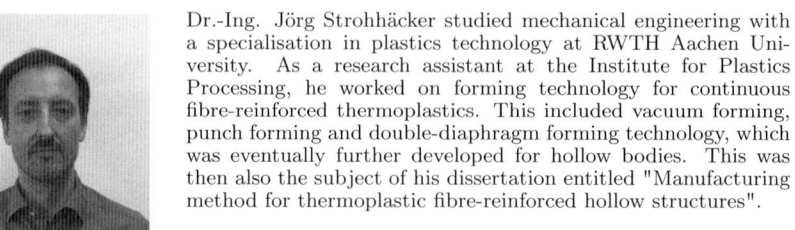

From 2008 to 2011, Jörg Strohhäcker headed the "Fibre-reinforced plastics" department at the Institute for Plastics Processing. In 2011, he then started as Programme Manager "Low Pressure Systems" at the company xperion Energy Environment GmbH at the location in Kassel. At the end of 2012, Dr Strohhäcker was promoted to Head of Development. After the company merger with Hexagon Composites, he then took on global responsibility for the RD Engineering department. This includes the evaluation of new materials, new processes and innovative liner and composite designs for pressure vessels for natural gas and hydrogen storage. As part of this role, he then also worked for over two years at another site in Lincoln, Nebraska, USA, before returning to the Kassel site in 2020.

1 Challenges in the development of type-IV high-pressure vessels for high H_2 storage densities

J. Strohhäcker[1], C. Kaufhold[1], F. Richter[1], P. Judt[1], V. Strubel[2], P. Jatzlau[3], C. Beckmann[4], R. Schäuble[5]

[1] Hexagon Purus GmbH
[2] InnovationGreen
[3] RayScan Technologies GmbH
[4] Fraunhofer-Institut für Werkstoffmechanik IWM
[5] Fraunhofer-Institut für Mikrostruktur von Werkstoffen und Systemen IMWS

Abstract

As the climate crisis and the resulting energy transition move forward, the role of "green" hydrogen generated by renewable energies is becoming increasingly significant. To implement this future-proof hydrogen technology, an infrastructure is needed to provide the distribution and transport of the energy carrier. This can be realized by centralized and decentralized storage and transport tanks for the hydrogen, which reliably and safely store the gaseous hydrogen at high energy density over long periods of time. Within the hydrogen network HYPOS (Hydrogen Power Storage & Solutions East Germany e.V.), the H2-HD project consortium has set itself the goal of developing new strategies for designing high-pressure hydrogen tanks (type-IV) for operating pressures of up to 1.000 bar. These pressure vessels consist of a plastic-based liner reinforced with a load carrying CFRP layer. As energy storage devices, the pressure vessels are integral components for a wide variety of transport and storage applications in the hydrogen network and must comply with high safety standards in their design and usage. The subject of the investigation in the H2-HD project is the material concept of the pressure tanks and its behavior, especially at operating pressures of up to 1.000 bar. The aim is to research design methods of the CFRP laminate structure for thick-walled tank in order to develop technically and economically innovative designs. Another key aspect of the project is to investigate the mechanical material behavior of the load carrying CFRP structure in order to perform an evaluation of the effects of manufacturing-related imperfections. Since these can´t be completely eliminated during the manufacturing of the tanks using the wet winding process, the containers are currently subjected to high safety factors. A precise understanding of the effects resulting from these imperfections on the behavior of the material would lead to further improvements in design and material utilization. In order to investigate these fundamentals of material science and to ensure the manufacturing quality of the safety-relevant material structures during the process, different methods of CT scan technology are being evaluated in the project. The results of this analysis will be used to develop a model that can be used for probabilistic simulation of material behavior. In addition, one focus of the project is on the thermoplastic liner of the tank, for which, among other things, the suitability of a new manufacturing process is to be tested. These topics of the project are presented in the paper.

1.1 Introduction

As the climate crisis and the resulting energy transition progress, the role of "green" hydrogen produced by renewable energies is becoming increasingly important. In order to implement this future-proof hydrogen technology, an infrastructure is needed to provide the distribution and transport of the energy carrier. This can be realised by centralised and decentralised storage and transport tanks for the hydrogen, which reliably and safely store the gaseous hydrogen over long periods of time with a high energy density and as low a weight as possible. The H2-HD project consortium, consisting of the two Fraunhofer Institutes Institute for Microstructure of Materials and Systems (IMWS) and Institute for Mechanics of Materials (IWM), the company RayScan Technologies GmbH and the company Hexagon Purus GmbH, has set itself the goal within the hydrogen network HY-

POS (Hydrogen Power Storage Solutions East Germany e.V.) of developing new strategies for designing the high-pressure hydrogen tanks (type-IV) for operating pressures of up to 1.000 bar. In order to increase the efficiency of the high-pressure vessels, the project consortium is investigating the material-mechanical behaviour of the composite laminate on a micro-, meso- and macroscopic level, looking at the effects of manufacturing-related defects and damage. These pressure vessels consist of a plastic-based liner reinforced with a load-bearing CFRP jacket. As energy storage devices, the pressure vessels are an integral component for a wide variety of transport and storage applications in the hydrogen network and must comply with high safety standards in their design and use. The subject of the investigation in the H2-HD project is the material concept of the high-pressure vessels and their behaviour, especially at operating pressures of up to 1.000 bar. The aim here is to research the CFRP laminate structure for thick-walled containers by means of improved design methods in order to develop technically and economically innovative designs. Another essential aspect of the project is research into the material-mechanical behaviour of the load-bearing CFRP structure in order to evaluate the effects of manufacturing-related imperfections. As these cannot be completely ruled out during the manufacture of the container using the winding process, the containers are currently subjected to high safety factors. A precise understanding of the effects resulting from these imperfections on the application behaviour of the material would lead to a further improved design and material utilisation. To investigate these material science fundamentals and to ensure the manufacturing quality of the safety-relevant material structures during the process, various methods of CT scan technology are being evaluated in the project. The results of this analysis will be used to develop a model that can be used for probabilistic simulation of the material behaviour. In addition, one focus of the project is on the thermoplastic liner, for which, among other things, the suitability of a new manufacturing process is to be tested. These partial aspects of the project are presented below.

1.2 Main section

1.2.1 Optimisation of pressure vessel performance

Within the framework of the H2-HD project, the laminate structure consisting of carbon fibres and thermoset matrix was considered. Especially for the storage and transport of hydrogen, it was shown that a higher operating pressure is important for efficient use [RRS05]. However, an increase in pressure in this range requires a coordinated, optimised material and component concept. It has been shown that in order to improve the effectiveness of thick-walled fibre composite vessels, it is important to look closely at the stresses and strains within the layer structure [Zu12]. Due to the fibre composite material with its anisotropic material properties, this design is highly complex and requires the use of modern CAE methods, e.g. for the numerical prediction of the failure loads and the resulting optimisation of the layer structure [RHA13]. With the help of the stresses and failure loads occurring in the individual layers of the laminate structure, it was investigated how the thick-walled high-pressure laminates can be optimised by a suitable choice of fibre angles, their sequence and sequencing. The analysis was simulated numerically and selected results were confirmed by container tests. Depending on the fibre angle, the layer structure for high-pressure containers can be divided into so-called circumferential layers (fibre angle close to 90°), which are considered load-bearing in the composite because they absorb the failure-critical forces in the circumferential container direction. On the other hand, there are layers with a low fibre angle which, especially in the composite, absorb the forces that arise axially on the container due to the internal pressure and in the dome. Due to the design and process, there are also layers with fibre angles in between in the laminate structure. Theoretical simulations of the laminate behaviour were used to compare different laminate structures, taking into account numerical and analytical calculation methods. The total load of the laminates was calculated, whereby non-linear and thick-walled material behaviour was assumed in the calculation. It was shown that by redistributing the position of the circumferential layers in the laminate, an improvement in the stress load was achieved. By iteratively shifting circumferential layers in the laminate, the overall stresses in the laminate could be reduced, potentially resulting in an increase in burst pressure. This was demonstrated in the test on containers whose circumferential layers were distributed in packages in the laminate, whereby a package consisted of several circumferential layers. The number of individual circumferential layers in the packages was changed in the test. This involved iteratively redistributing the amount of circumferential layers from the outside of the laminate to the inside. This redistribution resulted in an improvement in bursting resistance the more redistributed. The tests showed that a limit was reached in the laminate for this redistribution, above which further inward redistribution led to no further increase in burst pressure. Figure 1.1 shows the distributions of the circumferential layers in the different laminate versions that were simulated and tested for container strength in the burst test. Figure 1.2 shows a steady increase in burst pressure with progressive redistribution up to a certain point, which seems to have been reached for the laminate at hand at laminate version 5. There, further redistribution does not lead to any improvement in the burst test.

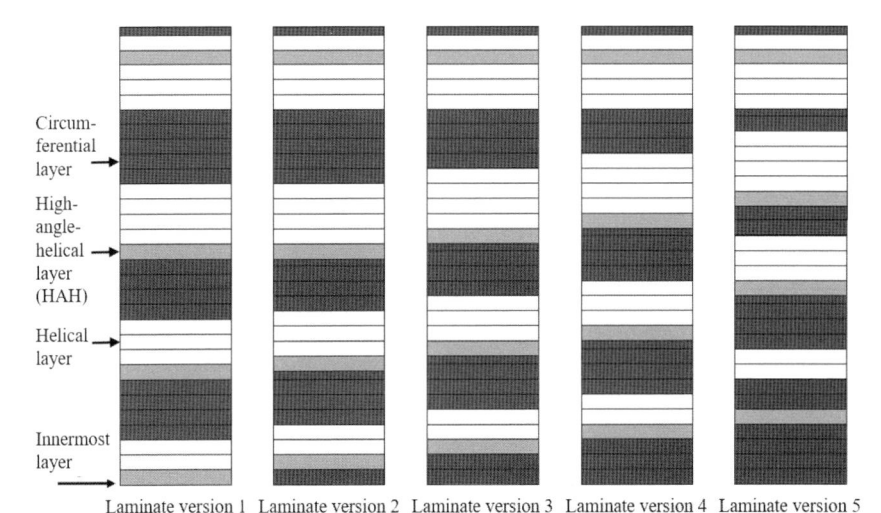

Figure 1.1: Laminate versions of a laminate structure in which further circumferential layers were distributed iteratively from the outside to the inside of the laminate with each version. Different layer types are marked accordingly by the shading ("HAH"="High angle helical" = layer with fibre angle between 75 ° and 90 °). The ply thicknesses shown are not to scale.

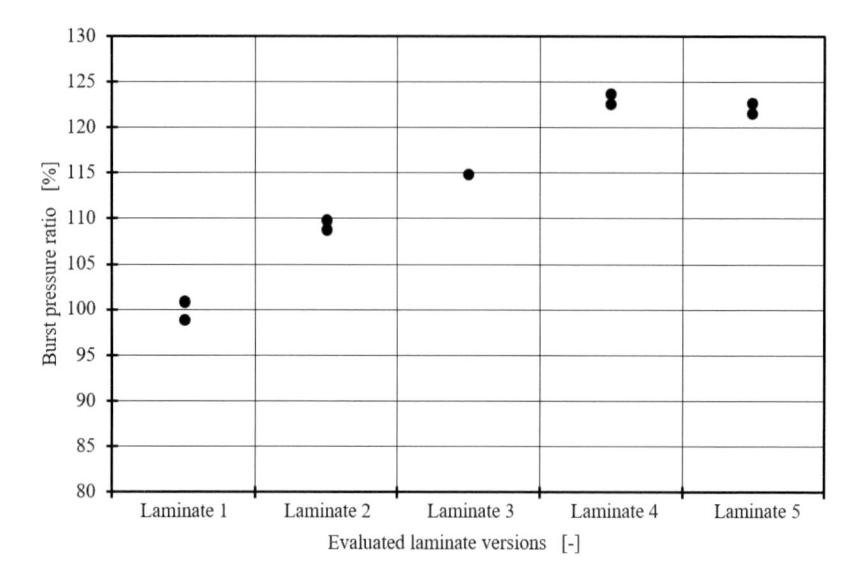

Figure 1.2: Results of the burst tests of the different laminate versions from Figure 1.1. The burst pressure increase in relation to laminate version 1 is shown.

The classical laminate theory (CLT), which is used in the calculation of thin-walled pressure vessels and according to whose theory the layers in the laminate structure are equally loaded, is not applicable to thick-walled laminates in the high-pressure hydrogen range [Sch07]. One explanation for the observed behaviour is that in the thick-walled laminate the inner layers have to bear higher loads than the layers that are further out. If there are then rather few and thin circumferential layer packages on the inside of the laminate structure, it could happen that these already reach their failure-critical load before the layers further on the outside can even utilise their full laminate strength. A failure of the inner circumferential layers then leads to a failure of the container. By redistributing the circumferential layers inwards in the laminate, this stress utilisation can be optimised. In another numerical study, more than 50 differently constructed high-pressure laminates were tested for their theoretical resistance to bursting failure. In the study, different laminates with circumferential layers, axial layers and intermediate layers were varied systematically as well as unsystematically in their order and sequencing. The aim here was to carry out a laminate analysis without taking into account regular design strategies that are conventionally adhered to due to geometric conditions or production-related reasons. Particularly noteworthy in the results of the study is the influence that can be exerted on the circumferential layers. As already mentioned, the circumferential layers are the most highly loaded layers due to the stress conditions in the container subjected to internal pressure. It has been shown that the load-bearing capacity of a laminate increases when high-angle helical (HAH) layers are placed in front of the circumferential layers or circumferential layer packages. These are layers with an angle between approx. 75° and 90°.

1.2.2 Liner-laminate connection

In the aforementioned H2-HD research project, the possibility of joining essential components of a type-IV high-pressure hydrogen tank was investigated: the composite laminate shell and the internal plastic liner, which serves as a permeation barrier [Ros18]. This bonding can be helpful when using the pressure vessels to compensate for the different thermal expansion coefficients of the two materials [Pro13]. In the study presented here, the feasibility of such a bond and its requirements were investigated. For this purpose, an intermediate layer was to be introduced between the fibre-matrix composite and the thermoplastic, which provides the appropriate adhesive strength and a mediating function between the strongly different joining partners [WKTM17]. For this purpose, different polymer-based materials were investigated as an intermediate layer. The study was carried out at the sample level, whereby the corresponding liner material based on polyamide, was pre-treated and then bonded with the epoxy resin, which is used as the matrix material. For the study of the properties, flexural specimens were made from the liner material, with dimensions of 150 x 20 mm^2. For reference, only the liner material sample was coated with the matrix resin. After pre-treatment, a reactive polymer film, for example, was applied to the material samples to be tested, which was then coated with the matrix resin to create a material combination that corresponds to that in the pressure vessel. The cured bending specimens were then subjected to a 180° bend, resulting in tensile stresses on the matrix side. For all specimens, the bending resulted in a cohesive fracture of the matrix layer. Prior to the cohesive fracture in the matrix layer, there was no adhesive detachment between liner material and epoxy resin or interlayer and epoxy resin. In the specimens without intermediary layer, fracture of the entire specimen and partial detachment of the epoxy resin was observed. The specimens with intermediary layer showed a clearly improved failure behaviour, whereby in some material combinations only a fracture in the matrix layer without large-scale detachment was observed. In this way, it was possible to demonstrate at sample level a way of creating a bond between the liner and the composite by using an intermediary film.

1.2.3 Analysis of CFRP pressure vessels using CT-scans

X-ray computed tomography (CT) scans are used for medical and industrial applications and provide a three-dimensional image/model that shows the differences in material density of the scanned object. In research and development, CT scans provide revealing insights into material structures. In the H2-HD research project, an H$_2$ transport container is being developed for operating pressures of up to 1.000 bar. CT scans are a useful support for the analysis of laminates in pressure vessels [Sco11, RJL+18] and can give a realistic assessment of the actual potential of laminate improvements and are therefore part of this project. During the project, several CT scans will be performed on laminate pieces, but also on complete vessels. Due to the dimensions of pressure vessels, especially those for H$_2$ transport applications, it is challenging and time consuming to perform a whole vessel scan with the aim of obtaining high resolution scan data. Furthermore, type-IV pressure vessels consist of individual parts with large differences in their material density, so that the selection of the correct scan parameters and the analysis of the scan results must be carried out with particular care.

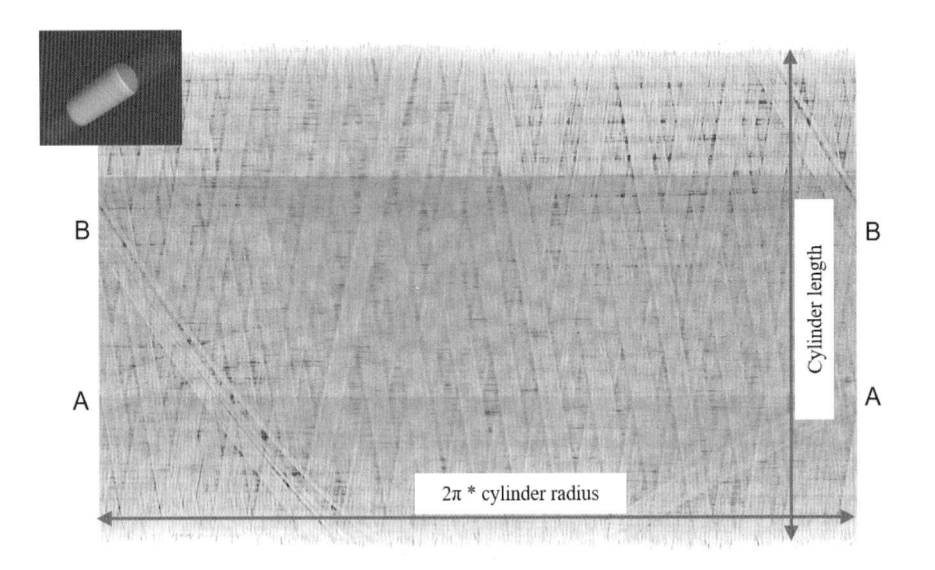

Figure 1.3: Unrolled image of the cylindrical section of a CNG vessel; pressure vessel CT scan with a resolution of 162 μm/voxel

Economical whole-vessel CT scans provide lower resolution images and are suitable for a preliminary assessment of the laminate quality as a whole. With an unrolled representation of the cylinder section of a total container CT scan, complete individual layers can be displayed and viewed, see Figure 1.3. The unrolled representation offers the possibility of obtaining a layer-by-layer overview of the laminate. Figure 1.3 shows a layer with a low winding angle. The underlying band of the combination layer (connection of layers with different fibre orientation) is still clearly visible. Higher image resolutions are obtained with a μ-CT scan (destructive, but high resolution) and the so-called region-of-interest (ROI) CT scan (non-destructive, but less detail than in μ-CT). Figure 1.4 shows a comparison between the image resolutions of a total vessel scan and a ROI CT scan as an example for the cylinder cross-section of a test vessel. The resolution of the ROI CT scan is higher by a factor of 4 and thus significantly more details of the laminate structure are visible. Pores in the total container scan are blurred and partly only show up as grey shading, whereas the pores in the ROI CT scan are clearly separated from fibres and resin.

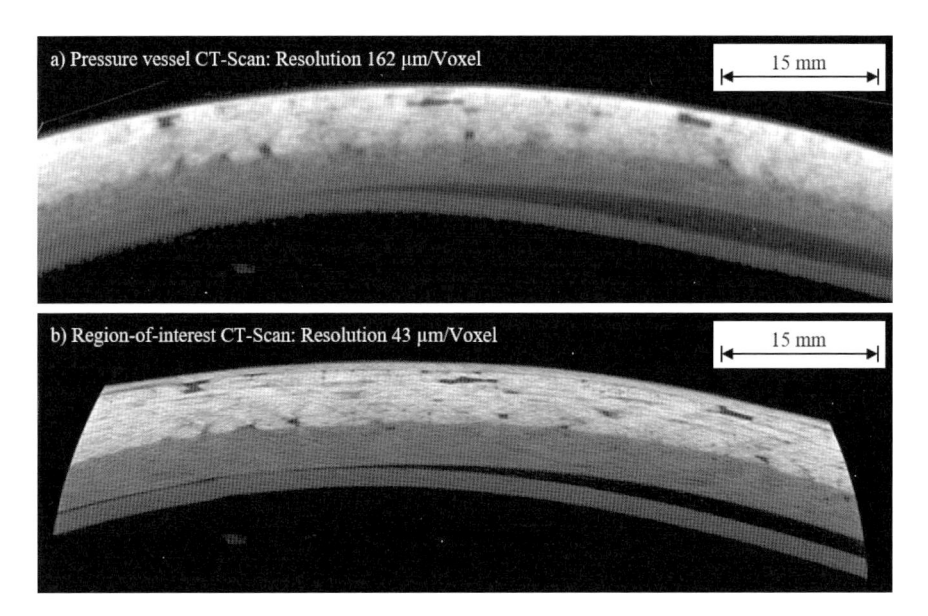

a) Pressure vessel CT-Scan: Resolution 162 µm/Voxel |← 15 mm →|

b) Region-of-interest CT-Scan: Resolution 43 µm/Voxel |← 15 mm →|

Figure 1.4: Comparison of the image quality of CT scans a) 162 µm and b) 43 µm voxel size; Cross-section of the cylinder of a test vessel

The unrolled representation of the cylindrical part of an H_2 test vessel in Figure 1.5 was created with a µ-CT scan and has approximately twice the resolution of an ROI CT scan.

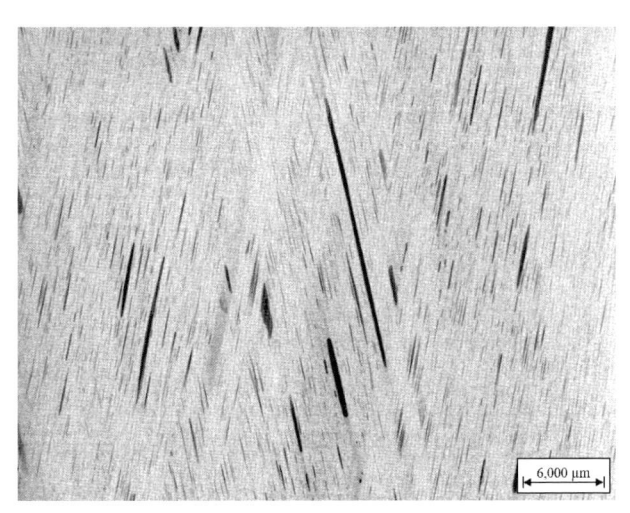

|← 6.000 µm →|

Figure 1.5: Unrolled image of a laminate piece from the cylinder area of an H_2 test vessel; µ-CT scan with a resolution of 24 µm/voxel

ROI CT scans offer the possibility to compare different states of the container laminate at selected positions, e.g. that of a fresh, unloaded state and after a loading history has been applied to analyse state changes due to loading. Suspected locations for the development of laminate damage are areas with imperfections in the laminate and are therefore investigated in detail in this project. Typical imperfections in CFRP laminates are e.g. pores of different shapes and sizes, fibre ondulations and areas of resin accumulation and have already been analysed in depth by various researchers [KCKK11, Kie11, Hei19]. Part of the project is to characterise these imperfections in terms of their occurrence, size and position in the laminate using porosity and fibre orientation analyses of the CT scan data. Figure 1.6 shows a fibre orientation analysis procedure developed in this project, which is used to determine the distribution of fibre orientation in the winding laminate. Laminates produced with the winding method typically have the winding angle with positive and negative signs in each layer (angle composite) and are limited to winding angles between 0° and ±90°. As a result, the fibre orientations determined according to Figure 6 extend over an angular range of 180°. In order to interpret the data meaningfully, the orientation of the image stack with respect to the container coordinate system must be known. Figure 1.7 shows a typical distribution determined on a 10 x 10 mm² CFRP section for a laminate section with the thickness of 2 mm. In addition to the cumulative representation, e.g. for the determination of distributions for specific winding layers in the laminate, the representation as a spatially resolved distribution over the laminate depth r is also possible.

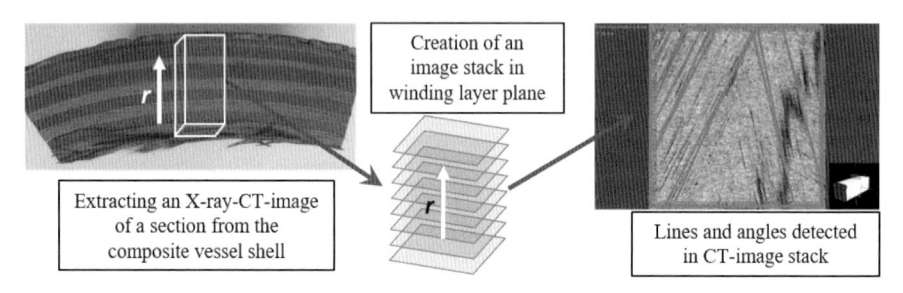

Figure 1.6: Procedure for fibre orientation analysis on CT scan data

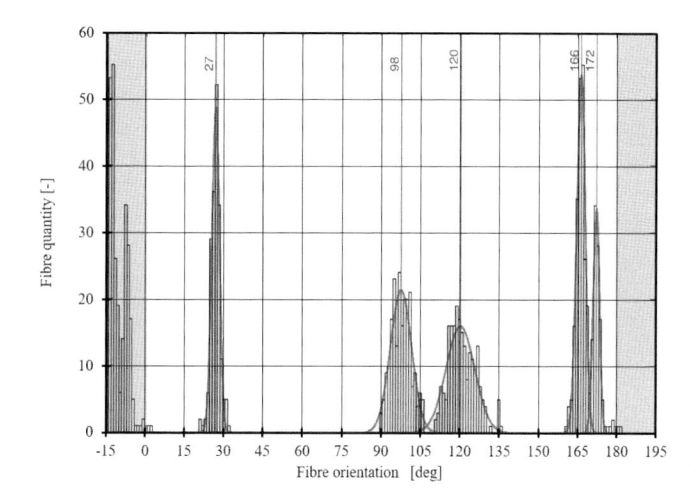

Figure 1.7: Typical distribution of fibre orientations for a section of a winding laminate

1.2.4 Simulation strategy for CFRP pressure vessels

The analysis results of the CT scan data are incorporated into a 3D simulation model with statistically distributed imperfections and a damage description according to Kachanov and Lemaitre [Kac58, Lem71]. The aim of the simulation is to evaluate the effect of different distributions of imperfections on container efficiency and their relevance for improvements in laminate quality, and thus to identify potential for improvement. Both the imperfections and the development of material damage are represented by reduced laminate stiffnesses and strengths (puck characteristics [Puc96]) in the macroscopic container model. For the calibration of the simulation model, tensile and Woehler tests of wound CFRP coupon specimens were carried out. Figure 1.8 shows the Woehler curve determined on CFRP coupon samples (UD, fibres transverse to the load direction) in double logarithmic representation.

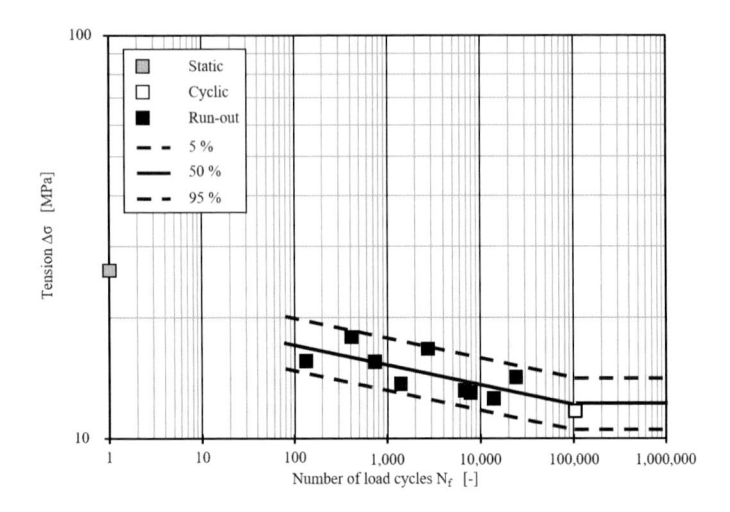

Figure 1.8: Double logarithmic representation of the CFRP Wöhler curve

The reduction of material properties by pores and ondulations are determined with analytical and numerical models on the angular composite level, which are parameterised according to the analysed CT scan data. Figure 1.9 shows six characteristic pores with length-to-diameter ratios between 6.25 and 25 embedded in a macromodel of the angular composite. Imposing 13 different strain-controlled load cases and assuming periodic boundary conditions results in a failure envelope depending on the winding angle for each pore model.

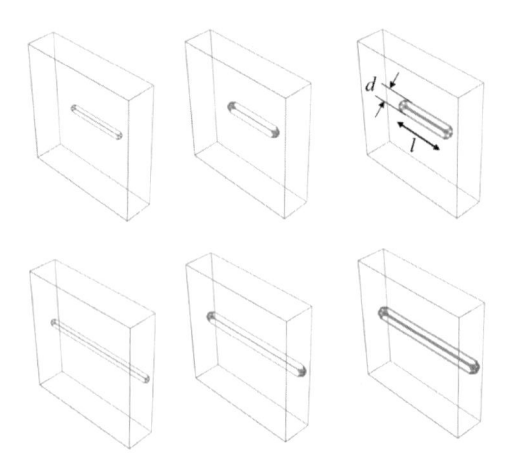

Figure 1.9: Six analysed macro models with pores in l/d ratio between 6.25 and 25

For the statistical analysis of the influence of process parameters in the winding process with regard to the mechanical properties of the pressure vessel, vessels with a simplified test laminate structure were fitted with strain gauges and various load scenarios were tested. Figure 1.10 shows the normalised measured values of the strain gages applied in the dome area (Figure 1.10 a) positions P-1 to P-3) and on the cylinder (Figure 1.10 b) Positions P-4 to P-6) during a compression test and burst test. For the majority of the measured values, a proportional relationship between vessel pressure and the maximum principal strain is shown. For quantitative evaluation of the measurements, local structural stiffnesses according to Equation (1.1) are determined.

$$C_S = \frac{p}{\epsilon_1} \quad (1.1)$$

The local structural stiffnesses determined at several positions in the cylinder show a relative standard deviation of 9.7 %. The differences of the local structural stiffnesses in the dome area are even more significant with a relative standard deviation of 30.5 %.

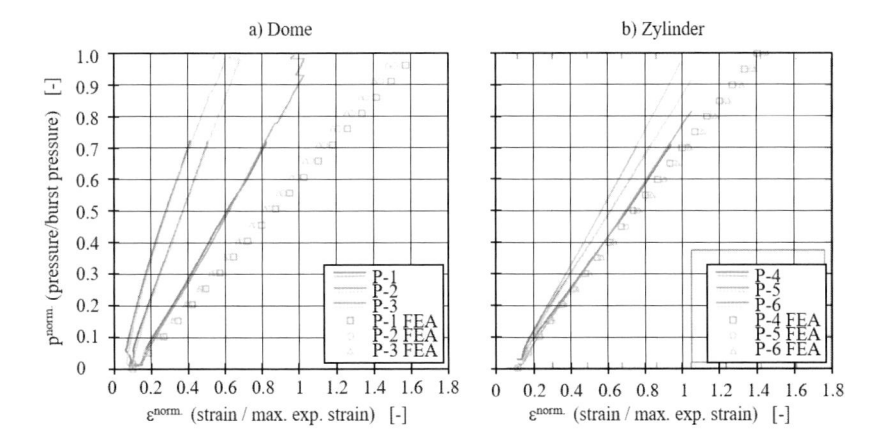

Figure 1.10: Normalised representation of the main expansions during a pressure (dark colours) and burst test (light colours) at measuring positions in the a) dome and b) cylinder of the container; results of the simulation are marked with FEA.

This is contrasted with the principal strains of a simplified container model without imperfections, which were determined at positions corresponding to the strain gage positions in the experiment and are marked with FEA in Figure 1.10. The structural stiffnesses of the container model have a relative standard deviation of 2.3 % in the cylinder area and 3.3 % in the dome area and are clearly below the measured scatter. In the remaining project period, various distributions of imperfections will be systematically analysed in order to further improve the simulation model on the one hand and to understand the scatter of the measured strains more precisely on the other.

1.2.5 Liner production using in-situ polymerisation of caprolactam

The state of the art for the production of thermoplastic liners, which serve as the gastight core of type-IV pressure vessels, is the blow moulding of a liner or liners produced by a process combination of extrusion, injection moulding and welding. These processes are industrially established and the high tooling costs are tolerable due to the long tool life and high discharge rates for the moderate to large volumes. The long delivery times (of several months) of the tools are known and must be taken into account accordingly in the project plan. However, especially at the beginning of a project, these established manufacturing methods bring some challenges for the container development. For example, due to the long delivery time of the liner tools and the subsequent sampling, the laminate development must primarily be based on numerical calculations. An additional limitation of the blow moulding process is the maximum liner size or length that can be produced, which depends, among other things, on the melt strength of the plastic. Especially for gas transport (see Figure 1.11), very large containers and thus liners are required. Here, the limits of blow moulding are already being exhausted today.

Figure 1.11: For the transport of gases, containers are equipped with pressure vessels

A promising alternative could be the production of liners by in-situ polymerisation of caprolactam. For this process, caprolactam is mixed in two separate storage tanks, one with catalyst and one with activator. Both containers are preheated to about 100-140 °C before the respective component mixtures are then mixed together and poured into a mould at about 140-180 °C [NN19], where the mixture completely polymerises to polyamide 6 within a short time (Figure 1.12).

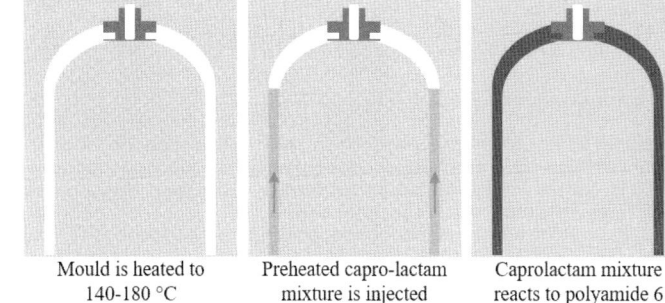

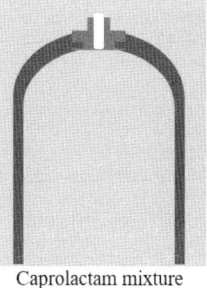

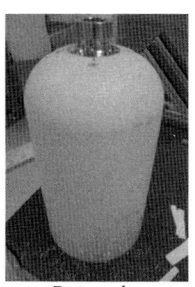

| Mould is heated to 140-180 °C | Preheated capro-lactam mixture is injected | Caprolactam mixture reacts to polyamide 6 | Part can be demoulded after a few minutes |

Figure 1.12: Liner casting process

Polymerisation and crystallisation occur in parallel in this process, as the reaction temperature is very close to the crystallisation temperature [VACB17]. At the processing temperature, the caprolactam is very fluid at about 10 mPas [TSKI17]. Accordingly, it can be filled into a mould with low pressure. In the simplest case, even gravimetric mould filling can be carried out. Due to the comparatively low pressure during the filling process, simple aluminium moulds or even silicone moulds can be used. The latter can be produced within a few days and are therefore particularly interesting for the production of first prototypes. Within the framework of the above-mentioned H2-HD project, the prototype production of liner halves by means of gravimetric filling of a silicone tool was investigated. Based on a liner model, a master mould was milled at the processor's site, which was used to produce a silicone mould. The mould was designed in such a way that the metallic boss part, which serves to hold the liner in the winding machine and for the subsequent assembly of valves, can be inserted and cast around during the subsequent prototype production. Due to the resulting form closure, the boss part is firmly connected to the liner. For the casting process, the mould including the boss part is heated and then completely filled with the preheated, low-viscosity caprolactam mixture within one minute. The finished component can be demoulded after just a few minutes. The produced liner half has a diameter of 280 mm, a length of approx. 500 mm and a wall thickness of approx. 3-4 mm. The pure plastic weight is approx. 1.8 kg. The tolerances achieved in the silicone mould cannot be compared to the tolerances achievable in metallic moulds of the established liner manufacturing methods. Due to the inherent weight of the silicone mould, there are, for example, sink effects and slight deformations of the mould. Among other things, this leads to wall thickness fluctuations between 70 and 170 % of the nominal wall thickness. Using sheet material, the weldability of the cast polyamide in hot plate welding was investigated in advance and successfully demonstrated. The liner halves were welded to complete liners with the previously determined welding parameters, tested for tightness and wrapped with the 700 bar laminate developed in the project. After the laminate had cured, the containers were subjected to a hydraulic burst test. It could be proven that the cast polyamide liner withstands the loads during the burst test and is only destroyed as a result of the laminate failure. The cast polyamide used is only heat stabilised and shows brittle behaviour. The Charpy impact strength is in the range of 2 kJ/m^2 both at room temperature and at -40 °C, which does not meet the requirements for a liner for type-IV pressure vessels. In principle, it is possible to add an impact modifier to the material mixture, but this can lead to new challenges due to an increase in viscosity and unknown compatibility with the silicone mould, for example.

1.3 Conclusions and outlook

Through the laminate calculations and tests carried out in the H2-HD project, it could be shown that in the design process of a thick-walled laminate, different factors can exert a strong influence on the container performance. In particular, the topics that can be influenced by the sequence and redistribution of layers in the container were presented. In the case of the distribution of circumferential layers, it was shown that these can achieve better container performance if the circumferential layers are distributed more concentrated inwards in the laminate. The theory was supported by the results of the burst tests. In the design of high angle helical plies, the influence of the order and sequencing of these plies was shown, especially when they are used before circumferential plies and can actively cause stress redistribution in the laminate. The implementation of such studies at container level can be envisaged here. In the investigation of the bonding of liner and composite material, the feasibility could be shown how a bonding of the components in the pressure vessel can be designed by using an intermediate layer. An implementation of this investigation from the sample level to the vessel level is aimed at in the further course of the project. Furthermore, the suitability of CT scans as a non-destructive testing and analysis method for pressure vessels was investigated and used for the development of an efficient H2 laminate. The analysis of the CT scan data reveals the process-related scatter in the area of porosity and fibre orientation and provides qualitative criteria for optimising the winding process with regard to porosity and fibre waviness. In the course of the project, the porosity could be greatly reduced in relevant areas of the container. Beyond the process optimisation, the findings from the analysis of the CT scan data are incorporated into the simulation strategy of the container model. Reduction factors were determined for the determined scatter of the fibre angles and the characteristic pores were evaluated with a macro model with regard to their strength. Further experimental investigations on CFRP structures at container and sample level were carried out and provide information on the scattering behaviour of the wound laminate. The findings from the various laminate analyses are combined in a container model that takes into account the statistically distributed imperfections in the laminate. With this simulation strategy, different distributions of imperfections can be mapped and evaluated in order to be able to show the further optimisation potential of the laminate. Another sub-project objective was to demonstrate the feasibility in principle of prototypical liner production by means of in-situ polymerisation. For small series and components with high demands on dimensional accuracy, an aluminium mould is probably recommended, but this could not be investigated within the scope of the project. The use of a silicone mould offers the advantage that it can be produced very quickly and at low cost. However, these time advantages in production also entail losses in the form of limited service life of the silicone moulds of approx. 20 components. Nevertheless, this process is of particular interest for the first iteration steps in laminate development, as it can shorten the time-to-market. Overall, the work presented here demonstrates the potential of high-pressure H2 vessels as a subject of current research. By optimising and better understanding efficiency, cost potential and safety, hydrogen technology can also be made more future-proof in order to use it to pursue the technological and climate policy goals of our time.

Acknowledgements

This publication is a scientific result of the research project "H2-HD", which was carried out within HYPOS e.V.. The research project entitled "H2-HD Plastic Hybrid High-Pressure Tank Systems for High H2 Storage Densities" was funded by the Federal Ministry of Education and Research (BMBF) (grant no. 03ZZ0743A-D), through the Project Management Organisation Jülich/Forschungszentrum Jülich GmbH, for which the project consortium would like to express its gratitude.

References

[Hei19] HEIECK, F.: *Qualitätsbewertung von Faser-Kunststoff-Verbunden mittels optischer Texturanalyse auf 3D-Preformoberflächen*. University of Stuttgart, Dissertation, 2019

[Kac58] KACHANOV, L. M.: Time of the rupture process under creep conditions. *Proceedings of the USSR Academy of Science*. 1958

[KCKK11] KREIKEMEIER, J.; CHRUPALLA, D.; KHATTAB, I. A.; KRAUSE, D.: Experimentelle und numerische Untersuchungen von CFK mit herstellungsbedingten Fehlstellen. *Proceedings of 10th Magdeburger Maschinenbau-Tage*. 2011

[Kie11] KIEFEL, D.: *Quantitative Porositätscharakterisierung von CFK-Werkstoffen mit der Mikro-Computertomographie*. Technical University of Munich, Dissertation, 2011

[Lem71] LEMAITRE, J.: Evaluation of dissipation and damage in metals submitted to dynamic loading. *Proceedings of the 1st ICM Conference*. 1971

[NN19] N.N.: *Datenblatt: AD-NYLON*. Heilbronn: Brueggemann Group, 2019

[Pro13] PROEBSTER, M.: *Elastisch kleben: aus der Praxis für die Praxis*. Wiesbaden: Springer Vieweg, 2013

[Puc96] PUCK, A.: *Festigkeitsanalyse von Faser-Matrix-Laminaten*. München: Carl Hanser Verlag, 1996

[RHA13] ROH, H. S.; HUA, T. Q.; ALUWALIA, R. K.: Optimization of carbon fiber usage in type 4 hydrogen storage tanks for fuel cell automobiles. *International Journal of Hydrogen Energy* 38 (2013) 29,

[RJL+18] ROJEK, J.; JOANNES, S. MAVROGORDATO, M.; LAIARINANDRASANA, L.; BUNSELL, A.; THIONNET, A.: Modelling the effect of porosity on the mechanical properties of unidirectional composites. The case of thick-walled pressure vessel. *Proceedings of the 18th ECCM Conference*. 2018

[Ros18] ROSEN, P. A.: *Beitrag zur Optimierung von Wasserstoffdruckbehältern - Volkswagen Aktiengesellschaft AutoUni Schriftenreihe Band 113*. Wiesbaden: Springer, 2018

[RRS05] RANA, M. D.; RAWLS, G. B.; SIROSH, N.: Codes and standards for gaseous hydrogen vessels. *Proceedings of the ASME Pressure Vessels and Piping Conference*. 2005

[Sch07] SCHÜRMANN, H.: *Konstruieren mit Faser-Kunststoff-Verbunden, 2. Auflage*. Berlin: Springer Vieweg, 2007

[Sco11] SCOTT, A. E.: *Analysis of a hybrid composite pressure vessel using multi-scale computed tomography techniques*. University of Southampton, Dissertation, 2011

[TSKI17] TAKI, K.; SHOJI, N.; KOBAYASHI, M.; ITO, H.: A kinetic model of viscosity development for in situ ring-opening anionic polymerization of -caprolactam. *Microsystem Technologies* 23 (2017),

[VACB17] VICARD, C.; ALMEIDA, O.; CANTEREL, A.; BERNHART, G.: Experimental study of polymerization and crystallization kinetics of polyamide 6 obtained by anionic ring opening polymerization of -caprolactam. *Polymer* 132 (2017) 6,

[WKTM17] WÜNSCHE, M.; KATHARINA, H.; TEUTENBERG, D.; MESCHUT, G.: Faserverstärkte Kunststoffe strukturell kleben. *adhaesion KLEBEN & DICHTEN* 61 (2017) 6,

[Zu12] ZU, L.: *Design and Optimization of Filament Wound Composite Pressure Vessels*. TU Delft, Dissertation, 2012

Abbreviations

Notation	Description
CAE	Computer-aided engineering
CFRP	Carbon fibre reinforced plastic
CLT	Classic laminate theory
CNG	Compressed natural gas
CT	Computer tomography
FEM	Finite element method
SG	Strain gauge
H2	Hydrogen
HAH	High angle helical
ROI	Region of interest
UD	unidirectional

2 Detection of the laminate thickness increase during the production of type-IV hydrogen pressure vessels in the wet filament winding process

Ch. Hopmann[1], R. Müller[1], J. Fuchs[1], D. Schneider[1], K. Fischer[1]
[1] Institute for Plastics Processing (IKV) at RWTH Aachen University

Abstract

The detection of the laminate thickness increase during wet filament winding is an important part for quality assurance in order to determine whether or not a sufficient resin mass is present in the laminate for fibre impregnation. It is pointed out that an optical working principle is preferred for the measurement system and its resolution must be significantly lower than the expected layer height of approximately 0.2 mm. Two inline measurement systems, based on the principles of laser-triangulation and shadow-casting, are presented which both meet those requirements.

Experimental trials were done in order to analyse the effects of process parameters and layer sequence on the laminate thickness increase during winding. While the laminate thickness increases strongly with higher settings for the resin loading of the fibre band, a slightly negative correlation was found in terms of fibre band tension. Furthermore, differences in the stacking sequence of hoop and helical layers of a given laminate seem to have no influence on its overall thickness. In order to investigate the differences in measured laminate thickness while winding and the cured laminate, a pressure vessel was wound with a part of a representative laminate. The layer thicknesses of the cured vessel were determined via microscopical cross-section analyses and compared with the data of both measurement systems. As expected, the layer thicknesses decrease about up to approximately 10 % because of the compaction effects during winding and curing.

It can be stated that both measurements systems provide highly relevant information in terms of resin mass in and on the laminate. Disadvantageous for further research purposes on the compaction effects during filament winding is their lacking ability to detect the position of the uppermost fibre layer. However, this problem could be avoided by winding with very low resin loading.

2.1 Introduction

Due to steadily increasing restrictions on pollutant emissions from conventional vehicles, alternative drive systems are becoming more and more important. The use of hydrogen as an energy carrier for existing mobility concepts is the focus of development projects [Ros18]. Currently, hydrogen is mainly stored in type-IV pressure vessels, which consist of a plastic liner made of Polyamide 6 (PA6) or Polyethylene (PE), a metal boss part to hold the valve and a composite shell to absorb the pressure forces [KET18]. Due to the high operating pressures of up to 700 bar (passenger car) or 350 bar (bus & truck), the highest importance is attached to pressure vessel safety [KET18]. The law requires the pressure vessels to be oversized by a factor of 2.25, resulting in minimum burst pressures of 1,575 bar and 787.5 bar respectively [NN19]. In order to meet these minimum burst pressures in a given batch, the pressure vessels are further oversized in the design in order to take process influences and fluctuations in the mechanical material properties into account [HAP+10, Ros18]. Since the material costs of the carbon fibre account for 50-70 % of the total manufacturing costs of a pressure vessel, there is a high potential for cost savings from reducing the second oversizing [HAP+10].

To increase the competitiveness of fuel cell-based vehicles, the DELFIN project, funded by the German Federal Ministry of Transport and Digital Infrastructure (BMVI), was created to research alternative materials and manufacturing processes in the production of type-IV pressure vessels for hydrogen storage. The project consortium (listed in section "Acknowledgements") with representatives from the entire value chain of composite pressure vessels and in particular several end users in the form of automotive OEMs achieved a reduction in costs and weight of a representative type-IV pressure vessel of at least 10 %. As part of the research project, the Institute of Plastics Processing (IKV) carried out simulative investigations on the mechanical and thermal behaviour of the vessel materials and their interfaces under realistic cyclic pressure loading. For this purpose, a physically based zero- and one-dimensional simulation methodology was developed for calculating the temperature development in type-IV pressure vessels with an optional one- or two-way coupling with structural-mechanical finite element models, which is based on only two model parameters and takes the real gas behaviour into account. The joint investigation of mechanical and thermal behaviour is particularly relevant for materials in hydrogen storage systems, since, according to internal project investigations, significant temperature fluctuations between -17 °C and +73 °C can occur in individual pressure vessel areas during refuelling.

The IKV also developed camera- and sensor-based measurement systems as well as adapted image processing algorithms for in-line quality monitoring of the wet filament winding process in order to detect process-side deviations of the fibre band deposition and laminate thickness increase to use them for a realistic reconstruction of the pressure vessel in the form of a digital twin. This virtual reconstruction enables the iterative optimisation of pressure vessel designs and winding programmes with regard to material efficiency in a very short time.

In previous publications, possibilities for recording the fibre band width, orientation and position in the wet winding process on the basis of an infrared camera have already been published [HOF+20, HMRL+21]. In this paper, two measuring systems for monitoring the laminate thickness increase are presented together with first investigations on the influence of winding process parameters on the laminate thickness increase. While the to presented systems can be used as stand-alone quality assurance systems, they are also an important part of the multi-system measurement environment for the digital twin.

2.2 Methodology for the determination of laminate thickness increase

The determination of the laminate thickness during the winding process provides information about the resin mass absorbed by the fibre band. After an initial calibration on the winding machine being used, the measurement results give basic information about sufficient or insufficient impregnation of the laminate, whereby no statement can be made about the impregnation at filament level. Besides that, fluctuations in laminate thickness increase can alter the resulting fibre volume content (FVC) of the manufactured laminate, thus influencing its mechanical properties [OYC+18, VM04]. For this reason, processing companies occasionally carry out manual measurements with circumferential tape measures when manufacturing pipes or pressure vessels in filament winding processes. However, the process must be interrupted for this procedure, which can lead to inhomogeneous fibre band impregnation in addition to a longer production time. Another disadvantage is that the measurements are dependent on the person carrying out the test and can lead to contamination of the measuring medium with resin.

For these reasons, high importance is attached to optical and thus non-contact measuring systems for recording laminate thickness. For example, Dynexa GmbH Co. KG, Laudenbach, Germany, developed a measuring method for determining the diameter of a wound tube by using an industrial camera and an image processing algorithm. In the future, the information generated by the measuring system will be transferred to the plant control system so that the machine can intelligently change the resin loading of the fibre band via the doctor blade gap in the impregnation bath [Kop20]. In the presented form, the measuring system requires the detection of the top and bottom side of the object, which could make it unsuitable for particularly large components, such as pressure vessels from the bus & truck sector because its cost-intensive telecentric lense needs to fit the outer diameter of the wound part. An alternative could be the use of two individual cameras. In addition, a bright background is necessary to create sufficient contrast in the pictures. In this publication, two measuring methods are investigated regarding their potential for fulfilling the requirements and for a sufficient monitoring. They only require the measurement of one side of the part and can be used independently of fluctuating lighting situations.

2.3 Specification of the measurement system

In order to enable an in-line detection of laminate thickness changes in the filament winding process, various requirements are made on the measuring system. On the one hand, the resolution of the system must be sufficiently high to be able to distinguish between individual layers with an expected thickness of 0.2-0.6 mm, which was evaluated via preliminary microscopical cross-section analyses. Therefore, a limit for the minimal resolution is set 50 % lower at 0.1 mm. In addition, the measurement must be carried out without contact in order to avoid influencing the top layer of resin and at the same time to eliminate resin contamination of the measuring device and operator dependency. Because of the fact that changes in laminate thickness occur only once per layer, a minimal required measurement frequency of 1 Hz is sufficient for the measurement task. Limitations regarding weight and size of the systems depend highly on the winding machine being used. In general, compact and light systems could be advantageous regarding flexibility in mounting positions.

2.4 Development of the measurement strategy

2.4.1 Machine technology and materials

The tests were carried out on a robot-based winding system from Hille Engineering GmbH Co. KG, Roetgen, Germany. The system consists of a linear axis on which the winding core can be rotated and moved in translational direction and a laying head mounted on a 6-axis KR300 industrial robot from Kuka AG, Augsburg, Germany. The compact laying head contains up to four fibre bobbins, an impregnation unit based on the roller impregnation principle (d = 200 mm, ϕ = 100-110 °) and a deposition roller, see Figure 2.1. In addition to an individual roving tension control, the fibre band tension can be modified via a fibre band tension control.

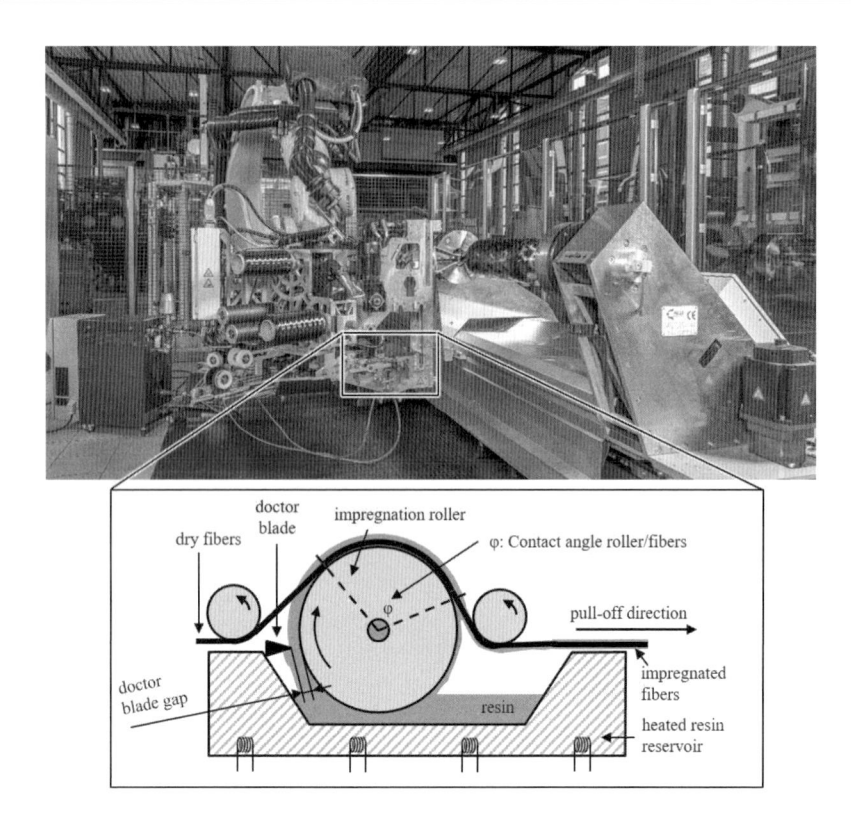

Figure 2.1: Robot-based winding machine at IKV and schematic view of the roller impregnation unit

The winding core shown in Figure 2.2 consists of a thermoplastic liner made of PA6 with the material designation Durethan BC550Z DUSXBL from LANXESS AG, Cologne, Germany, with metallic connecting elements (bosses).

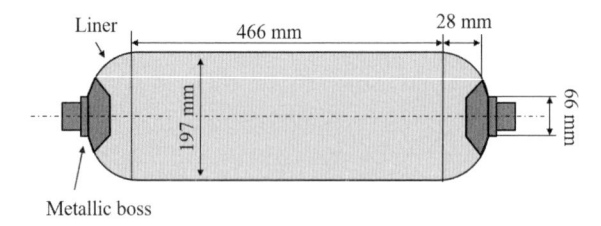

Figure 2.2: Geometry of the liner in a subscaled version of a representative type-IV pressure vessel

In the tests, four carbon fibre rovings (ITS 50) from Teijin Carbon Europe GmbH, Wuppertal, Germany, with a titre of 1,600 tex are used and formed into a fibre ribbon. The resin system used is a low-viscosity epoxy resin with the designation EPIKOTE Resin 04976 from Hexion Inc, Columbus, USA. Fibre and resin components represent a typical material combination for the production of type-IV pressure vessels. In order to assess the suitability of the measuring systems presented below for recording the laminate thickness increase, no curing of the wound pressure vessels is necessary. For this reason, only the resin component of the epoxy resin system is used, so that the fibre-resin mixture can be removed from the liner after winding and the liner can be reused. In order to still achieve the viscosity of the resin system of approx. 240 mPas at a processing temperature of 40 °C, the temperature of the resin component is set to 65 °C in the tests.

2.4.2 Measurement systems

Laser triangulation sensor

The principle of distance measurement using laser triangulation is based on the reflection of a laser beam. A triangulation sensor projects a laser beam perpendicularly onto the surface of the measuring body. In the reference state, the reflected beam falls back into the sensor via a lens and onto a specific position on a surface made of photodiodes. As soon as the position of the measuring body changes relative to the sensor, the point of incidence of the reflected laser beam on the photodiode surface varies according to trigonometric relationships, from which the change in distance can be calculated. The measurement principle of the sensor as well as the integration in the filament winding machine is shown in Figure 2.3.

For the experiments, a laser triangulation sensor with the designation optoNCDT ILD1900-25 from Micro-Epsilon Messtechnik GmbH & Co. KG, Ortenburg, Germany, is used because of its capability to fade out unwanted reflections of the part surface and its integrated software for data analysis. The measuring range of 25 mm is sufficiently large to map the entire winding process up to the final laminate thickness of 20 mm. The measuring rate of up to 10 kHz ensures a high data density. Anyhow, preliminary investigations showed highest signal intensities for a measuring rate of 250 Hz. The optimal exposure time was determined by the sensor itself to be 3,994.7 µs. The resolution of the sensor of about 5 µm is sufficiently low to distinguish different laminate layers with a thickness of approximately 0.2 mm. The small installation space for the sensor with around 5x5x5 cm^3 offers high flexibility regarding mounting positions. Therefore, the laser triangulation sensor meets all the requirements to fulfil the measurement task stated in section 2.3. The disadvantage of the sensor is the punctual measurement principle. Anyhow, for industrial purposes it is either possible to implement several sensors due to its relatively low costs in the range of a lower 4-digit number or to install line triangulation sensors which can consist of an array of several thousand single measurement points.

Regarding the data evaluation, the measured distance between sensor and laminate surface is output in the form of a csv-file and can thus be evaluated immediately. The software of the sensor also offers the possibility to show realtime measurement values.

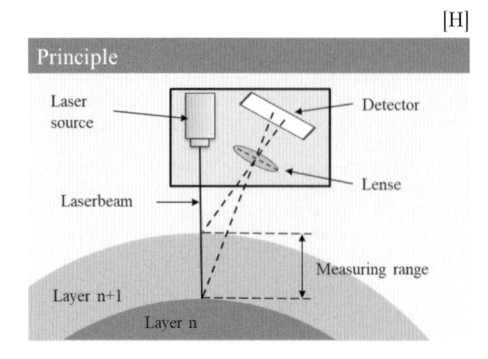

Figure 2.3: Measurement principle and machine implementation of the laser triangulation sensor

Shadow casting system

The second measuring system investigated is based on the detection of laminate thickness changes according to the shadow casting principle. The system is set up with a light source that is aimed directly at an industrial camera. The object to be measured, in this case the surface of the pressure vessel, is partially placed in the light beam path so that a section of the shadow is visible in the recorded image. In the process, the diameter of the winding body changes, which increases the area of the shadow in the image. After calibration, this increase of the shadow area can be transferred into a thickness increase given in millimetres.

An industrial camera with the designation VLXT-50M.I from the Baumer Group, Frauenfeld, Switzerland, with a telecentric lens of the model OPT-TS0142-228-AL from MaxxVision GmbH, Stuttgart, Germany, is used for the investigations. The telecentric lens allows only parallel light beams to fall on the camera sensor, thus eliminating optical distortion effects. As a light source, a diffuse LED spotlight with the designation DTL-1010-BE from MBJ Imaging GmbH, Hamburg, Germany, is used and, according to the above-mentioned principle, installed in the optical axis of the camera. The measurement principle and the integration in the winding machine is shown in Figure 2.4. An exposure time of 5,000 µs was selected as settings for imaging with regard to optimal contrast conditions. The measuring rate is limited by the maximum computing power of the computer and is approx. 12 Hz in this application. With the used components, the resolution of the measurement system was found to be about 0.04 mm, which is about five times lower than the expected thickness of an individual laminate layer with 0.2 mm. Therefore, the shadow casting system as well as the laser triangulation sensor meet the requirements of the measurement task stated in section 2.3. The complete shadow casting systems requires an installation space of around 20x20x20 cm^3 each for camera and light source and is therefore significantly larger than the laser triangulation sensor. The costs for the measurement systems in a range of a middle 4-digit number are slightly higher than the laser triangulation sensor.

The data evaluation is done on the basis of a self-developed Python™ programme, which determines the first white pixel on a vertical line in the recorded picture and thus the position of the winding body surface according to the threshold principle. After calibrating the system with an object of known thickness, pixels can be assigned a thickness in millimetres. Laminate thickness changes can subsequently be output in pixel values and converted to millimetres.

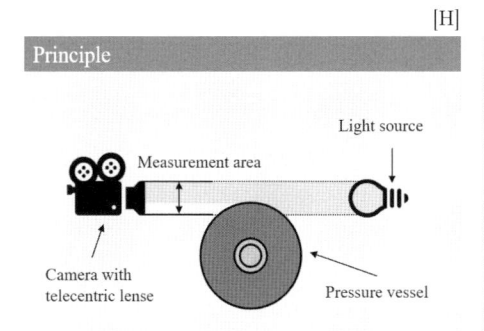

Figure 2.4: Measurement principle and machine implementation of the shadow casting system

2.5 In-line determination of the laminate thickness increase

Since both the laser triangulation sensor and the shadow casting system were found to be suitable for solving the measurement task with regard to the requirements stated in section 2.3, trials were done regarding the in-process determination of the laminate thickness increase during wet filament winding of a pressure vessel. In the following, the influence of different process settings as well as the layer sequence in the laminate was analysed by using the measurement systems.

2.5.1 Resin loading of the fibre band

In order to interpret the results on laminate thickness increase generated later, precise knowledge of the resin loading of the fibre band as a function of all subsequently varied process parameters is crucial. Therefore, the weight of the fibre band, consisting of four impregnated rovings, was measured after leaving the laying head by cutting out a metre in a triple determination. The fibre weight of 1,600 tex per roving was subtracted so that the sole weight of the resin is given. The tests were carried out with the resin component of the epoxy resin system at a temperature of 65 °C. The resin loading of the fibre band is adjusted indirectly via the doctor blade gap g in the impregnation unit. The measurement results are shown in Figure 2.5. As expected, the resin loading increases with higher doctor gaps. For example, at a roving tension of $F = 4$ N, the gravimetric resin content of the fibre band increases from 33.64 % (FVC = 56.2 %) at a doctor blade gap of $g = 100$ µm to 55.20 % (FVC = 34.5 %) at $g = 300$ µm. In addition, a lower resin loading of the fibre band is shown at higher roving tension forces. Thus, the measured resin loading at $F = 16$ N is between 2.6 % ($g = 100$ µm) and 19.0 % ($g = 150$ µm) below the recorded values at $F = 4$ N.

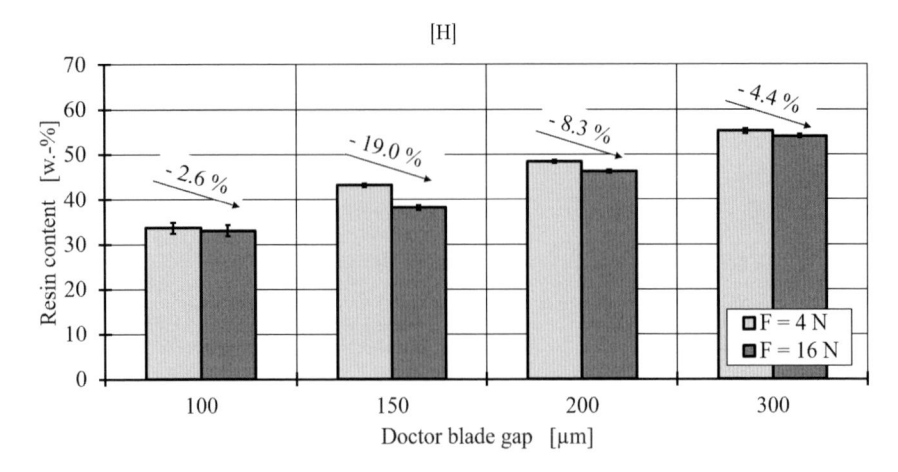

Figure 2.5: Resin loading of the fibre band in dependence of doctor blade gap g and roving tension force F

2.5.2 Laminate thickness increase at different positions on the pressure vessel

The measuring systems presented in section 2.4 were installed in various positions on the pressure vessel as shown in Figure 2.6. Since measurements of the laminate thickness in the industrial environment can only be carried out by using a circumferential tape measure (see section 2.2), it is possible for the first time with the optical systems to carry out measurements in the pole cap area and in the transition between the cylinder and pole cap area of the pressure vessel. A change in laminate thickness only occurs in the latter two measuring positions when helical layers are wound.

Table 2.1 shows the measured laminate thickness at the aforementioned positions after winding a representative laminate section with four circumferential layers (Ho) with a winding angle of 87.96 °, one helical layer (He) with 22.10 ° and four further circumferential layers (4Ho-1He-4Ho) with a roving tension force of $F = 16$ N and a doctor blade gap of $g = 200$ µm. In the cylindrical area, the laminate thickness measured with the laser triangulation sensor of approx. 3.8 ± 0.02 mm is significantly higher than in the pole cap and transition area with approx. 1.77 ± 0.04 mm and approx. 0.94 ± 0.04 mm respectively, as only the circumferential layers are measured in this position, see Table 2.1. The laminate thickness determined by hand with a circumferential tape measure is slightly smaller than that of the optical systems at approx. 3.68 mm, which may be due to a sinking of the measuring tape in the uppermost resin layer. It can be seen that measurements with the shadow casting system lead to lower values compared to the laser triangulation sensor at the identical measuring position in the cylindrical area and that these values simultaneously contain a higher standard deviation.

The characteristic of helical layers results in the fact that at the reversal point in the pole cap area there is extensive overlapping of neighbouring fibre bands, while in the cylindrical area these are mainly laid down next to each other. This overlap leads to higher laminate thickness increases which are reflected in the measured values in the pole cap area. At the reversal point of the helix layer, the thickness is 1.77-2.09 mm, while values of 0.83-0.94 mm are determined in the transition area.

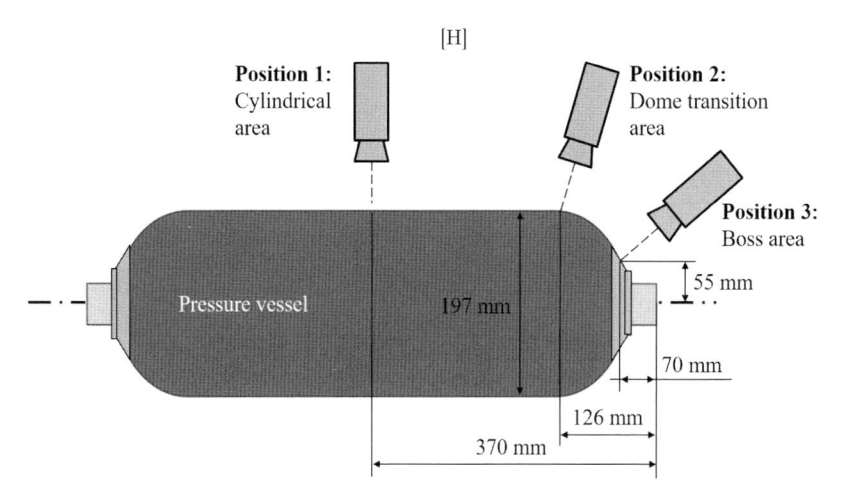

Figure 2.6: Measuring positions at the pressure vessel

Position	LTS	SCS	Manual measurement
Cylindrical area	3.78 ± 0.02	3.69 ± 0.07	3.68
Transition area	0.94 ± 0.04	0.83 ± 0.07	-
Dome area	1.77 ± 0.04	2.09 ± 0.13	-

Table 2.1: Measurement of laminate thickness at different positions of the pressure vessel (Laminate lay-up: 4Ho-1He-4Ho, F = 16 N, g = 200 µm)

2.5.3 Influence of resin loading on laminate thickness increase

Figure 2.7 shows the determined laminate thicknesses with both measuring systems after winding a representative laminate structure (4Ho-1He-4Ho) with different settings for the doctor blade gap (g). It can be seen that with a higher setting for the doctor blade gap, the laminate thickness increases. A laminate thickness of 3.01± 0.02 mm is measured with the laser triangulation sensor at g = 100 µm and 4.67±0.02 mm at a setting of g = 300 µm. The identical effect is seen in the measurements with the shadow casting system. This result can be explained with the findings from section 2.5.1. A higher resin loading of the fibre band leads to a greater laminate thickness increase per layer deposited.

A comparison of the results of both measuring methods shows that the values determined for the shadow casting system are about 9-18 µm below the values of the laser triangulation sensor at each measuring point. Furthermore, with a standard deviation of approx. 0.07 mm, the shadow casting system has a three to four times higher scatter in measured values.

Since both methods measure the surface of the uppermost resin layer on the winding body and not the position of the underlying carbon fibres, no statement can be made about the fibre volume content in the laminate or its gradient in the thickness direction.

[H]

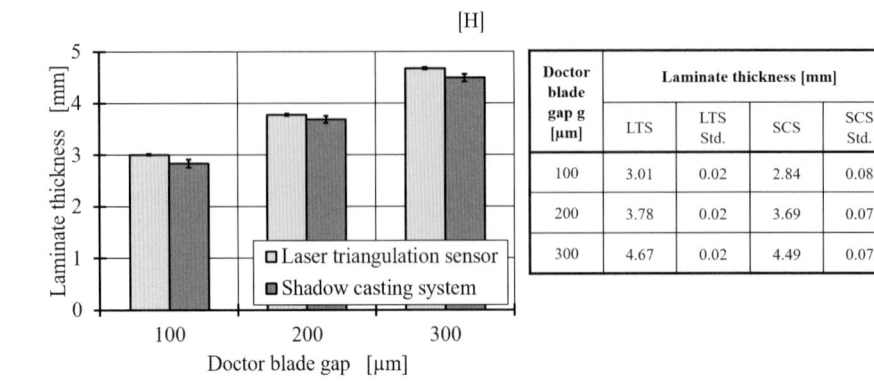

Doctor blade gap g [μm]	Laminate thickness [mm]			
	LTS	LTS Std.	SCS	SCS Std.
100	3.01	0.02	2.84	0.08
200	3.78	0.02	3.69	0.07
300	4.67	0.02	4.49	0.07

Figure 2.7: Influence of resin loading on laminate thickness increase for a fibre band tension force of F = 16 N

2.5.4 Influence of fibre band tension on laminate thickness increase

A change in the fibre band tension affects the compression of the laminate and thus the fibre volume content [LM20]. The fibres sink into the laminate, displacing resin and air pockets. The latter are manifested in the real process by the milky appearance of the uppermost resin layer on the winding body surface.

As an extension to Figure 2.7, Figure 2.8 shows measurements of the laminate thickness increase after winding the representative laminate (4Ho-1He-4Ho) with a roving tension force of F = 4 N. Contrary to expectations, measurements with the laser-triangulationsensor for a setting of the doctor blade gap of g = 200 μm and 300 μm at F = 4 N show a lower laminate thickness of 3.6±0.02 mm and 4.1±0.02 mm respectively compared to 3.8±0.02 mm and 4.7±0.02 mm at F = 16 N. Only at the lowest doctor blade gap setting of g = 100 μm does a lower roving tension force result in a higher laminate thickness. Analogous to the results from section 2.5.3, measurements with the shadow casting system lead to lower values compared to the laser triangulation sensor with a simultaneously higher standard deviation.

Since the volume of the fibres in the wound laminate remains constant regardless of the varied process settings, the result determined here can only be explained by a different volume of enclosed air or resin. It is known that air pockets are displaced to the outside at higher fibre band tensions and remain trapped in the resin mass[LM20]. However, the characteristically milky appearance of the uppermost resin layer could not be detected in these tests. The presence of air inclusions in the laminate must be investigated in future using microscopical cross-section analyses on the test plan set up here. From the investigations in section 2.5.1, a roving tension force of F = 16 N results in a generally lower resin loading of the fibre band than at F = 4 N, which is a possible explanation for the results determined here.

[H]

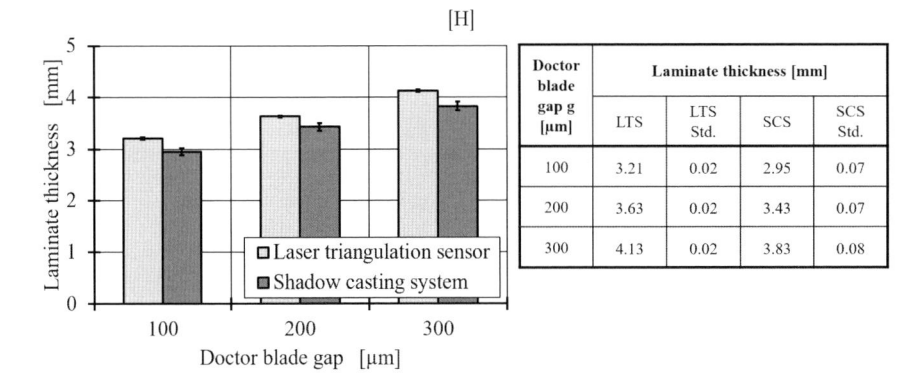

Doctor blade gap g [μm]	Laminate thickness [mm]			
	LTS	LTS Std.	SCS	SCS Std.
100	3.21	0.02	2.95	0.07
200	3.63	0.02	3.43	0.07
300	4.13	0.02	3.83	0.08

Figure 2.8: Influence of resin loading on laminate thickness increase for a fibre band tension force of $F = 4$ N

2.5.5 Influence of layer sequence on laminate thickness increase

In the cured laminate of type-IV pressure vessels from the wet filament winding process, helical layers with approx. 0.6 mm have a layer thickness about twice as high as circumferential layers with approx. 0.25 mm. The trajectory of flat helical layers with a winding angle of 22.1° in the present vessel geometry has a comparatively low curvature in the cylindrical area, resulting in low compaction forces. It is therefore assumed that helical layers are only compacted by the deposition of overlying circumferential layers with correspondingly high radial forces. In this section it will be analysed, whether or not the difference in layer thickness mentioned above is already visible during the winding process.

Figure 2.9 shows the thickness of two representative laminates after winding as a function of the doctor blade gap g. While the number of layers remains identical, only the position of the helical layer was varied so that it is either compacted (laminate 4Ho-1He-4Ho) or uncompacted (laminate 8Ho-1He). The laminate thickness was determined with both measurement systems.

It can be noted that for each of the different doctor blade settings the difference between both laminate sequences is negligible.This shows that the layer sequence of the wound laminate has no significant influence on its final thickness with a constant doctor blade gap for the analysed laminates. One explanation for this can be found in the approximately identical volume of fibre and resin that is deposited on the winding body. No statement can be made about the volume of air inclusions. The different layer thickness of circumferential layers (approx. 0.25 mm) and helical layers (approx. 0.6 mm) in cured laminates known from preliminary investigations suggests that there could be a difference between the thickness of the uppermost fibre layer in the results presented here. However, since the measuring systems presented here can only detect the position of the uppermost resin layer, this hypothesis cannot be tested with the present technique.

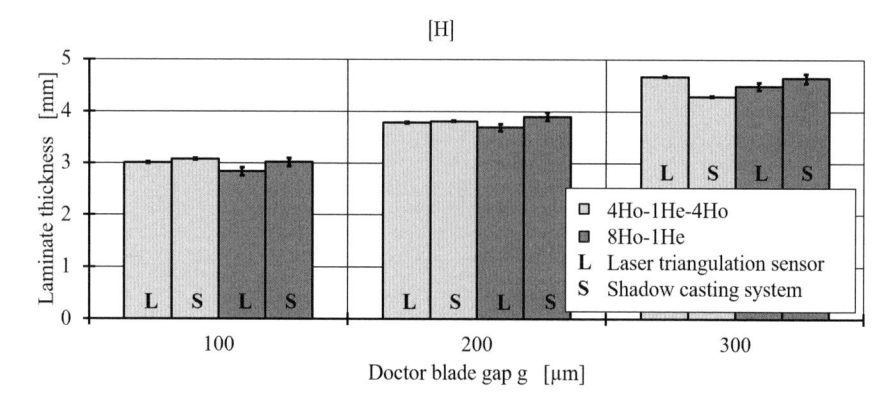

Figure 2.9: Influence of layer sequence on laminat thickness increase

2.5.6 Influence of the curing process on the laminate thickness

To investigate the change in the measured layer thicknesses due to the curing process, a pressure vessel was bewound with a part of a representative laminate structure (8Ho-1He-2Ho-1He-2Ho-1He) and cured. The winding angles for the different layers are sensitive information and therefore not mentioned here. In contrast to the previous investigations, the complete epoxy resin system, consisting of resin, curing agent and catalyst components in a gravimetric ratio of 100:80:0.1, with the designation EPIKOTE™ 04976 from Hexion Inc., Columbus, USA, was used. The wound vessel was cured at a constant temperature of 100 °C for a period of eight hours, while rotating at around 40 RPM.

Figure 2.10 shows the cross-section microscopy of the cured laminate, where different layers can be identified.The laminate thickness of the according steps in the winding program was determined with the measurement systems during winding and with the microscope in the cured laminate, see Table 2.2.In previous trials, constant settings for the roving tension force of $F = 16$ N and the doctor blade gap of $g = 150$ µm led to insufficient impregnation and thus to detachment of the lowest circumferential layers during test specimen conditioning. By winding the first four circumferential layers with a lower tension force of $F = 8$ N and an increased doctor blade gap of $g = 200$ µm, this problem could be reduced, although air pockets are still visible in the micrograph in Figure 2.10.

Figure 2.10 shows the measured laminate thicknesses after winding of different layers. Similar to the previous investigations, the measured values of the shadow casting system are slightly below the values of the laser triangulation sensor. In the microscopical cross-section analysis, the corresponding winding layers in the cured composite were identified and the laminate thickness up to the upper limit of this layer was determined with the evaluation software of the microscope. The laminate thicknesses thus determined are between 0.05 mm and 0.57 mm below the thickness measured during winding.

The lower laminate thickness in the cured laminate can be described by the superimposed effect of compacting and curing. It can be assumed that the thickness of the first four circumferential layers of 1.39 mm or 1.13 mm respectively is reduced by the deposition of the overlying layers and that excess resin and air pockets are displaced to the outside in radial direction. Due to the decrease in viscosity of the resin as a result of the temperature increase in the curing process, further displacement and thus a decrease in layer thickness can occur.

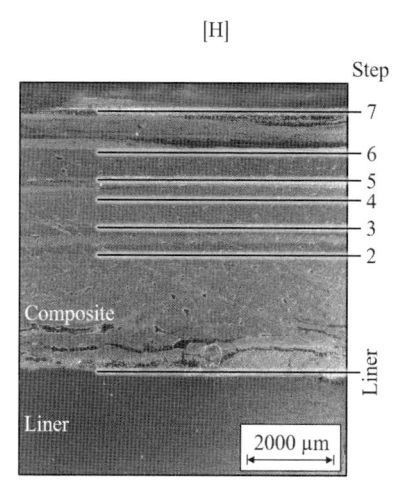

Figure 2.10: Detected layers in the cured laminate via cross-section microscopy

Winding step	Winding parameter (Layers, F, g)	LTS	SCS	Average	Microscope
1	4 Ho, 8 N, 200 µm	1.39 ±0.02	1.13 ±0.07	1.26	-
2	4 Ho, 16 N, 150 µm	2.69 ±0.02	2.65 ±0.07	2.67	2.62
3	1 He, 16 N, 150 µm	3.61 ±0.02	3.39 ±0.07	3.50	3.04
4	2 Ho, 16 N, 150 µm	4.14 ±0.02	3.91 ±0.07	4.03	4.11
5	1 He, 16 N, 150 µm	5.02 ±0.02	4.90 ±0.07	4.96	4.49
6	2 Ho, 16 N, 150 µm	5.60 ±0.02	5.27 ±0.07	5.44	5.14
7	1 He, 16 N, 150 µm	6.48 ±0.02	6.18 ±0.07	6.33	6.12

Table 2.2: Detected layer thicknesses in the cured laminate via in-line measurement systems and cross-section microscopy (LTS: Laser triangulation sensor, SCS: Shadow casting system)

2.6 Conclusions and outlook

After definition of the requirements for measurement systems in order to determine the laminate thickness increase during wet filament winding. Both laser triangulation sensor and shadow casting system meet the requirements and can therefore be considered suitable for the measurement task. In general, the laser triangulation sensor offers higher resolutions, lower costs and easy data evaluation. On the downside, the sensor is only able to provide punctual measurements. The shadow casting system on the other hand is able to evaluate large areas of about 100 mm, depending on the diameter of the telecentric lense used, at the same time.

Trials were done on the effect of different process settings, such as resin loading and fibre band tension, on the laminate thickness increase during wet filament winding. It was shown, that there is a high dependence of thickness increase and the doctor blade gap, which can be explained by the accompanying higher resin loading of the fibre band. Contrary to the expectations, an increase in roving tension force leads to higher laminate thicknesses. This effect is only visible for doctor blade gaps exceeding g = 200 µm.

Due to the low curvature of helix layers in the cylindrical area of a pressure vessel compared to hoop layers, it is estimated that the compaction of helix layers happens after winding further layers on top. In this paper, the laminate thickness of a compacted and uncompacted helix layer was measured and found to be almost equal. But, since the measurement systems are only able to determine the surface position of the uppermost resin layer, there could be differences in the position of the fibre layers in the laminate. This hypothesis has to be evaluated in microscopical cross-section analyses.

In order to investigate the change in laminate thicknesses after curing of a composite, a part of a representative laminate was measured while winding and evaluated via microscopical cross-section analyses after curing. As expected, the layer thicknesses decreased. No statement can be made about whether the compaction happened due to the radial compression force of the layers wound on top or during curing. Since the measurement systems are only able to determine the height of the resin and not the fibre layer, it could be also possible that the fibre layer itself does not compact at all during winding and curing.

Both measurement systems were found to provide useful information for quality assurance during wet filament winding. For further insights with regard to research purposes, it is planned to investigate correlation methods and alternative measurement techniques.

Acknowledgements

The research presented in this publication received financial support from the German Bundesministerium für Verkehr und Digitale Infrastruktur (BMVI) (No. 03B10104D). The authors extend their gratitude to BMVI, Projektträger Jülich (PTJ) and Now GmbH as well as to the project partners (Bundesanstalt für Materialforschung und -prüfung (BAM), BMW AG, cellcentric GmbH Co. KG, Elkamet Kunststofftechnik GmbH, Ford-Werke GmbH, ISATEC GmbH, NPROXX Jülich GmbH, Teijin Carbon Europe GmbH).

References

[HAP+10] HUA, T.; AHLUWALIA, R.; PENG, J.-K.; KROMER, M.; LASHER, S. AND MCKENNEY, K.; LAW, K.; SINHA, J.: *Technical Assessment of Compressed Hydrogen Storage Tank Systems for Automotive Applications*. U.S. Department of Energy, 2010

[HMRL+21] HOPMANN, C.; MAGURA, N.; ROZO LOPEZ, N.; SCHNEIDER, D.; FISCHER, K.: Detection and evaluation of the fibers' deposition parameters during wet filament winding. *Polymer Engineering and Science* (2021), p. 1–15

[HOF+20] HOPMANN, C.; OTREMBA, F.; FISCHER, K.; MAGURA, N.; WIPPERFÜRTH, J.; BECKER, F.; ITALIANO, F.; GIRELLI, A.; SCHNEIDER, D.: New methods for testing and quality assurance of high-performance FRP. *Proceedings of the 30th International Colloquium Plastics Technology*. 2020

[KET18] KLELL, M.; EICHLSEDER, H.; TRATTNER, A.: *Wasserstoff in der Fahrzeugtechnik: Erzeugung, Speicherung, Anwendung*. Wiesbaden: Springer Vieweg, 2018

[Kop20] KOPPENBERG, C.: Digital quality assurance of the wet filament winding process with optical measurement technology. *Proceedings of the AZL - Advanced Thermoset Production Workgroup*. 2020

[LM20] LASN, K.; MULELID, M.: The effect of processing on the microstructure of hoop-wound composite cylinders. *Journal of Composite Materials* 54(26) (2020), p. 3981–3997

[NN19] N.N.: *Regelung Nr. 134 der Wirtschaftskommission der Vereinten Nationen für Europa (UNECE) — Einheitliche Bestimmungen für die Genehmigung von Kraftfahrzeugen und Kraftfahrzeugbauteilen hinsichtlich der sicherheitsrelevanten Eigenschaften von mit Wasserstoff und Brennstoffzellen betriebenen Fahrzeugen (HFCV)*. Brüssel: Europäische Union, 2019

[OYC+18] OEZASLAN, E.; YETGIN, A.; COSKUN, V.; ACAR, B.; OLGAR, T.: The effect of layer-by-layer thickness and fiber volume fraction variation on the machanical performance of a pressure vessel. *Proceedings of the Asme International Mechanical Engineering Congress and Exposition*. 2018

[Ros18] ROSEN, P. A.: *Beitrag zur Optimierung von Wasserstoffdruckbehältern - Volkswagen Aktiengesellschaft AutoUni Schriftenreihe Band 113*. Wiesbaden: Springer, 2018

[VM04] VARGAS, G.; MIRAVETE, A.: Influence of the filament winding process variables on the mechanical behavior of a composite pressure vessel. *WIT High Performance structures and Materials II* 76 (2004),

Symbols

Symbol	Unit	Description
d	mm	Diameter
F	N	Roving pretension force
g	m	Doctor blade gap

Abbreviations

Notation	Description
FVC	Fibre volume content
He	Helix layer
Ho	Hoop layer
LTS	Laser triangulation sensor
PA6	Polyamide 6
PE	Polyethylene
SCS	Shadow casting system

3 Analysis of the thermal and mechanical stress of type-IV pressure vessels during hydrogen cycle tests

Ch. Hopmann[1], T. Gebhart[1], H. Çelik[1], K. Fischer[1]
[1]Institute for Plastics Processing (IKV) at RWTH Aachen University

Abstract

The establishing of hydrogen as a reliable energy carrier is closely linked to the performance and safety level of the storage components in particular. During operation, the storage systems such as composite over-wrapped pressure vessels (COPVs) are exposed to complex physical, mechanical and thermal loads. Since the mechanical and physical properties of the used materials are strongly temperature-dependent, thermal influences must be taken into account during the vessel design. Hydrogen pressure vessels for mobile applications have to perform safely in a wide temperature and filling pressure range, typically $[-40, 85]_{°C} \times [20, 875]_{barg}$. In order to be able to analyse the influence of the vessel geometry, in particular the length-to-diameter ratio, as well as filling conditions on the temperature distribution in the fluid, cyclic hydrogen tests were performed in accordance with ANSI/CSA HGV2 and UN GTR N0.13 / ECE R134. Two types of vessels were investigated showing the same inner and outer diameter but different length. It is observed that the length to diameter ratio substantially influences the temperature distribution within the fluid for room as well as elevated ambient temperatures, namely 23 °C and 50 °C at comparable filling conditions (-40 °C gas inlet temperature, 875 barg pressure at the end of filling, 3 min filling duration). Furthermore, the gas mass flow influences the temperature distribution within the fluid and shows higher spatial variations at higher gas mass flow.

The gas temperature during the cyclic tests is analysed using a measuring device that allows a temperature measurement at eight vertically and horizontally distributed positions. The sensor positions can be adjusted both axially and radially depending on the vessel geometry. This eliminates systematic measurement deviations resulting from the use of different thermocouples. Furthermore, fibre-optic measurements were performed on the vessels surface to analyse the axial and radial evolution of strain during cyclic loads. It is found that the radial strain increases gradually from the vessels pole to the vessel centre in the cylindrical region. A ratio of axial to radial stress of 2.6 was determined. At an internal pressure of 875 barg, an average radial strain of 0.55 % is achieved on the vessel surface, which is in agreement with digital image correlation analyses found in the literature.

3.1 Introduction

Hydrogen technology will play a key role in the field of energy transition in the future, in particular in transport and the intermediary storage of renewable energy without direct use. Composite over-wrapped pressure vessels for compressed gas storage are, along with cryogenic storage, the most developed technology where storage efficiency is critical [BWB17]. In the case of pressure vessels of type-IV, in which both thermoset composites and thermoplastics are used, the materials temperature, among other factors, has a high influence on their mechanical properties [TKE02, ICKY01, LKK18, MJB19]. In particular, the different coefficients of thermal expansion of the materials used can lead to the formation of gaps between the liner and the boss or composite, which can result in a significantly different load condition of the components compared to a fully adherent liner. Furthermore, the permeation properties are dependent on temperature [Sch07, CJ19]. High temperature gradients occur in the pressure vessel wall, especially during rapid filling of COPVs [MBAI+17, MBM+19, SBMAI15]. For mechanical analyses of such vessels under different ambient and filling conditions, the consideration of temperature influences is therefore an important aspect, especially with regard to the lifetime analysis, where cyclical loads are a major requirement of the analyses. It is known that the resulting spatial temperature distribution in a vessel depends on the injector geometry, the injector orientation, the initial/ambient temperature, the direction of filling, e. g. horizontally or vertically filling, as well as the vessel geometry [BABM18, MAMO16, MBM+19]. Furthermore, extensive analyses on the effect of variable filling temperature profiles can be found [BBB+17]. In filling protocols like SAE J2601, homogeneity of the gas temperature is a central assumption. However, depending on the filling procedure and the vessel geometry, this prerequisite is not fulfilled without restrictions [BBB+17]. Moreover, in the majority of analyses found in the literature, only vertical stratification effects were considered [MWTK12, AMM+14, SBMAI15, OAMM15, BBB+17, MAMO16, BBB+17, MBM+19].

In order to analyse the influence of the vessel geometry in particular, type-IV hydrogen pressure vessels were manufactured in the DELFIN project, which differ only in the length of the cylindrical part. The pole area and the laminate layup remained unchanged. A measuring apparatus was developed to analyse both horizontal and vertical stratification effects. Furthermore, the temperature distribution must be measured at different positions relatively to the vessel axis in order to detect zone formation as already found in computational fluid dynamic (CFD) analyses [MBM+19]. Hydrogen cycle tests in accordance with ANSI/CSA HGV2 and UN GTR N0.13 / ECE R134 were carried out to analyse the effect of both the vessel geometry and the gas mass flow on the resulting spatial temperature distribution. The vessel deformation, which represents an important criterion for assessing the mechanically induced stress, is realised with the help of fibre-optic measurements on the surface of the vessel, presented in Section 3.2.

3.2 Materials and methods

The instrumentation used for hydrogen cycle experiments is shown in Figure 3.1. Fullscale vessels have a length of 1500 mm, subscale vessels a length of 650 mm with an outer diameter of approximately 250 mm. The left part of the figure shows the positions of the thermocouples on the surface of the vessel.

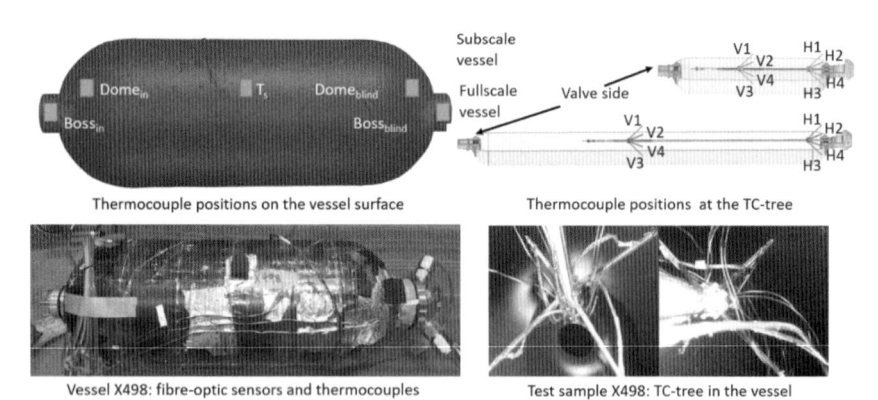

Figure 3.1: Visualisation of the used instrumentation for gas temperature and surface temperature measurement as well as the fibre-optic sensors for strain determination during hydrogen cycle tests.

In the lower left part the test sample X498, a subscale vessel, is shown. This is equipped with thermocouples on the surface of the vessel, gas temperature measurement equipment and fibre-optic sensors. The fibre-optic sensors are applied in five radial segments and four 90° offset axial segments in the cylindrical area of the vessel. A cyanoacrylate-based adhesive M Bond 200, Vishay Precision Group, United States was used for application. The right part of Figure 3.1 shows the designation of the measuring points in the sub- and fullscale vessels. The thermocouples are positioned in the first half and the rear section of the vessel, as seen from the valve side. The measuring points with odd sequence numbers are located in the vertical vessel plane, the measuring points with even sequence numbers in the horizontal plane. This configuration allows to detect both horizontal and vertical stratification effects. The 90° arrangement around the central lance can also detect the gas wobbling during the filling process. In addition, the measuring apparatus consists of an adapter with a pressure-resistant sensor cable bushing in which the measuring apparatus is screwed in. The measurement setup for full- and subscale vessels is designed to be interchangeable. The rear measuring points are placed as close as possible to the inner wall of the vessel, the front ones as close as possible to the presumed cooler area in the case of fullscale vessels. The distance between the thermocouples and the vessel wall is approximately 5 mm. All thermocouples used are type-K, 1.5 mm, class 1, with an expanded measurement uncertainty of 1 K (standard uncertainty times expansion factor k=2). The results of verification measurements are shown in Table 3.1.

Thermocouple	Cold in °C	RT in °C	Hot in °C
H1	-38.3	21.0	94.7
H2	-38.4	21.0	94.7
H3	-38.1	21.0	94.7
H4	-38.3	21.0	94.7
V1	-38.5	20.8	94.5
V2	-38.5	20.8	94.6
V3	-38.4	20.9	94.6
V4	-38.0	21.2	94.9
T279/M404 (reference)	-37.2	21.5	95.6

Table 3.1: Results of the thermocouple verification of the gas temperature measurement device at three different temperatures.

The fibre-optic measurements were performed using an OdiSI 6104 interrogator, Luna Innovations Roanoke, United States. Polyimide coated single-mode optical fibres with an outer diameter of 125 µm were used. The data acquisition was carried out with a spatial resolution of 0.65 mm. The system (including the interrogator, control unit and sensor) accuracy in the strain range $\pm 15.000\,\mu\varepsilon$ is given as $\pm 30\,\mu\varepsilon$, the measurement uncertainty as $\pm 7\,\mu\varepsilon$ [Inn16]. The measurement uncertainty of the composite of sensor fibre, adhesive and test sample is unknown at this time. Two vessels were tested in two separate test chambers for each test. The fibre-optic sensors as well as the thermocouples on the outside of the vessel were placed on both vessels. The gas temperature measurement was performed on one of the vessels.

In Figure 3.2 a schematic drawing of the hydrogen cycle test rig is shown. Two test chambers, into which the vessels are placed, are located in a burst-proof chamber. The chambers are used for permeation measurements and are connected to a mass spectrometer. Permeation measurements can be performed sequentially. The vessels are positioned at an 3.3° angle to the bottom of the chamber. The ambient temperature is recorded above the vessels in the test chamber. The vessels are fed with hydrogen via a supply line. A flow measurement is carried out directly at the hydrogen supply source by means of a mass flow meter. After the flow measurement, the hydrogen is cooled and cleaned. At a distance of 600 cm from the vessels a pressure measurement is carried out, at 295 cm a temperature measurement of the inflowing gas. The supply lines are sheathed with 20 mm Armaflex insulation, Armacell GmbH, Münster, Germany. The measurement setup is symmetrical, i.e. both the supply line length and the test sample volume are the same. The recording of the measurement data (temperature and fibre-optic strain measurement) was carried out with 1 Hz. The measurement uncertainties of the sensors used are shown in Table 3.2.

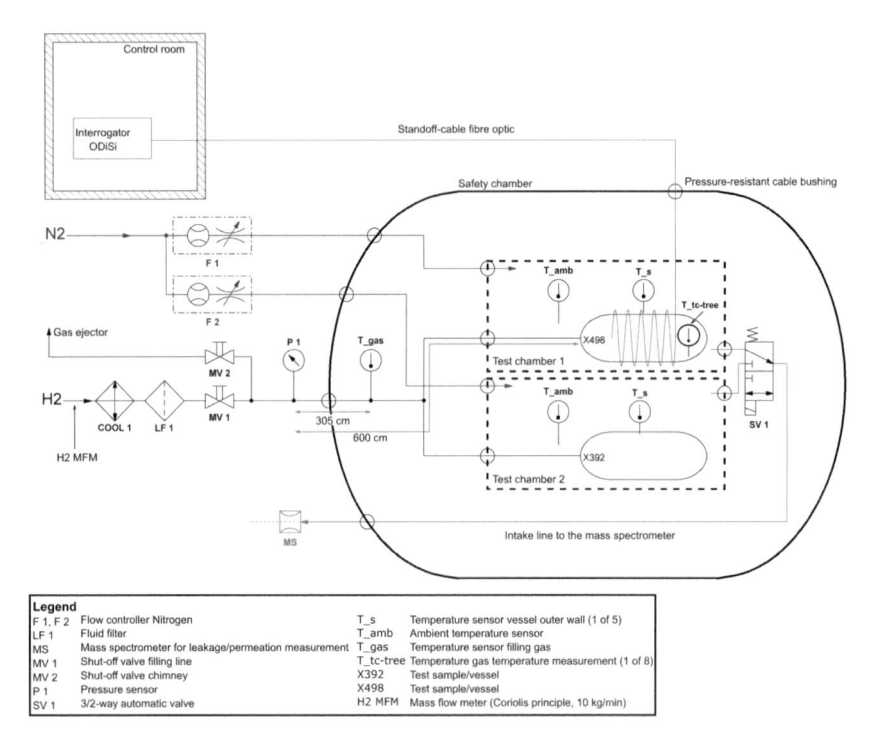

Figure 3.2: Schematic drawing of the hydrogen test stand used for parallel hydrogen cycle testing of two vessels in chamber one and two. The drawing is kindly provided by at TesTneT Engineering GmbH.

Measurand	Range of measurement	Unit	Uncertainty
Mass flow	0 to 10	kg/min	0.2 % (from measured value)
Pressure	0 to 105	MPa	0.5 % (from final value)
Temperature	-200 to 1250	°C	±1 K

Table 3.2: Measurement uncertainties of the used sensors. The uncertainty is given as expanded uncertainty (standard uncertainty times expansion factor k=2).

3.3 Results and discussion of the experimental investigations

In the following, the results of the temperature measurements in Section 3.3.1 are presented, those of the fibre-optic measurements in Section 3.3.2.

3.3.1 Gas temperature measurements

Figure 3.3 shows the gas temperature during filling for different filling sequences.

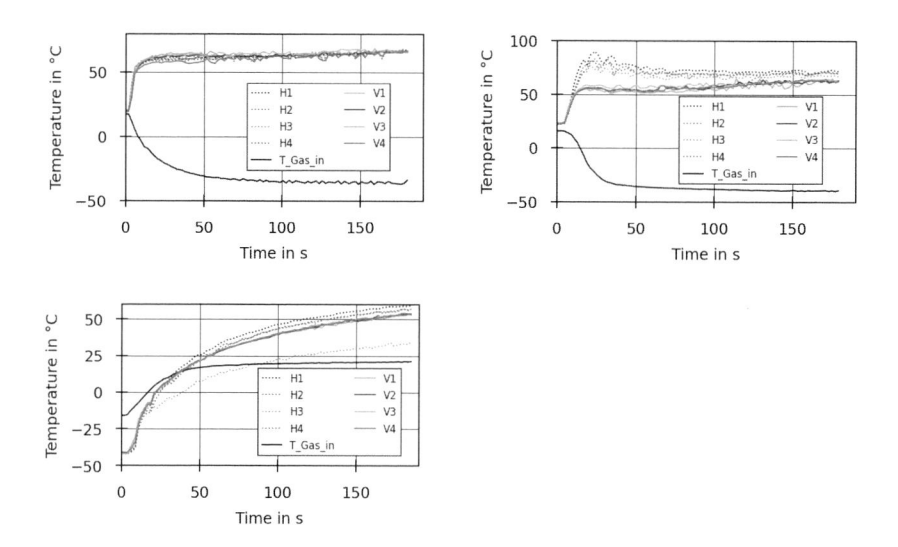

Figure 3.3: Results of the gas temperature development during hydrogen cycle tests for one filling cycle. Subscale vessel, 23 °C ambient temperature, test in accordance with ANSI/CSA HGV2 (upper left). Fullscale vessel, 23 °C ambient temperature (upper right), −40 °C ambient temperature (lower left), test in accordance with UN GTR N0.13 / ECE R134. For the purpose of clarity, only the measured values without measurement uncertainty are shown.

The room temperature (RT) and warm cycles 50 °C of the fullscale tests according to UN GTR No. 13 / ECE R 134 provide for a filling pressure of 875 barg (125 % nominal working pressure (NWP)), the cold cycles of 560 barg (80 % NWP). The subscale tests according to ANSI/CSA HGV2 were carried out at RT and a filling pressure of 875 barg (125 % NWP). A clear difference in the resulting gas temperature can be observed for the RT cycles. For the subscale vessels, an almost homogeneous temperature distribution can be seen at the front and rear measuring points Figure 3.3 (upper left). The temperature differences reach their maximum of 7 K after approximately 20 s of filling and decrease towards the end of filling. In the case of the fullscale vessels at RT, a clear difference of over 25 K can be observed after approximately 20 s of filling from the front to the rear measuring points, which decreases to approximately 10 K by the end of filling, Figure 3.3 (upper right). Furthermore, the fullscale measurement data at the front measurement points (V1 to V4) indicate a wobbling of the gas mass in the vertical plane. This becomes clear through the repeatedly overlapping temperature curves of the measurement points V2 and V3. In the case of the cold cycle, a temperature distribution with approx. 10 K spread can be seen, except for the rear lower measuring position, Figure 3.3 (lower left). The results for fullscale RT cycles agree well with numerical CFD analyses, shown in Figure 3.4.

Horizontal plane   Vertical plane

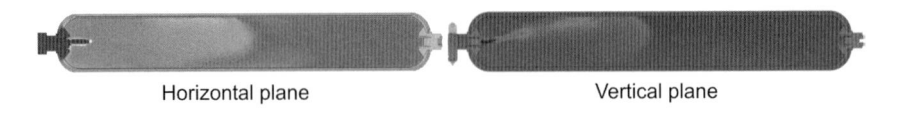

Figure 3.4: Schematic representation of the spatially inhomogeneous temperature distribution during filling using the example of a fullscale vessel. Results of CFD analyses kindly provided by Isatec GmbH.

In the horizontal section, the wobbling of the fluid is indicated by spatially different temperature zones. In the vertical section, the zone formation is clearly visible, which was determined experimentally for RT cycles of the fullscale vessel. The analysis results were provided by Isatec GmbH, a consortium partner of the DELFIN project. The 3D analyses were performed using the CFD solver FloEFD, Siemens, Munich, Germany. The experimentally determined gas inlet temperature and the pressure curve for an RT cycle were used as boundary conditions. Temperature- and pressure-dependent real gas properties as well as adaptive re-discretisation were used for the calculations.

In addition to the different length to diameter ratio, there is a difference in the gas mass flow rate of the fullscale and subscale vessels. It is almost twice as high in the case of fullscale vessels in the case of RT cycles due to the higher internal volume for the same filling time, depicted in Figure 3.5 (left). The gradual steps in the gas mass flow from approx. 50 s and 50 s in case of the subcale and fullscale vessels, respectively, are due to the control of the system.

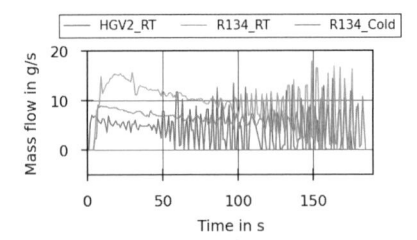

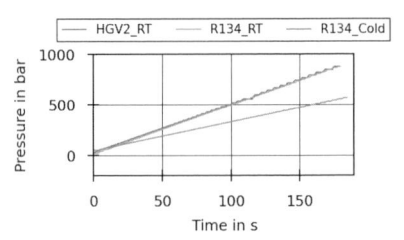

Figure 3.5: Results of the gas mass flow and pressure development during hydrogen cycle tests for one filling cycle. For the purpose of clarity, only the measured values without measurement uncertainty are shown.

The different gas mass flow seems to favour the formation of temperature zones in the case of fullscale vessels. Due to the higher gas mass flow, the inflowing gas seems to hit the inner liner wall with higher energy and hence is deflected. Thus, two temperature zones are formed. As the filling time increases, the entire gas mass seems to start moving which leads to better mixing. This is confirmed by the decrease in the temperature difference of the front and rear measuring points with increasing filling time. Furthermore, the gas mass flow decreases progressively over the filling time. In the case of cold cycles, the gas mass flow is significantly lower than in the RT cycles due to the lower filling pressure for the same filling duration, shown in Figure 3.5 (right). In the fullscale cold cycles, the lower gas mass flow leads to a more homogeneous temperature distribution. The different length to diameter ratio, on the other hand, seems to influence the dispersion of the gas temperature. This becomes clear when comparing the subscale RT and the fullscale cold cycles, shown in Figure 3.3.

After UN GTR No. 13 / ECE R 134 the cyclic test is started cold cycles ($\leq -40\,°C$ ambient temperature). In the first cold cycle, both tested fullscale vessels experienced liner failure in the area of both poles at an internal pressure of approx. 150 barg, visualised in Figure 3.6 (left).

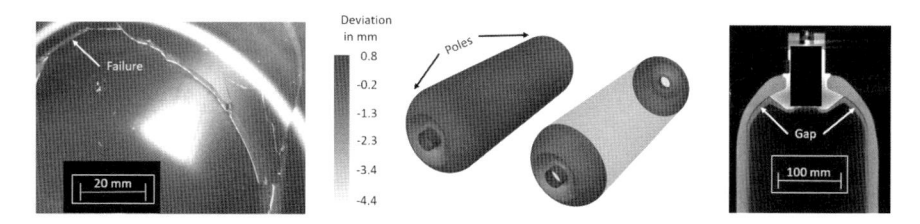

Figure 3.6: Failure characteristics of a liner during the first cold cycle in cycle tests according to UN GTR No. 13 / ECE R 134 (left), a target/actual comparison of the liner geometry after blow moulding (middle) and a CT section of a fullscale vessel (right). The CT-image is kindly provided by Bundesanstalt für Materialforschung und -prüfung (BAM).

The test was immediately terminated due to the high leakage rate. The fracture pattern macroscopically indicates a predominantly brittle failure of the liner material. The cracks extend around the entire pole area and are located in the area of the liner-boss contact area. A target/actual comparison of the liner outer geometry by means of 3D measurements (ATOS5, GOM GmbH, Germany), shown in Figure 3.6 (middle), indicates high actual deviations in the pole area of the liner after blow moulding. This applies to both full- and subscale vessels. In the preparation of the winding process, the liners and bosses are joined, which leads to a pre-tension in the liner pole area. This pre-stress is partially relieved by relaxation during the curing of the composite winding at elevated temperature due to the viscoelastic properties of the Polyamide 6 (PA6) liner material. Furthermore, computed tomography (CT) examinations were carried out with a Helix-CT scanner (BAM, Germany) with 600 kV high-power X-ray tube and 400 µm resolution. CT scans of the vessels, visualised in Figure 3.6 (right), show a gap between liner and boss as well as liner and composite.

A gap between the liner and the boss or carbon fibre winding was also detected by means of internal pressure-less CT analyses. Hence the liner can come loose at a pressure difference of 20 barg. This very likely led to liner failure in combination with the low temperatures which causes the liner to shrink and an associated reduction in the ductility of the liner material behaviour. Therefore the test sequences of the R134 cycle tests were changed in the following tests. Warm and RT cycles were carried out first to give the liner material time to close a possible gap through retardation and due to the increased temperature. Furthermore, the minimum pressure was increased from 20 barg to 50 barg to prevent the liner from detaching from the fibre composite winding and the bosses. The cold cycles were completed without any failures. This shows in particular the high relevance of the temperature influence, especially the low temperatures, on the failure behaviour of type-IV pressure vessels. The order of the temperature steps in the gas pressure cycle tests according to UN GTR No. 13 / ECE R 134 is therefore essential to detect a possible liner failure. Regarding first filling, the focus should be placed on cold temperatures in the vessel design.

3.3.2 Fibre-optic strain measurements

Figure 3.7 shows the results of the fibre-optic strain measurement at 875 barg internal pressure in the axial direction for a subscale vessel.

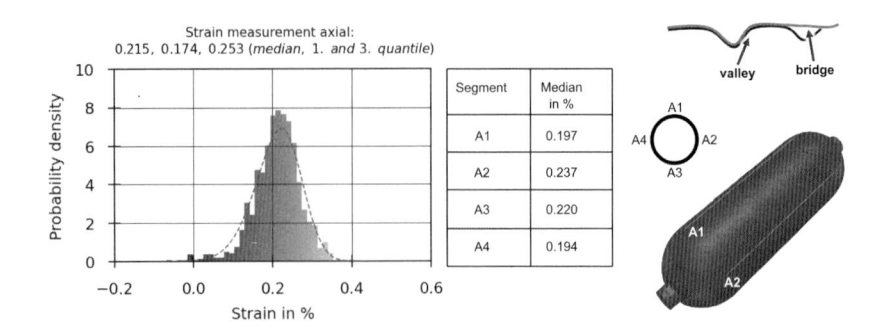

Figure 3.7: Results of the axial strain measurement during hydrogen cycle tests using fibre-optic sensors. The boundary conditions are 875 barg pressure and 23 °C ambient temperature. Test in accordance with ANSI/CSA HGV2. The strain is shown for all axial segments in normalised form as well as the probability distribution function assuming a skewed normal distribution.

Four sensor segments were applied at 90° intervals and evaluated. Segments A1 and A3 are located in the vertical plane of the vessel, segments A2 and A4 in the horizontal plane. The resulting probability density of the strain measurement values over the measurement results of all segments follows approximately a mildly left-skewed normal distribution. The median of the axial strain, which almost corresponds to the mode, is 0.215 % strain. The median distance of the fist quantile is 0.041 %, that of the third quantile 0.038 %. The median differs from segment to segment. Segments A1 and A4 with 0.197 % and 0.194 % have a lower median than segments A2 and A3 with 0.237 % and 0.220 %. The deviations lie within the first and third quantiles when considering all segments. However, if we consider the injector alignment (segment A4), the temperature influence caused by the impact of the inflowing gas on the vessel wall seems to be a conceivable possibility. Especially since 3D measurements (ATOS5, GOM GmbH, Germany) of the vessels show rotational symmetry with respect to the vessel axis. In addition, the measurements show compression at globally positive strain on the vessel outer surface due to the vessel expansion in axial in radial direction as well as very high strain maxima with partly more than 0.4 % strain. Since the last layer is a circumferential layer, the sensor crosses the fibre bands of the vessel in an angle of approximately 90° which leads to a high degree of ondulation of the sensors. As a result the sensors can experience compression in areas in local minima of the vessel surface, referred to here as valley. Furthermore, the ondulation of the sensor leads to micro-bending, which can lead to unrealistic outliers. This is underlined by a high not a number (NaN) content of 65 % of the measured values. NaN values occur at positions where the ODiSI software cannot find a plausible cross-correlation of the reference to the current measured frequency shift and thus no information can be output for the evaluation point.

Figure 3.8 shows the individual values of the strain measurement at 875 barg over the sensor path length. Both the strain minima and maxima appear to be outliers. This is supported by the fact that the strain minima and maxima always occur in a region of high NaN density. The choice of the first and third quantile as the limit for probable measurement results seems plausible. However, it must be mentioned that these are arbitrarily chosen limits. An influence of the surface waviness on the probability of the occurrence of outliers must be further investigated.

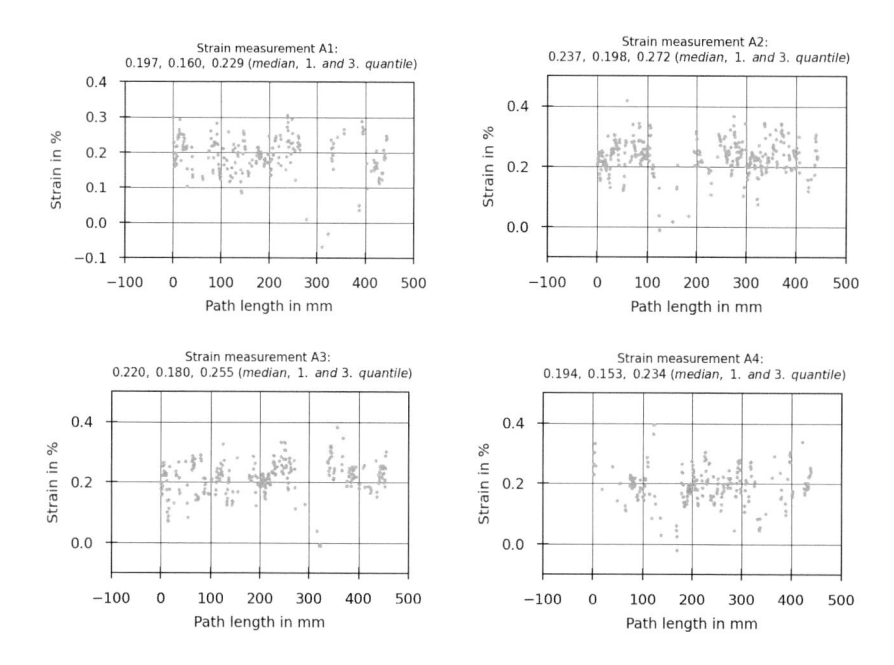

Figure 3.8: Results of the axial strain measurement during hydrogen cycle tests using fibre-optic sensors. The boundary conditions are 875 barg pressure and 23 °C ambient temperature. Test in accordance with ANSI/CSA HGV2. The strain is shown for each sensor segment over the path length. For the purpose of clarity, only the measured values without measurement uncertainty are shown.

Figure 3.9 shows the results of the fibre-optic strain measurement at 875 barg internal pressure in the radial direction for a subscale vessel. Five sensor segments were applied at 90° intervals and evaluated. Segments R1 and R5 are located in the transition region of the cylindrical and pole part of the vessel. Segment R3 is in the middle of the vessel and segments A2 and A4 between the already mentioned segments. The resulting probability density of the strain measurement values over the measurement results of all segments follows a mildly left-skewed normal distribution. The median of the radial strain, which almost corresponds to the mode, is 0.556 % strain. Thus there is a factor of 2.6 between axial and radial strain. The median distance of the fist quantile is 0.03 %, that of the third quantile 0.028 %. The quantile distance is hence smaller than for the axial measurements. The median of the segments increases from the poles towards the centre of the vessel. The difference in strain medians from R1 and R5 to R3 is 0.045 and 0.074 %, respectively, which is outside the first and third quantiles. Thus, the difference can be assumed to be significant. A higher radial vessel deformation in the cylindrical area in the centre of the vessel seems to be a plausible measurement result.

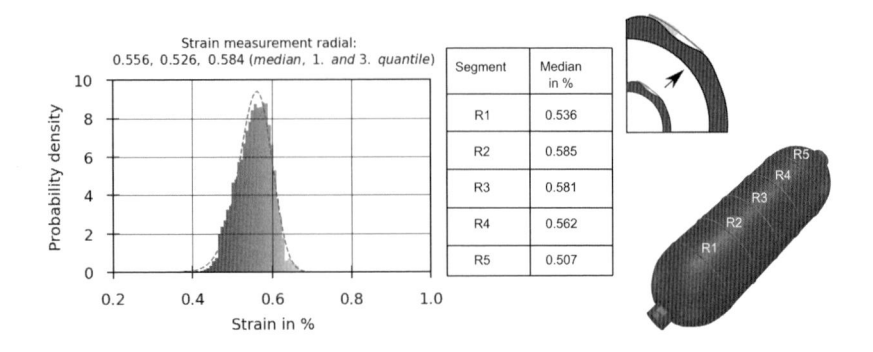

Figure 3.9: Results of the radial strain measurement during hydrogen cycle tests using fibre-optic sensors. The boundary conditions are 875 barg pressure and 23 °C ambient temperature. Test in accordance with ANSI/CSA HGV2. The strain is shown for all axial segments in normalised form as well as the probability distribution function assuming a skewed normal distribution.

Figure 3.10 shows the values of individual segments of the radial strain measurement at 875 barg over the sensor path length. A wave-like pattern of the strain values can be seen. This can be explained on the one hand by the fact that the last circumferential layer does not coincide exactly with 90° winding angle as the sensor fibre. This means that the sensor fibre does not follow a fibre band exactly and thus there is also a slight ondulation of the sensor. The local maxima and minima of the strain values always occur at the transitions of the waves. Therefore, these points seem to be outliers. The wave-like pattern can also be explained by the varying thickness of the adhesive layer. In areas of local minima of the vessel surface, it is higher, which may favour a certain damping by the adhesive between the sensor fibre and the vessel surface. The amplitude of the waves is about 0.05 % strain, corresponding to the first and third quantile distance. A strain at the vessel surface, which lies between the median and the third quantile, is therefore considered plausible. A significantly lower number of outliers also corresponds to a significantly lower NaN share of 5 %.

It is shown that fibre-optic sensors attached to the surface of the vessel can determine realistic strain values, which agree well with the measurement data found in the literature and determined by means of digital image correlation [NAB+20]. Outliers can be caused by the surface waviness of the vessels and must be reliably distinguished from local extremes induced by a damage event of the vessel by means of an evaluation routine in the future. Fibre-optic sensors also represent an alternative to pneumatic tests for structural health monitoring for vessels that have already been certified, since there is no intervention in the manufacturing process as if the fibre were embedded directly within the laminate. Furthermore, fibre-optic measurements can provide important reference data for the verification of the numerical structural design.

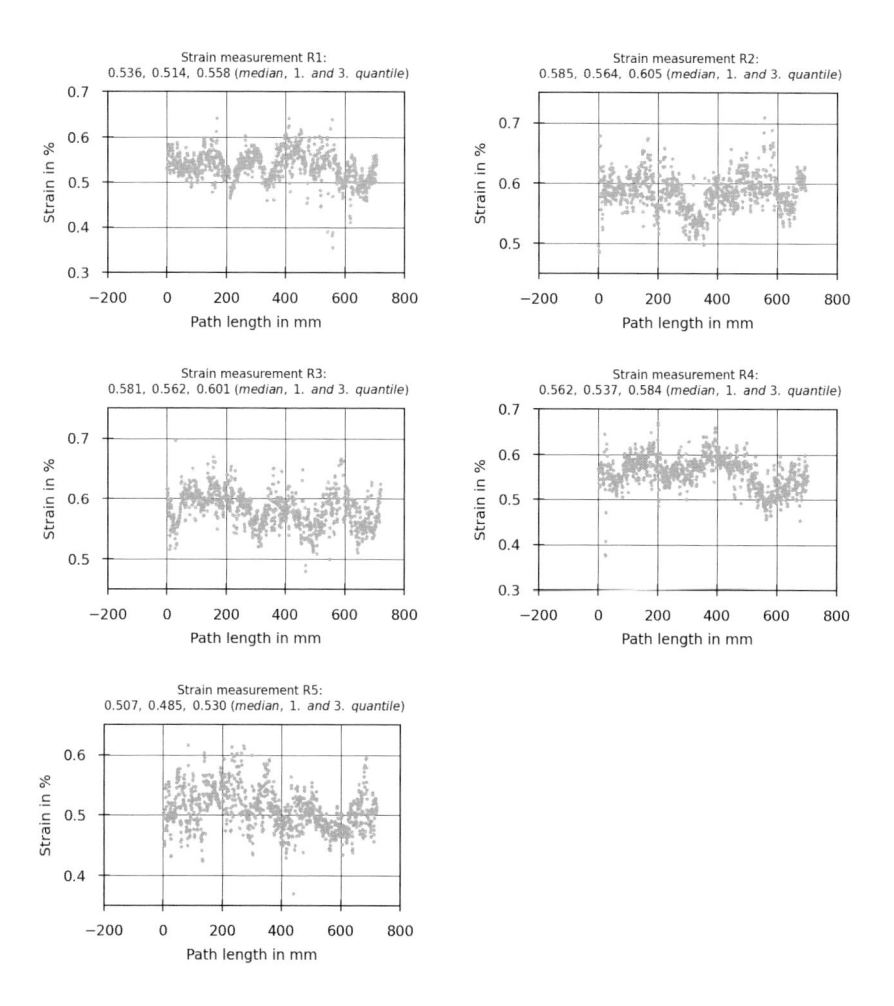

Figure 3.10: Results of the radial strain measurement during hydrogen cycle tests using fibre-optic sensors. The boundary conditions are 875 barg pressure and 23 °C ambient temperature. Test in accordance with ANSI/CSA HGV2. The strain is shown for each sensor segment over the path length. For the purpose of clarity, only the measured values without measurement uncertainty are shown.

3.4 Conclusions and outlook

Gas temperature measurements were carried out on both subscale and fullscale vessels, which differed only in the vessel length. Significant differences in the gas temperature were found. In the case of small vessels with a length to diameter ratio of 2.6, a homogeneous temperature distribution over almost the entire filling time was determined on the basis of eight horizontally and vertically distributed measuring points. Differences of [10;25] K were observed for fullscale vessels with a length to diameter ratio of 6. Both a zone formation and a wobbling of the gas mass in the horizontal vessel plane could be detected. Depending on the gas mass flow, the temperature differences can be reduced in the case of cold filling according to R134. This should be repeated for RT as well as warm fillings in order to determine a causal relationship between gas mass flow and maximum temperature differences or zone formation. During the first cold cycles of tests according to R134, two of two fullscale vessels showed liner failure in both pole regions. A detachment of the liner due to thermally induced shrinkage seems to be a plausible cause. A gap between the liner and the boss or carbon fibre winding was also detected by means of internal pressure-less CT analyses. This means that the liner can come loose at a pressure difference of 20 barg. Therefore, a firm connection between the liner and boss or carbon fibre winding cannot be assumed. The low temperatures in combination with a sudden increase in internal pressure due to the filling led to a macroscopically brittle fracture pattern. Low ambient temperatures in combination with a target/actual deviation of the liner outer geometry can lead to a total failure of the vessels and therefore should be accounted for in vessel design. Furthermore, fibre-optic strain measurements were carried out on the surface of the vessel. A ratio of 2.6 between radial and axial strain was determined. A temperature influence is suspected for the axial measurements. An increase in strain from the pole areas towards the centre of the vessel was observed. Fibre-optic measurements thus provide highly spatially resolved measurement results in order to be able to assess the vessel deformation during cyclic tests. An influence of the surface waviness of the vessels on the probability of the occurrence of outliers and the influence of adhesive thickness must be further investigated. Fibre-optic sensors should be further used for spatially high fidelity gas temperature measurements.

Acknowledgements

The research presented in this publication received financial support from the German Bundesministerium für Verkehr und Digitale Infrastruktur (BMVI) (No. 03B10104D). The authors extend their gratitude to BMVI, Projektträger Jülich (PTJ) and Now GmbH as well as to the project partners (Bundesanstalt für Materialforschung und -prüfung (BAM), BMW AG, cellcentric GmbH Co. KG, Elkamet Kunststofftechnik GmbH, Ford-Werke GmbH, ISATEC GmbH, NPROXX Jülich GmbH, Teijin Carbon Europe GmbH).

References

[AMM+14] Acosta, B.; Moretto, P.; de Miguel, N.; Ortiz, R.; Harskamp, F.; Bonato, C.; Miguel, N.; Ortiz Cebolla, R.; Harskamp, F.; Bonato, C.: JRC reference data from experiments of on-board hydrogen tanks fast filling. *International Journal of Hydrogen Energy* 39 (2014) 35, p. 20531–20537

[BABM18] Bourgeois, T.; Ammouri, F.; Baraldi, D.; Moretto, P.: The temperature evolution in compressed gas filling processes: A review. *International Journal of Hydrogen Energy* (2018), p. 2268–2292

[BBB+17] Bourgeois, T.; Brachmann, T.; Barth, F.; Ammouri, F.; Baraldi, D.; Melideo, D.; Acosta-Iborra, B.; Zaepffel, D.; Saury, D.; Lemonnier, D.: Optimization of hydrogen vehicle refuelling requirements. *International Journal of Hydrogen Energy* 42 (2017) 19, p. 13789–13809

[BWB17] Barthelemy, H.; Weber, M.; Barbier, F.: Hydrogen storage: Recent improvements and industrial perspectives. *International Journal of Hydrogen Energy* 42 (2017) 11, p. 7254–7262

[CJ19] Craster, B.; Jones, T. G.: Permeation of a range of species through polymer layers under varying conditions of temperature and pressure: In situ measurement methods. *Polymers* 11 (2019) 6, p. 1–21

[ICKY01] Im, K.-H.; Cha, C.-S.; Kim, S.-K.; Yang, I.-Y.: Effects of temperature on impact damages in CFRP composite laminates. *Composites: Part B* 32 (2001) 8, p. 669–682

[Inn16] Innovations, L.: *ODiSI-B Sensor Strain Gage Factor Uncertainty*. Luna Innovations, 2016

[LKK18] Lüders, C.; Krause, D.; Kreikemeier, J.: Fatigue damage model for fibre-reinforced polymers at different temperatures considering stress ratio effects. *Journal of Composite Materials* 52 (2018) 29, p. 4023–4050

[MAMO16] de Miguel, N.; Acosta, B.; Moretto, P.; Ortiz Cebolla, R.: Influence of the gas injector configuration on the temperature evolution during refueling of on-board hydrogen tanks. *Int. J. Hydrogen Energy* 41 (2016) 42, p. 19447–19454

[MBAI+17] Melideo, D.; Baraldi, D.; Acosta-Iborra, B.; Ortiz Cebolla, R.; Moretto, P.: CFD simulations of filling and emptying of hydrogen tanks. *International Journal of Hydrogen Energy* 42 (2017) 11, p. 7304–7313

[MBM+19] Melideo, D.; Baraldi, D.; Miguel, N.; Acosta, B.; De Miguel Echevarria, N.; Acosta Iborra, B.: Effects of some key-parameters on the thermal stratification in hydrogen tanks during the filling process. *International Journal of Hydrogen Energy* 44 (2019) 26, p. 13569–13582

[MJB19] Mahl, M.; Jelich, C.; Baier, H.: On the temperature-dependent non-isosensitive mechanical behavior of polyethylene in a hydrogen pressure vessel. *Procedia Manufacturing* 30 (2019), p. 475–482

[MWTK12] Monde, M.; Woodfield, P.; Takano, T.; Kosaka, M.: Estimation of temperature change in practical hydrogen pressure tanks being filled at high pressures of 35 and 70 MPa. *International Journal of Hydrogen Energy* 37 (2012) 7, p. 5723–5734

[NAB+20] Nebe, M.; Asijee, T. J.; Braun, C.; van Campen, J. M.; Walther, F.: Experimental and analytical analysis on the stacking sequence of composite pressure vessels. *Composite Structures* 247 (2020),

[OAMM15] ORTIZ CEBOLLA, R.; ACOSTA, B.; MIGUEL, N.; MORETTO, P.: Effect of precooled inlet gas temperature and mass flow rate on final state of charge during hydrogen vehicle refueling. *International Journal of Hydrogen Energy* 35 (2015),

[SBMAI15] SIMONOVSKI, I.; BARALDI, D.; MELIDEO, D.; ACOSTA-IBORRA, B.: Thermal simulations of a hydrogen storage tank during fast filling. *International Journal of Hydrogen Energy* 40 (2015) 36, p. 12560–12571

[Sch07] SCHULTHEISS, D.: *Permeation Barrier for Lightweight Liquid Hydrogen Tanks*. Universität Augsburg, Dissertation, 2007

[TKE02] TSCHOEGL, N. W.; KNAUSS, W. G.; EMRI, I.: The Effect of Temperature and Pressure on the Mechanical Properties of Thermo- and/or Piezorheologically Simple Polymeric Materials in Thermodynamic Equilibrium – A Critical Review. *Mechanics of Time-Dependent Materials* 6 (2002) 1, p. 53–99

Abbreviations

Notation	Description
BAM	Bundesanstalt für Materialforschung und -prüfung
CT	Computed tomography
COPVs	Composite over-wrapped pressure vessels
NaN	Not a number
PA6	Polyamide 6
RT	Room temperature

Manufacture of high-precision optical lenses

Moderator: Dipl.-Ing. Arne Schmidt, Röhm GmbH

Content
---

1 Plastics in optical application: Opportunities and challenges   305

A. Schmidt[1]
[1] Röhm GmbH

2 Automatic thermal mould design for optical parts in compression moulding   315

Ch. Hopmann[1], D. Fritsche[1], T. Hohlweck[1]
[1] Institute for Plastics Processing (IKV) at RWTH Aachen University

3 Development of a coupled thermo-optical process optimisation routine for plastic lenses in laser applications   330

Ch. Hopmann[1], C. Holly[3], J. Stollenwerk[2,3], B. Liu[1], J. Gerads[1], J. Hofmann[3]
[1] Institute for Plastics Processing (IKV) at RWTH Aachen University
[2] Fraunhofer Institute for Laser Technology ILT, Aachen, 52074 Aachen, Germany
[3] Chair for Technology of Optical Systems (TOS) at RWTH Aachen University

Dipl.-Ing. Arne Schmidt

Dipl.-Ing. Arne Schmidt is Head of Technical Marketing Automotive at Röhm GmbH. He holds a Plastics Engineering degree from the University of Applied Sciences Darmstadt and a Mechanical Engineering degree from the Technical University of Chemnitz. Followed by employment as research assistant at the University of Darmstadt in the field of Material Science. Finally, he joined the former Degussa AG in the business unit Methacrylates. In his actual role as Head of Technical Marketing Automotive, he translates the trends and requirements of the automotive industry into acrylic resin products.

1 Plastics in optical application: Opportunities and challenges

A. Schmidt[1]
[1] Röhm GmbH

Abstract

Optical plastics have been known and used for about 100 years. The LED as a point light source has significantly increased the use of plastics for optical parts. As a result of these new market needs, various developments were necessary. The precise optical characterization of plastics for the design and simulation of optics, as well as various developments in the field of injection moulding technology.

The optical characterisation of plastics follows well-known standards, which are used for inorganic glass since decades. These standards are now adapted to plastics and consider the temperature-dependent property profile of plastics. Plastics can be processed cost-efficiently and in large quantities into complex optics by using injection moulding. Due to their organic polymer structure, plastics are subjected to various physicochemical processes, which influence the optical quality of the moulded parts. During injection moulding the molecular weight can be reduced, additives converted or destroyed and thermooxidation can lead to the formation of chromophore substances. These conversion processes mainly depend on the processing parameters and the used polymer formulation.

Current simulation systems can analyse the processing parameters in terms of geometry and tolerances of optics. Simulation systems that can predict the resulting optical properties such as transmission, yellowness value, birefringence are not available.

1.1 History and market overview

1.1.1 History

Compared to the more than 3000 years of history of producing and processing inorganic glass, the approximately 100 years of experience in producing and processing transparent plastics is a very short period of time.

In Berlin in 1839, pharmacist Eduard Simon observed the polymerisation of styrene to polystyrene, a factor which led to the invention of one of the first transparent plastics. Despite this discovery, it took almost 100 years before commercial use began in 1931, when I.G. Farben-Werke started polystyrene production. However, the first optically transparent film material was celluloid, a product used as a carrier for photographic films, which was used from 1887.

Since then, plastics development has progressed rapidly. The first technical transparent thermoplastic that provided access to many optical applications – such as precise optical lenses, magnifying glasses, aircraft canopies, etc. – was polymethyl methacrylate (PMMA), invented in Darmstadt, Germany, in 1933 by chemist Dr. Otto Röhm and his team. PMMA is better known under the brand name PLEXIGLAS®, which is still used today.

Over the following years, other plastics that could be used to manufacture optical components were developed. In 1936, BASF in Ludwigshafen began production of styrene acrylonitrile (SAN), while in 1953, both Bayer AG and General Electric discovered polycarbonate (PC) simultaneously. One year later, in 1954, Hoechst AG added cyclo olefin copolymer (COC) to its portfolio. In 1965, Union Carbide introduced polysulfones (PSU) to the market.Transparent polyamides followed three years later, following the discovery of a method, which could reduce the formation of crystalline structures in polyamides which cause an opaque appearance. In 1970, polyethylene terephthalate (PET) was processed into transparent parts for the first time.

1.1.2 Definition of "optical parts"

Optics (Greek: optike: "study of the visible", optiko: "belonging to seeing", opsis: "seeing") is defined as a branch of physics that studies the behaviour and properties of light and its interaction with materials. The following definition can be used to specify "plastic optics":

A shaped part made of an organic polymer material which transmits, directs or scatters light in a predefined manner.

1.1.3 Market overview for optical plastics

Today, different plastics are used to manufacture optical applications. Figure 1.1 comprises these plastics that have become established for optical applications, and divides them into standard, technical and high-temperature materials in the "pyramid of plastics".

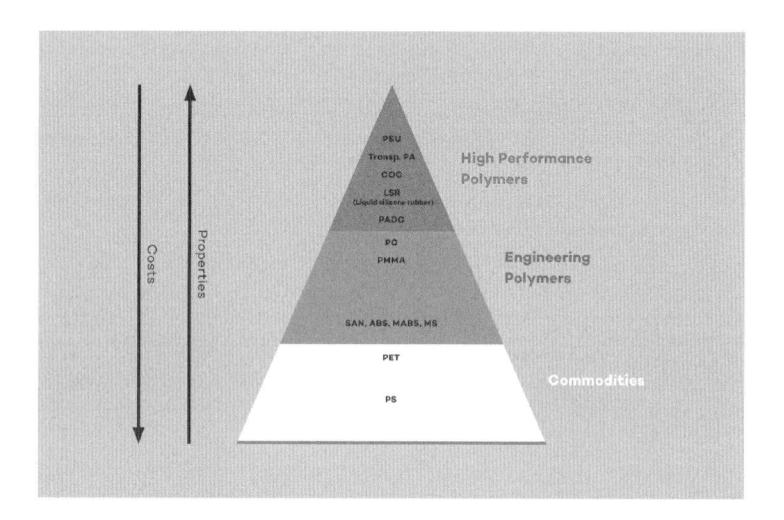

Figure 1.1: Classification of transparent plastics

Apart from liquid silicone rubber (LSR) and polyallydiglycol carbonate (PADC), Figure 1.1 contains only thermoplastics. However, these non-thermoplastic materials are also used in the field of plastic optics. PADC, for example, is a common material used in plastic eyeglass lenses under the well-known brand name "CR-39" (developed in 1945 by Columbia Southern Chemical Company, CR-39 for "Columbia Resin 39"). LSR materials, as well as epoxy-based thermoset moulding compounds, are used to create primary lenses of high-power LED, for example. LSR is also used in solar concentrator systems which need excellent thermal properties and UV stability.

1.1.4 Raw material suppliers for optical polymers

PMMA	PC	PS	ABS / SAN / ASA / MABS	PSU	Transp. PA	COC
Röhm	Covestro	BASF	BASF	Amoco	EMS Chemie	Mitsui
Trinseo	Dow	Dow	Lanxess/ Ineos	BASF	Evonik	Topas Advanced Polymers
Chi Mei	GE	Nova	Polimeri			Zeon Corporation
LG MMA	Idemitsu	Total	Dow			
Lucite/ Mitsubishi	Sabic	Others	Sabic			
Asahi	Teijin		Sinopec			
Sumitomo	Asahi		Chi Mei			
	Honam		LG			
	Mitsubishi		Cheil			
			Ineos			

Table 1.1: Suppliers of transparent plastics

1.2 Optical data

The following section summarises and explains the most relevant optical parameters for polymer raw materials that can be used to manufacture optical components.

1.2.1 Refractive index

The refractive index (n) is a dimensionless number, which describes the speed of light in vacuum (C_O) in relation to the speed of the light in a medium (C_M) [NN99]. Dutch mathematician *Willebrord Snellius* (1580-1626) discovered the fundamental approach for this, hence why it is known as "Snell's law":

$$n = \frac{C_0}{C_M} = \frac{sin(\alpha)}{sin(\beta)} \qquad (1.1)$$

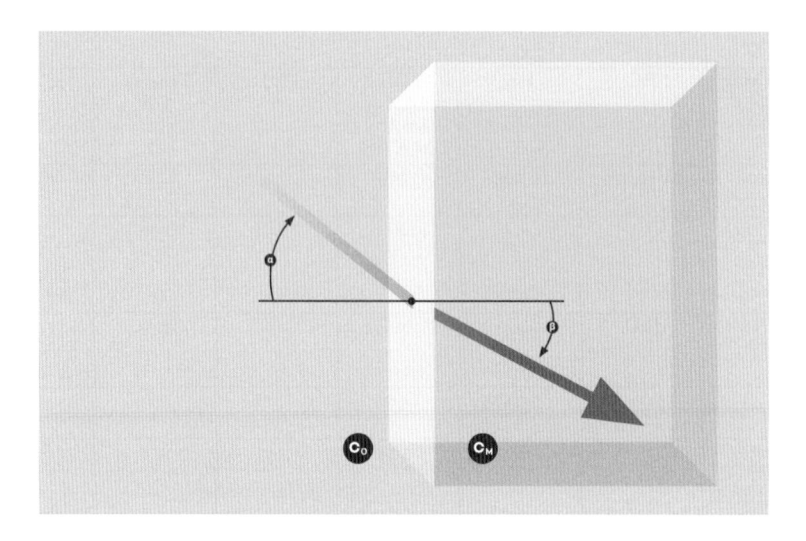

Figure 1.2: Refractive index

1.2.2 Refractive index and Abbe number

The refractive index of all optical materials depends on the wavelength of the incoming light. The Abbe number (V_D) describes the change of refractive index versus wavelength [NN98]. This dimensionless number provides a scale for the optical dispersion of a material. A high Abbe number implies a low dispersion.

$$V_D = \frac{n_{587.5618 \text{ nm}} - 1}{n_{486.1327 \text{ nm}} - n_{656.2725 \text{ nm}}} \quad (1.2)$$

1.2.3 Refractive index and temperature

For every optical polymer, the refractive index is a function of the temperature. In a certain temperature range, the refractive index displays linear behaviour with temperature. During this phase, the material is in a completely frozen, amorphous state. At a defined temperature, the mobility of polymer chain segments increases and the linear relation of the refractive index is lost. This behaviour is shown in Figure 1.3, based on an example of a standard PMMA grade (PLEXIGLAS® 8N):

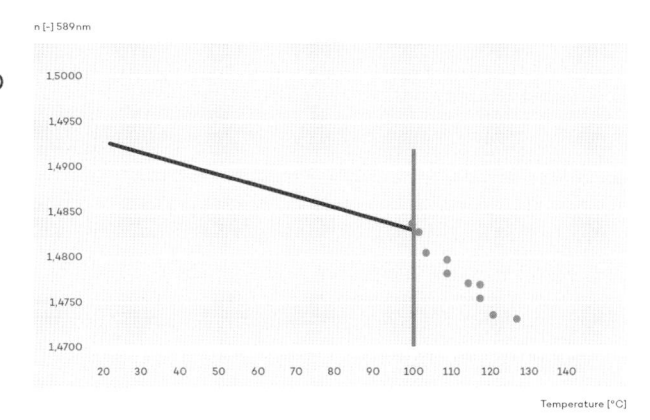

Figure 1.3: Refractive index of PLEXIGLAS® 8N at 589 nm as a function of temperature

1.2.4 Transmittance

The spectral transmittance (λ, D) is used to simplify the evaluation of the light transmittance for different wavelengths λ in finished transparent plate materials of different thicknesses D. It considers both the absolute transmittance (τ_i) of the material and the Fresnel reflections of both interfaces/surfaces. The measurement is based on ISO 13468 [NN97].

The overall transmittance of optical systems with multiple layers of different refractive indices (n) is affected by each transition of light at the interface from one optical layer to another. As the transmittance depends on the length of the optical path, indicating the thickness is essential. Transmittance values cannot be compared without specifying the thickness of the material. A high wall thickness is recommended for optical measurements, as optical effects only become visible here.

1.2.5 Optical loss absorption coefficient

Simulating optical systems requires knowledge of the absorption properties within the material with arbitrarily designed interfaces and thus different optical path lengths. This information is only indirectly contained in the transmittance . The transmittance specification is always associated with a thickness D and does not represent a pure material parameter. This behaviour is illustrated in Figure?? using the example of different absorption data for PMMA and a modified PMMA product.

1.2.6 Birefringence

Upon entering a material, light can be split in the two directions of polarisation. Some materials show a tendency for a variance in the refractive index for the different directions of polarisation. To observe birefringence, the optical material needs to show a tendency for that effect. For polymers, mechanical stress and intrinsic orientations will amplify this effect.

1.3 Performance of injection moulded optics

The optical properties of raw materials used to manufacture precise optical components can be determined by using several standards, e. g.:

- VDI/VDE 5596: "Optical Design for manufacturing - illumination optics, non-imaging optics and freeform optics - Optical material parameters"
- DIN 5033: "Colorimetry"
- DIN 5036-3: "Radiometric and photometric properties of materials; methods of measurement for photometric and spectral radiometric characteristics"

Manufacturers of plastic raw materials use these standards to characterise the optical properties of their products. However, the raw materials need to be processed through injection moulding, which has a significant influence on the optical properties of the moulded parts. During this processing step, the polymers are subjected to various physicochemical processes, which influence the optical quality of the moulded parts, e. g.:

- Thermo-Oxidation, which can lead to the formation of chromophore substances
- Polymer chain degradation, which can lead to lower molecular weight
- Degradation of additives such as UV-stabilisers and absorbers, scattering particles, thermal stabilisers
- Internal stress which can lead to birefringence

In section 1.2.1, the correlation of injection moulding and spectral transmittance is explained using the example of PMMA for a light guide.

1.4 Experimental setup

To analyse the influence of injection moulding parameters on the spectral transmittance of PMMA light guides, Röhm GmbH, Darmstadt/Germany, has conducted studies using the following processing equipment:

- Injection moulding machine: KraussMaffei KM250CX from KraussMaffei, Munich, Germany, 45 mm standard thermoplast screw, open nozzle
- Mould for light guide: optical length 950 mm, cross section: 10 mm x 10 mm, high-gloss optical finish. cold runner (see Figure 1.4)

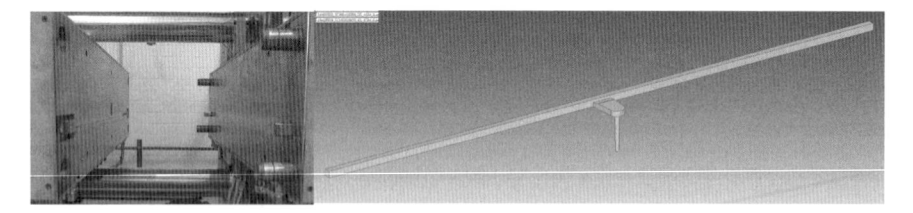

Figure 1.4: Light guide mould and light guide

For optical measurements, Röhm developed a "long path transmittance" spectrophotometer (see Figure 6) with a 5 nm measurement frequency step [NN21]. A halogen lamp with a spectrum between 360 nm and 2400 nm is used as a light-source. The test setup is validated for specimens up to 1000 mm. The device has an accuracy of $+/-$ 1 % in spectral transmittance.

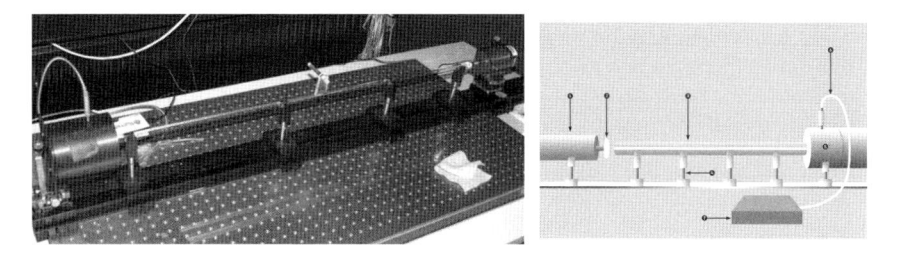

Figure 1.5: Long-path transmittance spectrophotometer

The light guides are moulded using PLEXIGLAS® 8N clear. The process parameters follow a design of experiment (DoE) schedule. Pre-drying of pellets for 4 hours at 80 °C in a dry-air dryer, while specimens are conditioned for 24 hours at 23 °C and 50 % relative humidity prior to testing. A minimum of 5 specimens are tested per inspection batch.

1.4.1 Results

The Yellowness Index for Standard Illuminant D65/10° (YI) [NN97] is used to compare all results. This index describes the change in color of a test sample from clear or white to yellow, which in turn has an impact on the optical quality of precise optical lenses. Table 2 shows the main process parameters (highest and lowest setting) in relation to the YI and its variance.

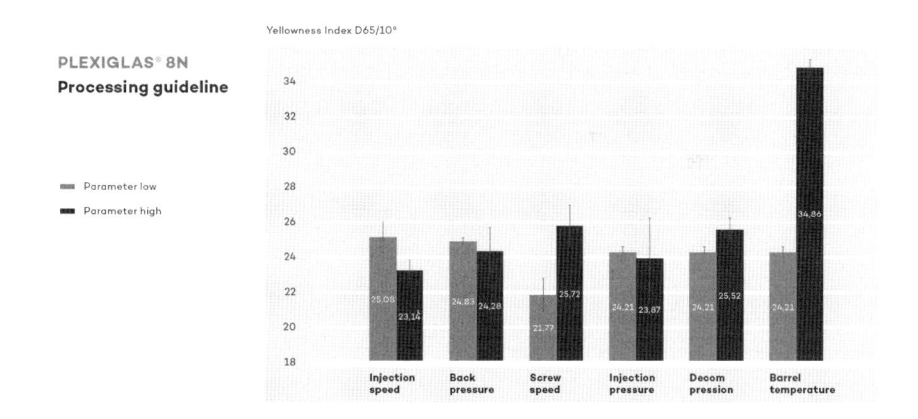

Figure 1.6: Results of optical quality of light guides in relation to process parameters

During the studies, the barrel temperature and screw rotation speed were identified as being the significant parameters that caused changes in the YI for a standard injection moulding process using a standard PMMA. For the PMMA polymer, it can be concluded that the thermo-oxidative reactions in the barrel are the primary root cause for the formation of chromophore substances, which are responsible for a higher Yellowness Index of the parts. During the process setup of a new mould for PMMA optics, the machine operator should therefore pay special attention to the plastification process in order to achieve the best results.

The results of this study are only valid for standard PMMA resins. By using special grades with different additive packages, the mechanism of polymer degradation may change and cause a different result. The polymer degradation mechanism hast to be determined in detail for each polymer and polymer formulation.

1.5 Conclusions

This report introduces the established optical polymers and their technical classification, as well as some manufacturers. It also explains the main optical parameters and measurement methods used to determine the properties of the raw materials. These are common methods and based on ISO standards. The optical material properties are determined by the manufacturers of the raw materials and can be found in the technical data documentation of the products.

The nature of polymers results in a strongly temperature-dependent property-profile, a factor which is also reflected in the optical properties. For the design of optical components, most raw material manufacturers provide the corresponding material data, considering this behaviour.

Considering the required processing steps to turn a raw material into an optical plastic component, it has to be taken into account that the optical properties can be changed by using injection moulding or extrusion processes. The relation between optical properties and processing conditions is illustrated using a long light guide and changes of the Yellowness Index at different process parameters. The results clearly show that not only the precise replication of the mould has to be considered during injection moulding, but also the physicochemical transformation processes at the polymer chain during processing. These processes are specific to each polymer, a fact which indicates that these processes and their impact on optical properties need to be determined for each polymer separately.

Currently, there are no validated simulation systems available that can predict the processing influence on the optical properties of components.

References

[NN97] N.N.: *DIN EN ISO 13468-1:1997-01: Plastics - Determination of the total Luminous Transmittance of Transparent Materials - Part 1: Single-Beam Instrument.* Berlin: Beuth Verlag, 1997

[NN98] N.N.: *ISO 7944:1998: Optics and Optical Instruments - Reference Wavelengths.* Berlin: Beuth Verlag, 1998

[NN99] N.N.: *DIN EN ISO 489:1999-08: Plastics - Determination of the refractive index.* Berlin: Beuth Verlag, 1999

[NN21] N.N.: *VDI/VDE 5596 - BLATT 3 - ENTWURF: 2021-10: Optical design for Manufacturing - Illumination Optics, Non-imaging and Freeform Optics - Optical Material Parameters.* Verein Deutscher Ingenieure, 2021

Symbols

Symbol	Unit	Description
α	°	Angle of incidence
β	°	Angle of reflection
τ	−	spectral transmittance
λ	mm	Wavelengths
C_0	m/s	Speed of light in vacuum
C_M	m/s	Speed of the light in a medium
D	mm	Thicknesses
n	−	Refractive index
V_D	−	Abbe number

Abbreviations

Notation	Description
ABS	Acrylonitrile butadiene styrene
ASA	Acrylonitrile styrene acrylate
COC	cyclo olefin copolymer
DoE	design of experiment
LED	Light-emitting diode
LSR	Liquid Silicone Rubber
MABS	Methyl methacrylate-acrylonitrile-butadiene-styrene
PA	Polyamid
PADC	Polyallydiglycol carbonate
PC	Polycarbonate
PET	Polyethylene terephthalate
PMMA	Polymethyl methacrylate
PSU	Polysulfones
SAN	Styrene acrylonitrile
UV	Ultraviolet
YI	Yellowness Index for Standard Illuminant D65/10°

2 Automatic thermal mould design for optical parts in compression moulding

Ch. Hopmann[1], D. Fritsche[1], T. Hohlweck[1]
[1] Institute for Plastics Processing (IKV) at RWTH Aachen University

Abstract

Optical plastic parts produced by injection moulding have high demands on the resulting part properties. The cooling conditions after the injection phase determine the residual stress and warpage of the part. As stresses and the shape of the part influence the refractive index both have to be minimised. This can be achieved with adaptive cooling, since the thermal history mainly influences these properties. Setting up the thermal mould design manually for such high-quality applications requires profound knowledge and experience about the cooling and shrinkage behaviour as well as the resulting residual stresses of thick-walled components. A novel approach, which locally evaluates the thermal process of the moulded part and derives a cooling channel layout, can be an automated and objective alternative.

The used methodology is the inverse thermal mould design, which calculates the optimal heat balance in the mould based on the thermal properties of the part. A homogeneous shrinkage behaviour can be derived from these temperature distributions. The evaluation of the optimal cooling conditions within the part assumes an appropriate quality function. Due to the variety of materials, geometries and morphologies that result from injection moulding, a validated quality function for thin-walled parts has to be verified for different shapes. Therefore, different quality functions rating different state variables within the part are used in a numerical optimisation and the temperature and density results are compared. Two suitable approaches are used. One approach considers the temperature deviation during the glass transition and the other approach takes the influence of the cooling rate into account. The process environment is a compression moulded optical lens with different varying wall thickness. Due to the nature of this process the relevant pressure in the melt is calculated with a thermomechanical model.

2.1 Introduction of the influence of thermal mould design on the geometrical precision

Material-related component shrinkage is one of the greatest challenges of modern plastics processing. This accounts especially for injection moulded components, which are characterised by a high functional density and increasingly complex component geometries. The shrinkage is affected by a lot of different influencing factors such as processing conditions, humidity, post-shrinkage conditions, etc. [MHMS14]. These effects superimpose each other, which leads to a locally different part shrinkage that needs to be anticipated during the mould design phase. The overall challenge is, that customers often have high demands on surface appearance and geometric stability.

From all the aforementioned parameters, the tempering system of the mould has one of the highest influences on the resulting part quality [HMMM18]. In contrast to the processing parameters, the tempering system is very difficult to be changed after the mould has already been manufactured. In the case of complex injection moulded parts, it is often not possible to design an optimal tempering system only based on the experience of the mould designer. Developing highly automatised and reproducible method for the thermal mould design has a great potential for a higher moulding precision. Additionally, cost and time savings can be achieved. The usual iterations of the injection mould before the start of production can be highly reduced. In addition to the resulting part warpage, inner tensions need to be considered in the case of optical components as they have a significant influence on the resulting optical quality. Several different approaches have been developed in recent years to automate the thermal mould design step [FKP21], which will be presented in the further course of this paper.

With the rise of additive manufacturing technologies, the thermal mould design becomes even more important. While the cavity geometry can be edited before post-curing of the additively manufactured mould, the enclosed tempering channel design cannot be adapted. Nevertheless, the idea of conformal cooling channels that extract the heat exactly at the location where it is necessary becomes more and more compelling for industrial applications. The closer the tempering channel gets to the cavity surface, the higher the resulting gradients are, which makes these layouts prone to errors within the thermal mould design. Precise calculations are necessary to create a suitable thermal mould design.

2.2 Methods for a systematic cooling channel design

Many different research groups are working in the field of automatic generation of cooling channels. An extensive literature review has recently been performed by Feng et al. [FKP21]. They cluster the scientific approaches into five categories:

- Experimental based design
- Design and optimisation based on the conformal cooling line
- Optimisation using expert algorithms
- Modular / parametrical design of conformal cooling channels
- Solid modelling based on topology optimisation

The first category is a hands-on approach, where different designs are created via experience and then validated via simulations [HYLH16, HHY16, LMLX16]. This method is costly due its iterative nature and the result strongly depends on the experience of the engineer. In the second category, cooling channels are geometrically optimised by keeping a constant distance to the moulded part. This is a very interesting approach as it is very quick in terms of calculation time [ASLG+10, WYW15]. Nevertheless, it does not consider different wall thicknesses or the varying cooling demand for complex geometries. In the category of "Optimisation using expert algorithms" several sophisticated mathematical approaches can be summed up [LLLH12, LLM05]. However, expert knowledge is necessary to be able to process these algorithms. Modular design of conformal cooling channels is a category, where the shape of the cooling channels is parameterised (e. g. the diameter and / or the length) [AY07, AY11]. These approaches generate very quick results, but do not necessarily cover all possible solutions and therefore, do not guarantee an absolute minimum of warpage. In the category of topology optimisation, several approaches are summed up. These improve e. g. the local heat conductance of the steel to improve the heat balance of the mould. These approaches take the tempering channel design as a given constant [HF01, Shi19]. As it can be seen the different approaches have all their advantages and disadvantages.

At the Institute for Plastics Processing (IKV) the methodology of the inverse thermal mould design has been focused in recent years. The advantage is that no starting tempering layout has to be designed and therefore in the future, no specific knowledge of the optimal layout has to be available. The approach is called inverse thermal mould design because a theoretical optimal state of the moulded part is assumed at the end of the cooling phase and respective heat fluxes in the mould are iteratively calculated so that this optimal state can be realised. The whole process is based on a mathematical optimisation, which uses a quality function to evaluate the thermal quality of the part. Based on the calculated heat fluxes in the mould steel an adapted tempering channel layout can be derived.

This method is also used in the design of cooling channels for optical applications [HGH21]. For injection moulded optical lenses the challenge is to produce a stress-free lens to avoid an inhomogeneous refractive index. As these optical lenses have high wall thicknesses due to their physical requirements, a long cooling time results. Consequently, thermally induced stresses occur during cooling and varying wall thicknesses increase the problem. The analysed production process in this contribution is called compression moulding. The manufacturing of optical components is much slower than regular injection moulding and therefore, compression moulding is used to maximise the surface quality and the contact time for optimal cooling efficiency. In this method additional external loads are applied by a decreasing cavity volume. This process can result in shorter cooling times. However, due to pressure on the fluid and solid areas of the part, further complex stresses can occur depending on the geometry. These process conditions change the demands of the thermal mould design. However, the temperature control of the mould through a cooling channel system still has a significant impact on achieving a uniform stress profile. Therefore, a modified method will be used.

Within this contribution, a target function based on classical thin-walled components is modified and a new definition of the quality function is derived based on theoretical aspects and the part quality.

2.3 New approach for the inverse thermal mould design

The methodology of the inverse thermal mould design consists of a thermal optimisation of the mould. Certain steps are necessary to perform a practical calculation of the optimal heat balance in the mould and extract a locally high-resolution cooling channel layout. The overall methodology is shown in the following Figure 2.1.

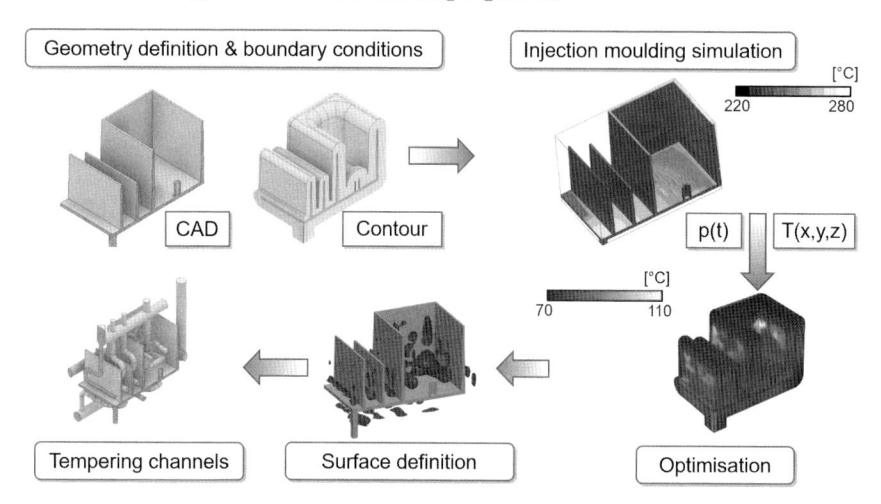

Figure 2.1: Methodology of the inverse thermal mould design

At first, the geometry and the necessary offsets are designed. To avoid the time-consuming thermal optimisation of a complete mould in later steps, a contour close to the part is designed, which represents the mould volume (see also Figure 2.2). The temperatures on the mould contour can be adapted automatically by the algorithm with respect to the evaluation of a quality function. Furthermore, an inner offset is defined as a second evaluation area. The general model setup is visualised exemplarily in Figure 2.1. In the next step, an injection moulding simulation without cooling channels is performed to calculate the cooling over the flow length during the filling phase. The temperature and pressure distribution calculated by the process simulation is then transferred to the optimisation software, whereby only the cooling phase is considered. In the thermal optimisation phase, an optimal temperature distribution in the mould is calculated. Based on this optimal temperature distribution, surfaces can be extracted that localise the optimal tempering channel position. Finally, a tempering channel layout is designed.

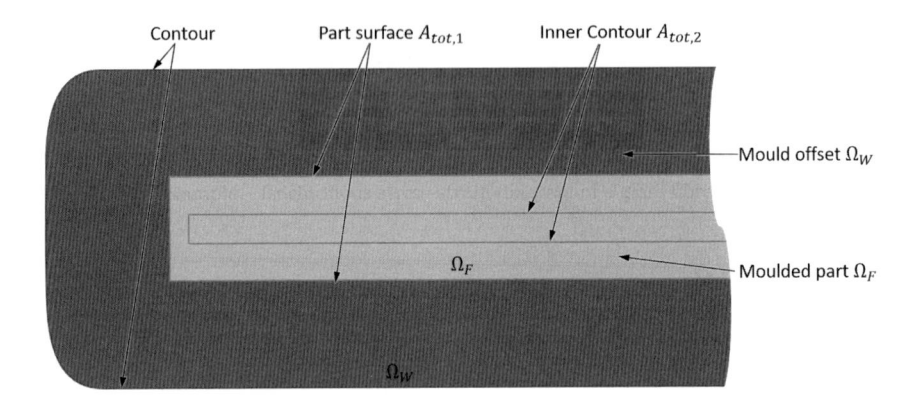

Figure 2.2: Detailed view of the derived contour

The executing optimisation algorithm requires a quality function that evaluates objectively the quality of the part. Several approaches have already been defined by different researchers [ALGGS13, Nik18], which all have different advantages and disadvantages. *Agazzi et al.* use a temperature-dependent approach that does not consider the compressibility of the molten plastic [ALGGS13]. Whereas *Nikoleizig* extends the quality function of *Agazzi et al.* and adds the evaluation of the density. The overall aim of the quality function described by *Nikoleizig* can qualitatively be described as [Nik18]:

$$Q(t_c) = min[T_P - T_{ejec}] + min[\bar{\rho}_p - \rho_p] \qquad (2.1)$$

The aim of the quality function is to minimise the difference between the local temperature in the part T_P and a given ejection temperature T_{ejec}. At the same time, the local density shall not vary from the overall averaged density at the end of the cooling phase t_c.

The disadvantages in this approach is that the density can only be evaluated at the end of the cooling phase, as an integration over every time step is numerically not feasible. Consequently, this approach does not consider the cooling history of the material, which is decisive for the material's morphology. To integrate the cooling history, a time integral over the density during the cooling phase would be necessary, which adds up the local differences in density over the time. This theoretically necessary time integral cannot be realised in the optimisation software COMSOL Multiphysics, Comsol AB, Stockholm, Sweden, because it leads to impractical calculation times. Tests to quantify the calculation time have been aborted due to their length and blockage of computing resources.

However, the density is purely temperature-dependent at the end of cooling because the pressure is at ambient level everywhere in the cavity. Consequently, this does not give any further information about the morphology compared to a purely temperature-dependent approach. These disadvantages lead to the idea of a new approach for the description of the quality function.

For the new approach, the temperature evaluation at the end of cooling is kept, because this term ensures a dimensional stability of the material. The direction of the optimisation towards a homogeneous temperature distribution is set. This temperature term is rated negatively if the local temperature is too far from the demoulding temperature. Otherwise, the term is rated positively. Consequently, the ejection temperature needs to be predefined and limits the process optimisation towards a short cooling and cycle time. However, the goal of homogeneous temperature distribution and rapid cooling are opposed to each other and cannot be optimised simultaneously without a comparison.

A possible evaluation criterion is the cooling rate. The cooling rate especially at the solidification temperature is determining the morphology of the plastic [MHMS14]. Homogenising this parameter should lead to less part warpage and more homogeneous properties of the moulded part. Especially, the crystallisation rate depends on the local cooling rate inside the material. Thus, the shrinkage can be influenced by integration of the cooling rate in the quality function. Due to the higher packing density, a higher crystallisation rate leads to higher local shrinkage [MHMS14]. Homogenising this parameter should improve the local shrinkage over the part volume.

In general, injection moulded parts experience different cooling rates over the wall thickness. Close to the cavity wall, a high cooling rate can be observed due to the contact of the hot melt with the cold mould. In the middle of the part, the cooling rate is significantly lower due to the isolation of the plastic boundary layer [MHMS14]. As this physical effect cannot be changed, it is important to generate homogeneous properties inside constant layers around the midplane of the part geometry. If the properties are not symmetric around the midplane, a lever effect is generated and warpage can be expected. These aspects lead to the newly formulated quality function [HH21]:

$$Q(T_K) = \sum_{i=1}^{m} \left(\frac{T_{ejec} - T_P(\vec{x_i}, t_c; T_K)}{T_{ejec}} \right)^2 * \frac{A_{elem,i}}{A_{tot,1}} * w_i + \sum_{j=1}^{k} \left(\left(\frac{\overline{\dot{T}(T)} - \dot{T}(\vec{x_j}, T_t)}{\overline{\dot{T}}} \right)^2 + \left(\frac{\bar{t_t} - t_t(\vec{x_j}, T_t)}{\bar{t_t}} \right) \right) * \frac{A_{elem,j}}{A_{tot,2}} * w_j \quad (2.2)$$

The first term of the equation is very similar to the previous quality function from *Nikoleizig* [Nik18]. It evaluates the temperature difference in the part T_P to the demoulding temperature T_{ejec} at the end of cooling t_c. In addition, the term is normalised based on the demoulding temperature T_{ejec}. Furthermore, the mesh influence is considered by the variable $A_{elem,i}$. This fraction ensures that every temperature node is only considered by the share of its respective element on the whole evaluation surface or volume $A_{tot,1}$.

The second row of the equation evaluates the morphology of the moulded part. The first term adds up the differences of the local cooling rates compared to an averaged cooling rate $\overline{\dot{T}(T)}$. The second term evaluates the solidification time. These two terms become very small in the case of a homogeneous cooling rate at a similar solidification time. Both terms are evaluated on the $A_{tot,2}$ (see Figure 2.2). $A_{tot,1}$ corresponds to the part surface. $A_{tot,2}$ is an offset surface inside the part. The inner offset is necessary, because at the interface of the mould and the melt different cooling rates appear in steel and plastic, which lead to numerical instabilities in a gradient-based optimisation. Based on the evaluation on these two surfaces, the optimisation algorithm subtracts a temperature distribution on the mould surface such that the shown function becomes minimal.

Both terms of Equation 2.2 are weighted with the factor w_j or w_i to achieve the best optimisation results. A sensitivity analysis of these weighting factors shows a higher priority of the temperature distribution for parts with a wall thickness of 1 to 2 mm, as the thermal conductivity of thermoplastics is relatively low [Hoh21]. Consequently, it has to be determined, if a different wall thickness also achieves a better result according to the quality function (see Equation 2.2) with the same weighting factors.

The new quality function has already been applied and validated on two different demonstrator geometries:

- a box geometry used for previous investigations [HSS18]
- a small housing demonstrator for a RaspberryPi used by the IKV for the K show in 2019 [SHR+19].

The results have shown that at least similar results to the quality function by *Nikoleizig* can be achieved for both geometries. These two parts have a common wall-thicknesses of 1 – 2 mm. In the next step of this contribution, other components that are manufactured in the injection moulding process are investigated. Optical components such as lenses require high wall thicknesses to achieve the desired optical properties. In the following chapter, an adaption of the methodology to this process variant in form of a quality function, which is adapted to this process, is explained. Eventually the influence of this new function on the optimisation algorithm is compared to new quality function for thin-walled parts.

2.4 Adaption to optical demonstrator

The use case is an indicator lens taken from an automotive application provided by Hella KG, Lippstadt, Germany, to ensure practical validity. Due to precision constraints, compression moulding is typically used. This features some wall thickness jumps on the upper side and an elongated lens on the lower half. In the compression moulding process, the two parts of the mould move towards each other and ensure a constant contact of the mould to the plastic. A very high surface quality is mandatory to ensure the correct light transmission [LDC+07]. For this complex process, the thermal optimisation needs to be extended by thermomechanical calculations to precisely calculate the pressure distribution inside the part. Calculating this pressure distribution is essential to predict the local specific volume correctly, which depends on the pressure and temperature distribution (pvT-behaviour). The precise description of the pvT-behaviour is necessary for a precise estimate of the expected part warpage. Additionally, the calculated pressure is simultaneously used to determine the pressure dependent heat transfer coefficient and heat conductivity locally. Usually, the mechanical compression replaces the holding pressure phase.

The temperature-dependent thermomechanical calculation in combination with the multiple iterations for the optimisation increases the calculation time significantly, which is why a two-dimensional approach is chosen [Hom21]. After a validation of this approach a 3D calculation might be feasible. In this contribution, a symmetric plane in the longitudinal direction is used for the calculations. The dimensions and the location of the cut plane are visualised in Figure 2.3.

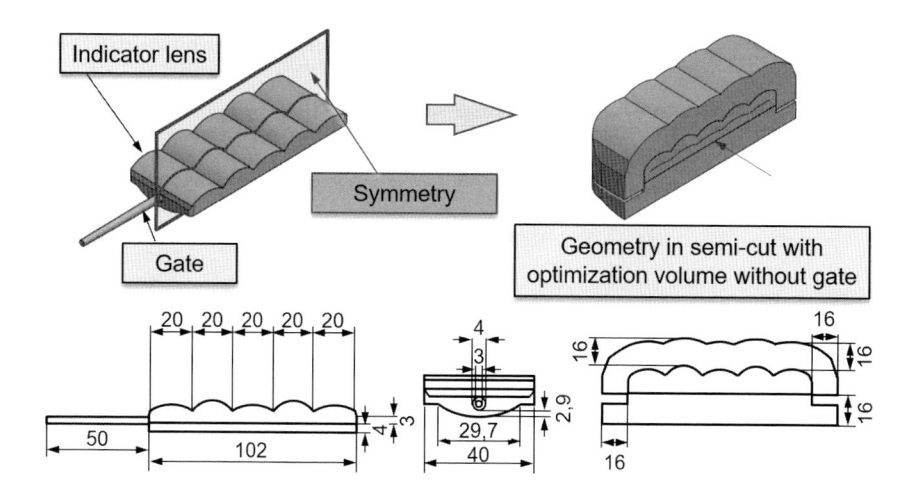

Figure 2.3: Dimensions of indicator lens and optimisation volume [Hom21]

The mould volume is split into two parts. The upper part (nozzle side) is fixed mechanically and the lower part is defined as the movable mould with a linear movement of 0.16 mm/s stopping at a compression of 0.5 mm of the plastic melt ensuring the mechanical compression.

In current work, only a temperature-dependent simulation has been used to evaluate the quality of the moulded part [Hom21]. The approach is based on two different criterions. To achieve a stable part geometry, the whole part volume needs to be below the glass transition temperature. Only when this condition is achieved, the physical shrinkage can be minimised. The part should achieve a temperature distribution that is as homogeneous as possible. Therefore, the surface temperature of the part is chosen as an optimisation objective, as better results can be achieved for an evaluated temperature field with similar values. The quality function based on these theoretical aspects is defined as follows:

$$Q(T_K) = \sum_{l=1}^{n} \int_{A_{Pi}} \left(\frac{T_g(p) - T_P}{T_g(p)} \right)^2 dA_{Pi} + \int_{A_P} \left(\frac{\bar{T} - T_P}{\bar{T}} \right)^2 dA_P \qquad (2.3)$$

The first term integrates the normalised difference between the part temperature T_P and glass transition temperature T_g over A_{Pi}, which is the midplane of the part. The aim of the approach is to cool down to the glass transition temperature as quickly as possible everywhere in the part. The midplane is the location that is the furthest away from the mould surface and will therefore solidify last. Differences of the temperature to the glass transition temperature are weighted negatively. Furthermore, the homogeneity of the temperature distribution on the part surface A_P is evaluated by the normalised difference between the part temperature and the average temperature $\bar{T}$.

The material used for this application is a Plexiglas 7N from Roehm GmbH, Darmstadt, Germany. The glass transition temperature of this material is 113.85 °C. The mechanical behaviour of the material is modelled thermoelastically with the temperature dependent tensile modulus (see Table 2.1).

Temperature [°C]	19	36	556	76	86	96	300
Young's modulus [MPa]	3.410	2.950	2.470	2.000	1.740	1.410	900

Table 2.1: Values of the young's modulus [Hom21]

In the following, a comparison of the purely temperature-dependent approach (see Equation 2.3) compared to the approach, which considers the morphology (see Equation 2.2) is performed on the two-dimensional cutting plane in the middle of the part (see Figure 2.3). In the following the approaches are referred to as temperature optimisation/approach and morphology optimisation/approach. The aim is to evaluate, whether the temperature optimisation specifically developed for thick-walled optical parts achieves better results than the morphology optimisation that has proven successful for thin-walled parts. However, the morphology approach also has the potential to determine a better solution by considering the significantly relevant cooling rate and normalised temperature values. The morphology approach is calculated with weighing factors emphasising the impact of the temperature distribution. The temperatures and specified process parameters are depicted in Table 2.2. The thermal optimisation starts at the end of filling and considers the whole compression and cooling phase.

Parameter	Value		Unit
Approaches	Morphology	Temperature	/
Temperature weighting w_i	100	/	[-]
Morphology weighting w_j	1	/	[-]
Process time	362		[s]
Stroke length	0,5		[mm]
Stroke time	3		[s]
Starting temperature steel T_W	75		[°C]
Starting temperature plastic T_P	240		[°C]
Maximum iterations	2.000		[-]

Table 2.2: Process parameters and weighting factors

The methodology of the inverse thermal mould design is stopped after the thermal optimisation step (see Figure 2.2) and a comparison of the results of the two optimisations is performed.

In the following Figure, the calculated temperature distribution in the moulded part at the end of cooling is shown.

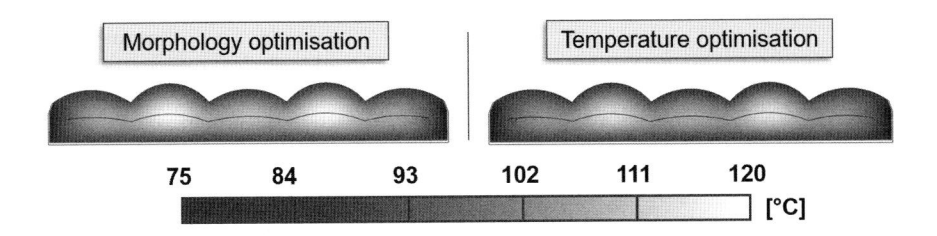

Figure 2.4: Temperature distribution in the part at the end of cooling

It can be seen that the temperature distribution with the morphology optimisation is wider than for the temperature optimisation. The surface is much colder because the optimiser focuses on the demoulding temperature of 85 °C. The temperature approach does not have a distinctive temperature to aim on but focuses on a homogenisation of the surface temperature. This result is expected, as the morphology optimisation not only focuses on the temperature distribution, but also on the solidification. In the following Table, the surface temperatures and their respective standard deviations for the two objective functions are shown.

	Average surface temperature	Std. Deviation
Morphology optimisation	84,23 °C	2,45 °C
Temperature optimisation	83,57 °C	2,23 °C

Table 2.3: Resulting surface temperatures of the two different optimisations

The aim of the morphological approach is to achieve a homogeneous 85 °C on the surface, whereas the temperature dependent approach tries to achieve a temperature of 110 °C, which is the glass transition temperature at ambient pressure on the midplane. The morphological approach is there closer to the required 85 °C. In the following Figure, the temperature distribution on the midsection is visualised (Figure 2.5).

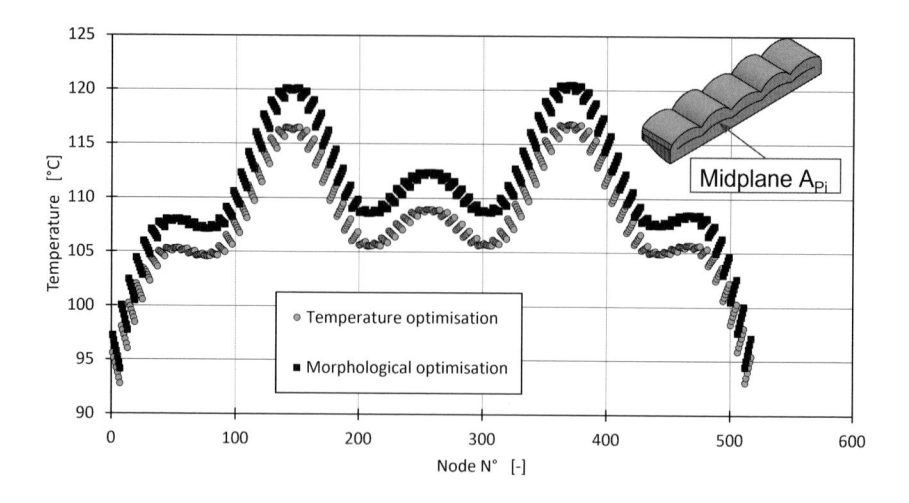

Figure 2.5: Temperature distribution on midsection line

It can be seen that temperature is fluctuating depending on the effective wall thickness at the respective nodes. The midplane at both ends of the part is significantly colder as this area can be cooled down from three sides at the edge of the indicator lens. The course of the two temperature curves is very comparable. The standard deviations are similar, but a general offset of 3 °C can be seen.

In the next step, the morphology needs to be evaluated, which influences the overall expected warpage. In order to compare the results in a meaningful way, previous work has considered the value of the objective function [HGH21, Nik18]. This is not useful in this case, because the formulation of the two objective functions (see Equation 2.2 and Equation 2.3) is fundamentally different. Thus, different numerical values result, which are comparable in the same model. Different approaches cannot be compared. For a good moulding quality, a homogeneous course of the density over time is important. The local pressure and temperature influence the local shrinkage, which is represented by the density. However, it is not the exact value of the density that is relevant, but only the fluctuation of the density over the moulded part from the mean value. For this reason, the standard deviation of the density at each time step of the cooling phase as well as the average moulded part surface temperature at the time of demoulding are used as central evaluation criteria. This course of the density over time is shown in the following Figure 2.6.

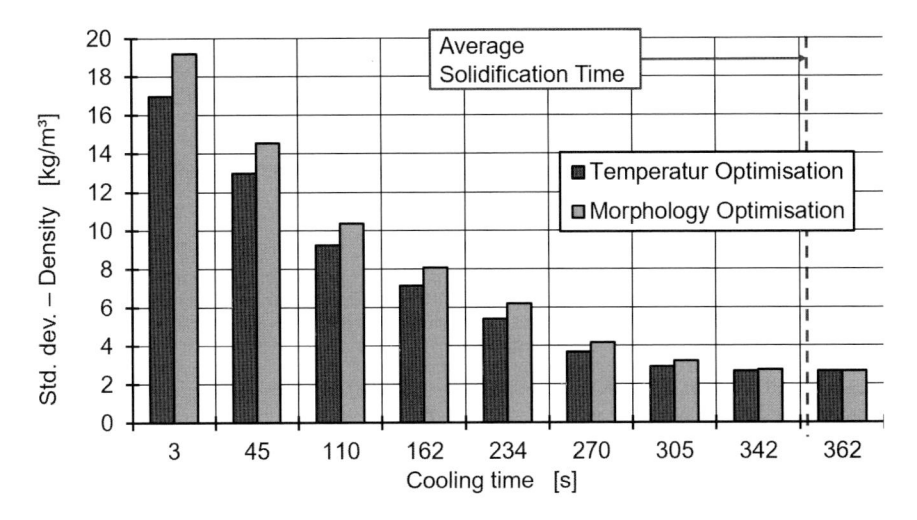

[Bar chart: Std. dev. – Density [kg/m³] vs Cooling time [s], comparing Temperatur Optimisation and Morphology Optimisation across cooling times 3, 45, 110, 162, 234, 270, 305, 342, 362. Legend box labeled "Average Solidification Time".]

Figure 2.6: Comparison of the standard deviation of the density

It can be seen that, the differences in the deviation of the density are very small. The fluctuation compared to the overall value of the density is around 2 %. The temperature-dependent approach shows a lower variation of the density over time compared to the morphological approach, which equalises over time and reaches a similar distribution at the end of cooling. In order to be able to clearly assess the influence of these slightly deviating density standard deviations, a simulation and a comparison of the distortion with the two derived cooling channel layouts will be advisable in the future.

2.5 Discussion

The shown results are not expected, because the definitions of the two optimisation functions are quite different. Although the evaluation and normalisation of temperature is similar, it was expected that the consideration of freezing time and cooling rate would have a significant impact. The course of the temperature is very similar. An offset of a few degrees can be found, but it doesn't show big differences in the standard deviation, which is relevant for the part quality and warpage. The temperature-dependent approach has a very different focus compared to the morphological approach. A reason for the quite similar results might be the low thermal conductivity of the plastic. The low thermal conductivity, amplified by the high wall thickness, lowers the optimisation potential, as any local influence of the optimised temperature distribution is damped in the centre of the moulded part. As the cooling time is comparably long, an inhomogeneous temperature distribution in the mould might be evened out. The temperature is the driving quality feature in this process and has a higher influence on the objective function leading to the shown similar results.

For the compression moulding process, both approaches lead to realistic results for the inverse thermal mould design. Currently, the temperature-dependent approach guarantees a slightly lower variation of the density and still ensures the achievement of the final temperature. This approach is therefore preferred in the case of thick-walled geometries. However, different weighting factors might be beneficial due to the high wall thickness to achieve better results with the quality function according to Equation 2.2.

The performed optimisations are numerically very challenging as they combine a complex thermomechanical calculation with a numerical optimisation and a high amount of calculations have to be conducted. Currently, an elastic material model is used to calculate the resulting pressures due to the mechanical compression of the melt. The melt is highly viscous and the validity of this approach needs to be ensured in future practical trials.

2.6 Outlook

In this contribution, two different optimisation approaches have been tested in order to check their influence on the quality of compression moulded optical parts. To correctly calculate the pressure distribution, the thermal approach by *Nikoleizig* was extended to a thermomechanical approach. The calculation of the mechanical stresses is challenging for the numerical stability of the optimisation. Therefore, a reduction of the problem on two dimensions was performed.

In further research, the interdependencies between the thermal and the mechanical calculations should be investigated. Currently, an elastic material model is used. The melt is highly viscous why an elastoplastic approach will be checked in the future. As this extended mechanical approach adds more complexity, calculation times need to be monitored closely. Furthermore, practical trials will be performed to evaluate the resulting quality of produced parts. They will be compared with parts from series production from Hella, which are produced with a conventional mould. On the one hand, geometric aspects such as part warpage will be checked. On the other hand, the optical quality such as the light distribution will be part of the investigation.

Acknowledgments

The depicted research has been funded by the Deutsche Forschungsgemeinschaft (DFG) as part of the Collaborative Research Centres CRC 1120. We would like to extend our thanks to the DFG.

References

[ALGGS13] AGAZZI, A.; LE GOFF, R.; GARCIA, D.; SOBOTKA, V.: MCOOL® : Optimal cooling system design in injection molding process. *Society of Plastics Engineers* (2013), p. 261–264

[ASLG+10] AGAZZI, A.; SOBOTKA, V.; LE GOFF, R.; GARCIA, D.; JARNY, Y.: A Methodology for the Design of Effective Cooling System in Injection Moulding. *International Journal of Material Forming* 3 (2010), p. 13–16

[AY07] AU, K. M.; YU, K. M.: A scaffolding architecture for conformal cooling design in rapid plastic injection moulding. *The International Journal of Advanced Manufacturing Technology* 34 (2007), p. 496–515

[AY11] AU, K. M.; YU, K. M.: A scaffolding architecture for conformal cooling design in rapid plastic injection moulding. *Computer-Aided Design* 43 (2011) 8, p. 989–1000

[FKP21] FENG, S.; KAMAT, A. M.; PEI, Y.: Design and fabrication of conformal cooling channels in molds: Review and progress updates. *Applied Thermal Engineering* 171 (2021) 121082,

[HF01] HUANG, J.; FADEL, G. M.: Bi-Objective Optimization Design of Heterogeneous Injection Mold Cooling Systems. *Journal of Mechanical Design* 123 (2001) 2, p. 226–239

[HGH21] HOPMANN, C.; GERADS, J.; HOHLWECK, T.: Investigation of an inverse thermal injection mould design methodology in dependence of the part geometry. *International Journal of Material Forming* 14 (2021) 2, p. 309–321

[HH21] HOHLWECK, T.; HOPMANN, C.: Thermal Optimisation of Injection Moulds by Solving an Inverse Heat Conduction Problem. *Enhanced Material, Parts Optimization and Process Intensification*. Cham, 2021

[HHY16] HU, P.; HE, B.; YING, L.: Numerical investigation on cooling performance of hot stamping tool with various channel designs. *Applied Thermal Engineering* 96 (2016), p. 338–351

[HMMM18] HOPMANN, C.; MENGES, G.; MICHAELI, W.; MOHREN, P.: *Spritzgießwerkzeuge*. München: Hanser, 2018

[Hoh21] HOHLWECK, T.: *Inverse thermische Optimierung zur wissensbasierten thermischen Spritzgießwerkzeugauslegung*. RWTH Aachen, dissertation, 2021

[Hom21] HOMBERG, A.: *Übertragung der Methodik zur inversen Kühlkanalauslegung auf spritzgeprägte Kunststoffoptiken mit Wanddickensprüngen*. RWTH Aachen, Master's Thesis, 2021 – supervisor: J. Gerads

[HSS18] HOPMANN, C.; SCHMITZ, M.; SCHNEPPE, T.: Digitisation of mould development in precision injection moulding. *Internationales Kolloquium Kunststofftechnik*. 2018

[HYLH16] HE, B.; YING, L.; LI, X.; HU, P.: Optimal design of longitudinal conformal cooling channels in hot stamping tools. *Applied Thermal Engineering* 106 (2016), p. 1176–1189

[LDC+07] LIN, Y. J.; DIAS, P.; CHUM, S.; HILTNER, A.; BAER, E.: Surface roughness and light transmission of biaxially oriented polypropylene films. *Polymer Engineering Science* 47 (2007) 10, p. 1658–1665

[LLLH12] LI, C. G.; LI, C. L.; LIU, Y.; HUANG, Y.: A new C-space method to automate the layout design of injection mould cooling system. *Computer-Aided Design* 44 (2012) 9, p. 811–823

[LLM05]   LI, C. L.; LI, C. G.; MOK, A.: Automatic layout design of plastic injection mould cooling system. *Computer-Aided Design* 37 (2005) 7, p. 645–662

[LMLX16] LI, H.; MEI, Y.; LIN, B.; XIAO, H. Q.: Design and Optimization of Conformal Cooling System of an Injection Molding Chimney. *Materials Science Forum* 850 (2016), p. 679–686

[MHMS14] MENGES, G.; HABERSTROH, E.; MICHAELI, W.; SCHMACHTENBERG, E.: *Menges Werkstoffkunde Kunststoffe*. München: Carl Hanser Fachbuchverlag, 2014

[Nik18]   NIKOLEIZIG, P.: *Inverse thermische Spritzgießwerkzeugauslegung auf Basis des lokalen Kühlbedarfs*. RWTH Aachen, dissertation, 2018 – supervisor: C. Hopmann

[Shi19]   SHIN, K.-H.: A method for representation and analysis of conformal cooling channels in molds made of functionally graded tool steel/Cu materials. *Journal of Mechanical Science and Technology* 33 (2019) 4, p. 1743–1750

[SHR+19] SCHMITZ, M.; HOPMANN, C.; RÖBIG, M.; PELZER, L.; TOPMÖLLER, B.; WURZBACHER, S.: Jenseits menschlicher Fähigkeiten. *Kunststoffe* 109 (2019), p. 142–145

[WYW15]   WANG, Y.; YU, K.-M.; WANG, C. C.: Spiral and conformal cooling in plastic injection molding. *Computer-Aided Design* 63 (2015), p. 1–11

Symbols

Symbol	Unit	Description
ρ_p	kg/m^3	Part density
w_i	−	Weight factor temperature term
w_j	−	Weight factor morphology term
$A_{elem,i}$	m^2	Element surface
$A_{elem,j}$	m^2	Element surface inner offset
A_{Pi}	m^2	Midplane surface
A_P	m^2	Part surfacet
$A_{tot,1}$	m^2	Total part surface
$A_{tot,2}$	m^2	Total inner offset surface
p	Pa	Pressure
Q	−	Quality value
T	K/s	Temperature
$\dot{T}$	K/s	Temperature rate
T_{ejec}	K	Ejection temperature
T_g	K	Glass transition temperature
T_P	K	Part temperature
T_t	K	Solidification temperature
t_c	s	Cooling time
t_t	s	Solidification time
x	mm	Position vector

Abbreviations

Notation	Description
IKV	Institute for Plastics Processing

3 Development of a coupled thermo-optical process optimisation routine for plastic lenses in laser applications

Ch. Hopmann[1], C. Holly[3], J. Stollenwerk[2,3], B. Liu[1], J. Gerads[1], J. Hofmann[3]
[1] Institute for Plastics Processing (IKV) at RWTH Aachen University
[2] Fraunhofer Institute for Laser Technology ILT, Aachen, 52074 Aachen, Germany
[3] Chair for Technology of Optical Systems (TOS) at RWTH Aachen University

Abstract

For high power laser applications, the use of plastic lenses is difficult to realise due to the high temperature dependence of the optical properties, the low temperature resistance and high thermal expansion. For this reason, the IKV in cooperation with the Chair of Technology of Optical Systems (TOS), RWTH Aachen, Germany, developed a design routine that couples thermo and optical process optimisation in order to increase the effective design of plastic optics for laser applications and reduce cost-intensive tool iterations. Therefore, this contribution presents the coupled thermo-optical process optimisation routine and provides a practical validation of the developed method. The results of the characterisation of the required optical properties in this system will be also introduced. To validate the effectiveness of the coupled routine, a planar-convex lens, which usually requires very high geometrical accuracy in practical applications, is used as an exemplary geometry. After the process optimisation and simulation, the deformed geometry was very close to the desired lens geometry in thermally unloaded state, with an average difference between the two geometries of 0.36 μm, compared to 0.04 mm prior to optimisation.

3.1 Introduction

Due to increasing cost pressure in the field of laser material processing, there are growing efforts to replace glass optics by plastic optics. In addition to increased freedom in design and weight reduction, plastic optics offer the possibility of cost-effective production by means of injection moulding. Even though modern optical plastics absorb only a small portion of the laser light (<3 %/cm) [Bon19], a significant temperature change in the optic is induced by the absorbance. The locally changing temperature distribution leads to a change in the focus position, which is called thermal lensing. Because of the change in focus, the intensity on the work piece is decreased, which might reduce the machining quality or even lead to an abortion of the machining process. This currently restricts the use of plastic optics to low laser powers (<1 W). In this contribution the compensation of a given lens geometry for thermal effects for one specific operation state including the applied laser power and the environmental circumstances is demonstrated. A basic requirement for the compensation of thermal effects is an exact knowledge of the material parameters, especially the optical parameters like for instance the refractive index. Besides, the fabrication aspects of the lens production using injection moulding are considered. As optical elements are sensitive to geometric variations in the order of the wavelength of the laser light, a simulative compensation for production errors like e. g. shrinkage is performed. Within one integrated coupling, all fabrication and optical aspects are digitally considered, so that that expensive and time-consuming iterations of the tooling can be reduced.

The optimisation routine is divided into two areas: the thermo-optical design and the process design. In the thermo-optical design, an optical layout is examined by means of thermo mechanics on the basis of the material data and environmental conditions. The lens geometry is then further optimised until the thermally loaded lens meets the optical requirements, e. g. spot diameter. The unloaded lens geometry that is determined for the best optical performance in this process is then transferred to the process design. Here, the process and the mould geometry are optimised, so that the desired geometry can be met.

This contribution is structured as follows: Following this introduction, the process simulation is introduced in Section 3.2. In Section 3.3, the practical experiments and measurement methods of moulding accuracy as well as optical properties will be described. Section 3.4 presents the validation of the process simulation with help of practical results. In Section 3.5, the developed coupled thermo-optical process optimisation routine will be introduced and Section 3.6 wraps up this contribution with the conclusion and an outlook on future work.

3.2 Process simulation

In this contribution, the software Moldex3D 2020 provided by CoreTech System Co., Ltd., Zhubei, Taiwan, was used to simulate the injection moulding process in order to predict the warpage and shrinkage of the lens with corresponding machine setting parameters and material properties. A plano-convex lens geometry to be optimised is depicted in Figure 3.1. The lens shown is an asphere made of acrylic (PLEXIGLAS® 7N manufactured by Röhm GmbH, Darmstadt, Germany, with a back fcal length (BFL) of 304.02 mm[URL20c].

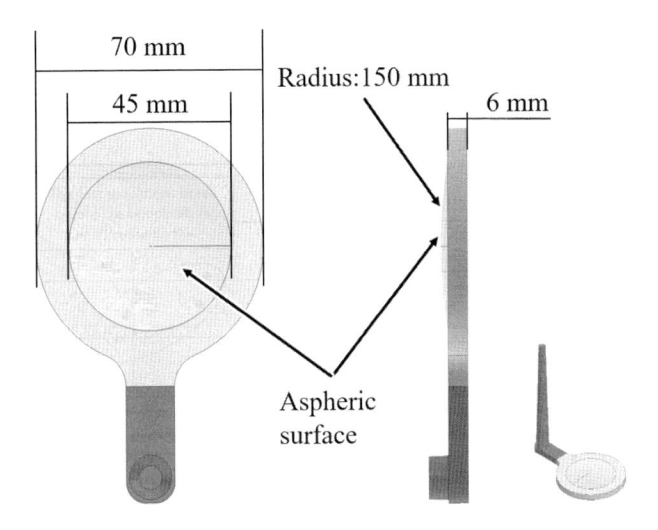

Figure 3.1: Original lens geometry

The ability of the process simulation to accurately predict the deformation of the lens is also a pre-requisite for coupled optimisation. Therefore, the process simulation will be introduced in this chapter.

3.2.1 Implementation of process simulation

The process simulation needs to be pre-processed before the execution of the optimisation routine, which includes the preparation of the model, the selection of materials and the determination of process parameters.

The CAD model of the mould and the corresponding simulation model of the plano-convex lens with cooling channels are shown in Figure 3.2. Then the prepared CAD model needs to be meshed, which was made using the boundary layer mesh (BLM) provided by the Moldex3D software. The supporting mesh types include tetrahedral elements and multiple-layer BLM to approach flow and heat behaviours near the mould cavity surface [URL20b].

To model pvT-behaviour and viscosity the Tait law and the Cross-WFL is used, which is applied to determine the shrinkage and flow behaviour. The relevant parameters for these models are provided by the Moldex3D material database [URL20b].

As shrinkage is the focus of the process design, it will be evaluated during every phase of the injection moulding process, such as injection, packing and cooling phase as well as the phase from ejection to ambient condition. In addition, in the warpage analysis, the influence of in-mould constraint effect and flow-induced residual stress are also considered [URL20b].

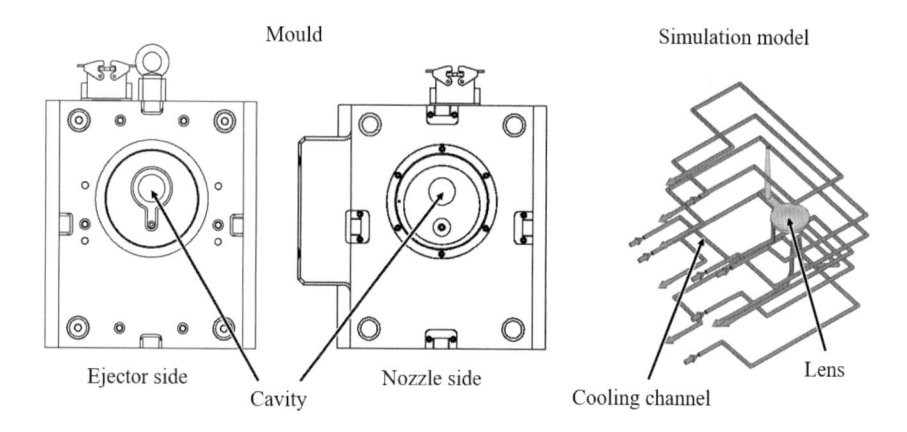

Figure 3.2: The mould and the corresponding simulation model of the plano-convex lens

3.2.2 Statistical evaluation of process parameter

Moldex3D provides a DoE analysis module to automatically to perform DoEs and support of the evaluation. In this module, the Taguchi method was adopted, which uses statistical methods to conduct experiments and analyse the production process. Depending on the purpose of the optimisation, it may first be necessary to determine the quality factor and the machine setting parameters [KFLB15]. The quality factor represents the objective of the optimisation, which in this case is to minimise volumetric shrinkage. The machine setting parameters are the variables to be optimised. Three machine setting parameters were selected based on a preceding literature research: melt temperature, maximum packing pressure and cooling time [JP05, LK01]. After the machine setting parameters have been confirmed, their appropriate values, which are also called levels, were determined based on the material information and experiences (see Table 3.1) [URL20c, URL20a]. A full factorial experimental design with totally 27 runs was chosen for the Taguchi method.

Machine Setting parameter	Unit	Level 1	Level 2	Level 3
Melt temperature	[°C]	230	240	250
Maximum packing pressure	[MPa]	50	75	100
Cooling time	[s]	200	250	300

Table 3.1: Machine setting parameter windows

In the Taguchi method, quality response and S/N (signal to noise) ratio are applied to evaluate the DoE results. The quality response is the average of the quality factor from different runs at each level of each machine setting parameter in the Taguchi array. The S/N ratio integrates the effect of the average quality responses and their standard deviations. As the value of S/N is normalised, a higher S/N ratio is always better regardless of the characteristics of the quality factor. The variation of S/N ratio shows, how dominating this machine setting parameter is over the quality factor between different levels. If the variation is insignificant, it means that this machine setting parameter has less effect on this quality factor [KFLB15, URL20b].

Table 3.2 shows the corresponding S/N ratio of the chosen machine setting parameters and their effects, which are the absolute value range of the S/N ratio. Compared to the melt temperature and cooling time, maximum packing pressure has a large effect on the volumetric shrinkage. The best level of each machine setting parameter is summarised according to its S/N ratio in this table. A best and a worst machine setting parameter set is determined based on the DoE analysis results (see Table 3.3) and their simulative total displacement are shown in Figure 3.3. The maximal total displacement of the best process parameter was 0.16 mm, which is already near to the minimal volumetric shrinkage (0.15 mm) of the worst machine setting parameter set. This result also verified the validity of the Taguchi DoE analysis.

Level	S/N ratio [dB]		
	Melt temperature	Maximum packing pressure	Cooling time
1	-5.16	-5.19	-5.15
2	-5.14	-5.15	-5.15
3	-5.13	-5.09	-5.15
Effect	0.03	0.1	0.01
Best level	3	3	1,2,3

Table 3.2: S/N ratio of each level of the chosen machine setting parameters and their effects on the quality factor

Parameter set	Best parameter set	Worst parameter set
Injection speed [cm^3/s]	35.4	35.4
Holding pressure [MPa]	100	50
Coolant temperature [°C]	50	50
Cooling time [s]	300	300
Melt temperature [°C]	250	230
Holding pressure time [t]	70	70

Table 3.3: selected machine setting parameter sets for the further practical validation

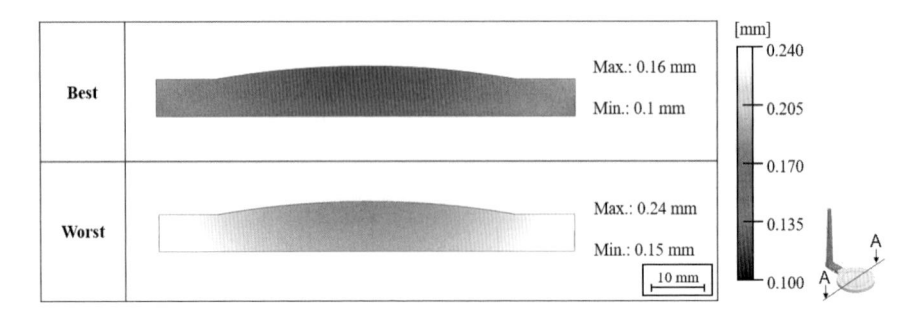

		Max.: 0.16 mm	[mm]
Best		Min.: 0.1 mm	0.240 — 0.205
Worst		Max.: 0.24 mm Min.: 0.15 mm 10 mm	0.170 — 0.135 — 0.100

Figure 3.3: Simulative volumetric shrinkage of the determined machine setting parameter sets

The best and worst machine setting parameter set were selected for the further practical trials in order to evaluate the simulation results, which will be discussed in detail in the next chapter.

3.3 Practical experiments

To validate real world usage the results of the simulations have to be validated in practical trials. For this purpose, plastic lenses are produced and examined at the IKV using the existing lens mould.

3.3.1 Process parameter

The practical experiments are carried out on an all-electric injection moulding machine of the type e-motion 160 from ENGEL AUSTRIA GmbH, Schwertberg, Austria. The integrat 40 temperature control unit from gwk Gesellschaft Wärme Kältetechnik mbH, Meinerzhagen, Germany, is used as cooling system. The same parameter stages that were investigated in the process simulation are used for the experimental set-up (see Table 3.1 and Table 3.3). In order to reduce the practical test plan, the results from the simulation are taken into account and the best and worst parameter combinations were examined (see Table 3.3). In each case, 10 process cycles were first carried out in order to achieve thermal stability of the process without the time required for start-up exceeding the time required for production, and then 10 test specimens were taken for the measurements.

3.3.2 Measurement of the moulding accuracy

The surface of the plastic lenses are measured using a chromatic white light sensor on a Zeiss O-Inspect-442 from Carl Zeiss IQS Deutschland GmbH, Oberkochen, Germany. The measuring accuracy of the O-Inspect is 1.6 $\mu m + x/250$ with $x =$ measuring distance in mm. For this purpose, two measurements are carried out on the aspherical and flat surfaces: one in the direction of flow and one perpendicular to the flow direction (Figure 3.4). The results are recorded in coordinates and to enable a comparison of the surfaces, the results are calibrated, so that a variation in the height through manual placement is eliminated.

The corresponding results of the measurement are shown in Section 3.4, which are used to compare to the simulation results.

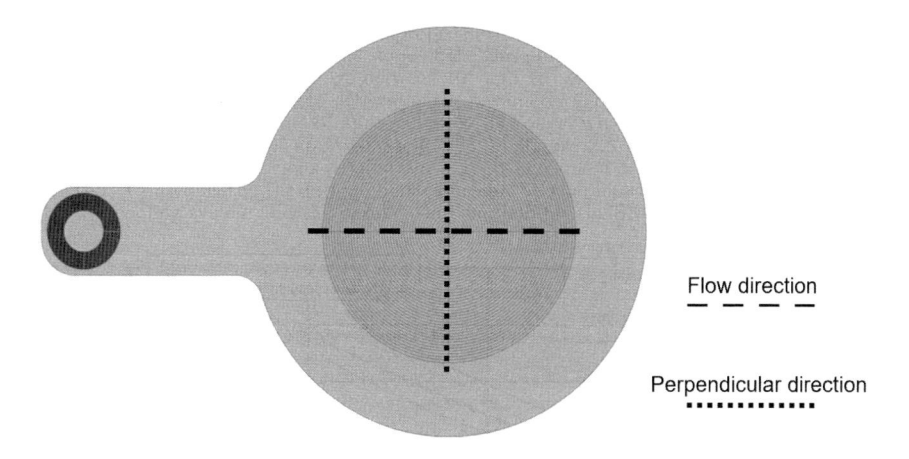

Figure 3.4: Measurement position on the lens surface and an example for the results

3.3.3 Measurement of the optical properties

A pre-requisite for a successful optimisation of the lens is a precise knowledge of the exact operation state and the material parameters. In order to investigate the influence of potential deviations between the material parameters present in the experiment and those underlying the simulations, a sensitivity analysis has been carried out [HBSL20]. It is shown that the optimisation is most sensitive to the following material parameters: refractive index, thermo-optic coefficient and absorption coefficient. As the manufacturer does not specify those parameters for the used laser wavelength (1070 nm), we carried out a material characterisation. For the measurements of the refractive index and the thermo-optic coefficient an abbe-refractometer is used.

For the measurements of the thermo-optic coefficient the temperature has been changed in between 24 °C and 40 °C. The results of the characterisation of the thermo-optic coefficient are shown in Figure 3.5 (left). The depicted measurement errors result from an error propagation of the uncertainties of the single measurements of the refractive index. It might be seen that the measured values for different lenses fluctuate more than the depicted measurement errors. The mean value for the thermo-optic coefficient is $(-157.4 \pm 14.4) \cdot 10^{-5} K^{-1}$. It can be assumed that the deviations result from slight variations of the fabrication parameters even though the same machining parameters have been chosen during the injection moulding. The values for the refractive index are slightly changing from lens to lens, but also changing within the geometry of the lens. Therefore, the mean value of the refractive index in the centre position was determined to 1.4837 ± 0.0004 for the optimisation.

Furthermore, the absorption coefficient of the lenses was determined. As a precise direct measurement of the absorption is difficult to achieve and requires expensive equipment, an indirect measurement by tracking the temperature of the lens by thermography was carried out. The maximum temperature in the centre position of the lens is compared with the results of FEM simulations. By tuning the absorption coefficient in the simulations such that the temperature in the simulation matches the experimentally measured temperature, the absorption coefficient is determined at 10 W. By comparing experimental and simulation results for different powers, using the pre-determined absorption coefficient, we assure that the absorption coefficient is generally valid. The results for different laser powers are depicted in Figure 3.5 (right). Averaging over the different lenses, an absorption coefficient of $(1.463 \pm 0.051)\%/cm$ for $1070\ nm$ is determined.

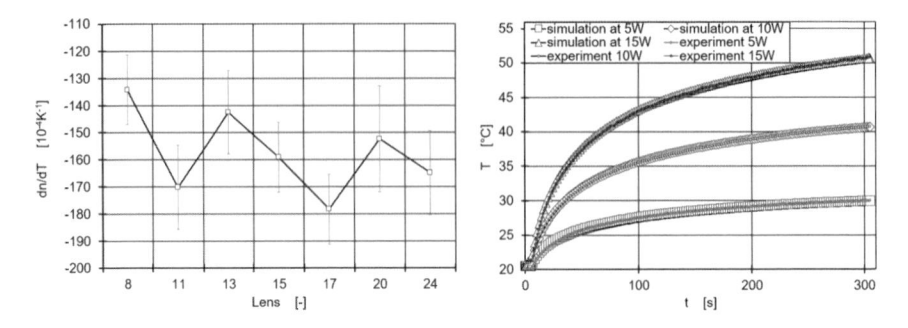

Figure 3.5: Thermo-optic coefficient & Determination of the absorption coefficient

3.4 Validation of the simulation

In this section, the comparison of the simulation results with the measurement results will be presented. Figure 3.6 and Figure 3.7 shows the comparison between the measured and simulated results for the aspheric surface respectively the planar surface in the flow direction. Due to the good repeatability of the measurements, only one set of data has been chosen for comparison with the simulated data. Compared to the planar surface, the measurement of the aspheric surface agrees better with the simulation results. However, there is still some deviation between them, and the actual measured shrinkage of the lens is greater compared to the simulated results. For the planar surface, there is a relatively large difference between the simulated and measured results. For the worst machine setting parameter set, the simulated and actual shrinkage directions do not agree and such an inconsistency can make a large difference in the optical properties of the lens. The simulated and measured results for best machine setting parameter set on the planar surface are relatively consistent, but there are still differences in the values. These discrepancies are mainly caused by the limitations of the simulation itself and uncontrollable factors during the actual experiment. However, it is difficult to quantify these differences simply by a direct comparison of measured and simulated values. Therefore, they are further compared here with using mean absolute error (MAE).

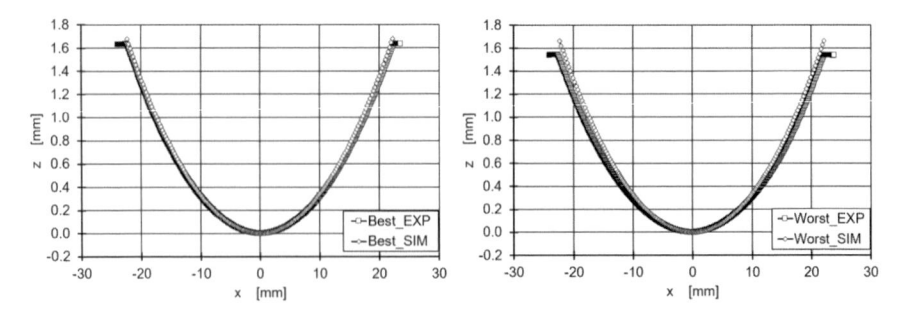

Figure 3.6: Comparison of the measurement and simulation results of aspherical in Flow-direction

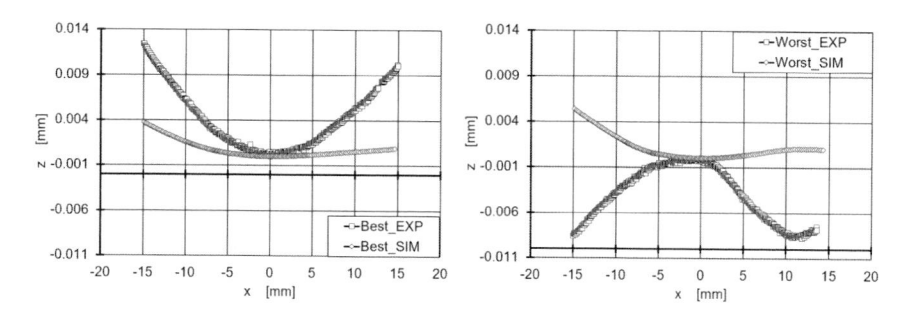

Figure 3.7: Comparison of the measurement and simulation results of planar in Flow-direction

The MAE values for the two machine setting parameter sets in Table 3.3 for aspheric surface as well as planar surface in all directions are summarised in Figure 3.8. For aspheric surfaces, the results for best machine setting parameter set is relatively stable, and the differences in flow direction and across direction are not significant, which is beneficial for optical properties of the lens. On the other hand, the worst parameter set has relatively large differences in different directions, which also indicates that the simulated and actual results of this set differ the most. For planar surface, the results for best parameter set is similar to those for aspheric surfaces. The results for worst parameter set is also worse. Therefore, the best machine setting parameter set, which has the best agreement between simulated and measured results in comparison with the other parameter sets, was chosen as the parameter set for the further optimisation process.

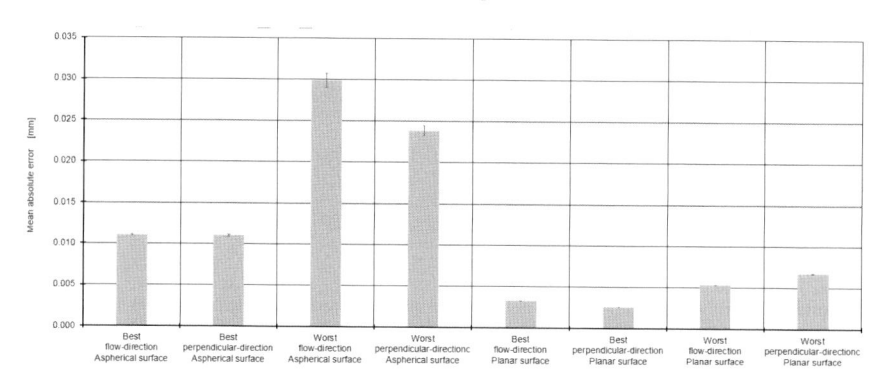

Figure 3.8: Mean absolute error (MAE) between measurement and simulation results

3.5 Development of the coupled optimisation routine

Figure 3.9 illustrates the developed coupled optimisation routine, which consists of thermo-optical optimisation and process optimisation.

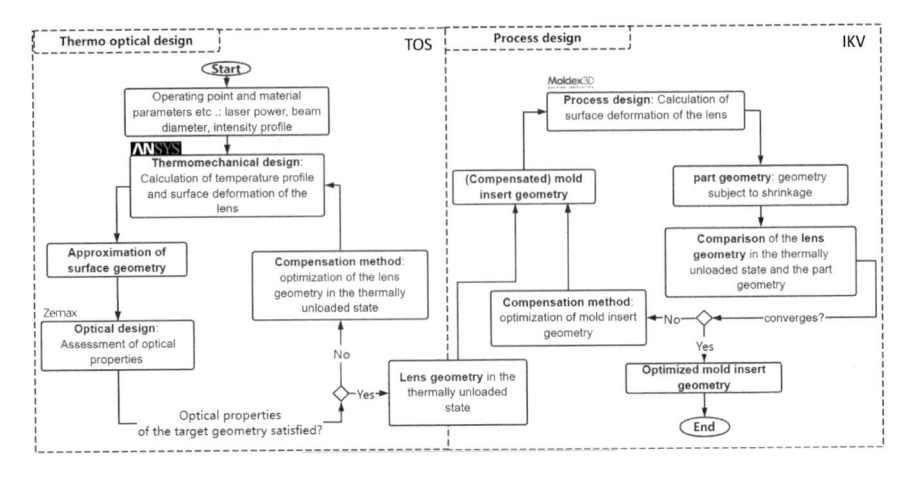

Figure 3.9: Coupled optimisation routine

3.5.1 Thermo-optical optimisation

With the knowledge of the operation state and the material parameters, the temperature profile and the deformation of the lens are calculated within FEM-simulations. For this purpose, the FEM environment Ansys Workbench from Ansys Inc., Canonsburg (PA), USA, is used. Subsequently, the FE datasets, which exist as point clouds, are approximated with continuously differentiable functions by using the TOP Simulation software developed at TOS and finally transferred to the optics simulation program Zemax Optical Studio from Zemax LLC, Kirkland (WA), USA. By using user defined DLLs the thermal effects are included into the setup. Next, if the beam size and the back focal length deviate too much from the initial values of the geometry without thermal effects, the lens geometry is optimised. For this purpose, the radius, the conical constant and the aspherical coefficients are varied for the curved lens side so that the spot size is minimal. Afterwards the optimised geometry without thermal deformations is again imported in the FEM environment and that procedure is iterated until the shape of the lens geometry and thus the focal length of the lens converges. The optimisation was stopped when the deviation in the back focal length (BFL) is less than 1 mm.

Contributing to the focus shift is the deformation of the lens geometry due to the non-zero expansion coefficient and the change in refractive index with temperature. As the thermal lens due to the refractive index change scales with the total lens thickness, the lens thickness is kept constant during the optimisation. Because of the already present tooling, the edge thickness of the lens is restricted to six millimetres. In order to have enough degrees of freedom during the optical optimisation, the diameter of the curved part of the first lens surface is reduced to 15 mm.

The results of the iterative optimisation are depicted in Table 3.4, according to which the aspherical surface of the lens can be described. As the focal length of the lens and the change of focal length caused by the thermal effects are of the same order, there is no significant focusing by the optics in the initial state anymore. Therefore, for compensation a greater bending of the lens surfaces is needed. Contributing to the focus shift is the deformation of the lens geometry due to the non-zero expansion coefficient and the change in refractive index with temperature. As the thermal lens due to the refractive index change scales with the total lens thickness, the lens thickness is kept constant during the optimisation. Because of the already present tooling, the edge thickness of the lens is restricted to six millimetres. In order to have enough degrees of freedom during the optical optimisation, the diameter of the curved part of the first lens surface is reduced to 15 mm. Regarding the results of the back focal length after three iterations, we observed that with each iteration the back focal length gets closer to the target value of 304.02 mm. The final optimised lens geometry after iteration 3 are displayed in Figure 3.10. Now, even with thermal effects the lens achieves diffraction limited focusing at a BFL of 304.72 mm.

Iteration	0	1	2	3
$R_1 [mm]$	150.000	4.302	4.289	1.236
$\alpha_2 [mm^{-2}]$	$1.452 \cdot 10^{-5}$	-0.10697	-0.10697	-0.39485
$\alpha_4 [mm^{-4}]$	$-2.174 \cdot 10^{-8}$	$-1.572 \cdot 10^{-3}$	$-1.572 \cdot 10^{-3}$	$3.942 \cdot 10^{-3}$
$\alpha_6 [mm^{-6}]$	$-1.504 \cdot 10^{-13}$	$-3.303 \cdot 10^{-5}$	$-3.303 \cdot 10^{-5}$	$-5.996 \cdot 10^{-5}$
$\alpha_8 [mm^{-8}]$	$-5.615 \cdot 10^{-17}$	$4.614 \cdot 10^{-7}$	$4.614 \cdot 10^{-7}$	$6.278 \cdot 10^{-7}$
κ	$-1.51 \cdot 10^{-9}$	-2.359	-2.363	-1.068
BFL*$[mm]$	na	316.23	302.92	304.72

*during laser emission in equilibrium state

Table 3.4: Results of the thermo-optical optimisation

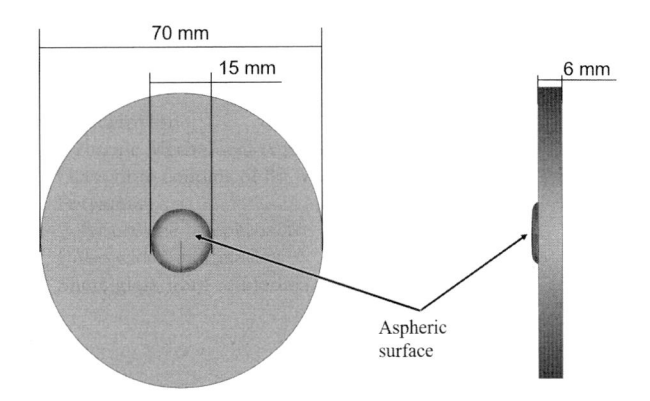

Figure 3.10: Optimised lens geometry

3.5.2 Process optimisation

After the lens geometry in unloaded state (Figure 3.10) has been determined by the thermo-optical optimisation, the process optimisation can be carried out. First, the values of relevant machine setting parameters for the injection moulding simulations are determined based on a DoE. At the same time, the CAD model of the desired geometry and its corresponding mould information such as the cooling channel should be prepared for the injection moulding simulation. After the simulation, a deformed geometry described in scattered points is exported for the developed Rapid Semi-Automatic Iterative Compensation (RSIC) method, by which every single scattered point should be compensated, in order to achieve global compensation. This process is carried out iteratively until convergence of the optimisation process is achieved.

Before performing the simulation, the original CAD model, which is also the optimisation target of the RSIC method, needs to be meshed in the simulation software. The surface mesh of the model is described by a stl-file, which can represent a 3-dimensional surface geometry in triangular elements [KJM97]. Once the mesh has been successfully generated and the other parameters such as material and machine setting parameters have been set, the simulation can be carried out.

After each iteration of the simulation, the warpage and shrinkage of the part first needs to be evaluated in order to test the compensation effect of the mould cavity. The average Euclidean distance D (see Equation 3.1) for evaluating the difference between these two surface meshes is used, where the sum of all vertices corresponding to the target surface mesh and the deformed surface mesh were calculated and then averaged.

$$D = \frac{1}{n}\sum_{i=0}^{n}\sqrt{(x_{target,i}-x_{deform,i})^2+(y_{target,i}-y_{deform,i})^2+(z_{target,i}-z_{deform,i})^2}$$

(3.1)

This way, the average distance between all the corresponding vertexes of the two surface meshes can be determined. A smaller D-Value means that the two meshes are more similar, which also means that the optimisation is more effective. In the entire optimisation process, the D-Value is also used to determine the number of iterations. After each iteration of the simulation, the current D-Value is used and compared with the previous D-Value until there is convergence, which is D-Value not getting any smaller than last iteration. If after an iterative simulation the D-Value does not converge, the compensation process is started. The basic idea of this procedure is adding the deviation of the injection-moulded part from the current mould cavity to the target geometry to form a compensated mould cavity for the next iteration. The specific method of compensation (see Equation 3.2 and Equation 3.3) is shown below, where m represents the current iteration; i. e. when m is 1, compensated coordinates of the corresponding vertexes are derived through Equation 3.2 and the compensated stl-file is then generated directly. This compensated stl-file is imported into the second simulation as a surface mesh to continue the simulation. If m is greater than 1, the compensation process is carried out via Equation 3.3 and the generated stl-file is also used by the next simulation.

$$(x_i,y_i,z_i)_{compensation,m+1} = (x_i,y_i,z_i)_{target} \cdot 2 - (x_i,y_i,z_i)_{deform,m} \quad (m=1) \quad (3.2)$$

$$(x_i,y_i,z_i)_{compensation,m+1} = (x_i,y_i,z_i)_{target} + (x_i,y_i,z_i)_{compensation,m} - (x_i,y_i,z_i)_{deform,m} \quad (m>1)$$

(3.3)

By modelling the entire compensation process programmatically, a rapid analysis and compensation process is achieved. However, the overall computation time is determined by the number of elements of the surface mesh to be analysed. For a surface mesh with 199,900 elements, the overall analysis and compensation time is approximately 40 s.

3.5.3 Coupled optimisation routine

In this application, the RSIC method focused on the aspherical surface and its corresponding planar surface, which both have high requirements of geometrical accuracy. The error indicators shown in Figure 3.11 represents the deviation of the D-Value, which is the average Euclidean distance between the vertexes of the target surface and deformed surface. As shown in Figure 3.11. Regardless of the aspherical surface or the planar surface, their D-Values in the optimisation process have the same the change trend.

When the mould cavity was not compensated, the D-Values of the aspherical and planar surface were 40 μm respectively 38 μm, and their deviation were 13 μm respectively 14 μm. After the first compensation, the D-Values of both surfaces and their deviations were significantly reduced, which also means that after the first compensation, the deformed lens was much closer to its target geometry. On this basis, a second compensation was carried out. Both D-Values continued to decrease, as expected by using the iterative compensation method. After three compensations, the D-Values of the both surface converged at 0.34 μm and 0.31 μm, there corresponding deviations were 0.11 μm and 0.12 μm. Compared to the results without compensation of the mould cavity, the volumetric shrinkage with three iterative compensations of the mould cavity were greatly optimised. After the 4th compensation, the D-Value of the both surfaces increased slightly again, which is probably that the step size of the optimal machine setting parameter set was not small enough and it missed the global minimum. Therefore, the mould cavity used by the third compensation was selected as the optimised mould cavity, which should be manufactured by mould manufacturer in a next step. However, beforehand, the optimisation results need to be verified again: the deformed geometry of the lens by third compensation, which should be very similar to lens geometry in the unloaded state, was exported for the further thermo-optical validation. The decision was made to compare the two geometries in the unloaded state, because if the optical results match well in the unloaded state, they will also do during laser irradiation. A comparison of the optical results of the compensated geometry after the simulated moulding process with the original geometry of the thermally optimised lens shows that the back focal length is only slightly changing from 183.644 mm to 183.635 mm. This again proves that the compensation has been successful.

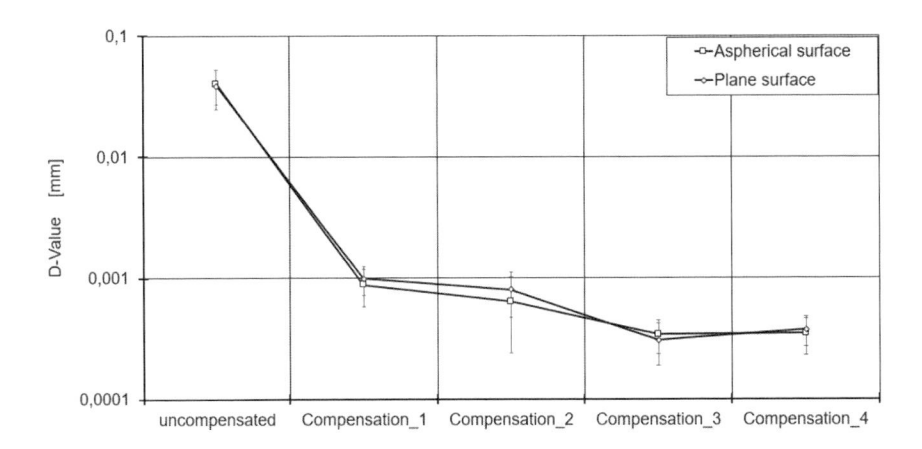

Figure 3.11: Results (D-Value) of the iterative compensation process

3.6 Conclusions and outlook

The contribution showed that realistic estimates can be made using injection moulding simulation with suitable process parameters. This enables the design of a mould for the production of plastic optics. With the coupling of the process design and the thermo-optical design, the optics can be optimised to such an extent that they are theoretically suitable for use in laser applications. The compensation results of the example model show that the proposed method can achieve simulative optimisation with very high accuracy of 0.34 μm. The RSIC method can be applied on any position on the surface mesh, which also makes it very suitable for the compensation of complex structures.

Going forward, this project plans to design and manufacture the optimised lens geometry for laser applications. In practical experiments, the beam acoustics behind the lens will be measured and compared with the simulation results. In addition, it will be systematically investigated which manufacturing parameters have the greatest influence on the optical parameters.

Acknowledgements

The research project 20797N of the Forschungsvereinigung Kunststoffverarbeitung has been sponsored as part of the "industrielle Gemeinschaftsforschung und -entwicklung (IGF)" by the German Bundesministerium für Wirtschaft und Energie (BMWi) due to an enactment of the German Bundestag through the AiF. We would like to extend our thanks to all organisations mentioned. SimpaTec GmbH supplied the used injection moulding software in the experiments. We wish to express our sincere gratitude for the sponsorship and the support. In addition, we also would like to thank the Chair of Technology of Optical Systems for the kindly cooperation.

References

[Bon19]   BONHOFF, T.: *Multiphysikalische Simulation und Kompensation thermooptischer Effekte in Optiken für Laseranwendungen; 1. Auflage.* RWTH Aachen University, dissertation, 2019

[HBSL20]   HOFMANN, J.; BONHOFF, T.; STOLLENWERK, J.; LOOSEN, P.: Auslegungsmethodik für Kunststoffoptiken in Laseranwendungen. *[121. Jahrestagung der Deutschen Gesellschaft für angewandte Optik e. V., DGaO, 2020-06-02 - 2020-06-06, Bremen, Germany].* Jun 2020

[JP05]   JIALING, W.; PENGFEI, W.: The simulation and optimization of aspheric plastic lens injection molding. *Journal of Wuhan University of Technology-Mater. Sci. Ed.* 20 (2005) 2, p. 86–89

[KFLB15]   KEMMLER, S.; FUCHS, A.; LEOPOLD, T.; BERTSCHE, B.: Comparison of Taguchi Method and Robust Design Optimization (RDO): by application of a functional adaptive simulation model for the robust product-optimization of an adjuster unit (2015),

[KJM97]   KAI, C.; JACOB, G. G. K.; MEI, T.: Interface between CAD and Rapid Prototyping systems. Part 2: LMI — An improved interface. *The International Journal of Advanced Manufacturing Technology* 13 (1997), p. 571–576

[LK01]   LU, X.; KHIM, L. S.: A statistical experimental study of the injection molding of optical lenses. *Journal of Materials Processing Technology* 113 (2001) 1-3, p. 189–195

[URL20a]   CAMPUSPLASTICS: *PLEXIGLAS® 7N.* URL: https://www.campusplastics.com/campus/en/datasheet/PLEXIGLAS%C2%AE+7N/R%C3%B6hm+GmbH/21/a7aaeaa5/SI, 01.12.2020

[URL20b]   CORETECH SYSTEM CO., L.: *Moldex3D Help 2020.* URL: http://support.moldex3d.com/2020/en/r1.html, 01.12.2020

[URL20c]   RÖHM, G.: *PLEXIGLAS® 7N.* URL: https://www.plexiglas-polymers.com/en/plexiglas-7n, 01.12.2020

Symbols

Symbol	Unit	Description
A	mm^2	Querschnittsfläche
B	mm	Breite
D	mm	Average Euclidean distance
N	–	Sum of vertices
x	mm	Value in x-direction in coordinate system
y	mm	Value in y-direction in coordinate system
z	mm	Value in z-direction in coordinate system
m	–	Number of current iteration

Abbreviations

Notation	Description
BFL	Back focal length
BLM	Boundary layer mesh
CAD	Computer-Aided Design
DoE	Design of experiment
FE	Finite Element
FEM	Finite Element Method
MAE	Mean absolute error
PMMA	Polymethyl methacrylate
RSIC	Rapid Semi-Automatic iterative Compensation
stl	Standard triangle elements
S/N	Signal to noise

Dimensioning methods for failure mechanisms of technical thermoplastics

Moderator: Dr. Matthew Beaumont, Siemens AG

--- Content ---

1 Investigation of creep on torsion-loaded polymer components of circuit breakers   347

D. Finck[1], W. Erven[1], M. Beaumont[1]
[1] Siemens AG

2 Accelerated fatigue characterisation of technical thermoplastics by temperature controlled fatigue tests at high frequencies   357

Ch. Hopmann[1], R. Dahlmann[1], H. Çelik[1]
[1] Institute for Plastics Processing (IKV) at RWTH Aachen University

3 Influence of fibre curvature on the interfacial shear stresses   372

Ch. Hopmann[1] R. Dahlmann[1], F. Di Battista[1]
[1] Institute for Plastics Processing (IKV) at RWTH Aachen University

Dr. Matthew Beaumont

Matthew's interest in the use of advanced composites in bicycles as a teenager led him to study materials science and specialise in carbon fibre composite materials at MIT in Cambridge, Massachusetts, USA. After taking a year to gain experience in low-volume manufacturing of composite sports equipment, he continued his studies with a masters degree and doctorate in Aerospace Engineering from Purdue University in West Lafayette, Indiana, USA, with his focus on manufacturing and structural analysis of fibre-reinforced composites.

His professional life began in Laupheim, Germany at the Airbus subsidiary Aircabin, where he was a process development engineer for the production of glass fibre composite cabin interior components for Airbus and Dornier aircraft. After transferring to Airbus in Hamburg, he gained management experience as the project lead for the creation of the Airbus Americas engineering centre in Mobile, Alabama. With the successful begin of operations at the centre, Matthew took an opportunity at Airbus corporate research, then known as EADS Innovation Works, leading the composites research in Ottobrunn, Germany.

Following a desire to branch out beyond the aerospace industry, he joined GE Global Research, taking the position of Lab Manager for Composites Manufacturing and Automation Technology in Garching. After successfully leading the development of automated fibre placement technology, preforming and infusion processes and development of automation solutions various GE businesses, he was asked to join GE Additive at its formation to initiate and lead GE Additive's Customer Experience Center in Garching. After three exciting years in the nascent AM industry, in 2020 he joined Siemens Corporate Technology in Munich as the executive head of the global technology field for Materials Design and Manufacturing Technologies.

1 Investigation of creep on torsion-loaded polymer components of circuit breakers

D. Finck[1], W. Erven[1], M. Beaumont[1]
[1]*Siemens AG*

Abstract

A methodology for modelling the creep behaviour of bulk moulding compounds is presented. The time-dependent material behaviour was determined in creep tests and the critical strain limit was determined using the isochronous stress-strain relationship of the material. In order to model the creep behaviour at the component level, the Norton-Bailey creep law was fitted on the basis of the experimental creep data and applied to a connecting element of a circuit breaker in a structural simulation. Due to discrepancies between the simulation and the component-level experiments, the material model was recalibrated using reverse engineering, resulting in more accurate creep simulations at the component level.

1.1 Introduction

The design of thermoplastic and thermoset polymer components that are sensitive to creep will be addressed in this paper. SIEMENS produces low voltage circuit breakers, which must carry and switch high currents reliably and provide protection against faulty electrical circuits. A very important part for closing the electrical circuit is a spring-loaded thermosetting connecting element, which is put into focus in this paper. The connecting element has to resist and transfer the spring force. Also temperatures of up to $110\,°C$ have to be withstood. A thermosetting material was put to choice, as the component has to be an electrical isolator with high arc resistance, must be cost efficient and demonstrate high thermo-mechanical stability. Especially the last mentioned requirement is of special interest for the following investigations, since the installation situation induces creep of the polymer component. In order to take the creep behaviour into account in the mechanical design, the method according to Menges, which has been established for polymer components exposed to creep, is applied [MHMS11]. Here, a strain limit is assigned to the material. The material strain must not be exceeded by the sum of the different mechanical deformation mechanisms, such as creep, plastic deformation and elastic deformation. It is explained how current FE modelling can be integrated into this design process.

1.2 SIEMENS circuit breakers

SIEMENS produces circuit breakers in a variety of power classes with a wide range of application scenarios. Typical functions a circuit breaker must fulfill in the application are, for example, the opening and closing of electrical circuits, detecting and switching off of excess current, detecting and switching off short-circuit currents. In all these scenarios, safe galvanic isolation must be ensured [Erv06].

A thermosetting material solution is particularly suitable for the connecting element, as high forces have to be switched while being affected by temperature over time. In addition, high dimensional stability, dielectric strength as well as fire resistance requirements are required. Creep calculation is required as the lifetime of the devices can be as long as 10 up to 25 years. Experience has shown that all these points can be achieved cost-effectively with bulk moulding compounds (BMC). Particular part analysed in this paper is the circuit breaker SIEMENS 3VA2, which is shown in Figure 1.1 [Erv06].

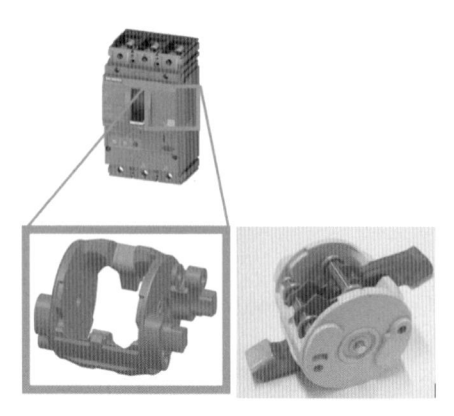

Figure 1.1: SIEMENS Circuit Breaker 3VA2 with thermosetting connecting element in detail [Erv06]

1.3 Production BMC polymer components

BMC usually consists of a fast curing thermosetting matrix and reinforcement fibres. There are different fibre length and materials commercially available. Also, different fillers to raise the Young's modulus and reduce shrinkage are common additives. With rising fibre length, the anisotropy of the produced component is more characteristic [Sch07].

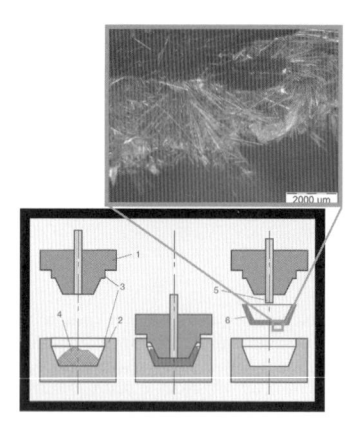

Figure 1.2: Schematic BMC press process [URL21c, Sie19]

One of the manufacturing processes of BMC is compression moulding. The kneadable mass is placed in a heated steel mould, pressed and the part is cured for several seconds or minutes afterwards. The process is shown schematic in Figure 1.2. A microscopic image of a broken BMC specimen is giving an indication about the internal material structure.

1.4 Design method for components exposed to the risk of creep according to Menges

Menges method refers to a critical strain limit to be assigned to polymeric materials, up to which no damage to the material is to be expected under load. If this strain limit is exceeded, cracks and further damage can form and expand in the material. Menges mentions critical strain limits, which are based on long-term experiences and can be used to derive corresponding design limits of common unfilled thermoplastics [MHMS11]. Originally, Menges only intended the application for strongly creeping thermoplastics, but since the thermoset system presented here also shows a significant creep tendency, the application is adopted [Ehr06]. Menges suggests to investigate more exotic polymers or those with fibre fillings experimentally [MHMS11]. Obtaining the critical strain limit can be a challenging task. Menges refers to the methodology according to Oberbach, which uses knock-down factors, which can depend on the loading situation, internal polymer structure, quasi-static material tests at different temperatures etc. to estimate the critical strain limit [Obe81]. Another mentionable methodology for this task could also be acoustic emission testing, where microphones attached e.g. to a tensile specimen, are used to detect the beginning internal fracture or matrix debonding during testing and therefore the critical strain limit [TPW17].

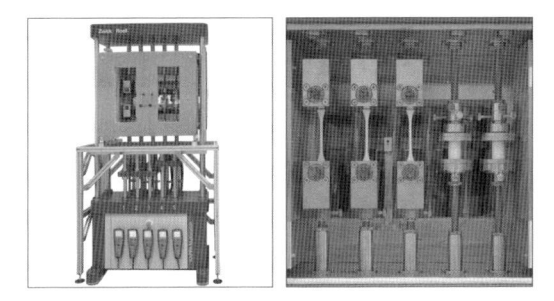

Figure 1.3: Example given - Creep test setup - Machine Zwick Kappa 5x 10 kN [URL21d]

Alternatively, creep tests can be conducted to obtain the limiting design strain. These tests are necessary anyway, since the later described creep modulus or Norton-Bailey curve fitting has to be considered. Detecting a critical strain limit in such tests can be difficult, since it is not easy to observe the beginning of internal material fracture. The tertiary creep stage, where cracks form, which significantly reduce the stiffness of the specimen and therefore accelerate the creeping are the only available signal. It should be noted therefore, that in such complex tests, the service conditions of the polymer component should be reproduced as accurately as possible, e.g. including the maximum or average service temperature. Determination of critical strain from creep tests can be done in an isochronous stress-strain diagram. The transition from the linear Hookean behaviour into flattening of the stress-strain graphs indicates the location of the critical strain. Taprogge and Menges stated an example of such an investigation (ref. Figure 1.4) [MT74]. For component development, however, the critical or design strain should still be assigned with a safety factor. The safety factor to be used depends on the application scenario. Menges also provides suggestions for possible safety factors. [MHMS11]

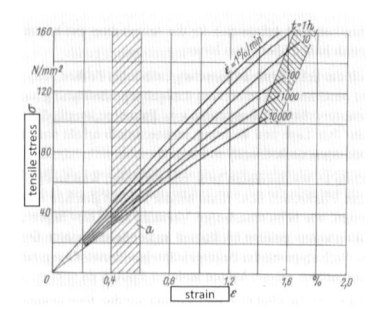

Figure 1.4: Example – Isochronous stress-strain diagram of a glass fibre reinforced UP-resin [MT74]

Figure 1.4 displays the isochronous stress-strain diagram of a glass fibre reinforced thermosetting UP-resin. In the graphs shown, the transition from the linear stress-strain segment is difficult to identify, which is why α marks a transition region. The previously mentioned critical strain can be located in this region. For a more precise determination of the critical strain, microscopic observations for crack detection or cyclic tests for the detection of permanent deformations can be performed [MT74]. Conclusively, the last mentioned method is the most promising to analyse a thermosetting material, since Odenbach did not cover these type of polymers and an acoustic emission setup was not available [Obe81]. Therefore, an isochronous stress-strain diagram was obtained from a conducted three-point bending test. Isochronous diagrams refer to time-constant curves. The axes were normalised with the last scale value of the diagram.

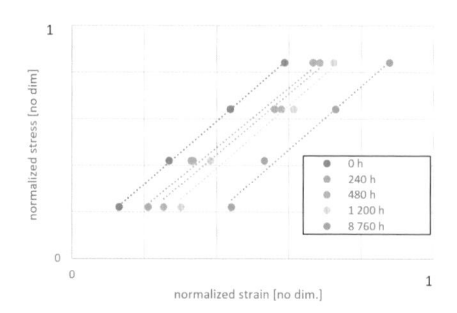

Figure 1.5: Normalised Isochronous stress-strain graph of three-point creep tests over time of used BMC

Unfortunately, the chosen stresses were not high enough to detect the transition from the linear stress-strain behaviour (compare Figure 1.4 and Figure 1.5). Therefore, the design strain can be temporarily and conservatively found to the highest strain of the $8760\,h$ isochronous curve. However, future test series are planned to find the critical strain limit of the material even more precisely.

1.5 Simulative evaluation with creep modelling according to the Norton-Bailey creep law

Continuous component deformation under load due to creep can result in load redistributions and new stress fields. There are different strategies to consider this process in the component design. A common method so far, and also recommended by Menges, is to use a creep modulus instead of the material Young's modulus in the simulation/manual calculation [MHMS11]. The creep modulus represents a reduced Young's modulus, which is intended to represent the deformation of the component due to creep at the time of design. The problem with this approach is that only one point in time in the component's life can be investigated with a computation. In addition, the use of a knocked-down Young's modulus does not correctly represent the actual physical process, since there is no softening of the material due to creep. In complex simulations with structural loads occurring in time sequences, this modelling can produce very inaccurate results. Furthermore, a softened Young's modulus is assigned to the entire component, even though there may be large variation in local creep strain within the component. A "too soft" modelling is not always conservative, for instance in a complex assembly, where a stiff component has to bear higher loads than a soft one, due to load transfers within a surrounding structure.

In the present study, it was decided that creep modelling with a creep strain that actually increases as a function of time and stress, should be done instead. It has to be pointed out that the following calculation is done with some simplifications. The main one is to use a homogeneous isotropic material model. All common simulation tools have implemented approaches, some of them more than one, to do this task. As a standard, the Norton-Bailey creep data approximation formula has found acceptance in material representation in CAx software. The Norton-Bailey creep data approximation formula represents a polynomial formula, which is composed of three material constants A, B and D and two variables σ and t [MD07].

$$\varepsilon_{creep}(\sigma, t) = A \cdot \sigma^B \cdot t^D \tag{1.1}$$

The three material constants A, B and D are iteratively adjusted to match the creep measurement data as accurately as possible. Commonly used tools for this purpose are spreadsheet programs and the least squares method. Creep test measurements and a Norton-Bailey curve-fitting of the investigated BMC material is shown in Figure 1.6.

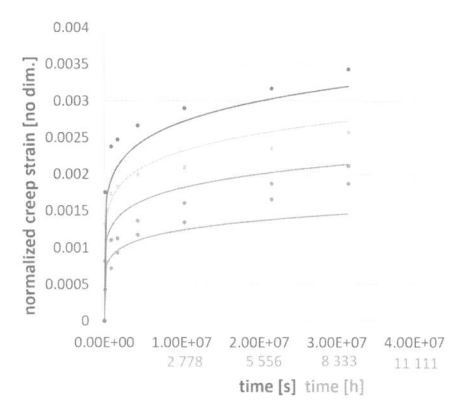

Figure 1.6: Creep strains of three-point creep tests over time with Norton-Bailey fitting of used BMC with standardised y-scale

The input of the creep data in the FE software SIEMENS NX is shown in Figure 1.7.

Figure 1.7: Norton-Bailey creep law and input mask in the material card - Software: SIEMENS NX 1980 [URL21a]

By implementing the Norton-Bailey creep law in the material card, the simulation is able to calculate element-wise creep strains based on the element stresses and the overall simulated time. The stress distribution in a connecting element and the corresponding creep strains are shown in Figure 1.8. Creep strains build up depending on the local stress over time. As there were only marginal stress redistributions due to the mounting situation, the creep strain and stress distribution are almost the same. The time steps for the steady-state transient implicit simulation must have a temporally reasonable sequence. For example, the torsional stress due to the spring occurs at first. The circuit breaker then heats up in operation and is then subjected to this state for a longer period of time.

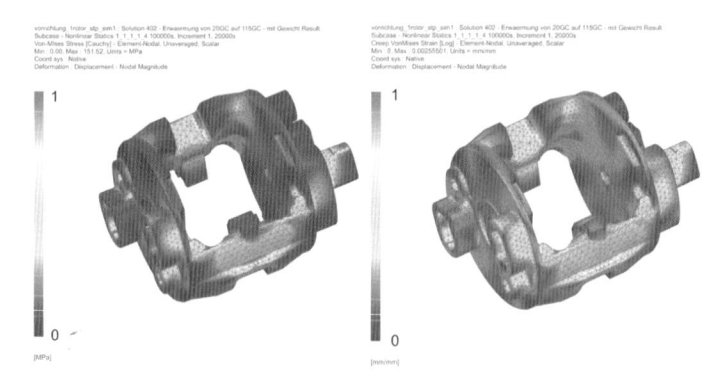

Figure 1.8: Simulation of thermosetting connecting element in SIEMENS NX 1980 at 20000 s with standardised scales - left: Von-Mises-stress [MPa] - right: Creep strain (no dim)

This modelling can not only predict the expected total material strains, but also consider, for example, the spring force loss on the contact, which closes the electrical circuit.

1.6 Advanced approaches of simulative creep modelling – reverse engineering of Norton-Bailey creep constants

It was observed that a discrepancy exists between simulated deformation curves of the circuit breakers over time and experimental comparison tests (compare Figure 1.9). Simulations were performed using the Norton-Bailey creep law fitted to coupon measurement data, as described in section 1.5.

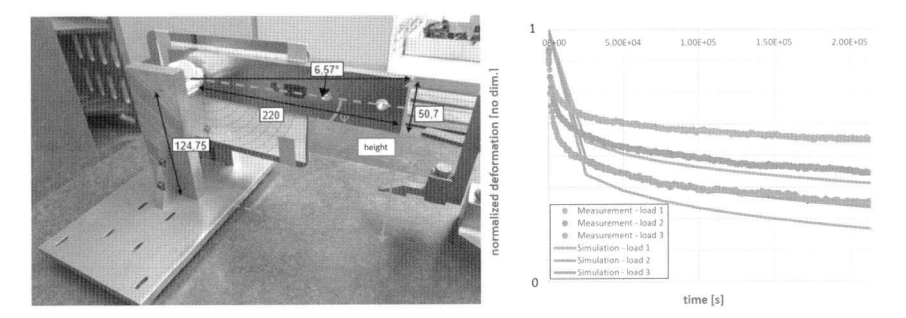

Figure 1.9: Experimental creep testing and simulation of circuit breakers - left: Test setup - right: Deformation-time graph of simulated and measurement data

A possible approach to increase the accuracy of the prediction could be to better account for the anisotropy of the material in the simulation. Such simulative approaches have already been described in other studies [Pfl01, MO10, FHG16]. However, since the anisotropic modelling of short-fiber-reinforced thermoset materials is very complex and also raises a wide range of questions with regard to accuracy, a different approach was chosen here for better simulative capture of the circuit breaker creep. One of the difficulties is the exact prediction of the fibre orientations, which can vary considerably depending on the exact placement of the preform in the press mold. Experiments were performed on circuit breakers subjected to torsional loading. Subsequently the software Siemens HEEDS was used to adjust the Norton-Bailey material card characteristic values in the simulation so that the creep experiments were properly fitted in the simulation [URL21b]. The software iteratively enters different values into the variable fields of the material card and seeks to achieve the most optimal curve fitting possible using optimisation algorithms by performing as few iterations as possible (compare Figure 1.10). The measured values of three load levels on the circuit breakers could be optimally approximated in sequence. Thus, the circuit-breaker-specific material card parameters for a creep simulation with the Norton-Bailey creep law were finally found.

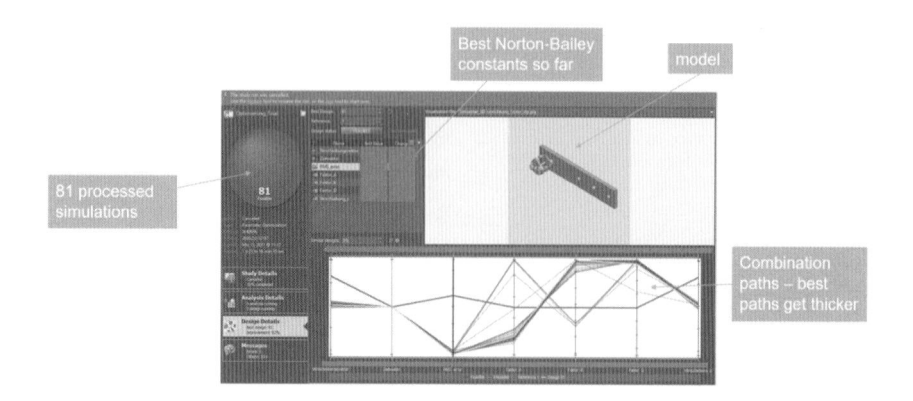

Figure 1.10: Screenshot optimisation software Siemens HEEDS – used for finding of optimal Norton-Bailey creep law regression constants fitting

The reverse-engineering approach offers the opportunity to perform a material characterisation with relatively little effort for a very complex problem. The disadvantage, however, is that the material properties are strongly dependent on the component and do not consider the anisotropic material behaviour in detail. These disadvantages can be tolerated as long as no significant geometry changes take place on the component and respect is shown to the anisotropic-isotropic modelling accuracy. The material properties generated in this way allow the component to be simulatively integrated into a complex circuit assembly with the correct input-output force response over time. In addition, local strain peaks can be evaluated and taken into account in the design of the component.

1.7 Conclusions and outlook

The studies shown, provide information on the design of thermosetting BMC components exposed to creep using Menges method [MHMS11]. The polymer connecting element from a SIEMENS power circuit breaker was used as a practical example, as it is exposed to constant loading which leads to significant creep strain over the course of the service life. Menges strain-based design methodology continues to be a standard in today's consideration of polymer components subjected to creep. The integration of modern simulative modelling techniques into this approach is possible and has been partially presented here. An insight was also given into how reverse-engineering can be used to develop new approaches in the simulatively modelling creep of anisotropic materials. Advantageous for this method is that no coupon characteristics are required. A disadvantage is that a circuit breaker component must already be available before the simulation is conducted. Further inaccuracies are to be expected if the circuit breaker undergoes a significant geometry update. How large the mismatch is for future power circuit breaker generations and whether a better/acceptable simulation accuracy can be achieved, compared to complex anisotropic creep modelling, is not yet investigated.

References

[Ehr06] Ehrenstein, Georg W.: *Faserverbund-Kunststoffe Werkstoffe - Verarbeitung - Eigenschaften 2., völlig überarbeitete Auflage.* München, Wien: Carl Hanser Verlag, 2006

[Erv06] Erven, W.: Kunststoffe in Schaltgeräten: Fachtagung Kunststoffe und SIMULATION. 2006

[FHG16] Fliegener, S.; Hohe, J.; Gumbsch, P.: The creep behavior of long fiber reinforced thermoplastics examined by microstructural simulations. *Composites Science and Technology* 131 (2016), p. 1–11

[MD07] Miravalles, M.; Dharmawan, I.: *The creep behaviour of adhesives: A numerical and experimental investigation.* CHALMERS UNIVERSITY OF TECHNOLOGY, Göteborg, Sweden, Master's thesis, 2007

[MHMS11] Menges, G.; Haberstroh, E.; Michaeli, W.; Schmachtenberg, E.: *Menges Werkstoffkunde Kunststoffe.* München: Carl Hanser Verlag, 2011

[MO10] Matsuda, T.; Ohno, N.: Predicting the elastic-viscoplastic and creep behaviour of polymer matrix composites using the homogenization theory. *Creep and Fatigue in Polymer Matrix Composites* (2010), p. 113–148

[MT74] Menges, G.; Taprogge, R.: *Kunststoff-Konstruktionen - Rechenbeispiele.* Düsseldorf: VDI Verlag, 1974

[Obe81] Oberbach, K.: Berechnung von Kunststoff-Bauteilen, Berechnungsmethoden und zulässige Werkstoffanstrengungen Konstruieren mit Kunststoffen Bd. 91 (1981), p. 181–196

[Pfl01] Pflamm, T.: *Auslegung und Dimensionierung von kurzfaserverstärkten Spritzgussbauteilen.* TU Darmstadt, Dissertation, 2001

[Sch07] Schürmann, H.: *Konstruieren mit Faser-Kunststoff-Verbunden - 2. bearbeitete und erweiterte Auflage.* Berlin: Springer Verlag, 2007

[Sie19] Siemens AG: Microscopic image broken BMC specimen (2019),

[TPW17] Trauth, A.; Pinter, P.; Weidenmann, K.: Investigation of Quasi-Static and Dynamic Material Properties of a Structural Sheet Molding Compound Combined with Acoustic Emission Damage Analysis. *Journal of Composite Science* 1 (2017) 2: 18,

[URL21a] N.N.: *SIEMENS DIGITAL INDUSTRIES SOFTWARE - PRODUCTS - NX.* URL: https://www.plm.automation.siemens.com/global/de/products/nx/, 04.10.2021

[URL21b] N.N.: *Siemens HEEDS.* URL: https://www.plm.automation.siemens.com/global/de/products/simcenter/simcenter-heeds.html, 21.10.2021

[URL21c] N.N.: *Thermoset Compression Moulding.* URL: https://www.aareplast.ch/en/thermoset-processes/thermoset-compression-moulding., 08.10.2021

[URL21d] N.N.: *ZwickRoell - Zeitstandsprüfmaschinen.* URL: https://www.zwickroell.com/fileadmin/content/Files/SharePoint/user_upload/PI_DE/88_959_Kappa_Multistation_PI_D.pdf, 01.10.2021

Symbols

Symbol	Unit	Description
A	–	empirical material constant
B	–	empirical material constant
D	mm	empirical material constant
t	s	time
ε_{creep}	–	creep strain
σ	MPa	stress

Abbreviations

Notation	Description
BMC	Bulk Moulding Compound
CAx	Computer Aided x
UP	Unsaturated Polyester

2 Accelerated fatigue characterisation of technical thermoplastics by temperature controlled fatigue tests at high frequencies

Ch. Hopmann[1], R. Dahlmann[1], H. Çelik[1]
[1]*Institute for Plastics Processing (IKV) at RWTH Aachen University*

Abstract

To accelerate the fatigue characterisation a novel adaptive cooling system was developed. It could be proven that with this novel cooling system the temperature could be controlled at one predefined level even at high frequencies. As a result, the disadvantageous influence of the temperature increase could be prevented and thus a characterisation of the fatigue life can be carried out at high frequencies. The quasi-isothermal tests resulted in higher fatigue endurance, but the characteristics of fatigue behaviour remained unchanged. Therefore, with the use of the Basquin-Equation, the slope of the stress-cycle-line (SN-line) could be determined during the quasi-isothermal experiments and transferred to the 5 Hz experiments by parallel shifting. The shift is determined by conventional fatigue tests at 5 Hz for the highest and the medium load level. By combining the conventional and the quasi-isothermal tests and simultaneously slipping the experiments at 5 Hz for one million load cycles, the characterisation duration could be reduced by a factor of around 5.5.

2.1 Introduction

Short glass fibre reinforced thermoplastics (SGFRT) are established engineering materials due to their good mechanical properties, the possibility of component and function integration and the economical production using injection moulding. Driven by higher demands on product requirements, considering cost and resource efficiency, and increasing component complexity, short glass fibre reinforced plastics are increasingly used in semi-structural applications with static as well as dynamic-cyclic loadings. During the development process of such structural components, decisions about manufacturing methods, tools and the materials commonly must be made at a very early stage, which means that the responsibility for the economical sustainability of the respective part is very high at this point. For this reason, the development process is enhanced by simulative analyses regarding the processing and structural behaviour in order to obtain a reliable estimation of the fulfilment level of the targeted objectives. The accuracy of the simulation results is highly influenced by both, the model quality, and the input data for the model calibration. The material models must be able to correctly approximate the mechanical characteristics. Therefore, a deep understanding of the material and the specification of the necessary model parameters are required. The mechanical properties of SGFRT are strongly dependent on the fibre configuration at the micro-level (fibre content, fibre orientation and fibre length distribution) and the embedding thermoplastic matrix. Hence, the material behaviour is anisotropic and shows time dependency, through which the load history influences the deformation and the stress state within the composite material. There are already established material models and integrative calculation methods that can represent the anisotropic and time-dependent material behaviour of short-fibre composites under quasi-static and long-term loads in numerical simulation models with a high degree of accuracy [Bra06, Haa18]. The determination of the necessary material parameters can also be carried out using effort-optimised testing methods, for instance by applying the principle of time-temperature shift [Kü12]. However, proving fatigue performance using simulation methods is still a challenge due to the incomplete understanding of the fatigue process of SGFRT compared to quasi-static or long-term behaviour. The interfering and non-linear material effects lead to a complex behaviour under dynamic-cyclic loading. Also, the influence and interactions of the material micro-configuration (fibre content and orientation) and the applied loads in terms of stress level/ratio, frequency and temperature need to be investigated further.

Therefore, if a failure prediction under fatigue load is required, very often extensive and cost-intensive component tests are carried out to ensure the durability.

Nevertheless, there are already the first design methods for initial fatigue estimations considering the local fibre orientation. For this purpose, Wöhler tests are carried out in analogy to isotropic materials to characterise the tolerable fatigue-stress for different loading conditions. Due to the anisotropic mechanical behaviour the characterisation has to be carried out for different material orientations. Based on these orientation-dependent experimental fatigue strength data, a generally applicable and orientation-independent Master-Wöhler curve can be derived, which allows first approximations for the fatigue assessment of SGFRT materials [Kaw04]. To take the orientation effects into account significantly more experimental tests are required to determine the fatigue life, which makes the already expensive and time-consuming characterisation even more costly. This paper introduces a novel testing methodology to accelerate the fatigue characterisation of short glass fibre reinforced plastics. This improvement is achieved by significantly increasing the test frequency during the fatigue experiments. Under the assumption that the increase in frequency must not lead to a change in the active failure mechanisms (as discussed in chapter 2.2), comparable experimental data can be obtained.

2.2 Fatigue mechanisms in short fibre reinforced thermoplastics

The heterogeneous microstructure of SGFRT, in particular the fibre configuration and the fibre-matrix interaction, leads to complex material behaviour. In addition, if a dynamic-cyclic load characterised by a constant mean load and a high-frequency oscillating fraction is applied on SFRTs, the material response and the fatigue process are influenced by many different effects. For dimensioning against fatigue, the material behaviour is characterised in Wöhler experiments using standardised test specimens and determining the number of oscillations N_B that leads to specimen failure [Hai06]. The stress form (load ratio, frequency, oscillation form etc.) is kept constant throughout the entire characterisation. Depending on the application requirements either force-controlled or displacement-controlled tests are performed, analogous to tests with constant load exposures (relaxation/retardation) [RV07]. A specific decrease in material stiffness or exceeding of a defined material damping can also be selected as a failure criterion in addition to the specimen fracture [RV07].

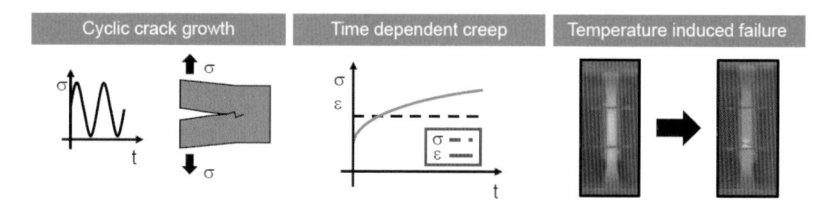

Figure 2.1: Mechanism effecting the failure procedure during fatigue

However, the resulting SN-line only provides information about the final fatigue life, but there is no differentiation of the failure mechanisms and types. The degradation of mechanical properties due to progressive material damage are not considered. The fibre/matrix interactions on the micro-level are also reflected in the damage behaviour on the macro-level for short glass fibre reinforced plastics, resulting in different failure modes (compare Figure 2.1). If an alternating load is applied that is significantly below the quasi-static failure load, fatigue is characterised by a slow crack growth process and the time of failure is determined by the cyclic fracture [MF15, SKSK91, PKEG20]. When SFRTs are subjected to high stresses close to the quasi-static strength, yielding occurs and the active damage mechanism is plasticity. To differentiate between cyclic fracture and plastic yielding in the case of dynamic-cyclic loading, long-term tests at an equally high, but constant load can be performed [PKEG20, KKG16]. The decisive parameter is the time of failure. If the dynamic-cyclic loading results in a higher life span, the dominant failure mechanism is characterised by plastic yielding since the duration of the stress load is the driving variable here. Once the time of failure due to fatigue falls below the maximum achievable failure duration compared the static long-term tests, fatigue damaging due to fracture accumulation predominates [KKG16]. Furthermore, the damage caused by fatigue is also influenced by creep of the thermoplastic matrix [KKG16, Sch17]. Based on fatigue characterisations at different load conditions and frequencies, it is shown that the crack growth scales with the applied number of cycles and the creep with the load duration [KKG16, PG21]. In addition, the visco-elastic properties of the matrix and the low thermal conductivity lead to self-heating. The applied stress level and frequency have a direct influence on the dissipated power and hence the temperature within the material [BK09, HKF+20]. As a result of continuous load and the associated temperature increase of the material, thermally induced yielding occurs [RHB19]. Since in this work the acceleration of fatigue tests by increasing the test frequency from $5\,Hz$ to $60\,Hz$ is investigated, it must be ensured that the active failure mechanisms and the interactions are not altered to allow an equivalent fatigue prediction. By increasing the test frequency, higher strain rates occur during dynamic-cyclic loading, which is why the thermoplastic matrix can exhibit a stiffer material response. However, the influence on the micro-mechanical stress state and thus on the fatigue behaviour is estimated to be negligible, since the frequency range is approximately within one decade. As previously discussed, when exposed to continuous dynamic cyclic loading, thermoplastics can heat up, which can negatively affect the mechanical properties and thus the fatigue life. For this reason, the thermodynamic interactions and the different heat transport mechanisms are discussed in more detail in the next chapter.

2.3 Influence of hysteresis on the material temperature

Due to their molecular structure, which is characterised by a polymeric structure, the mechanical properties of thermoplastics are dependent on the stress level, the stress duration and, particularly, the temperature [BBO+13, MHMS11]. Typical semi-crystalline representatives of technical thermoplastics are polyamide (PA) or polybutylene terephthalate (PBT), which are distinguished by their mechanical properties (modulus of elasticity, strength, impact resistance). The glass transition temperature T_G of these materials is in a range of about $50\,°C$ to $70\,°C$, which, when exceeded, causes a significant change in the mechanical properties [BBO+13, MHMS11, OM12]. On the one hand, the stiffness is reduced and on the other hand, the viscous properties become more dominant. The latter is particularly important for dynamic-cyclic loads. With each cycle, kinetic energy is introduced into the material. Depending on the visco-elastic properties (storage modulus E_S and loss modulus E_L, a fraction of this energy is dissipated during loading/unloading and converted into thermic energy. This is a result of the internal friction effects between the macromolecules [OM12, WS12]. If the stress is plotted against strain, a hysteresis loop is formed in the case of dynamic-cyclic loading (compare Figure 2.2). The dissipated energy w_D corresponds to the enclosed area and can be described as a function according to Equation 2.1 depending on the loss modulus E_L and strain amplitude ε_0. For the derivation of this relationship, the time-dependent parameters stress $\sigma(t)$ and strain $\varepsilon(t)$ can be approximated with idealised sinusoidal considerations [BCR15, MHMS11, OM12, WS12].

$$w_D = \oint \sigma(t)d\varepsilon = \int_0^{2\pi/\omega} \sigma(t)\frac{d\varepsilon(t)}{dt}dt = \pi\varepsilon_0^2 E_L \qquad (2.1)$$

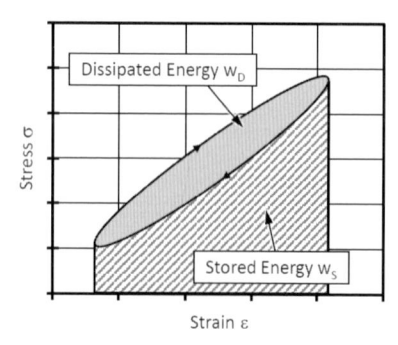

Figure 2.2: Hysteresis behaviour during dynamic-cyclic loading

The resulting temperature is also influenced by interactions with the environment. To describe the temperature development within the material, the first law of thermodynamics, which considers the conservation of energy, can be used. Every closed system has an internal energy u, which can be changed by the interaction with the environment [KS19]. The interaction is described by the supply and removal of work w and heat q.

$$du = dq + dw \qquad (2.2)$$

With the definition for the free internal energy $\psi = u - Ts$ and some transformation steps according to [NK12], the Equation 2.3 to describe the temperature evolution over time is derived. The resulting temperature depends on the heat flux $div(\vec{q})$ across the system boundaries, the thermal power supply due to the hysteresis dissipative power $\dot{w}_D$ caused by the visco-elastic material properties, the thermoelastic effects due to $T\frac{\partial \sigma}{\partial T}\dot{\varepsilon}_e$ and the changes of the internal variables V_k. Thereby, the dissipated power $\dot{w}_D$ is the product of the dissipated energy per cycle w_D and the load frequency f [BCR15]. The internal variables represent e.g. defects, crystalline micro-structures, and configurations of micro-cracks [LC94].

$$\rho c \dot{T} = -div(\vec{q}) + \dot{w}_D + T\frac{\partial \sigma}{\partial T}\dot{\varepsilon}_e + \dot{V}_k \frac{\rho T \partial^2 \psi}{\partial T \partial V_k} - \frac{\rho \partial \psi}{\partial V_k}\dot{V}_k \qquad (2.3)$$

However, the thermo-elastic effect and the changes of the internal variables can be neglected compared to the other variables, which is why the material temperature development over time depends mainly on the internal dissipative power $\dot{w}_D$ and the heat flux $div(\vec{q})$ [SMLS+17]. Energy input and loss across system boundaries is achieved by conduction, convection, and radiation. Whereby, the dominant heat transport mechanism is thermal convection [JMLSC13, SMLS+17]. Convection is the heat transfer mechanism which conducts thermal energy between a surface and a fluid. The temperature difference is thereby determined by the convection rate α, the surface temperature T_S and fluid temperature T_a. The convection rate α is largely influenced by the fluid velocity.

$$\frac{\dot{Q}_S}{A} = \dot{q_S}'' = \alpha \cdot (T_s - T_\alpha) \qquad (2.4)$$

The thermal conductivity is responsible for the temperature distribution within the cross-section. In anisotropic materials, the thermal conductivity of the compound results from the interaction of the constituents fibre and matrix, analogous to the mechanical properties [CW96, FLM19, FM03, HDEO09]. If the dissipated power cannot be transferred to the environment, the temperature in the material will increase and alter the mechanical response [DMJQ10, PG21, SMLS+20, SMLS+17].

2.4 Conventional fatigue characterisation

For the fatigue investigations, a glass fibre reinforced polybutylene terephthalate with 30 $wt. - \%$ (PBT-GF30) Pocan B3235 provided by Lanxess AG, Cologne, Germany, is selected, since this material has high industrial relevance, and the mechanical properties are not dependent on the absorbed water content. The specimen plates with the dimension of 115 mm x 115 mm x 2 mm are injection moulded. An Ergotech 80/420-310 injection moulding machine from Sumitomo Demag Plastics Machinery GmbH, Schwaig, Germany, is used for the plate production. The injection moulding process is set according to manufacturer's recommended parameters. The melt distributor with the sprue is removed after the ejection of the part to avoid warpage due to its wall thickness variations.

The ratio between surface and volume is particularly important for temperature development in fatigue tests, which is why a plate thickness of 2 mm is chosen. Additionally, thin plate provides a thinner core layer. The core layer contains predominantly fibres that are oriented transversal to the flow direction of the melt. The shear zones, on the other hand, have an orientation longitudinal to the flow direction. For the fatigue experiments, test specimens of the type 1BA according to the DIN EN ISO 527 are cut out with a CNC machine. The specimen's test area has a cross-section of 10 mm^2. A single specimen is extracted from each plate in longitudinal direction, so that possible local differences regarding the fibre orientation over the plate have no influences on the obtained test results. Therefore, the fibre configuration on the micro level over all specimens is comparable and the repeatability of the tests is ensured. For the fatigue characterisation, a servo-hydraulic testing machine of the type Schenck VHF7 from Instron Schenk Testing Systems GmbH, Darmstadt, Germany is used. The force measurement is conducted with a 5 kN load cell. The testing machine is equipped with a control and data acquisition system by means of a digital control system of the type testControl II from Zwick GmbH Co. KG, Ulm, Germany. The data acquisition system can record data with up to 17 kHz, which is important to obtain a sufficiently high sampling rate even at higher test frequencies. Hence, hysteresis response can be measured with high accuracy. The specimen strain is measured with a high-precision extensometer of the type EXA20 1.25x for small elongations up to ±1.25 mm from Sandner-Messtechnik GmbH, Biebesheim, Germany. The gauge length is 20 mm. Hence, strains up to ±6.25% can be measured. A mounting frame is used to ensure that the extensometer is always positioned at the exact same way. Hence, there is only minimal fluctuation regarding the specimen clamping over all tests.

A temperature chamber with a working range of $-40\,°C$ to $250\,°C$ from Brabender Realtest GmbH, Moers, Germany keeps the ambient temperature constant. The actual temperature of the specimen during the dynamic-cyclic tests and the resulting self-heating is recorded by a pyrometer of the type of CSmicro LT22H from optris GmbH, Berlin, Germany. This pyrometer can measure temperatures from $-50\,°C$ to $1030\,°C$. The emission rate of the tested material was verified with temperature step tests and comparing the measured temperature with the internal thermoelement of the sensor. The fatigue tests are carried out force-controlled for a stress ratio of $R \approx 0$ in the tensile range. To prevent the specimens from being exposed to compression load due to irregularities in the control system of the testing machine a lower force of 20 N is set for all fatigue tests. This lower load equates to a minimum material stress of 2 N/mm^2, which is not expected to have a negative impact on the damage mechanisms and fatigue results. The conventional fatigue tests are conducted at an ambient temperature of 23 °C. After the specimen has been positioned, the temperature chamber is closed, and it is waited for the specimen temperature to reach equilibrium with the environment. In this paper, the fatigue strength in the range of 10^4 and 10^6 cycles is investigated as a typical range for fatigue [Hai06]. If the stress is plotted over the number of cycles to failure, a linear relation for the stress-cycle diagram for double logarithmic scaling is obtained for many materials. Mathematically, this relation can be described by Basquin Equation 2.5. It provides a relationship between the stress and the number of cycles to failure N using a position parameter C and a slope factor k.

$$N = C \cdot \sigma^{-k} \qquad (2.5)$$

Three load levels for σ_{max} of 78.0 N/mm^2, 68.5 N/mm^2 and 62.0 N/mm^2 are chosen to determine the fatigue behaviour in the range of 10^4 to 10^6 cycles. Each load level is assessed with two to three tests.

As previously discussed, the main objective is to accelerate fatigue characterisation by increasing the load frequency to obtain faster information about the behaviour of the material. Therefore, initially tests are carried out at a frequency of 5 Hz to determine a reference behaviour of the material. Afterwards, the tests are repeated at a frequency of 60 Hz, which means that the same number of load cycles is applied to the test specimen in a time that is 12 times shorter than in the 5 Hz experiments. The fatigue data for both frequencies are shown in Figure 2.3 and the stress levels are normalised to the quasi-static tensile strength of 1223.7N/mm^2. For the 5 Hz fatigue tests, the obtained cycle range for the selected stress range is 10^4 to 10^6 load cycles. Furthermore, the progression is approximately linear, which means that the fatigue life for the 5 Hz tests can be described with sufficient accuracy using the Basquin Equation. In comparison, the 60 Hz fatigue tests exhibit a different behaviour. On the one hand, the cyclic fatigue life is significantly lower. For the upper and medium load levels, the fatigue life is decreased by a half decade to a full decade. For the lowest load level, ca. $4.3 \cdot 10^5$ to $6.7 \cdot 10^5$ cycles are achieved, which is a factor of about two less compared to the 5 Hz tests. On the other hand, the course of the fatigue-cycle relationship is also characterised by a regressive dependency, whereby the course cannot be approximated by the Basquin Equation properly.

This behaviour can be explained by using the temperature data. Figure 2.4 shows the averaged and maximum temperatures reached for the conducted experiments. If the maximum temperature is close to the average temperature, it can be assumed that a quasi-static temperature was reached during the dynamic-cyclic loading. If there is a large deviation, it can be assumed that a progressive temperature increase was present, which is why the material has failed due to temperature rise. The temperature measurements shows that there is only a small increase in temperature at a frequency of 5 Hz. Furthermore, the difference between averaged and maximum temperature is minor. Especially in the cyclic ranges above 10^5 cycles there is almost no temperature increase. This can be explained by the linear material behaviour at the corresponding load levels. As soon as the material behaviour becomes non-linear, the stress leads to the plastic range of the material, which is why increased dissipation takes place. However, for the 60 Hz tests it can be stated that for all loads a progressive temperature increase occurs in the experiments. Despite the temperature of 23 °C in the temperature chamber, all averaged temperatures are clearly above 30 °C for the 60 Hz characterisation. Furthermore, the maximum temperatures also

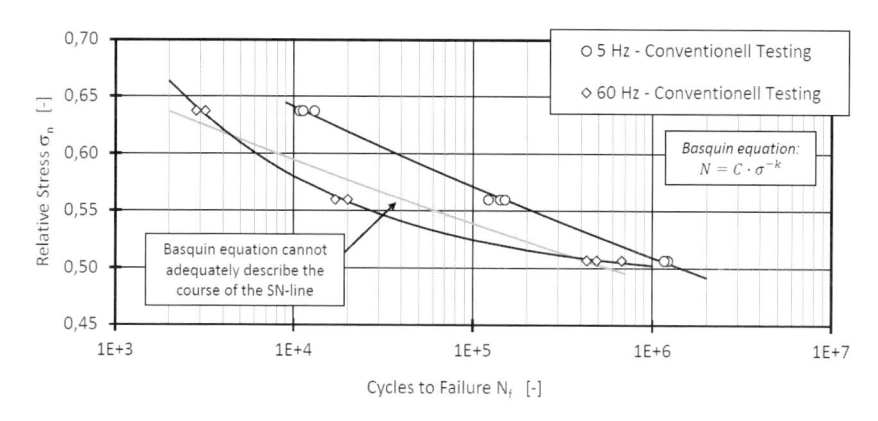

Figure 2.3: Comparison of fatigue characterisation at the frequencies of $5\,Hz$ and $60\,Hz$

deviate from the averaged temperatures by up to a maximum of ca. $30\,°C$. Many of the test specimens reach maximum temperatures in the range of $40\,°C$ to $65\,°C$.

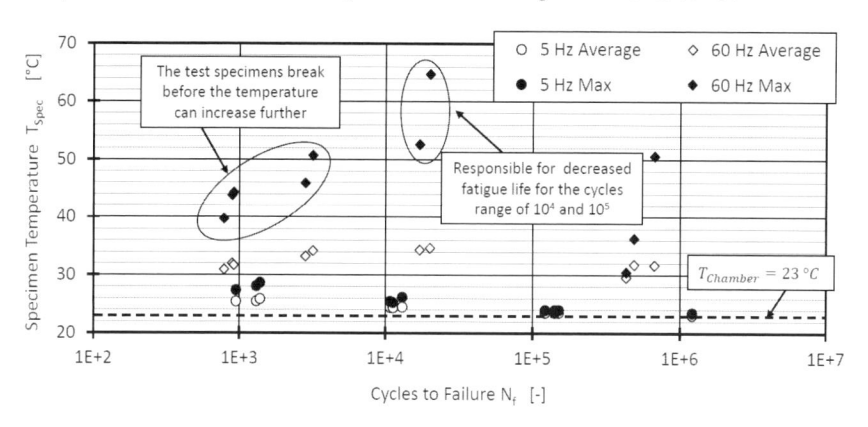

Figure 2.4: Frequency dependent temperature increase during fatigue characterisation

The temperature measurements can also be used to explain the significantly reduced fatigue life for the relative stress level of $\sigma_n = 0.56$. For this load, significantly higher temperatures can be achieved than for the other stress levels. Since the increase in temperature also leads to a reduction in strength, the fatigue life is lower in this range. For the higher load levels, the stress duration is not sufficient enough for the specimen to heat up further, which is why the temperature dependent strength reduction is less distinctive.

Due to the temperature increase resulting from the high dissipative power at high frequencies and the low thermal conductivity of the material, the damage mechanisms within the material are altered and the obtained information is not comparable with the $5\,Hz$ fatigue tests. Therefore, no accelerated fatigue characterisation is possible by increasing the test frequency only.

2.5 Development of an adaptive cooling system for quasi-isothermal fatigue characterisation

In order to keep the temperature development within reasonable bounds despite high dissipation at high frequencies, an adaptive cooling system is developed. The experimental setup is shown in Figure 2.5. Two ventilator packs are utilised to variably increase the convective heat flux around the specimen. To direct the air exactly onto the specimen, the ventilators are equipped with air routing parts. The vents are positioned in such a way that the specimen can be uniformly blown from the sides. An control system based on an arduiono was developed to control the rotational velocity of the ventilators. The temperature of the pyrometer is taken as the input signal and the power of the ventilators is adjusted according to the required reference temperature of the specimen via pulse-width modulation (PWM). Since the inertia effects of the ventilators and the high loop repetition rate of the arduino system, an on-off controller is implemented for the ventilator-power adaption. As soon as the reference temperature is exceeded, the ventilators are activated with full power. If the temperature is lower than the reference temperature, the ventilators are no longer supplied with any power. However, they still continue to rotate due to the mechanical inertia. Thus, the convective heat flow is not abruptly discontinued. Due to the high frequency of the arduino, a steady temperature control results. Hence, using the developed adaptive cooling system quasi-isothermal fatigue tests can be performed.

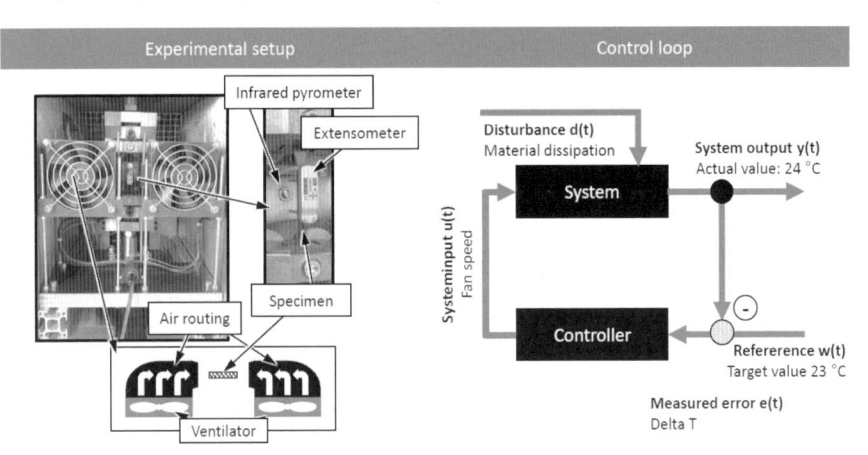

Figure 2.5: The developed novel adaptive cooling system

Compared to the previously described conventional tests, there are a few additional parameters to adjust according to the test parameters. The test frequencies and load levels used are the same as for conventional fatigue tests. However, the temperature development is not only determined by the convection rate α, but also by the ambient temperature T_a. Therefore, in order to keep the material temperature constant throughout the test, it must be ensured that the operating range of the cooling system corresponds to the possible temperature increase of the specimen. Hence, the ambient temperature is set to a level that the ventilation power of at least 20 % is achieved at the beginning of each fatigue test. This results in a chamber temperature in the range of $15°C$ - $21°C$ depending on the load level. However, the setting for the chamber temperature depends on the material and must be verified iteratively for each load level and test frequency in advance. In order to check the functionality of the adaptive cooling system, a test is performed at $90\,Hz$ and at an ambient temperature of $15.5°C$ (compare Figure 2.6). At the beginning of the experiment, it can be observed that the temperature of the test specimen is increased from $15.5°C$ to $23.0°C$ within a short time. As soon as the set temperature is reached, the ventilators start to work, and the material temperature is maintained within a range of $\pm1°C$. This corresponds to a class 1 temperature tolerance according to DIN 291. It can also be observed that the ventilator power fluctuates over the duration of the experiment due to the temperature variation in the chamber. Due to the thickness of the specimens being $2\,mm$, it can be assumed that the temperature distribution within the specimens is not strongly deviating from the measured surface temperature. Hence, the developed adaptive cooling system enables quasi-isothermal fatigue tests that suppress the temperature increase and consequently do not change the damage mechanisms through temperature change despite higher frequencies.

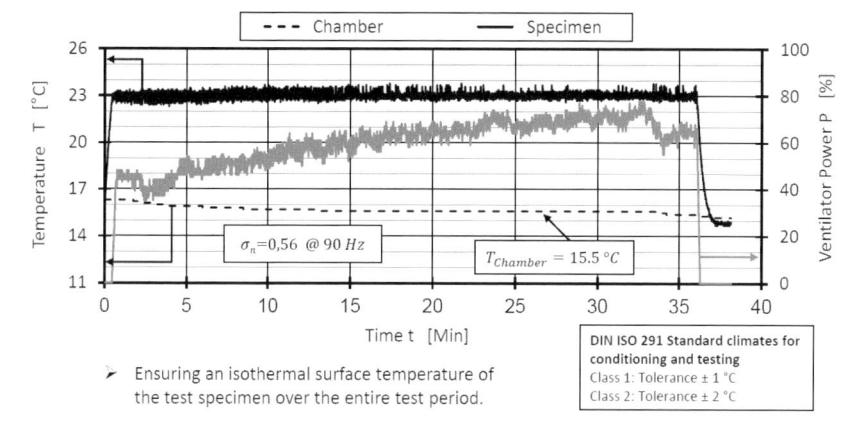

Figure 2.6: Ensuring the quasi-isothermal fatigue experiments by adjusting the ventilation power and the ambient temperature

2.6 Quasi-isothermal testing to accelerate the fatigue characterisation

With the developed novel adaptive cooling system, it is possible to keep the specimen temperature constant over the entire test duration. Therefore, the fatigue characterisation is repeated at 60 Hz and at the same load levels as before. The achieved fatigue data of the quasi-isothermal experiments are compared with the previous results in Figure 2.7.

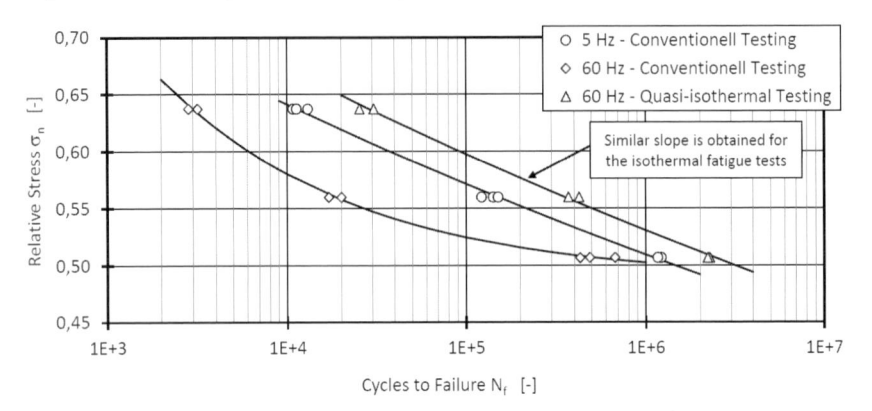

Figure 2.7: Comparing quasi-isothermal fatigue results with conventional experiments

The fatigue life of the conventional tests at 5 Hz and 60 Hz is exceeded at all load levels. There is a difference of a factor of two to three between the achieved load cycles compared to the 5 Hz experiments. Furthermore, it is noticeable that the obtained results, analogous to the 5 Hz experiments, form a straight line in the SN-diagram, which allows the Basquin Equation to be applied. In addition, it is also evident that the slope of the Basquin correlation is almost identical to the 5 Hz experiments ($k_{CT5Hz} = 19.95$ and $k_{IT60Hz} = 19.32$). To understand why quasi-isothermal fatigue tests lead to higher cycles to failure, it is necessary to consider the damage mechanisms of SGFRT-materials. As discussed previously, material fatigue is caused by cyclic crack growth on the one hand and visco-elastic effects (retardation) at the micro level on the other hand. If the frequency is increased significantly, the material has less time to retard per cycle than using a lower frequency. Consequently, the micro-mechanical stress state of the material is less affected by creep after the same number of applied cycles, indicating that the material exhibits less material damage. This means that a higher number of cycles can be withstood by the material. However, since the slope of the SN-line is approximately identical to the 5 Hz experiments, it can be assumed that the damage mechanisms are not significantly altered by the higher frequency.

The advantage of the fact that the SN-lines are shifted parallel to each other can be used to generate the fatigue life data for 5 Hz faster. The characterisation of fatigue life for the range 10^4 to 10^6 on three load levels and three tests per load level requires approximately 185 $hours$ of pure testing time. In addition, there is idle time between the individual tests. Since short glass fibre reinforced thermoplastics demonstrate anisotropic mechanical properties, additional testing must be carried out transversely to the flow direction. As a result, about ca. one whole month of continuous testing is required to characterise a material in this cycle range and frequency. Considering a three times longer cyclic life at 60 Hz than at 5 Hz, the same characterisation, including all repetitions, takes only approximately 16 $hours$ of testing. Since the characteristics of the 60 Hz tests are comparable to the conventional testing, it is reasonable to combine the conventional and the quasi-isothermal testing at both frequencies. Especially since the 10^6 cycles have the greatest influence on the total test duration. Therefore, it is proposed that a full fatigue data set at $60\,Hz$ is first generated in order to determine the slope of the Basquin Equation. Subsequently, only the load levels for the two cycles range between 10^4 and 10^5 are characterised for $5\,Hz$. These data points for $5\,Hz$ are used to determine the parameter C of the Basquin Equation. This results in a total testing time of approximately $34\,hours$, which equals a reduction of the testing time by a factor of approximately 5.5. In Figure 2.8 the new combined SN line is compared with the previous results and a good agreement with the $5\,Hz$ tests can be achieved. Consequently, the fatigue behaviour can be characterised faster with the novel adaptive cooling system and the parallel shift method for the SN-line.

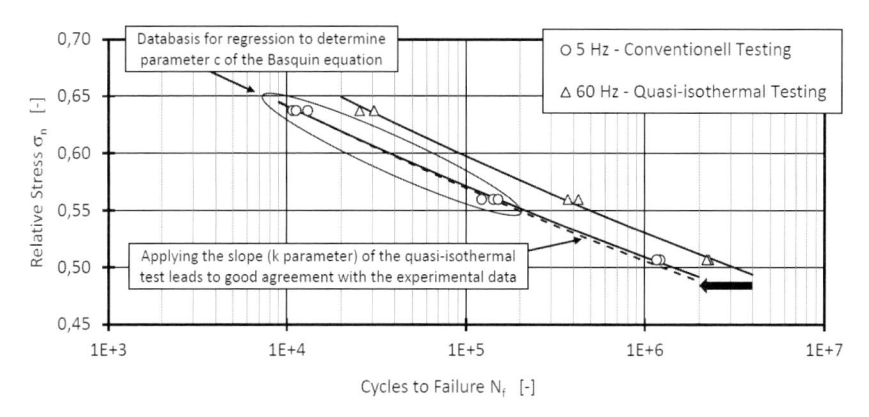

Figure 2.8: Parallel shift method for accelerated fatigue characterisation of SFRT Materials

2.7 Conclusions and outlook

In this work, a newly developed method was presented so that the fatigue characterisation for short fibre reinforced thermoplastics is accelerated by several times. It was first demonstrated that increasing the frequency alone is not sufficient. The increased frequency led to an increase in the dissipated power, and thus the material heated up due to the dynamic-cyclic stress and caused prematurely failure. For this reason, a novel adaptive cooling system was developed and validated to ensure that the material temperature is maintained at a constant level even at high test frequencies. The validation of the developed cooling system showed that the temperature remains within a deviation of $\pm 1^\circ C$ over the entire test period and thus corresponded to a class 1 temperature control. Using this novel cooling system, it was possible to carry out fatigue tests at $60\,Hz$, which were characterised by a quasi-isothermal temperature control at $23^\circ C$. Based on these test results, the parameters of the Basquin Equation were determined. Eventually, it was shown that the slope for the conventional $5\,Hz$ experiments and the slope of the quasi-isothermal experiments at $60\,Hz$ are almost identical. Therefore, a parallel shift can be applied to the SN line. The offset is determined by finding the C-parameters using conventional test data at $5\,Hz$ and the load levels for 10^4 and 10^5 cycles. Using this methodology, the duration of the experimental characterisation can be decreased by a factor of approximately 5.5. In subsequent investigations it is necessary to investigate whether this methodology can also be applied for transversely oriented specimens. Furthermore, it must be determined if the method can also be applied to other thermoplastic material types. Another question is, if dynamic mechanical analyses (DMA) can provide information whether the method is applicable in principle and if a maximum frequency for the accelerated fatigue characterisation can be determined.

Acknowledgments

The research project (20556N) of the Forschungsvereinigung Kunststoffverarbeitung was sponsored as part of the „Industrielle Gemeinschaftsforschung und -entwicklung (IGF)" by the German Bundesministerium für Wirtschaft und Klimaschutz (BMWK) due to an enactment of the German Bundestag through the AiF. The investigated material was provided by Lanxess AG. We would like to extend our thanks to all organizations mentioned.

References

[BBO+13] BAUR, E.; BRINKMANN, S.; OSSWALD, T.; RUDOLPH, N.; SCHMACHTENBERG, E.: *Saechtling Kunststoff Taschenbuch.* München: Carl Hanser Verlag, 2013

[BCR15] BENAARBIA, A.; CHRYSOCHOOS, A.; ROBERT, G.: Thermomechanical behavior of PA6.6 composites subjected to low cycle fatigue. *Composites Part B: Engineering* 76 (2015), p. 52–64

[BK09] BERNASCONI, A.; KULIN, R.: Effect of frequency upon fatigue strength of a short glass fiber reinforced polyamide 6: A superposition method based on cyclic creep parameters. *Polymer Composites* 30 (2009) 2, p. 154–161

[Bra06] BRANDT, M.: *CAE Methoden für die verbesserte Auslegung themoplastischer Spritzgussbauteile.* RWTH Aachen University, Dissertation, 2006. ISBN: 978-3-86130-850-8

[CW96] CHEN, C.-H.; WANG, Y.-C.: Effective thermal conductivity of misoriented short-fiber reinforced thermoplastics. *Mechanics of Materials* 23 (1996) 3, p. 217–228

[DMJQ10] DE MONTE, M.; MOOSBRUGGER, E.; JASCHEK, K.; QUARESIMIN, M.: Multiaxial fatigue of a short glass fibre reinforced polyamide 6.6 – Fatigue and fracture behaviour. *International Journal of Fatigue* 32 (2010) 1, p. 17–28

[FLM19] FU, S.; LAUKE, B.; MAI, Y.: *Science and Engineering of Short Fibre-Reinforced Polymer Composites*: Elsevier Science, 2019

[FM03] FU, S.-Y.; MAI, Y.-W.: Thermal conductivity of misaligned short-fiber-reinforced polymer composites 88 (2003) 6, p. 1497–1505

[Haa18] HAAG, J. V.: *Modellierung des zeitabhängigen thermomechanischen Steifigkeitsverhaltens spritzgegossener kurz- und langfaserverstärkter Thermoplaste.* RWTH Aachen University, Dissertation, 2018. ISBN: 978-3-89653-428-6

[Hai06] HAIBACH, E.: *Betriebsfestigkeit - Verfahren und Daten zur Bauteilberechnung.* Berlin Heidelberg: Springer-Verlag, 2006

[HDEO09] HEINLE, C.; DRUMMER, D.; EHRENSTEIN, G. W.; OSSWALD, T.: Thermally conductive modified polymers- Part 1: Experimental analysis and modelling description of a process dependent part property. *Zeitschrift Kunststofftechnik/Journal of Plastics Technology* (2009) 5, p. 338–358

[HKF+20] HOPMANN, C.; KORTE, W.; FISCHER, K.; SCHMITZ, M.; ONKEN, J.; VERWAAYEN, S.; ÇELIK, H.: Integrative simulation methods for optimised injection moulding products. *30th International Colloquium Plastics Technology* (2020),

[JMLSC13] JEGOU, L.; MARCO, Y.; LE SAUX, V.; CALLOCH, S.: Fast prediction of the Wöhler curve from heat build-up measurements on Short Fiber Reinforced Plastic. *International Journal of Fatigue* 47 (2013), p. 259–267

[Kaw04] KAWAI, M.: A phenomenological model for off-axis fatigue behavior of unidirectional polymer matrix composites under different stress ratios. *Composites: Part A* 35 (2004), p. 955–963

[KKG16] KANTERS, M. J. W.; KUROKAWA, T.; GOVAERT, L. E.: Competition between plasticity-controlled and crack-growth controlled failure in static and cyclic fatigue of thermoplastic polymer systems. *Polymer Testing* 50 (2016), p. 101–110

[KS19] KAISER, W.; SCHLACHTER, W.: *Energie in der Kunststofftechnik: Grundlagen und Anwendungen für Ingenieure.* München: Carl Hanser Verlag, 2019

[Kü12]   KÜSTERS, K.: *Modellierung des thermo-mechansichen Langzeitverhaltens von Thermoplasten*. RWTH Aachen University, Dissertation, 2012. ISBN : 978-3-86130-513-2

[LC94]   LEMAITRE, J.; CHABOCHE, J.-L.: *Mechanics of solid materials*. Cambridge: Cambridge University Press, 1994

[MF15]   MORTAZAVIAN, S.; FATEMI, A.: Fatigue behavior and modeling of short fiber reinforced polymer composites: A literature review. *International Journal of Fatigue* 70 (2015), p. 297–321

[MHMS11]   MENGES, G.; HABERSTROH, E.; MICHAELI, W.; SCHMACHTENBERG, E.: *Menges Werkstoffkunde Kunststoffe*. München: Carl Hanser Verlag, 2011

[NK12]   NADERI, M.; KHONSARI, M. M.: Thermodynamic analysis of fatige failure in a composite laminate. *Mechanics of Materials* 46 (2012), p. 113–122

[OM12]   OSSWALD, T. A.; MENGES, G.: *Material Science of Polymers for Engineers*. München: Carl Hanser Verlag, 2012

[PG21]   PASTUKHOV, L. V.; GOVAERT, L. E.: Crack-growth controlled failure of short fibre reinforced thermoplastics: Influence of fibre orientation. *International Journal of Fatigue* 143 (2021),

[PKEG20]   PASTUKHOV, L. V.; KANTERS, M. J. W.; ENGELS, T. A. P.; GOVAERT, L. E.: Influence of fiber orientation, temperature and relative humidity on the long-term performance of short glass fiber reinforced polyamide 6. *Journal of Applied Polymer Science* (2020), p. 50382

[RHB19]   RÖSLER, J.; HARDERS, H.; BÄKER, M.: *Mechanisches Verhalten der Werkstoffe*. Wiesbaden: Springer Fachmedien Wiesbaden, 2019

[RV07]   RADAJ, D.; VORMWALD, M.: *Ermüdungsfestigkeit - Grundlagen für Ingenieure*. Berlin Heidelberg: Springer-Verlag, 2007

[Sch17]   SCHÖNEICH, M.: *Charakterisierung und Modellierung viskoelastischer Eigenschaften von kurzglasfaserverstärkten Thermoplasten mit Faser-Matrix Interphase*. Saarland University, Dissertation, 2017

[SKSK91]   SATO, N.; KURAUCHI, T.; SATO, S.; KAMIGAITO, O.: Microfailure behaviour of randomly dispersed short fibre reinforced thermoplastic composites obtained by direct SEM observation. *Journal of Materials Science* 26 (1991) 14, p. 3891–3898

[SMLS+17]   SERRANO, L.; MARCO, Y.; LE SAUX, V.; ROBERT, G.; CHARRIER, P.: Fast prediction of the fatigue behavior of short-fiber-reinforced thermoplastics based on heat build-up measurements: application to heterogeneous cases. *Continuum Mechanics and Thermodynamics* 29 (2017) 5, p. 1113–1133

[SMLS+20]   SANTHARAM, P.; MARCO, Y.; LE SAUX, V.; LE SAUX, M.; ROBERT, G.; RAOULT, I.; GUÉVENOUX, C.; TAVEAU, D.; CHARRIER, P.: Fatigue criteria for short fiber-reinforced thermoplastic validated over various fiber orientations, load ratios and environmental conditions. *International Journal of Fatigue* 135 (2020),

[WS12]   WARD, I.; SWEENEY, J.: *Mechanical Properties of Solid Polymers*: Wiley, 2012

Symbols

Symbol	Unit	Description
A	mm^2	Cross-section
E_L	N/mm^2	Loss modulus
E_S	N/mm^2	Storage modulus
f	Hz	frequency
k	–	Basquin parameter, slope
N	–	Number of cycles to failure
R	–	Stress ratio
T	$°C$	Temperature
V	–	Internal
w	J	Work
C	–	Basquin parameter, position
q	J	Heat
α	–	Convection rate
w_D	J	Dissipated energy
ε	$\%$	Strain
σ	N/mm^2	Stress

Abbreviations

Notation	Description
DMA	Dynamic Mechanical Analyses
GF30	Glass fibre content of 30 wt.-%
PA	Polyamide
PBT	Polybutylene terephthalate
PWM	Pulse-width modulation
SGFRT	Short glass fibre reinforced thermoplastics

3 Influence of fibre curvature on the interfacial shear stresses

Ch. Hopmann[1] R. Dahlmann[1], F. Di Battista[1]
[1]*Institute for Plastics Processing (IKV) at RWTH Aachen University*

Abstract

The interphase between the fibre and the matrix in long glass fibre reinforced polypropylene components can be identified as a weak spot of the material combination due to the non-polarity of both components. Therefore an extensive simulative study is made to investigate the influence of different fibre lengths, fibre orientations, and fibre curvatures on the shear stress uptake of the interphase. The results show that with reduced fibre lengths the shear stresses on the fibre are higher which implies that shorter fibres will debond earlier from the matrix than longer fibres. The same effect can be observed when the fibre orientation approaches the direction of the local tensile field. A curvature of the fibre for values higher than 90° leads to a shift of the maximum shear stress from the fibre ends to the surface areas which are inclined by 45° degrees to the tensile direction.

3.1 Introduction

The material class of long fibre reinforced thermoplastics (LFRT) combine excellent mechanical properties with a high geometrical degree of freedom and a good processability. Thus, LFRTs are gaining more and more relevance for structural applications [NLH+19]. The mechanical properties of LFRTs depend on the local configuration of the fibres and the adhesive bond between the fibre and the matrix. The main influencing parameters are the fibre orientation distribution (FOD), the fibre volume content and the fibre length distribution (FLD) [TV96, TV97]. For the prediction of mechanical properties of fibre reinforced materials in dependence of the mentioned parameters homogenisation models based on the theories of *Eshelby* and *Mori-Tanaka* or unit cell approaches can be used [Esh57, MT73, TL99].

Additionally, the long fibres within the component tend to curve during the injection-moulding process [BGY+21, VGMH+10] which also effects the local mechanical properties of LFRTs. Generally, a curved fibre leads to a constant local reorientation of the fibre. Since the fibres can take up only forces in fibre direction this constant local reorientation causes a smaller reinforcement contribution of the fibre to the composite. In [BN08] a comparison of a unit cell homogenisation approach of an single fibre and a fibre bundle model with an analytical homogenisation approach for a discontinuous glass fibre reinforced Polypropylene shows that the curvature of the fibre leads to a significant stiffness reduction in fibre direction but only a small stiffness increase perpendicular to the fibre direction. *Drach et al.* compares the the overall material stiffness in the case of an continuous sinusoidal fibre for different crimp ratios of the fibre. Especially the stiffness value in fibre orientation direction decreases significantly with increasing crimp ratios [DKS16].

The main researches focus especially on the stiffness prediction but the effect of the fibre curvature on the strength and failure behaviour of LFRTs can be considered unexplored. Thus, the scope of this article is an extensive study of different fibre geometries and alignments to investigate the influence on the initial debonding behaviour of fibres within the matrix.

3.2 Simulation model

3.2.1 Material modelling

The investigated material combination is Polypropylene (PP) reinforced with glass fibres. This material combination is widely used due to its low costs of the raw materials but also good mechanical properties. Since both ingredients are materials with a low chemical potential, the adhesive bond between the two components is weak by nature [EB12, Tho07]. Thus, the interphase can be identified as one of the weak spots of this particular material combination.

To be able to analyse the adhesive zone between the PP matrix and the glass fibre it is modelled as interface by defining a cohesive interaction between the adjacent surfaces of the fibre and the matrix. The PP matrix is modelled as an elasto-plastic material whereas the glass fibre is considered ideally linear elastic. The defined material values can be taken from the Table 3.1. All mentioned components are modelled without taking failure into account.

Material	Property	Symbol	Value	Unit	Source
PP (matrix)	Young's modulus	E_m	1564	MPa	Measured
	Yield stress	σ_Y	6	MPa	
	Poisson's coefficient	ν_m	0.4	-	[Sch07]
Glass	Young's modulus	E_f	74000	MPa	[URL22]
	Poisson's coefficient	ν_f	0.2	-	
Interphase	Stiffness	K_n, K_s, K_t	5000	MPa/mm	[GM02]

Table 3.1: material properties

3.2.2 Geometrical modelling

To investigate the influence of the varying fibre geometries a simulation model is built based on a cuboid matrix and a single fibre which is positioned in the middle of the matrix. The dimensions of the matrix are fixed for each simulation and are chosen to be much higher than the fibre dimensions to create a homogeneous strain field and neglect edge effects. Contrary to the fixed matrix, the fibre dimensions are varied by three parameters: the length, the orientation and the curvature. All three parameters can widely vary within LFRT composites. Therefore a wide spectrum is set for the simulative investigations (see Table 3.2).

Parameter	Symbol	Values	Unit
Length	L	1, 3, 5, 7	mm
Orientation	ϕ	0, 30, 45, 60, 90	°
Curvature	α	2, 60, 90, 120, 180	°

Table 3.2: parameter overview

The orientation angle is defined by the tensile direction and the straight line drawn between the two fibre ends. So an orientation angle of 0° describes a fibre which lies parallel to the tensile direction and an orientation angle of 90° relates to a fibre which lies perpendicular to the tensile direction. The curvature angle represents the angle of a circular segment and is defined by the two planes which go through the two surfaces of the ends of the fibres.

For example, a fibre curvature angle of 90° represents a quarter circle and a fibre curvature angle of 180° represents a half circle. Figure 3.1 shows a graphical interpretation of the two angles for the example of an orientation angle $\phi = 30°$ and a curvature angle $\alpha = 90°$.

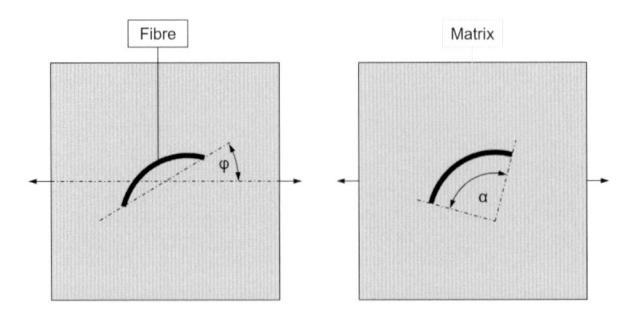

Figure 3.1: Graphical interpretation of the orientation angle (left) and the curvature angle (right)

3.3 Interfacial shear stresses in dependence of fibre length, orientation and curvature

The dependence of the different fibre geometries and alignments on the interfacial shear stresses is detected by comparing the curves of the shear stresses on the convexly shaped outer side of the fibre. At this side the shear stresses are mostly on a higher level than on the concavely shaped side of the fibre. Especially with an orientation angle of 0°, the difference between the convex and concave side of the fibre is higher with increasing degree of curvature, because the curvature of the fibre shields the matrix area within the convex region from the applied strain field (Figure 3.2).

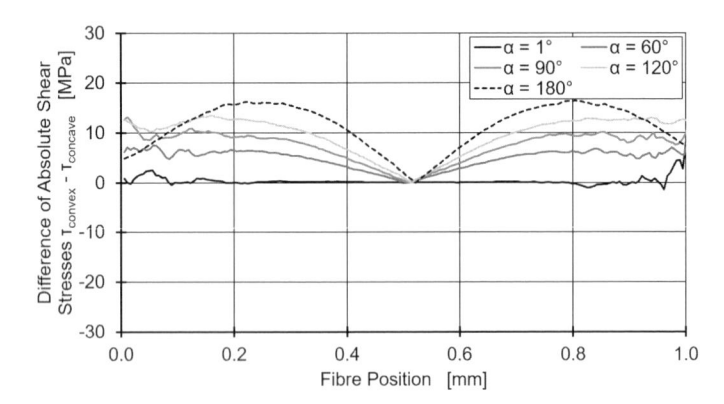

Figure 3.2: Comparison of the shear stresses on the concavely and convexly shaped fibre side for a fibre length L = 1 mm and a fibre orientation angle $\phi = 0°$

3.3.1 Influence of fibre length

The influence of different fibre lengths is shown in Figure 3.3 as an example by comparing the shear stress curves for the different fibre lengths at a constant fibre orientation angle of 0° and a constant fibre curvature angle of 90°. The shear stress is plotted against the local orientation angle of the fibre with respect to the tensile direction, which is spanned by the local perpendicular fibre cross-section and the tensile direction. The graph shows that the maximum shear stress is at the fibre ends, then decreases rapidly within a short section of the fibre and then drops to a value of 0 MPa towards the centre of the fibre. The further course is shown via the point-mirrored course of the first half of the fibre. All shear stress curves show a qualitatively comparable course, whereby the level of shear stress is higher with reduced fibre length. One explanation is the smaller outer surface of the shorter fibres. This results in a shorter distance over which the forces can be transferred from the matrix to the fibre. Furthermore, for the same angle of curvature of the fibre, the radius of curvature is smaller for shorter fibres, which results in a stronger curvature and thus a more unfavourable position of the fibre.

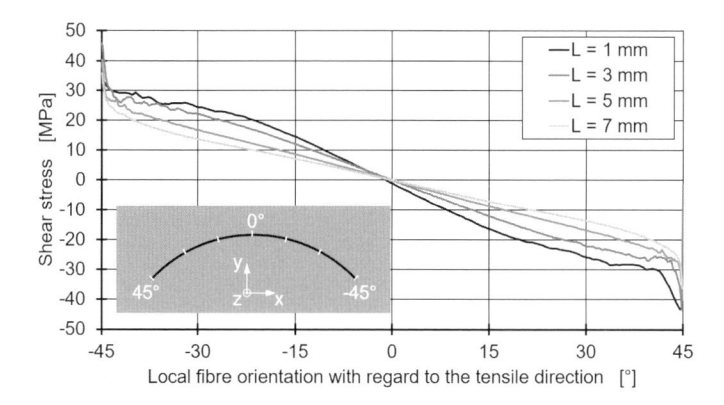

Figure 3.3: shear stress in dependence of varying fibre orientations and lengths

3.3.2 Influence of fibre orientation

A comparison of the shear stress curves for different fibre orientation angles can be seen in Figure 3.4 as an example for a constant fibre length of 5 mm and a constant fibre curvature angle of 90°. In addition to the already described curves for an orientation of 0°, two other types of shear stress curves are mainly caused by the reorientation. The orientation angles of 30°, 45° and 60° show comparable curves between -15° and 90° local fibre orientation. The highest absolute shear stress value is at the right fibre end. From this maximum value, the shear stress decreases to a local minimum and then increases to a positive value up to the left fibre end with the exception of the 60° orientation course. In the course of 60° orientation, on the other hand, the shear stress increases to 0° after the local minimum near the local orientation angle of 45° and then decreases again towards the left end of the fibre. Basically, it can be stated that the maximum shear stress is present at the fibre ends. The disoriented fibres have higher values than the ideally oriented fibre. Due to the reorientation of the fibres and the curvature angle of 90°, the fibre ends are closer to the ideal fibre orientation of 0°, which means that a higher stiffness jump between fibre and matrix can be expected at the fibre end, which results in a faster transfer of forces from the matrix to the fibre and thus in higher shear stress values. The local minimum for the graphs is at the point disoriented by 45° to the tensile direction. In ductile materials, the

45° planes represent shear stress maxima, which favours the slippage of matrix from the fibre at these points.

A different course can be seen for the fibre oriented perpendicular to the tensile direction. There, the shear stress shows almost continuously negative values with two shear stress minima. The shear stress level for this combination is lower than for the other configurations. Due to the 90° disorientation, the fibre is in the most unfavourable orientation for force transmission, which is why a perpendicular peeling of the matrix from the fibre is to be expected here rather than a detachment by parallel sliding of the matrix.

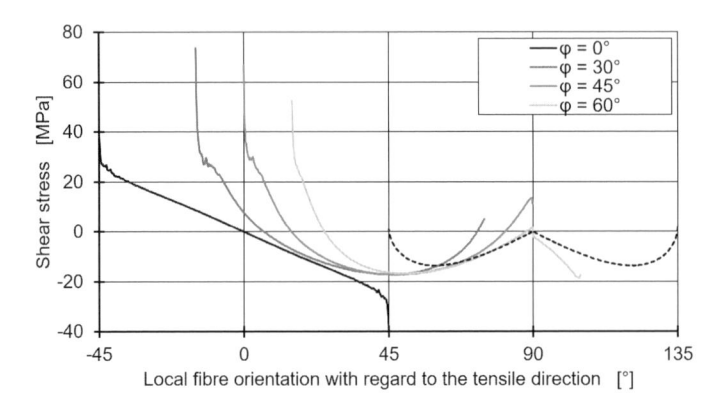

Figure 3.4: shear stress in dependence of varying fibre orientations

3.3.3 Influence of fibre curvature

For the consideration of the influence of curvature, two graphs are shown in Figure 3.5, which contain the shear stress curves for different angles of curvature for a constant fibre length of 5 mm and a constant fibre orientation angle of 0°. On the left side, the curves are plotted against the position on the fibre in order to be able to evaluate the values at the same fibre position. The right side shows the same curves over the local fibre orientation angle. It can be noted that up to a fibre curvature of 90°, the fibre ends represent the location of maximum shear stress, with the level decreasing with increasing curvature. This is due to the fact that the curvature moves the fibres further away from their ideal straight alignment and thus the stiffness jump at the fibre ends becomes smaller. For the fibres with a curvature angle of more than 90°, the shear stress maximum again shifts from the fibre ends towards the centre into the areas that are inclined by +45° or -45° to the tensile direction.

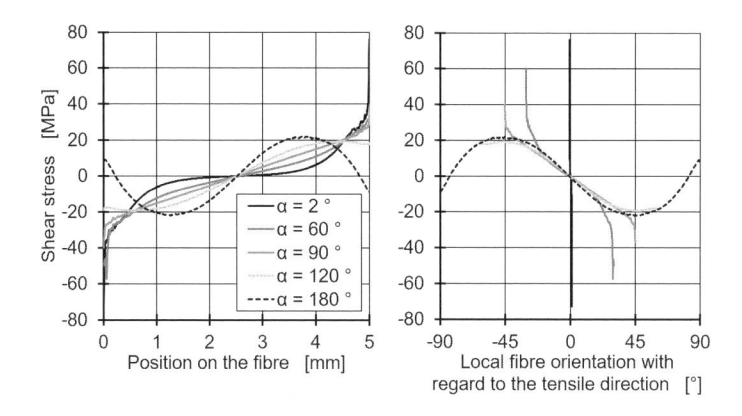

Figure 3.5: shear stress in dependence of varying fibre curvatures

3.3.4 Maximum stress on the fibre as indicator of the force transfer on the fibre

For an overall evaluation of the effects, Figure 3.6 shows the maximum stress that occurs in the fibre as a function of the fibre orientation angle and the fibre curvature angle. Since the stress is induced in the fibre via the force transfer from the matrix, this quantity serves as a good indicator for the fibre effectiveness.

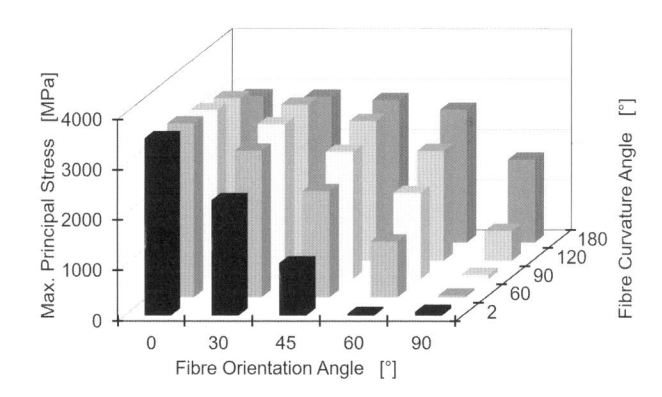

Figure 3.6: Maximum principal stress in the fibre for varying fibre orientations and curvatures (L = 5 mm)

For a fibre orientation angle of 0°, the straightest fibre has the highest stress value of 3520 MPa. As the curvature increases, the maximum induced stress value decreases to a value of 3225 MPa for a fibre curvature angle of 120°. In comparison, the drop in maximum stress for a fibre with a fibre curvature angle of 180° is significantly lower at 2902 MPa. The comparable stress level for the 0° oriented fibre can be explained by the fact that these fibres have a fibre area ideally oriented in the direction of tension. In addition to this, there are still large fibre parts around this maximum in both directions, through which force transmission is made possible (Figure 3.7). The curvature does lead to a deterioration in force transmission, which especially becomes noticeable at a curvature angle of 180°.

However, the angle of curvature means that for the orientation angles considered, there is always a fibre component ideally oriented in the direction of tension. This leads to the fact that, in contrast to the other curvatures, the level of the maximum induced stress is comparably high for the fibres curved by 180° with the exception of the fibre oriented by 90° to the tensile direction. However, the orientations cause the fibre portion to the right of the stress maximum to decrease with increasing orientation angle, which is why the stress level becomes smaller due to the smaller transmission area. For the 90° disoriented fibre, there are also parts in the tensile direction, but since these are the fibre ends, no tensile stress can be introduced there. Therefore, in this case the stress maximum shifts slightly towards the centre of the fibre. Due to the unfavourable position, the stress value of 1680 MPa is almost 1000 MPa below the maximum value for a 60° disoriented fibre.

The effect that the force transmission is lower in the fibres oriented perpendicular to the tensile direction increases with decreasing fibre curvature. Due to a straighter course of the fibre, this orientation state leads to larger fibre portions being close to the worst orientation of 90° for force transmission and thus almost no tensile forces can be taken up by the fibre. From a curvature angle of 90° and lower, the maximum stress values are therefore below 100 MPa.

Considering the remaining parameter combinations with a fibre orientation angle between 30° and 90° and a fibre curvature angle between 2° and 120°, the maximum stress induced in the fibre increases with decreasing orientation angle and increasing curvature angle. On the one hand, higher curvatures lead to fibre components being ideally aligned over a wider orientation space. On the other hand, at lower orientation angles, fibres are closer to the ideal alignment, resulting in an improved alignment for taking up tensile stresses.

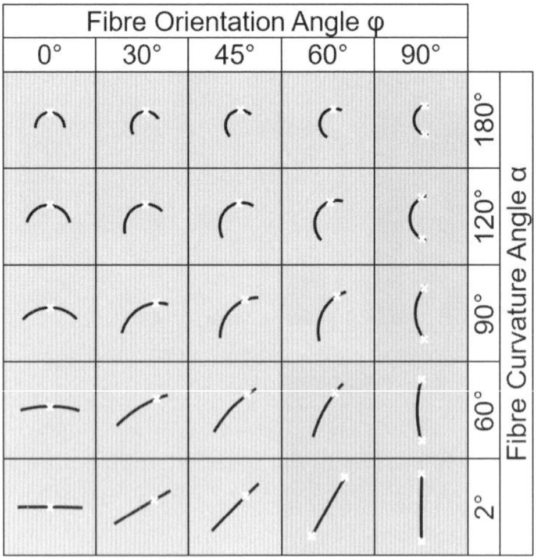

Figure 3.7: Maximum principal stress locations in the fibre for varying fibre orientations and curvatures (L = 5 mm)

3.4 Conclusions and outlook

A comprehensive numerical study of the effects of varying fibre lengths, fibre orientations and fibre curvatures on the interphase shear stress in a homogeneous idealised strain field was executed. In principle, shorter fibres lead to a higher shear stress level than longer fibres; moreover, the highest shear stress values are usually observed at the fibre ends, especially if these are oriented in the tensile direction. If the fibre ends deviate significantly from this optimal orientation, the maximum shear stress shifts to the fibre areas that are 45° disoriented to the tensile direction, whereby the maximum values are lower in this case. Therefore, it can be stated that the fibre ends or the 45° disoriented fibre areas can be identified as the starting point of a possible fibre-matrix separation.

Further investigations should focus on the consideration of fibre-fibre interactions, as these also influence the local stress fields. By considering real fibre volume contents, for example, the influence of adjacent fibres can be taken into account in the investigations. Furthermore, the integration of material damage is an important pillar to gain further insights into possible damage mechanisms as well as the damage process in LFRT components.

Acknowledgements

The depicted research has been funded by the Deutsche Forschungsgemeinschaft (DFG) within the project HO 4776/56-1. The authors would like to extend our thanks to the DFG.

References

[BGY+21] BECHARA, A.; GORIS, S.; YANEV, A.; BRANDS, D.; OSSWALD, T.: Novel modeling approach for fiber breakage during molding of long fiber-reinforced thermoplastics. *Physics of Fluids* 33 (2021) 7, p. 1–11

[BN08] BAPANAPALLI, S.; NGUYEN, B. N.: Prediction of Elastic Properties for Curved Fiber Polymer Composites. *Polymers and Polymer Composites* 29 (2008) 5, p. 544–550

[DKS16] DRACH, B.; KUKSENKO, D.; SEVOSTIANOV, I.: Effect of a curved fiber on the overall material stiffness. *International Journal of Solids and Structures* 100-101 (2016), p. 211–222

[EB12] ETCHEVERRY, M.; BARBOSA, S. E.: Glass fiber reinforced polypropylene mechanical properties enhancement by adhesion improvement. *Materials* 5 (2012) 6, p. 1084–1113

[Esh57] ESHELBY, J. D.: The determination of the elastic field of an ellipsoidal inclusion, and related problems. *Proceedings of the Royal Society of London. Series A. Mathematical and Physical Sciences* 241 (1957), p. 376–396

[GM02] GAO, S. L.; MÄDER, E.: Characterisation of interphase nanoscale property variations in glass fibre reinforced polypropylene and epoxy resin composites. *Composites - Part A: Applied Science and Manufacturing* 33 (2002) 4, p. 559–576

[MT73] MORI, T.; TANAKA, K.: Average stress in matrix and average elastic energy of materials with misfitting inclusions. *ACTA METALLURGICA* 21 (1973), p. 571–574

[NLH+19] NING, H.; LU, N.; HASSEN, A. A.; CHAWLA, K.; SELIM, M.; PILLAY, S.: A review of Long fibre-reinforced thermoplastic or long fibre thermoplastic (LFT) composites 6608 (2019),

[Sch07] SCHÜRMANN, H.: *Konstruieren mit Faser-Kunststoff-Verbunden - 2. bearbeitete und erweiterte Auflage.* Heidelberg: Springer Berlin, 2007

[Tho07] THOMASON, J. L.: Interfaces and interfacial effects in glass reinforced thermoplastics. *Proceedings of the 28th Risø International Conference on Materials Science* (2007) April, p. 75–92

[TL99] TUCKER, C. L.; LIANG, E.: Stiffness predictions for unidirectional short-fiber composites: Review and evaluation. *Composites Science and Technology* 59 (1999), p. 655–671

[TV96] THOMASON, J. L.; VLUG, M. A.: Influence of fibre length and concentration on the properties of glass fibre-reinforced polypropylene: 1. Tensile and flexural modulus. *Composites Part A: Applied Science and Manufacturing* 27 (1996) 6, p. 477–484

[TV97] THOMASON, J. L.; VLUG, M. A.: Influence of fibre length and concentration on the properties of glass fibre-reinforced polypropylene: 4. Impact properties. *Composites Part A: Applied Science and Manufacturing* 28 (1997) 3, p. 277–288

[URL22] N.N.: *E-Glass Fiber, Generic.* URL: https://www.matweb.com/search/DataSheet.aspx?MatGUID=d9c18047c49147a2a7c0b0bb1743e812, 14.06.2022

[VGMH+10] VÉLEZ-GARCIA, G. M.; MAZAHIR, S.; HOFMANN, J.; WAPPEROM, P.; BAIRD, D.; ZINK-SHARP, A.; KUNC, V.: Improvement in orientation measurement for short and long fiber injection modled composites. *Society of Plastics Engineers - 10th Annual Automotive Composites Conference and Exhibition 2010, ACCE 2010* (2010) January, p. 799–809

Symbols

Symbol	Unit	Description
α	°	Fibre Curvature Angle
E_f	MPa	Young's Modulus of the Fibre
E_m	MPa	Young's Modulus of the Fibre
$K_{n,s,t}$	MPa/mm	Stiffness Parameter of the Interfacial Contact
L	mm	Fibre Length
ν_f	–	Poisson's Coefficient of the Fibre
ν_m	–	Poisson's Coefficient of the Matrix
ϕ	°	Fibre Orientation Angle
σ_y	MPa	Yield Strength
τ	MPa	Shear Stress

Abbreviations

Notation	Description
FLD	Fibre Length Distribution
FOD	Fibre Orientation Distribution
LFRT	Long Fibre Reinforced Thermoplastics
PP	Polypropylene

Resilient control of the injection moulding process

Moderator: Dr.-Ing. Thomas Walther, ARBURG GmbH + Co. KG

Content
---

1 Paths to the self-optimising injection moulding machine 385

T. Walther[1]
[1] ARBURG GmbH + Co. KG

2 Geometry dependent injection - simulative approach to realise constant flow front velocity and shear rates at the flow front in injection moulding 392

Ch. Hopmann[1], T. Köbel[1]
[1] Institute for Plastics Processing (IKV) at RWTH Aachen University

3 Increased process stability in the injection moulding of post-consumer recyclates using cavity pressure for process control 406

Ch. Hopmann[1], K. Hornberg[1]
[1] Institute for Plastics Processing (IKV) at RWTH Aachen University

Dr.-Ing. Thomas Walther

Dr.-Ing. Thomas Walther studied mechanical engineering at the University Stuttgart. After completing his doctorate on the subject of dynamic mould temperature control in injection moulding, he joined ARBURG in 2000 to work in application technology development. Since 2009, he has been head of application technology and since 2021 he has been responsible for application and process development.

1 Paths to the self-optimising injection moulding machine

T. Walther[1]
[1] ARBURG GmbH + Co. KG

Abstract

If an injection moulding production is compared with a production environment in which metal is processed, you will find significant differences in the workflows and processes. In metal processing, the processes for the processing machines are usually created in advance, optimised and transferred directly to the machines. Orders can then be repeated almost as often as desired. In an injection moulding production, processes for new moulds are created directly at the machine by an application engineer. The quality of the components in a series process is continuously checked. If necessary, the process must be corrected during series production. The mostly individual injection moulds combined with a very wide range of plastics that are processed require this individual process finding. There is a desire for technical support for more efficient process finding and process support. Different approaches are briefly presented below.

1.1 Tasks of a machine control for an injection moulding machine

The essential key component of an injection moulding machine is the machine control. The control specifies the cycle sequence of the injection moulding process (see Figure 1.1). The single process steps must be parameterised and controlled. The core competence is the injection step. With a controlled forward movement of the screw the molten plastic is injected into the cavity of the mould. By cooling the plastic solidifies in the mould to a rigid plastic part. After opening the mould the part is ejected or taken out. In order to achieve the melt, the material has to be plasticised. The required amount is metered by rotating the screw under the influence of the specified back-pressure. The temperature conditions in the barrel are controlled and monitored. Beside the moulding process the mould movements are also coordinated in the cycle sequence. This includes the movements of the mould halves, the primary mould axis as well as different auxiliary axes. These axes include ejector movement, core pulls, shot-off nozzles, rotating units or further special mould functions.

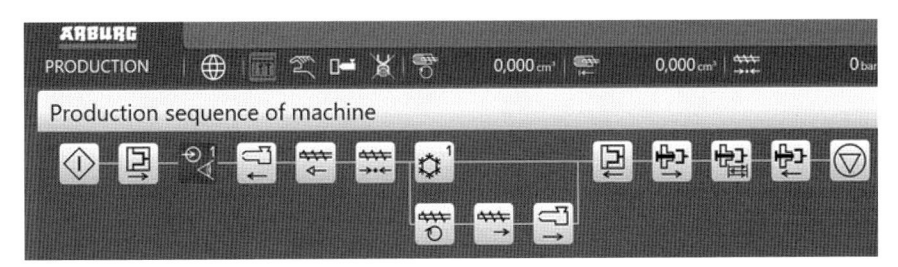

Figure 1.1: Cycle sequence of an ARBURG injection moulding machine

Depending on the machine type and version, these movements are electrically, hydraulically or pneumatically driven. Depending on the requirements, the movements vary from simple switching functions up to highly dynamic controlled ones. In addition the machine control has to fulfil several basic functions. Among them is the monitoring of movements, fill levels, the regulation of different temperature control circuits as well as the implementation of a safety circuit conception. If the machine is equipped with a robot system this functionality is also integrated in the machine control. The different actuators are monitored by a large number of sensors and regulated based on the signals of these sensors, the main function of the control. In addition the process is monitored to prove the uniformity of the process. Therefore, process key figures are calculated, monitored and stored. The proof of quality of the produced parts is derived out of these process key figures. With the progress of digitalisation this data management becomes increasingly important. During the sampling of new moulds, the process is parameterised and adjusted by a specialist so that the plastic parts can be produced in the required quality. In this step the acceptable process window is determined, which is important later if process adjustments during the production become necessary. Increasingly, data management becomes important, as well on the machine and in relation to third-party systems, requiring extensive interface options for external data evaluation and analysis.

1.2 Goals and challenges in process setting

A process sequence must be set up so that injection moulded parts can be produced. The challenge and profession here is to establish an uniform, stable process on the injection moulding machine with the individual mould and the specified material with which good parts can be produced. This is the mission of the application engineer, who makes use of the possibilities of the machine control system and the resulting process key parameters. However, the application engineer does not know how accurate and uniform the consistency of the raw material is that is fed to the machine via the material hopper. Nor does the machine know how the melt behaves in the mould and how the condition of the mould changes during the ongoing process. In the majority of cases, there are no sensors installed in the system that could provide information to the machine control system, neither in the plasticising unit nor in the mould. Figuratively speaking the machine respectively the application engineer is confronted to two black boxes (see Figure 1.2).

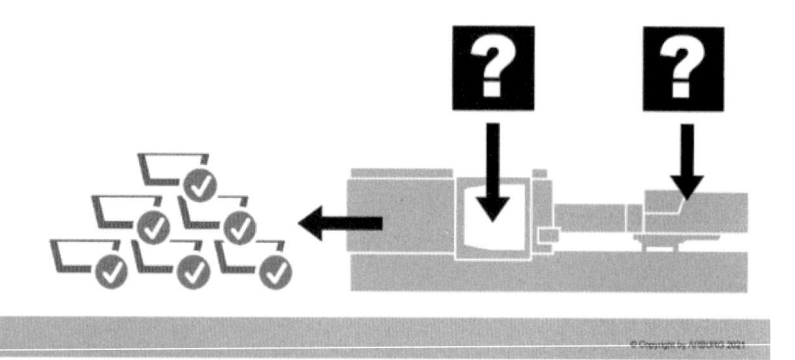

Figure 1.2: Challenge: Plasticising unit and mould as black box

The conditions of these black boxes are not detectable in the standard cases. If additional sensors are implemented, the conditions can be detected even partially. By processing virgin material, the material manufacturer guarantees that the properties of the polymer are highly consistent. Nevertheless, it happens that narrow process windows arise, so that even if material properties are within the specified tolerances, processes have to be re-adjusted in order to be able to maintain the required component qualities. The challenges arise when recycled material is used. Greater variations in material properties are to be expected. The same situation can apply, if bio-polymers or bio-based materials are used. Inevitably, these conditions lead to resulting variations in mould filling and thus in part quality. Furthermore, environmental conditions can also influence the stability of injection moulding processes, temperature fluctuations at different times of the year or day as well as fluctuations of the humidity. For this purpose, the control system is parameterised and set by an application engineer. Data management is becoming increasingly important. A comprehensive interface for external online data evaluation and data analysis is also required.

1.3 Lifting potentials

The irreversible and necessary trend towards the use of alternatives to the virgin petrochemical based plastics requires the further development of the injection process. Though this requirement is not new. There were and are different, partly combined approaches how more transparency can be brought into these black boxes. The basic idea is that a process should be as robust and stable as possible, and ideally the machine should also be able to react independently to irregularities or even, as mentioned in the title, to optimise the process independently. Some known approaches are briefly described below.

1.3.1 Additional sensors

Higher quality and better control can be achieved, if additional sensors can be used beyond those installed in the machine. Both in the injection mould and in the plasticising unit, there is experience of how signals from sensors can be used profitably. Pressure or temperature sensors in the cavity of injection moulds help to achieve more transparency of the conditions in the cavity with the corresponding signals. In the cavity, the moulded part is influenced by the holding pressure when the part is cooled down from melt temperature to demoulding temperature. The direct, local recording of the conditions in the cavity enables these signals to be used without distortion. Machine influences as well as cross-machine model influences are equalised. Based on a step response of the temperature signal at the end of the flow path the position of the melt front can be detected. So the signal can easily used as an independent switch-over signal to switch between injection step and the holding pressure step. This option can be ordered for most of the machines for many years. The efficiency of this option depends on the position, where the sensor can be placed in the cavity. Good results can be achieved when the mould has only one cavity.

Using multi-cavity moulds with a hotrunner, each cavity with a sensor, there are various possibilities to improve the uniform filling of the cavities. Either the machine control or external control systems can balance the filling of the cavities by adjusting the temperature in the zones of the hotrunner. The uniformity of cavity filling over the number of cavities is positively influenced and as a consequence the improved uniformity of the part weights. The use of cavity pressure signals for process monitoring and thus quality monitoring is often used for technical components. In addition, there are control concepts that use the internal pressure as the basis for a reference curve control. The measured internal pressure curve of cycles with verifiable good component is specified as a set point curve for the following processes (see Figure 1.3). The control system regulates the holding pressure process in such a way that the same conditions are reproduced in the cavity shot by shot as in the reference process. Irregularities, for example from batch fluctuations in the melt, or even creeping wear issues in plasticising components can thus be compensated for in an automated manner to a certain degree.

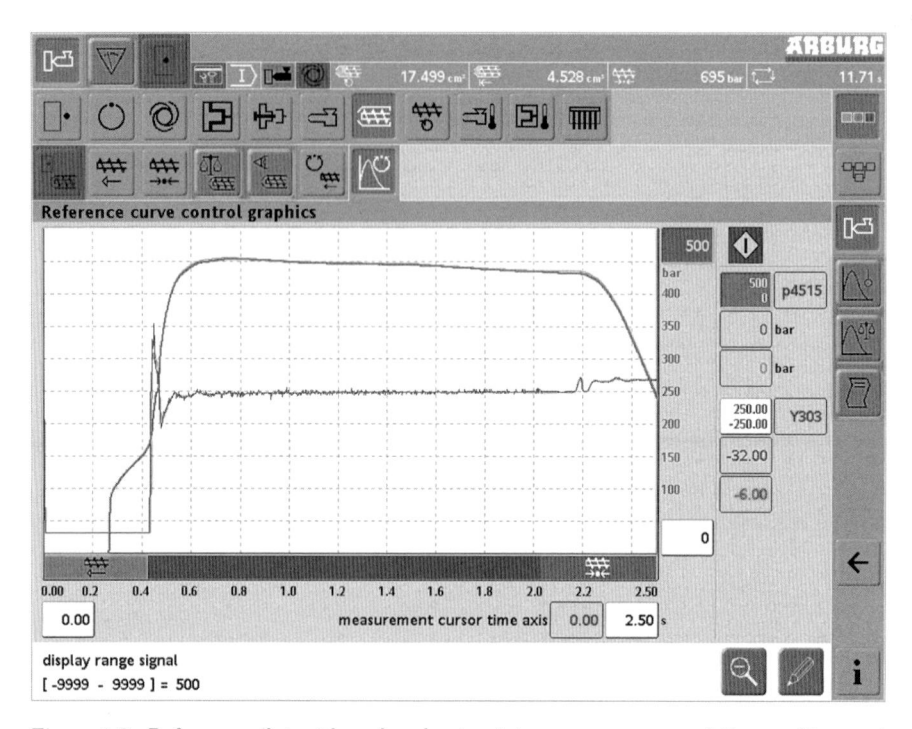

Figure 1.3: Reference pilot with real and set point pressure curve and the resulting real respectively set control behaviour

As in the mould, the conditions in the plasticising cylinder can also be detected and analysed via additional melt pressure sensors. Changes in the injection pressure under constant conditions can be indicative of viscosity fluctuations and can be controlled accordingly. By using additional sensor technology, special attention must be paid to the correct functionality of the signal chain. It must be ensured at all times that the sensors are mounted correctly and operate without interference. The connection from the sensor to the amplifier has to be made with the correct cables. The appropriate amplifiers must also be correctly calibrated and the configuration data in the machine control system must correspond to the correct specifications. The consequence of incorrectly transmitted signals is a misinterpretation of the values, in this case a control system can have counterproductive effects. The sensors of the injection moulding machine are usually placed indirectly and thus protected against premature wear. The signals are often automatically re-calibrated cyclically.

1.3.2 Process models

Process models for the forming process were already successfully established in the 1990s. On the basis of DOE (Design of experiments), different test settings are carried out around a robust parameter setting. Multiple linear regression conceptions are used to calculate mathematical relationships between machine setting variables and the resulting characteristics of the moulded parts. In addition to the measurement results determined for the components in the respective tests, process parameters from machine and mould sensors are also included in the analysis. The result is a process model that makes transparent with which parameter settings the component quality can be specifically influenced.

Technically, the method has proven itself. The model created provides an exact process image. The relationships between setting parameters and component quality are transparent. In the final stage, the developed regression model can be loaded into the machine control system. The dimensioning of the components is displayed and editable on the control surface. If quality losses are detected, the application engineer can directly adjust the dimensions of the part on the control surface. The relevant machine parameters are automatically adjusted in the background via the stored process model. In practice, the procedure usually turns out to be too time-consuming. The execution of the test plan as the basis of the data structure including the measurement of the components with a statistically reliable sample size means a big workload. Especially with complex part geometries, the measurement effort can be correspondingly high, so that measurement results are available days after the tests. For these reasons, the method has not become widely accepted for normal sampling. The model determination must be carried out anew for each component. In the field, the method is used for the analysis of long-lasting problem cases, but rarely for everyday process set-ups.

1.3.3 Adaptive control systems

Another approach that has found its way into the series of various machine control systems in different forms is an adaptive control system for improving process consistency. The aim of this type of system is to improve process stability and shot weight consistency. The initial situation is usually a parameter setting as a reference, with which an optimum component quality is achieved. The reference process is represented by characteristic process values and signal characteristics. The focus is on the injection pressure curve over the screw path during the injection process. During injection, the current pressure curve is continuously compared with the reference cycle. If deviations occur in the injection pressure curve that influence the filling level of the cavity, re-adjustments are made during the current process based on the stored process knowledge. Both the switch-over point and the follow-up pressure curve are automatically adjusted if necessary. This keeps the mould filling constant, minimises shot weight fluctuations and increases the process stability. Most systems are a software solution without additional sensors in the mould or in the machine.

1.3.4 Development of AI based process models

With the progress of digitisation, further possibilities arise. Larger volumes of data can be captured and analysed much faster and at higher resolution. The data sources are also becoming more diverse. In the past, only those process parameters and curves of those sensors were available that were directly related to the process. Primarily, this was the data relating to the screw movement. Today, much more information from the control system is available digitally and can be used to describe the process. This is, for example, continuous information from force or displacement transducers from various machine axes or signals from the converter for the various servo drives. The torque required for the current metering process can be used for process analysis, as can the force profile required for ejecting the component. In conjunction with the associated quality data and the logging of process adjustments, a comprehensive database can be generated for analysis methods. Known, rule-based methods reach their limits more quickly due to the complexity of the database. AI methods can be used to determine individual process models. The great opportunity, however, is to learn overarching relationships from many different process models. These insights can then be available for process creation and process optimisation of unknown components and lead to good results more quickly.

1.3.5 Use of digital component data as a basis for process adjustment

Another promising approach is to use digital part data in conjunction with results of a mould filling simulation as a data basis for the process adjustment (see Figure 1.4). Nowadays, various digital information and knowledge about the injection moulding parts are usually available at the time when new moulds are on the machine the first time. The parts are usually designed as a 3-D model using appropriate CAD systems. The mould filling process and the ideal cooling of the component are calculated in simulation calculations and the resulting part properties are derived. In addition to an optimised component geometry, the simulation results are used to obtain specifications for mould design. The goal of this approach is to use this existent, digitally processable prior knowledge for process adjustment and process optimisation.

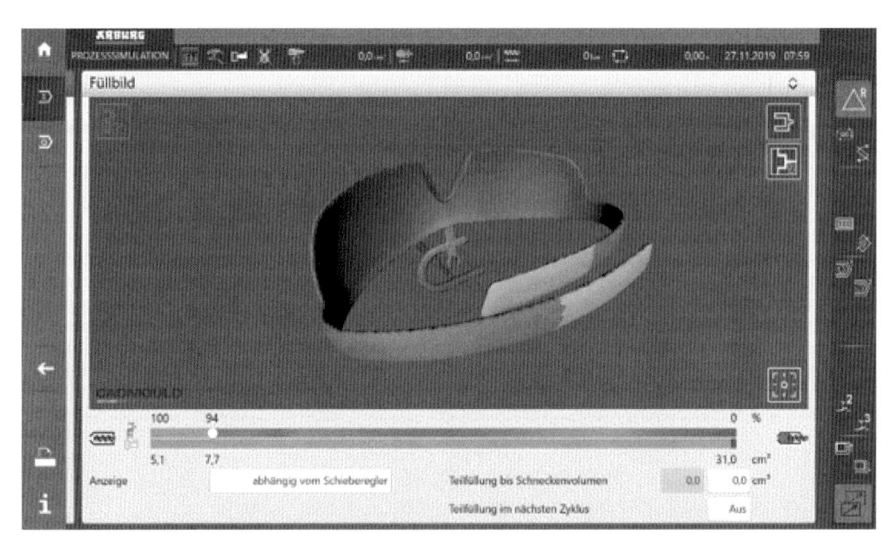

Figure 1.4: Display of simulation results on the injection moulding machine

The process boundary conditions for optimum components have been determined from the simulation calculations. This results in an injection moulding process setting that includes melt temperatures, mould temperatures and the parameters for the filling profile. The simulation results are transferred directly to the machine control system via an interface. Within the control system, the appropriate parameter suggestion for this mould is determined and proposed with the corresponding stored expert knowledge. The application engineer can start the sampling with a pre-configured process setting tailored to the component. In addition, the simulated mould filling process is graphically displayed on the machine control in a three-dimensional view. The mould filling of the cavity can be displayed step by step. In this way, the application engineer can "experience" the peculiarities of the component and mould before the first shot, identify the critical points and derive his approach strategy to set up the process.

1.4 Summary

The identified black boxes for the injection moulding machine are still the major challenges and obstacles for a self-optimising injection moulding machine. The mostly individual injection moulds in combination with a very wide field of plastics, always present individual situations, where well trained specialists are needed for process finding. With the use of additional sensors, there are various concepts with which process control can be facilitated. However, the use of these sensors is always countered by the need for control and maintenance of the sensors, which must be organised very carefully and also carried out by qualified personnel. If the signals are not reliable, the concepts based on them are invalid. Process control based on elaborated process models works technically very well. The general effort for model creation was too high in the past. The AI-based methods advancing with digitisation are very promising. However, there is a need for overarching collaboration and opening in order to be able to make the necessary data volumes available. In practice, the protection of the individual know-how often stands in the way of overriding interests. A promising approach is the merging of simulation and production. The individuality of components and moulds are taken into account in this approach. Process knowledge that can be generated through simulation flows directly into process finding and optimisation and will accelerate these processes in the future. Adaptive control concepts will also continue to develop, especially to compensate fluctuations during production. Here, too, digitisation and the further development of data acquisition and fast processing will improve the performance of the concepts. The big challenge is the definition of the references that always underlie these systems. The self-optimising injection moulding machine has not yet been implemented, but the paths taken are heading in the right direction.

2 Geometry dependent injection - simulative approach to realise constant flow front velocity and shear rates at the flow front in injection moulding

Ch. Hopmann[1], T. Köbel[1]
[1] Institute for Plastics Processing (IKV) at RWTH Aachen University

Abstract

In the injection moulding process the injection volume rate influences the formation of orientations, residual stresses as well as crystalline structures and consequently the mechanical properties and the dimensions of the part. In addition, the external surface properties are decisively influenced by the injection volume rate. To ensure that precision parts meet the quality requirements, it is necessary to profile the injection volume rate as a function of the component geometry. So far the iterative adjustment of the injection volume rate on the injection moulding machine, based on expert knowledge, is state of the art. This process is time-consuming and costly, especially for complex components.

In order to make this process more systematic and efficient, an optimisation routine for profiling of the injection volume rate is developed. The aim of the routine is the analysis of a suitable injection volume rate profile dependent on the part geometry and several boundary conditions. Two possible strategies are examined: Achieving a constant flow front velocity and achieving a constant shear rate at the flow front. Based on an injection moulding simulation with a constant injection volume flow profile, the flow front area and the wall thickness over the filling level are derived. This data serves as input to the optimisation routine, which outputs an optimised injection volume flow profile after specifying various boundary conditions, such as the selection of the strategy, the injection time and the number of profile stages. The simulative validation of the optimised injection profile confirms that this enables a significant equalisation of the flow front velocity or the shear rate at the flow front over the flow path.

2.1 Recommendations for the injection phase dependent on part geometry

Depending on the requirements of the process and the part quality, there are different criteria that are decisive in the injection phase. In general there are basic recommendations for the injection volume rate depending on the geometrical aspects of the part [Bic12, Gru05]:

- Fast filling of the sprue
- Reduction of the injection volume rate when the melt front passes the gate
- Fast filling of the part
- Reduction of the injection volume rate shortly before reaching volumetric filling

In the sprue, the melt should be injected quickly to achieve short injection times and to avoid a cooling of the melt front (see Figure 2.1). When the gate is reached, the injection volume rate should be reduced in order to avoid excessive shear rates and thus material degradation. For economical production, the injection flow rate should be as high as possible in order to keep the cycle time low. In addition, a high injection volume flow can minimise the cooling down of the melt front over the flow path, as the dissipation heat compensates the cooling down of the melt during injection. Upwards, the injection flow rate is limited by material degradation due to excessive shear and pressure losses [Bic12, Gru05, HKH21]. A frequently published approach is to profile the injection velocity in order to keep the flow front velocity constant [Bou89, Gru05, Lei71, SMK98, Thi77, YCLG16]. At constant wall thickness along the flow path a constant flow front velocity results in a constant mean shear rate over the component thickness at the flow front. This results in an uniform orientation of the molecules along the flow path [Lei71, Thi77]. In the case of cross-sectional changes, the injection volume flow has to be adjusted proportionally to the change in the cross-sectional area in order to achieve a constant flow front velocity in accordance with the continuity equation. Towards the end of the flow path, a reduction of the injection volume rate is recommended in order to avoid material damage by burners and overmoulding of the cavity in the compression phase [Bic12, Gru05].

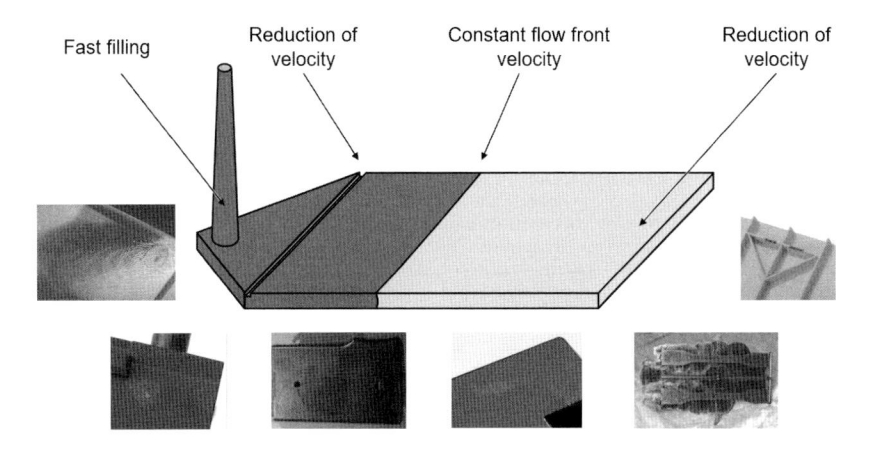

Figure 2.1: Recommended adaption of injection volume rate dependent on part geometry and possible part failures resulting from a badly chosen injection volume rate [Sti77, NN21]

2.2 Development of the simulative optimisation routine

The setting of an injection volume rate profile on the injection moulding machine is still a very iterative process, which is based on the one hand on the expert knowledge of the setter and on the other hand on the recommendations of the material manufacturer. This procedure is time-consuming and expensive. To make this approach more methodical and efficient, injection moulding simulation is to be used to calculate an ideal injection volume rate profile in advance. Therefore, a suitable optimisation routine is developed.

2.2.1 Aim of the optimisation routine

As explained in chapter 2.1, a constant flow front velocity with uniform wall thickness leads to an uniform orientation along the flow path. As this does not apply to components with a variation of the wall thickness, a second strategy is pursued: Keeping the shear rate constant at the flow front. In addition some boundary conditions have to be fulfilled:

- Achieving a constant flow front velocity or a constant shear rate at the flow front along the flow path
- Specification of number of profile steps
- Specification of filling time
- Efficient process without trial-and-error-steps

2.2.2 Workflow of the optimisation routine

The scheme of the workflow in order to determine the optimised injection volume rate profile is shown in Figure 2.2. In a first step, a simulation with a constant injection volume rate is performed. This equals the classical setup without any volume rate profile. In the simulation, the flow front velocity is calculated by default for each node at the fill level at which this node is part of the flow front. Thus, there are many different flow front velocities at each filling level step, which are assigned to all nodes that are part of the flow front at that point in time. Using the software Matlab from MathWorks Inc. in Nattick (MA), USA, the different flow front velocities of the elements are averaged over the filling level so that the mean flow front velocity results for each filling level respectively time step. In the next step, the melt front area above the filling level is calculated from the mean flow front velocity at a constant injection volume profile. Assuming incompressible flow, this can be calculated from the quotient of the constant injection volume flow rate and the mean flow front velocity at the respective point in time. Due to the compressible behaviour of thermoplastic melt the assumption of an incompressible flow is defective. As the melt volume in the cavity is compressed, the resulting melt volume rate at the flow front is in general less than the melt volume rate, that is injected at the sprue. Therefore, the real melt volume rate at the flow front and thus the calculated melt front area can be slightly overestimated. However, the more critical aspect is the abrupt change of the melt front area along the flow path, e. g. because of a change of wall thickness. These time steps, that are the switch over points of the injection volume profile, are not influenced by the assumption of incompressible flow as they are considered in the mean flow front velocity.

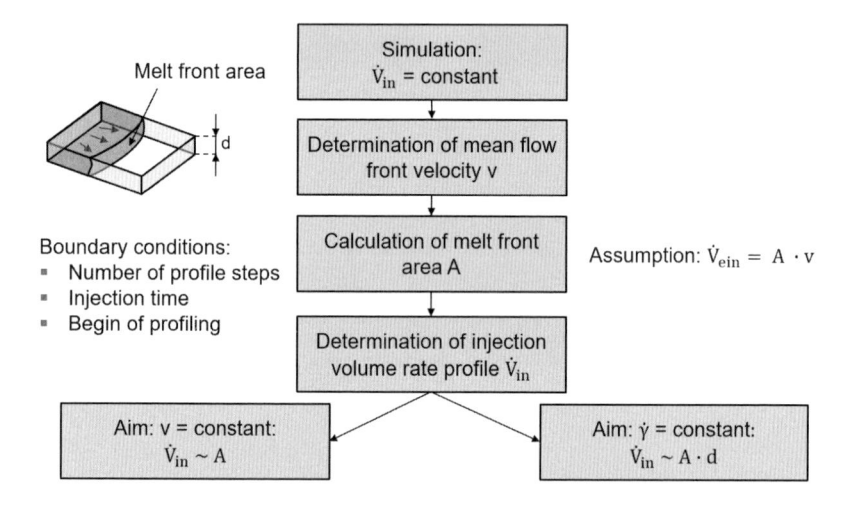

Figure 2.2: Worfklow of the optimisation routine

2.3 Simulative validation

In the following chapter, the optimisation routine is presented on the basis of the simulation results.

2.3.1 Part geometry

For the simulative investigation, a part geometry consisting of cone-shaped sprue, manifold, film gate and subsequent plate is selected. The plate itself consists of five areas in which the wall thickness varies alternately between 2 and 3 mm (Figure 2.3). The used material is a Polypropylene (PP) of the type Sabic PP 579S from Sabic, Riad, Saudi Arabia. The necessary material data for the simulation are used from Cadmould material database.

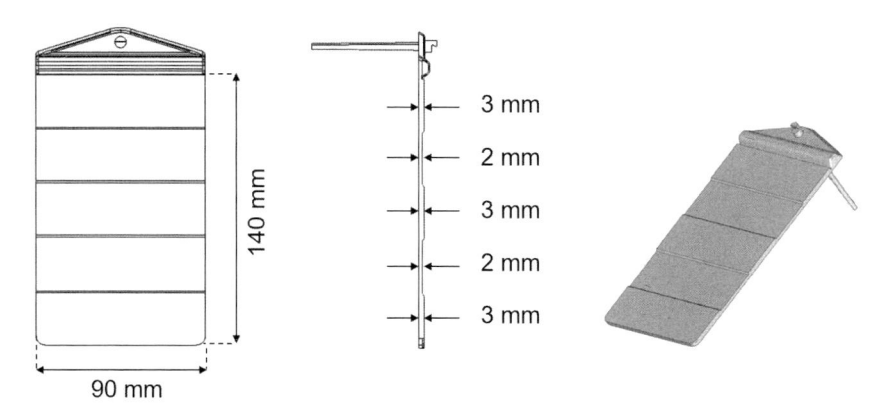

Figure 2.3: Used part geometry with variation of thickness

2.3.2 Flow front velocity and shear rate at constant injection volume rate

The process simulation software Cadmould from simcon kunststofftechnische Software GmbH, Würselen, Germany, is used to calculate the injection moulding process. As explained in chapter 2.2.2, the basis for the optimisation routine is a simulation of the injection moulding process with a constant injection volume rate. For the present case an injection volume rate of 23.5 ccm/s respectively a filling time of app. 1.7 seconds is applied. The flow front velocities of all nodes that are part of the flow front at a certain level are shown in Figure 2.4 (dark grey curve). After smoothing, the mean flow front velocity results above the filling level (black curve). At a filling level of 3 % the cone-shaped sprue is filled. Due to the small cross-sectional area in the sprue, a very high flow front velocity occurs. Because of the increase of the cross-sectional area in the subsequent manifold, the flow front velocity decreases. The high deviation of the flow front velocities of the different nodes at a filling level of approx. 27 % can be attributed to the nodes in the thin film gate. As the flow front does not pass through all nodes in the film section at the same time, the increase in the mean flow front velocity in this area is only moderate. Due to the reduction of the wall thickness from 3 to 2 mm at a filling level of app. 40 %, the average flow front velocity increases accordingly by approx. 33 %. The same effect appears when the wall thickness increases, where the flow front velocity drops accordingly by 33 %. Regarding the plate-shaped geometry, the increase in mean flow front velocity at the end of the filling process is surprising. As the middle of the flow front slightly advances at the end of the filling, the melt reaches the end of the flow path there first. Due to the still unchanged injection volume rate, the flow front velocity of the still moving flow front therefore increases significantly. Comparing the resulting flow front velocity from Figure 2.4 with the recommendations from Figure 2.1, it can be concluded that a profiled injection volume flow is necessary.

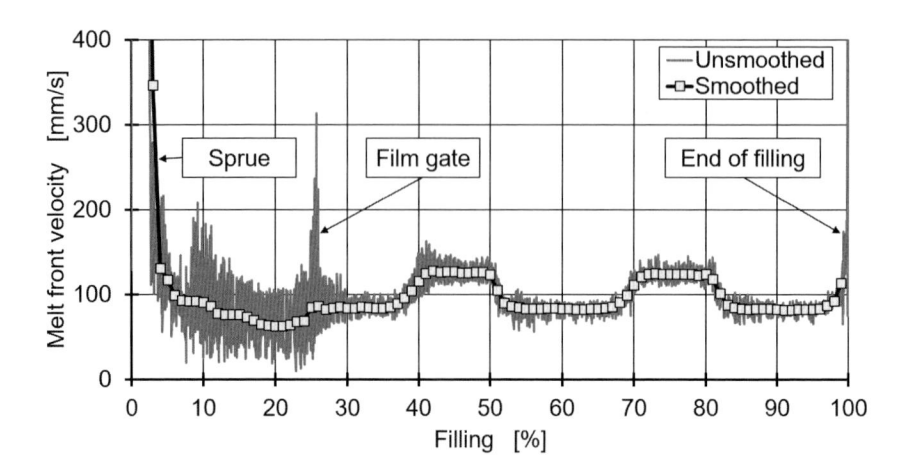

Figure 2.4: Resulting flow front velocity over filling level at constant injection volume rate of 23,5 ccm/s

Figure 2.5 shows the mean thickness above the filling (black curve). The mean thickness is calculated from the mean of all wall thicknesses of those nodes that are part of the flow front at a certain filling level. It correlates well with the geometry. The mean shear rate over the thickness can be calculated by the quotient of the mean flow front velocity and the mean thickness (grey curve). As the flow front velocity increases by a factor of approx. 1.5 and the thickness decreases by a factor of 1.5 during the wall thickness jump from 3 to 2 mm, the mean shear rate increases by a factor of 2.25 and the other way around.

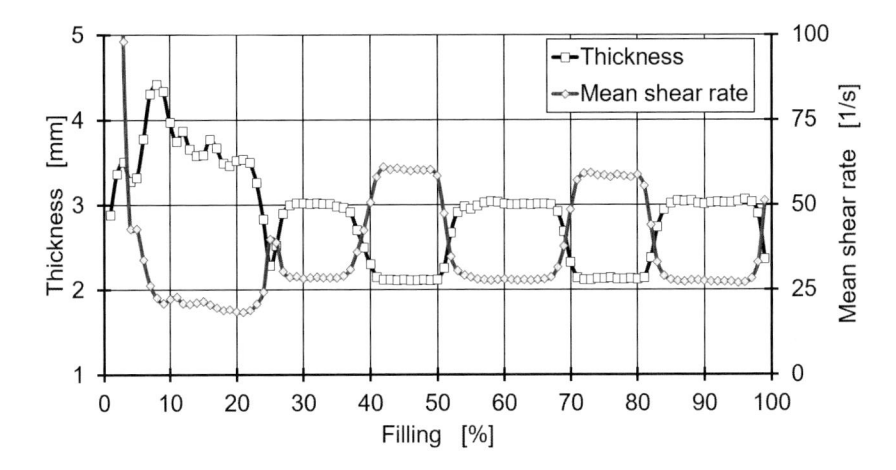

Figure 2.5: Thickness and resulting mean shear rate over filing level at constant injection volume rate of 23.5 ccm/s

However, the mean shear rate over the thickness at the flow front does not necessarily determine the orientations in the outer layers. The local shear rates are relevant for the orientations in the outer layers. In order to analyse whether the mean shear rate over the thickness at the flow front correlates with the shear rates in the outer layers, the ratio of the maximum shear rates at the flow front in the area with a wall thickness of 2 and 3 mm are analysed. The ratio of the maximum shear rates at the flow front, which occur in the range of the wall thickness of 2 and 3 mm, is 2.28 (Figure 2.6). This means that the ratio of the mean shear rates over the wall thickness at the flow front also indicates the ratio of the maximum shear rates in the outer layers at the flow front.

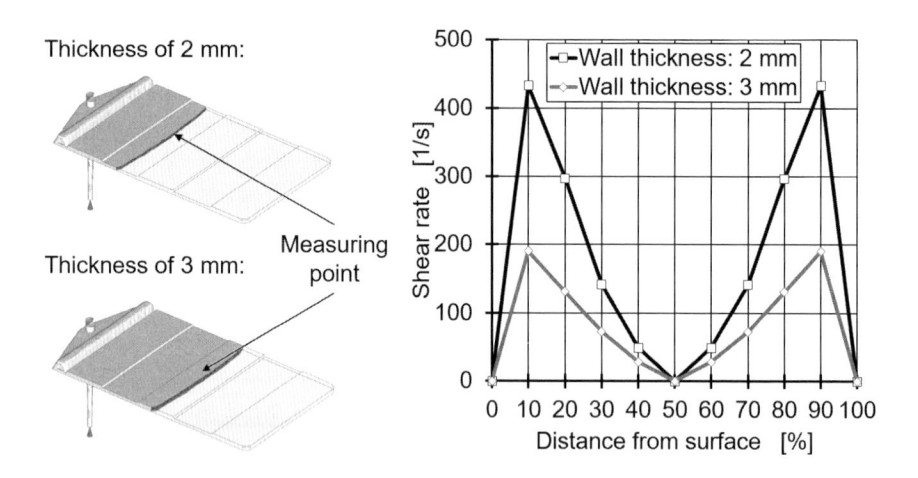

Figure 2.6: Shear rate over thickness at flow front in the sections with a thickness of 2 and 3 mm at constant injection volume rate of 23.5 ccm/s

2.3.3 Optimisation of the injection volume rate

According to Figure 2.2, the flow front area is calculated as quotient of the constant volume rate and the resulting flow front velocity over the filling level (Figure 2.7, black curve). While the flow front area is very small in the cone-shaped sprue, it increases continuously afterwards in the sprue manifold. From a filling level of 27 %, the plate itself is filled and the different wall thickness ranges can be clearly seen. Not to be expected with this geometry is the significant drop in the flow front area at the end of the filling. The flow front does not reach the end of the panel at the same time, as the middle of the flow front is slightly hurrying ahead. For this reason, the flow front area is successively reduced.

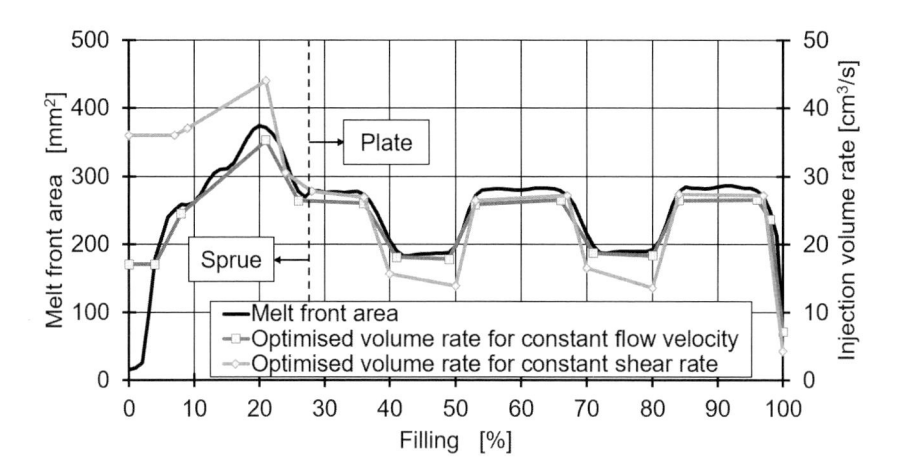

Figure 2.7: Calculated melt front area and optimised injection volume rates in order to gain the flow front velocity respectively shear rate constant

In order to keep the flow front velocity constant, the optimised injection volume rate profile has to be proportional to the flow front area. Using the software Matlab, the curve of the flow front area is smoothed and the points from the curve, at which a defined change in gradient is exceeded, are determined. At these levels, steps are set for the volume rate profile. The smaller the critical change in gradient is defined, the more profile steps are generated in the volume flow profile. As boundary conditions 15 profile steps, profiling from 3 % (to avoid profiling of the cone-shaped sprue) and an unchanged injection time compared to the constant injection volume rate are chosen. Based on these conditions, an automatic injection volume rate profile is calculated, which should result in a constant flow front velocity (Figure 2.7, dark grey curve).

To achieve a constant shear rate at the flow front, the injection volume rate has to be proportional to the product of flow front area and wall thickness. If the same boundary conditions are specified, the injection volume rate profile shown in Figure 2.7 (light grey curve) results. In the area of the sprue manifold, the greater thickness results in a higher injection volume rate compared to the injection volume rate for a constant flow front velocity. In the thin sections with a wall thickness of 2 mm, on the other hand, a lower injection volume rate results in order to keep the shear rate constant. The switching points between the profile stages are not yet ideally selected by the algorithm as there is a change of injection volume rate despite constant melt front area and thickness in the thin sections, e. g. at the section between 40 and 50 %.

In a second injection moulding simulation, the injection is then carried out with the optimised profile in order to gain a constant flow front velocity. In a third simulation, the optimised profile in order to gain a constant shear rate at the flow front is applied. The resulting melt front velocities are shown in Figure 2.8. The flow front velocity at constant injection volume rate deviates up to 201 % from the mean value after passing the film gate at a filling of 27 % (Figure 2.8, black curve). The injection profile, optimised to gain a constant flow front velocity, results in a maximum deviation of the flow front velocity of only 51 % (Figure 2.8, dark grey curve).

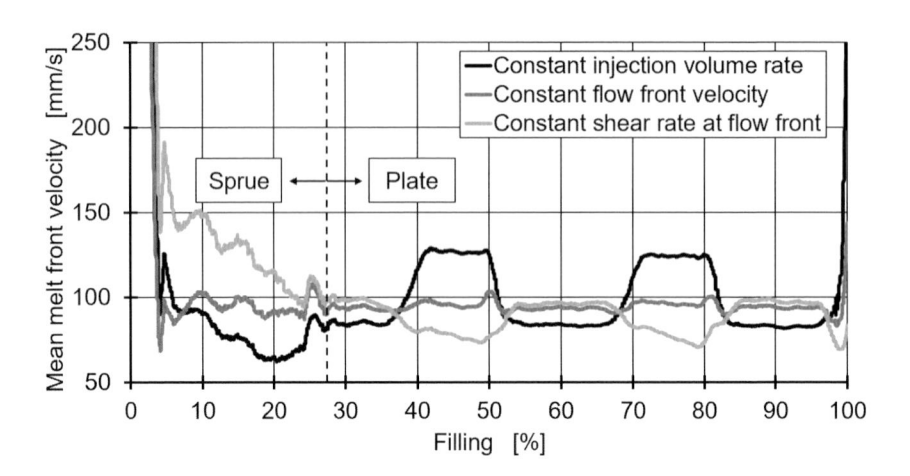

Figure 2.8: Mean melt front velocity resulting from a constant injection volume rate and the two optimised volume rate profiles from Figure 2.7

Figure 2.9 confirms this result: The local flow front velocities are homogenised by the optimised injection volume rate profile. The shear rate-optimised injection profile leads to higher flow front velocities in the sprue manifold and lower flow front velocities in the sections of thin wall thickness (Figure 2.8, light grey curve).

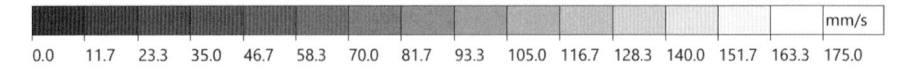

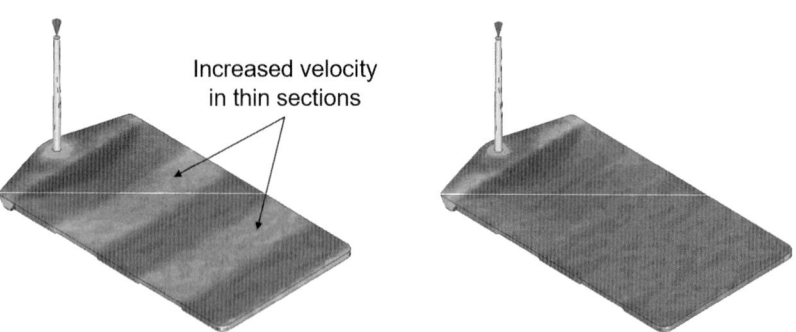

Figure 2.9: Local melt front velocities resulting from a constant injection volume rate of 23.5 ccm/s (left) and the flow front optimised volume rate profile from Figure 2.8, dark grey curve (right)

Figure 2.10 shows the mean shear rate at the flow front, resulting from the constant and the two optimised injection volume rate profiles. The mean melt front shear rate at constant injection volume rate deviates up to 317 % from the mean value after passing the film gate at a filling of 27 % (Figure 2.10, black curve). The injection profile, optimised to gain a constant flow front velocity, leads still to a maximum deviation of 124 % (Figure 2.10, dark grey curve), while the shear optimised injection volume rate profile results in a maximum deviation of 44 % (Figure 2.10, light grey curve). In the simulation no significant quality differences between the three strategies can be detected with regard to macroscopic properties, such as shrinkage and warpage. In the simulation programme, the resulting orientation of the polymer molecules or the resulting morphology are not simulated, which is why a possible influence on the macroscopic properties cannot be mapped. Therefore, experimental investigations are necessary to analyse the resulting morphology and the corresponding part quality.

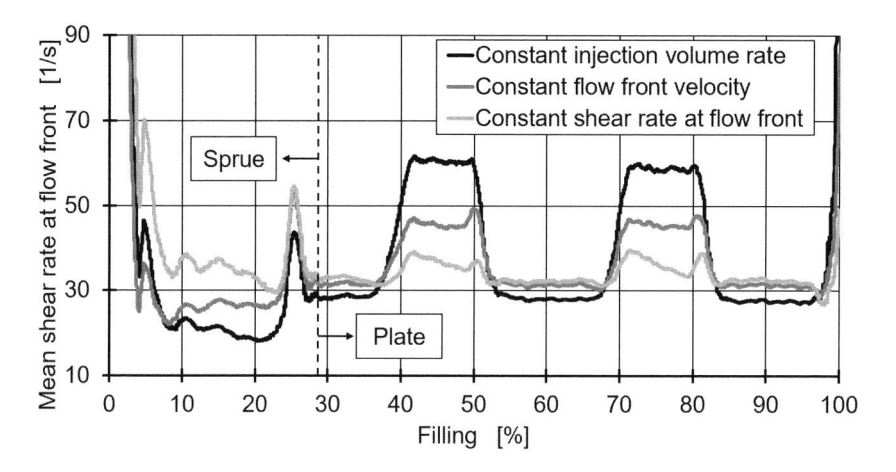

Figure 2.10: Mean shear rates at flow front resulting from a constant injection volume rate and the two optimised volume rate profiles from Figure 2.7

2.4 Transfer to the practice

A requirement for the use of the optimised injection volume rate profile in practice is the exact mapping of the correlation between the injection volume rate profile set on the machine and the resulting volume rate into the mould cavity. There are several reasons, why the volume rate flowing into the cavity deviates from the set injection volume rate at the injection moulding machine:

- Leakage of the non-return valve when accelerating at the begin of injection
- Compression of the melt in the screw antechamber
- Deformation of screw, cylinder and mould
- Inertia respectively dynamics of the screw control system

As the non-return valve is not (completely) closed at the beginning of the injection process, there is a leakage volume flow over the valve. Likewise, small leakage volume flows are possible even if the non-return valve is closed. Another reason for the deviation is the compression of the melt in the screw antechamber. Due to the high pressures, deformations can also occur, such as compression of the screw or expansion of the barrel. Furthermore, the inertia and the control behaviour of the translatory drive of the screw influence the actual injection speed.

Optimised volume rate profile from simulation

- Optimisation routine
  - Part geometry
  - Material
  - Mould

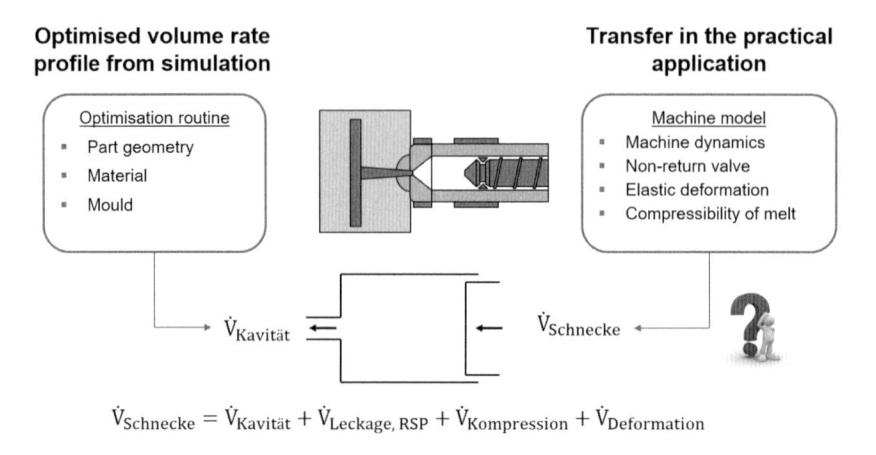

Transfer in the practical application

- Machine model
  - Machine dynamics
  - Non-return valve
  - Elastic deformation
  - Compressibility of melt

$$\dot{V}_{Schnecke} = \dot{V}_{Kavität} + \dot{V}_{Leckage, RSP} + \dot{V}_{Kompression} + \dot{V}_{Deformation}$$

Figure 2.11: Concept of the machine model to calculate the injection volume rate profile to be set at the machine

To take these effects into account, a machine model is currently being developed and validated by experimental tests. In this way, it should be possible to convert the injection volume flow profile obtained from the optimisation routine into the injection profile to be set on the machine. Subsequently, the different injection strategies will be validated in practice to assess whether a constant flow front velocity or a constant shear rate at the flow front positively influences the component properties and can increase the final part qualities.

2.5 Conclusions and outlook

In order to make the iterative adjustment process of a suitable injection volume rate profile more efficient and objective, a simulative optimisation routine was developed. Two strategies are investigated for this purpose: Achieving a constant flow front velocity at the flow front and achieving a constant shear rate at the flow front. This routine makes it possible to derive an optimised injection volume flow profile by means of a single simulation depending on the component geometry after selecting the desired strategy and further boundary conditions (number of profile stages, injection time, start of profiling). The injection volume flow profiles obtained from the optimisation routine were validated by simulation. As a result, a more uniform flow front velocity or shear rate at the flow front along the flow path can be achieved for the investigated component with thickness variations.

Currently, a machine model is being developed by which the optimised volume rate profile can be transformed into the injection volume rate profile to be set on the machine. Only this model makes it possible to accelerate the actual set-up process on the injection moulding machine and to make it more objectively. In a further step, it will be investigated to what extent the injection strategies can improve the actual component quality. For this purpose, the strategies are also to be tested using complex components from industry.

Acknowledgment

The research project 20935 N of the Forschungsvereinigung Institute for Plastics Processing has been sponsored as part of the "Industrielle Gemeinschaftsforschung und -entwicklung (IGF)" by the German Bundesministerium für Wirtschaft und Klimaschutz (BMWK) due to an enactment of the German Bundestag through the AiF. We would like to extend our thanks to all organizations mentioned.

References

[Bic12] BICHLER, M.: *Prozessgrößen beim Spritzgießen - Analyse und Optimierung*. Beuth Praxis. Berlin, Vienna, Zurich: Beuth Verlag GmbH, 2012

[Bou89] BOURDON, K.: *Computer aided set-up of injection moulding machines*. RWTH Aachen University, Dissertation, 1989

[Gru05] GRUBER, J.-M.: *Process control based in cavity pressure for the injection moulding process of thermoplastics*. RWTH Aachen University, Dissertation, 2005

[HKH21] HOPMANN, C.; KÖBEL, T.; HORNBERG, K.: Optimal einspritzen - aber wie?. *Kunststoffe* 111 (2021) 8, p. 42–45

[Lei71] LEIBFRIED, D.: *Untersuchungen zum Werkzeugfüllvorgang beim Spritzgiessen von thermoplastischen Kunststoffen*. RWTH Aachen University, Dissertation, 1971

[NN21] N.N.: *Spritzgussfehler bei technischen Thermoplasten*. BASF, 2021

[SMK98] SPEIGHT, R. G.; MONRO, A. J.; KHASSAPOV, A.: Benefits of velocity phase profiling for injection molding. *SPE Annual Technical Conference Paper*. Atlanta, 1998

[Sti77] STITZ, S.: *Analyse der Formteilbildung beim Spritzgießen von Plastomeren als Grundlage für die Prozeßsteuerung*. RWTH Aachen University, Dissertation, 1977

[Thi77] THIENEL, P.: *Der Formfüllvorgang beim Spritzgießen von Thermoplasten*. RWTH Aachen University, Dissertation, 1977

[YCLG16] YANG, Y.; CHEN, X.; LU, N.; GAO, F.: *Injection molding process control, monitoring, and optimization*. München, Wien: Carl Hanser Verlag and Hanser Publishers, 2016

Abbreviations

Notation	Description
PP	Polypropylene

3 Increased process stability in the injection moulding of post-consumer recyclates using cavity pressure for process control

Ch. Hopmann[1], K. Hornberg[1]
[1]Institute for Plastics Processing (IKV) at RWTH Aachen University

Abstract

In injection moulding, the process is continuously affected by disturbances that influence the part quality. As a result, adjustments to the process parameters are constantly necessary so that the quality requirements can be maintained. For example, when material viscosity changes, the switchover time and holding pressure level have to be adjusted to avoid switchover marks in the part or burr formation. By processing recyclates, especially post-consumer recyclate, the viscosity changes of batches are very pronounced. As the batches of recycled material are still comparatively small, frequent process adjustments are necessary. In addition, there are incompatible ingredients in the material, resulting generally in major process fluctuations. This paper presents a cavity pressure control as an alternative concept to conventional process control, which is intended to compensate the effects of disturbing influences.

Injection moulding trials are performed by processing two different batches of post-consumer material in order to evaluate the compensation potential of a cavity pressure control method. In addition, the process settings are changed to reflect thermal influences and dosing fluctuations as additional disturbance variables. The trials are carried out comparatively with cavity pressure control and with conventional process control. The part quality is evaluated by weight and mechanical part properties.

The evaluation of the part weight shows a significantly higher process consistency with cavity pressure control. The effects of batch fluctuations on the part weight can be reduced by more than 70 %. In the case of thermal changes, the process consistency is even more than 80 % higher compared to the conventional process. Although, the mechanical properties were more than 1 MPa higher with conventional process control, which could result from a lower pressure level of the experimental plan for cavity pressure control.

In conclusion, the results clearly show the potential of process control based on process variables. By phase-unifying controlling of the cavity pressure, batch fluctuations as well as thermal influences can be compensated. The quality can be adjusted specifically via the holding pressure level. This approach significantly increases the process consistency when processing post-consumer recyclates, enabling the material to be used for new applications. This enables the increased use of post-consumer waste for new products and supports compliance with climate targets.

3.1 Introduction

With a total processing volume of 14.23 million tonnes in Germany, post-consumer recyclate (PCR) accounts for 1.02 million tonnes, a share of about 7 %, and continues to increase by about 4.1 % annually [LSH20]. The processability of the plastic waste must be increased for a recyclable plastics industry. This is increasingly demanded by the society as well as by the European policy (European Green Deal) [NN19, RAK17].

Polypropylene (PP) is the most commonly used plastic for mass-produced articles in injection moulding, which accounts to the second largest share in PCR after PE [LSH20]. PCR-PP is produced by sorting at least 90 % of the household waste (yellow bag) from form-stable PP articles. The waste is shredded, washed, fine-sorted, dried and extruded to produce new plastic granulate [RAK17]. The material recycling of post-consumer plastics waste and its constituent of substances and materials is specified in DIN ISO/TR 17098 [NN14]. However, the ranges of the individual constituents are kept wide at several percent, causing the material properties to differ widely from batch to batch. Incompatible constituents include metals, glass and paper [NN14].

The use of recyclates leads to fluctuating processing conditions due to the different composition and history of the material batches [MK11, BB14]. The fluctuations in the resulting part quality grow with increasing recyclate use and lead to a decrease in the mechanical and optical part properties [ASUM03, MOY08]. Post-consumer recycling research is weak due to the high variance in recyclate composition, which makes it difficult to achieve reproducible test conditions [BHHM08, PH07, vCA+21]. Therefore, the reproducibility of the process is low by using PCR, which is caused by the low consistency of the PCR properties in addition to the usual disturbing influences such as thermal process changes.

Recent research has already shown that the use of a real-time process control concept can enable better processing and higher process consistency for the processing of recyclate [MDK+20]. However, the presented process control concept is limited to injection phase control. This article proposes an alternative process control concept based on the cavity pressure, which controls the process in a phase-unifying approach by specifying a cavity pressure reference for the complete cycle. The aim is to increase the process consistency when processing different materials in various environmental conditions in order to be able to produce parts of constant quality.

3.2 State of the art

The conventional injection moulding process consists of a velocity-controlled injection phase and a pressure-controlled holding pressure phase. The switchover between those phases is time or process variable dependent. The process is influenced by various disturbances, so the ideal switchover point between the injection and holding pressure phases has to be adapted cyclically to the current process conditions in order to prevent quality losses [KVW+10, Sch19, Ste07]. Already, some injection moulding machine manufacturers have developed procedures that adapt the switchover point and holding pressure phase to the current process conditions in order to compensate process changes [GM16, NN17, ST17]. However, these systems are individual solutions from separate manufacturers and continue to use discrete switchover, keeping the undefined process condition at the switchover point as a weakness of the process [KVW+10, HH20].

An approach was developed at the Institue for Plastic Processing (IKV) to solve the switchover problem, which uses available process knowledge by means of physically motivated models and applies cross-phase information for process control. The aim is to enable a continuous and reproducible injection and holding pressure process despite process disturbances. The cavity pressure is used as the only controlled variable for the entire filling process, enabling holistic process control. The advantage of a model-based control (MPC) approach is to predict the future process behaviour with a dynamic process model. Based on this prediction, the MPC uses a quality function to calculate the control output signal that leads to an optimal process behaviour. In contrast to conventional control methods, the controller parameterisation is not based on abstract controller parameters, but primarily on the backed-up process model, Figure 3.1.

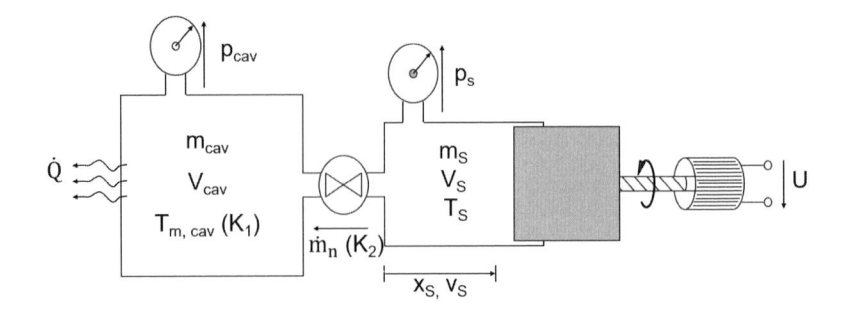

Figure 3.1: Process model for MPC

With the physical replacement model, consisting of two pressure vessels and a hydraulic throttle, it is possible to achieve very good control results in combined operation with an MPR.The parameter V_{cav}, which represents the cavity volume, has to be be adjusted when changing moulds. The parameter K_1 specifies the flow behaviour of the melt and is determined in an initial identification test. The parameter K_2 specifies the material-specific cooling behaviour of the melt and therefore allows a calculation of the actual melt temperature. The remaining parameters are determined in real time by measuring the corresponding process variables.

The reduction of the model to the most important process interrelationships ensures that the real-time demand for inline cavity pressure control can be fulfilled [SAV+19, Ste19]. The process control concept has been continuously evolved since 2013, resulting in a significant improvement of control performance in recent years (cf. [HRR+16, HRRZ13, RSH+14]).

In addition to a high control quality, the shape of the reference trajectory is decisive for achieving a high, reproducible part quality, as the resulting part quality can be specifically adjusted via the cavity pressure curve [HHV+21]. For the actual concept, the cavity pressure curve and the specified reference trajectory are sketched in Figure 3.2.

The injection process starts with a constant injection speed in order to ensure a fast filling of the cavity. In the compression phase, the injection speed is successively reduced so that the cavity pressure curve steadily approaches the reference trajectory. The so-called pvT-optimisation for the holding pressure phase ensures a constant part weight in the presence of disturbances. The pvT-behaviour of the used material is taken into account for the generation of a cavity pressure reference trajectory.

As there is a direct correlation between the moulded part weight and the specific volume, the first requirement is to achieve an identical specific volume when the 1-bar line is reached and thus ensure a constant local shrinkage. The second requirement is isochoric process control, which is characterised by the realisation of a constant specific volume throughout the entire holding pressure phase. The effect of the mass temperature changes as well as the mould temperature changes on the part weight are significantly lower using pvT-optimisation compared to conventional process control [Sch11].

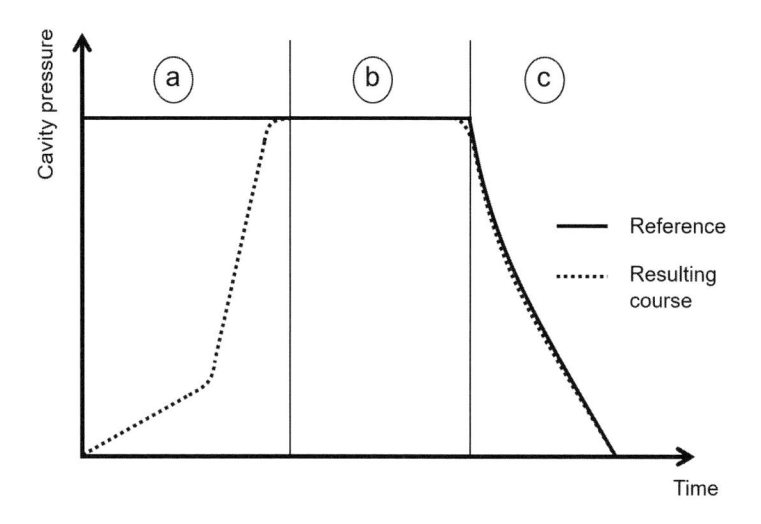

Figure 3.2: Cavity pressure reference and resulting cavity pressure curve for MPC

3.3 Differences between PCR batches in pvT-behavior

The pvT-behaviour of the material has to be determined for the use of MPC, so that the cavity pressure curve can be defined material-specifically for process control. For this purpose, the pvT-behaviour of two different PCR material batches, a polypropylene of the type Systalen PP-C44000 gr000, Systec Platics GmbH, Cologne, Germany, is measured with a pvT500 from Göttfert, Buchen, Germany. A comparison of the material data indicates whether an adjustment of the process is necessary for different material batches. The determined pvT-behaviour is shown in Figure 3.3.

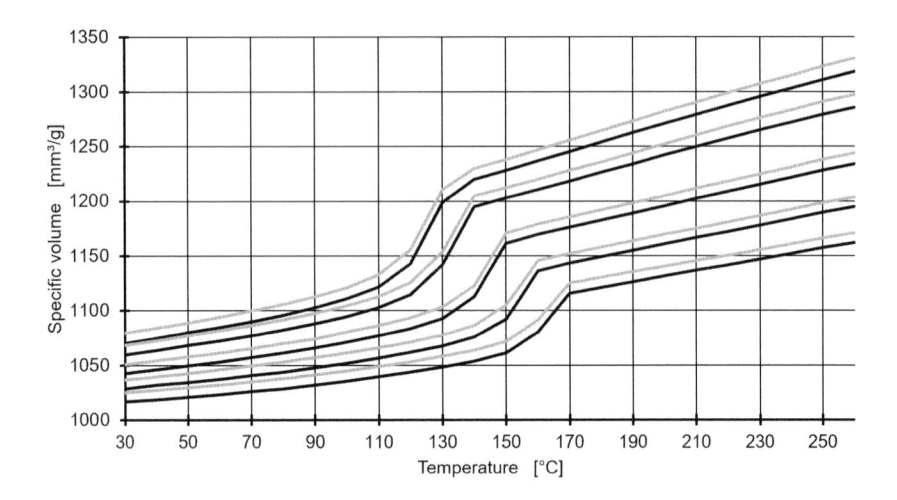

Figure 3.3: pvT-measurements of two different batches Systalen PP-C44000 gr000 with pressure levels 200, 400, 800, 1200 and 1600 bar

Batch 1 has a lower specific volume at the same temperature and pressure than batch 2. This results in a shorter effect period of the holding pressure, as the predefined specific volume is reached earlier. On the other hand, the melt temperature is higher at this stage, so the melt solidifies later. However, the differences between the material batches are negligible, as the measuring method itself has a significantly greater influence on the material behaviour difference between laboratory and real processing. The measuring accuracy of the pvT measuring cell as well as the cooling rate dependence of the pvT characteristics lead to a significant shift of the measured pvT-curves in comparison to the real process [WHSH19]. Due to the qualitatively identical pvT-data of both PCR batches, the same material data is used for process control with MPC.

3.4 Methods

An injection moulding machine of the type "Allrounder 520A 1500-400/400", manufactured by Arburg GmbH Co. KG, Loßburg, Germany, was used for the experiments. The machine is specified to process an external voltage signal at the inverters of the electric drive unit. The control algorithm was applied to an external real-time controller "PXI-8108" and setup in LabVIEW 2020, by National Instruments, Austin (TX), USA. The sampling time of the real-time controller was set to 8 ms. The part geometry and sensor positions are shown in Figure 3.4.

Combined cavity pressure and temperature sensor

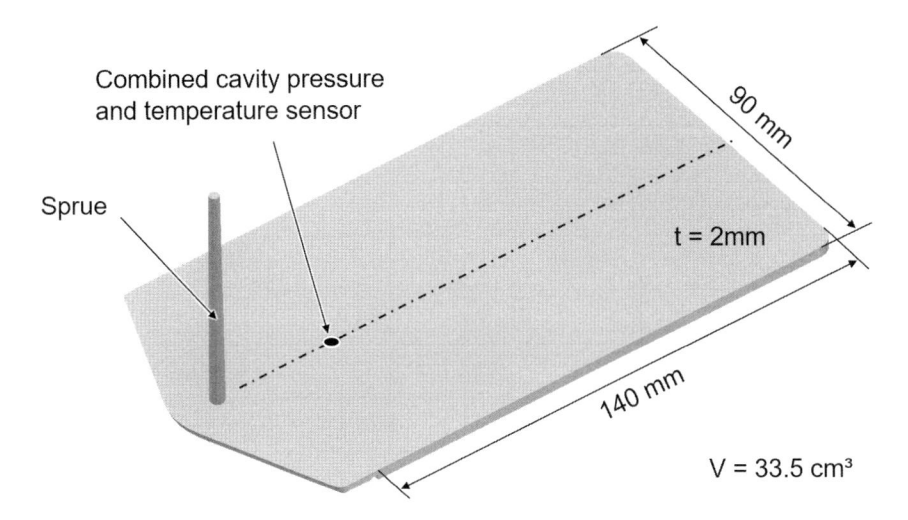

Sprue

90 mm

t = 2mm

140 mm

V = 33.5 cm³

Figure 3.4: Part geometry and sensor positions

The plate is 140 mm in length, 90 mm wide and 2 mm thick with a shot volume of 33.5 cm³. For process data acquisition in the cavity, a combined temperature and cavity pressure sensor near the sprue of the type "6190 CA" manufactured by Kistler Instrumente AG, Winterthur, Switzerland, was used.

The 2^{4-1} fractional design of experiments, including the central point, is given in Table 3.1. The varying process parameters are mould temperature, barrel temperature, holding pressure and dosing volume to replicate thermal and machine disturbances. This parameters replicate the main process disturbances and also have a high impact on the mechanical part properties [MK11]. The variation of cavity pressure level is investigated whether a specific adjustment of the part quality is possible despite the influence of disturbances. For conventional process control, 50 bar of holding pressure were added, as the pressure loss from screw antechamber to the sensor is about that amount. As control strategies, the conventional process control was compared to the model-based cavity pressure control as presented in Section 3.2. Additionally, two PCR batches have been processed with different viscosity and material composition. Batch 1 has a MFR of 14.2 g/10min and batch 2 a MFR of 10.5 g/10min. The experimental plan was run four times in total, by combining the material batches and control strategies with each other for all process parameter combinations.

	Mould Temperature [°C]	Barrel Temperature [°C]	Holding Pressure [bar]	Dosing Volume [cm³]	Process Control Strategy	Material Batch
CP	230	40	300	50		
1	240	50	325	52	1	CPC
2	240	50	275	48		
3	220	50	325	48		
4	220	50	275	52		
5	240	30	325	48	2	MPC
6	240	30	275	52		
7	220	30	325	52		
8	220	30	275	48		

Table 3.1: Level value for process parameters, control strategy and material batch

The part quality was defined by thin section analysis, part weight and mechanical tensile tests. The tension rods of the type "1BA" were milled out along and across the flow direction according to ISO 2818 for mechanical tensile tests [NN19]. The part weight and tensile strength were determined for five parts per trial setting.

3.5 Results

3.5.1 Thin section analysis of recycled plastic parts

The composition of material batches influences defects inside of injection moulded parts. Incompatible impurities such as metal, glass or paper can occur in PCR with a small percentage. As defects can reduce the mechanical properties of injection moulded parts, material inclusions and cutting depths (caused by contaminants) were investigated by light microscopic thin section images using a Leica DM4500M microscope, manufactured by Leica Biosystems, Wetzlar, Germany. The structural microscopic examinations were performed with a magnification of 200. Figure 3.5 shows the defects near and far from the sprue across the cross-section of the part.

The investigated factors, which are influencing the part structure, were material batch, control strategy and process settings. The process settings do not influence the occurrence of cutting marks and material inclusions close to the sprue, whereas far away from the sprue, significantly more defects can be seen with low holding pressure. With conventional process control, larger material inclusions can be seen, especially far from the sprue. In addition, the batch has an influence on the occurrence of part defects, depending on the amount of impurities contained within the batch. With cavity pressure control, no differences can be seen in the thin section images between both material batches. These statements were made on the basis of a relatively small sample size, as microscopic analysis is very time-consuming. Thus, the validity is limited here. In conclusion, it can be seen that impurities are visible in the parts made of PCR. As a result, the general process fluctuations are also more pronounced with PCR compared to virgin material.

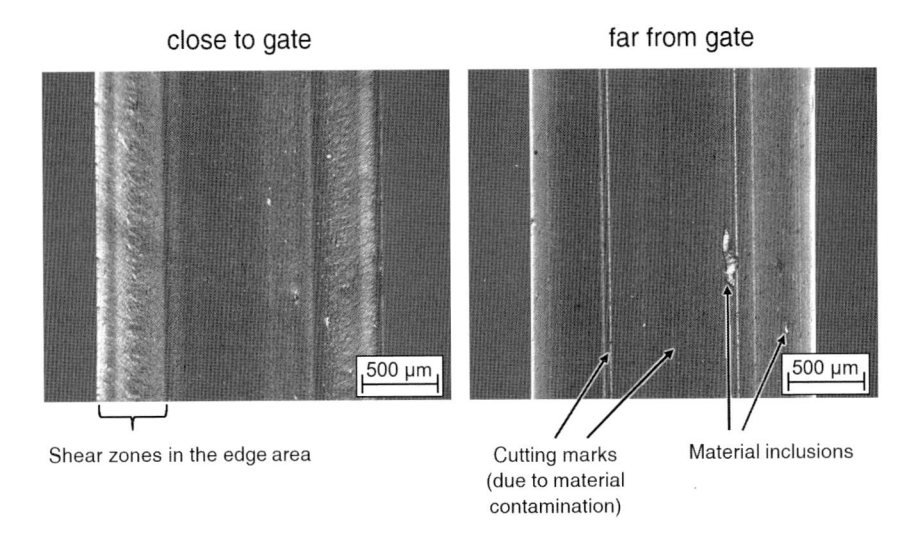

close to gate   **far from gate**

Shear zones in the edge area   Cutting marks (due to material contamination)   Material inclusions

Figure 3.5: Microscopic thin-section images close and far from the gate for high mould temperature and low holding pressure with conventional control for an exemplary part from batch 2

3.5.2 Influence of disturbances on process stability for different control strategies

The influence of process parameter changes on the part weight is shown in Figure 3.6 for the different process control strategies and material batches. It can be clearly seen that the reproducibility of the part quality can be significantly increased with MPC. A statistical regression analysis was performed to evaluate the experimental data in detail. The parts with cavity pressure control are generally lighter. A complete transfer of the experimental parameters between screw pressure and cavity pressure is not possible. With conventional process control, the cavity pressure drops during the holding pressure phase, whereas with cavity pressure control the cavity pressure is kept constant and thus the screw pressure increases. The difference in weight between the control strategies has a negligible influence on the variation of the part weight when changing the process settings, so that the significance of the test results is given.

For the conventional process control, an increase in part weight of about 0.2 g with an increase in mould temperature can be detected. The mould temperature has the greatest influence on the part weight before the holding pressure level, which is only slightly half as large. With cavity pressure control, the influence of the mould temperature on the part weight is reduced by more than 80 %, so that the pressure level in the holding pressure phase has at least three times the influence on the part weight than mould temperature. This means that the influence of a pressure change of 1 bar corresponds to a change in mould temperature of 3 °C. As the cavity pressure is the only main parameter influencing

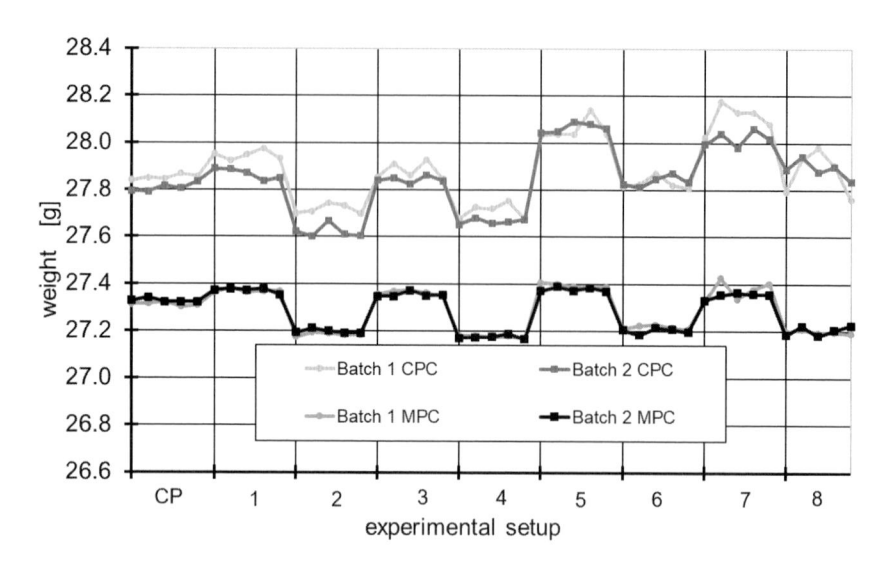

Figure 3.6: Part weight for conventional process control (Conv) and cavity pressure control (MPC) for different process settings and material batches

the part weight, it is possible to adjust the part weight specifically by defining the cavity pressure level, even if thermal changes occur in the process.

In addition, the process stability within a test setting indicates the ability to compensate random disturbances such as impurities in PCR material (cf. 3.5.1). The process stability is higher when cavity pressure control is used. The standard deviation of the weight is 10.2 mg on average, whereas the standard deviation with conventional process control is 27.5 mg. This means that cyclical fluctuations in material homogeneity due to impurities are more effectively compensated.

Furthermore, a change of material batch was investigated. The parts from batch 1 are heavier with conventional process control, as a higher pressure transmission is possible due to the lower material viscosity of batch 1. In comparison, the influence of the batch and thus the material viscosity can be reduced by more than 73 % by cavity pressure control. In Figure 3.6, almost no differences in part weight can be seen between the two material batches. Consequently, the results underline a higher process stability with cavity pressure control.

3.5.3 Mechanical analysis

The mechanical properties of the injection moulded parts are influenced by the control strategy as well as material batch and process settings. The tension rods along the flow direction show strong ductile failure, whereas the tensile rods crosswise to the flow direction show brittle failure. By comparing the mechanical properties along the flow direction at break, the tensile strength is two times greater using cavity pressure control, as well as the elongation at break. The tensile strength is 11.1 MPa (cf. 6 MPa) and the elongation at break is 90 % (cf. 50 %). For the tension rods crosswise to the flow direction, no significant differences are discernible.

A comparison between the material batches shows that the tension rods from batch 1 have a higher tensile strength at break with a lower elongation at break. As a result, the tension rods exhibit a more brittle fracture behaviour.

The maximum tensile strength is one of the most important mechanical parameter besides tensile strength and elongation at break. The mean values and standard deviations of the maximum tensile strength lengthwise and crosswise to the flow direction are shown in Table 3.2 as an average over all test points.

Maximum Tensile Strength [MPa]	Conventional control, Batch 1	Conventional Control, Batch 2	Cavity Pressure Control, Batch 1	Cavity Pressure Control, Batch 2
Lengthwise				
Mean Value	30.05	30.46	28.97	29.22
Standard Deviation	0.31	0.46	0.35	0.41
Crosswise				
Mean value	26.14	26.62	25.97	26.15
Standard Deviation	0.36	0.31	0.39	0.32

Table 3.2: Maximum tensile strength lengthwise and crosswise to the flow direction for different batches and control strategies

The tension rods from batch 1 have lower maximum tensile strengths than the tension rods from batch 2, although the proportion of impurities is lower. Batch 1 has a proportion of 5.5 % impurities in comparison to batch 2 with a proportion of 2 %. Only the proportion of metallic impurities is with 0.22 % higher than batch 2, which has a proportion of 0.08 %. The metallic impurities might be taller and thus cause larger defects in the part. This hypothesis would have to be tested with an examination of further batches. However, the differences between the two batches in total are small with a deviation in maximum tensile strength of less than 0.5.

There are significantly greater differences between the two control strategies, as well as depending on the positioning of the test sample in the part. The maximum tensile strength is more than 3 greater in the flow direction than crosswise to flow direction. The part strength is anisotropic even though there are no fibres in the material. The tension rods crosswise to flow direction were milled away from the sprue, so that less holding pressure has an effect in this area and thus the part has lower strength. For verification purposes, test specimens could be tested in further trials close to the sprue crosswise to the flow direction in order to quantify their positional effect.

The conventional process control achieves overall higher tensile strengths, especially lengthwise to flow direction with a higher maximum tensile strength up to 1.24 and thus 4.2 %. This may have been caused by higher pressures in the holding pressure phase for this process control method. Consequently, the material is compressed more which results in a higher homogeneity of the tension rods, as described at microscopy analysis in the prior section. This argument is strengthened by the fact that the strength differences are less than 0.5 away from the sprue.

The test results of the individual experimental points are examined in detail in order to investigate the influence of the process parameters on the part strength depending on control strategy and material batch. The maximum tensile strengths lengthwise to flow direction are shown in Figure 3.7.

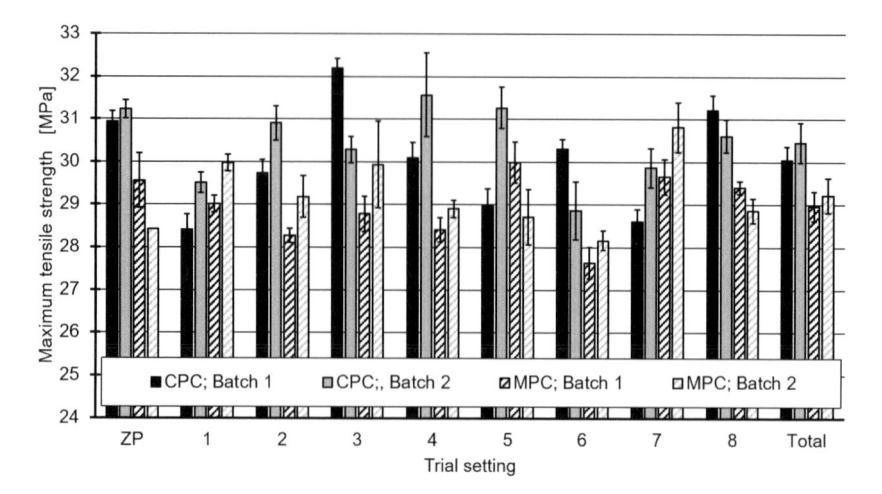

Figure 3.7: Maximum tensile strength along the flow direction for conventional process control (CPC) and cavity pressure control (MPC) and different batches

The control strategy has a major influence on the tensile strength of the parts. At high mould and cylinder temperatures in combination with low holding pressure (case 1), as well as at low temperatures and high holding pressure (case 2), the tensile strengths are higher with cavity pressure control than with conventional process control. The higher strength in case 1 might result from a better holding pressure transmission caused by the higher temperatures. In addition, the holding pressure duration increases with the pvT-optimised cavity pressure trajectory, so that homogeneous part cooling is possible. At low temperatures and high pressures (case 2), the decrease in cavity pressure in the holding pressure phase is more pronounced with conventional process control, and this could lead to an inhomogeneity in the part. These effects appear to be material independent.

The maximum tensile strength is the highest at the central point with conventional process control, which demonstrates that standard process temperatures from material data sheets are appropriate. With cavity pressure control, the tensile strengths are greatest at experimental setup 7 at low temperatures and high holding pressure.

The standard deviations of the experimental results are too large to statistically evaluate the influence of mould temperature, cylinder temperature, holding pressure level and dosing volume. The regression models have an accuracy of less than 40 %, so the significance of the process variables cannot be determined. Consequently, further tests with a larger sample size of, for example, ten tensile test specimens are necessary. Injection moulding of tensile bars also offers a way to reduce the variation, as the mechanical processing of tensile bars during milling can have a major influence on tensile strength as well.

3.6 Conclusions and outlook

This paper presents a cavity pressure control system designed to increase the stability of injection moulding processes. The use of PCR-PP and changing thermal process conditions were examined in injection moulding tests. The investigation of different PCR batches showed that only minor differences in the pvT-behaviour of the batches could be detected, compared to other inaccuracies. Microscopic transmitted light images showed defects in the part, which can lead to mechanical failure.

The processing of PCR-PP can be improved by using advanced cavity pressure process control methods. Process stability was assessed by evaluating the part weight. The results demonstrate that cavity pressure control can reduce batch-to-batch variations by more than 73 % compared to the conventional process control. Thermal fluctuations can be reduced by more than 80 %.

The mechanical part properties were evaluated with tensile tests. It was found that the tensile strength was up to 1.24 higher with the conventional process control. One reason could be the overall lower pressure load in the injection and holding pressure phase with cavity pressure control.

The standard deviations of all process settings were very high for all experimental points, preventing a statistical correlation between process setting and tensile strength. Therefore, a higher sample size is necessary for a higher significance of the results. Further experiments with additional material batches are required to evaluate the influence of the material composition on the tensile strength of injection moulded parts. It is assumed that a high metallic content has a negative influence on the tensile strength.

Acknowledgment

The studies presented in this report on the development of a cross-phase process control strategy (DFG Research Project No. 638619) are funded by the German Research Foundation (DFG). We extend our thanks to the DFG. We would also like to thank all the companies supporting this work by providing plastic material, machinery and further equipment.

References

[ASUM03] AURREKOETXEA, J.; SARRIONANDIA, M. A.; URRUTIBEASCOA, I.; MASPOCH, M.: Effects of injection moulding induced morphology on the fracture behaviour of virgin and recycled polypropylene. *Polymer* 44 (2003) 22, p. 6959–6964

[BB14] BHATTACHARYA, D.; BEPARI, B.: Feasibility study of recycled polypropylene through multi response optimization of injection moulding parameters using grey relational analysis. *Procedia Engineering* (2014) 97, p. 186–196

[BHHM08] BRACHET, P.; HØYDAL, L. T.; HINRICHSEN, E. L.; MELUM, F.: Modification of mechanical properties of recycled polypropylene from post-consumer containers. *Waste Management* 22 (2008) 12, p. 2456–2464

[GM16] GIESSAUF, J.; MAIER, C.: Wie reproduzierbar ist die Reproduzierbarkeit?. *Kunststoffe* 106 (2016) 6, p. 62–66

[HH20] HOPMANN, C.; HORNBERG, K.: The Switchover Problem and its Consequences: Phaseless Process Control Should Revolutionize the Switchover Process. *Kunststoffe International* 110 (2020) 2, p. 40–44

[HHV+21] HORNBERG, K.; HOPMANN, C.; VUKOVIC, M.; STEMMLER, S.; ABEL, D.: Auswirkungen zyklischer Prozesseinflüsse auf den Werkzeuginnendruckverlauf und die Bauteilqualität im Spritzgießprozess. *Zeitschrift Kunststofftechnik/ Journal of Plastics Technology* 17 (2021) 3, p. 179–203

[HRR+16] HOPMANN, C.; RESSMANN, A.; REITER, M.; STEMMLER, S.; ABEL, D.: A Self-optimising Injection Moulding Process with Model-Based Control System Parameterisation. *International journal of computer integrated manufacturing* 29 (2016) 11, p. 1190–1199

[HRRZ13] HOPMANN, C.; RESSMANN, A.; REITER, M. C.; ZÖLLER, D.: Strategy for Robust System Identification for Model Predictive Control of Cavity Pressure in an Injection Moulding Process. In: Schmitt, R.; Bosse, H. (Editor): *ISMTII 2013*. Aachen: Apprimus-Verl., 2013

[KVW+10] KAZMER, D. O.; VELUSAMY, S.; WESTERDALE, S.; JOHNSTON, S.; GAO, R. X.: A comparison of seven filling to packing switchover methods for injection molding. *Polymer Engineering & Science* 50 (2010) 10, p. 2031–2043

[LSH20] LINDNER, C.; SCHMITT, J.; HEIN, J.:. Stoffstrombild Kunststoffe in Deutschland2020

[MDK+20] MORITZER, E.; DEUSE, J.; KROLL, A.; RICHTER, R.; HOPP, M.; SCHMITT, J.; SCHULTE, L.; SCHRODT, A.; WITTKE, M.: Einsatz von maschinellem Lernen für die Rezyklat-Verarbeitung. In: Institut für wissenschaftliche Veröffentlichungen (Editor): *Jahresmagazin Kunststofftechnik 2020.* , 2020

[MK11] MEHAT, N. M.; KAMARUDDIN, S.: Investigating the Effects of Injection Molding Parameters on the Mechanical Properties of Recycled Plastic Parts Using the Taguchi Method. *Materials and Manufacturing Processes* (2011) 26,

[MOY08] MERAN, C.; OZTURK, O.; YUKSEL, M.: Examination of the possibility of recycling and utilizing recycled polyethylene and polypropylene. *Materials & Design* 29 (2008) 3, p. 701–705

[NN14] N.N.: *DIN ISO/TR 17098: Stoffliche Verwertung von Verpackungsmaterialien – Bericht über Substanzen und Materialien, die die stoffliche Verwertung behindern können*. Berlin: Beuth Verlag, 2014

[NN17] N.N.: *DE 10 2015 117 237 B3: Verfahren zur Bestimmung eines realen Volumens einer spritzgießfähigen Masse in einem Spritzgießprozess*. Patent, Deutsches Patent- und Markenamt, 13.04.2017

[NN19] N.N.: *DIN EN ISO 2818: Herstellung von Probekörpern durch mechanische Bearbeitung.* Berlin: Beuth Verlag, 2019

[PH07] PREMPHET, K.; HORANONT, P.: Improving performance of Polypropylene through combined use of aalcium carbonate and metallocene-produced impact modifier. *Polymer-Plastics Technology and Engineering* 40 (2007) 3, p. 235–247

[RAK17] RUDOLPH, N. S.; AUMANATE, C.; KIESEL, R.: *Understanding plastics recycling: Economic, ecological, and technical aspects of plastic waste handling.* Cincinnati and Munich: Hanser Publishers and Hanser Publications, 2017

[RSH+14] REITER, M.; STEMMLER, S.; HOPMANN, C.; RESSMANN, A.; ABEL, D.: Model Predictive Control of Cavity Pressure in an Injection Moulding Process. *IFAC Proceedings Volumes* 47 (2014) 3, p. 4358–4363

[SAV+19] STEMMLER, S.; AY, M.; VUKOVIC, M.; ABEL, D.; HEINISCH, J.; HOPMANN, C.: Cross-phase Model-based Predictive Cavity Pressure Control in Injection Molding. *2019 IEEE Conference on Control Technology and Applications (CCTA).* 2019

[Sch11] SCHREIBER, A.:. Regelung des Spritzgießprozesses auf Basis von Prozessgrößen und im Werkzeug ermittelter Materialdaten. 2011

[Sch19] SCHÖTZ, A.: *Abmusterung von Spritzgießwerkzeugen: Strukturierte und analytische Vorgehensweise.* München and © 2019: Hanser3., aktualisierte und erweiterte auflage edition, 2019

[ST17] SCHIFFERS, R.; TOPIC, N.: More Stability Increases Attractiveness: Optimized Process Control for Injection Molding of Thermoset Molding Compounds. *Kunststoffe International* 107 (2017) 8, p. 26–29

[Ste07] STEINKO, W.: *Optimierung von Spritzgießprozessen.* München, Wien: Carl Hanser Verlag, 2007

[Ste19] STEMMLER, S.: *Intelligente Regelungsstrategien als Schlüsseltechnologie selbstoptimierender Fertigungssysteme.* RWTH Aachen University, Dissertation, 2019

[vCA+21] VAN THODEN VELZEN, E. U.; CHU, S.; ALVARADO CHACON, F.; BROUWER, M. T.; MOLENVELD, K.: The impact of impurities on the mechanical properties of recycled polyethylene. *Packaging Technology and Science* 34 (2021) 4, p. 219–228

[WHSH19] WANG, J.; HOPMANN, C.; SCHMITZ, M.; HOHLWECK, T.: Influence of measurement processes on pressure-specific volume-temperature relationships of semi-crystalline polymer: Polypropylene. *Polymer Testing* 78 (2019), p. 105992

Symbols

Symbol	Unit	Description
K_1	–	melt flow behaviour
K_2	–	cooling behaviour
V_{cav}	cm	cavity volume

Abbreviations

Notation	Description
CPC	Conventional Process Control
MPC	Model Predictive Control
PCR	Post-Consumer Recyclate

> 31st International Colloquium Plastics Technology 2022 · Session 11

Long-fibre-reinforced plastic battery housing

Moderator: Dr.-Ing. Roman Bouffier, KAUTEX TEXTRON GmbH & Co. KG

---------------- Content ----------------

1 Lightweight, sustainable and safe: plastic battery housings for electric vehicles   423

R. Bouffier[1], N. Bergmann[1]
[1]*KAUTEX TEXTRON GmbH Co. KG*

2 Integration of thermoplastic endless fibre reinforcement into the LFT compression moulding process   436

F. Block[1], D. Schneider[1], K. Fischer[1], Ch. Hopmann[1]
[1]*Institute for Plastics Processing (IKV) at RWTH Aachen University*

3 Modelling and validation of continuous carbon-fibre reinforced Sheet Moulding Compound-based structural components   450

J. Neuhaus[1], H. Wang[2], D. Schneider[1], K. Fischer[1], Ch. Hopmann[1]
[1]*Institute for Plastics Processing (IKV) at RWTH Aachen University*
[2]*Aachen Center for Integrative Lightweight Production (AZL) of RWTH Aachen University*

Dr.-Ing. Roman Bouffier

Dr. Roman Bouffier studied Mechanical Engineering at the RWTH Aachen with a focus on plastics processing. After graduation in 2007 he worked as scientist in the field of continuous fiber reinforced thermoplastics and finished this period with a PhD in 2012. Thereafter he joined the managing board of the Institute of Plastics Processing at RWTH Aachen. As chief engineer Dr. Bouffier was responsible for the scientific staff. In 2015 he started at Kautex Textron in Bonn. Since 2016 he has been the Director Research at Kautex. In his position Dr. Bouffier was responsible for the development of pressurised tank systems for plug-in hybrid vehicles and high pressure vessels for hydrogen storage. Currently, he is also responsible for the material and process development for high voltage automotive battery systems.

1 Lightweight, sustainable and safe: plastic battery housings for electric vehicles

R. Bouffier[1], N. Bergmann[1]
[1]*KAUTEX TEXTRON GmbH Co. KG*

Abstract

Traction batteries are the new energy storage system in the automotive sector. Current battery systems mostly use metallic housings. As battery systems mature and electric vehicle volumes increase, plastic-based battery systems are coming more into focus. The ability to produce large series in a robust and functionally integrated way makes plastic battery housings a highly competitive solution. Kautex has therefore once again made it its business to replace costly metal constructions with such advantageous plastic solutions as with fuel tanks in the 1970s.
Due to the high mechanical requirements, long-fiber-reinforced materials such as SMC and LFT are particularly suitable for battery housings. The presentation provides an insight into the advantages of compression molding with long-fiber-reinforced plastics and the differentiation from short-fiber-reinforced thermoplastics on the one hand and continuous-fiber-reinforced plastics on the other. The essential requirements for traction battery systems, such as crash, crush and fire protection are described and corresponding solutions in plastic are presented. The solutions presented are currently under development with various OEMs worldwide, have successfully passed comprehensive trials and are on their way into series production.

1.1 Introduction

Electrically powered vehicles mark the beginning of a new era of energy and mobility. Powered by renewable energy, they are a central component of a smart and resource-conserving urban lifestyle. The market ramp-up has picked up speed - also thanks to global subsidies for the technology in all continents.
In 2021, around 1,100,000 e-vehicles were newly registered in Germany, including around 350,000 pure battery electric vehicles (BEVs) and 325,000 plug-in hybrid electric vehicles (PHEVs). This corresponds to year-on-year increases of around 83 % (BEV) and 62 % (PHEV), while at the same time new registrations of passenger cars with the classic gasoline and diesel powertrains declined by 28 % and 36 % respectively. This trend continues in 2022. By May 2022, 135,029 BEVs and 112,677 PHEVs had already been registered. This means that around 25 % of the newly registered vehicles were equipped with an electric powertrain [URL22].
A common feature of BEVs and PHEVs is that the vehicles' batteries can be charged externally via the power grid. Plug-in hybrids additionally combine an internal combustion engine and an electric powertrain. PHEVs feature batteries with a capacity of about 10-25 kWh and a range of about 10-80 km. BEVs, on the other hand, have a significantly higher electric range and battery capacity (25-120 kWh, 100-500 km) (Figure 1.1).

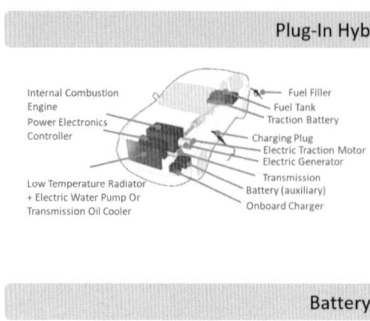

Plug-In Hybrid Electric Vehicle (PHEV) *(10-25 kWh Battery)*

- Electric motor assists the internal combustion engine (ICE). Under light load only electricity is consumed
- The battery replenishes itself via energy generated by the ICE as well as regenerative braking
- Compared to a standard hybrid, PHEVs may be plugged into an outlet or charging station to recharge the on-board battery
- PHEV has the ability to run solely on battery power alone, as well as just gas or a combination of both

Battery Electric Vehicle (BEV) *(25-120 kWh Battery)*

- Run exclusively on electricity via on-board batteries (no ICE)
- Charged by plugging into an outlet or charging station
- Increased usage of regenerative braking
- Adapted platforms for pure BEV
- Large battery packs of up to 5m² *[Kautex]*

Figure 1.1: Variation in architectures when moving to electrified propulsion systems

1.2 Structure of current battery systems

Battery systems in electrically powered vehicles are based on a suitable number of individual cells connected in series and parallel. The structure and shape of a complete battery system are defined not only by the technical requirements but also by the conditions in the vehicle. The battery design therefore depends on the available installation space in the vehicle. Typical examples of battery systems in electric vehicles are the Volkswagen ID.3 and the Mercedes-Benz EQC (Figure 1.2).

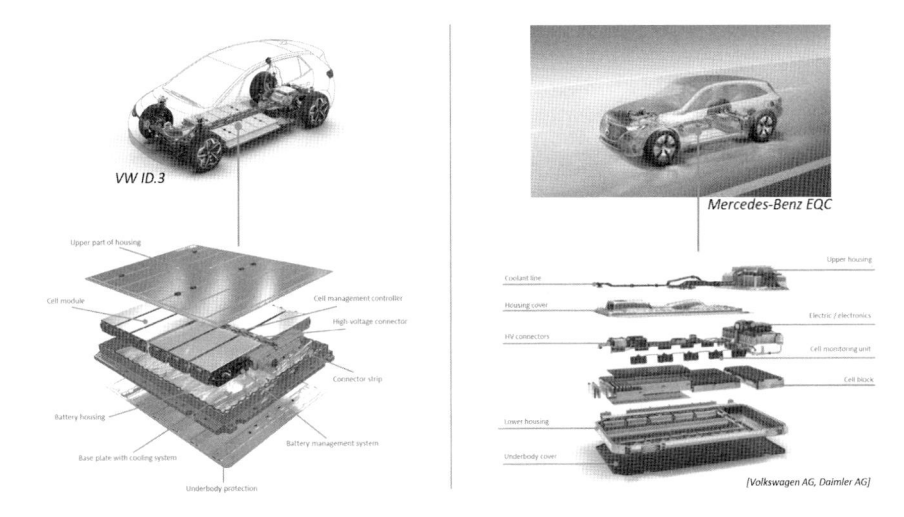

Figure 1.2: BEV Examples - VW ID.3 Mercedes-Benz EQC

Usually, a certain number of battery cells are mechanically and electrically combined into a so-called module. These modules have a rather low voltage level (approx. 60 V) and manageable masses of mostly < 50 kg. This facilitates assembly and increases safety during installation. Particularly in the case of pouch cells, the modules also often take over the task of mechanical bracing in order to reduce the above-described swelling of the cells and the resulting greatly accelerated aging.
Inside the high-voltage battery system of the Volkswagen ID.3, up to twelve battery modules based on 24 lithium-ion cells each are installed and interconnected. The number of modules that are combined to form a battery system is variable. The greater the range required by the customer, the more modules are installed in the battery system. With basically the same structure, up to 408 V are available in the system [URL20].
In the case of the Mercedes-Benz EQC, the battery consists of 384 cells, which are combined in six modules. The two front and rear modules each consist of 72 cells, while the two middle modules each consist of 48 cells [URL18].
The battery cells and modules are always surrounded by a battery housing, which performs numerous important functions such as crash safety, fire protection and sealing.

1.3 The battery housing and its functions

The most important task of a battery housing is to protect the cells and modules inside from critical influences and thus to permanently ensure safe operation for the vehicle occupants (Figure 1.3). As with any energy storage system, battery systems pose serious risks. Similar to a fuel tank, extremely high amounts of energy are stored in a very small space. An uncontrolled release of this energy must be prevented in normal operation and at least delayed in an emergency so that it is possible to exit the vehicle or rescue the occupants with sufficient time. What in the fuel tank area is fuel leakage and the associated fire risk, in the battery area is the thermal or mechanical overstressing of a battery cell and the associated thermal runaway. In this case, the battery cell enters an unstable state and releases the energy contained in it in the form of flames, electric arcs, escaping decomposition gases and particles.

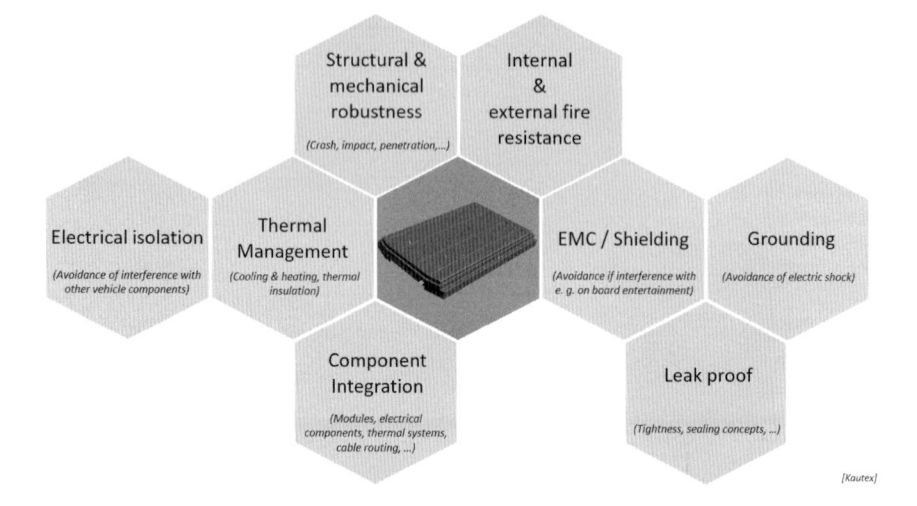

Figure 1.3: Requirements for battery housings

To prevent thermal runaway, a battery housing needs to perform the following subtasks:

- Protection of cells against external mechanical loads such as side crash, front and rear crash, impacts in the vehicle underbody, vehicle touchdown, etc.,
- Leak tightness to prevent from corrosion and short circuits,
- Protection of cells against excessive heat (external fire protection and sufficient cooling),
- Durable fixation of the components.

If prevention of thermal runaway is not possible, the battery housing must provide protection for the occupants through specific internal fire protective measures. Additionally, the battery housing is responsible for preventing (electrical) interference with other vehicle components. Various shielding solutions can be used for this purpose.
In the following, the protection against mechanical and thermal overload will be discussed in more detail.

1.3.1 Protection against mechanical overload

In the case of BEVs, the batteries required are usually so large that separate platforms have become established for these vehicles, usually referred to as skateboard platforms. These are characterized by the fact that almost the entire vehicle underbody is fitted with a flat battery. Such battery systems have masses of 400 to 800 kg and must be designed to be so rigid to protect the cells that they have a significant impact on the overall vehicle structure and vehicle dynamics. In addition, the battery systems are very exposed due to their size and positioning in the underbody and must therefore be particularly well protected.
Both in the event of a vehicle crash and in the event of the vehicle touching down, high loads act on the battery system. Due to the battery installation space, which usually extends between the axles to the rockers, the side pole impact is the most critical crash test for BEV battery systems. While in the front and rear crashes there are still sufficient energy absorption zones beyond the axles, in the side impact only the rockers lie between the impacting obstacle and the battery system. To avoid mechanical stress on the cells as far as possible in this case, battery systems are usually equipped with a solid crash frame

which, together with the rocker, must provide sufficiently high resistance to deformation. In addition, longitudinal and transverse beams are usually inserted inside the housing to reduce the deformation of the battery housing. A system test resulting from this load case is the so-called crush test. In this test, a quasi-static load of at least 100 kN is applied laterally and held for a defined time. The battery system must not be damaged in such a way that the internal modules or cells are damaged. Crush tests of battery housings are typically performed following the GB38031-2020 standard. For the tests, each specimen is placed individually on a test sled according to its orientation in the car and gently secured with clamping claws. A high contact area to the test bench needs to be ensured for the test specimen. The test specimen can thus be moved in one direction via a linear hydraulic cylinder, which drives the track-guided sled, while all other degrees of freedom are fixed. Opposite to the carriage, a test stamp is fixed to a support beam firmly anchored to the floor. During the test procedure, the force, displacement and contact foil signal are recorded. The test setup is shown in Figure 1.4.

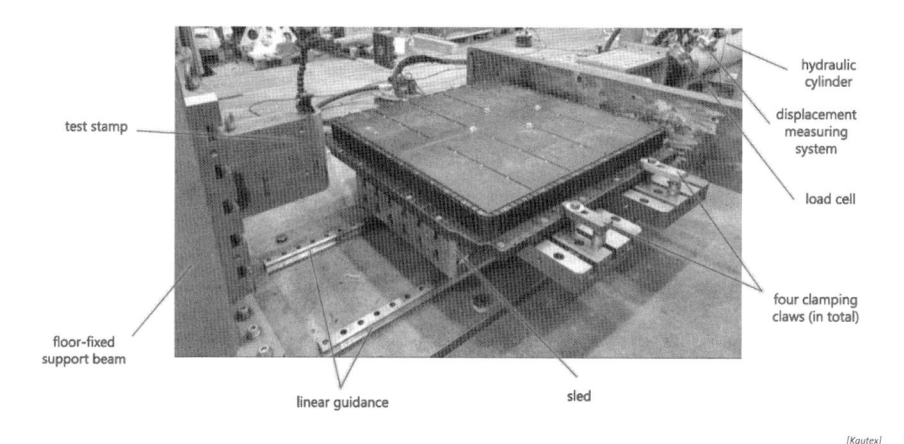

Figure 1.4: Crush Test Setup

For test execution, a horizontally or vertically arranged hydraulic cylinder with a certain stroke is used for load application. The cylinder, usually mounted on a clamping block, moves the test sled with the test specimen fixed so that it is pressed onto the opposite test stamp. The cylinder force is transmitted to the test specimen via the test stamp. A load cell and a displacement measuring system on the cylinder are used to measure the applied force and the deformation of the test specimen during the test. First, a certain pre-load is applied to the test specimen. Then, the test specimen is pressed against the test stamp with a constant travel ramp of about 0.5 mm/s until the maximum force of at least 100 kN is reached. The displacement reached at that maximum force is held for five minutes and finally reset, with the cylinder moving the sled with the test specimen away from the test stamp. Figure 1.5 shows exemplary images of a thermoplastic battery housing after the crush test.

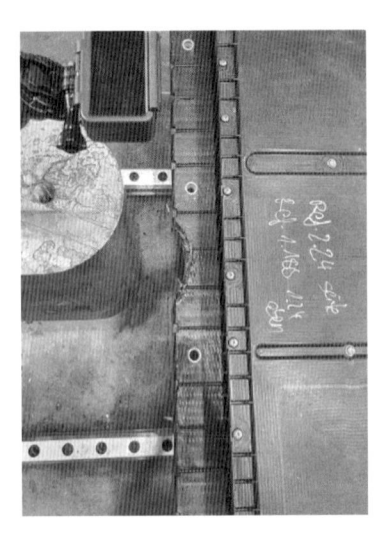

Figure 1.5: Specimens after crush test

The underbody of the vehicle is exposed to many external influences. The most critical case for the battery systems is probably the dynamic impact of the entire vehicle on a smaller object such as a stone or a bollard. Here, high energies act on small surfaces. As in no other spatial direction, there is a particular lack of package space for deformation and energy dissipation, since the so-called Z-height has a direct influence on customer comfort and the perceived vehicle design.
The exact conditions of this bottom impact vary so much from vehicle to vehicle depending on the type of use, vehicle mass and manufacturer requirements that no generally applicable test has yet been established at system level.

1.3.2 Protection against thermal overload

In addition to ensuring the structural mechanical stability of the overall system and protecting the battery cell modules from external influences, the primary tasks of a battery housing include in particular ensuring adequate safety for the vehicle occupants and the environment against hazards emanating from the battery cell modules themselves, i. e. from the interior of the housing. A particularly critical battery condition in terms of safety occurs when a thermal runaway (so-called internal fire) occurs in a cell of the system. This is a self-reinforcing reaction that can spread from one cell to the entire battery system (propagation), releasing large amounts of energy in the form of heat, hot gases and particle streams in a very short time. The battery enclosure must be able to withstand these loads for a sufficiently long time or be designed to suppress propagation so that safety requirements can be met.
Furthermore, the energy storage devices must be protected from a flame source located outside the battery enclosure (so-called external fire).
Apart from the technical tasks, there are further requirements to be fulfilled like a low carbon footprint of the product, a low weight and competitive cost. These aspects are discussed in the following.

1.4 Plastic based battery housings

In general, battery housings for BEVs and PHEVs represent very material-intensive and large components with numerous demanding functional and structural requirements. Today's battery housings usually consist of die-cast aluminum, complex welded aluminum profiles or deep-drawn structures made of sheet steel or aluminum. Weighing more than 500 kg, the traction battery and its housing often account for up to a third of the total weight of a electric vehicle with a long range. Its high weight therefore makes it an attractive target for weight reduction.

1.4.1 Lightweight potential and sustainability

As battery systems mature and volumes increase, plastic-based battery systems are increasingly coming into focus. The main advantages over metallic solutions include lower weight, high leak tightness and corrosion resistance as well as lower costs due to very high functional integration potential and manufacturing processes suitable for large-scale production. Overall plastics can not only improve the technical performance of battery systems but also make them greener and more cost competitive than their metal alternatives. Therefore, Kautex again decided to start a development journey and replace costly metal constructions for battery housing systems with advantageous plastic solutions. As early as the 1960s, Kautex successfully implemented such a technology shift by developing the world's first plastic-based fuel tank and established itself as the global market leader for innovative plastic fuel tank systems for the automotive industry (Figure 1.6).

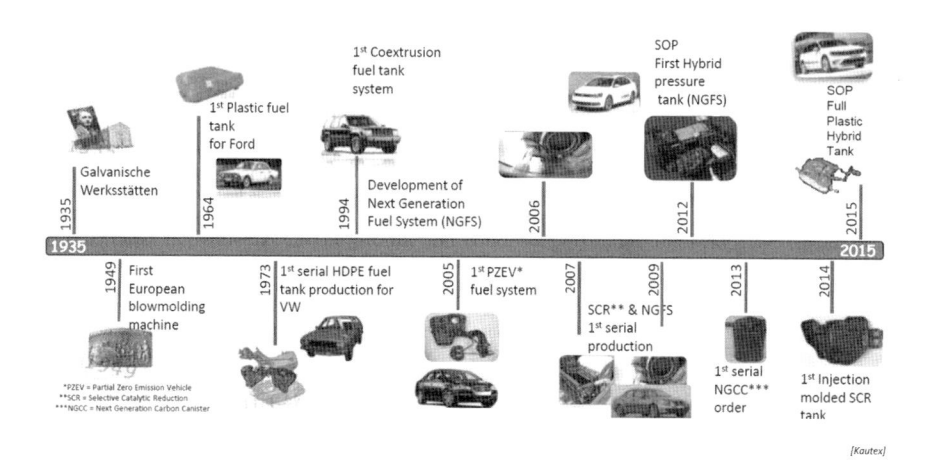

Figure 1.6: Kautex history in fuel tank technology development

With the Pentatonic system, Kautex is developing lightweight, adaptable battery system solutions for hybrid and fully electric vehicles based on fiber-reinforced plastics (FRP). Results of a life cycle analysis for an exemplary battery housing system (1600 x 1300 x 200 mm) show the particular advantages of plastic-based battery housing systems (Pentatonic) over conventional solutions made of steel or aluminum. In addition to a significant weight reduction (Pentatonic: 55 kg vs. aluminum: 61 kg vs. steel: 89 kg), the kg CO_2 equivalent during the manufacturing phase can be reduced by about 59 % and 22 %, respectively, compared to aluminum and steel. By using CO2-optimized polymers and recyclates, an even greater reduction can be achieved compared to a steel concept. A plastic-based battery housing system also enables significant CO2 savings during the use phase (Figure 1.7).

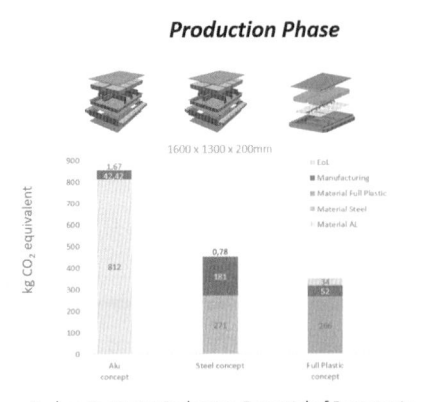

Production Phase

1600 x 1300 x 200mm
- EoL
- Manufacturing
- Material Full Plastic
- Material Steel
- Material Al

Carbon Footprint Reduction Potential of Pentatonic
→ 59 % compared to Aluminum concept
→ 22 % compared to Steel concept

Use Phase

Pentatonic Systems significantly reduce the carbon footprint in use phase:

Pentatonic vs. Aluminum concept: *
→ 93 kg less CO_2 over lifetime per vehicle
→ 93 Mio kg less CO_2 over lifetime per fleet

Pentatonic vs. Steel concept: *
→ 556 kg less CO_2 over lifetime per vehicle
→ 556 Mio kg less CO_2 over lifetime per fleet

*Assumption:
- Lifetime: 200,000 km
- 1,000,000 vehicles of one model of production lifetime
- Reduction of CO_2 consumption of 7.5 g CO_2/km per 100 kg weight reduction
- Part weight Battery System:
  - Aluminum concept: 89 kg
  - Steel concept: 61 kg
  - Pentatonic: 55 kg

[Kautex]

Figure 1.7: Carbon footprint of plastic-based battery housing systems versus conventional solutions made of aluminum and steel

1.4.2 Material and process selection

For the production of battery housings in large quantities, extrusion processes for long-fiber-reinforced materials such as thermoset sheet molding compounds (SMC) and thermoplastic molding compounds (LFT) are particularly suitable. SMC mostly consist of unsaturated polyester resins, unoriented glass fibers with a fiber length of 10-50 mm, and mineral fillers and additives. These molding compounds are processed by the standard compression molding process. The main applications of SMC are in the exterior of cars and trucks (e. g. exterior trim, cabin roofs). For the very cost-sensitive automotive industry, SMC represent an economical lightweight alternative to conventional steel or aluminum components. Compared with today's established textile-based manufacturing processes for FRP components (e. g. RTM, wet pressing; up to 50 % waste, limited automation, limited geometric complexity, long cycle times), the main arguments in favor of using SMC are its high cost-effectiveness due to low material costs, the possibility of a fully automated manufacturing process, very high material utilization (hardly any waste) and the possibility of forming complex three-dimensional geometries and integrating features like metallic components. To further increase the lightweight potential of SMC, there are current efforts to develop new SMC material combinations. The potential of these so-called high-performance SMC materials (HP-SMC) is still not fully usable, as prediction of fiber orientation in the component through process simulation tools is still not mature enough. Current developments and solutions in these areas are discussed in detail in the third article.

The specific advantages of SMC for the production of battery housing systems include high mechanical properties and, in particular, high temperature and fire resistance. On the other hand, challenges arise with regard to recyclability and slightly longer cycle times.

In addition to SMC, the extrusion of long-fiber-reinforced thermoplastics (LFT) has become established. In LFT extrusion, a molten LFT charge is put into a mold, compressed and

cooled under high pressure. The short cycle times of less than one minute and the possibility of material recycling enable sustainable, efficient and thus cost-effective component production. Compared to injection molding of short-fiber-reinforced thermoplastics, less fiber damage during plasticizing and mold filling usually results in a higher average fiber length and better mechanical properties in the part.
One process variant that is particularly well established in the automotive sector is the Direct-LFT process (D-LFT). In this process, the reinforcing fibers and the thermoplastic matrix are dosed separately and only combined in the plasticizing process to form an LFT molding compound, enabling the flexible use of cost-effective raw materials (reinforcing fibers, matrix, additives).
The process technology for the D-LFT process comprises two compounding steps in two specially optimized extruders. The matrix polymer is compounded in a normal twin-screw extruder. To ensure homogeneity of the polymer melt, high screw speeds and shear elements are used. The polymer melt is then fed through a die into a second extruder, which serves as a fiber feeding and mixing unit. In this twin-screw extruder, the fiber rovings are continuously fed directly from the bobin. A special design of screw and barrel impregnates and cuts the fibers. All extruders operate continuously to produce long fiber extrudate that is cut to the component-specific weight.
After the LFT charge (220 °C for PP, 280 °C for PA6) is placed in the lower die, the press is closed. The time period before the first contact of the cavity of the upper die with the molding compound should be kept as short as possible to avoid uneven cooling as well as freezing effects on the cold cavity surface.
After contact of the upper mold with the molding compound, the molten material flows in the cavity during the closing movement. The cavity is sealed during the pressing process by means of dip edges. Depending on the flow path lengths and the degree of deformation, the reinforcing fibers are oriented. This results in locally different fiber orientations and thus, compared to metals, anisotropic properties in the component.
In view of the challenging product requirements (large-size component geometries, highest mechanical demands, cost sensitivity), Kautex has been working for several years on further development of the D-LFT technology. As a result, it has been possible to implement a fully automated integration solution of continuous fiber-reinforced thermoplastic semi-finished products for further component reinforcement. In this process, known as E-LFT, thermoplastic semi-finished products are first conditioned by means of a tempering concept optimized for the respective component complexity and then inserted into the mold together with the D-LFT charge and compression molded. In this way, load-path-optimized and geometrically highly complex LFT components with highest mechanical properties can be produced fully automatically in short cycle times (Figure 1.8).

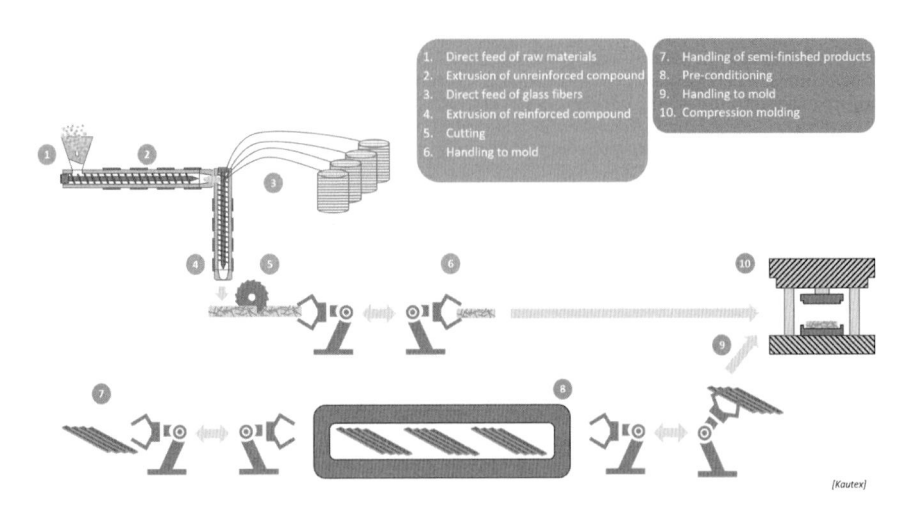

Figure 1.8: Endlessfiber-LFT Compression Molding

1.5 First global full scale off tool plastic battery housing

To underscore the potential of E-LFT technology for the production of heavy-duty battery housing systems, Kautex developed customized materials and CAE methods in cooperation with the company Lanxess and implemented a plastic prototype battery housing based on the highly relevant all-electric series platform of the VW ID.3. This holistic development and the merging of the competencies of materials, process and product enabled a rapid implementation of a marketable plastic battery housing.

During the transformation of the metallic design into a plastic solution, all relevant vehicle- and system-specific boundary conditions and restrictions, such as installation space, position of the battery modules and electrical and hydraulic interfaces, were taken into account in order to make the prototypes usable for vehicle testing and validation at overall system level by incorporating them into the vehicle architecture. In this way, directly usable results can be made possible for customers. Figure 1.9 shows a comparison of the metallic design (VW) and the plastic-based solution implemented by Kautex/Lanxess.

Figure 1.9: ID.3 Battery housing – Comparison of metallic (left, VW) and plastic-based design (right, Kautex/Lanxess)

During development, all functional requirements such as structural integrity, fastening, sealing, cooling as well as internal fire and EMC were considered. To validate the component properties, a comprehensive Design Verification Plan (DVP) was run for all battery housing components (lower shell, upper shell, underbody protection), analyzing mechanical properties (crush, crash, bottom impact) as well as thermal and chemical resistance, among others.

1.6 Conclusions

Overall FRP have a great potential to improve safety, weight performance, sustainability and cost competitiveness of battery housings. The possibility to create a function integrated one shot battery tray without the need to weld, glue or bolt and at the same time locally reinforce the structure to meet all mechanical requirements is a complex engineering task but also a clear advantage versus metal based solutions. Challenges like immature process simulation and the integration of multifunctional composite materials are still to be improved and will unleash further potential of the mentioned processes and materials. These topics are currently under investigation at the IKV and will be presented in the 2 following papers.

References

[URL18] SCHAAL, S.: *Mercedes EQC: Die Batterie im Detail.* URL: https://edison.media/erklaeren/mercedes-eqc-die-batterie-im-detail/23004130.html, 12.06.2022

[URL20] N N: *Die Hochvolt-Batterie des VW ID.3.* URL: https://www.incoming-mobility.com/die-hochvolt-batterie-des-vw-id-3/, 12.06.2022

[URL22] N N: *Neuzulassungen - Zahlen des Jahres 2021 im Überblick.* URL: https://www.kba.de/DE/Statistik/Fahrzeuge/Neuzulassungen/ neuzulassungen_node.html, 12.06.2022

Abbreviations

Notation	Description
BEV	Battery Electric Vehicle
CAE	Computer-aided Engineering
D-LFT	Direct Long Fibre-reinforced Thermoplastic Process
DVP	Design Verification Plan
E-LFT	Endless Long Fibre-reinforced Thermoplastic Process
EMC	Electromagnetic Compatibility
FRP	Fibre-reinforced Plastics
HP-SMC	High Performance Sheet Moulding Compound
IKV	Institute for Plastics Processing
LFT	Long Fibre-reinforced Thermoplastics
OEM	Original Equipment Manufacturer
PA6	Polyamide 6
PHEV	Plug-in Hybrid Electric Vehicles
PP	Polypropylene
RTM	Resin Transfer Molding
SMC	Sheet Moulding Compound
VW	Volkswagen

2 Integration of thermoplastic endless fibre reinforcement into the LFT compression moulding process

F. Block[1], D. Schneider[1], K. Fischer[1], Ch. Hopmann[1]
[1] *Institute for Plastics Processing (IKV) at RWTH Aachen University*

Abstract

In this paper, the basic investigations regarding the material characterisation of long fibre reinforced thermoplastics (LFT) for the application as battery housing are discussed. The mechanical requirements of battery housing of EV are addressed as the basis of material investigations of thermoplastic hybrid components consisting of LFT and continuous fibre reinforcements (organosheets and UD-tape laminates). The bottom impact test and the crush test are two exemplary component test procedures for battery housings on which following tests of the demonstrator part are based. The bottom impact test and the crush test are examined in more detail [KKB+21]. To analyse the material behaviour against the background of the requirement for crush behaviour, bending tests are carried out at specimen level. For this purpose, components are manufactured using the back-compression moulding process with varying continuous fibre reinforcement. As an approximation to the bottom impact test, the instrumented puncture tests is carried out at specimen level for different continuous fibre reinforcements of the LFT. The bending behaviour of hybrid components consisting of LFT and UD-tape-based laminates shows the lowest standard deviation at a long heating time and thus a complete melting of the UD-tape based laminates. Reducing the heating time of the laminates in back-compression moulding causes a significant increase in the standard deviation. To energy absorption capacity based on instrumented puncture tests shows the advantage of a single-sided organosheet reinforcement in comparison to a UD-tape reinforcement or no reinforcement of the LFT.

2.1 Introduction

Fibre-reinforced plastics (FRP), which are used as a material mainly due to their high lightweight potential. Of great importance here are the high specific strengths and stiffness of FRP as well as the possibility of designing components to withstand extreme loads. In order to reduce carbon dioxide emissions through lower energy consumption in the entire transport sector, material lightweight construction is increasingly used e.g. through the application of hybrid FRP. The recently accelerating mobility turnaround holds numerous potentials for novel, innovative plastics-based solution concepts. This is particularly evident in the current reorientation of individual transport towards alternative drive concepts. One of the underlying issues here addresses the sector of battery-powered electric vehicles [EGH+12, Mal18]. Increasing customer demands on vehicle performance, especially with regard to a significant increase in range, require new concepts for battery systems. The integration of high-capacity and thus high-weight battery systems in existing vehicle concepts is mainly carried out in the vehicle floor. Among other things, this results in high mechanical demands on the battery housing, which, in addition to the weight of the battery cells, must also absorb loads acting in a crash while keeping its own weight as low as possible. Long-fibre reinforced thermoplastics (LFT) and thermoset sheet moulding compounds (SMC) offer the potential to combine long and continuous fibre FRP for large-area components such as battery housings. [LLNA20, TP20, HNF+20].
As part of the public founded research project LightMat Battery Housing (EFRE-0801511) the Institut for Plastics Processing and its project partners Kautex Textron GmbH Co. KG, Fraunhofer Institute for Production Technology (IPT), PART Engineering GmbH and FRIMO Lotte GmbH are developing functionalised unidirectional fibre-reinforced semi-

finished products for the large-scale production of plastic battery housings in electromobility.

2.2 Potentials of combining LFT with continuous fibre reinforced plastic

2.2.1 Applications of LFT

In application areas that require a high number of components, the extrusion and pressing of long fibre reinforced thermoplastic composites (LFT) has been of great importance for years. With this process, lightweight components can be produced economically in medium and large series (10,000 to over 100,000 units/year) [Mal18, Sch07]. In the LFT process, the molten LFT semi-finished product is placed in a mould, pressed and cooled to demoulding temperature (Figure 2.1) [Ehr06, NM04]. The short cycle times combined with a high system availability as well as the recycling possibility enables an efficient and thus cost-effective component production [TP20].

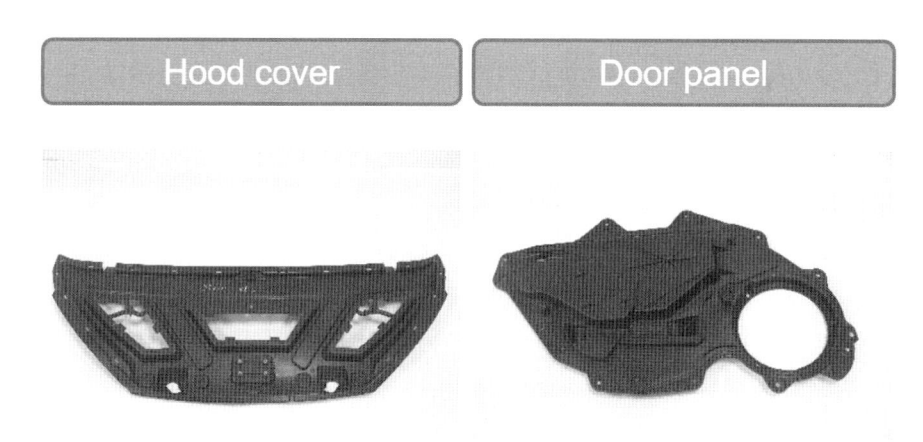

Figure 2.1: LFT components in automotive applications [URL12, URL10]

Compared to injection moulding of short-fibre reinforced thermoplastics, significantly higher average fibre lengths (> 4 mm) and therefore increased mechanical properties can be achieved due to the lower fibre damage during plasticising and compression moulding [Sch07]. For this reason, extrusion of LFT is used as standard for the production of large-area and ribbed components [Sch07, TP20]. Typical applications are mainly underbody panels, spare wheel recesses and instrument panel supports in the automotive industry (Figure 2.1) but also applications in the electrical industry and mechanical engineering (e.g. housings, workpiece carriers and pump bases) in the construction industry (e.g. formwork and scaffolding bases) and in sports articles and furniture [Mal18, URL21, TP20].

2.2.2 Hybridisation of LFT components with UD-tape based laminates and organosheets

The properties of LFT components are determined to a decisive extent by the processing and process control during extrusion. Thus, the achievable mechanical properties are lim-

ited due to the fibre contents (max. 40 wt.%) and fibre lengths (4.5 - 50 mm) compared to continuous fibre reinforced TP-FRP components [Sch07]. These limitations result from the requirement for the lowest possible flow resistance to ensure sufficient flow of the LFT compound in the mould. Moreover, due to the moderate surface quality, e.g. due to freezing effects and fibre markings on the component surface, LFT components cannot be used in the visible area [HGBE05].
However, LFT components are not suitable for the application of highly stressed, large-area components. The structural, mechanical requirements for a battery housing based on LFT can be achieved by hybridisation with continuous fibre-reinforced TP-FRP. [Mal18, KKB+21]. Organosheets are already being used in the economical production of large-series components made of TP-FRP as reinforcement of injected mouldeded parts.
In addition to organic sheets, unidirectional tapes (UD-tapes) are increasingly used as local reinforcing elements due to their high specific strength [URL20]. Applications suitable for large-scale production are being found as local reinforcement elements in, for example, injection-moulded components in sectors such as automotive engineering [URL20]. By means of automated tape laying processes, UD-tapes can be produced into individualised laminate structures in a load-path-compatible manner at a high production speed [JPB17]. A solution is the process chain of hybridising LFT components with continuous fibre-reinforced TP-FRP semi-finished products. In the publicly funded LightMat Battery Housing project, such a process chain was used.

The hybridisation of continuous fibre-reinforded semi-finished products (UD-tapes and organosheets) and LFT combines the properties of the high energiy absorption capacity of LFT with the high achievable stiffness of UD-tapes [Mal18]. The hybridisation is intended to absorb and distribute mechanical loads, for example from a crash. Focus of this paper are the investigations to generate material data for structural simulations of crush and bottom impact test. By using the same matrix materials for the long and endless fibre reinforced FRPs a materially bonded connection can be achieved. In the design of large-area components structural simulations are used to investigate the mechanical requirements. For the validation of the structural simulations of the project Partner Part Engineering GmbH, material data of hybrid components under environmental conditions are required.
Mechanical tests such as bending tests are used to characterise the hybrid materials with respond to the load cases in a crush and bottom impact test, material data in different load directions and material combinations. Furthermore, puncture tests at specimen level can serve as an indicator for the component behaviour in bottom impact cases. For the validation of structural simulations, simple load types such as the bending test can be performed at the component level. In this publication, results of bending and puncture tests on LFT hybrid components are discussed.

2.3 Combination of compression moulding and continuous fibre-reinforced semi-finished products to produce hybrid components

For the investigations of hybrid components from LFT back compression moulding with continuous fibre-reinforced semi-finished products at the IKV, a press of the type VSEFE 3000/2000/330 b Performance from LWB Steinl GmbH Co. KG, Altdorf, is used. A single-screw extruder of the type KMH 60 S from Kannegiesser KMH Kunststofftechnik GmbH, Vlotho, is used for plasticising the LFT granulates. An LFT in granulate form is used at the IKV for the component production. For the following investigations, a polyamide 6 with 40 wt.% glass fibre reinforcement (BASF Ultramid B3WG8 LFX bk23215 - PA6-LFG40) with a fibre lenght of 6 mm from the company BASF, Ludwigshafen is used. The granulate is gradually heated up to the melt temperature of $T_m = 300$ °C. The LFT compound is ejected and cut from the filled screw cavity. A plate mould with the dimensions 200 x 400 mm^2 is used for the compression moulding process (Figure 2.2).

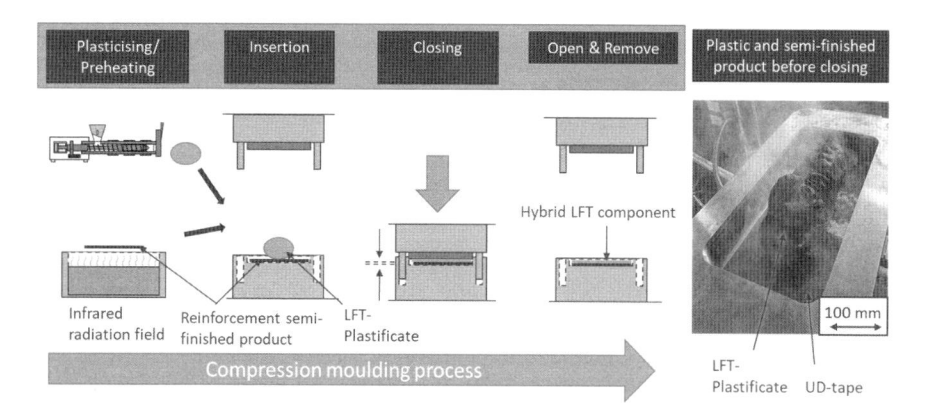

Figure 2.2: LFT compression moulding process for hybrid component production using and UD-tape based laminates and organosheets at IKV Aachen

For the compression moulding process of the LFT components, UD-tape based laminates and organosheets are used. The production of the tape-based laminates was carried out with the help of the project partner Fraunhofer IPT. For this purpose, PA6/GF UD-tapes were produced in a first step using the tape line available at the IKV. Therefore E glass fibres of the type StarRov 440 2400 895 QS with a fibre fineness of 2400 tex from Johns Manville Corp., Denvillle, USA, and Durethan BM240H2.0 from Lanxess AG, Cologne, are used. The UD-tapes have a fibre volume content (FVC) of 45 % and a thickness of 0.25 mm. For further processing in the insitu tape laying process at the Fraunhofer IPT, the UD-tapes are slitted to a width of 25 mm. The PA6/GF organosheets of the type Tepex dynalite 102-RG600(x)/47 % are purchased from Lanxess AG, Cologne and cut to size at the IKV. Parallel to the extrusion of the LFT material, the UD-tape based laminates and the organosheets are molten by means of an IR emitter field. In order to optimise and thus reduce the handling time, the molten laminates and organosheets are placed into the mould immediately before the LFT compound is placed. After manual placement of the pre-heated UD-tape based laminates, the molten LFT compound is placed centered on the inserted semi-finished products. In order to avoid rapid cooling of the semi-finished products after insertion into the temperature-controlled mould and to maintain the flowability of the LFT compound, a process point was selected that includes a relatively high mould temperature of $T_{mould} = 110$ °C. After a cooling time of $t_{cool} = 240$ s the hybridised plate is demoulded. The residual heat of the LFT mass may achieve a sufficient heat input in the joining zone. To analyse the influence of lower cycle times in the joining zone, the UD-tape-based laminates are provided in three heating stages. Laminates with no preheating time $t_h = 0$ s, with $t_h = 30$ s and with $t_h = 90$ s are used.

2.4 Test procedures and evaluation methodology

To determine the material data base for the structural simulation, the bending properties of the hybrid structures are examined. Therefore bending tests are carried out according to DIN EN ISO 14125 [NN11]. For the analysis of the test specimen a 4-point bending test is selected, since the force application between the two support points implies a constant bending moment to the specimen. For the investigation of the influence of varying pre heating configurations of UD-tape-beased laminates, test specimens with varied pre-heating times prior to the compression moulding process at constant LFT processing temperatures are analysed (see Table 2.1).

Test specimens				
-	Layup	4-point bending (UD-tape up)	4-point bending (UD-tape down)	puncture test
1	1 mm UD-tape 4 mm LFT	5 specimen	5 specimen	-
2	4 mm LFT	-	-	7 specimen
3	5 mm LFT	-	-	7 specimen
4	0.5 mm Organosheet 4 mm LFT 0.5 mm Organosheet	-	-	7 specimen
5	1 mm Organosheet 4 mm LFT	-	-	7 specimen
6	0.5 mm UD-tape 4 mm LFT 0.5 mm UD-tape	-	-	7 specimen
7	1 mm UD-tape 4 mm LFT	-	-	7 specimen

Table 2.1: Layer setup and resulting test specimens for the 4-point bending and puncture tests of the hybrid components produced at the IKV

Due to the chosen test setup, the upper side is subjected to compression and the lower side to tension. Therefore the bending tests are carried out on two configurations. The results of the bending analysis is used by the structural simulation as input parameters. Accordingly, five bending test specimens each in the configuration in wich the UD-tape is placed on the side of high tension (UD-tape down) and the configuration the UD-tape is placed on the side of high pressure (UD-tape up) are analysed. The laminate and therefore fibre orientation is placed in the longitudinal direction of the specimen. Thus, the joining zone (Figure 2.3, right) is subjected to varying shear stresses (Figure 2.3, left) assuming a linear transverse force progression with half of the edge fibre stress. Refering to the standard [NN11], test specimens of class II with a size of 80 x 15 x 4 mm^3 are used, which are explicitly to be chosen for hybrid materials. A water-cooled diamond saw is used to prepare the specimens in order to keep the heat input into the material as low as possible. The support width of 66 mm results in an A/D ratio of 16.67, whereby this value indicates a measure between the lever arm resulting from the support width and the thickness of the body to be tested. The test is carried out using a multifunctional testing machine of the type Z150 and load cell of the type XforceK with a nominal load of 150 kN from ZwickRoell GmbH Co. KG, Ulm, Germany. The test travel is measured via the crosshead movement [NN11]. The test parameters for the 4-point bending tests are shown in Table 2.2.

Test parameters		
Parameter	Unit	Value
Pressure fin radius	[mm]	5
Supporting pin radius	[mm]	5
span	[mm]	66
Preload force	[N]	2
Testing speed	[mm/min]	2
Temperature	[°C]	26

Table 2.2: Test parameters used in the 4-point bending test of the hybrid LFT-UD-tape specimens

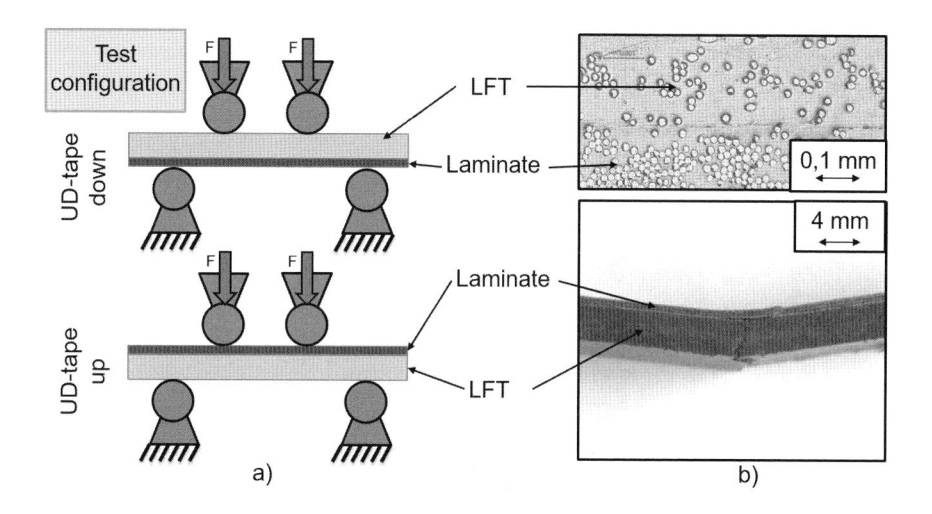

Figure 2.3: a) Test setup of the 4-point bending tests of the hybrid components LFT/UD-tape b) Fracture and micro-graph of a hybrid panel before and after the test

As an alternative to crash tests on test specimen level, hybrid components consisting of LFT and continuous fibre reinforcements (UD-tape-based laminates and organsheets) are analysed in puncture tests in accordance with DIN EN ISO 6603-2 [NN02]. For this purpose, the test specimen (60 x 60 mm^2) is placed in the CEAST 9350 drop tower from Instron GmbH, Darmstadt, Germany.
Different layer setups are chosen for the investigation to compare the influence of the thickness of the LFT material component and with the test specimens reinforced with continuous fibres on one and both sides. Therefore in addition to the hybridized test specimen, two different thicknesses of the LFT specimens are analysed. The focus of the investigation in the puncture test is the influence of the layer structures on the energy absorption capacity. The test specimen layup is shown in Table 2.1. The PA6/GF organosheets of the type Tepex dynalite 102-RG600(x)/47 % from Lanxess AG, Cologne, are used in two different thicknesses of 0.5 mm and 1 mm. The UD-tapes based laminates provided by the Fraunhofer IPT, Aachen, are used in the configuration of 0.5 mm thickness and the laminate structure [0°, 90°] and 1 mm thickness and the laminate structure [0°, 90°]$_s$ (symmetrical).

The thickness of the LFT layer in the hybrid test specimens is kept constant at 4 mm and the single-sided application of 1 mm thickness and the double-sided application of 0.5 mm of the TP-FRP semi-finished products results in a total thickness of 5 mm. The wall thickness of the LFT components without endless fibre reinforcement is varied from 4 mm to 5 mm to match the single LFT thickness and the total thickness of the hybrid components. The samples of the single-sided fibre reinforcement are positioned with the LFT on the impact side to counteract any tensile stress occurring on the mandrel exit side and thus exploit the energy absorption potential.

2.5 Results and discussion

2.5.1 Bending behaviour of the hybrid structures

Figure 2.5 shows the results of the 4-point bending tests of the UD-Tape-LFT plates based test specimen. Here, the bending strength of the respective test configurations UD-tape down and UD-tape up are shown with varying preheating times of the tape laminates (compare Table 2.1).
The test point UD-tape down with cold-applied laminates shows a bending strength of 140.1 N/mm^2 with a scatter of 29.1 N/mm^2. However, the median is 112.6 N/mm^2. The failure behaviour of the test point UD-tape down cold shows an initial failure of the joint interface for each sample. Following this, a subsequent increase in force is observed until the failure of the LFT layer. Furthermore, a high standard deviation of the tested samples of this test point can be seen. By increasing the heating time of the tape laminates to $t_h = 30$ s, the bending strength is significantly increased. At UD-tape down with $t_h = 30$ s the bending strength is at 238.0 N/mm^2 with a scatter of 32,7 N/mm^2. In addition to the increase in strength, there is a further increase in deviation. If the heating time is increased to $t_h = 90$ s, no further increase in bending strength can be achieved. The resulting strength of the test point is 206.5 N/mm^2, including the outlier, at a further reduction of the deviation to 14.5 N/mm^2. Therefore it can be seen, that the further increased heating time t_h doesn't result in a higher bending strength but in the significantly reduction of the scatter of the bending strength.

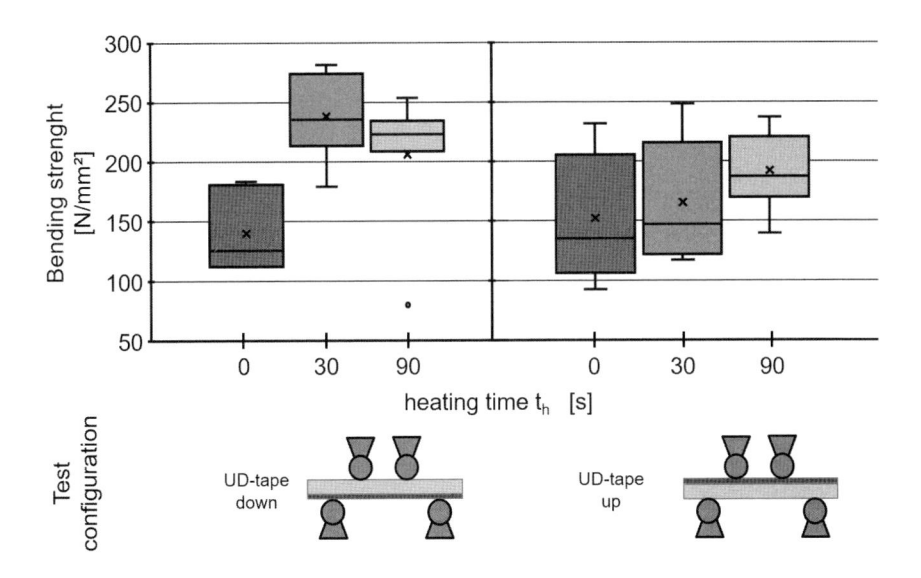

Figure 2.4: Influence of the pre-heating time of the laminate on the bending behaviour of the hybrid plates

For the test configuration UD-tape up, Figure 2.4 shows a slightly higher bending strengths for the process point of heating time $t_h = 30$ s (146.91 N/mm^2) compared to $t_h = 0$ s (135.39 N/mm^2). For the heating time of $t_h = 90$ s the bending strength increases even further to 187,45 N/mm^2. It is noticeable that the test configuration UD-tape up shows a significantly higher deviation in all process points of the preheating time compared to UD-tape down, which results in a lower reproducibility of the results. The results can be used as validation parameters of the structure simulation wich is carried out by the project partner PART Engineering GmbH, Bergisch Gladbach. For a series production, a high pre-heating time is selected to ensure the homogeneous melting of the UD-tape-based laminates and organosheets to produce reproducible components.

2.5.2 Energy absorption capacity of hybrid structures

Figure 2.5 shows the results of the puncture tests with varying wall thickness and endless fibre reinforcement configuration. The calculated energy absorption until failure, induced by a force drop, is plotted over the plate configurations as a measure of energy absorption.

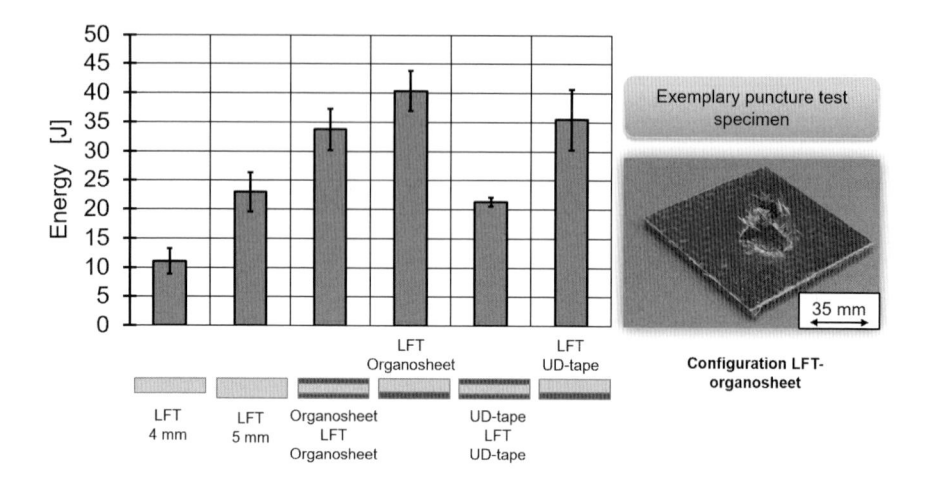

Figure 2.5: Energy absorption capacity of hybrid components of varying semi-finished reinforcements

The LFT panels with 4 mm wall thickness show as expected the lowest energy absorption capability. This result is to be expected due to the lower thickness, but supports the evaluation of the increase in energy absorption of the hybridised components. If the test point with 5 mm wall thickness is considered, it already shows an higher energy absorption potential of 22.9 J. The test specimen of the double-sided UD-tape laminate reinforcement show an energy absorption in the range of the non-reinforced PA6 with 5 mm wall thickness. In contrast to the UD-tape reinforcement the organosheet reinforcement of the LFT has a significantly higher energy absorption of 33.7 J. The one-sided reinforcement in the hybrid structure of 4 mm thick LFT and 1 mm thick UD-tape laminate achieves an energy absorption capacity of 35.5 J. The energy absorption capacity is thus in the range of the double-sided organosheet reinforcement. The one-sided reinforcement of the LFT panels with a 1 mm thick organosheet shows the highest energy absorption of 40.4 J. This results in a clear increase in the energy absorption capacity. This results in a significantly higher absorption capacity of the hybrid components with organosheet reinforcement. For use in the vehicle under body in a battery housing, the one-sided organosheet reinforcement is therefore the right layer configuration with regard to a possible impact load case based on the puncture test results. For the application of the battery housing, an insertion of organosheets in the underbody structure can be benifitial to the possible load case of an bottom impact test.

2.6 Conclusions and outlook

The requirements for a plastic battery housing in the field of electromobility are complex. TP-FRP offers high potentials in terms of fuctionalisation and hybridisation. The compression moulding process is highly suitable for series production of large area components, such as battery housings. Furthermore, the aspects of recyclability and high mechanical properties of the continuous fibre reinforcements should be mentioned.
D-LFT processed parts show improved mechanical properties compared to injection moulded parts due to the potential of resulting in lower fibre length degradation through the process. The concept of back compression moulding is used to introduce local continuous fibre reinforcements into the hybrid component. Bending tests as well as puncture tests of the

hybrid LFT components were carried out to analyse the mechanical properties due to the requirements for a battery housing in electric vehicles and for the validation of the structural simulation at PART Engineering GmbH. The bending tests of the hybrid components show a higher reproducability at high heating times. Furthermore, in addition to UD-tapes, organosheets could also be hybridised as single or double-sided reinforcement. The puncture tests carried out on the basis of different hybrid setups show a high energy absorption capacity of the configuration of an organosheet top layer used compared to UD-tape reinforcement and organosheets applied on both sides. For certain load cases at component level, such as instrumented puncture tests, organosheet reinforcements can thus achieve an increase in energy absorption at critical points while the overall wall thickness remains the same. It was shown that a disproportionate increase in the bending force could be achieved when the wall thickness was increased, but that the potential of continuous fibre reinforcement is primarily apparent in flat components. These results are incorporated into the structural simulation in the research project LightMat Battery Housing (EFRE-0801509) and thus provide the basis for designing a plastic battery housing as a series demonstrator. In the research project, a scaled series demonstrator has been designed at Kautex Textron GmbH Co. KG, Bonn, with the support of all project partners and is produced in a D-LFT process at Kautex Textron, as shown in Figure 2.6.

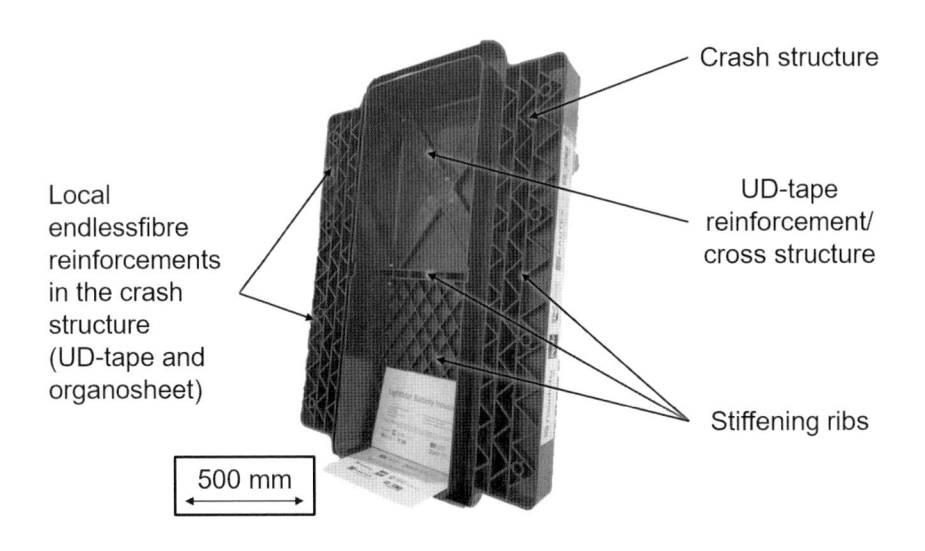

Figure 2.6: Scaled series demonstrator for a plastic battery housing consisting of D-LFT with endlessfibre reinforcements organosheets and UD-tape bases laminates

The scaled series demonstrator includes further functionalisation options. In addition to the local reinforcement of the LFT for targeted load distribution in the event of a crash, topics such as fire resistance, electrical insulation are addressed. The back-compression moulding process offers the opportunity to integrate functionalities such as metallic inserts and fire-resistant materials. To increase electromagnetic compatibility, EMC sheets can be processed in the back-compression moulding. For passenger protection the possibility of a functionalisation with heat-resistant materials is also being investigated due to thermal runaway of the high capacity battery packs. Furthermore the influence of the functionalisation options addressed are analysed by means of mechanical, electrical and thermic investigations.

Acknowledgements

This research "Development of functionalised, unidirectional fibre-reinforced semi-finished products for the large-scale production of novel, highly stressed plastic battery housings for electric vehicles" (EFRE-0801509) was funded by the European Regional Development Fund (ERDF). We would like to take this opportunity to express our thanks for the funding and support. Furthermore, we would like to express our gratitude to the project partners Fraunhofer Institute for Production Technology (IPT) Aachen, Frimo Lotte GmbH, Lotte, Kautex Textron GmbH Co. KG, Bonn, PART Engineering GmbH, Bergisch Gladbach.

References

[EGH+12] EMERSON, D.; GRAUER, D.; HANGS, B.; REIF, M.; HENNING, F.; MARTSMAN, A.; TAGE, S.: Using unidirectional glass tapes to improve impact performance of thermoplastic composites in automotive applications. *Unleashing the power of design : 12th Annual Automotive Composites Conference and Exhibition.* Troy, Michigan, USA, 2012

[Ehr06] EHRENSTEIN, GEORG W.: *Faserverbund-Kunststoffe Werkstoffe - Verarbeitung - Eigenschaften 2., völlig überarbeitete Auflage.* München, Wien: Carl Hanser Verlag, 2006

[HGBE05] HENNING, F.; GEIGER O.; BRÜSSEL, R.; ERNST, H.: Sichtbauteile aus langfaserverstärkten Thermoplasten. *Kunststoffe 3* (2005), p. 118–122

[HNF+20] HOPMANN, C.; NEUHAUS, J.; FISCHER, K.; SCHNEIDER, D.; GONCALVES, R.: Metamodelling of the Correlations of Preform and Part Performance for Preform Optimisation in Sheet Moulding Compound Processing. *Journal of Composites Science* 4 (2020) 3,

[JPB17] JANSSEN, H.; PETERS, T.; BRECHER, C.: Efficient Production of Tailored Structural Thermoplastic Composite Parts by Combining Tape Placement and 3d Printing. *Procedia CIRP, Volume 66* (2017), p. 91–95

[KKB+21] KOTAK, B.; KOTAK, Y.; BRADE, K.; KUBJATKO, T.; SCHWEIGER, H.-G.: Battery Crush Test Procedures in Standards and Regulation: Need for Augmentation and Harmonisation. *Batteries* 7 (2021), p. 1–28

[LLNA20] LENGSFELD, H.; LACALLE, J.; NEUMEYER, T.; ALTSTAEDT, V.: *Faserverbundwerkstoffe: Prepregs und ihre Verarbeitung.* München: Carl Hanser Verlag, 2020

[Mal18] MALNATI, P.: Hybrid thermoplastics give load floor impact strength - Project leads to development of new compression process for selective application of D-LFT on UD tape laminates. *CompositesWorld 12* (2018), p. 44–47

[NM04] NEITZEL, M.; MITSCHANG, P.: *Handbuch Verbundwerkstoffe. Werkstoffe, Verarbeitung, Anwendung.* München: Carl Hanser Verlag, 2004

[NN02] N.N.: *Kunststoffe - Bestimmung des Durchstoßverhaltens von festen Kunststoffen - Teil 2: Intrumentalisierter Schlagversuch (ISO 6603:2000; Deutsche Fassung EN ISO 6603:2000).* Berlin: Beuth Verlag, 2002

[NN11] N.N.: *Faserverstärkte Kunststoffe. Bestimmung der Biegeeigenschaften (ISO 14125:1998 + Cor.1:2001 + Amd.1:2011); Deutsche Fassung EN ISO 14125:1998 + AC:2002 + A1:2011.* Berlin: Beuth Verlag, 2011

[Sch07] SCHEMME, M.: Langfaserverstärkte Thermoplaste (LFT) - Entwicklungsstand und Perspektiven. *10. Internationale AVK-Tagung* (2007),

[TP20] TIWARI, A., B. D. P. D.; PEREIRA, C.: Thermoplastics in electric-vehicle battery applications - A lightweight and efficient EV battery concept. *Procedia PIAE, Volume 2020* (2020), p. 349–364

[URL10] MALNATI, P.: *UBSs: Coming to North America.* URL: https://www.compositesworld.com/articles/ubss-coming-to-north-america, 14.04.2022

[URL12] MALNATI, P.: *Spare wheel well: Functional integration.* URL: https://www.compositesworld.com/articles/spare-wheel-well-functional-integration, 14.04.2022

[URL20] Droege, M.: *Auf die Bremse steigen - mit Kunststoff.* URL: https://www.kunststoff-magazin.de/fvk-werkstoffe/auf-die-bremse-steigen—mit-kunststoff-634330.htm, 05.04.2022

[URL21] N.N.: *LFT Langfaserverstärkte Thermoplaste.* URL: https://www.coperion.com/de/industrien/kunststoffe/lft-langfaserverstaerkte-thermoplaste, 05.09.2021

Symbols

Symbol	Unit	Description
t_h	s	Heating Time
T_m	$°C$	Melt Temperature
T_{mould}	$°C$	Mould Temperature

Abbreviations

Notation	Description
EMC	Electromagnetic Compatibility
EV	Electric Vehicles
FRP	Fibre-reinforced Plastics
IKV	Institute for Plastics Processing
IPT	Institute for Production Technology
LFT	Long fibre-reinforced Thermoplastics
PA6	Polyamide 6
TP	Thermoplastic
TP-FRP	Thermoplastic Fibre-reinforced Plastics
UD-tape	Unidirectional Fibre-reinforced Tape

3 Modelling and validation of continuous carbon-fibre reinforced Sheet Moulding Compound-based structural components

J. Neuhaus[1], H. Wang[2], D. Schneider[1], K. Fischer[1], Ch. Hopmann[1]
[1]*Institute for Plastics Processing (IKV) at RWTH Aachen University*
[2]*Aachen Center for Integrative Lightweight Production (AZL) of RWTH Aachen University*

Abstract

In the light of climate change and further increasing demands on cost optimisation, the importance of functional integration and resulting decrease in material use is increasing continuously in the manufacturing field. In composite processing, Sheet Moulding Compounds (SMC) enable the one-shot production of large and geometrically complex parts. Further process-integrated hybridisation with load-optimised reinforcement structures based on unidirectional carbon or glass fibres further increase the viability of SMC in applications with high demands on mechanical properties, such is the case in large scale traction battery housings. Load specific optimisation of these structures however necessitates simulation approaches which enable a high number of iterations, and thus must be fast in both model set-up and calculation. While methods for simulating the mechanical behaviour of fibre-reinforced parts with local reinforcements exist, these either make use of part discretisation's unique to a reinforcement structure or induce limitations on viable part geometry complexities. In this paper, an alternative approach is proposed which enables the use of part discretisation's independent of the reinforcement geometry and is based on mapping of reinforcement structures directly onto existing part discretisations already used in both structural and process simulation of non-reinforced SMC parts. The approach enables the accurate prediction of the influence of local reinforcement structures on both local and global deformation of a non-planar part under three-point bending loads. By using mesh structures consisting of pre-cured carbon-fibre towpregs, global maximum part deformation is reduced by 17.02 % (measurement) and 21.37 % (simulation), respectively.

3.1 Introduction

In recent years and following the break-through of automotive applications of electrical traction systems, large-scale battery housing applications based on long fibre reinforced polymers materials such as long-fibre reinforced thermoplastics (LFT) and thermosets (Sheet Moulding compounds - SMC) have been consistently increasing in usage. In comparison to the metal-based solutions initially used in this field, these materials offer superior lightweight properties and low tooling and investment costs due to their single-shot compression moulding processing capabilites. In particular Sheet Moulding Compounds (SMC) have a long and successful history in enabling the production of both large scale and geometrically complex parts in a wide range of high-priced and cost-sensitive applications alike, ranging from aerospace engineering (e.g. large-area cladding elements in the passenger compartment), rail and road vehicle construction (e.g. seat shells as well as interior and exterior cladding elements and mentioned traction battery housings) to the energy and construction sectors (e.g. electrical insulators, cryogenic insulators, shower and bath tubs). In the field of traction battery housings, demands on mechanical properties are especially harsh. E.g., the chinese industry standard GB 38031-2020 requires crushing resistance of up to 100 kN in both frontal and perpendicular direction, to be retained for 10 minutes [NN20]. Approaches to increase mechanical properties for SMC-based semistructural and structural parts have been increasingly demonstrated in recent years. One of these approaches is the hybridisation of SMCs by incorporating continuous glass and carbon fibre reinforcements. The use of textiles and unidirectional reinforcements alike have been demonstrated in the past [BFDH18, Bü18, TW18]. Mechanical simulation approaches of these hybrid materials typically make use of part discretisations unique to one reinforcement structure geometry [FHS+17]. While the approach is highly suitable for simulating reinforcements of defined load introduction points, as presented by *Fette et al.*, simulating planar reinforced SMC with varying optimisation geometries (as is the case e.g. implementing a geometry optimization) would necessitate the use of individual part discretisations. *Mehl et al.* implemented a bi-directional algorithm enabling the optimization of local continuous reinforcements in anisotropic fibre reinforced parts. By using a part discretization based on evenly spaced square elements, placement of the reinforcements is approximated [MSM+21]. While the approach is successfully implemented for a cantilever geometry, scaling of the approach to complex parts may be challenging as the number of necessary elements would increase significantly. Therefore, an alternative 2.5D-based approach for simulation of locally reinforced SMC components is proposed in this study, which enables the use of part discretisation's independent of the reinforcement geometry. Furthermore, the approach is based on part discretisation's identical to those used in both process and structural simulation and as such, no further restrictions on geometrical part complexities are induced. Simulated deformation behaviour of both reinforced and non-reinforced SMC component is validated by using cantilever and 3-point bending tests on component level using a novel measurement cell of the also Aachen-based project partner Aachen Center for Integrated Lightweight Production (AZL), Aachen.

3.2 Manufacturing of hybridised and non-hybridised SMC parts

The evaluation of the deformation behaviours is carried out on a tray geometry with dimensions of 697 x 300 x 50 mm^3 (Figure 3.1), using HUP16/40 of Polynt GmbH, Miehlen, Germany, as base material.

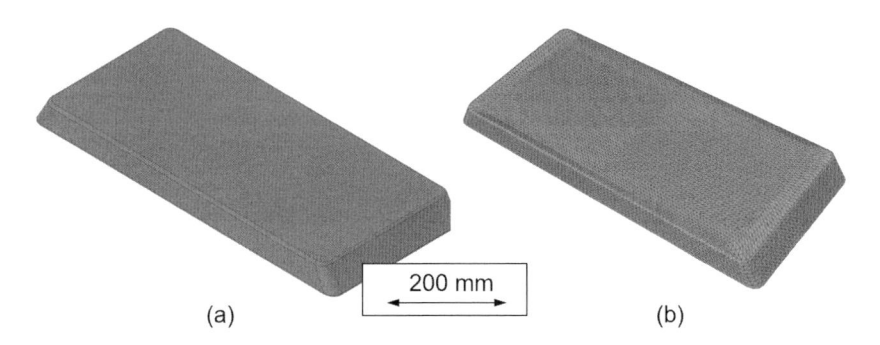

Figure 3.1: (a) Part geometry, (b) Discretisation used in process and structural simulation (load case: 3-point bending)

Cycle time neutral manufacturing of reinforced SMC can be conducted both with pre-cured and uncured towpreg structures. However, uncured towpreg structures exhibit a high rate of deformability, which reduces placement and orientation accuracy in the final part and whose prediction may greatly increase process simulation complexity (Figure 3.2). To counter these drawbacks, pre-cured towpreg structures are manufactured and used in this study. Towpreg structures of SIGRAPREG C TP50/11-4.4/255-E420/39, SGL Carbon, Wiesbaden, Germany, are manually placed in grid patterns with 60 mm distance in 0/90°-Orientation and cured by infrared radiation for 5 minutes. The SMC charges are cut to a size of 80 % of the projected mould surface using a cutting template and placed between two towpreg structures onto the mould surface, immediately prior to initialisation of the compression moulding cycle (Figure 3.3). Additional processing parameters used are summarised in Table 3.2. Target values for part thicknesses are 3 mm (side walls) and 3.6 mm (plane), which is determined by charge weight present.

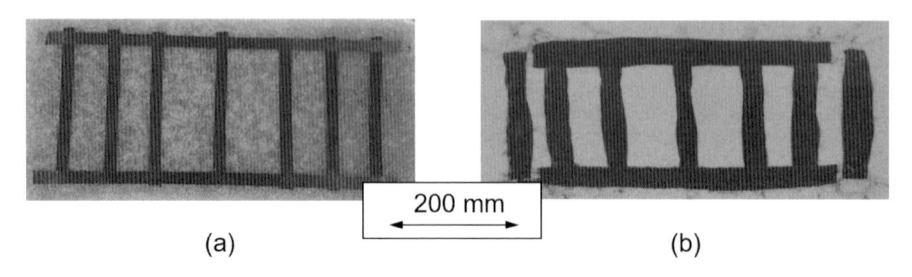

Figure 3.2: Placement of uncured towpreg structures before (a) and after (b) compression moulding cycle

Figure 3.3: Placement of towpreg structures and SMC charge on compression tool (left to right) [AZL]

Parameter	Unit	Value
Compression force	[kN]	2,100
Temperature lower mold	[°C]	135
Temperature upper mold	[°C]	150
Compression speed	[mm/s]	5
Curing time	[min]	5
Time between initial mould contact and compression initiation	[s]	45

Table 3.1: Processing parameters used in Production of both hybridized an non-hybridized SMC parts

3.3 Testing of hybridized parts

Testing of hybridized parts is conducted on a novel test cell developed by AZL Aachen (Figure 3.4) [FEWN21]. This test cell is capable of evaluating local and global part stiffnesses under defined loads. Loads are applied by a servoelectric cylinder manufactured by Festo SE Co. KG, Esslingen am Neckar, Germany. For initial validation of the material parameters used, cantilever testing is conducted (Figure 3.4 (c)). A load of 150 N is applied in 530 mm distance relative to the clamping fixture. For further validation of the material parameters and the simulation approach, a 3-point bending load is applied to the manufactured parts. Support is realised by two cylindrical supports with 50 mm diameter and a distance of 150 mm relative to the parts transversal symmetry plane, respectively. A load of 540 N is applied to the parts central point with a spherical force applicator with a diameter of 50 mm. Movement speed is 0.1 mm/s. Part deformations and surface strains are detected by an optical system of the type ATOS 5 from GOM GmbH, Braunschweig, Germany. Part manufacturing is conducted in triple repetition for cantilever testing, unreinforced and reinforced 3-point bending evaluation respectively.

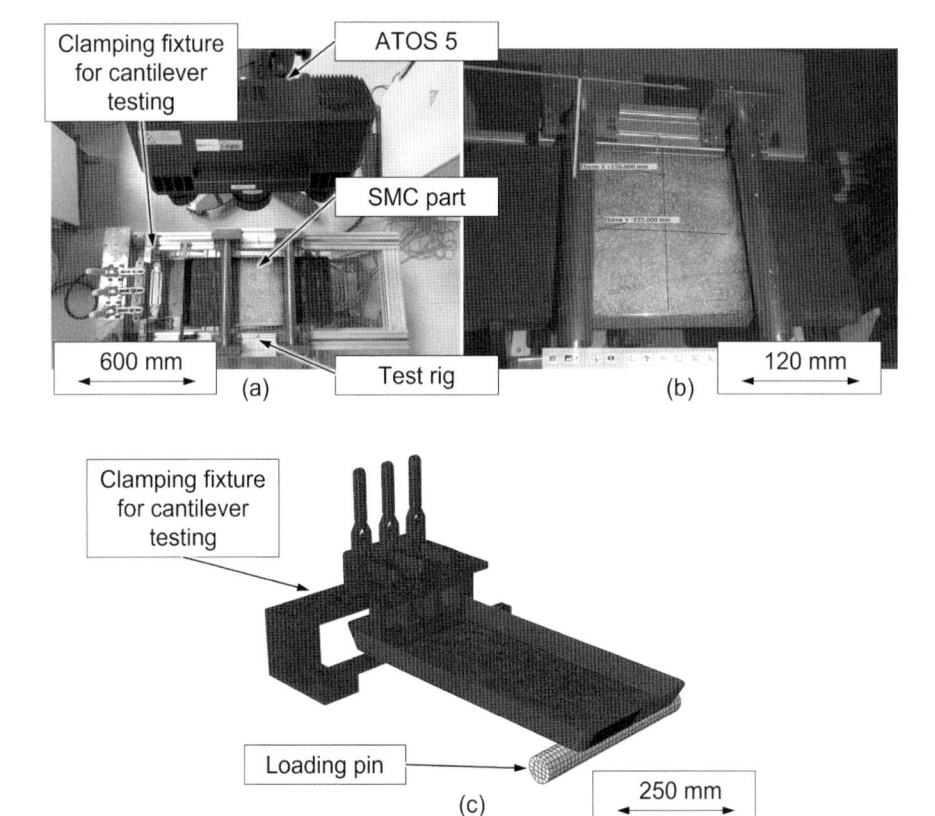

Figure 3.4: (a) Test cell configuration for SMC part evaluation under 3-point bending load. (b) Close-up of evaluation zone under 3-point bending load. (c) Schematic of test cell in cantilever load configuration

3.4 Digital twin of hybridised and non-hybridised SMC part

Prediction of part performance requires, as previously mentioned, the coupling of process and structural simulation. Procedures implemented are presented in the following subchapters. Discretisation of the geometry into a structured shell mesh was conducted in Abaqus 2021, Dassault Systèmes, Vélizy-Villacoublay, France. For initial evaluation of the part deformation behavior and, subsequently, material parameters used the geometry is discretised with 27,010 symmetrical S3R-elements. These comparatively small elements are used to minimise nonlinearities in the clamping zone. Simulation of 3-point bending behavior is conducted using a discretisation based on 13,104 symmetrical S3R-elements (Figure 3.1(b)). This shell element type was chosen based on compatibility with the process simulation procedure, and subsequently, it also applied in definition of towpreg positions, mechanics and final structural simulation.

3.4.1 Process simulation

Calculation of the fibre orientation probability distribution function (FOD) resulting from SMC flow during compression moulding was also conducted using Express 6.0, M-Base Engineering + Software GmbH, Aachen, Germany. This software was developed in close collaboration with the IKV, Aachen, Germany, and it has been used for a 2.5D process simulation of thermoplastic and thermoset compression moulding alike. It is based on the control volume approach as described by Osswald, which has been shown to accurately predict the filling pattern in compression moulding of thin geometries under the assumption of planar flow [HNF+20, Oss87, Sem98, Spe91].

$$\frac{\partial}{\partial x}(S\frac{\partial p}{\partial x}) + \frac{\partial}{\partial y}(S\frac{\partial p}{\partial y}) - \dot{h} = 0 \qquad (3.1)$$

The flow conductivity S is derived from the flow gap height h and the shear and temperature-dependent viscosity η :

$$S = \frac{h^3}{12\eta} \qquad (3.2)$$

From the pressure distribution, the gap-wise average velocities U and V are derived, from which the fibre orientation probability distribution function (FOD) is subsequently calculated. [HNF+20, Oss87, TF83]

$$\overline{U} = -\frac{S}{h}\frac{\partial p}{\partial x} \qquad (3.3)$$

$$\overline{V} = -\frac{S}{h}\frac{\partial p}{\partial y} \qquad (3.4)$$

Anisotropy of viscosity is taken into account by simulating the filling of the geometry with five shell geometries stacked on top of each other, for which the momentary temperature change due to conductive heat transfer is considered individually. From the layer-wise average velocities, fibre orientations are determined for each layer using the orientation model described by *Folgar and Tucker* [TF84]. This model includes a phenomenological diffusion term, with which fibre–fibre interactions are taken into account by the fibre interaction coefficient C_I [HNF+20, TF84].

$$\frac{\partial \psi}{\partial t} = C_I \frac{\partial^2 \psi}{\partial \Phi^2} - \frac{\partial}{\partial \Phi}(\psi(-sin(\Phi)cos(\Phi)\frac{\partial v_x}{\partial x} - sin^2(\Phi)\frac{\partial v_x}{\partial y} + cos^2\Phi\frac{\partial v_x}{\partial x}$$
$$+ sin(\Phi)cos(\Phi)\frac{\partial v_y}{\partial y})) \quad (3.5)$$

FOD were transferred to MATLAB 2020b, Mathworks, Natick, MA, USA for calculation of the mechanical properties of each element and definition of the towpreg prositions. The mechanical properties of the SMC fraction are calculated by methods described by *Advani and Tucker*, which are suitable for use in thin-walled compression moulding [AT87]. The calculation of mechanical properties is divided into three successive steps. Initially, the fourth-order stiffness tensor is calculated under an assumption of unidirectional fibre orientation in the SMC using the governing equations of *Halpin and Tsai* [HK76]. Then, the FOD-dependent stiffness tensors T of each element are calculated by orientation tensor averaging [AT87]. For a closer description of the process, one is referred to *Hopmann, Neuhaus et al.* [HNF+20]. Material parameters used are summarized in Table 3.2, and were derived from SMC composition (Unsaturated polyester and glassfibre-based class-A SMC0400 of Menzolit). In investigations conducted prior, resulting stiffnesses were shown to be in high agreement on test coupon level. Suitability will be further validated on component level under both cantilever and three-point bending load.

Parameter	Unit	Value
Thermal conductivity	[W/mK]	0.555
Heat transfer coefficient (Tool/SMC)	[W/m²K]	2,000
Fibre weight fraction	[-]	0.4
Fibre interaction coefficient	[-]	0.07
Elastic modulus glass fibre	[GPa]	73
Elastic modulus SMC-Matrix (UP)	[GPa]	6.25
Poisson ratio fibre	[-]	0.22
Poisson ratio matrix [Pir04]	[-]	0.33
Fibre aspect ratio (length/diameter)	[-]	3,000
Initial fibre orientation	[-]	isotropic

Table 3.2: Parameters for process and structural simulation. Source [Kre11] if not stated otherwise

3.4.2 Mapping to structural simulation

The position of each individual towpreg is defined as a vector with starting and ending point as well as orientation. Towpreg width after compression moulding is evaluated by caliper gauge at 11.5 mm (thickness at 0.42 mm, respectively). Using this width, a virtual volume is spanned around each towpreg vector. The overlap percentage of each towpreg volume and discretisation element is subsequently calculated and the local thickness of the towpreg in each element defined accordingly (50 % overlap -> 50 % local thickness). A visual representation of the local thickness distribution can be found in Figure 3.5, darker colours indicating a higher percentage of total overlap. Stiffness tensor components of the towpregs not included in the data sheet provided (e.g. poisson ratio) are calculated with Compositor, IKV Aachen. The material parameters used in structural simulation are shown in Table 3.3.

Parameter	Unit	Value
Density	[g/cm^3]	1.518
Elastic modulus in fibre direction	[GPa]	123.475
Elastic modulus perpendicular to fibre direction	[GPa]	9.024
Fibre weight Fraction	[-]	0.6285
Elastic modulus glass fibre	[GPa]	73
Poisson ratio perpendicular to fibre orientation	[-]	0.2705

Table 3.3: Mechanical properties of the towpregs

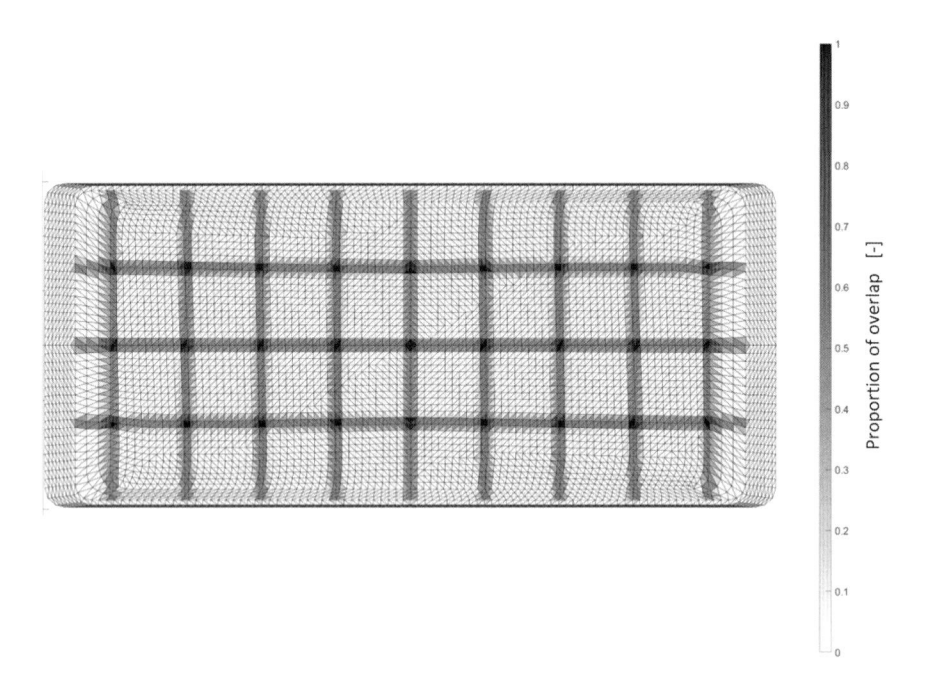

Figure 3.5: Towpreg overlap fraction of geometry discretisation

Finally, the resulting stiffness tensor components are exported to an Abaqus 2021 input-file (.inp) as individually defined materials for each element using an automated script. The 5 (no towpreg overlap), 7 (one towpreg overlapping on part top and bottom) or 9 (two towpregs on part top and bottom) layers respectively were treated as plies of a composite shell section, which was also automatically created using the script. A schematic representation is shown in Figure 3.6.

	Ply 1
	Ply 2
Ply 1	Ply 3
Ply 2	Ply 4
Ply 3	Ply 5
Ply 4	Ply 6
Ply 5	Ply 7
	Ply 8
	Ply 9

(Layouts shown: (a) 5 plies, (b) 7 plies, (c) 9 plies)

Figure 3.6: Schematic representation of the composite layup of each element with none (a) two (b) and four (c) towpreg overlaps present. The SMC fraction is shown in light gray, whereas towpreg plies are shown in darker gray

As SMC and towpreg stiffness matrices are calculated and defined in local coordinate systems, these were also supplied to the .inp file for the SMC and towpreg fractions of each element. The digital twin of the test cell in 3-point bending configuration is depicted in Figure 3.7. Interaction of the parts was defined as frictionless.

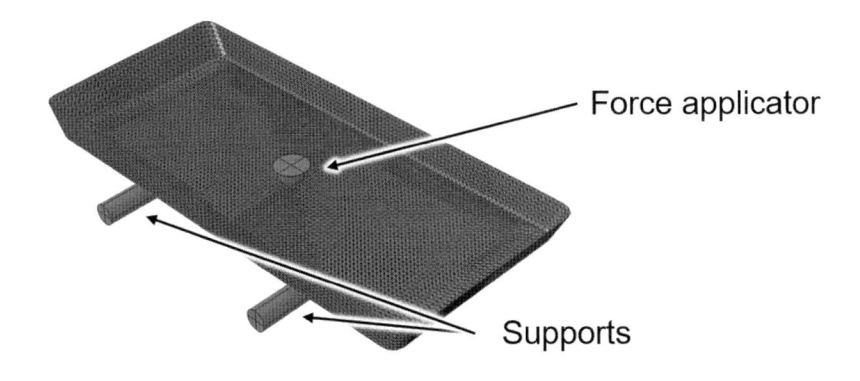

Figure 3.7: Digital twin of the test cell in 3-point bending configuration

3.5 Results – validation of material parameters under cantilever load

In Figure 3.8, part deformations evaluated by the test cell and simulation for the cantilever load are compared. Deformation of part 1 closely matches the simulation. This part exhibits part thicknesses most closely matching the simulation (charge weight 1784.2 g), while part 2 (1744.4 g) and part 3 (1793.6 g) deviate in both charge weight, part thickness and resulting deformation. Deviations in charge weight can be traced back to deviation in surface weight of the SMC used. Part 2 exhibits a buckling behavior close to the clamp caused by the lower charge weight, which increases total deformation. Correlations between part deformation behavior and part weight/thickness are therefore complex but can be accurately simulated when taken into account. Furthermore, suitability of the material parameters used is shown.

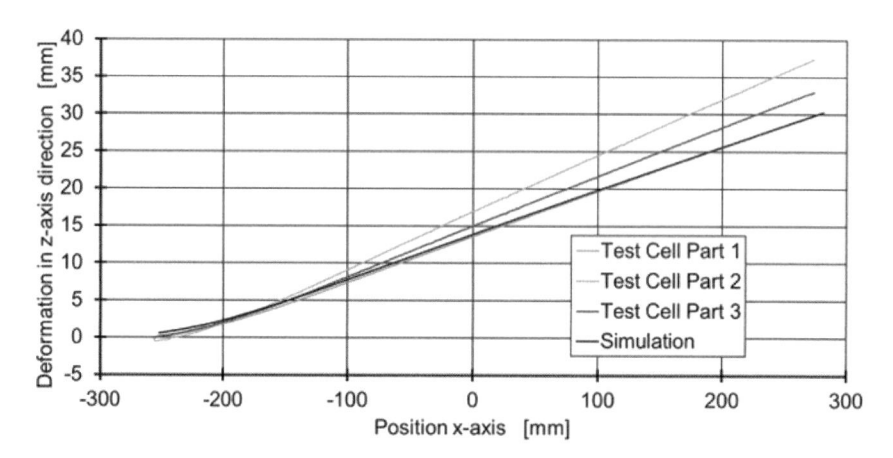

Figure 3.8: Comparison of deformations evaluated by test cell and simulation for non-reinforced (a) and reinforced SMC part (b)

3.6 Results – validation of the simulation approach under 3-point bending

In Figure 3.9, part deformations evaluated by the test cell and simulation are of both reinforced and non-reinforced pars is compared. In general the total deformation of the part and curvature of the bending line are accurately predicted for both the reinforced and non-reinforced SMC parts. The deformation of the part due to hybridisation is reduced by 17.02 % (test cell) and 21.37 % (simulation) respectively. Discrepancies may be traced back to several reasons. Firstly, differences in part positioning, coordinate system definition, friction coefficients and resulting buckling behaviour may occur between the test cell and simulation procedure as shown during cantilever testing, which may also result in the negative deformations seen during test cell analysis. Secondly, thicknesses of the parts evaluated vary due to variations in SMC surface weight, with reinforced part 1 most closely matching the thickness of the digital twin. Thirdly, while matrix stiffness of the SMC may be accurate under predominantly tensile loads as shown during cantilever testing, the side walls are loaded predominantly with compressive forces during 3-point bending. Thus, tensile and compressive modulus of the matrix may differ (Figure 3.9 (a)).

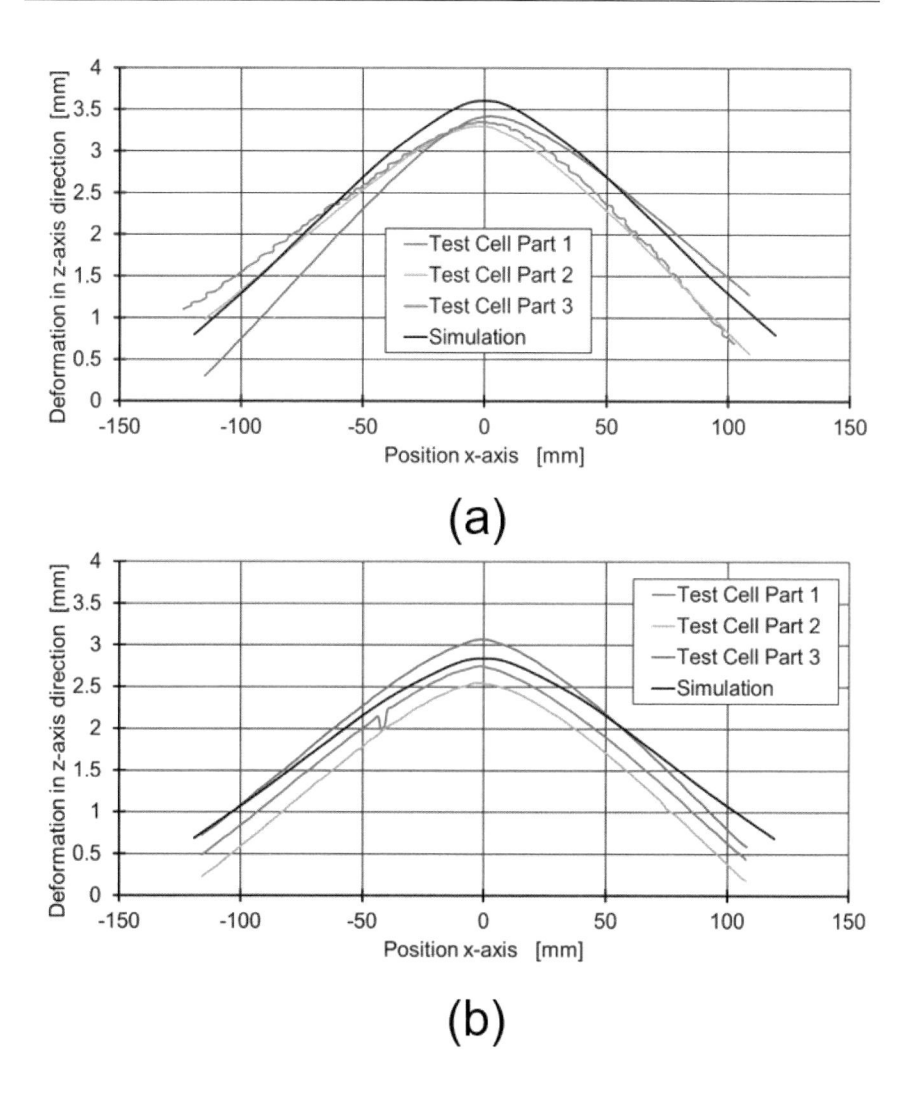

Figure 3.9: Comparison of deformations evaluated by test cell and simulation for non-reinforced (a) and reinforced SMC part (b)

In Figure 3.10, the tensions present in the towpregs (plies 9 and 7, (a) and (b) respectively) and surface layer (ply 5 - (c)) of the SMC fraction are shown. Tensions in the fibre orientation direction of the towpregs are significantly higher than those present in the SMC fraction, showing the effect of load displacement inside the hybridised part to the more stiffer towpreg structures. Contact elements close to the 3-point bending support structures exhibit compression loads of 6 MPa maximum. Loads in the surface layer of the SMC fraction are homogenously distributed. Tensions orthogonal to the towpreg fibre orientation and shear tensions are shown in Figure 3.11. Both are significantly lower than tensions in fibre direction, resulting from the the significantly lower stiffness of the towpregs in non-orientation load directions (see Table 3.3).

(a)

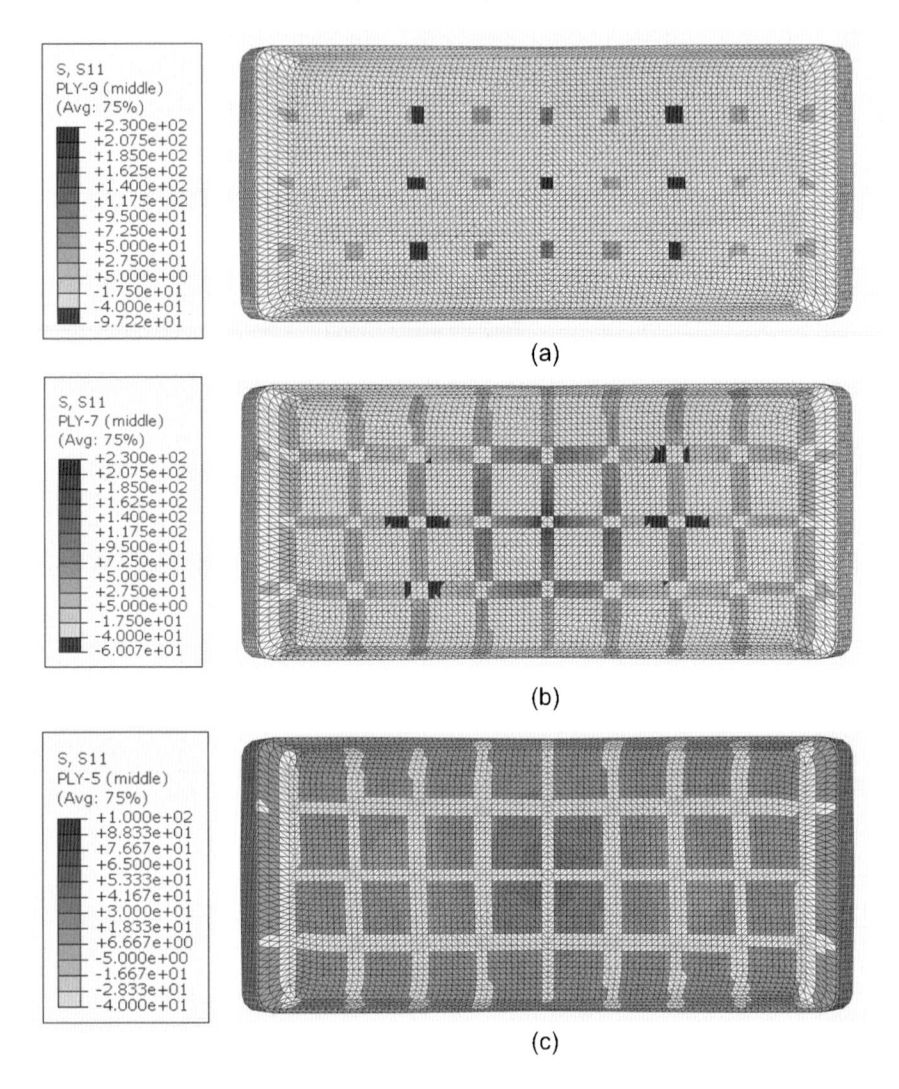

(b)

(c)

Figure 3.10: Tensions in main fibre orientation in composite plies 9 (a), 7 (b) and 5 (c) in [MPa]

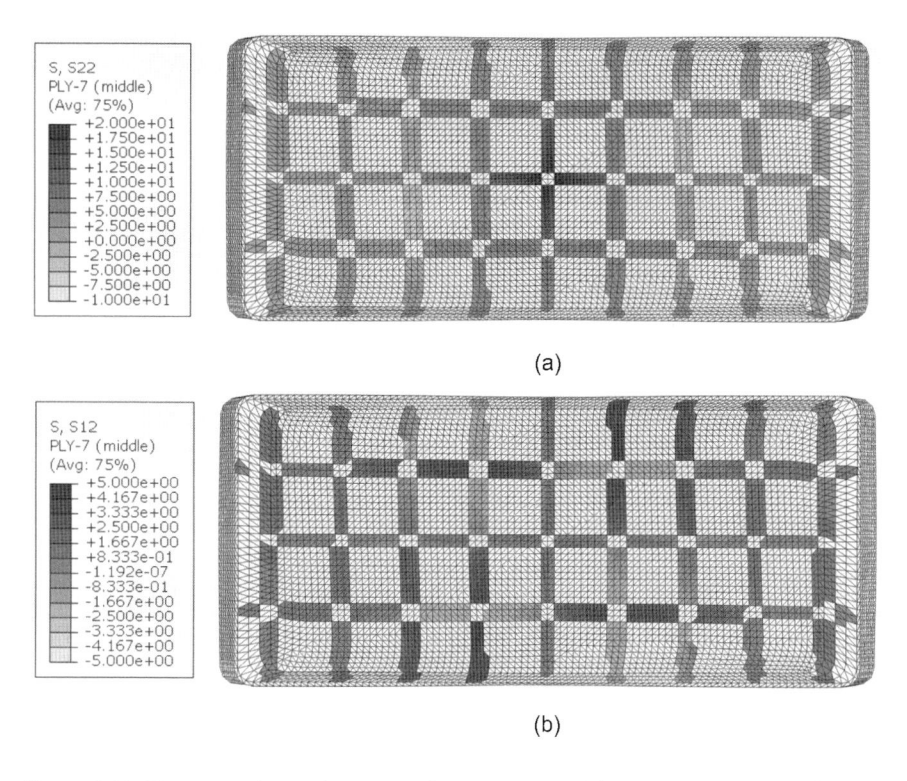

Figure 3.11: Tension orthogonal to main fibre orientation (a) and shear tension (b) of ply 7 in [MPa]

3.7 Conclusions and outlook

SMC materials offer highly competitive manufacturing properties in the field of large-scale traction battery housings, namely low cycle times and low investment costs. Mechanical demands on these parts however are harsh, necessitating cost efficient local reinforcements of highly stressed part areas. In this paper, a procedure for process simulation, mapping and subsequent structural simulation of a locally reinforced SMC part have been presented. The resulting digital twin is compared to measurement results obtained by a novel test cell by AZL Aachen. The procedure enables the prediction of part deformation and bending line curvature under 3-point bending loads. Stiffness increases due to reinforcement structures in comparison to non-reinforced structured and load distributions are accurately predicted, enabling the effective design of these parts, and therefore minimisation of testing and part design iterations. In the future, the accuracy of the procedure will be further increased by incorporating both friction effects and the use of compressive matrix stiffnesses. The use of a finer mesh may also increase prediction accuracy and will be refined in subsequent work. Furthermore, more complex geometries and load cases such as torsion may also be incorporated in both the test cell and the digital twin, further increasing the potential in part design of the procedure presented (Figure 3.12). The investigations set out in this report received financial support from the European Regional Development Fund (No.: EFRE-0801121), to whom we extend our thanks.

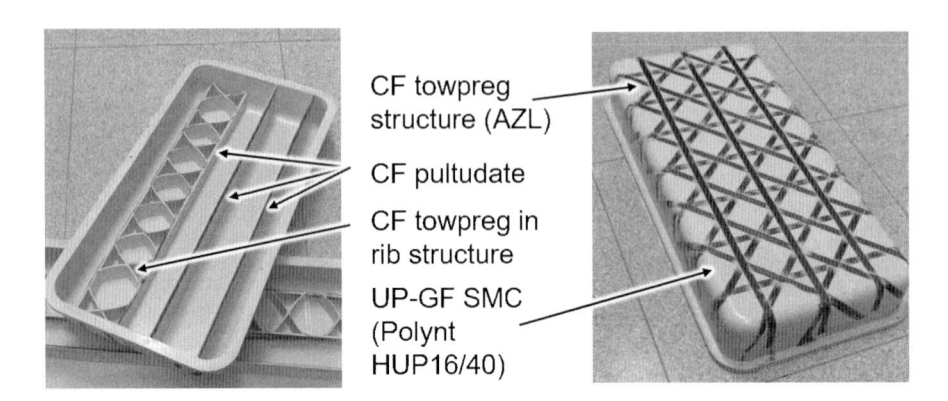

Figure 3.12: HybridSMC-demonstrator with multiple hybridisation concepts

Acknowledgements

This research "Development of a holistic methodology for the use of innovative hybrid materials for the cost-efficient series production of functionally integrated composite lightweight components based on sheet moulding compounds (SMC) and thermoset injection moulding compounds" (EFRE-0801121) was funded by the European Regional Development Fund (ERDF). We would like to take this opportunity to express our gratitude for the funding and support. Furthermore, we would like to express our gratitude to the project partners Aachen Center for Integrative Lightweight Production (AZL) of RWTH Aachen University, Aachen, BYK-Chemie GmbH, Wesel, Kautex Textron GmbH Co. KG, Bonn, M-Base Engineering + Software GmbH, Aachen, SimpaTec Simulation Technology Consulting GmbH, Aachen.

References

[AT87] ADVANI, S. G.; TUCKER, C. L.: The Use of Tensors to Describe and Predict Fiber Orientation in Short Fiber Composites. *Journal of Rheology* 31 (1987) 8, p. 751–784

[BFDH18] BÜTTEMEYER, H.; FETTE, M.; DREWES, S.; HERRMANN, A.: Vertical Deformations of Continuous Carbon Fibre Textiles in Hybrid Sheet Moulding Compound Processing. 09 2018

[Bü18] BÜCHELER, D.: *Locally Continuous-fiber Reinforced Sheet Molding Compound.* Karlsruher Institut für Technologie (KIT), Dissertation, 2018

[FEWN21] FISCHER, K.; EMONTS, M.; WANG, H.; NEUHAUS, J.: Continuous-discontinuous sheet moulding compounds – Effect of hybridisation on mechanical material properties. *AVK Composites Report* 3 (2021), p. 6–7

[FHS+17] FETTE, M.; HENTSCHEL, M.; SANTAFE, J. G.; WILLE, T.; BÜTTEMEYER, H.; SCHIEBEL, P.: New Methods for Computing and Developing Hybrid Sheet Molding Compound Structures for Aviation Industry. *Procedia CIRP* 66 (2017), p. 45–50

[HK76] HALPIN, J. C.; KARDOS, J. L.: The Halpin-Tsai equations: A review. *Polymer Engineering Science* 16 (1976) 5, p. 344–352

[HNF+20] HOPMANN, C.; NEUHAUS, J.; FISCHER, K.; SCHNEIDER, D.; GONCALVES, R.: Metamodelling of the Correlations of Preform and Part Performance for Preform Optimisation in Sheet Moulding Compound Processing. *Journal of Composites Science* 4 (2020) 3,

[Kre11] KREMER, C.: *Vorhersage der Oberflächenwelligkeit von Bauteilen aus Sheet Moulding Compound durch die Simulation der prozessinduzierten Eigenspannung.* RWTH Aachen University, Dissertation, 2011

[MSM+21] MEHL, K.; SCHMEER, S.; MOTSCH-EICHMANN, N.; BAUER, P.; MÜLLER, I.; HAUSMANN, J.: Structural optimization of locally continuous fiber-reinforcements for short fiber-reinforced plastics. *Journal of Composites Science* 5 (2021) 5, p. 1–16

[NN20] N.N.: *Electric vehicle traction battery safety requirements.* Standard, Beijing, China, 2020

[Oss87] OSSWALD, T.: *Numerical Methods for Compression Mold Filling Simulation.* University of Illinois, Dissertation, 1987

[Pir04] PIRY, M.: *Mechanical dimensioning of SMC-components and characterisation of the relevant material properties.* RWTH Aachen University, Dissertation, 2004

[Sem98] SEMMLER, E.: *Simulation of the Mechanical and Thermomechanical Behavior of Thermoplastic Fiber-reinforced Compression Moulded Parts.* RWTH Aachen University, Dissertation, 1998

[Spe91] SPECKER, O.: *Compression Moulding of SMC: Computer Simulations for Computer-Aided Lay-Out of the Process and for Determination of the Component Properties.* RWTH Aachen University, Dissertation, 1991

[TF83] TUCKER, C. L.; FOLGAR, F.: A model of compression mould filling. *Polymer Engineering Science* 23 (1983) 2, p. 69–73

[TF84] TUCKER, C. L. I.; FOLGAR, F.: Orientation Behavior of Fibers in Concentrated Suspensions. *Journal of Reinforced Plastics and Composites* 3 (1984) April, p. 98–119

[TW18]　Trauth, A.; Weidenmann, K. A.: Continuous-discontinuous sheet moulding compounds – Effect of hybridisation on mechanical material properties. *Composite Structures* 202 (2018), p. 1087–1098

Symbols

Symbol	Unit	Description
ψ	–	Fibre Orientation Distribution
C_I	–	Fibre Interaction Coefficient
h	mm	Flow Gap Hight
p	Pa	Pressurey
S	mm^2	Flow Conductivity
v	mm/s	Flow Speed

Abbreviations

Notation	Description
SMC	Sheet Moulding Compound
FOD	Fibre Orientation Distribution

Joining of innovative and sustainable material combinations

Moderator: Dipl.-Ing. Michael Dietrich, WEGENER International GmbH

---- Content ----

1   IR-Welding of PET-foam, a sustainable way?   469

M. Dietrich[1]
[1] WEGENER International GmbH

2   Welding of polypropylene post-consumer recycled material   478

Ch. Hopmann[1], R. Dahlmann[1], M. Weihermüller[1]
[1] Institute for Plastics Processing (IKV) at RWTH Aachen University

3   Welding of TPE-TP composites: Development of strategies to produce aging-resistant welds with reduced anisotropy   492

Ch. Hopmann[1], R. Dahlmann[1], P. Knupe-Wolfgang[1], S. Bölle[1], Ch. Kley[1], J. Wipperfürth[1]
[1] Institute for Plastics Processing (IKV) at RWTH Aachen University

Dipl.-Ing. Michael Dietrich

Michael Dietrich studied mechanical engineering at RWTH Aachen University, specialising in design engineering. In 1999 he completed his studies with a diploma thesis at the Institute for Plastics Processing (IKV). This was followed by a position at the company Projekt Automation GmbH in Cologne, Germany, with a focus on automation and robotics. In 2001, he moved to Jouhsen-Bündgens Maschinenbau GmbH in Stolberg, Germany, where he worked mainly in the field of upsetting cutting of wires. In 2004, he took over the function of technical manager at the company Wegener International GmbH, Eschweiler, Germany, the market leader in the field of butt welding and bending of thermoplastic sheets and plates. Since 2008, he has been the technical managing director of Wegener International. Mr Dietrich is also a member of the Sponsoring Association of the SKZ Würzburg, Germany, and the Sponsoring Association of the IKV. Furthermore, he is a member of the German Association for Welding and Allied Processes (DVS).

1 IR-Welding of PET-foam, a sustainable way?

M. Dietrich[1]
[1] WEGENER International GmbH

Abstract

The question of sustainability has long transcended the boundaries of private consumption. The need to assess our industrial processes from an economic, ergonomic and ecological point of view has long since been extended to include the aspect of sustainability. The motivation of this article is not so much to provide proof of the sustainability of a specific process, but rather to show by means of an example how the combination of the most modern materials technology in conjunction with supposedly old process technology can be an approach to solutions for current and future tasks. Last but not least, the complexity of the topic becomes just as clear as the never-ending need for further technological development.

1.1 Introduction

With the development and growing use of thermoplastics, the joining techniques also inevitably developed to increase the complexity of the parts and the dimensions or for economical or applicational reasons. Joining processes or joints can be subdivided into material-locking, form- or force-fitting joints and can be reversible or irreversible. Particular attention in this paper is paid to welding technology a material-locking joining techniques from a sustainability point of view. The reasons for this are explained in more detail in the chapter on sustainability. A variety of welding techniques have been developed in recent decades to meet product requirements. In turn, a wide variety of attributes can be found for classification. Why thermoplastics can be permanently welded and which mechanisms come into play is described very impressively by *Grewell and Benatar* using a model in which squeeze flow and diffusion are the supporting mechanisms that allow molecular chains to pass through the boundary layers and thus enable the molecular chains of the joining partners to become entangled [GB07]. The extent to which secondary forces, such as Van der Waals forces, can become effective at the molecular level is not described. The decisive factor for a weld, regardless of the welding process, is ultimately to determine the required welding parameters of pressure and temperature as a function of time [HM15, EHE08]. Furthermore, *Grewell and Benatar* describe that the completeness of the "healing", i.e. the tangling of the molecular chains corresponding to the virgin material, has a decisive influence on the quality of the weld [GB07]. An essential distinguishing feature between the welding processes is the generation of heat for the plasticisation of the joining zone. Here, for example, a distinction can be made between internal friction, as in ultrasonic or vibration welding, and external heat supply, as in hot gas welding or infrared (IR) welding. The boundary conditions are now used to select the best possible welding process. In the present example, the large-area welding of polyethylene terephthalate (PET) foam sheets into blocks, all processes with internal heat generation can be excluded for technical reasons. Since the product geometry does not allow continuous welding processes, the only remaining methods are heated-tool butt welding and Infrared(IR)-welding. However, both processes still have system-related disadvantages. The high melting temperature of PET requires a welding temperature between 260°C and 290°C [NN84]. This means that the polytetrafluoroethylene (PTFE)-based heating element coatings available on the market are permanently operated at their thermal load limit, which results in premature degradation of the PTFE components. Also, especially with large (and thus heavy) heating elements, the so-called changeover time, i.e. the time between the detachment of the heating element from the product to the contact of the joining surfaces, is technically very long. Since the surface temperature drops rapidly during this time, the associated loss of energy must be compensated by an extended plasticising phase or an increased process temperature.

In addition, the welding cycle is extending, which has a negative impact on its economic efficiency. These disadvantages do not occur with IR-welding due to the process.

1.2 IR-Welding

Electromagnetic radiation in the range between 780 µm and 1 mm wavelength is defined as IR radiation according to DIN 5031-7 (withdrawn) [NN84]. There are different international definitions regarding the division of this range. It directly follows the range of visible light; colloquially, the transitional range is often referred to as red light. According to DIN 5031, the wave range of IR radiation is divided into four frequency or wavelength ranges:

IR-A (NIR = Near Infrared) Wave Length 0.78 µm - 1.4 µm:
Short-wave IR radiation is usually generated by quartz tube emitters. These work at surface temperatures between 1800°C and 2400°C and a wavelength range between 1.0 µm and 1.4 µm. A major advantage of these emitter types is the extremely short reaction time of approx. 1 s [URL21j]. A disadvantage, on the other hand, is the high distance to the plastic surface required due to the high surface temperature and the resulting high scattering range of the emitter.

IR-B (NIR = Near Infrared) Wave Length 1,4 µm – 3,0 µm:
In this so-called long-wave range of the NIR, emitters are realised whose wavelength is above 1.4 µm and which operate at surface temperatures between 1400°C and 1800°C. Their response time is between 1 s and 2 s. The emitters can be used in the long-wave range. [URL21j]. With regard to their possible applications for welding thermoplastics, the same applies as for emitters operating in the IR-A range. The still very high surface temperature of the emitters in relation to the melting and decomposition temperatures of the thermoplastics also requires a relatively large distance to the joining surface and thus also a wide dispersion [Tro08].

IR-C (MIR = Mid range Infrared) Wave Length 3,0 µm – 50,0 µm:
Various designs of long-wave radiators in this wavelength range have become established on the market. For example, open metal wire grids fixed on a thermal insulator are offered [URL21l], but also coils protected by quartz tubes. Wire filament emitters embedded in ceramics are also produced. Their wavelength range lies between 2.4 µm and 2.7 µm with surface temperatures between 800°C and 950°C. With appropriate power controls, the wavelength range of this type of emitter can be increased up to 9.6 m. Their reaction time is between 1 min and 4 min [URL21f, URL21g]. The so-called dark radiators emit even longer wavelengths. These are usually heating plates made of steel, rarely with a ceramic coating, which can reach surface temperatures of up to 650°C. Due to the low surface temperatures, the distance to the joining surface must be kept very small. This process is also referred to as non-contact hot plate welding. [HM15]

IR-C (FIR = Far Infrared) Wave Length 50,0 µm – 1,0 mm [URL21n]:
This frequency range is only listed here for the sake of completeness. It does not play a significant role in technical heating. However, the terahertz radiation to be assigned to this wavelength range is used, for example, in full-body scanners at airports. Interesting investigations have also been carried out in the recent past on the non-destructive testing of plastic weld seams [URL21n].

1.3 PET, a development

The existence of polyethylene terephthalate (PET), probably only came to the attention of broader sections of the population with the growing spread of the PET bottles. Its successful history began in Germany in the early 90s of the last century and continues to this day. The soft-drink-market in the USA, however, demanded the introduction of the PET bottle already at the end of the 70s [URL17]. What is probably less well known is that PET was thus given a second life. During the Second World War, this material was developed and patented in Great Britain as a fibre material and later further developed by Imperial Chemical Industries (ICI, today Akzo Nobel, Amsterdam, Netherland) and Du Pont, Wilmington, Delaware, USA, for the USA. Older people may remember the first successful life of PET, especially in the 1960s and 1970s in Germany under the brand names Trevira and Diolen for synthetic fibres in textiles [DEEH12]. According to *statista*, the global consumption of PET in 2020 was 27 Mto [URL21o]. The platform marketsandmarkets estimates the PET-foam market at US$ 316 million in 2020 and forecasts growth to US$ 448 million in 2025 [URL21m]. Assuming a raw material price of 1.00 €/kg and an average dollar exchange rate of 0,88 €/$ for 2020, this would lead to a consumption of approx. 278 kto/a [URL21k, URL21p]. By comparing these two very rough estimations, we find that the demand for PET-foam only accounts for one percent of the total production of PET, but at approx. 278 kto/a it is still representing a significant quantity. By also taking into account that PET foam can currently be produced a 100% from bottle flakes, i.e. pure regranulate, without any loss of strength, and high availability by different manufacturers, this example already gives an idea of the potential of this material within the circular economy.

Physics of PET:

PET is a semi-crystalline polyester from the group of thermoplastic polycondensates. It's versatility is made possible, among other things, by the use of suitable nucleating agents which increase the crystallisation rate to such an extent that degrees of crystallisation of up to 40% can be achieved, but also by the incorporation of suitable co-monomers to reduce the crystallinity, so that crystal-clear products such as the well-known PET bottle can be produced [DEEH12].

Consumer goods/ producer goods:

The use of PET continues to be the main focus in the packaging industry. In 2010, the distribution was as follows:

- Fibres 65%
- Packaging 29%
- Vapourised films 4%
- Specialities 2%

Due to the dynamics of this market described above, current figures are likely to be different. Basically, however, a large proportion of the products are to be found in the low-cost sector or in articles with a short service life. The use of PET in the area of technical products or products with a longer service life is also diverse but still in absolute development. Due to its good sliding properties, PET is used, among other things, in gear construction for gear wheels, cams, couplings, etc., as an unreinforced but also as a glass fibre-reinforced material. In recent years, since 2008 to be exact, PET has conquered another area of application. In that year, WEGENER International GmbH, Eschweiler, Germany, delivered the world's first block welding machine for the welding production of blocks from extruded PET foam sheets. Whereas until then the cores of wind turbine rotor blades were made of balsa wood, foamed SAN or PVC, PET foam is now the material of choice for this application [URL21a, URL21b, URL21d, URL21i]. However, more and more market areas are being opened up by PET or rPET (recycled PET) foams. Even in the insulation sector, low-density foams made of PET could soon be found.

1.4 Sustainability

A definition of sustainability is difficult. The BUND (Bund für Umwelt- und Naturschutz Deutschland = Association for protection of environment and nature Germany) speaks of three components: Efficiency, consistency and sufficiency [URL21c]. The Dictionary of the Standard High German language offers this definition, among others: "Prinzip, nach dem nicht mehr verbraucht werden darf, als jeweils nachwachsen, sich regenerieren, künftig wieder bereitgestellt werden kann" (=Principle according to which no more may be consumed than can be regrown, regenerated, provided again in the future at any one time) [URL21e]. Grober refers to Hans Carl von Carlowitz, who in 1713, as head of the Freiberg Mining Authority, wrote the Sylvicultura oecomomica and in it formulated the principle of sustainability for the first time, at least from a forestry perspective [Gro13]. In addition, two definitions from the Brundtland Report of 1987 on sustainable development should be mentioned: 1) "Sustainable development is development that meets the needs of the present without compromising the ability of future generations to meet their own needs." [NN87]. 2) "In essence, sustainable development is a process of change in which the exploitation of resources, the direction of investments, the orientation of technological development; and institutional change are all in harmony and enhance both current and future potential to meet human needs and aspirations" [NN87]. It should be indisputable that these goals cannot be achieved under the given boundary conditions without the use of renewable energy generation. The contribution of wind energy should be emphasised, especially considering the next generation of offshore wind turbines, such as the Haliade-X from General Electric Company, Bosten Massachusetts, USA, whose prototype in Rotterdam has already set a record with 312 MWh of electricity generation in 24 hours, without having been switched to the maximum 14 MW [URL20]. In this respect, we can perhaps give a first cautious partial answer to the question whether the welding of recycled PET-foam is sustainable in nature; the final product seems to serve sustainability.

Recycling:

It should also be indisputable that without recycling, the principles of sustainability can hardly be met, especially under the postulate of at least approximately maintaining our standard of living, although it may be debatable whether these are needs in the sense of the Brundtland Report. The fact that recycling already works very well in the area of PET bottles, at least in Germany, is shown, among other things, by a report from 2016 with the reference year 2015 by the GVM (Gesellschaft für Verpackungsmarktforschung - Society for Packaging Market Research), which found a recycling rate of 97.9% for PET bottles with a deposit refund system and remarkable 93.5% for all PET bottles. This refers to the recycling feed rate defined in the EU specifications and VerpackV [NN16].

Downcycling/ Upcycling:

Following the definitions of Chandler and Werther, upcycling in the following refers to material recycling that leads to a new product with a significantly longer product life. Accordingly, closed cycles (bottle-to-bottle) are not considered. All other recycling processes up to thermal recovery are also considered downcycling in this article, following Chandler and Werther [PMRG19]. Last but not least, the increased demand for products made from recycled materials has led to many products made of rPET in the last decade, such as school satchels, shoulder bags, clothes, car components, etc. [URL21h] Above all, however, the production of PET foam for wind converters from so-called bottle flakes can be seen as an upcycling process, since the lifespan of a wind converter should supposedly be longer than that of a lemonade bottle. With regard to our question about the sustainability of the welding process, we are thus one step further, because PET core material is already being produced 100% from regranulate, and some manufacturers even offer core material made exclusively from rPET [URL21b, URL21d, URL21h]. With the material to be welded from 100% recyclate, we are probably not contradicting the concept of sustainability.

1.5 IR-Welding of 100% recycled PET-foam

The product wind converter blade is made of PET foam inside. This is because it is the most efficient way to make it. Efficiency is one of the main building blocks of sustainability. A major reason for welding the extruded PET foam sheets is the anisotropy created by the extrusion of the foam. The load-oriented definition of the core material also makes this possible, but the direction of extrusion in the component must be rotated to achieve this. To put it simply: The previous height becomes the length afterwards. Another reason for welding is the stabilising effect of the weld seams in the product. In the later component, the weld seams form a kind of grid or parallel planes of solid, made from compacted unfoamed material and thus has higher strength. This might be interpreted as an analogy to sandwich structures. This structure increases the anisotropy according to the load and thus foam material can be saved, thus also fulfilling the aspect of sufficiency. Finally, it makes the process very flexible in terms of possible product heights. The extrusion process can be operated with maximum efficiency without considering the customer-specific product height. What remains unconsidered here is the welding to the final product, as used, for example, in a rotor blade made from the Haliade-X converter described above, which has a length of 107 m. Now that the welding process itself fulfils sustainability criteria, the question of the specific welding process remains. So far, it has only been possible to deduce why only a heated element butt welding process or an IR welding process can be considered for this task. Functionally, both processes fulfil the technical requirements. The weaknesses of contact heating lie on the one hand in the non-stick coating of the heating element, which, depending on the PET composition, has a very short service life and requires frequent re-coating. This currently produces waste materials that cannot be recycled due to the process. Another disadvantage, as already mentioned, is the extended changeover time due to the detachment process of the heating element from the product; this leads to a disadvantage. Due to the nature of the process, these disadvantages do not occur with IR welding. However, the wavelength of the emitters as well as the absorption maximum of the PET foam must be subject to special consideration. Different materials show different absorption spectra. Therefore, the emitter wavelength must be adjusted to the absorption spectrum of the material depending on the process. However, completely different boundary conditions must be taken into account when welding foam than when welding solid materials. The background is as follows: For a high-quality butt weld, a defined layer thickness must be plasticised. This should be done as quickly as possible to keep the total heat input to a minimum. In the case of solid materials, the "depth effect" of the IR emitter can be used here, as plasticising is not only carried out via heat conduction from the surface, but at the same time heat is also generated directly in the depth by absorption of the IR radiation. By definition, this depth effect is stronger the further apart the absorption maximum of the material and the wavelength of the emitter are [EHE08, Tro08]. In the case of foams, however, this effect is usually counterproductive, because absorption generally takes place at a lower level due to the low density. Under certain circumstances, this leads to such a softening of the foam areas close to the joining surface that considerably more material is permanently compressed under the joining pressure than in the otherwise comparable hot plate welding process. While in IR welding with dark radiators the welding losses, i.e. the length losses, can be limited to less than 1 mm due to the complete melting of the foam including irreversible compression, these losses are a power of ten higher in IR welding with short-wave IR radiators and thus fulfil the sustainability conditions to a lesser extent than the dark radiators. It is therefore important to select the wavelength of the IR emitter used close to the absorption maximum of the material to be welded, as is the case with dark emitters. In summary, the sustainability of the process has been positive so far.

Energetic View:

So far, it has not been possible to carry out a comparative energetic analysis of dark radiator technology in this area of application in comparison to the HS-technology. Since the major disadvantage of dark radiators is their reaction time, the radiator must be kept permanently at working temperature with cycle times between 30 s and 60 s. Currently, however, the proportion of heating time is only between 15% and 25% of the cycle time, which results in considerable potential for energy savings.

1.6 Conclusions and outlook

The question of whether a process is sustainable or not can hardly be answered in absolute terms. Relatively speaking, answers can be found, but the quality of the answers depends crucially on the quality of the question. Using the example of joining extruded foam sheets made of recycled PET, it could be shown that in the question of sustainable ways of sustainable industrial products, it makes perfect sense to look for solutions not only in the latest process technology in combination with "high-tech materials", but also that supposedly traditional "low-tech processes" can offer solutions in combination with supposed "waste" as a starting material. The rethinking associated with this is probably an essential part of the path to greater sustainability. If one accepts the aspect of sustainability as an essential part of the industrial challenges in the twenty-first century, the solutions will be found with the corresponding openness to the new and the knowledge of the existing.

References

[DEEH12] DOMINGHAUS, H.; ELSNER, P.; EYERER, P.; HIRTH, T.: *Kunststoffe - Eigenschaften und Anwendungen*. Heidelberg: Springer Verlag, 2012

[EHE08] EYERER, P.; HIRTH, T.; ELSNER, P.: *Polymer Engineering - Technologien und Praxis*. Berlin, Heidelberg: Springer Verlag, 2008

[GB07] GREWELL, D.; BENATAR, A.: Welding of Plastics: Fundamentals and New Developments.. *International Polymer Processing* 22 (2007) 1, p. 43–60

[Gro13] GROBER, U.: Urtexte – Carlowitz und die Quellen unseres Nachhaltigkeitsbegriffs. *Natur und Landschaft* 88 (2013) 2, p. 46–48

[HM15] HOPMANN, C.; MICHAELI, W.: *Einführung in die Kunststoffverarbeitung*. München: Carl Hanser Verlag, 2015

[NN84] N.N.: *DIN 5031-7: Optical radiation physics and illumination engineering; terms for wavebands (withdrawn)*. Berlin: Beuth Verlag, 1984

[NN87] N.N.: *Our Common Future: Report of the World Commission on Environment and Development*. : The World Commission on Environment and Development, 1987

[NN16] N.N.: *Aufkommen und Verwertung von PET-Getränkeflaschen in Deutschland 2015*: Forum PET in der IK Industrievereinigung Kunststoffverpackungen e.V., 2016

[PMRG19] PIRES, A.; MARTINHO, G.; RODRIGUES, S.; GOMES, M. I.: *Sustainable Solid Waste Collection and Management*. : Springer International Publishing AG, 2019

[Tro08] TROUGHTON, M.: *Handbook of Plastics Joining - A Practical Guide*. United States of America: William Andrew Inc., 2008

[URL17] N.N.: *Forum PET in der IK - Industrievereinigung Kunststoffverpackungen e.V.: Von der Küche bis zum Kleid: recyceltes PET ist ein wertvoller Rohstoff für alle Fälle*. URL: https://www.forum-pet.de/media/archiv/detail.php?nID=67, 14.10.2021

[URL20] N.N.: *General Electric: Turning Point: Powerful Haliade-X Offshore Wind Turbine Sets New Record, Passes Crucial Milestone, GE News*. URL: URL:https://www.ge.com/news/reports/turning-point-powerful-haliade-x-offshore-wind-turbine-sets-new-record-passes-crucial, 14.10.2021

[URL21a] N.N.: *Airex: Foam and balsa sandwich solutions for renewable energy, 3A*. URL: https://www.3accorematerials.com/en/markets-and-products/renewable-energy, 15.10.2021

[URL21b] N.N.: *Armacell: Wind - ArmaFORM PET*. URL: https://local.armacell.com/en/armapet/markets/wind/, 15.10.2021

[URL21c] N.N.: *BUND - BUND für Naturschutz und Umwelt in Deutschland: Suffizienz – was ist das? Eine Definition*. URL: https://www.bund.net/ressourcentechnik/suffizienz/suffizienz-was-ist-das/, 15.10.2021

[URL21d] N.N.: *Diab: Diab's core materials for wind market*. URL: https://www.diabgroup.com/markets/wind/, 15.10.2021

[URL21e] N.N.: *Duden: Nachhaltigkeit*. URL: https://www.duden.de/rechtschreibung/Nachhaltigkeit, 15.10.2021

[URL21f] N.N.: *Elstein-Werk M.Steinmetz GmbH & Co KG: SHTS / Elstein Infrarotstrahler*. URL: https://www.elstein.com/de/produkte/einzelstrahler/flaechenstrahler/shts/, 14.10.2021

[URL21g] N.N.: *Elstein-Werk M.Steinmetz GmbH & Co KG: SHTS, Emission und Absorption.* URL: https://www.elstein.com/de/anwendungen/technikcenter/grundlagen/emission-und-absorption, accessed 14.10.2021

[URL21h] N.N.: *Forum PET in der IK - Industrievereinigung Kunststoffverpackungen e.V.: PET - Eine Zeitreise.* URL: https://www.forum-pet.de/material/pet-flasche, 14.10.2021

[URL21i] N.N.: *Gurit: Structural Core Materials.* URL: https://www.gurit.com/en/our-business/composite-materials/structural-core-materials, 15.10.2021

[URL21j] N.N.: *Heraeus Noblelight: Infrarot Zwillingsrohrstrahler Goldene 8.* URL: https://www.heraeus.com/de/hng/products_and_solutions/infrared_emitters_and_systems/golden_8_infrared_emitters/twin_tube_emitters_golden_8.html, 14.10.2021

[URL21k] N.N.: *kiweb: POLYMERPREISE - Kunststoff Information.* URL: https://www.kiweb.de/Default.aspx?pageid=399docid=247898query=(BAT:pet%20AND %20BAT:preis), 14.10.2021

[URL21l] N.N.: *Leister: Beratung Krelus Infrarottechnologie von Leister.* URL: https://www.leister.com/de/Services/Industrial-Consulting/Consulting-Infrared-Technology, 14.10.2021

[URL21m] N.N.: *marketsandmarkets: PET Foam Market Global Forecast to 2025 / MarketsandMarkets.* URL: https://www.marketsandmarkets.com/Market-Reports/pet-foam-market-19032949.html, 14.10.2021

[URL21n] N.N.: *SKZ: Terahertz-Technologie.* URL: https://www.skz.de/forschung/aktuelles/terahertz-technologie, 14.10.2021

[URL21o] N.N.: *Statista: Tiseo, I.: PET demand worldwide 2030.* URL: https://www.statista.com/statistics/1128658/polyethylene-terephthalate-demand-worldwide/, 14.10.2021

[URL21p] N.N.: *Statista: Wechselkurs US-Dollar gegenüber Euro - Jährliche Entwicklung.* URL: https://de.statista.com/statistik/daten/studie/254757/umfrage/wechselkurs-des-us-dollars-gegenueber-dem-euro-jahresmittelwerte/, 17.10.2021

Symbols

Abbreviations

Notation	Description
FIR	Far infrared
IR	Infrared
MIR	Mid range infrared
NIR	Near infrared
PET	Polyethylene terephthalate
PTFE	Polytetrafluoroethylene

2 Welding of polypropylene post-consumer recycled material

Ch. Hopmann[1], R. Dahlmann[1], M. Weihermüller[1]
[1]Institute for Plastics Processing (IKV) at RWTH Aachen University

Abstract

In order to increase confidence in the use of post-consumer materials (PCRs) for applications with high value-added potential, such as injection moulding applications with subsequent joining processes (e.g. welding) an investigation was carried out into the welding behaviour of a PP-PCR in ultrasonic and hot plate welding. In the ultrasonic welding tests, the welding force and the welding amplitude were varied in three steps. During the hot plate welding trials, the joining force, the temperature of the hot plate and the heating time were varied in two steps with central point. The material was welded to itself as well as to virgin material in order to consider possible combinations in a structural unit. The results were evaluated with regard to the achievable short-time tensile strength.

During the investigation, the PP-PCR was successfully welded to itself as well as to virgin PP in each of the two processes. The maximum welding factors were in both processes at a similar level to those of the virgin material. However, the maximum weld strength achievable in hot plate welding (17.2 MPa) is significantly higher than that in ultrasonic welding (10.9 MPa). In addition, the PCR reacted significantly more sensitively to the variation of the considered process parameters in ultrasonic welding than in hot plate welding, which indicates a more stable processing and welding behaviour in the hot plate welding process. It was also remarkable that in both processes the weld strengths of the PCR material showed lower standard deviations than the virgin material.

2.1 Introduction

Over the past decades, plastics have found their way into almost all areas of life and, due to their broad spectrum of properties, make an important contribution to overcoming a wide range of social challenges. However, the success of polymer materials and the associated market expansion has led to increasing risks for humans and nature [BHM19]. In particular, the disposal and handling of the waste produced represents a major challenge [BHM19]. In addition, the current level of reuse of plastic waste leaves an enormous economic potential unused [NN16].

However, the substitution of virgin material by PCR is not possible without further investigation, as the property profile and the associated processing and subsequent processing behaviour as well as the resulting part properties of PCR can differ from virgin material [BPC17, JM03, LKF06, VCC+20]. For example, the length of the molecular chains or the number of cross-branches can change due to the previous history or thermal load in the recycling process and influence the mechanical properties [BA06, BPC17]. Furthermore, contamination of the material with foreign substances (e.g. other polymer types, organic and inorganic components, metals or paper) cannot be ruled out [HCF20, URL19]. Furthermore, the property profile of PCR is subject to variations depending on the input materials and individual history of the components. Due to these differences and fluctuations, the design process as well as the processing methods and parameters must be adapted [RKA20]. So far, however, there are only insufficient findings and recommendations for action to compensate the changed property profile of PCR and the influences on process control, which inhibits its widespread use on the market. In particular, the effects on subsequent processing properties, such as the welding behaviour, have hardly been considered so far. However, it cannot be ruled out that the altered polymer structure (e.g. chain length, number of cross-branches, molecular weight distribution) or foreign substances have an impact on the welding process. For example, foreign substances can negatively influence the sound propagation during ultrasonic welding or the interdiffusion of the polymer chains in the joining zone. Furthermore, the changed polymer structure and morphology could have an influence on the welding process via the thermal and caloric material properties. For example, the energy input during welding must be adapted to the melting and decomposition temperatures in order to prevent degradation of the material. For this reason, the welding behaviour of a commercially available PCR polypropylene (PP) with two different welding method is investigated and compared to a virgin PP in this paper. In order to enable a fast industrial application of the results and to support an increase in the application quantities of PCRs, established series welding processes of the plastics processing industry such as ultrasonic and hot plate welding are considered first. The consideration of welding processes with different mechanisms of heat input (heat conduction, internal friction) serves to identify a welding process with the help of which a stable welding process and reproducible welding results can be achieved despite the changed property profile.

2.2 Materials and welding processes

PCRs are produced from household waste (yellow bag). The waste is sorted, shredded, washed, finely sorted, dried and then extruded to produce new granulate [RKA20]. The recycling of post-consumer plastic waste and its material composition are specified in DIN ISO/TR 17098 [NN14]. The welding behaviour of recyclates (especially PCR) has hardly been considered so far. There are only very few papers in the literature that have addressed this topic. One example is the study by *Ghasemi et al.* In the context of the investigation, the weldability of a PCR polyamide 6 in the vibration welding process was examined. A drastic reduction in the weld strength of up to 70% was found, which was attributed to impurities with PP [GMBK15].

Ultrasonic welding is one of the established serial welding processes in the plastics processing industry and is used in almost all branches of industry, such as the automotive sector, medical technology or the packaging sector [Ehr04, Pot04, Rot09]. In addition to the wide range of applications, the process is characterised by extremely short welding times, partially less than one second, as well as relatively low investment costs, high automation potential and low energy consumption compared to other welding processes [Nat10, NN08, Rot09].

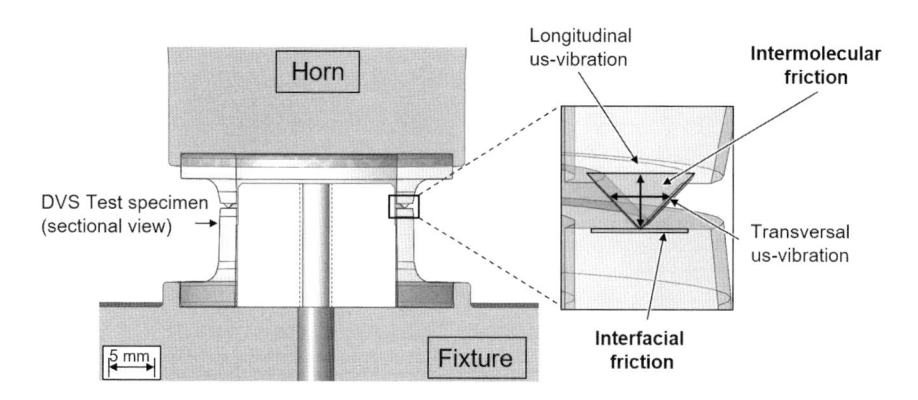

Figure 2.1: Heating mechanisms during ultrasonic welding of plastics

Ultrasonic welding is a one-step welding process, as the heating and joining process is not separated in time. In ultrasonic welding, the joining partners are subjected to mechanical vibrations (longitudinal) at a frequency of typically 20 - 70 kHz and an amplitude in the range of 10 - 40 µm via the welding tool, the so-called horn, and welded under pressure [Ehr04, Pot04]. The heating mechanism is based on the one hand on the interfacial friction of the two joining partners and on the other hand on friction and damping processes at the molecular level, see Figure 2.1 [Pot72]. Neglecting the interfacial friction, the heat development during ultrasonic welding can be described by Equation 2.1 [GB07]:

$$Q = \frac{\omega * A_0^2 * E''}{2} \qquad (2.1)$$

Q is the internal heat generation [W/m^3], ω is the welding frequency [radius/sec], A$_0$ is the amplitude and E" is the loss modulus of the welded material [GB07].

Besides ultrasonic welding, hot plate welding is one of the most widespread welding methods and represents a simple, reliable and economical way of producing high-strength weld seams [GBP03]. The joining partners are heated by contact (heat conduction) or by thermal radiation by means of a heated heating element and then joined under pressure [Pot04]. Within the scope of this study, only classical hot plate welding based on heat conduction is considered. There are almost no restrictions on the process with regard to the product geometries and sizes as well as the plastics that can be welded. Hot plate welding is a two-stage welding process, as the heating and joining processes are separated from each other in terms of time. Thus it enables the joining of different plastics, if physically possible, as the heating temperatures and times can be set separately for the two joining partners [Pot04]. A disadvantage of the process are the comparatively long cycle times. The process sequence begins after the joining partners have been placed in the fixtures with the contact of the joining surfaces on the heating element, see Figure 2.2. The process is divided into the heating, changeover and joining phases. In the heating phase, a distinction is also made between the initial matching phase under pressure and subsequent pressure less heating.

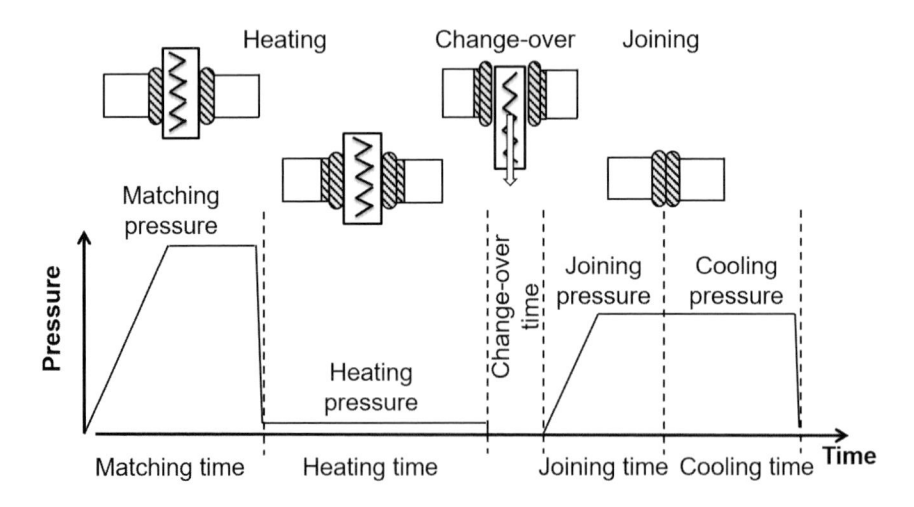

Figure 2.2: Schematic overview of the process sequence for hot plate welding

2.3 Approach, Material, Equipment and Methods

In order to investigate the weldability of PCR with the different welding techniques respectively the different mechanism of heat input, welding tests and subsequently short-time tensile tests are carried out with a commercially available PP-PCR and compared with the results of test specimens made of virgin material. Furthermore, the recycled material is also welded with virgin material in order to explore possible combinations in a structural unit. Ultrasonic welding was chosen because it is a process that reacts very sensitively to changes in the material properties. In addition, hot plate welding is considered a relatively robust process.

Investigated Materials

The post-consumer recycled material selected is a commercially available polypropylene (PP) of the type Systalen PP-C44000 from Systec Plastics GmbH, Cologne, Germany. The material consists of 100% post-consumer packaging and is particularly suitable for injection moulding applications. As the virgin material a polypropylene (PP) type 579S from Sabic, Riyadh, Saudi Arabia is used. Some material properties of the test materials can be found in Table 2.1 below.

Properties	Unit	PP-C44000	PP 579S
Density	[g/cm^3]	0.92	0.91
Tensile strength	[MPa]	25 (EN ISO 527-1/2)	35 (ASTM D638)
Modulus of elasticity	[MPa]	1200 (EN ISO 527-1/2)	1900 (ASTM D790 A)
MFR (230°C/ 2.16 Kg)	[g/10 min]	12	47

Table 2.1: Properties of investigates materials [NN19, NN20]

Used Equipment and Process Parameters for Specimen Manufacturing

For carrying out the ultrasonic welding trials the DVS test specimens with energy director according to DVS guideline 2216-1 [NN12] are used as test specimens. The joining partners for the subsequent welding processes are produced by injection moulding on an e-motion 160/440 injection moulding machine from Engel Austria GmbH, Schwertberg, Austria.

The ultrasonic welding tests are carried out with a welding frequency of 20 kHz on a HiQ Dialog 20/6200 from Herrmann Ultraschalltechnik GmbH Co.KG, Karlsbad, Germany. The welding tests are performed with the specification of a joining path of 0.7 mm (path-controlled) in order to ensure complete melting of the energy director (height 0.6 mm) and complete welding. During the tests, the amplitude and the joining force are varied in three steps. The variation of the welding parameters can be taken from Table 2.2.

Parameters	Unit	Variation								
Joining force	[N]	300			400			500		
Amplitude	[µm]	30	35	38	30	35	38	30	35	38
Experimental setup	[-]	U1	U2	U3	U4	U5	U6	U7	U8	U9

Table 2.2: Experimental setup for ultrasonic welding

The trigger force was kept constant at 100 N as well as the holding time (0,1 s). The holding force was adjusted to the respective level of the welding force.

The hot plate welding tests were carried out on injection moulded plate specimens (80 x 80 x 2 mm^3). The tests were carried out on a hot plate welding machine type SM0500 from Wegener International GmbH, Eschweiler, Germany. Within the scope of the investigation, the heating element temperature, the heating time and the joining pressure are varied in two stages with a central point (cp), see Table 2.3.

Parameters	Heat element temperature	Heating time	Joining pressure	Experimental setup
Unit	[°C]	[s]	[MPa]	[-]
Variation	200	30	0.08	HP1
			0.12	HP2
		50	0.08	HP3
			0.12	HP4
	220	30	0.08	HP5
			0.12	HP6
		50	0.08	HP7
			0.12	HP8
	210	40	0.10	CP

Table 2.3: Experimental setup for hot plate welding

In accordance with DIN EN ISO 527-2, test specimens with a width of 15 mm are sawn from the butt-welded plates. To exclude edge effects, the edge areas with a width of 10 mm are removed. In order to generate targeted failure in the weld seam, the test specimens are provided with a drill hole with a diameter of d = 3 mm placed in the centre of the weld seam in accordance with DVS Guideline 2203-2 [NN10].

Test Methods

To determine the mechanical properties (tensile strength) of the welded specimens, short-time tensile tests are carried out on a universal testing machine of type Z10 from Zwick GmbH Co KG, Ulm, Germany. Per test point 8 specimens are tested. In order to convert the determined tensile strength into a breaking stress, microscopic images of the weld seams are taken by microscope VHX-5000 from Keyence, Osaka, Japan, and the weld seam width is measured.

2.4 Results

In the following, the results from the short-term tensile tests of the ultrasonic test specimens made of virgin material and the test specimens made of recycled material are first presented and compared. All test specimens examined failed in the weld seam.

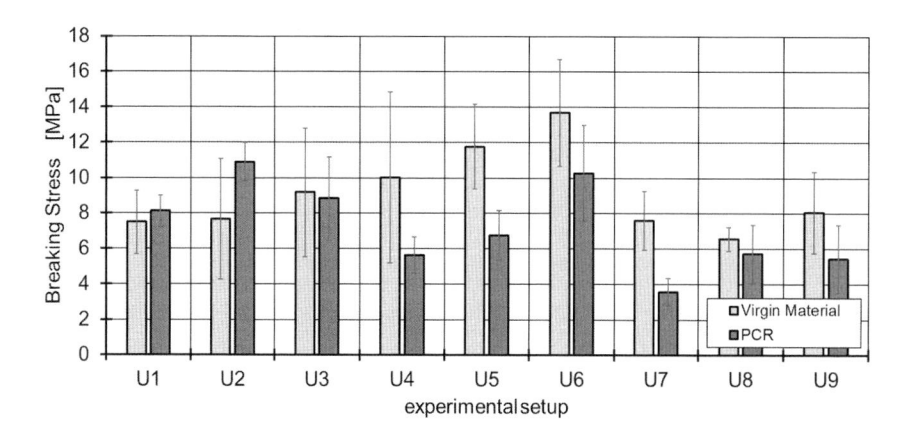

Figure 2.3: Breaking stress of the weld seam in dependency of the investigated welding parameters in ultrasonic welding

As can be seen in Figure 2.3, the breaking stress of the welded test specimens made of virgin material tends to increase slightly with increasing amplitude by a joining force of 300 N (process point 1, 2, 3). This trend continues with increasing joining force from 300 N to 400 N and leads to a maximum breaking stress of 13.7 MPa at an amplitude of 38 µm and a joining force of 400 N (process point 6). With a further increase in the joining force to 500 N, a clear drop in the breaking stress occurs. Furthermore, the results at a joining force of 500 N are almost independent of the amplitudes considered, so although the highest breaking stress is achieved at the largest amplitude (38 µm), the strength level of the other two process point (7 and 8) are in the range of the standard deviation.

The increasing breaking stress with increasing amplitude can be explained by a higher energy input into the test specimen. Since the heat generated in the material is quadratically dependent on the strain or compression of the molecules caused by the welding amplitude [GB07]. Due to the increasing energy input, higher temperatures can be expected in the welding process and thus an improved possibility for interdiffusion of the molecular chains. The decrease of the breaking stress with a further increase of the joining force could be due to a greater deformation of the tip of the energy director and the resulting larger contact area respectively the reduced sound focusing in the interface of the two joining partners, which reduces the heat development in the material. Another explanation could be the increasing and accelerated expulsion of the melt in the joining seam area with increasing welding force. The accelerated flow movement results in a stronger orientation of the molecular chains along the flow direction and thus transverse to the load direction.

When looking at the breaking stress of the recycled test specimens, the previously clearly established correlation between increasing amplitude and increasing breaking force no longer seems to be as pronounced as with the test specimens made of virgin material. For all selected joining force levels, the lowest breaking stress is found at the smallest amplitude (30 µm, process point 1, 4 and 7). However, only in the middle force range of 400 N is a clear increase in the breaking stress with increasing amplitude over all amplitude levels (30 µm, 35 µm, 38 µm). At a joining force of 300 N and 500 N, respectively, the highest breaking stress is achieved at a mean amplitude of 35 µm (process point 2 and 8). The highest breaking stress of 10.9 MPa of the investigated recycled specimens is achieved at a joining force of 300 N and an amplitude of 35 µm (process point 2). Furthermore, a high breaking stress of 10.2 MPa is detected at a joining force of 400 N and an amplitude of 38 µm and thus at the same test point at which the highest breaking stresses of the test specimens made of virgin material (13.7 MPa) were detected. However, the process point is associated with a significantly higher standard deviation than with a joining force of 300 N and an amplitude of 35 µm. Overall, when switching from virgin to recycled material, there seems to be a shift of the optimal process window, with regard to the maximum achievable breaking stress, towards lower joining forces. One explanation for this could be the lower stiffness of the PCR. Due to the lower stiffness, the specimen respectively the energy director is more compressed and deformed with the same welding force than the test specimens made of virgin material. This means on the one hand, that the specified joining distance is reached more quickly and a smaller amount of energy is introduced into the weld seam during the welding process. On the other hand, a higher deformation of the tip of the energy director lets to a bigger contact area between the welding partners, which reduces the focussing of sound and hinders the heat development in the material, as already mentioned. In addition, the breaking stress of the recycled specimens is more dependent on the welding parameters considered (amplitude, joining force). Thus, the maximum and minimum values of the breaking stress achieved for the recycled test specimens are up to 68% (related to the maximum value) different in the considered test area compared to 52% for the test specimens made of virgin material. This may indicate a smaller optimal process window when welding recycled materials in the ultrasonic welding process compared to virgin materials. But it is also noticeable, that at almost all process points, with the exception of a joining force of 500 N and an amplitude of 35 µm (process point 8), a lower standard deviation is recorded when welding recycled material compared to virgin material. Furthermore, it is surprising that the PCR material has a comparable or even a slightly higher welding factor (max. breaking stress related to the base material strength). The maximum welding factor of the virgin material is 0.39 in the parameter sets considered and 0.44 for the recycled material, which speaks for a comparable weldability of the PCR material to the virgin material in the ultrasonic welding process.

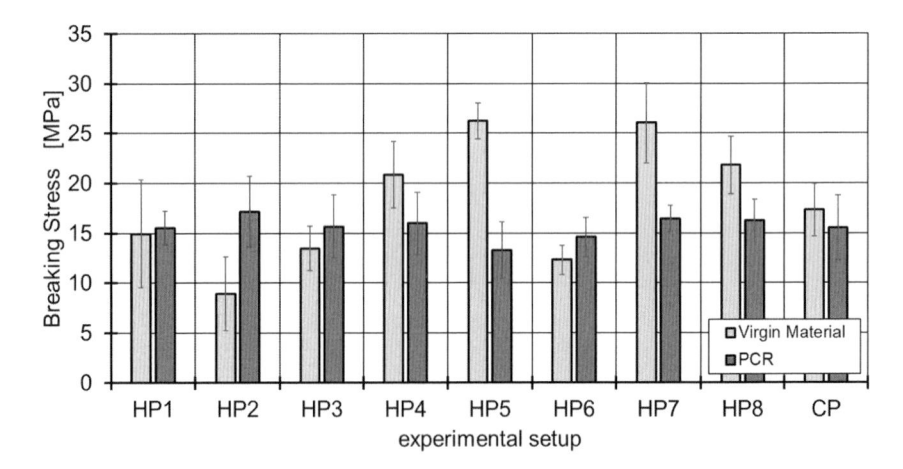

Figure 2.4: Breaking stress of the weld seam in dependency of the investigated welding parameters in Hot plat welding

When considering the results of the test specimens joined by hot plate welding, it is interesting to note that the influence of the parameter variation in the experimental setup under consideration has a significantly lower effect on the achievable breaking stresses of the recycled material than on those of the virgin material, see Figure 2.4. In the case of the virgin material, the difference between the maximum and minimum breaking stress is

66% (based on the maximum value) compared to 23% for the recycled material. For the test specimens made of virgin material, an increase in the energy input (temperature and time) leads to an increase in the achievable breaking stresses at low (see process points 3, 5 and 7, except process point 1) as well as at higher joining pressures (see process points 2, 4 and 8, except process point 6). The maximum breaking stress of the virgin material is 26.2 MPa and is achieved at a heating element temperature of 220°C, a heating time of 50 s and a joining pressure of 0,08 MPa (process point 7). The trend towards higher breaking stresses with increasing energy input is not observed with PCR. The maximum breaking strength (17,2 MPa) of the PCR is achieved at the lowest energy input with a heating element temperature of 200°C and a heating time of 30 s (process point 2). Related to the tensile strengths of the unwelded base materials, this leads to very similar welding factors. In the case of the virgin material, a maximum welding factor of 0.75 is achieved, and in the case of the recycled material, 0.69. That points again towards a comparable weldability of the two materials considered. In combination with the clearly low effects of the parameter variation on the achievable breaking stresses of the PCR, this indicates a good welding behaviour of the PCR and a wide process window in the hot plate welding process. As with the ultrasonically welded specimens, the recycled material has on average slightly smaller standard deviations compared to the virgin material.

In addition to the welding tests with both joining partners made of virgin material or recycled material, welding tests with both processes (ultrasonic and hot plate welding) were also carried out with one joining partner made of PCR and one joining partner made of virgin material (VM), to investigate possible combinations of the two materials (virgin and PCR) in a structural unit. In the case of ultrasonic welding, the positioning of the test specimens in relation to each other was varied as well. Accordingly, tests were carried out in which the joining partner made of virgin material was positioned above the joining partner made of recycled material (VM up - PCR down) and vice versa (RM up - PCR down).

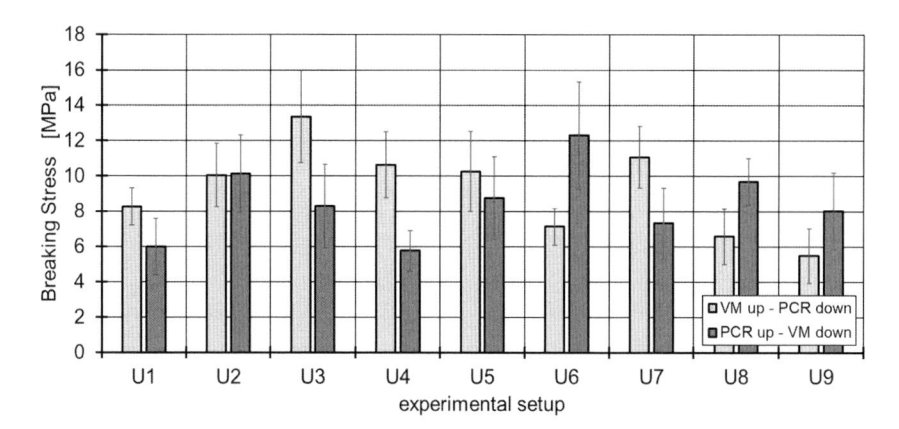

Figure 2.5: Breaking stress of the weld seam of the material combination in dependency of the investigated welding parameters in ultrasonic welding

All in all, no unambiguous statement can be made about the influence of the varied welding parameters in the examined process window, as can been seen in Figure 2.5. Although the stress at breaking of the material combination VM up - RM down clearly increases with increasing amplitude at a joining force of 300 N, a trend in the opposite direction can be seen when the joining force is increased to 400 N or 500 N, which also could possibly be connected to the lower stiffness of the recycled material. The maximum breaking stress of the material combination VM up - RM down is 13.4 MP, achieved at a joining force of 300 N and an amplitude of 38 µm (process point 3) and is thus at the level of the welded test specimens made of pure virgin material.

For the material combination RM up - VM down, a maximum breaking stress of 12.3 MPa is achieved at a joining force of 400 N and an amplitude of 38 µm and is thus at the same level as the results of the test specimens made of recycled material. This could indicate that the maximum achievable strength level of the material combination is rather dependent on the material of the upper joining partner. A possible explanation could be the sound conductivity and the resulting energy input into the joining zone. However, due to the partly small differences in strength and the existing standard deviations, these statements must be viewed very critically. Nevertheless, it can be stated that the strength level of the material combinations lies between the levels of the fracture strengths of the pure recycled material and the virgin material and thus a basic welding suitability for the ultrasonic welding process is given.

The results of the material combination in hot plate welding are presented below, see Figure 2.6:

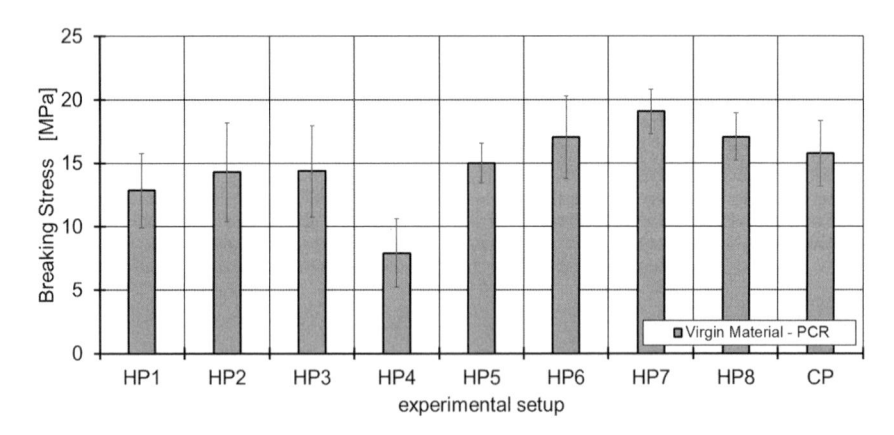

Figure 2.6: Breaking stress of the weld seam of the material combination in dependency of the investigated welding parameters in hot plate welding

As with the hot plate welding tests with the pure virgin material, an increase in the energy input tends to lead to an increase in the achievable fracture strength at low joining pressure (process points 1, 3, 5 and 7) as well as at the higher joining pressure (process points 2, 6 and 8, except 4). The highest fracture strength is thus again achieved at process point 7 (heating element temperature 220°C, heating time 50 s, joining pressure 0,08 N/mm2) with a value of 19.1 MPa. The average strength level is approximately in the range of the pure PCR welds and is thus, as expected, oriented towards the weaker joining partner. The standard deviations are around the standard deviations of the pure virgin material and thus above the level of the pure PCR. However, it can also be stated here that the material combination is as well suitable for hot plate welding in principle and even higher weld strengths can be achieved than in ultrasonic welding.

2.5 Conclusions and outlook

In the course of the investigation, it was found, that the recycled PP post-consumer material under consideration can be joined with itself as well as with PP virgin material in the ultrasonic and hot plate welding process. In the ultrasonic welding process, the weld seam strength of PCR in the short-time tensile test reacts much more sensitively to the variation of the process parameters considered compared to the virgin material. This is especially noticeable in the significantly greater difference between the maximum and minimum weld strengths that can be reached. For the virgin material the difference is 52% (related to the maximum value) and for the PCR 68%. Nevertheless, slightly higher weld factors were achieved for the PCR specimens (0.44) compared to the virgin material (0.39) in the considered process window and this with lower standard deviations on average. In combination with the difference between the achievable weld strengths, this indicates a smaller optimal process window for processing PCR in the ultrasonic welding process compared to virgin material. With regard to the maximum achievable weld strengths, there appears to be a slight shift in the optimal process window towards lower joining forces when switching from the selected virgin material to the PCR. A different picture appears when considering the achievable weld seam strength in hot plate welding. The difference between the maximum and minimum weld strengths in the investigated process window was significantly smaller for the PCR (22%) than for the virgin material (66%). However, with a value of 0.75, slightly higher welding factors were obtained with the virgin material compared to the PCR (0.69). It was noticeable that, as in the ultrasonic welding tests, there were on average smaller standard deviations in the weld strength of the PCR. In summary, the results indicate that the PP-PCR under consideration exhibits good welding behaviour in the hot plate welding process, whereby a slight shift towards lower energy inputs (heating element temperature and heating time) occurs with regard to the maximum achievable weld strength. As already mentioned at the beginning, the combinations of the materials (PCR and Virgin) could also be successfully welded in both processes. In ultrasonic welding, the achievable strength level showed a slight dependence on the upper joining partner, which is particularly responsible for the sound propagation into the joining zone and thus for the heating process. Therefore, slightly higher weld strengths could be achieved if the upper joining partner is made of virgin material. Overall, the strength level of the material combinations lies between the results from pure virgin material and pure PCR. In the case of hot plate welding, however, the achievable strengths are more oriented towards the results from pure PCR test specimens and accordingly the weaker joining partner. Whereby the highest weld seam strength in hot plate welding with a value of 19.1 MPa is clearly above the results in ultrasonic welding (13.4 MPa). This again indicates that although both processes are basically suitable for processing the PP-PCR under consideration, but based on the present results, the heating element process in particular is to be recommended. With regard to the welding factors achieved, it can also be stated that the investigated PCR has comparable welding properties to the virgin material and is therefore not only suitable for the downcycling process, but also for use in applications with a higher value-added potential. However, it would be interesting to take a deeper look at the recycled material in order to be able to make well-founded statements about the welding behaviour and the further processing properties. For example, it would be useful to determine the molecular weight distribution before and after injection moulding and after welding, especially in the weld seam, to get an indication of the narrower standard deviation. Although recycled materials often have low molecular weights, due to the previous thermal stresses, but the distribution of the molecular weight could be narrower and thus could lead to a more uniform heating and flow behaviour and consequently to a more homogeneous welding behaviour. In addition, for a safe and reliable use of the material, the ageing behaviour, the influence of media on the weld strength as well as different load cases, such as shear, bending, impact or tightness, should be investigated depending on the target application. Furthermore, it would be interesting to investigate other welding processes in order to be able to determine a possible sensitivity or preference of the material with regard to the heat input mechanism used. For example, in the case of radiation-based welding processes (laser welding, infrared welding), any foreign substances contained in the material could lead to locally different optical properties (absorption, transmission, reflection) and thus to locally varying heating and welding behaviour along the weld seam. Only when well-founded knowledge of the existing welding processes is available is it possible to select the optimal welding process for series applications of recycled PP post-consumer materials.

References

[BA06] BASFAR, A.; ALI, K.: Natural weathering test for films of various formulations of low density polyethylene (LDPE) and linear low density polyethylene (LLDPE). *Polymer Degradation and Stability* (2006) 91, p. 437–443

[BHM19] BAUR, E.; HARSCH, G.; MONEKE, M.: *Werkstoff-Führer Kunststoffe Eigenschaften - Prüfung - Kennwerte.* 11. Auflage. Munich: Carl Hanser Verlag, 2019

[BPC17] BARBOSA, L.; PIAIA, M.; CENI, G.: Analysis of Impact and Tensile Properties of Recyceld Polypropylene. *International Journal of Materials Engineering* 7 (2017) 6, p. 117–120

[Ehr04] EHRENSTEIN, G.: *Handbuch Kunststoff-Verbindungstechnik.* Munich: Carl Hanser Verlag, 2004

[GB07] GREWELL, D.; BENATAR, A.: Welding of Plastics: Fundamentals and New Developments.. *International Polymer Processing* 22 (2007) 1, p. 43–60

[GBP03] GREWELL, D.; BENATAR, A.; PARK, J.: *Plastics and Composites Welding Handbook.* Munich: Hanser Gardener Publications, 2003

[GMBK15] GHASEMI, H.; MIRZADEH, H.; BATES, P.; KAMAL, M.: Characterization of recycled polyamide 6: Effect of polypropylene and inorganic contaminants on mechanical properties. *Polymer Testing* 42 (2015), p. 69–78

[HCF20] HORODYTSKA, O.; CABANER, A.; FULLANA, A.: Non-Intentionally Added Substances (NIAS) in recyceld plastics. *Chemosphere* 251 (2020),

[JM03] JANSSON, A.; MOLLER, K.: Degradation of post-consumer polypropylene materials exposed to simulated recycling - mechanical properties. *Polymer, Degradation and Stability* 82 (2003), p. 37–46

[LKF06] LUZURIAGA, S.; KOVAROVA, J.; FORTELNY, I.: Degradation of pre-aged polymers exposed to simulated recycling: Properties and thermal stability. *Polymer Degradation and Stability* 82 (2006), p. 1226–1232

[Nat10] NATROP, J.: Systematische Auswahl eines anforderungsgerechten Fügeverfahrens an Praxisbeispielen. *Umdruck zur IKV-Fachtagung: Fügen von Kunststoffen - Technologien für erfolgreiche Verbindungslösungen* (2010),

[NN08] N.N.: Ultraschallschweißen - Runter mit den Nebenzeiten. *Joining Plastics* 2 (2008) 3, p. 180

[NN10] N.N.: *DVS-Richtlinie DVS 2203-2: Prüfen von Schweißverbindungen an Tafeln und Rohren aus thermoplastischen Kunststoffen - Zugversuch.* Düsseldorf: DVS Media GmbH, 2010

[NN12] N.N.: *DVS-Richtlinie 2216-1: Ultraschallschweißen von Kunststoffserienteilen - Prozessbeschreibung, Maschinen und Geräte, Einflussgrößen, Konstruktion, Qualitätssicherung.* Düsseldorf: DVS Media GmbH, 2012

[NN14] N.N.: *DIN ISO/TR 17098: Stoffliche Verwertung von Verpackungsmaterialien - Bericht über Substanzen und Materialien, die die stoffliche Verwertung behindern könnnen.* Berlin: Beuth Verlag, 2014

[NN16] N.N.: *The New Plastics Economy: Rethinking the future of plastics.* : World Economic Forum, Ellen MacArthur Foundation and McKinsey Company, 2016

[NN19] N.N.: *Sabic PP 579S.* : Product information, SABIC, 2019

[NN20] N.N.: *Systalen PP-C44000 GR000.* : Product information, Systec Plastics GmbH, Cologne, 2020

[Pot72]   POTENTE, H.: *Untersuchung der Schweissbarkeit Thermoplastischer Kunststoffe mit Ultraschall.* RWTH Aachen University, Dissertation, 1972

[Pot04]   POTENTE, H.: *Fügen von Kunststoffen - Grundlagen, Verfahren, Anwendungen.* Munich: Carl Hanser Verlag, 2004

[RKA20]   RUDOLPH, N.; KIESEL, R.; AUMNATE, C.: *Einführung Kunststoffrecycling - Ökonomische, Ökologische und technische Aspekte der Kunststoffabfallverwertung.* Munich: Carl Hanser Verlag, 2020

[Rot09]   ROTHEISER, J.: *Joining of Plastics - Handbook for Designers and Engineers.* Munich: Carl Hanser Verlag, 2009

[URL19]   N.N.: *Ettlinger: Schmelzefilter mit Mikroperforation für PET.* URL: https://www.k-aktuell.de/technologie/ettlinger-schmelzefilter-mit-mikroperforation-fuer-pet-recycling-65715/, 15.09.2021

[VCC+20] VELZEN, E.; CHU, S.; CHACON, F.; BROUWER, M.; MOLENVELD, K.: The impact of impurities on the mechanical properties of recycled polyethylene. *Packaging Technology and Science* 34 (2020) 4, p. 219–228

Symbols

Symbol	Unit	Description
A_0	m	Amplitude
E''	MPa	Loss modulus
ω	$radius/sec$	Welding frequency
Q	W/m^3	Loss modulus

Abbreviations

Notation	Description
DVS	Deutscher Verband für Schweißen und verwandte Verfahren
EU	European Union
PCR	Post-consumer recyclate
PEEK	Polyether ether ketone
PMMA	Polymethyl methacrylate
PP	Polypropylene
RM	Recycled material
VM	Virgin material

3 Welding of TPE-TP composites: Development of strategies to produce aging-resistant welds with reduced anisotropy

Ch. Hopmann[1], R. Dahlmann[1], P. Knupe-Wolfgang[1], S. Bölle[1], Ch. Kley[1], J. Wipperfürth[1]

[1] Institute for Plastics Processing (IKV) at RWTH Aachen University

Abstract

Thermoplastics and thermoplastic elastomer composite copolymers (TP-TPE) are often multi-injection moulded by overmoulding the TP-component with TPE. During processing a high anisotropy of the part that influences mechanical properties such as elongation and strength evolves. Compared to multicomponent injection moulding welding may have advantages, since the re-melting of the TPEs in a welding process offers the possibility of resetting the orientations and thus reducing the anisotropy of the mechanical properties in the area of the weld seam. In addition, welding of TP-TPE components has a higher economic potential regarding complex joints. For this reason, the aim of this study is to investigate the weldability of TP-TPE components and to investigate the resulting mechanical properties. At the example of laser transmission welding it can be shown that TP and TPE components can be well welded. However, a significant reduction of the anisotropy cannot be observed in standard laser transmission welding.

3.1 Introduction

The demand for plastics with pronounced rubber-like properties, such as high elasticity, stretchability and good recovery behaviour, is steadily increasing. With annual global growth rates of over 6%, thermoplastic elastomers (TPE) are among the most important innovative polymer materials [Ell13]. In 2015, the main field of application for TPEs was in the transportation industry (of which 72% was in the automotive industry), at around 60%, and this is where they will continue to be seen in the future [BDS11, Ell15, Ell16, Geb12]. Other areas of application are in the electrical/electronics (14 %), construction (11 %), medical, food and hygiene (8 %), consumer, leisure and sports (3 %), and other sectors (3 %) [Ell16]. In the future, as a result of active development in TPE materials, e.g. development of "high-performance TPE", "super-TPE-V", "ready-to-use TPV" or "medical-grade TPE" [Fri08, Gan15, Kne12, Ose08], a further substitution of classical elastomers by TPE can be assumed. An increase in the global processing volume of TPE by approx. 30 % is predicted for the period from 2015 to 2020 [Ell16].

Like thermoplastics (TP), thermoplastic elastomers (TPE) are melted for processing and do not need to be vulcanized after shaping. Compared to processing elastomers, this results in lower energy consumption and shorter cycle times. The resulting lower production costs are an incentive to use TPEs over classic elastomers. Due to their good adhesion properties to polyolefins and through modification to almost all technical TPEs, TPEs are particularly suitable for multicomponent technology. In this way, components made of soft TPE and hard TP can be manufactured with a high degree of functional integration and the assembly effort can be reduced. Further advantages of TPE are the lower density compared to elastomers, the recyclability, and the high freedom of design [KS15, Ose06, RS13, Ven05]. On the other hand, there are some disadvantages that must be taken into account in the application and design of products made of TPE. These include the low temperature resistance, the plasticity of the material, the low media resistance compared to rubber compounds made from special elastomers, and the high material costs [KS15, Ose06, RS13, Ven05].

Due to the advantages and disadvantages, TPE and TPE-TP composites are used wherever the lower properties of TPE compared to elastomers do not impair the functionality of the component. Areas of application include the automotive sector, household appliances and medical technology. Specific applications include car body seals, bellows and roller bellows, dosing systems, children's dishes, seals for freshness boxes, respiratory masks or medical membranes [Gan15, Geb12, RS13].

The majority of applications for TPE-TP composites are exposed to permanent temperature and media influence by water, oil or medical preparate. The bond between TPE and TP represents a potential weak point in the component. For an economical use of TPE-TP composites, the resistance of the joint to temperature and media is essential. The manufacturing process plays a decisive role here. The standard for TPE-TP composites is multicomponent injection moulding, in which a bond is achieved by overmoulding or gating the TP component with TPE [DGN07, ML04, OS99, SWSH14, WH09]. A disadvantage of multi-component injection moulding is the resulting anisotropy of the mechanical properties (e.g. strength and elongation) due to the high shear stress on the melt during the process. [BKW+16, Hop16, RS13].

The directional dependence of the mechanical properties can lead to a weak point of the composite and thus to premature failure if it is frequently loaded perpendicular to the orientation. Furthermore, for many applications of TPE-TP composites, such as seals or handles, there is no clear preferred loading direction, so the reduced mechanical properties perpendicular to the orientation have a detrimental effect on the application. It must therefore be designed against failure perpendicular to the orientation, which results in over-dimensioning and thus the unneccassary use of material.

Alternative manufacturing processes to multicomponent injection moulding are adhesive bonding and welding processes. Compared to multicomponent injection moulding, the re-melting of the TPEs in a welding process also offers the possibility of resetting the orientations and thus reducing the anisotropy of the mechanical properties in the area of the weld seam. Therefore, the aim of this study is to develop process control strategies for laser transmission welding of TPE with TP to produce aging-resistant welds with reduced anisotropy of the mechanical properties. The melting and cooling in the welding process shall be used specifically to reduce the anisotropy. The process control is to be designed in such a way that an aging-resistant weld is obtained to ensure the economic viability of the application.

3.2 Processing of TPE

The term TPE covers an entire family of materials that can be divided into two groups based on their molecular structure: Block copolymers and polymer blends (see Figure 3.1) [Ven07].

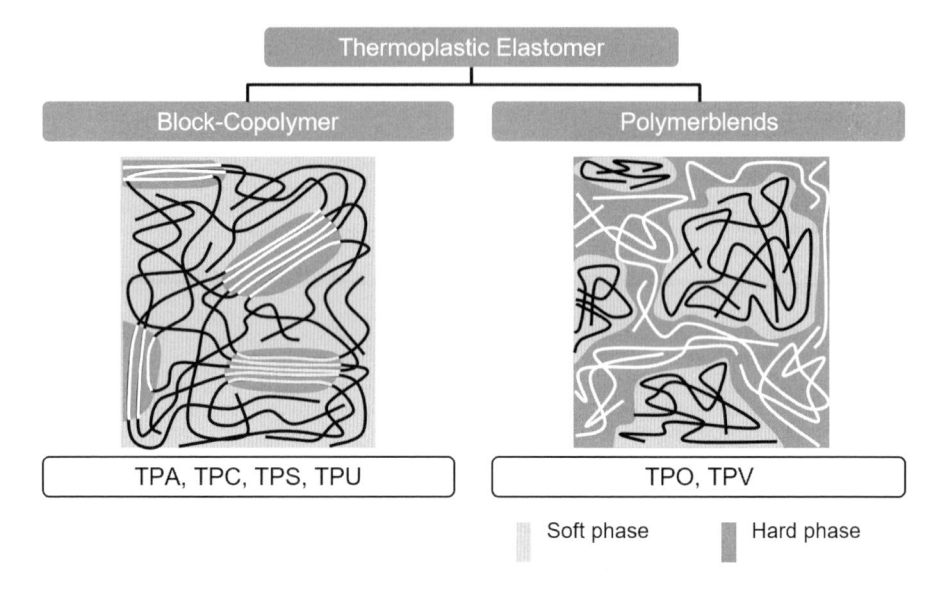

Figure 3.1: Classification of thermoplastic elastomers

In the group of block copolymers, soft (elastomeric) and hard (thermoplastic) segments are chemically bonded in one molecule. The different segments are incompatible with each other. Local demixing leads to the formation of hard segments, which act as physical cross-linking points in the soft matrix. By heating the material, the hard domains are melted and TPE can be processed like TP. The hard segments of the block copolymer thus determine the processing properties, while the soft segments of the block copolymer determine the elastic properties of the material [Nag04]. The main representatives of the block copolymers are the styrene block copolymers (TPE-S), the polyurethanes (TPE-U), the polyether esters (TPE-E) and the polyether amides (TPE-A) [Ven07]. In the context of this paper, polymer blends consist of a thermoplastic matrix material (hard phase) in which uncrosslinked or crosslinked elastomer particles (soft phase) are embedded. This includes thermoplastic polyolefins with uncrosslinked (TPE-O) and crosslinked elastomer phase (TPE-V). TPE materials qualitatively exhibit the same characteristic mechanical material behaviour, but depending on their composition (e.g. ratio of soft to hard phase) their behaviour differs quantitatively [Zys93]. TPEs, especially TPVs, have a pronounced anisotropic mechanical material behaviour after processing in the injection moulding process [LS94, Hop16, KKC87, TM00, WZN+03].

During injection moulding, the processed material is exposed to high forces and cooling gradients. These result in high shear and elongation of the plastic melt, causing the polymer chains and fillers to align (see Figure 3.2.

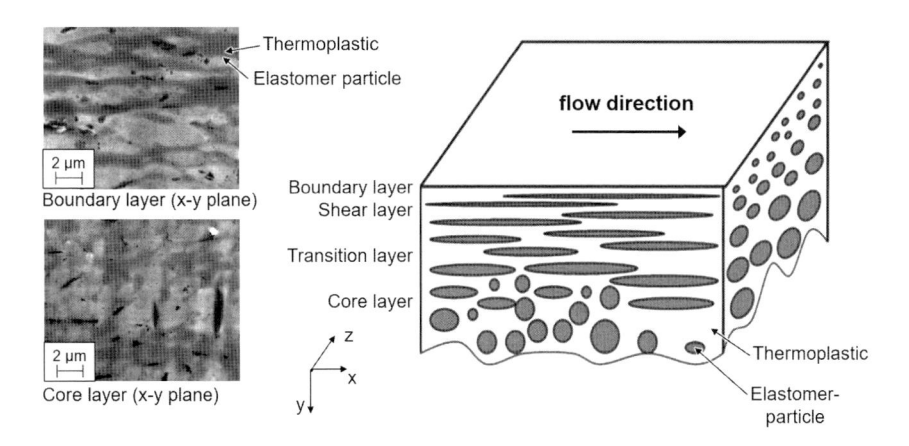

Figure 3.2: Morphology of injection moulded TPV parts

Friction between the molecules also leads to shear heating and a structurally induced reduction in viscosity. The temperature differences between the melt and the mould lead to very high cooling gradients, especially in the edge layer of the moulded part, while in the centre of the moulded part the solidification of the melt is slower. Depending on the polymer, the temperature and deformation gradients, an inhomogeneous and anisotropic morphology is formed by relaxation, solidification and crystallization processes. In the case of polymer blends, the elastomer particles in the surface layer, i.e. at the surface of the moulded part, are strongly deformed in the direction of flow. At the transition to the centre of the moulded part, the deformation of the elastomer particles decreases due to the low velocity and cooling gradients during the injection moulding process. In the centre of the moulded part, the particles are oriented [HB15, HBK16, Hop16]. In the case of block copolymers, the elastomer and thermoplastic segments are also oriented in the flow direction during the injection moulding process [BKW+16, CH04, YK93].

3.3 Welding of thermoplastic elastomers

Compared with multicomponent injection moulding, the re-melting of the TPEs in a welding process also offers the possibility of resetting the orientations and thus reducing the anisotropy of the mechanical properties in the area of the weld seam. With the adhesive joining processes, a more flexible adaptation of the manufacturing process and the plant technology to new geometries or applications is possible. In addition, complex joining geometries can be implemented, which in multi-component injection moulding can only be realised with very expensive special tools. In addition, the downstream joining process enables more targeted influence to be exerted on the joint properties. Compared to adhesive bonding, the advantages of welding include shorter cycle times, no need for surface pre-treatment, and higher media and temperature resistance [DB02, Hab01, Hab90].

Challenges in welding TPE-TP composites arise from the different mechanical, thermal and rheological properties of the two materials as well as the areas of strong orientation in the edge region of TPE moulded parts after the primary forming process. Both represent a significant influence on the strength and residual stresses resulting from the welding process, which exert a strong influence on the aging behaviour of the joints. Furthermore, the welding process is influenced by the speed of heating and cooling, the joining pressure and the single- or multi-stage nature of the process.

In practice, raw material and machine manufacturers generally only provide information on the general weldability of materials and on possible welding processes that can be used for TPE in addition to TP [URL18d, URL18c, URL18a, URL18b]. Furthermore, basic investigations have already been carried out on the plastic series welding processes hot plate, ultrasonic, vibration, high frequency and laser transmission welding, which deal with the weldability of TPE with itself as well as with TP [AAAK16, AOD13, RLK15, RS05, Sch02, Tro08, WBM12]. Both block copolymers and polymer blends were considered as materials on which the maximum achievable weld strengths were determined in tensile and tensile shear tests for the different welding processes. For hot plate welding [Sch02] and laser transmission welding [RLK15], high short-term welding factors (ratio of the tensile strength of the weld to the tensile strength of the base material) of up to approx. 0.8 and 0.75, respectively, could be achieved with respect to the TPE base material. Due to the complex influencing variables and interactions, a reduction in the anisotropy of the mechanical properties and an increase in the aging resistance of welded TPE-TP joints is not possible on the basis of the current state of research and development.

3.4 Materials and test procedure

The aim of the investigations is to verify the influence of the downstream laser welding process on the orientation of the TPEs. The change in orientation caused by the welding process is to be determined with the aid of short-term tensile tests on type 1BA tensile test specimens of the base materials and on welded specimens. The orientation is to be inferred from the changing tensile strengths and elongations at break. Plates measuring 80 mm x 80 mm x 2 mm were produced for the three materials on the injection moulding machine Arburg Allrounder 520A, Arburg GmbH und Co. KG, Germany. Tensile test specimens of type 1BA as well as rectangular specimens for the laser transmission welding tests, which have been cut out of these (see Figure 3.3).

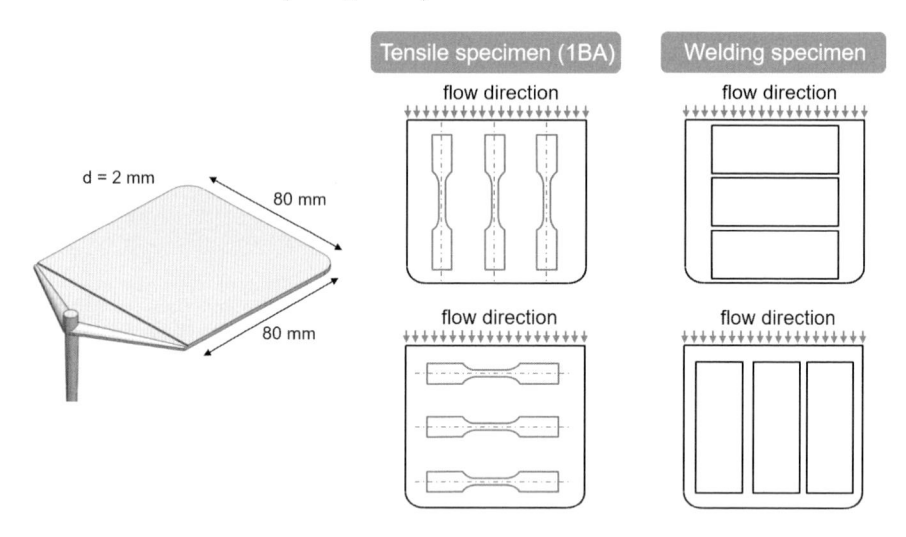

Figure 3.3: Production of tensile and welding specimen by injection moulding

The specimens are taken parallel to the direction of flow as well as perpendicular to the direction of flow, so that different molecular orientations are present within the specimens. To determine the tensile strength and elongation at break of the materials, short-time tensile tests were carried out on the tensile testing machine Z100, Zwick Roell GmbH Co. KG, Germany in accordance with DIN EN ISO 527-1. Using the type 1BA tensile test specimens, the above mechanical properties have been determined for the base materials for orientations in and transverse to the direction of loading. A laser system of type LDF 400-40 from Laserline GmbH, Germany, which can be operated in the power range from 40 W to 400 W and has a wavelength of 940 nm, has been used to produce the weld specimens by laser transmission welding. The line optics used have an irradiation area of 27 mm x 1.5 mm at the focus.

To determine suitable test points, a suitable process window for the materials used was determined in the first step (see Figure 3.4). For this purpose, the irradiation time and laser power have been varied and the weld seam and the welding process have been evaluated. From the suitable points, five process points were selected for each material with which test specimens were produced. The process points were selected in such a way that both the laser power and the irradiation time could be varied in two stages without any decomposition occurring during the welding process.

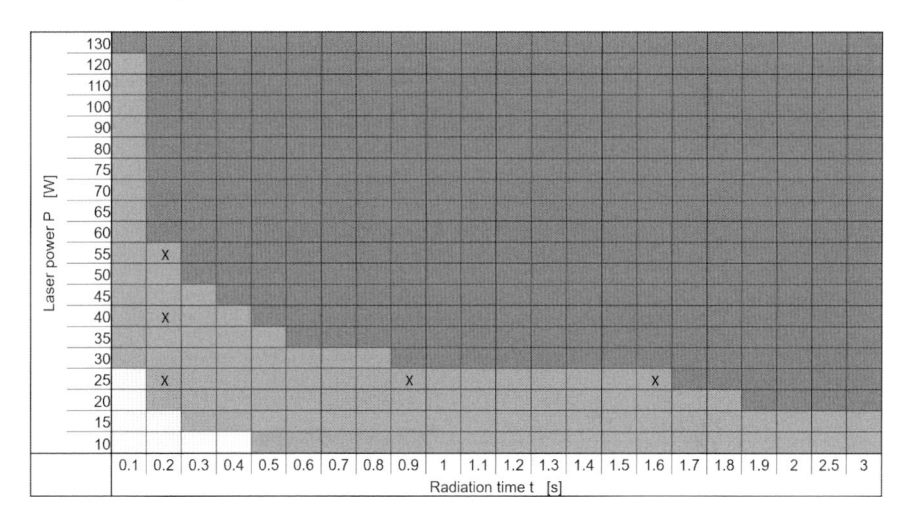

Figure 3.4: Determination of suitable welding parameter

For each selected process point, ten test specimens are produced for the tensile shear tests. In addition, the test specimens are produced in two variations. For one variation, the orientations set in the injection moulding process run longitudinally and once transversely to the flow direction during welding (see Figure 3.5). The type of clamping for the tensile shear tests is also shown in Figure 5. Specimens of the same material and the same orientation are always welded together.

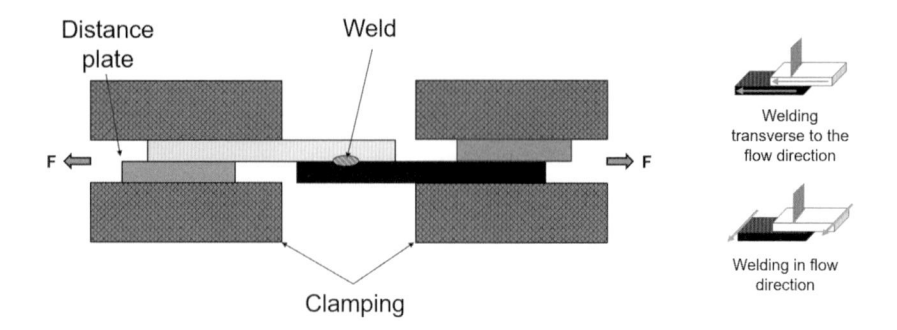

Figure 3.5: Principle of tensile-shear-test

By evaluating the tensile shear tests, it has been verified whether the orientations produced in the injection moulding process could be cancelled out by the welding process. The tests were carried out on the machine Z010TH, Zwick GmbH Co. KG, Ulm, Germany. The results of the tests are described and explained in the next chapter.

Two materials have been used within the scope of the investigations. The Thermolast K TF7AAC Shore A 70 material from KRAIBURG TPE Co.KG GmbH, Waldkraiburg, Germany, serves as the representative for the block polymers. To test the hypothesis for polymer blends, the material Alfafter XL A70L 2GP0050 Shore A70 from Albis Plastic GmbH, Düsseldorf, Germany, is used. For each material, additional tensile test specimens (1BA) and weld test specimens containing 0.3% carbon black have been prepared. Compounding was carried out by a co-rotating intermeshing twin screw extruder type ZSK 26 Mc from Coperion GmbH, Stuttgart, Germany.

3.5 Results

3.5.1 Characterisation of the injection moulded test specimen

Figure 3.6 shows the results for the tensile tests. The tension achieved and the strains at the maximum force achieved are always compared for the two orientations longitudinal and transverse to the tensile direction.

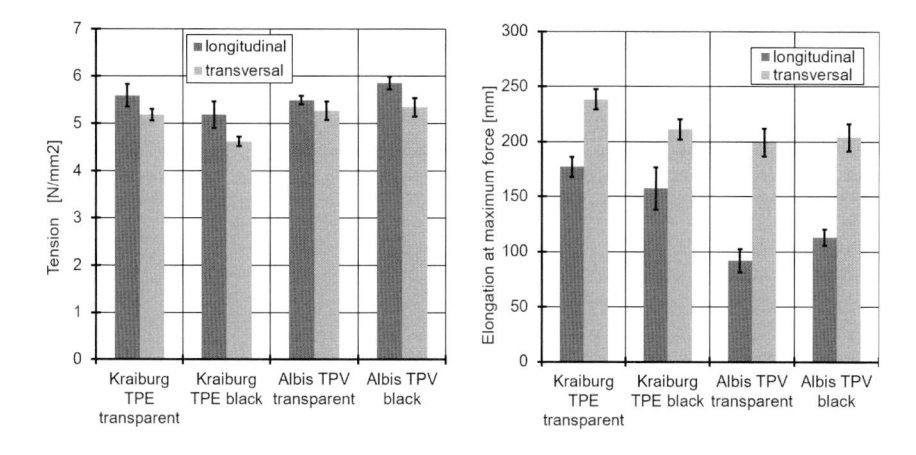

Figure 3.6: Results of tensile tests

For both the block copolymer from KRAIBURG TPE Co.KG GmbH and the polymer blend from Albis Plastic GmbH, the tension achieved for the specimens oriented in the tensile direction is higher than for the specimens oriented transverse to the tensile direction. The differences in the tension are in the range of 10%. This shows that the orientations introduced in the injection moulding process have a clear influence on the strength of the material. The reason for this is that the orientation causes the polymer chains to lie largely in the direction of the load, which means that the stronger covalent bonds have to be broken for the material to fail. In contrast, in the specimens oriented transverse to the tensile direction, the polymer chains also lie for the most part transverse to the tensile load. This means that only the secondary valence forces have to be broken for the material to fail.

The right-hand diagram in Figure 3.6 shows the elongation when the maximum force is reached. For all material combinations, the elongation of the specimens oriented transversely to the tensile direction is higher than the elongation achieved for the specimens oriented longitudinally to the tensile direction. The strain differences are between 30% and 100%.

3.5.2 Characterisation of the weld

Figure 3.7 shows the results of the tensile tests for the welded test specimens with the block copolymer from KRAIBURG TPE Co.KG GmbH.

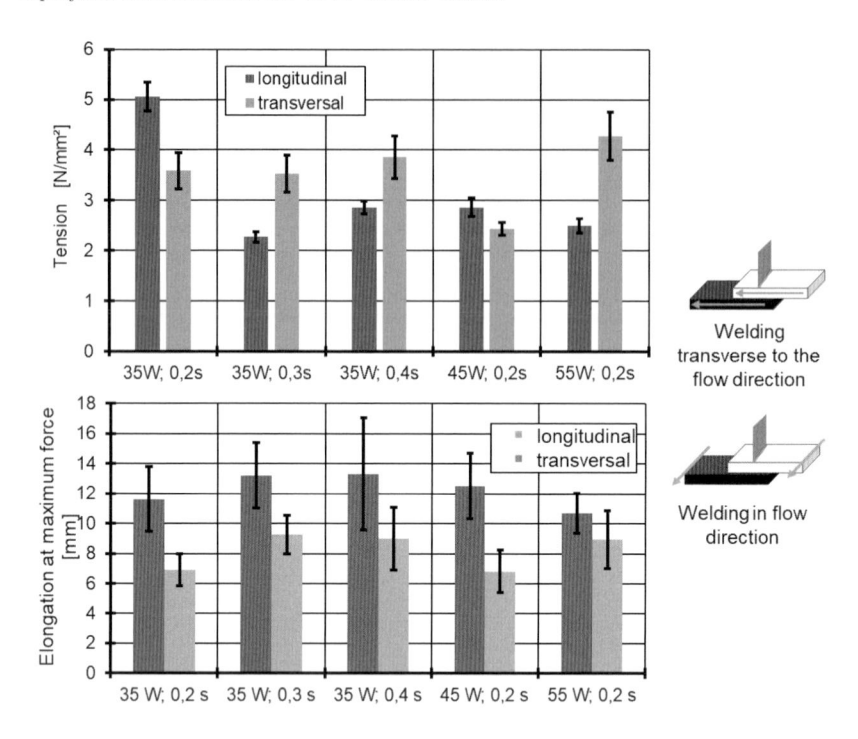

Figure 3.7: Results of tensile-shear-tests for the block-copolymer (Kraiburg)

The specimens oriented transverse to the flow direction during the welding process and thus in the tensile direction achieve higher tension for all process points. As before, this is due to the existing pre-orientation caused by the injection moulding process. The differences in tension are in the range of 10%. This behaviour is also reflected in the elongations achieved at maximum force (see Figure 3.7 bottom). There, the specimens oriented longitudinally to the flow direction during the welding process and thus transverse to the tensile direction achieve the greater strains at maximum force for all process points. The strain differences are between 30% and 100%. This is also due to the existing pre-orientation caused by the injection moulding process. Since the differences in tensions and the strains achieved at maximum force are still of the same order of magnitude after the welding process, it can be concluded that the welding process did not neutralise the orientations introduced in the injection moulding process. If no orientations were left in the material as a result of the welding process, both specimen variants would have to have similar maximum forces and achieved strains.

Figure 3.8 shows the results for the tensile shear tests for the polymer blend used from Albis Plastic GmbH.

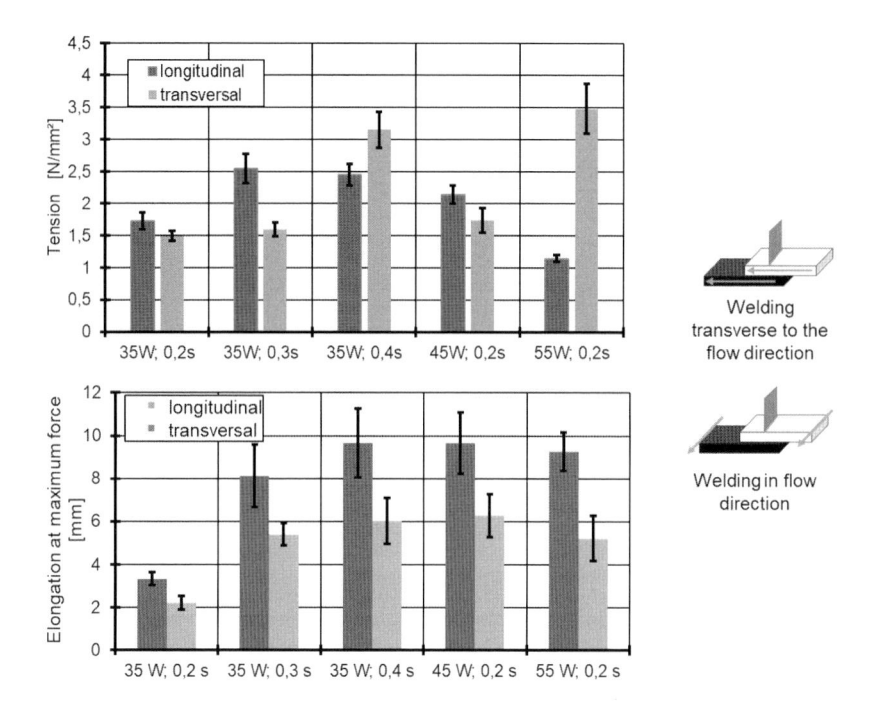

Figure 3.8: Results of tensile-shear-tests for the block-copolymer (Albis)

Even for these welded specimens, there are still differences in the tension and elongations achieved at maximum force after the welding tests. Thus, also for this group of materials, the welding process was not able to dissolve the orientations introduced in the injection moulding process.

3.6 Conclusions and outlook

TP-TPE components are typically processed in multi-injection moulding, which results in a high anisotropy of the mechanical properties. In addition, complex geometries are cost intensive since special tools are required. A potential alternative to multi-injection moulding is welding of single components. On the one hand welding is suitable to join complex parts, which leads to reduction of costs, on the other hand the re-melting of the components can lead to a reduction of the anisotropy. For this reason, in this study the weldability and the resulting mechanical properties of TP-TPE components were determined from laser-transmission welding. Two materials, the Thermolast K TF7AAC Shore A 70 as representative for block polymers as well as the Alfafter XL A70L 2GP0050 Shore A70 as representative for polymer blends are characterised. For the production of the test specimen for tensile and tensile-shear tests plates are injection moulded. From this, test specimens are prepared longitudinally and transversally to the flow direction. In tensile tests the resulting anisotropy could be shown. The results serve as benchmark to the following tensile-shear tests. The tensile-shear test specimens are produced from injection moulding followed by welding the specimens in laser-transmission welding. For laser transmission welding a suitable process window is first determined, to prevent a bad adhesion or a decomposition of the material by varying radiation time and laser power. For both materials also the welded parts show anisotropic effects. The difference is nearly in the same order of magnitude as the anisotropy of the base material itself. It can be concluded that the determined process window is able to produce a good adhesion of the materials but does not reduce the anisotropy of the welded parts.

Acknowledgement

The research project 21296 N of the Forschungsvereinigung Institute for Plastics Processing has been sponsored as part of the "Industrielle Gemeinschaftsforschung und -entwicklung (IGF)" by the German Bundesministerium für Wirtschaft und Klimaschutz (BMWK) due to an enactment of the German Bundestag through the AiF. We would like to extend our thanks to all organizations mentioned.

References

[AAAK16] AL-ALI ALMA'ADEED, M.; KRUPA, I.: *Polyolefin Compounds and Materials – Fundamentals and Industrial Applications*. Cham, Heidelberg, New York, Dordrecht, London: Springer International Publishing AG Switzerland, 2016

[AOD13] ADEN, M.; OTTO, G.; DUWE, C.: Irradiation Strategy for Laser Transmission Welding of Thermoplastics Using High Brilliance Laser Source. *International Polymer Processing* 3 (2013) 28, p. 300–305

[BDS11] BUBAK, B.; DECKERS, E.; SCHRAUWEN, R.: TPVs in automotive. *TPE Magazine* 2 (2011) 1, p. 28–33

[BKW+16] BRUBERT, J.; KRAJEWSKI, S.; WENDEL, H.; NAIR, S.; STASIAK, J.; MOGGRIDGE, G. D.: Hemocompatibility of styrenic block copolymers for use in prosthetic heart valves. *Journal of Material Science: Materials in Medicine* 3 (2016) 27, p. 23–43

[CH04] CASTELLETTO, V.; HAMLEY, I. W.: Morphologies of block copolymer melts. *Current Opinion in Solid State and Material Science* 6 (2004) 8, p. 426–438

[DB02] DILTHEY, U.; BRANDENBURG, A.: *Schweißtechnische Fertigungsverfahren 3 – Gestaltung und Festigkeit von Schweißkonstruktionen*. Berlin, Heidelberg: Springer-Verlag, 2002

[DGN07] DOLANSKY, T.; GEHRINGER, M.; NEUMEIER, H.: *TPE FIBEL – Grundlagen Spritzguß.*. Ratingen: Dr. Gupta Verlag, 2007

[Ell13] ELLIS, P.: Global TPE markets - forecasts to 2018. *TPE Magazine International* 4 (2013) 5, p. 238–239

[Ell15] ELLIS, P.: The future of high performance thermoplastic elastomers. *Proceedings of thermoplastic elastomers world summit*. Barcelona, Spain, 2015

[Ell16] ELLIS, P.: Die Zukunft der Hochleistungselastomere bis 2020. *Gummi Fasern Kunststoffe* 12 (2016) 69, p. 767–766

[Fri08] FRITZ, H.: Innovative Thermoplastische Vulkanisate (TPV). *Umdruck zur Fachtagung des VDI: Internationales VDI-Forum "TPE 2008"*. Neu-Ulm, 2008

[Gan15] GANDERT, E.: Thermoplastische Elastomere auf der Fakuma 2015 - Marktchancen für TPE. *Plastverarbeiter* 12 (2015) 66, p. 42–182

[Geb12] GEBERT, K.: TPE in Automotive. *Umdruck zur Deutschen Kautschuk Tagung*. Nürnberg, 2012

[Hab90] HABENICHT, G.: *Kleben – Grundlagen, Technologie, Anwendungen*. Berlin, Heidelberg: Springer-Verlag, 1990

[Hab01] HABENICHT, G.: *Kleben – erfolgreich und fehlerfrei*. Wiesbaden: Vieweg Sohn Verlagsgesellschaft mbH, 2001

[HB15] HOPMANN, C.; BRUNS, P.: Investigations on the process induced morphology of thermoplastic vulcanizates. *Umdruck zur Deutschen Kautschuk-Tagung/ International Rubber Conference*. Nürnberg, 2015

[HBK16] HOPMANN, C.; BRUNS, P.; KAMMER, S.: Charakterisierung des anisotropen Werkstoffverhaltens von spritzgegossenen TPV-Bauteilen. *KGK Kautschuk Gummi Kusntstoffe* 10 (2016) 69, p. 51–57

[Hop16] HOPMANN, C.: *Strukturuntersuchungen zum anisotropen Werkststoffverhalten von thermoplastischen Vulkanisaten und dessen Beschreibung innerhalb einer Prozesssimulation*. Abschlussbericht zum IGF-Forschungsvorhaben Nr. 18076N, 2016

[KKC87]   KARGER-KOCIS, J.; CSIKAI, I.: Skin-Core Morphology and Failure of Injection-Molded Specimens of Impact-Modified Polypropylene Blends. *Polymer Engineering Science* 4 (1987) 27,

[Kne12]   KNEISSL, B.: TPE mit besonderen Eigenschaften. *Umdruck zur Deutschen Kautschuktagung.* Nürnberg, 2012

[KS15]   KAISER, W.; SOMMER, F.: *Kunststoffchemie für Ingenieure: Von der Synthese bis zur Anwendung.* München: Carl-Hanser-Verlag, 2015

[LS94]   LAVEBRATT, H.; STENBERG, B.: Anisotropy in injection-molded Ethylene-Propylene-Diene Rubbers. *Polymer Engineering Science* 1 (1994) 24,

[ML04]   MICHAELI, W.; LETTOWSKY, C.: Zukunftssicherung durch Verfahrensintegration. *Kunststoffe* 94 (2004) 5, p. 20–24

[Nag04]   NAGDI, K.: *Gummi-Werkstoffe: Ein Ratgeber für Anwender.* Ratingen: Gupta-Verlag, 2004

[OS99]   OSEN, E.; SCHKUHR, M.: Thermoplastische Elastomere (TPE). *Kunststoffe* 89 (1999) 10, p. 176–182

[Ose06]   OSEN, E.: Thermoplastische Elastomere - neue Materialien, neue Anwendungen. *Handbuch zum Seminar: Thermoplastische Elastomere.* Würzburg, 2006

[Ose08]   OSEN, E.: Thermoplastische Elastomere - Neue Werkstoffe erweitern das Produktspektrum. *Umdruck zur Fachtagung "Thermoplastische Elastomere".* Würzburg, 2008

[RLK15]   RUOTSALAINEN, S.; LAAKSO, P.; KUJANPÄÄ, V.: Laser Welding of Transparent Polymers by Using Quasi-simultaneous Beam Off-setting Scanning Technique. *Physics Procedia* 78 (2015) 15, p. 272–284

[RS05]   RAULIE, R.; SMITH, D.: Radio Frequency Welding of Thermoplastic Vulcanizates. *Proceedings of the 63th Annual Technical Conference (ANTEC).* Boston, 2005

[RS13]   RÖTHEMEYER, F.; SOMMER, F.: *Kautschuktechnologie: Werkstoffe-Verarbeitung-Produkte.* München, Wien: Carl-Hanser-Verlag, 2013

[Sch02]   SCHMACHTENBERG, E.: *Untersuchungen zur Schweißeignung thermoplastischer Elastomere (TPE).* Abschlussbericht zum DFG-Forschungsvorhaben Schm 682/18-1, 2002

[SWSH14] SEITZ, V.; WINTERMANTEL, E.; SCHÖNEBERGER, M.; HOFSTETTER, M.: Thermoplastische Elastomere (TPE). *Kunststoffe* 114 (2014) 11, p. 72–77

[TM00]   TANG, H.; MARTIN, D.: Characterization of the near-surface crystalline structure and morphology in injection molded TPO. *Proceedings of the 58th Annual Technical Conference (ANTEC).* Orlando, 2000

[Tro08]   TROUGHTON, M.: *Handbook of Plastics Joining - A Pracitcal Guide.* Norwich: Plastics Design Library, 2008

[URL18a] N.N.: *DUPONT HYTREL THERMOPLASTIC POLYESTER ELASTOMERS - DESIGN GUIDE.* URL: http://www.dupont.com/content/dam /dupont/products-and-services/plastics-polymers-and-resins/thermoplastics/ documents/Hytrel/Design Guide for Hytrel.pdf, 03.09.2021

[URL18b] N.N.: *FRIMO - INFRAROT verbindet Kunststoffe.* URL: www.frimo.com/de/produkte/frimo-technologien/fuegen-kleben/infrarot schweissanlagen.html?file=files/frimo/content/mediathek/competence_ sheets/FRIMO_Infrarot_1013_DE_low.pdf, 03.09.2021

[URL18c] N.N.: *Thermoplastische Polyurethan-Elastomere (TPU) - Elastollan® -Verarbeitungshinweise.* URL: http://www.polyurethanes.basf.de/pu/solutions/elastollan/de/function /conversions:/publish/content/group/Arbeitsgebiete_und_Produkte /Thermoplastische_Spezialelastomere/Infomaterial/elastollan_verarbeitung_de.pdf, 03.09.2021

[URL18d] N.N.: *THERMPOLAST Eigenschaften Verarbeitung.* URL: http://www.kraiburg-tpe.com/assets/files/uploads/thermolast_technik_d_1560.pdf, 03.09.2021

[Ven05] VENNEMANN, N.: TPE-Werkstoffe im Überblick – Eigenschaften, Entwicklungstrends und neue Produkte. *GAK - Gummi Fasern Kunststoffe* 7 (2005), p. 430–431

[Ven07] VENNEMANN, N.: TPE - Werkstoffe im Überblick.. *Umdruck zur Tagung der VDI-Gesellschaft Kunststofftechnik: Haftung und Materialverbunde.* Düsseldorf, 2007

[WBM12] WU, C.-Y.; BENATAR, A.; MOKHTARZADEH, A.: .: Comparison of Ultrasonic Welding and Vibration Welding of Thermoplastic Polyolefin. *Welding in the World* 56 (2012) 1, p. 69–75

[WH09] WINTERMANTEL, E.; HA, S.-H.: *Medizintechnik - Life Science Engineering.* Berlin, Heidelberg: Springer-Verlag, 2009

[WZN+03] WANG, Y.; ZHANG, Q.; NA, B.; DU, R.; FU, Q.; SHEN, K.: Dependence of impact strength of the fracture propagation direction in dynamic packing injection molded PP/EPDM blends. *Polymer* 43 (2003), p. 4261–4271

[YK93] YAMAOKA, I.; KIMURA, M.: Effect of morphology on mechanical properties of a SBS triblock copolymer. *Polymer* 34 (1993), p. 4399–4409

[Zys93] ZYSK, T.: *Zum statischen und dynamischen Werkstoffverhalten von Thermoplastischen Elastomeren.* Friedrich-Alexander-Universität Erlangen-Nürnberg, Dissertation, 1993

Abbreviations

Notation	Description
TP	Thermoplastics
TPE	Thermoplastic elastomer
TPE-S	Thermoplastic elastomers based on styrene block copolymers
TPE-U	Thermoplastic elastomers based on polyurethans
TPE-E	Thermoplastic elastomers based on polyesther esters
TPE-A	Thermoplastic elastomers based on polyether amides
TPE-V	Thermoplastic polyolefines with crosslinked elastomer phase
TPE-O	Thermoplastic polyolefines with uncrosslinked elastomer phase

Development of resilient injection moulded products

Moderator: Dr.-Ing. Björn Fink, Roche Diagnostics GmbH

---------- Content ----------

1 Resilient injection moulding products – proof via process-validation and a possible leaner validation approach   509

B. Fink[1]
[1] Roche Diagnostics GmbH

2 Strategic selection of parameters for the validation of injection moulding processes in medical devices   519

Ch. Hopmann[1], M. Schöll[1]
[1] Institute for Plastics Processing (IKV) at RWTH Aachen University

3 Analysis of the testing influence on the tensile shear strength of microform-fit back-moulded plastic/metal hybrid components   533

Ch. Hopmann[1], P. Wagner[1]
[1] Institute for Plastics Processing (IKV) at RWTH Aachen University

Dr.-Ing. Björn Fink

Dr.-Ing. Björn Fink started his study of mechanical engineering at the RWTH Aachen in 1997. After completing his Vor-Diplom Björn turned to plastics engineering. With the Diplom in 2004 he finished his studies and joined the Institute of plastic processing as research assistant. In his role in the extrusion department Björn researched mainly the impact of different visco-elastic material models of polymer melts on simulation results regarding the extrudate geometry after leaving the extrusion die.

After successfully completing his doctoral thesis Björn worked for the Victrex Europa GmbH as Technical Service Engineering supporting multiple projects to develop application made of VICTREX PEEK across all industries. He then took over the position of the Automotive Technology Manager. In this role he identified future potential PEEK-application in the Automotive Industry.

In 2015 Björn joined Roche as group leader of the primary packaging group. The group is responsible for the validation of the manufacturing processes at external manufactures of the packaging components. The components are mainly injected or extrusion blow moulded. Since 2020 Björn is leading the Packaging Engineering department.

1 Resilient injection moulding products – proof via process-validation and a possible leaner validation approach

B. Fink[1]
[1] Roche Diagnostics GmbH

Abstract

In the medical industry it is of highest priority not to put defective products on the market. Usually, the validation of manufacturing processes is a major part of achieving this goal. Validation is a way to show that the processes are capable of guaranteeing the specified component quality. In general the proof comes with extensive testing and costs. Therefore, it is analysed whether the validation efforts of similar processes via injection moulding can be reduced. The production of the same product with different colour masterbatches was analysed. The melt and component properties of two colour variants of a component of a medical device are investigated in order to work out possibilities for streamlining the validation. Differences in the heating and cooling properties are observed between the colour variants, but their relevance for validation is considered to be low. With these results multiple options to reduce validations efforts are available without affecting the product quality.

1.1 Introduction

The term resilience is taken from psychology. In psychology a resilient person is being described as a person, which is able to successfully re-align after smaller and bigger events, which puts her or his life at disorder. In technology resilience may therefore be described as the capability of a technical system to adjust to perturbations. A good example is a football. During use the football is highly stressed and deformed. However, the football always manages to return to his original shape and properties after each kick.

Hence, resilient injection moulding products therefore cannot move out of their equilibrium no matter how big the disturbance will be or will be able to move back to their equilibrium after disturbance. One aspect of resilient products is that the products can be manufactured according to the specification. During early development work this is usually proven by feasibility studies. During industrialisation or ramp-up phase it is mandatory in almost all industries to proof that with the developed process settings the product can be manufactured million and million times according to its specification. In the medical industry the proof of robust manufacturing is done via process validation. After a brief discussion of the validation concept in the medical industry a possible leaner validation approach to a current high effort validation is shown.

1.2 Validation in the medical industry

The medical industry is an industry with a vast product portfolio, i.e. drugs and implants including packaging, syringes, surgery tools and diagnostic analysers including consumables and reagents and their packaging to name a few. All these products have in common that the performance of these products are very critical for the patient. Thus, the authorities want to make sure that all these products are performing according to the product specification. In principal this can be shown by a 100% control. This approach is on one hand very labour and cost intensive and one the other hand not always possible. Therefore, the authorities want to see answers to the question, if the manufacturer is capable to produce safe and fully operational products. The answer is almost all of the time a fully validated manufacturing process in combination with a compliant and working Quality Management System. This is the foundation to ensure that defective products or products with the wrong material are not being put on the market.

The standards Code of Federal Regulations Title 21 Part 820 by the U.S. Food Drug Administration (FDA) and the DIN EN ISO 13485 describe the manufacturing and the Quality Assurance of medical products [NN16, NN17].Key elements are a validation-process with established methods to proof robust manufacturing of capable products. The validation procedure must be documented and it is mandatory to re-validate the processes periodically. The Global Harmonization Task Force (GHTF), which consists of members of multiple national health agencies recommends an approach with three steps: Installation qualification (IQ), Operational qualification (OQ) and Process Qualification (PQ) [Gue20, NN04, Pet05].It is now common practice that a Design Qualification (DQ) step is performed prior to the IQ. The following section is a brief summary of the validation steps. For more details, please see for example [AB17, RL19, LBS+18, The17, ZSZ+17].

Design Qualification: At the end of this step it is shown on a theoretical basis or virtual assessment that the chosen technical equipment is fit to manufacture the product according to the specification. A first process risk assessment is performed.

Installation Qualification: The aim of the IQ is to proof that all components defined in the manufacturing specification are installed and fully operational on the final manufacturing site. For injection moulding equipment it is usually shown that the components of the injection moulding machine like plastification-, clamping- and control-unit work accordingly. The same needs to be shown for components of the mould like ejection-system, hot runner system, and cooling channels [VRA12].

Operational Qualification: Usually the main work in the OQ is to define the process limits and show that the process is robust within these limits. The foundation for the work is to identify the Critical Quality Attributes (CQA) and the Critical Process Parameters (CPP). With a Design of Experiment (DoE) the limits for the CPP are defined. From the limits a process point is derived and it is shown in a short-term production that the cpk is above 1.33 for defined CQA. This procedure is adapted to the injection moulding process.

Process Qualification: Building on the knowledge of the OQ during the PQ a long-term process robustness is usually shown. Typically, three production batches are manufactured and the cpk needs to be above 1.33. The PQ can only be performed by the operating plant. The risk assessment should be finalised with the experience from the three production lots at the end of the PQ.

It is common knowledge and a crucial part of Good Manufacturing Practice (GMP) to build a Master Qualification Plan (MQP) at the beginning of each validation process. Within the MQP it must be described in detail, which technical systems are validated. It is also part of the MQP to define, which dimensions need a cpk above 1.33 and how to proceed with multiple cavities during the validation process. The MQP is the preferred tool to show the authorities why and which validation strategy was chosen. The report of each validation step then proofs that you have worked according to the MQP [Bü13].

1.3 Possible validation strategy as OEM

As described above the FDA's Code of Federal Regulations Title 21 Part 820 and the DIN ISO 13485 are the framework for the validation of manufacturing processes. The common approach nowadays is the four Qualification phases DQ, IQ, OQ and PQ. Additionally, the framework asks the manufacturers to describe the plan and the rational and then show that the validation was done according to the plan.

The freedom, which comes with this approach, is much appreciated because in the supply chain there are also different roles with diverse aims for the validation. Roche Diagnostics does not mould polymer parts itself. Roche Diagnostics sources the polymer parts within its supplier network and uses the polymer parts to manufacture the medical product. Being the manufacture Roche Diagnostics must proof that the sourced parts do not lead to defective products on the market. Therefore, Roche provides a framework for different polymer parts. The following shows the framework for a random injected-moulded part of one reagent carrier. The amount of qualification and validation effort may be adjusted for every part on the basis of a risk assessment.

In the DQ-Phase the mould- and validation-requirements are finalised after a joined discussion with the supplier. With this information the supplier is able to set-up the MQP. In the IQ, the acceptance of the mould by the supplier is approved. The approval is typically done on the basis, that the supplier has checked the correct setup of the mould, all auxiliary equipment and sensors.

The approval of the initial sample test report is a core part of the OQ. With the approval of the initial sample test report the supplier may start the short-term process qualification. The final report is approved by the OEM. To proof that the produced parts with the new mould work in the final device, application specific testing is performed and documented at the OEM internally.

When all tests are passed the supplier performs the preliminary process capability analysis and process validation under real process conditions according to the requirements of the OEM. With the approval of the PQ-test report, the PQ-phase and validation of the injection moulding process is closed.

The key for the supplier and OEM is to define the Validation-plan and align the validation-processes at the supplier and at the OEM. The responsibility of planning the DoE, of performing the testing and of doing the short-term qualification and of doing the preliminary process capability analysis and process validation sits with the supplier. The framework for creating the DoE can be approved in a technical meeting or supplier audit when needed. Besides being compliant to the international standards it should be clear to the supplier and OEM that a proper validation is the key-element to provide the foundation for a robust production of polymer parts for many years.

1.4 Leaner validation approach with multiple coloured closures

From commercial perspective the supplier and the OEM are driven to reduce cost and efforts for the validation without putting the quality of the product in jeopardy. There are many ideas on how to reduce the validation efforts using new technologies like enhanced simulation techniques and novel temperature and pressure sensors.

In packaging for technical devices coloured components are a powerful and easy way to prevent the user from mixing up products. I.e. the liquid in bottles with the same size and outer appearance can be differentiated simply by different colours of the cap of the bottle. This principle is also used in this example leading to a snap cap in multiple colours using different master batches Figure 1.1.

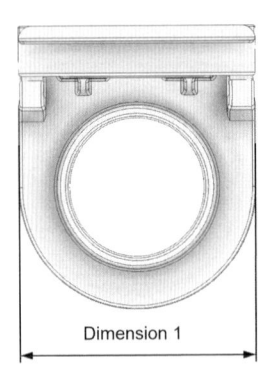

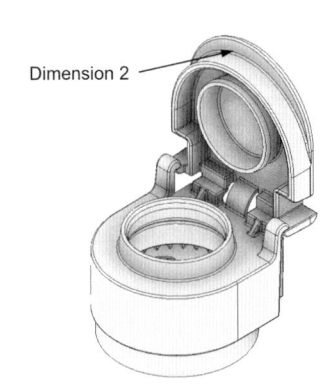

Figure 1.1: Snap Cap design

According to the validation framework each snap cap with a different colour is treated as one product and hence for each coloured snap cap the above described validation process must be applied. In order to reduce cost and efforts a possible leaner approach is investigated by equivalence testing [Kle21]. Table 3.1 shows the different material composition used to compare the white and caramel snap caps. The studies with two batches of the base material will support the definition of the equivalence limits whereas the experiments with the serial material using the white and caramel master batch will provide knowledge on the equivalence of the serial materials.

Name	Base Material	Masterbatch
base Material batch 1	Polypropylen - batch 1	-
base Material batch 2	Polypropylen - batch 2	-
Serial Material white	Polypropylen - batch 1	Specified% masterbatch white
Serial Material caramel	Polypropylen - batch 1	Specified% masterbatch white

Table 1.1: Materials to be investigated

First the impact of the master batches on the melt properties is studied by an experimental injection-moulding process. The experimental injection moulding process has the advantage that it provides more extensive measuring methods that the current serial production of the snap-caps. The melt properties are characterised during the injection- and cooling-phase comparing process-parameters and geometrical dimensions of the component. For this, preliminary tests are conducted using a full factorial experiment design with variations of the master batches and several process parameters. The master batch influence is then further investigated in the equivalence tests according to two one sided tests (TOST). Finally, the equivalence of the serial part, the snap cap, is investigated.

1.4.1 Impact of master batch on the melt properties

Figure 1.2 shows the experimental set-up and the variables measured during the process and on the moulded part. Between cylinder and mould a nozzle with temperature and pressure sensors is introduced. The used mould is equipped with multiple internal temperature and pressure sensors. With the measured data from all sensors it is possible to calculate the temperature increase, the injection work and the filling and cooling behaviour of the melt. All data are used to assess the melt properties of the materials and moulding settings. Some dimensions of the moulded part like wall thickness, part weight and warpage are measured and assessed as well [Kle21].

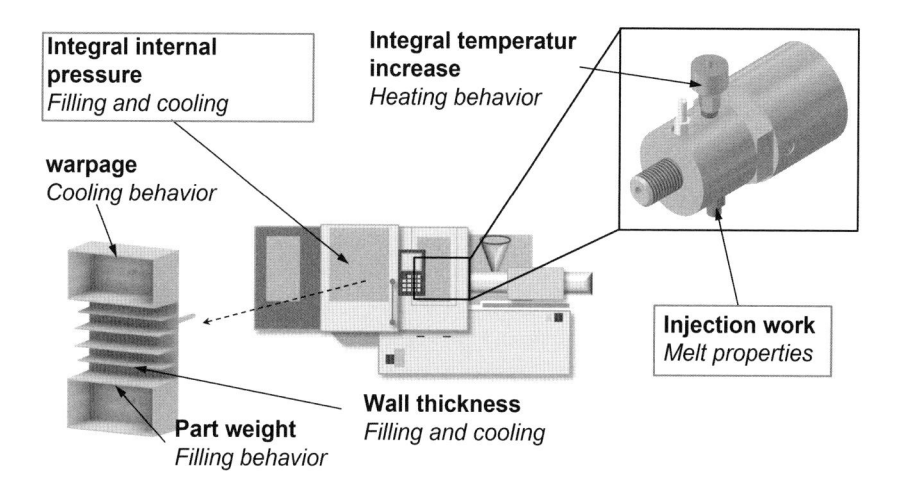

Figure 1.2: Experimental set-up for equivalence study

Figure 1.3 shows the injection work depending on the used material. Comparing the two batches of base material the injection work of batch 2 is 0.2 bar·cm^3 lower compared to batch 1. Most likely there is a slight difference in viscosity of the two batches, which leads to the different injection work. To define the equivalence limits of the TOST equivalence test a two-sample t-test with a level of significance of 5% is carried out with the measured data from the two batches. With the resulting 95% confidence interval, the outer limit is assessed as the maximum batch fluctuation, which is included in the definition of the equivalence limits. For the injection work, the outer limit of the 95% confidence interval of the difference between batch 1 and 2 is 0.2 bar·cm^3. From this the equivalence limits are defined at an average value of 0.25 bar·cm^3. The confidence interval of the differences between the master batches ranges from 0.05 bar·cm^3 to 0.10 bar·cm^3 and is thus within the equivalence limits of ± 0.25 bar·cm^3. This means that the injection work of the materials with both master batches is equivalent.

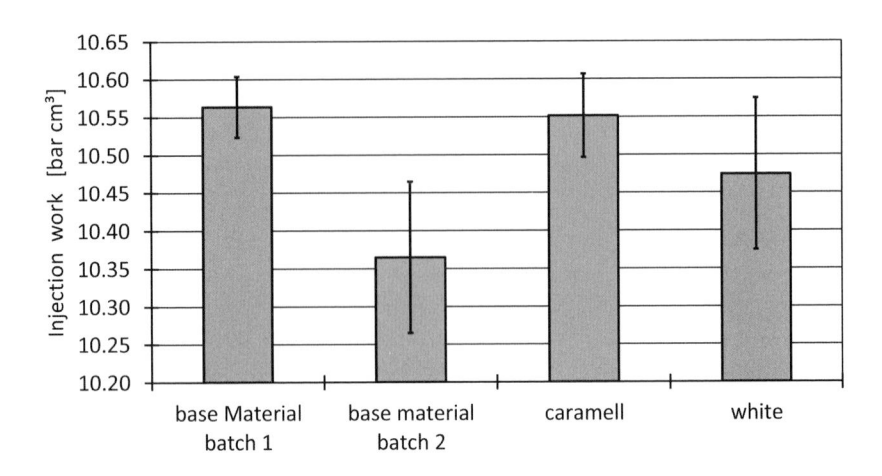

Figure 1.3: Material study - injection work

Figure 1.4 shows the holding pressure integral as a function of the used materials. For batch 2, a holding pressure integral that is 8.4 bar·s greater than for batch 1 is observed. This is also consistent with the theory that batch 2 has a lower viscosity than batch 1. The outer limit of the 95% confidence interval for the difference between batches 1 and 2 is 10.5 bar·s. From this, as well as from the changes due to normal temperature fluctuations (8.2 bar·s difference), the equivalence limits of the holding pressure integral are defined at ± 9.35 bar·s.

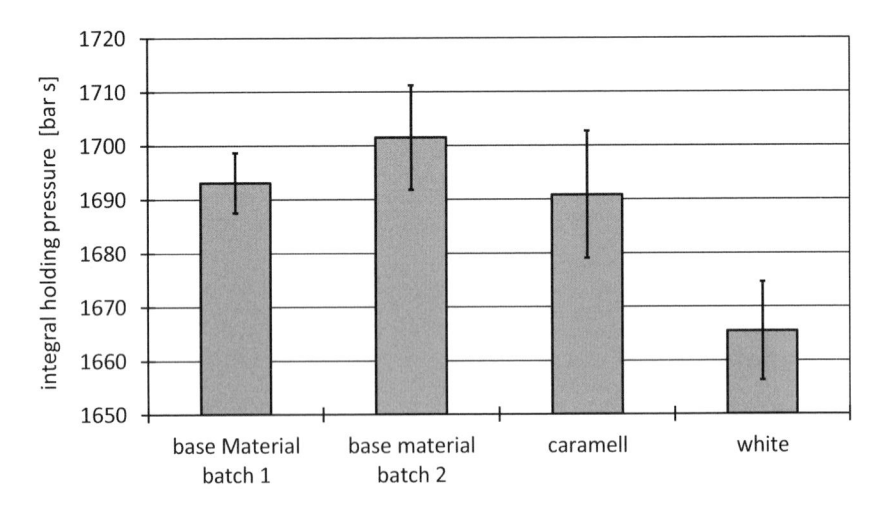

Figure 1.4: Material study - holding pressure

In the case of the white master batch, a holding pressure integral that is about 25 bar·s lower is observed than in the case of the caramel-coloured master batch. The result of the TOST equivalence test shows, that the confidence interval for the differences between the two materials is outside the equivalence limits of ± 9.4 bar·s. The difference between the materials with the master batches is up to three times greater than the defined technical equivalence. The holding pressure integral of the master batch is therefore not equivalent. As an equivalence of the injection work is observed, the explanation for this cannot be a change in viscosity. The difference may be caused by a different cooling behaviour of the master batch, which could lead to faster cooling of the material with the white master batch, when it comes into contact with the mould. The melt solidifies more quickly and thus the holding pressure may have a less significant effect, resulting in a lower holding pressure integral.

1.4.2 Impact of the master batches on the dimensions and properties of the snap cap

The impact assessment is done with the serial process on a serial multi-cavity mould. 1000 injection moulding cycles are performed and from each cavity 25 parts were taken randomly and defined dimensions and part properties are measured. As an example the results are discussed with one dimension and one part-property from one cavity.

As depicted in Figure 1.5 dimension 1 is smaller of the part with the white master batch than the dimension 1 with the caramel master batch. The difference is smaller than the difference between the base material batches and the standard deviation. The 95% confidence interval of the difference ranges from -0.005 mm to 0.003 mm, which means that the difference between the materials with the master batches lies between these values with a 5% probability of error. The confidence interval lies within the equivalence limits of ± 0.017 mm. Dimension 1 is therefore equivalent for the materials with the white and caramel master batches. Equivalence is also determined for the other cavities.

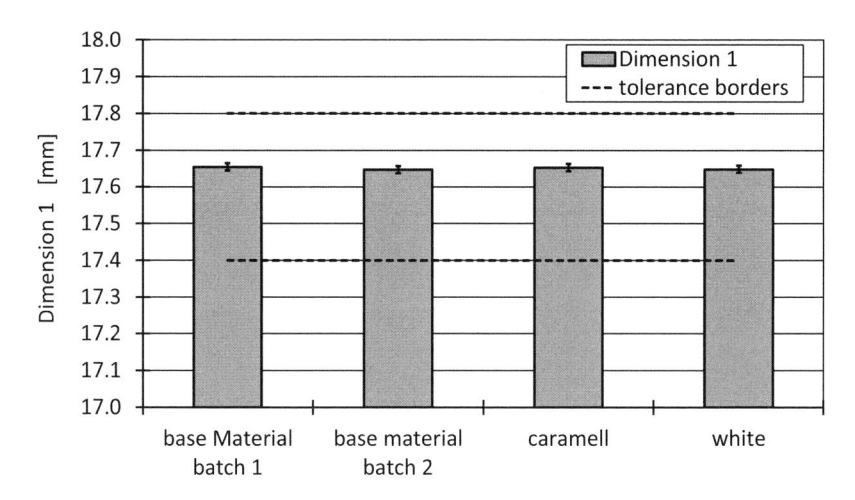

Figure 1.5: Equivalence study on serial part - dimension 1

Figure 1.6 shows the results of the equivalence study for a part property (dimension 2). Dimension 2 observed with parts moulded with batch 1 is 2.8 N greater than with parts moulded with batch 2, which is greater than the mean standard deviation of 0.79 N and is therefore significant. The mean values of the master batches are within the tolerance range from 7 N to 16 N. The mean 95% confidence interval of the differences in all cavities ranges from 0.43 N to 3.05 N. The equivalence limits are set for the mean value of the batch fluctuation and 10% share of the tolerance range of 1.75 N.

Dimension 2 is influenced by a few other dimensions of the part. Differences between the material mixtures in these dimensions could lead to differences in dimension 2 as well. For parts manufactured with batch 1 and batch 2 there are no significant differences in the said dimensions measured. The observed difference in this attribute cannot therefore be explained by a dimensional difference only. Instead, a lower surface roughness or a lower modulus of elasticity could lead to the fact that less force is required. From previous investigations it is also known that the storage time between manufacturing the snap caps and the testing of the part property has a major influence on the measurement result. Adding different master batches adds to the complexity because different additives have an impact on the surface roughness and on the material properties like the tensile modulus, which will have an influence on this particular attribute of the snap cap.

As an example for one cavity the 95% confidence interval of the difference ranges from 0 N to 0.33 N and is therefore within the equivalence limits of ± 1.75 N. The parts taken from this cavity are equivalent regarding dimension 2. Equivalence is also determined for the other cavities.

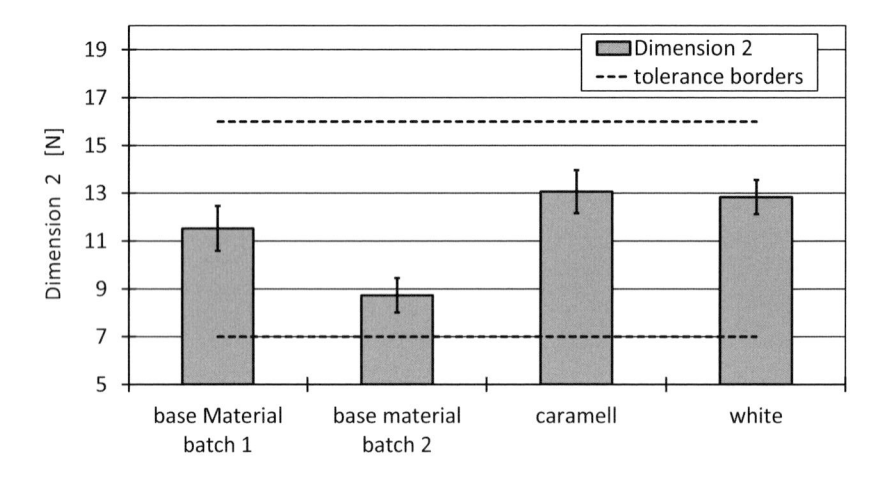

Figure 1.6: Equivalence study on serial part - dimension 1

1.4.3 Possible next steps

In order to decrease the validation efforts for multi-coloured components an equivalence study is a valuable tool, where the analysis of the melt properties is supporting the equivalence study with real parts. However, it is not always possible to perform an equivalence study. A risk assessment may be performed in these rare cases.

The successful equivalence study gives several opportunities to cut down the validation effort. One option is to perform the application specific testing at the OEM with parts manufactured with one colour during the OQ. Another option is to reduce the amount of PQ-runs at the moulder by manufacturing only 3 PQ-batches (6 PQ-batches before) by switching from one colour to the other after each run.

1.4.4 Conclusions

In the medical industry it is mandatory to proof robust manufacturing processes via qualification and or validation. Together with a working Quality Management System these are the key factors to produce resilient injection moulding parts. The regulatory framework for validating manufacturing processes gives a lot of freedom. At the beginning of every validation effort a MQP should be set-up, which describes in detail what steps and criteria must be met during the validation process. This approach gives also the room to reduce time and cost and still being compliant.

A possible less effort approach for coloured parts is described. On basis of extensive studies on finished parts it can be shown that for the parts with different colours there is equivalency. With the proven equivalency there are different options to reduce validation efforts.

References

[AB17] ANES, J.; BEAUMONT, J.: Process Validation for Injection Molding Medical Components. *Plastics Engineering* 73 (2017) 5, p. 38–43

[Bü13] BÜRKLE, E.; KARLINGER, P. W. P.: *Reinraumtechnik in der Spritzgießverarbeitung*. München: Carl Hanser Verlag, 2013

[Gue20] GUERRA, M.: *Validation of the injection molding process to produce the primary packaging of tissue valves*. Politecnico di Torino, Master thesis, 2020 – supervisor: Prof. Silvia Spriano and Arnaldo Giannetti and Carlo Guala

[Kle21] KLEIN, E.: *Möglichkeiten der Rationalisierung der Validierung von Spritzgießprozessen in der Medizintechnik bei der Verarbeitung unterschiedlicher Farb-Masterbatches*. RWTH Aachen University, unreleased Master thesis, 2021 – supervisor: Matthias Schöll

[LBS+18] LUSARDI, G. AND BROWN, R. S. B. H. M.; BROWN, D.; SCULLY, S.; BUTLER, D.; VALLEY, E. AND THERRIEN, M.: *"Part Process" Transferability Using Machine Independent Variable (MIV) Methodology for Multiple Machines*. White Paper, 2018

[NN04] N.N.: *Quality Management Systems - Process Validation Guidance*. White Paper, 2004

[NN16] N.N.: *DIN EN ISO 13485 - Medizinprodukte-Qualitätsmanagementsysteme - Anforderungen für regulatorische Zwecke*. Berlin: Beuth Verlag, 2016

[NN17] N.N.: *21 CFR Part 820 (Quality System Regulation)*. Code of Federal Regulations, N.N., 2017

[Pet05] PETRETICH, C.: *Development of a validation methof for thermoplastic injection molding processes for the contract medical device manufacturer*. Bowling Green State University, Dissertation, 2005

[RL19] RAZENBÖCK, J.; LHOTA, C.: Konforme dynamische Prozesse - Assistenzsysteme in die Validierungsstrategie einbinden. *Medplast* 13 (2019), p. 18–21

[The17] THERRIEN, M.: 'Part Process' Development and Validation for Multiple Machines. *Medical Product Outsourcing Magazine* (2017), p. 156–161

[VRA12] VESELOV, V.; ROYTMAN, H.; ALQUIER, L.: Medical Device Regulations for Process Validation: Review of FDA, GHTF, and GAMP requirements. *Journal of Validation Technology* (2012), p. 82–92

[ZSZ+17] ZHAO, Y.; SHENG, K.; ZHANG, X.; HENGYI, Y.; MIAO, R.: Process Validation and Revalidation in Medical Device Production. *Procedia Engineering* 174 (2017), p. 686–692

2 Strategic selection of parameters for the validation of injection moulding processes in medical devices

Ch. Hopmann[1], M. Schöll[1]
[1] Institute for Plastics Processing (IKV) at RWTH Aachen University

Abstract

In medical technology, all process steps whose results are not fully checked have to be validated for the manufacturing of products. For the validation of injection moulding processes, a process window is usually defined on the basis of machine setting parameters. The disadvantage of the setting parameters is that they do not guarantee reproducible process conditions and correlate only to a limited extent with the component quality due to fluctuations. Therefore, the suitability of process parameters for validation is analysed. In order to keep the effort for validation as low as possible, it is investigated, whether the limits of the process window of the process parameters can also be readjusted by specific variation of the machine setting parameters without changing the correlation between process parameters and part quality.

Investigations show that the influence of process variations on the correlation between process parameters and quality characteristics can be simulated by the targeted variation of machine setting parameters. However, for individual quality characteristics, differences in the correlation factors of the individual process parameters occur. A possible reason for this may be that the material behaviour is influenced differently by the varied machine setting parameters than by the fluctuations.

2.1 Introduction

In medical technology, the expenditure from the product idea to series-ready production is particularly high, as extensive tests and documentation are required for the validation of all process steps in production. Validation is required by various standards and guidelines for processes whose quality cannot be completely verified afterwards. Validation serves as proof that both product and process are within planned specifications. As the injection moulding process is a mass production method, a 100 % control of the individual components often does not make economic sense. In this case, the validation has to show that a process window has been identified in which the quality of the manufactured components meets the product specifications. In the subsequent production, the validated process window has not to be left. Up to now, machine setting parameters have been used to determine the process window of the injection moulding process. However, these do not represent the process conditions and cannot identify fluctuations in the process, such as the processing of different material batches. Therefore, the possibility of validating an injection moulding process with the help of process parameters will be investigated within this contribution.

2.2 State of the art

In the state of the art, the requirements for validation as well as the established procedure for validation are presented first and challenges are identified. Finally, different process parameters that describe the process behaviour are discussed.

2.2.1 Validation of injection moulding processes in medical technology

The manufacture and quality assurance of medical devices is regulated by the U.S. Food Drug Administration (FDA) in the Code of Federal Regulations Title 21 Part 820 and by the DIN EN ISO 13485 standard [NN16, NN17].The functionality and safety of a product have to be verified by a 100 percent inspection. If this is not possible, the manufacturing process of the product has to be validated. Due to the high quantities and partly non-destructive testing of the functionality, injection moulding processes are generally validated [AB17, RL19].According to 21 CFR Part 820 and ISO 13485, the validation has to show with a high degree of certainty and through established methods that the manufacturing process is capable of producing safe and functional products. The validation procedure has to be documented and the process capability has to be checked regularly. A precise procedure is not defined in any of the regulations. The Global Harmonization Task Force (GHTF), a consortium of several national medical regulatory authorities, such as the FDA, recommends a three-step procedure that is also used for injection moulding processes [Gue20, NN04, Pet05]: Installation Qualification (IQ), Operational Qualification (OQ) and Process Qualification (PQ).

The Installation Qualification outlines that all required machinery, equipment and systems for manufacturing are installed and fully functional at the manufacturing site according to the manufacturer's requirements [NN04, VRA12].

Operational Qualification shows the parameter limits within which the process is still capable of producing safe and functional products (worst-case testing). If not already known from previous investigations, it is first examined which process parameters have an influence on the Critical Quality Attributes (CQA). The Critical Process Parameters (CPP) determined in this way are validated in the further procedure [AB17]. In injection moulding processes, machine parameters such as the set cylinder and mould temperature, holding pressure height and duration or speed of injection are generally validated as CPP [Gue20, Pet05].

In Process Qualification, the long-term stability of the process under series production conditions is examined [NN04]. The long-term process capability (cp, cpk) at the process point determined in the OQ is demonstrated in three production batches and should generally be cp, cpk $\geq$ 1.33 [ZSZ+17].

All phases of the process validation have to be documented. There are no specifications for the exact sample size in the respective validation phase. The sample size has to be chosen by the producer and statistically justified in a rationale [NN17]. In the event of technical changes to the process or observed changes to the CQA, a re-validation of the process has to be carried out [VRA12]. In the running series process, it has to be regularly checked that the process is within the defined limits [NN04].

2.2.2 Approaches to shorten the validation procedure

A complete process validation is time-consuming and cost-intensive. If an injection mould is used on several, different injection moulding machines, there are existing approaches to carry out a shortened validation on each individual injection moulding machine instead of a complete process validation. For this purpose, the complete validation (IQ, OQ, PQ) is first carried out on one injection moulding machine. Machine-independent parameters such as pressures, temperatures and screw movements are measured and recorded in the validation protocol. The chosen parameters should describe the properties of the melt (temperature, viscosity, pressure and cooling behaviour) as well as the process operations [The17]. Hence, the validation is based on the material properties and the process operations is independent of the used injection moulding machine [Gro00, LBS+18]. By using the same mould on another injection moulding machine, the machine settings are now selected in such a way that the process variables match the process variables of the fully validated machine noted in the validation protocol. The process stability is demonstrated in a shortened, four-hour PQ with a smaller test scope. This can result in time, personnel, material and cost savings, but there is an increased measurement effort for temperatures, pressures and movements during the injection moulding process [LBS+18].

2.2.3 Process variables in injection moulding

The interfering influences on the injection moulding process can result in rejects being produced even despite a stable process point. Due to changing process conditions, such as fluctuating material properties, it is not sufficient to reproduce only machine parameters. Instead, the setting parameters have to be adapted to the changed process conditions [Ham04, HW15].

One possibility is using process indicators. Process indicators reflect the current process conditions and are usually formed from several process parameters or process parameter curves. In addition to the process conditions, the indicators correlate to the quality criteria. Indicators are usually dimensionless, as they are only a measure of the changed process conditions and not a physical quantity [Mus00]. In the following, the process key figures that will be considered in the further course of this contribution will be explained:

The injection pressure integral PI_{Inj} is a process indicator with high relevance, as it correlates directly with the flowability, because the pressure in the screw ante-chamber depends on the material-, mould- and machine-specific parameters [LBS+83]. It is formed as the integral of the injection pressure p_M measured with the machine M during the injection phase Inj (see Equation 2.1).

$$PI_{Inj} = \int_{t_0}^{t_1} p_M(t)dt \qquad (2.1)$$

Another indicator is the cavity pressure integral during the injection phase PI_{InjCav}. It is calculated as the integral of the cavity pressure p_{Cav} during the injection phase (see Equation 2.2).

$$PI_{InjCav} = \int_{t_0}^{t_1} p_{Cav}(t)dt \qquad (2.2)$$

It correlates with fluctuations during the filling of the cavity. These can result, for example, from different closing behaviour of the non-return valve, deposits on the flow channel surfaces or fluctuations in the switchover point.

A further indicator is the integral temperature increase at the sensor position in the injection phase TI_{Inj}. It is formed from the integral of the mould wall temperature T_{Cav} in the cavity Cav during the injection phase (see Equation 2.3).

$$TI_{Inj} = \int_{t_0}^{t_1} T_{Cav}(t)dt \qquad (2.3)$$

This correlates directly with the mould and melt temperature and is therefore a measure of the shrinkage potential of the plastic.

Another indicator is the averaged cavity pressure integral during the holding pressure phase PI_{PacCav}. This is calculated from the integral of the cavity pressure p_{Cav} during the holding pressure phase Pac (see Equation 2.4).

$$PI_{PacCav} = \frac{1}{t_{Pac}} \int_{t_0}^{t_1} p_{Cav}(t)dt \qquad (2.4)$$

Like the injection pressure integral, this key figure is also dependent on the flowability of the melt. As the pressure gradient increases with higher viscosity, the cavity pressure decreases under otherwise identical conditions [Zöl03]. In addition, the process parameter also takes the mould temperature into account. As the cooling rate depends directly on the cavity temperature, this effect can also be seen in the cavity pressure. With a warmer mould temperature, the difference between melt and mould temperature is smaller, so that the melt solidifies more slowly.

2.3 Objectives and procedure

Due to the disadvantages of the previous procedure for the validation of injection moulding processes in the example of medical technology, the possibility of validating the process with the help of process parameters is to be investigated. As it is not economical to deliberately introduce disturbances, such as fluctuations in the thermal equilibrium or fluctuations in the flowability, into the process, it will be analysed whether the disturbances can be readjusted through targeted variation of the machine setting parameters.

For this purpose, the influence of process fluctuations on the component quality is investigated in a first step. Therefore, process indicators are identified that correlate with the component quality in this process. In a further step, the influence of a variation of the machine setting parameters on the process parameters is investigated. Furthermore, the effect on the part quality is considered. As in the first step, the correlation between the process parameters and the part quality is analysed. Finally, it is assessed whether the influence of variations on the correlation between process parameters and part quality can be simulated by varying the machine setting parameters and whether the injection moulding process can be validated meaningfully with the help of process parameters.

2.4 Operating resources

The injection moulding tests were carried out with a hydraulic injection moulding machine (see Table 3.1). In order to be able to additionally investigate the influence of the fluctuations and the machine parameter variations on the melt quality, a special measuring nozzle was used. With this nozzle, the melt pressure can be measured via a pressure sensor. The melt temperature is measured with an immersion thermocouple that reaches 1 mm into the flow channel and an IR sensor that has a thermocouple on the detector and on the sensor surface.

Component	Type	Manufactor
Injection moulding machine	160-1000 CX	KraussMaffei Technologies GmbH, Munich, Germany
pT-sensor	6190CA	Kistler Instrumente AG, Winterthur, Switzerland
Melt pressure measuring chain	4021B30HAP1	Kistler Instrumente AG, Winterthur, Switzerland
IR sensor	NTS 2017-IR-BTS/STS	FOS Messtechnik GmbH, Schacht-Audorf, Germany
Immersion thermocouple	EST-71523-10-M5-1500-0	Therma Thermofühler GmbH, Lindlar, Germany
Scale	PLJ 700-3C	KERN & SOHN GmbH, Balingen-Frommern, Germany
Basic material	PP 579s	Saudi Basic Industries Corporation, Riad, Saudi-Arabia
Colour masterbatch	Polybatch L3116	Basell Polyolefine GmbH, Wesseling, Germany

Table 2.1: Used resources

Polypropylene is processed during the trials. In order to simulate both variations, different masterbatch contents are processed (see Table 3.1).

A box-like geometry with additional rib structures and openings serves as the test part. A combined pressure-temperature-sensor (pT-sensor) is installed in the corresponding injection mould. Figure 2.1 shows the test component and the sensor position. In addition, the component dimensions are marked, which will be used to assess the component quality. These are measured with a coordinate measuring machine. In addition to the dimensions, the component weight is also considered.

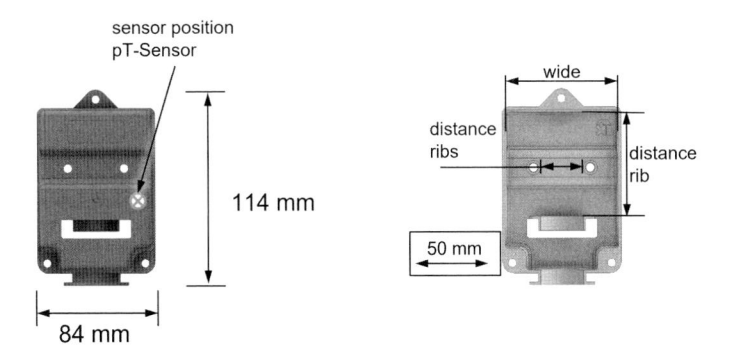

Figure 2.1: Test component with marked sensor position and depicted quality characteristics

2.5 Analysis for validation of the injection moulding process in medical technology

This section investigates whether the influence of disturbances on the correlation between process parameters and component quality can be simulated by varying the machine setting parameters. For this purpose, the effects of disturbances on the correlation are analysed first. Based on this, the correlation is explored with targeted variation of the machine setting parameters before the two approaches are compared with each other.

2.5.1 Analysis of the influence of process fluctuations on the injection moulding process

In a first step, the influence of fluctuations on the process parameters and the component quality is analysed. For this purpose, the process is deliberately manipulated by introducing fluctuations in individual test points. In this way, a change in viscosity is simulated by different masterbatch proportions. In addition, two different batches of the base material were processed. Disturbances of the thermal equilibrium are initiated by varying the mould temperature and the cylinder heating. The disturbances and their stages are summarised in Table 3.2).

Level [-]	Batch [-]	Master batch share [%]	Mould temperature [°C]	Cylinder heating [°C]
+	1	2	35	250
-	2	0	25	245

Table 2.2: Interferences introduced

The parameters are varied according to a full factorial experimental design. The fluctuations in the test points are meant to represent natural fluctuations, which also occur in production.

To assess the influence and connections between the individual variables, the Pearson correlation is calculated. This describes the linear correlation between two variables. The correlation factors can assume values between -1 and 1. A correlation of 1 / -1 describes a perfectly positive / negative linear relationship between the variables, whereas a correlation of 0 describes a non-linear relationship.

The influence of the fluctuations on the melt quality (e.g., temperature and pressure) can be measured with the help of the additional sensors in the measuring nozzle. Both the batch (0.73) and the masterbatch (0.51) can be identified with the maximum pressure in the nozzle. As both fluctuations affect the viscosity, different amounts of energy are required to inject the melt into the cavity.

The variation of the melt temperature correlates with the maximum melt temperature measured with the IR sensor (0.99). The variation of the mould temperature has no significant influence on the pressure and temperature in the nozzle.

The evaluation of the component weight shows that the standard deviation does not exceed 0.01 within a test point. Across the test points, the standard deviation of 0.17 is many times higher (see Figure 2.2).

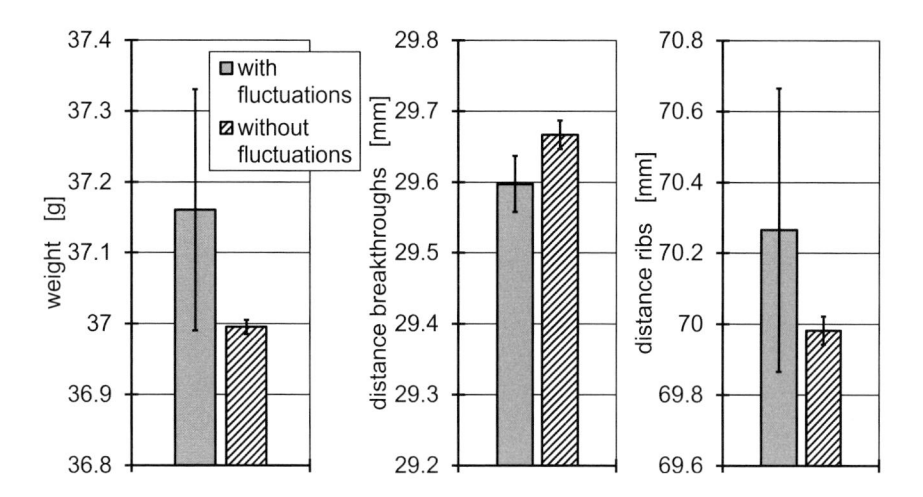

Figure 2.2: Influence of fluctuations on the standard deviation of the individual quality criteria

The same can be observed for Distance Ribs. The maximum standard deviation within a test point is 0.04, whereas a standard deviation of 0.4 is measured across all tests. These results make it clear that reproducing the machine setting parameters does not always result in the same component quality. Different process conditions change the material behaviour of the plastic, such as the pressure gradient or the shrinkage potential, so that the process has to be adapted to the existing conditions in order to achieve consistent quality.

Process parameters can be used to identify changes in the prevailing process conditions. Furthermore, they offer the advantage that they also correlate with the component quality due to their dependence on the production conditions. In the following, the correlations between process parameters and faults as well as component quality will be analysed in more detail. For this purpose, the process parameters from Section 2.2.3 will be considered.

The evaluation of the process parameters with regard to their correlation with the melt quality shows that they reflect the process conditions. Furthermore, the injection pressure integral (0.99) and the cavity pressure integral during the injection phase (0.42) correlate with the melt pressure in the nozzle. Additionally, the integral temperature increase TI_{Inj} correlates significantly (0.42) with the melt temperature of the plastic measured with the IR sensor.

Figure 2.3 shows the correlation of the process parameters with the distance of the breakthroughs. The injection pressure integral (-0.64) and the cavity pressure integral during the injection phase (-0.74) correlate strongly with the quality criteria.

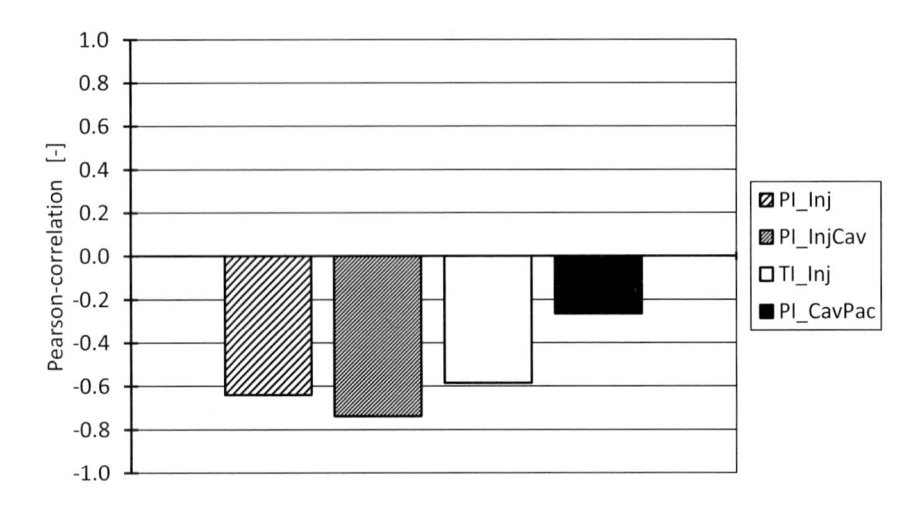

Figure 2.3: Correlation of the process variables with the component quality using the example of the spacing of the breakthroughs when exposed to disturbances

The integral temperature increase during the injection phase TI_{Inj} correlates significantly with the component dimension (-0.58). Only for the averaged cavity pressure during the holding pressure phase no statically significant correlation is identified.

Correlations can also be observed between the other quality characteristics, component weight, width and spacing of the rib, and the process parameters. This illustrates that the process parameters reflect the manufacturing conditions and consequently also correlate with the quality.

2.5.2 Influencing the process variables through targeted variation of the machine setting parameters

In the validation of injection moulding processes, no disturbances can be deliberately introduced into the process in order to determine the limits of the process parameters for the process window, as the effort involved becomes significantly greater. Therefore, in a second step, it is investigated whether the influence of natural fluctuations on the process parameters can be simulated through targeted variation of the machine setting parameters. For this purpose, the machine setting parameters injection speed, holding pressure, holding pressure time, switchover point, back pressure and speed were varied according to a screening test plan. The levels of the individual setting parameters are summarised in Table 2.3).

Level [-]	Holding pressure [bar]	Holding pressure time [s]	Injection speed [cm³/s]	Switchover point [cm³]	Back pressure [bar]	Rotation speed [r/min]
+	325	17	120	17.2	75	150
-	275	7	40	16.8	25	50

Table 2.3: Varied machine setting parameters

Figure 2.4 shows the influence of the machine setting parameters using the example of the mean cavity pressure during the holding pressure phase. The injection speed (0.48), the holding pressure (0.56) and the holding pressure time (-0.56) correlate strongly with the cavity pressure. The influences are statistically significant. The adjustment parameters switchover point, back pressure and rotation speed do not significantly influence the cavity pressure.

Due to a higher holding pressure, the plastic in the cavity is compressed more strongly with the same pressure gradient, so that more material is introduced into the cavity. Due to a longer holding pressure time, the drop in the cavity pressure curve is taken into account when calculating the key figure, so that the average cavity pressure decreases with a longer holding pressure time.

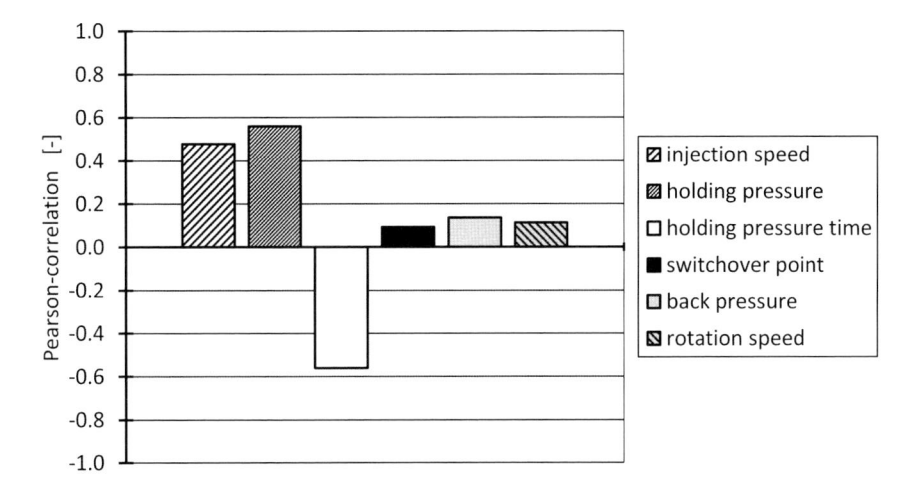

Figure 2.4: Influence of the variation of the machine setting parameters on the mean cavity pressure during the holding pressure phase

A significant correlation with the machine setting parameters can also be identified for the other process parameters. The injection pressure integral correlates significantly with the injection speed (0.91). This can be explained by the fact that the pressure increases when the melt is injected faster into the cavity. The influence of the other machine setting parameters on the cavity pressure is not significant.

The integral increase of the mould wall temperature only correlates significantly with the injection speed (0.87). Due to the faster injection speed, the material is sheared more strongly, which results in an increase in temperature. The higher melt temperature also results in a lower pressure gradient in the melt. Therefore, the injection mould cavity pressure integral also increases with the injection speed. The pressure integral is also slightly influenced by the changeover point (0.21), the back pressure (0.23) and the rotation speed. By varying the switchover point, the amount of plastic that is introduced into the cavity during the injection phase differs. The setting parameters rotation speed and back pressure influence the shear and thus the temperature of the melt during dosing. This is underlined by the slight but significant correlation of the back pressure (0.14) and the rotation speed (0.17) with the maximum melt temperature measured with the IR sensor.

The evaluation shows that the process variables are influenced by the targeted variation of the machine setting parameters. Furthermore, the process parameters correlate with the component quality. Figure 2.5 shows an example of the correlation of the process variables with the distance of the breakthroughs when the machine setting parameters are varied. The component dimension is most strongly reflected by the injection pressure integral (-0.51) and the cavity pressure integral during the injection phase (-0.61). Also the mean cavity pressure correlates significantly with the distance. The other quality characteristics also correlate with some of the process parameters. This shows that the correlations between process characteristics and part quality can also be established by varying the machine setting parameters.

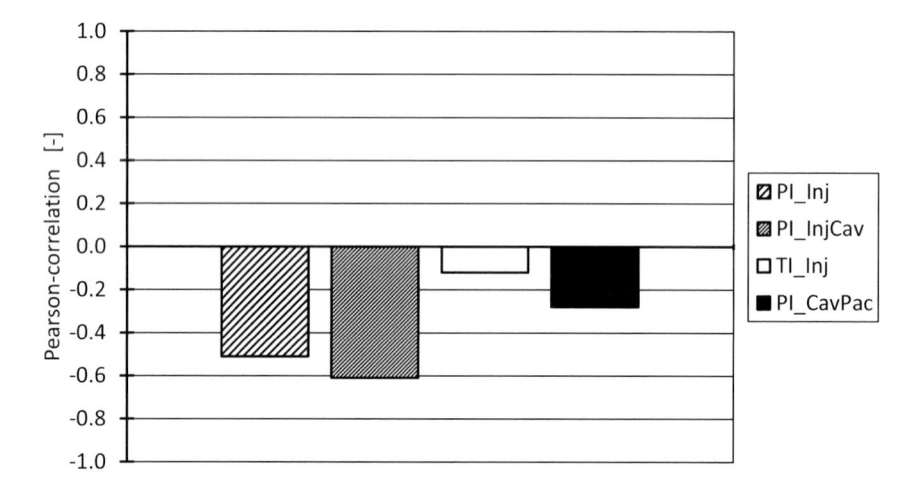

Figure 2.5: Correlation of the process variables with the distance of the breakthroughs when varying the machine setting parameters

2.6 Comparison of the correlation between the process variables and the component quality in case of process variations and in case of variation of the machine setting parameters

In order to be able to recreate the disturbances for validation using the variation of the machine setting parameters, this section analyses whether the individual quality characteristics correlate equally with the process parameters regardless of the fluctuations due to variations or machine setting parameters.

Figure 2.6 shows the differences in the correlation factors of the individual quality characteristics. It can be seen that the correlation factors for the quality criterion width are almost identical. The situation is similar for the component weight, but the correlation factor of the cavity pressure integral differs by 0.21. Also, for the distance rib, the factors for the correlation of the integral temperature increase and the mean cavity pressure during the holding pressure phase when introducing the disturbances agree with those when varying the machine setting parameters. Only the influence of the injection pressure integral deviates by -0.19. For the spacing of the breakthroughs, the correlation factors of the integral temperature increase differ. The other process parameters are equal.

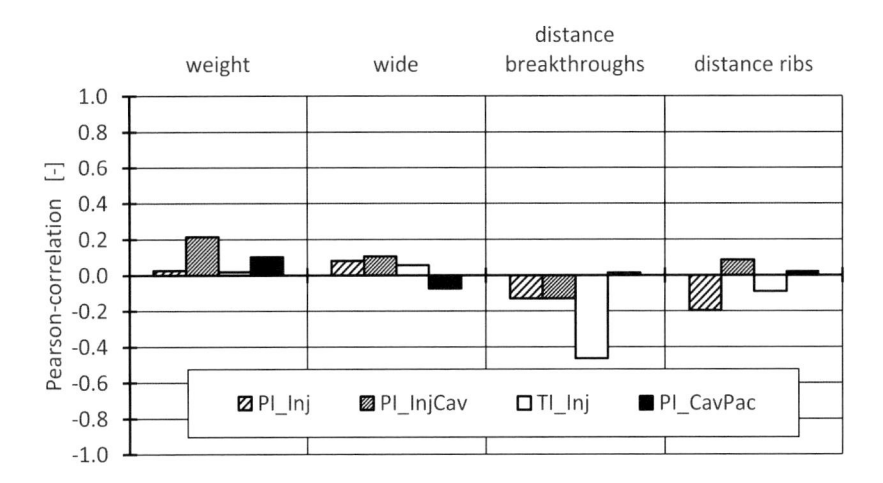

Figure 2.6: Comparison of the correlation of the process variables with the component quality when exposed to fluctuations and variation of the setting parameters

One possible reason for the partly significant differences may be that the variations in the two investigations influence the material behaviour differently in individual aspects. This can lead to differences in the correlations between individual process parameters and quality variables. Therefore, the influence of the variations on the material behaviour should be investigated simulatively in further steps. In this way, more in-depth analyses can be carried out.

2.7 Conclusions and outlook

The investigations show that by using process parameters the prevailing process conditions can be described. Furthermore, the correlation between part quality and process parameters was demonstrated, which highlights the advantage of process parameters over machine setting parameters.

It was also demonstrated that the melt quality is influenced by the specific variation of the machine setting parameters. Again, it became clear that the process parameters correlate with the quality of the manufactured components. A comparison of the correlations between process parameters and the quality criteria when fluctuations are applied and when the machine setting parameters are varied shows that most of the correlation factors almost match. However, individual differences are also apparent, which may be due to different material behaviour. In order to be able to make general statements about the possibility of validation with the help of process parameters, the material behaviour under the influence of variations should therefore be investigated further.

Furthermore, it should be analysed whether the relevant process parameters for the respective quality parameters can be predicted with the help of injection moulding simulations. Additionally, the prediction of the machine setting parameters, with which the corresponding process parameters can be influenced, should also be investigated in order to keep the amount of testing during validation as low as possible.

Acknowledgements

The research project 21127 N of the Forschungsvereinigung Kunststoffverarbeitung was sponsored as part of „industrielle Gemeinschaftsforschung und -entwicklung (IGF)" by the German Bundesministerium für Wirtschaft und Klimaschutz (BMWK) due to an enactment of the German Bundestag through the AiF. We would like to extend our thanks to all organizations mentioned.

References

[AB17] ANES, J.; BEAUMONT, J.: Process Validation for Injection Molding Medical Components. *Plastics Engineering* 73 (2017) 5, p. 38–43

[Gro00] GROLEAU, R.: Location Independetn PPAP Streamlined Global Manufacturing. *Proceedings of Automotive Conference, University of Michigan.* 2000

[Gue20] GUERRA, M.: *Validation of the injection molding process to produce the primary packaging of tissue valves.* Politecnico di Torino, Masterthesis, 2020 – supervisor: Prof. Silvia Spriano and Arnaldo Giannetti and Carlo Guala

[Ham04] HAMAN, S.: *Prozessnahes Qualitätsmanagement beim Spritzgießen.* TU Dresden, Dissertation, 2004

[HW15] HEINZLER, F.; WORTBERG, J.: Qualitätsregelung beim Spritzgießen - Teil 2: Adaptive, druckgeregelte Prozessführung. *Zeitschrift Kunststofftechnik* 11 (2015) 3, p. 157–179

[LBS+83] LAMPL, A.; BROWN, R.; SMITH, B.; HARRINGTON, M.; BROWN, D.; SCULLY, S.; BUTLER, D.: Kontrolle des Formfüllvorgangs beim Spritzgießen mit Hilfe der Einspritzarbeit. *Plastverarbeiter* 34 (1983) 10, p. 1105–1108

[LBS+18] LUSARDI, G. AND BROWN, R. S. B. H. M.; BROWN, D.; SCULLY, S.; BUTLER, D.; VALLEY, E. AND THERRIEN, M.: *"Part Process" Transferability Using Machine Independent Variable (MIV) Methodology for Multiple Machines.* White Paper, 2018

[Mus00] MUSTAFA, M.-H.: *Modellbasierte Ansätze zur Qualitätsregelung beim Kunststoffspritzgießen.* Universität-Gesamthochschule Essen, Dissertation, 2000

[NN04] N.N.: *Quality Management Systems - Process Validation Guidance.* White Paper, 2004

[NN16] N.N.: *DIN EN ISO 13485 - Medizinprodukte-Qualitätsmanagementsysteme - Anforderungen für regulatorische Zwecke.* Berlin: Beuth Verlag, 2016

[NN17] N.N.: *21 CFR Part 820 (Quality System Regulation).* Code of Federal Regulations, N.N., 2017

[Pet05] PETRETICH, C.: *Development of a validation methof for thermoplastic injection molding processes for the contract medical device manufacturer.* Bowling Green State University, Dissertation, 2005

[RL19] RAZENBÖCK, J.; LHOTA, C.: Konforme dynamische Prozesse - Assistenzsysteme in die Validierungsstrategie einbinden. *Medplast* 13 (2019), p. 18–21

[The17] THERRIEN, M.: 'Part Process' Development and Validation for Multiple Machines. *Medical Product Outsourcing Magazine* (2017), p. 156–161

[VRA12] VESELOV, V.; ROYTMAN, H.; ALQUIER, L.: Medical Device Regulations for Process Validation: Review of FDA, GHTF, and GAMP requirements. *Journal of Validation Technology* (2012), p. 82–92

[Zöl03] ZÖLLNER, O.: Grundlagen zur Schwindung von thermoplastischen Kunststoffen. *Bayer Magazin* (2003) 3,

[ZSZ+17] ZHAO, Y.; SHENG, K.; ZHANG, X.; HENGYI, Y.; MIAO, R.: Process Validation and Revalidation in Medical Device Production. *Procedia Engineering* 174 (2017), p. 686–692

Symbols

Symbol	Unit	Description
PI	$bar\,s$	Pressure integral
t_0	s	point in time 1
t_1	s	point in time 2
p	bar	pressure
t	s	Time
TI	$C\,s$	Temperature integral
T	C	Temperature

3 Analysis of the testing influence on the tensile shear strength of microform-fit back-moulded plastic/metal hybrid components

Ch. Hopmann[1], P. Wagner[1]
[1] Institute for Plastics Processing (IKV) at RWTH Aachen University

Abstract

In order to meet the high demands on tolerances and surface qualities, for example in automotive lighting applications, a hybrid part composed of a light metal (i. e. aluminium) and amorphous thermoplastics (i. e. polycarbonate) could be an alternative to established mono-materials. To remove the need for adhesion promoter systems, a micro-form-fitted connection is used to achieve a quasi full surface bond independent of the plastic. Therefore, undercut microstructures are created by laser ablation in the metal surface. The joining process of the metal and the plastic part is combined with the moulding process of the plastic component through back-injection moulding.

To meet the part requirements of plastic/metal hybrids, the bonding strength between the two materials is one of the most important characteristics. Due to different joining approaches, a uniform testing of those plastic/metal hybrids is an ambitious challenge. To validate different testing methods for the plastic/metal bond, two tensile shear test specimens are investigated. For both test standards (DIN EN 1465, ISO 19095) a stabilizing retainer is used to reduce bending stresses during the testing. DIN EN 1465 provides a much bigger test specimen in comparison to tensile shear specimens according to ISO 19095. Parameters influencing the bond between metal and plastic component as for example laser structure distances, laser parameters and pre-treatments are varied to validate the influence of the specimen size. Regardless of the chosen test specimen, the tensile shear strength of the hybrid joint increases with decreasing structural spacing. A pickling treatment removes structural irregularities, deepens the structures, and allows a deeper filling thus also leading to a higher bond strength. The achieved strength of the test specimens according to DIN EN 1465 tends to be higher than those according to ISO 19095, but comparability between the two testing standards can be achieved within the scope of the standard deviation.

3.1 Introduction

Automotive headlights are highly relevant vehicle components from a safety and marketing point of view. Lighting contributes significantly to road safety [KK20] and LED headlights in particular open up new design possibilities for conveying the vehicle and brand image and thus represent an important stylistic element of modern automobiles [DHJT13, HH08]. As a result, the visual and functional requirements for headlight assemblies are increasing.

Due to the small size of the light source in LED headlights, even small defects in the reflector surface can lead to imaging errors in the reflection. Therefore, tight manufacturing tolerances, excellent surface qualities and high dimensional stability even under elevated temperatures are necessary [LBQI03, URLb]. These requirements can rarely be met by a single material, or only at high production costs, which is why a hybrid structure of light metals as for example aluminium and amorphous thermoplastics offers a promising approach in new lighting applications. A light metal component made of a die-cast alloy, or a bent sheet metal part creates the support module, which enables tight tolerances over a wide temperature range due to its low thermal expansion. It also serves as a heat sink to dissipate the high temperatures generated in the light source [QGBI02, Rös14]. However, die-cast components have limitations in terms of surface quality and freedom of design [Abt20, FG20]. By back moulding the light metal component with amorphous thermoplastics, high tolerance requirements and the high surface qualities required for metallisation can be achieved without additional post-processing due to the lower shrinkage compared to semi-crystalline thermoplastics [LBQI03, QGBI02]. Furthermore, the use of the injection moulding process enables a wide range of possibilities for functional integration and a high freedom of design, as well as a high level of economic efficiency due to short cycle times with no need for post-processing [LBQI03].

For the appropriate use of plastic/metal hybrids, the bonding strength between the two materials is one of the most important characteristics. The achievable strength depends mainly on the bonding method [Roe14]. Whereas industrially, full-surface bonding agents or local injection points are often used, the IKV Aachen and project partners analyse a (quasi)full surface bond through laser structured micro-form-fit elements. These are created in the light metal component and enable a connection, which is independent of the moulding material.

Due to the different joining approaches, uniform testing of plastic/metal hybrids is an ambitious challenge. A test based on DIN EN 1465 with a modified tensile shear test specimen geometry for hybrid components is often used [NN09]. For the specific testing of plastic/metal hybrid components, a new and significantly smaller tensile shear test geometry was published in ISO 19095 in 2015 [NN15a, NN15b]. The use of the test specimen according to ISO 19095 is necessary to secure standardised testing and save material during the trial periods. However, up to now there has not been sufficient comparability of the strengths achievable by the different test specimens. Therefore, this study compares the different tensile shear strengths and therefore the bond strength of equally bonded test specimens to create a basis for assessment by following the process route presented in Figure 3.1. At first aluminium metal sheets are sized, cut and laser structured. Afterwards some specimens receive a pickling treatment. The metal component is than back-moulded and the produced test specimens are analysed through tensile shear tests.

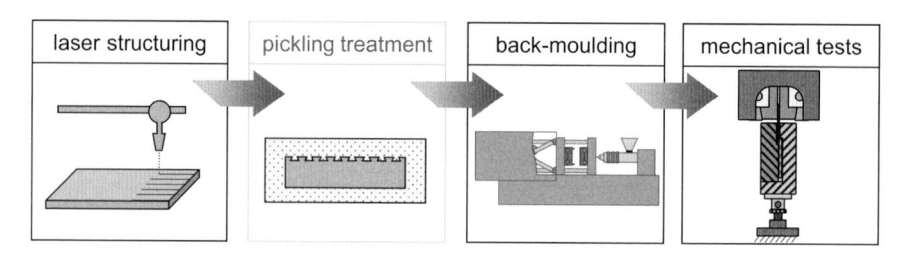

Figure 3.1: Process route to analyse the influence of plastic/metal hybrid test specimens on the resulting tensile shear strength

3.2 Joining of plastic/metal hybrid components

In the production of plastic/metal hybrids, In-Mould Assembly (IMA) and Post-Mould Assembly (PMA) processes can be distinguished. Both processes describe the joining of plastic and metal components. In the multi-stage post-mould process the two components are manufactured separately in an upstream process and then joined together to form a single component [Drö20]. In the in-mould process, the joining process is simultaneously to the moulding process of the plastic component [HFKE17]. In the post-mould process, a bond between the components is created with the help of adhesives, screws or thermal joining [Bre11]. The separate production of the semi-finished products allows greater component tolerances compared to the IMA process, as the metal component has no sealing function [Drö20]. The advantage of bolting or clamping is that no chemical similarity of the components is needed. [Fle21]. The substance to substance joining by adhesive also enables a bond of dissimilar substances, but a surface pre treatment is often necessary to ensure sufficient wetting by the adhesive [Pau13]. In contrast to the PMA, the cycle time can be significantly reduced through the IMA [Bre11, Drö20]. The upstream production step for the plastic component is omitted. The metal sheet inserted in the injection mould is bonded to the plastic melt by injection or back-moulding during the moulding process [Bre11].

3.3 Testing of plastic/metal hybrid components

A frequently used test method for determining the bond strength is the tensile shear test according to DIN EN 1465. It is formally used to analyse the component quality and ageing behaviour of bonded metal joints [RS15]. As shown in Figure 3.2, the specimen geometry can be partially modified for the investigation of plastic/metal hybrids. The metal sheet can be used as an inlay in the back-moulding process and the tensile shear strength of the plastic/metal hybrid joints can be determined quickly and easily. However, the tensile shear strength of the hybrid joint does not only depend on the joining surface, but is influenced by several factors The mechanical properties and the surface condition of the individual materials and test specimen geometry, operating conditions such as the joining process have a considerable influence on the load-bearing capacity of the joint [Ras86a, Ras86b].

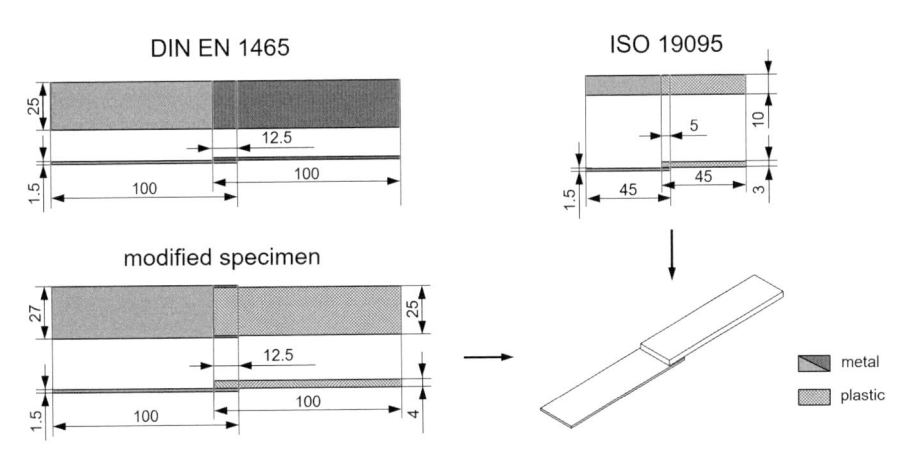

Figure 3.2: (Modified) tensile shear test specimen according to DIN EN 1465 and ISO 19095

In free-hanging tensile shear tests, as for example DIN EN 1465, the plastic component often breaks before the joint fails [PHW18]. The significance of the test with regard to the bond strength is very limited, because it is not certain, whether the maximum adhesion has been reached [PHW18, Ras86a]. In addition to the influence of the specimen geometry, there is also a considerable influence of other factors, such as the specimen restraint or the ductility of the components [RS15]. Due to the free-moving surfaces of the specimen and the displacement of the tensile forces, bending moments and notch effects are possible, which lead to uneven stress distributions [RS15]. Heterogeneous stress distributions can lead to stress peaks, which cause early failure [BHH+11]. This means that the significance of the strength of the bond under investigation is low. In addition, shear stresses and peel stresses in hybrid composites with adhesion promoters lead to lower tensile shear strengths due to adhesive and component elongation [RS15]. The bending and torsion forces, which would considerably reduce the informative value of the test according to DIN 1465, can be minimised by adapting the test methods and specimens [Hab16, RS15]. To prevent failure of the plastic part, there is an alternative tensile shear test method specifically for plastic/metal hybrid components, which is specified in ISO 19095 [NN15b, NN15c]. The size of the specimen as shown in Figure 3.2 is reduced and the specimen is not clamped between two specimen grips, but the joint is instead enclosed on the plastic side by a so-called 'retainer'. The retainer applies pressure to the plastic component at the end of the connection, avoiding stresses due to plastic expansion as in DIN EN 1465. In addition, the retainer counteracts the bending of the connection as a result of the bending moment, which reduces strength-reducing peel stresses.

3.4 Test specimens and test setup to analyse the tensile shear strength of back moulded plastic/metal hybrid components

3.4.1 Used materials

To analyse the tensile shear strength of plastic/light metal hybrid components, the rolled aluminium alloy $AlMg_3$ from Rosen Metall-Service, Aachen, Germany, is chosen, due to its weather resistance and frequent usage in the automotive sector [URLa]. The metal sheet is 1.5 mm thick and the specimens are cut with standard guillotine shears. The investigated moulding compound is the unfilled polycarbonate (PC) Makrolon 2205 black from Covestro AG, Leverkusen, Germany. PC is chosen because of its great optical surface properties, which allow metal coatings and appliances in automobile reflectors [RG15].

3.4.2 Laser structuring of the light metal components

A laser system from the Fraunhofer Institute for Laser Technology (ILT), Aachen, Germany, is used to create undercut microstructures. The laser system is an ytterbium fibre laser type ROFIN FL 020 C with a nominal power of 2000 W from Rofin-Sinar Laser GmbH, Hamburg, Germany. The laser has a beam radius of 20 µm and generates high-precision microstructures through mirror steering systems. Two modules, varioSCAN and intelliS-CAN from the manufacturer Scanlab GmbH, Puchheim, Germany, are used to guide and control the laser system. In addition to different line distances, the laser parameters are varied.

The aim of the micro structuring process is the creation of undercuts for the plastic melts resulting in a strong bond. Previous studies on the filling behaviour showed that with an isothermal process a structure opening width of less than 100 µm can result in the plastic not being able to flow completely into the deep undercuts [Bau20]. As the opening width increases, better mould filling is achieved. In addition, the burrs at the structure edges can negatively influence the flow behaviour of the plastic melt, which can hinder the filling degree of the microstructures [Str20].

To create sufficiently deep microstructures and to maximise the bond strength, the laser with a power of 1200 watts moves with 15 m/s. The laser beam width is 20 µm and as shown in Figure 3.3 for aluminium, one linear structure is formed by five lines with a distance of 30 µm in between and three repetitions. Previous tests showed that the laser structures influence the achievable bond strength. To validate the test specimens independent from the laser structuring process parameters, a reference laser structure 2 is analysed. For the adapted parameters the laser power is reduced to 900 W and the distance between the five single lines is increased to 50 µm.

Exemplary laser structured aluminium sheets and the build-up process of the microstructures are shown in Figure 3.3. Furthermore, the different structure distances investigated in this study are presented for the different test specimens.

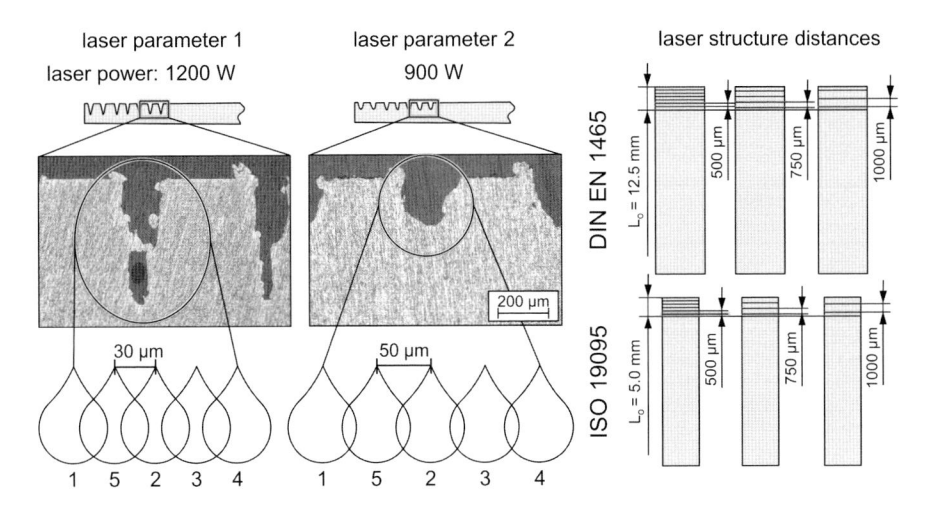

Figure 3.3: Build-up of the microstructures and investigated structure distances

3.4.3 Pickling treatment

In order to improve filling during the back-moulding process, the structure openings are partially enlarged with the help of a pickling treatment. The pickling treatment also increases the reproducibility of the specimens, because it reduces the surface roughness and removes excess material created by the laser structuring process. The pickling is carried out by JUBO Technologies GmbH, Wuppertal, Germany. A treatment shown in (Table 3.1) is implemented to specifically react with the aluminium substrate.

Process	Product	Parameter
Cleaning	TPID 10570-44	50 g/l; 50 °C; 6 min
Rinsing	Water	RT; 0.5 min
Acid Pickling	JUBOclean 5000 PE	50 g/l; 60 °C; 10 min
US-rinsing	Water	RT; 0.5 min
VE-rinsing	VE-water	RT; 0.5 min
Conversion	JUBOcoat 1056	10 g/l; RT; pH: 3.5; 2 min
VE-rinsing	VE-water	RT; 0.5 min
Drying	Circulating air	120 °C; 10 min

Table 3.1: Parameters for the pickling treatment of aluminium

3.4.4 Back-moulding

An intElect2 100/470-250 series injection moulding machine from Sumitomo (SHI) Demag Plastics Machinery GmbH, Schwaig, Germany, is used to back-mould the test specimens. The maximum clamping force of the machine is 1,00 t, the screw diameter is 30 mm, and the maximum injection pressure is 2,510 bar. A mould base with interchangeable inserts is used to produce the tensile shear specimens according to DIN EN 1465 and ISO 19095 as shown in Figure 3.2. Two water tempering units from the manufacturer Regloplas AG, St. Gallen, Switzerland, type Regloplas P 140, are used. The water temperature is set to 105 °C, which results in a mould temperature of 98 °C due to heat losses. Before processing, the material is dried at 120 °C for three hours in a type ERD 65 B dryer from Fasti GmbH, Ispringen, Germany. The selected setting parameters shown in (Table 3.2) vary due to the different specimen size and thickness in order to ensure demoulding without distortion.

Parameter	unit	DIN EN 1465	ISO 19095
Injection flow rate	[cm³/s]	40	30
Dosing volume	[cm³]	16	16
Switchover volume	[cm³]	14	14
Packing pressure	[bar]	650	850
Packing pressure time	[s]	6	6
Cooling time	[s]	40	15
Back pressure	[bar]	50	50
Mould temperature	[°C]	98	98
Melt temperature	[°C]	290	290

Table 3.2: Injection moulding processing parameters - tensile shear specimens

3.4.5 Analysing methods and machines

The filling of the microstructures is examined on the basis of micro-sections using a VHX-5000 light microscope by Keyence Deutschland GmbH, Neu-Isenburg, Germany. For the tensile shear tests a universal testing machine of the series Z010 of the manufacturer Zwick GmbH Co. KG, Ulm, Germany, is used. The maximum test load is 10 kN and the crosshead speed is freely adjustable between 0.0005 mm/min and 2,000 mm/min. As specified in ISO 19095, the testing speed is set to 10 mm/min [NN15c]. The evaluation of the measured values determined by the tensile testing machine is carried out with the testXpert II software developed by the same manufacturer. This test software calculates the corresponding tensile shear strength based on the overlap area and the measured values. Previous studies showed the significant influence of a stabilising element, a retainer, to improve the reliability of tensile shear tests. As shown in Figure 3.4, the retainer influences the course of the force-deformation curve and thus the resulting tensile-shear strength and the fracture behaviour. The achievable strength of the bond increases significantly independent of the underlaying test standard, because only very small bending deformations at the joint are geometrically possible in the retainer. Additionally, stress peaks, which can occur through bending, are reduced. Therefore, a retainer is used for the following tests allowing a comparability of the test specimens due to equal testing conditions.

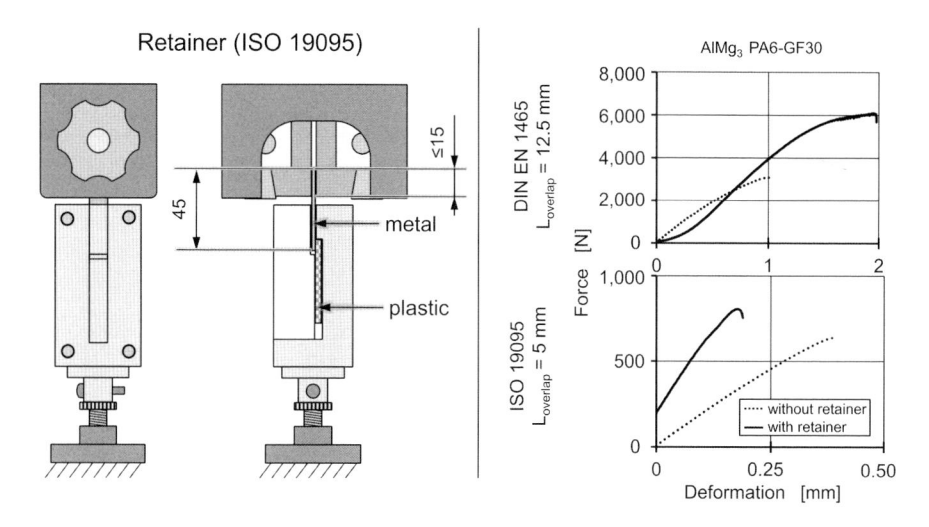

Figure 3.4: Construction layout and influence of the retainer used in tensile shear strength testing

3.4.6 Analysis of the test method influence on the resulting tensile shear strength

The comparison of the underlaying test standard is investigated through the use of different test specimens with equally pre-processed and structured aluminium sheets, which are back moulded with unfilled polycarbonate. To validate the results and to minimise the influence of fluctuations during the laser structuring process, pickled test specimens are analysed as well. In Figure 3.5 the influence of the chosen laser parameters on the tensile shear strength is presented. Focusing on the laser structuring influence, higher overall strengths can be achieved with laser parameter 2. As can be also seen in Figure 2.5, this is due to the larger structure openings and the associated better filling behaviour. As laser parameter 2 has larger structures, a higher strength can be achieved before pre-treatment than with laser parameter 1.

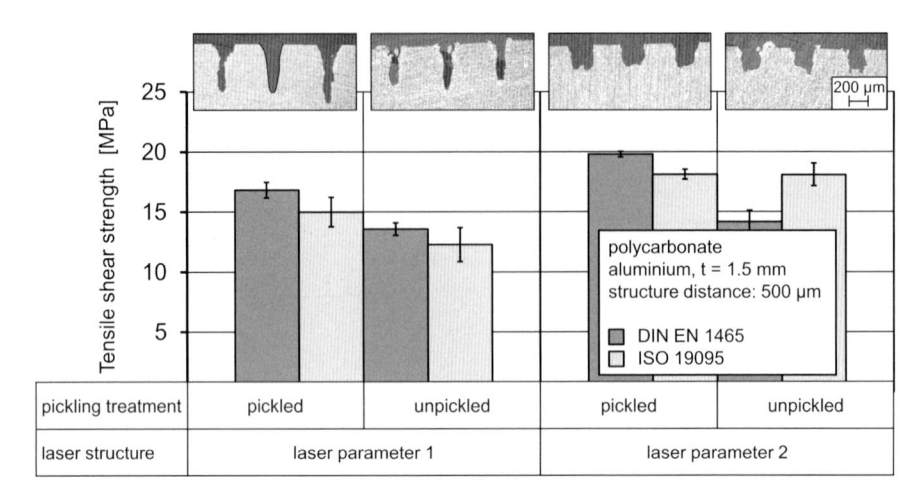

Figure 3.5: Influence of the laser parameters, pre-treatment, and test specimen on the resulting tensile shear strength

A comparison of the pickled samples shows that there is a similar relationship between the two test specimens regardless of the chosen laser parameters. For laser parameter 1 a tensile shear strength of 16.81 (± 0.65) MPa can be achieved with the bigger specimen and a slightly reduced strength of 14.99 (± 1.22) MPa with the specimen according to ISO 19095. Reducing the laser power and increasing the distance in between one laser line increases the tensile shear strength to 19.80 (±0,22) MPa (DIN EN 1465) and 18.12 (±0.43) MPa (ISO 19095). On average, the measured strength of the specimens according to DIN EN 1465 is therefore approx. 10°% higher than the tensile shear strength of the smaller test specimens. However, the standard deviation reduces the accuracy of the results and possible interpretations. It is noticeable that only the specimens structured with laser parameter 2, which were back moulded without pre-treatment show a higher strength using the smaller test specimen (ISO 19095). It can be assumed that without pickling treatment the irregularity of the microstructures led to an increase in strength in the small overlap area. After the pickling treatment, the effect was reversed.

Figure 3.6 shows the influence of the test specimen on the resulting tensile shear strength before and after the pickling treatment for the specimens structured with laser parameter 1. Furthermore, the tensile shear strength depending on the laser structuring distances is demonstrated. The highest tensile strength of 16.81 (± 0.65) MPa can be achieved with narrow structuring (500 µm) for the bigger test specimen (DIN EN 1465) after the pickling treatment. Reducing the specimen size and the overlap area results in a decrease to 14.99 (± 1,22) MPa. A similar behaviour can be observed for the unpickled and for the test specimens with a structure spacing of 750 µm. Using the test specimens according to ISO 19095 tends to result in a lower tensile shear strength. However, the difference is only slightly significant due to the measured standard deviation. Focusing on the broader spaced linear structures (1000 µm) the achievable strengths approach each other. The results after the pickling treatment show a tensile shear strength of 8.17 (± 0.55) MPa for specimens according to DIN EN 1465 and 8.40 (± 0.52) MPa for the smaller ISO 19095-specimens. The back-moulded specimens without pre-processing generate similar results with a difference of less than 1 MPa between the two test specimen sizes. It can be assumed that the transferability of the test results is more likely to be accurate for larger structural distances.

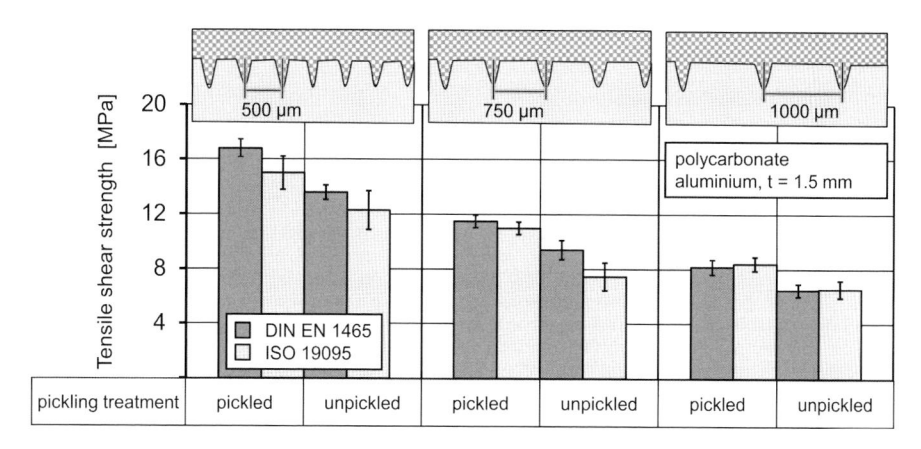

Figure 3.6: Influence of the structure distance, pre-treatment, and test specimen on the resulting tensile shear strength

With increasing space between the linear laser structures, the tensile shear strength decreases. This is independent of the used specimen and the pre-treatment. Fewer structures lead to less anchoring possibilities for the plastic component and thus lower bond strength. Even smaller spacing could further improve the tensile shear strength, but it would also lead to thermal challenges during the laser structuring process and would be uneconomical due to the process time extension. The pickling treatment positively effects the measured tensile shear strength regardless of the specimen size. As can be seen in Figure 3.7, the process is able to widen the structure openings and to remove excess material, burrs and small reclosures in the structure itself. This leads to a deeper filling in the back-moulding process and thus higher bond strength.

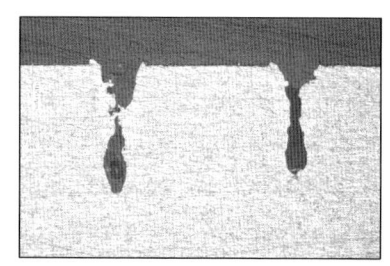

before pickling treatment   after pickling treatment

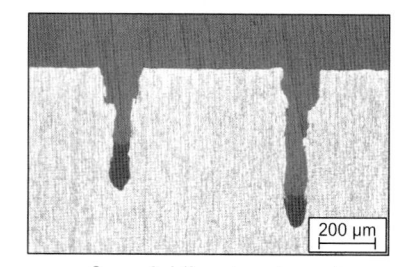

Figure 3.7: Influence of the pickling treatment on the microstructure geometry

Even though the structures cannot always be completely filled, the filling depth increased significantly in comparison to previous studies. With narrower opening widths, the plastic tends to shear off during the tensile shear tests. The deep filling and larger connection through broader structure openings strengthens the bond between the two components, because of the increase in anchoring possibilities. Higher bond strengths open up further application possibilities for plastic/metal hybrids.

3.5 Conclusions and outlook

The aim of the investigation was to compare and contrast the different possibilities to analyse plastic/metal hybrid components with two different (partially) standardised tensile shear test specimens. Therefore, equally pre-treated and laser structured tensile shear specimens were analysed. The experiments show that reducing the specimen size according to ISO 19095 slightly influences the resulting shear strength. A tendency towards a reduction in bond strength can be observed regardless of the used laser parameters and pre-treatments. Overall the results can be used to compare the different test specimens. Even though the results show deviations between the specimens, a transferability of the standards onto each other can be derived. Effects that influence the tensile shear strength of a plastic/metal hybrid component affect both test specimens equally. The bond strength increases with decreasing structure distances and after a pickling treatment. The results achieved are particularly important with automotive applications in mind, because the comparability of different test standards ensures that the industry-specific requirements can be met accordingly. For the specific use of plastic/metal hybrids in headlight units, the geometric structure and the structural arrangement of the microform-fit joint can be adapted to achieve a trade-off between cost-effectiveness and mechanical properties. In order to further prove the comparability of plastic/metal hybrid components test standards, investigations should be carried out with other plastic moulding compounds as well as other metal alloys. Additionally, the usage of different form closure elements and primers should be analysed to verify the results independent of the bond characteristic. To evaluate the plastic/metal hybrid bond for its application in automotive lightning technologies the influence of temperature changes on the bond strength and the media tightness of the microform-fit interlocking must be analysed.

Acknowledgements

The project "Development of a microform-fit plastic/light metal hybrid material composite with a Class A surface - Form-LIGHT" with the funding designation ERDF-0801465 is funded by the European Regional Development Fund (ERDF). We would like to take this opportunity to thank all the institutions for their support and encouragement.

References

[Abt20] ABTS, G.: *Kunststoff-Wissen für Einsteiger - Grundlagen, Eigenschaften und Recycling polymerer Werkstoffe*. München: Carl Hanser Verlag, 2020

[Bau20] BAUSER, M.: *Simulative Auslegung und praktische Validierung der Formfüllung von mikorstrukturierten Hybridformteilen im Spritzgießen*. RWTH Aachen University, Master thesis, 2020

[BHH+11] BYSKOV, J.; HOLM, A.; HOJSHOLT, R.; SÁ, P.; BALLING, P.: Testing the permeability and corrosion resistance of micro-mechanicalls interlocked joints. *Applied Physics A* 104 (2011) 3, p. 1–5

[Bre11] BRECHER, C.: *Integrative Produktionstechnik für Hochlohnländer*. Berlin Heidelberg: Springer-Verlag, 2011

[DHJT13] DECKER, D.; HAGE, M.; JERG, F.; TARTASCHUK, E.: Produktentstehungsprozess für Scheinwerfer und Heckleuchten. *ATZ - Automobiltechnische Zeitschrift* 115 (2013) 11, p. 888–893

[Drö20] DRÖDER, K.: *Prozesstechnologie zur Herstellung von FVK-Metall-Hybriden*. Forschungsbericht, 2020

[FG20] FIEBIG, S.; GLAUBITZ, F.: *Prozesstechnologie zur Herstellung von FVK-Metall-Hybriden*. Berlin, Heidelberg: Springer Verlag, 2020

[Fle21] FLEISCHER, J.: *Intrinsische Hybridverbunde für Leichtbautragstrukturen - Grundlagen der Fertigung, Charakterisierung und Auslegung*. Berlin: Springer Vieweg, 2021

[Hab16] HABENICHT, G.: *Kleben - erfolgreich und fehlerfrei*. Berlin: Springer Vieweg, 2016

[HFKE17] HOFFMANN, L.; FAISST, B.; KOSE, K.; EGGERS, F.: Herstellung und Charakterisierung hochfester Kunststoff-Metall-Hybride. *Lightweight Design* 10 (2017), p. 50–55

[HH08] HAMM, M.; HUHN, W.: Weltweit erster Voll-LED-Scheinwerfer im Audi R8. *ATZ - Automobiltechnische Zeitschrift* 110 (2008) 10, p. 894–901

[KK20] KOBBERT, J.; KHANH, T.: Objektive Bewertung des Sicherheitsbeitrags von heutigen Kfz-Scheinwerfern. *ATZ - Automobiltechnische Zeitschrift* 122 (2020) 6, p. 70–75

[LBQI03] LIESE, M.; BLUHM, R.; QUEISSER, J.; ICKES, G.: Thermoplaste für innovative Fahrzeugbeleuchtung. *ATZ - Automobiltechnische Zeitschrift* 105 (2003) 8, p. 700–706

[NN09] N.N.: *DIN EN 1465: Klebstoffe - Bestimmung der Zugscherfestigkeit von Überlappungsklebungen*. Berlin: Beuth Verlag, 2009

[NN15a] N.N.: *ISO 19095-1: Plastics - Evaluation of the adhesion interface performance in plastic-metal assemblies - Part 1: Guidelines for the approach*. Genf, Switzerland: ISO copyright office, 2015

[NN15b] N.N.: *ISO 19095-2: Plastics - Evaluation of the adhesion interface in plastic-metal assemblies - Part 2: Test specimens*. Genf, Switzerland: ISO copyright office, 2015

[NN15c] N.N.: *ISO 19095-3: Plastics - Evaluation of the adhesion interface in plastic-metal assemblies - Part 3: Test methods*. Genf, Switzerland: ISO copyright office, 2015

[Pau13] PAUL, H.: *Bewertung von langfaserverstärkten Kunststoff-Metall-Hybridverbunden auf der Basis des Verformungs- und Versagensverhaltens.* Karlsruher Institut für Technologie, Dissertation, 2013

[PHW18] POPOV, V.; HESS, M.; WILLERT, E.: *Handbuch der Kontaktmechanik - Exakte Lösungen axialsymmetrischer Kontaktprobleme.* Berlin: Springer Vieweg, 2018

[QGBI02] QUEISSER, J.; GEPRÄGS, M.; BLUHM, R.; ICKES, G.: Trends bei Automobil-Scheinwerfern. *Kunststoffe* 92 (2002) 3, p. 90–97

[Ras86a] RASCHE, M.: Problematik der Prüfung von Polymer-Metall-Verbunden im Zugscherversuch. *Adhäsion* 10 (1986), p. 10–18

[Ras86b] RASCHE, M.: *Qualitätsbestimmende Einflußgrößen bei Kunststoff-MetallKlebverbindungen.* Düsseldorf: Deutscher Verlag für Schweißtechnik (DVS) GmbH, 1986

[RG15] REINARTZ, K.; GLÄSSER, M.: Die Vorteile von Polycarbonat-Kunststoffen für die LED-Lichttechnik. *Elektronik Praxis* (2015),

[Rös14] RÖSNER, A.: *Zweistufige laserbasierte Prozesskette zur Herstellung von Kunststoff-MetallHybridbauteilen.* Fraunhofer Institute for Laser Technology ILT, Dissertation, 2014

[RS15] RASCHE, M.; SYPEREK, D.: Zugscherversuch nach DIN EN 1465 - Ergebnisse mit Vorsicht behandeln. *Adhäsion KLEBEN DICHTEN* 11 (2015), p. 26–31

[Str20] STRAETEN, K. V. D.: *Laserbasiertes Fügen von Kunststoff-Metall-Hybridverbindungen mittels selbstorganisierter Mikrostrukturen.* RWTH Aachen University, Dissertation, 2020

[URLa] N.N.: *3.3535 Datenblatt.* URL: https://facts.kloeckner.de/werkstoffe/aluminium/3-3535/, 04.10.2021

[URLb] N.N.: *Technik & Produkte.* URL: https://www.hella.com/techworld/de/Technik/Beleuchtung/Scheinwerfer-219/, 27.09.2021

Abbreviations

Notation	Description
IMA	In-Mould Assembly
PC	Polycarbonate
PMA	Post-Mould Assembly

Additive manufacturing of large components

Moderators: Dr.-Ing. Wolfgang Meyer, Volkswagen Aktiengesellschaft
Olaf-Björn Kölle, Volkswagen Aktiengesellschaft

Content

1 Automotive industry requirements for the additive manufacturing of large parts 549

W. Meyer[1], O. Kölle[1]
[1] *Volkswagen Aktiengesellschaft*

2 Investigation into the differences of mechanical properties of additive manufactured and injection moulded parts 558

Ch. Hopmann[1], R. Dahlmann[1], J. Austermann[1]
[1] *Institute for Plastics Processing (IKV) at RWTH Aachen University*

3 Expanding non-planar additive manufacturing with variable layer height to be used on a six-axis robot arm 577

Ch. Hopmann[1], R. Dahlmann[1], L. Pelzer[1]
[1] *Institute for Plastics Processing (IKV) at RWTH Aachen University*

Dr.-Ing. Wolfgang Meyer

Dr.-Ing. Wolfgang Meyer studied mechanical engineering, specialising in aeronautical and aerospace engineering. He earned his Diploma of Engineering at the Technical University Braunschweig in 1985. Since then, he worked for Volkswagen, taking up various technical and managerial positions. Among his many jobs within VW, he served as the head of production at Volkswagen Motor Polska, an engine factory in Polkowice, Poland, during its first four years of production from 1998 until 2002. After earning a doctorate in 2008 from Otto von Guericke University Magdeburg for work in the field of automobile production, he became head of body manufacturing at the Volkswagen Pilot Hall, a position he held for five years. Dr Meyer is currently head of Central Functions and Innovations at the Pre-Production Centre of Development.

Olaf-Björn Kölle

Olaf-Björn Kölle completed his apprenticeship as a plastics technician at Volkswagen. He then began working in the department of technical development in the production of plastic parts. Since the procurement of the first additive manufacturing system by Volkswagen in 1997, Mr Kölle has been responsible for the field of additive manufacturing at the Volkswagen Pre-Production Center as a technical expert concerning all questions regarding the process chain of the additive manufacturing of plastic parts.

1 Automotive industry requirements for the additive manufacturing of large parts

W. Meyer[1], O. Kölle[1]
[1] Volkswagen Aktiengesellschaft

Abstract

The high flexibility of additive manufacturing (AM) methods allows for an economical manufacturing of automotive prototype parts, speeding up the overall development process and decreasing risks. By the utilisation of AM methods parts can be tested and validated before a final design is defined. Despite this, the application of AM methods is currently limited to the early phases of the development cycle, such as geometric prototypes. In the scope of this report, the uses, advantages and requirements that are to be met by AM prototypes are illustrated with the help of example parts of successful developments. Additionally, current limitations of AM prototypes are analysed and an outlook to further research and development tasks to extend the use of AM prototypes is given.

1.1 Introduction

Tracking current publications on the industrial use of AM methods, one can gain the impression that these manufacturing methods have now become established on the market and can be used directly for a wide range of tasks. When it comes to the manufacture of automobiles in high-volume production, it is worth taking a closer look and identifying processes in which additive manufacturing methods can reasonably be introduced. It can be assumed that the operation of a high-volume passenger car production (defined as > 300 units/day $=$ 75 000 units/year was not a primary objective when these methods were developed [Sie99]. As shown in Figure 1.1, the benefits of additive manufacturing are in its freedom in design, its flexibility, and – because workpiece-specific tools and devices are unnecessary – the rapid and cost-efficient availability of the first product (job #1). Weaknesses manifest in terms of productivity and process stability. The fulfilment of these two parameters, though, is imperative for the installation of a high-volume industrial manufacturing system.

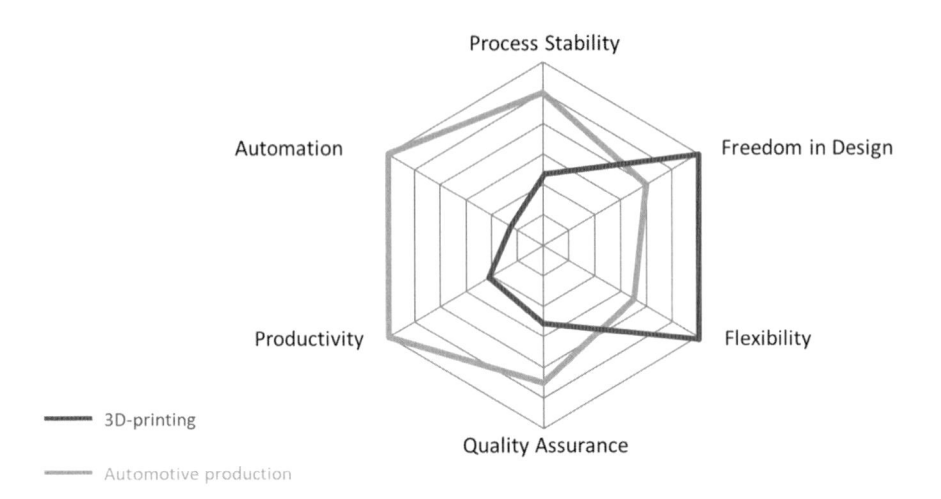

Figure 1.1: AM compared with conventional manufacturing methods in car production [SC19]

However, in analysing the entire life cycle of an automobile – extending from concept design to development, production, and sales, and to customer service – there are indeed sub-processes therein that warrant the use of additive manufacturing methods. To this end, this article analyses the prototype construction process, which is part of the development process. Among other things, prototype construction is characterised by:

- the need to provide test mules, whose properties do not yet need to correspond to those of the later series-production parts in some cases.
- required part specifications that are essential for certain tests and do not correspond to those of the later series-production parts.
- a product specification that changes frequently and is to some extent incomplete.

Using examples, we demonstrate below how these specific requirements for prototype construction were able to be met with the aid of additive manufacturing methods, thus enabling the development process as a whole to speed up.

1.2 Practical examples for the use of AM parts in the development process

The examples listed below are intended to provide a small glimpse into the various possibilities of AM methods that are currently available to solve certain tasks that emerge during the development process. In this context, rapid and cost-efficient part availability is of particular importance.

In the first example, various parts of a front end are shown, whose installation must be secured by means of physical parts (Figure 1.2). Experience has shown that fitment trials carried out using solely virtual methods sometimes do not provide sufficiently accurate conclusions in this context. Such is the case, e.g., when flexible parts (hoses, wiring) have to be joined. In addition, the possibilities of a human hand during the joining process, e.g., lightly pushing aside adjacent parts, cannot yet be simulated virtually. In the early development phase shown here, the parts that are provided by means of additive manufacturing methods are not yet required to fulfil the specifications of the later series-production parts, e.g., in terms of strength or heat resistance.

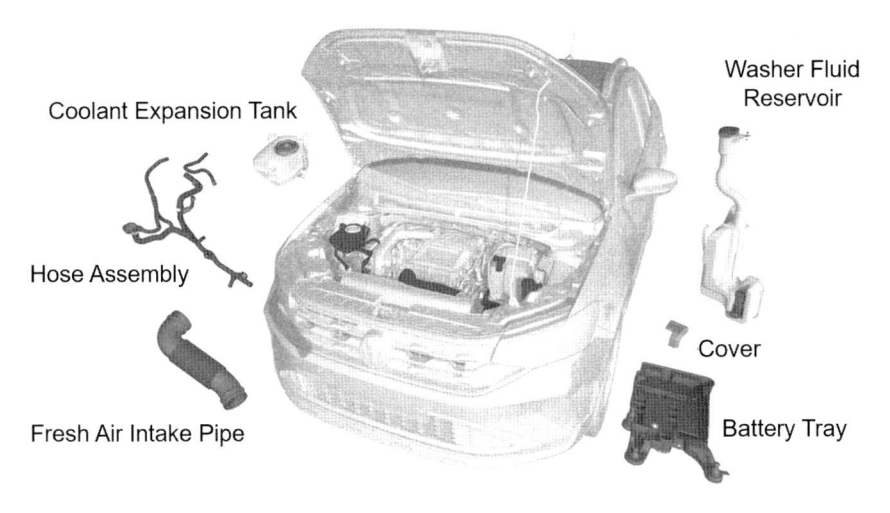

Figure 1.2: Front end and parts to be installed [source: Volkswagen AG]

In the next example, the development task is to ensure that the components of a transmission are sufficiently wetted with transmission fluid during various operating situations. To simulate this, the transmission must be installed in a transparent/translucent transmission case (Figure 1.3) to enable a visual observation and assessment of the transmission fluid flowing through that results during operation. Because the heat resistance of the transmission case produced by additive manufacturing is significantly below the typical operating temperatures of transmission fluid, a substitute fluid that is suitable for simulating the flow behaviour of transmission fluid at lower temperatures must be used for these tests. The part was manufactured using the PolyJet method on a Stratasys J850 system.

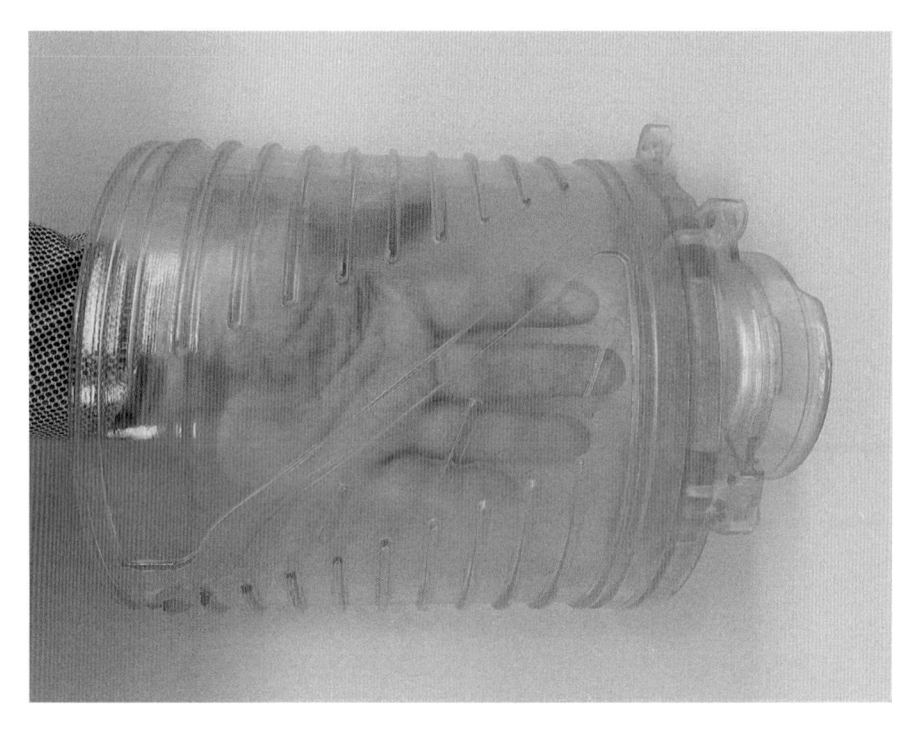

Figure 1.3: Transparent/translucent transmission case [source: Volkswagen AG]

Example 3 addresses a development task that can be solved very well with the aid of additive manufacturing. There are characteristic groups of parts that are subject to exceedingly frequent changes, owing to their dependencies with other parts, whose final manifestation is not ultimately defined until a relatively late point in time. To that end, Figure 1.4 shows various model phases of a cable holder used in a large number of vehicles. The cables to be held are subject to highly frequent changes, depending on the vehicle functions being implemented. With the aid of 3D-printed sample parts, the design engineer can very quickly perform cable-holding tests using the respective cable holder variant and thus, in a short time, arrive at an engineering result that meets the technical requirements.

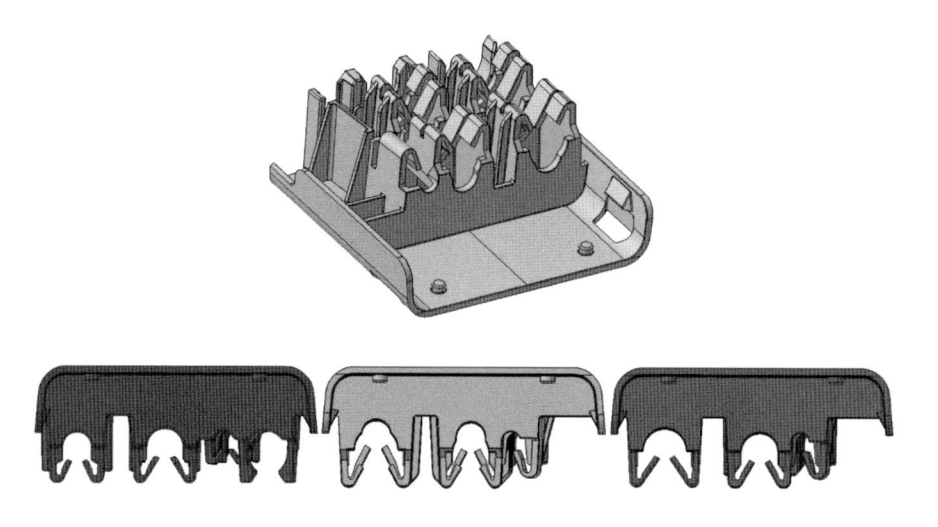

Figure 1.4: Various model phases of a cable holder [source: Volkswagen AG]

1.3 Tasks that cannot be covered by AM methods available on the market

The examples mentioned above have shown that the use of additive manufacturing methods can be helpful, particularly to solve tasks from the early development phase. In doing so, the aforementioned benefits of these methods, such as speed, flexibility, and freedom in design, are utilised. However, every development process typically ends with the issuance of the series-production release by the appropriate design engineer. For this release to be issued in good faith, the design engineer relies on results from tests carried out on parts manufactured from the specified series-production material using a method that is close to series production. For plastic parts, this would be an injection-moulded (IM) part, e.g., manufactured using a limited-production tool. The two mentioned parameters – use of series-production material and series-production method – cannot currently be implemented with the means of additive manufacturing, because:

- virtually all additive manufacturing methods available on the market use materials adapted to the manufacturing process used, which are also often offered exclusively by the system manufacturer.
- the respective additive manufacturing method differs significantly from the series-production method, because, e.g., the virtual model of the part is sliced into increments that are then joined successively to form the physical part. These differences in method lead in turn to different part properties; the anisotropy of 3D-printed parts is cited as an example of this.

Due to the mentioned limitations of additive manufacturing methods, it is therefore often necessary to construct prototype tools or limited-production tools to validate the series-production capability of newly developed parts, in order to thus manufacture test parts using a close-to-production method. Because we know – from our over 25 years of experience with the use of additive manufacturing in prototype construction and our daily, close collaboration with vehicle design engineers – that the large majority of questions that arise can be solved using 3D-printed parts, it seems naturally appealing to investigate whether additive manufacturing cannot also provide support during the series-production release process. For this reason, we have begun a joint research project with the Institute for Plastics Processing (IKV), which is explained below.

1.4 Joint development project with IKV

To derive reliable conclusions on the part properties of later series-production parts to be expected from the part properties of parts produced using additive manufacturing, one must develop a corresponding mapping function for this purpose. An approach to solving this problem is shown in Figure 1.5 .

```
┌─────────────────────────────────────────┐         ┌─────────────────────────────────────────┐
│       Additive Manufacturing            │         │         Injection Moulding              │
├─────────────────────────────────────────┤         ├─────────────────────────────────────────┤
│ Parameter ─┐                            │         │                            ┌─ Parameter │
│ Material   ├─► AM-Process ─► Part       │ ◄─ ? ─► │  Part       ◄─ IM-Process ─┤ Material   │
│ Geometry ──┘                Properties  │         │  Properties                └─ Geometry  │
└─────────────────────────────────────────┘         └─────────────────────────────────────────┘
                            │
                Objective 1: Predictability
         Prediction of the mechanical part properties
         of AM parts based on the materials and
                        parameters

                Objective 2: Transferability
         Establishment of a transferability of the properties of
         parts produced using additive manufacturing and
                those using injection molding
```

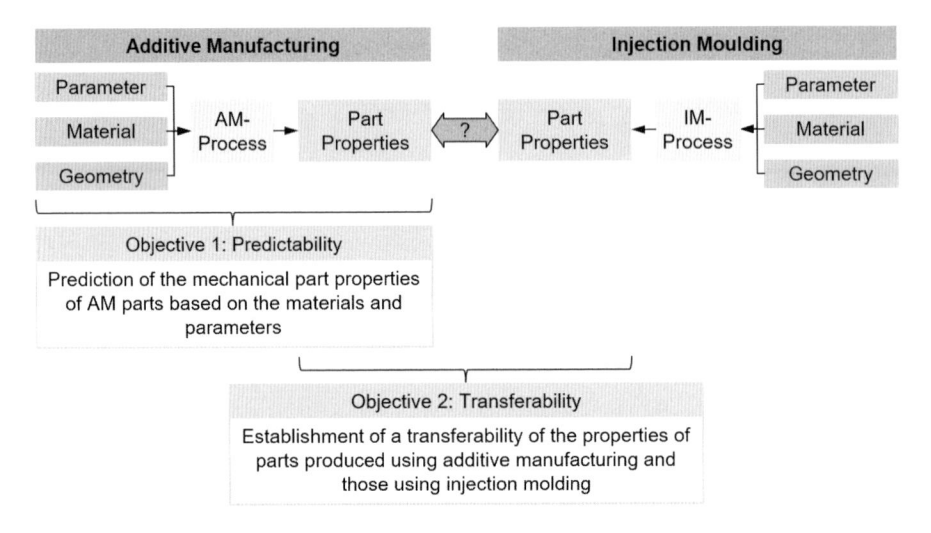

Figure 1.5: Starting points for developing a mapping function

Because the spectrum of additive manufacturing methods and the materials used is at this time quite wide and diverse, a mapping function of this kind can only be determined for a clearly-defined subsection of this spectrum. As part of a research project being worked on by IKV and Volkswagen AG, it was therefore determined that the same material must be used for both additive manufacturing and for injection moulding. Furthermore, the methods used must originate from the same method group. Therefore, additive manufacturing of the parts is carried out on the Yizumi Extruder system shown in Figure 1.6 (fused layer modelling, FLM), which is able to process the same materials as the injection-moulding system used for comparison.

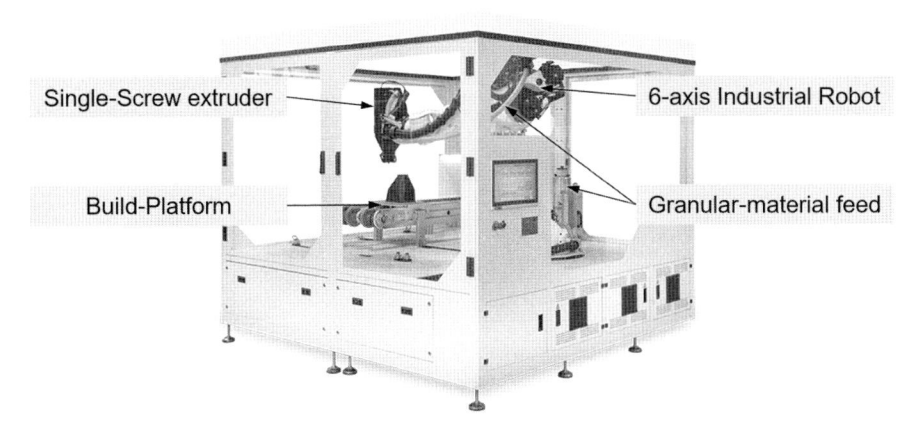

Figure 1.6: Yizumi Extruder 500-C-U-D + Automation Cell; manufacturing system used for the analyses

The research project is entitled:

- "Analyses on the predictive power of plastic parts produced using melt-based additive manufacturing" and is scheduled for completion by mid-year 2023.

1.5 Summary and outlook

This report analyses the use of additive manufacturing methods in the development process of automobiles. Using selected examples from prototype construction, it demonstrates that additive methods for the manufacture of prototype parts can be effectively utilised for certain task areas, especially from the early development phase, thus enabling the achievement of a significant speed-up of the overall development process.

In order to extend the utilisation of the time and cost benefits of additive manufacturing in the development process, a research project was initiated between Volkswagen AG and IKV. The objective of this project is to determine a predictable correlation between the part properties of a prototype part produced using additive manufacturing and those of the later series-production part, in order to thus facilitate doing away with the manufacture of limited-production tools used for series-production validation, and ultimately to realise significant time and cost benefits. Beyond the issues and current challenges listed as examples above, additive manufacturing can and does make a specific contribution toward the fulfilment of fundamental requirements for present and future automobile manufacture.

References

[SC19] STACHE, R.; CRULL, S.: Möglichkeiten und Herausforderungen durch neue 3D-Druckverfahren im Karosseriebau, Generative Fertigungsverfahren für die Karosserie. *Automotive Circle.* 2019

[Sie99] SIEGERT, K.: Entwicklungstendenzen in der Blechumformung. *Umformtechnik 2000 plus.* 1999

Abbreviations

Notation	Description
AM	Additive Manufacturing
FLM	Fused Layer Modelling
IM	Injection Moulding

2 Investigation into the differences of mechanical properties of additive manufactured and injection moulded parts

Ch. Hopmann[1], R. Dahlmann[1], J. Austermann[1]
[1] Institute for Plastics Processing (IKV) at RWTH Aachen University

Abstract

Screw-extrusion based additive manufacturing (AM) technologies offer an efficient way to manufacture prototype-parts, due to the possibility to use series materials as well as their tool-less, layer-wise manufacturing principle. However, this layer-wise manufacturing principle also results in anisotropic part properties, which differ from the mechanical properties of the final parts. This currently limits the use of AM-prototypes in the product development of technical injection moulded (IM) parts. Especially, the use of AM-prototypes in pre-series mechanical testing is limited. With the goal to further expand the use of AM-prototypes, fist the different mechanical behaviours of injection moulded and additively manufactured parts have to be assessed. In the scope of this work, the differences in mechanical properties of additively manufactured and injection moulded tensile-, 3-point bending- and notched impact test specimen manufactured from an injection moulding grade Polypropylene (PP) are investigated. Additionally, microscopic investigations are performed to analyse the material behyviour of tensile specimens manufactured using injection moulding and additive manufacturing. From the results, suitable applications and limitations of AM-prototype parts in pre-series testing are assessed.

2.1 Introduction

The field of additive manufacturing of polymers is comprised of a multitude of manufacturing techniques. All of these AM-techniques are defined by the joining of layers to directly produce a part from 3D data [NN18]. A specific subset of these are melt-based AM-processes, also referred to as material extrusion. Melt based AM-processes are characterised by the use of thermoplastic materials, which are plasticised in a movable plasticizing unit and subsequently deposited as a strand in layer-wise fashion, thus building up a part [Oss17, NN18]. Examples of melt based AM-processes are the Fused Filament Fabrication, in which thermoplastic mono-filament feedstocks are processed and screw extrusion based processes, which can employ any common thermoplastic pellets as feedstock. [GRS15, HL16, GKT19, HLD+20].

A well-established use of AM-processes is the rapid prototyping of parts during product development [GKT19]. Despite the steady gain of importance of additively manufactured end-use parts, nowadays prototyping remains the main application of additive manufacturing [NN19d, Scu20]. Especially with regards to the prototypical realization of injection moulding parts, the use screw extrusion based AM-processes shows a high potential [WRK+08]. This is based on two reasons. First, the strand- and layer-wise manufacturing principle allows for a tool-free part production, resulting in a time- and cost-effective manufacturing of prototype parts [HD03, FP16]. Furthermore, the direct manufacturing based on 3D-data allows for a quick iteration of development states, as in every production cycle a different geometry can be manufactured. Second, the utilisation of a screw-based plasticising principle enables the use of the series materials in a prototypical phase, as common thermoplastic pellets can be processed. This is a key advantage of screw-based AM-technologies compared to other AM-technologies, which are limited to the available feedstock materials. However, these cannot represent the properties of the series material. In addition to this, using large scale kinematics, such as six-axis industrial robots or gantry systems, provides the possibility to not only prototype small parts, but also parts with a size of 2 m x 1 m x 2 m or lager [LPN+21, HL16].

A promising concept is the use of melt-based AM-prototypes for mechanical pre-series testing. This approach to use AM-prototypes could contribute to the early evaluation of partial functions or functional structures at any stage of development. Consequently, design-flaws can be identified at an early stage and resolved in a cost-neutral manner, resulting in overall lower development cost. However, despite being able to manufacture large and complex prototype parts form the series material, the use of such AM-prototype parts is often limited to models or prototypes for the purpose of only testing geometrical features. This is based on the severe difference in part properties of injection moulded and additively manufactured parts. As AM-parts are manufactured in strand- layer-wise fashion, their part properties are also governed by this [CK17]. For example, the interlayer bonding strengths, in terms of tensile strength, commonly only achieve between 25 % and 60 % of the tensile strengths that can be achieved within a layer, i.e. in direction of deposited strands [CCMP16, HLD+20]. This distinct anisotropy of AM-parts differentiates them from the injection moulded counterparts they prototypically represent, making it challenging to infer the mechanical performance of IM-parts from tests conducted on prototype AM-parts. Despite its potential, so far there are no approaches in the literature for transferring the properties of melt-based additively manufactured and injection-moulded components. Although some comparisons of the properties of melt-based additively manufactured and injection-moulded components are presented, the properties of injection-moulded components only serve as a reference for AM-parts [AMO+02]. Based on this, the aim of a research project, which is currently carried out at the IKV in collaboration with Volkswagen Aktiengesellschaft, Wolfsburg, Germany, is to investigate the transferability of the properties of AM-prototypes and IM-parts. The underlying hypothesis is that the mechanical properties of injection moulded components can be inferred from the mechanical properties of melt-based AM-prototype using a suitable transfer function.

2.2 Material, equipment and methods

2.2.1 Approach

In order to better use additive manufactured parts in prototypical applications in the development of injection moulded parts, the differences in mechanical behaviour between AM- and IM-parts need to be better understood. While the motivation focuses on complex, medium to large size parts manufactured in melt based AM-processes, the fundamental differences in mechanical behaviour of AM- and IM-parts have to be established first, in order to identify suitable uses and limitations. Consequently, the aim of this investigation is to quantify the differences in mechanical performance of additively manufactured and injection moulded parts by means of mechanical testing, using standardised test methods. For this purpose, three test methods, tensile tests, 3-point bending tests and notched Charpy impact tests are carried out using injection moulded and additively manufactured specimen made from the same material. To account for the anisotropy of AM-parts two load-orientations are investigated. First, specimens are tested on which the load is introduced parallel to the deposited strands. Second, specimens are tested on which the load is applied orthogonal to the strand direction. Based on the results, the differences in mechanical behaviour of IM-an AM-parts is assessed and cases derived in which AM-prototypes can serve as a stand-in for IM-parts during development.

2.2.2 Investigated material

The presented investigations are carried out using the injection moulding grade polypropylene Finalloy EBP 820/2 C16, Total Research Technology Feluy, Feluy, Belgium. EBP 820/2 C16 is an elastomer-modified PP-compound with 20 % talc filler content. Talc is mineral filler, with plate like geometry, commonly used in PP as a nucleation agent and to reduce shrinkage during solidification. The material is used as a series material for automotive exterior parts, such as panelling and side skirts, making it suitable to be investigated within the context of additively manufactured prototype parts for pre-series investigation.

As per manufacturer recommendation, the material is dried for 2 hours at 80 °C before processing. [NN19c, NN19d]

2.2.3 Used equipment and process parameters for specimen manufacturing

Test specimens are manufactured both by AM- and IM to determine the difference in mechanical performance using the above described material. AM test specimens are manufactured using an Extruder-500-CU-D + Automation Cell screw-extrusion AM-system from Yizumi Germany GmbH, Alsdorf, Germany. The AM-System is composed of a single screw extruder plasticising unit and a six-axis positioning system. As plasticizing unit, a 16 mm screw diameter single-screw extruder is used. The plasticising unit is separated into three heating zones. Additionally, a forth heating element heats the plasticising unit's nozzle. As a kinematic a KR30 HA, Kuka Systems GmbH, Augsburg, Germany, six-axis industrial robot is used, allowing for fast and precise positioning in a large-volume work area. While principally allowing for non-planar manufacturing, all test specimens are manufactured in a layer-wise fashion. The parts are deposited on an unheated glass buildplate. To increase adhesion of the parts during manufacturing a PP-adhesive of type 64014, Tesa SE, Norderstedt, Germany is used as bonding agent. For manufacturing of the IM-specimens an injection moulding machine of the type Allrounder 520 A, Arburg GmbH + Co KG, Loßburg, Germany is used. The process parameters used for manufacturing of the AM- and IM-specimens are listed in Table 2.1.

Additive Manufacturing Parameters	
Extrusion width [mm]	2
Numbers of perimeter [mm]	2
Layer height [mm]	0.6
Printing speed [mm/s]	80
Barrel temperature (nozzle to hopper) [°C]	240; 240; 230; 220
Injection Moulding Parameters	
Barrel temperature (nozzle to hopper) [°C]	240; 240; 230; 220; 210; 200
Mould temperature [°C]]	40
Injection speed [cm^3/s]	120
Holding pressure [bar]	250
Holding pressure time [s]	6
Cooling time [s]	55

Table 2.1: AM- and IM-process parameters used for specimen manufacturing

The test specimens are not directly manufactured. AM test specimens are milled from specimen blanks taken from single-wall octagonal geometries as shown on the left side of Figure 2.1. The octagon has side length of 160 mm and height of 100 mm. The wall-thickness of the octagon is chosen to be 4 mm, corresponding to the width of two parallel strands. For better adhesion a brim is added to the bottom of the geometry. After manufacturing, the octagon is separated into plates at its edges. The sides before and after a layer-change, occurring at one of the edges of the octagon, are discarded to account for process instabilities which could impact the mechanical properties. In Figure 2.1 these sides are designated side H side and A. From the remaining plates specimen blanks in the direction of strand deposition, designated AM-P, as well as specimen orthogonal to the direction of strand deposition (AM-O) are taken, as shown in Figure 2.1. From each plate four blanks are cut. The specimen blanks are subsequently used for milling the geometries for tensile, bending and impact testing as is described in the following section.

Similar to the AM specimen blanks, the IM-blanks are cut from plates as shown on the right of Figure 2.1. The plates have a height of 174 mm and a width of 168 mm at a thickness of 4 mm. From each plate two specimen blanks are taken orthogonal to the direction of the melt flow, leading to the shown specimen designation IM-O. The lower part of the plate is not used to manufacture specimen due to ejector markings.

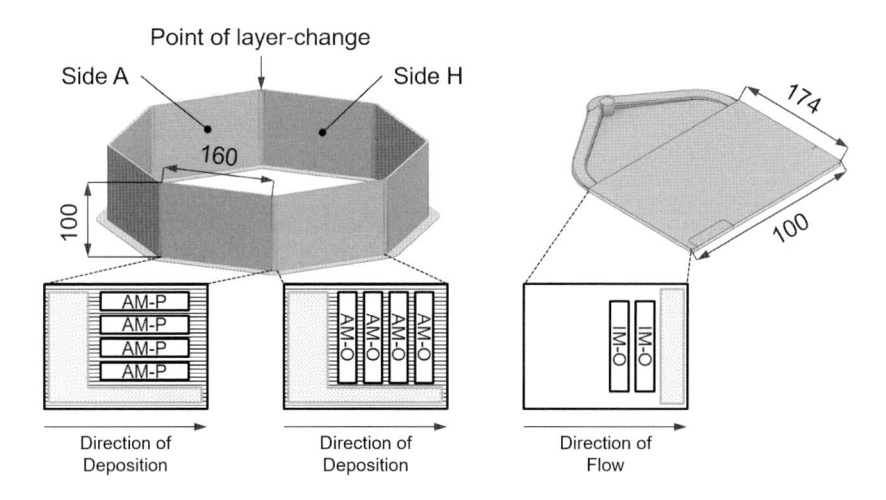

Figure 2.1: AM and IM test-geometries from which specimen blanks are taken

2.2.4 Test methods

For tensile testing, specimens of type 1BA as specified in DIN EN ISO 527-2 are machined from the AM- and IM-specimen blanks, as shown on the left side of Figure 2.2 [NN19b]. These are subsequently tested following DIN EN ISO 527 at room temperature using a universal testing machine of type Z100, Zwick GmbH Co. KG, Ulm, Germany, equipped with a 1kN load cell and an extensometer to determine the tensile strain. The tests are carried at a test speed of 1 mm/min at tensile strains below 0,25 % during which the tensile modulus is determined. When a tensile strain of 0,25 % is exceeded, a test speed of 20 mm/min is used until failure occurs. For each IM-O, AM-P, and AM-O ten valid tensile tests are carried out.

3-point bending test are carried out in accordance to DIN EN ISO 178 using the preferred test specimen type of length 80 mm, width 10 mm and thickness 4 mm, which are milled from IM-O, AM-P and AM-O specimen blanks as shown in the middle of Figure 2.2 [NN19a]. Testing is carried out using the same universal testing machine and load cell as for the tensile tests. To determine the flexural modulus a test speed of 2 mm/min is used until a flexural strain of 0,25 % is reached. From this point on, a test speed 10 mm/min is used until failure occurs or the test specimen slides of the bending supports. Ten valid tensile tests are carried out for each series.

Charpy impact tests are performed following DIN EN ISO 179-1 using notched 1eA standard specimens with a length 80 mm, width 10 mm and thickness 4 mm, machined from the test specimen blanks [NN10]. Exemplary Charpy impact test specimens are shown on the right of Figure 2.2. For each test series 10 valid tests are carried our using an impact test machine of type D-7900, Zwick GmbH Co. KG, Ulm, Germany. For IM-O a 2 J impact pendulum is used, while for AM-P a 7,5 J and for AM-O a 0,5 J impact pendulum is employed. During testing the notch impact strength as well as the mode of failure is recorded.

The specimen cross-section used for the calculation of the tensile and bending stresses is determined from the width and thickness of the respective test specimen. Width and thickness are measured the at three different locations of the gage section using a caliper and subsequently averaged. In case of AM-specimens the outer contour of the specimens is measured. The average gage section width and thickness IM-O, AM-P and AM-O tensile and 3-point bending specimen are documented in Table 2.2. Furthermore, the porous mesostructure of the AM-specimen is not taken into account for the determination of the stresses during testing. As such, the determined stresses and strains represent part properties rather then pure materials properties, as stress-redistribution effects as well as notch effects are included in the measurements.

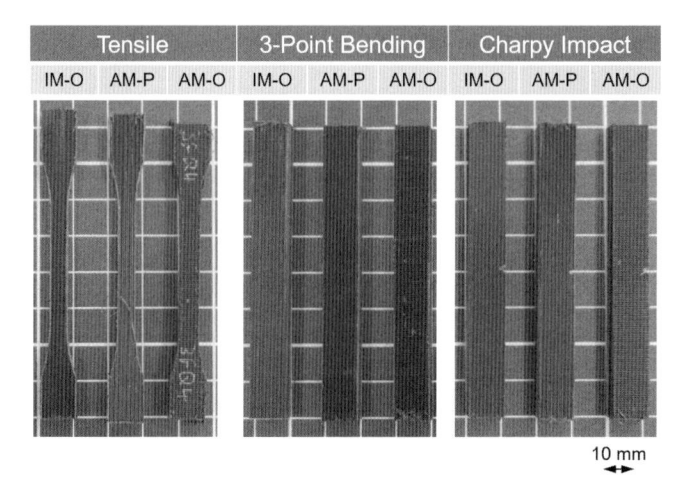

Figure 2.2: Exemplary specimens used for mechanical testing

	IM-O	AM-P	AM-O
Average tensile specimen gage thickness [mm]	3.56 ± 0.03	4.03 ± 0.09	4.07 ± 0.09
Average tensile specimen gage width [mm]	4.98 ± 0.01	5.00 ± 0.02	5.01 ± 0.04
Average bending specimen thickness [mm]	3.55 ± 0.02	4.02 ± 0.08	4.06 ± 0.06
Average bending specimen gage width [mm]	9.99 ± 0.01	9.95 ± 0.02	10.00 ± 0.02

Table 2.2: Average tensile and 3-point bending specimen thickness and width

To gain further insight into the differences in mechanical behaviour of AM and IM parts, the tensile test specimen are investigated using microscopy methods subsequent to tensile testing. Thin-sections of IM-O, AM-P and AM-O tensile test specimen are prepared and investigated using transmitted light microscopy. Additionally, field emission scanning electron microscopy (FESEM) is performed on tensile specimen fracture surfaces, providing additional information on the differences in specimen failure behaviour as well as on the influence of filler materials and their orientation on the mechanical behaviour of the tensile test specimen.

2.3 Results

2.3.1 Tensile tests

Exemplary stress-strain curves of the tensile tests performed on IM-O, AM-P and AM-O specimen are shown in Figure 2.3. For better readability the stress-strain response in small strain-rages is shown enlarged. The stress-strain curves of the injection moulded specimens and specimens, which are tested parallel to the deposition direction show a similar, ductile behaviour. For both IM-O and AM-P a yield strength can be identified at comparable strains, followed by an area of plastic deformation before fracture. However, AM-P specimens show a significantly higher elongation at break than IM-O. Contrary to that, AM-O specimen display brittle behaviour, reaching only low stresses before failure occurs at small strains. Based on this behaviour, in the following the maximum reached stress is used for comparison, corresponding to the yield stress for IM-O and AM-P and to the tensile strength for AM-O.

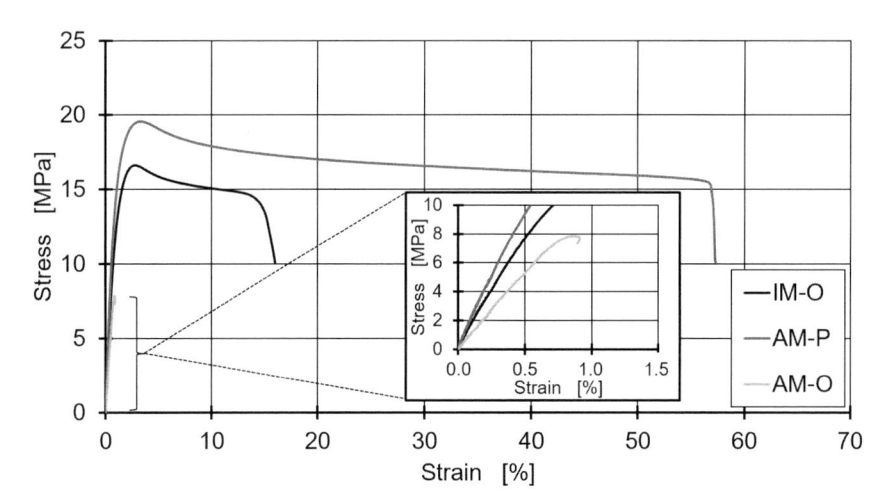

Figure 2.3: Exemplary tensile stress strain curves for the investigated series

In addition to the stress-strain curves, Figure 2.4 shows exemplary tensile specimens after testing on the left as well as their fracture surface on the right. Consistent with the behaviour seen in Figure 2.3, IM-O and AM-P show ductile failure marked by stress-whitening. Based on the fracture surface of IM-O and AM-P it can be inferred that failure occurs abruptly as soon as the ultimate tensile strength is reached. Yet, the uneven failure surface of AM-P indicates that failure starts by rupture of a single strand, followed by failure of the remaining strands and subsequent breaking of the whole specimen. AM-O specimens show failure between layers, as is commonly observed and thus expected from additively manufactured parts, due to a lower interlayer strength [LSBG02, BLSG04]. The fracture surfaces of the tensile specimens is further investigated by microscopic investigations, as detailed in subsection 2.3.4.

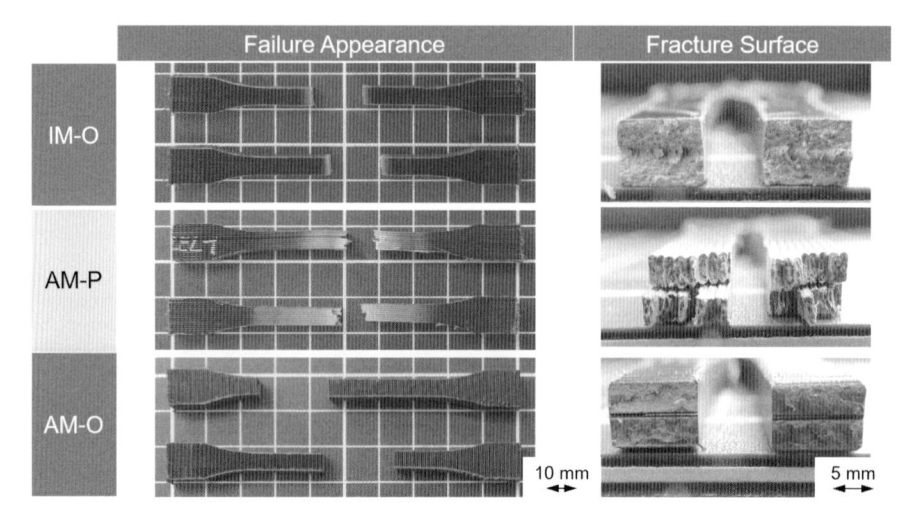

Figure 2.4: Tensile specimen failure appearance and fracture surface

The average maximum tensile stress determined from injection moulded and additively manufactured specimens is shown in Figure 2.5 a). The determined tensile modulus is shown in Figure 2.5 b). In addition to these determined values, data provided by the datasheet is shown as a dashed line [NN19c]. The data sheet value given in Figure 2.5 a) is specified as the yield strength.

From the comparison it becomes clear that the highest maximum tensile stress and tensile modulus is reached by additively manufactured parts, which are loaded parallel to the direction of strand deposition. For AM-P a maximum average tensile stress of 19 MPa ± 0.7 MPa at a yield strain of 3.5 % ± 0.1 % and a tensile modulus of 1,851 MPa ± 142 MPa is determined. This is followed by the maximum tensile stress and tensile modulus of the injection moulded specimens at 17 MPa ± 0.2 MPa at 2.7 % ± 0.1 % yield strain and 1,587 MPa ± 25 MPa, respectively. By far the lowest maximum stress is determined for AM-O and consequently a force application orthogonal to the strand direction at 7 MPa ± 0.8 MPa, at which the specimens break at the layer boundary at a strain of 0.7 % ± 0.1 %. Notably, the difference in tensile modulus compared to IM-O and AM-P is less pronounced at 1,059 MPa ± 140 MPa. None of the investigated series reaches the values specified by the manufacturer of a yield strength of 20 MPa at 5 % yield strain and tensile modulus of 1,950 MPa. This might be based on different testing conditions used by the manufacturer to determine the values given in the datasheet.

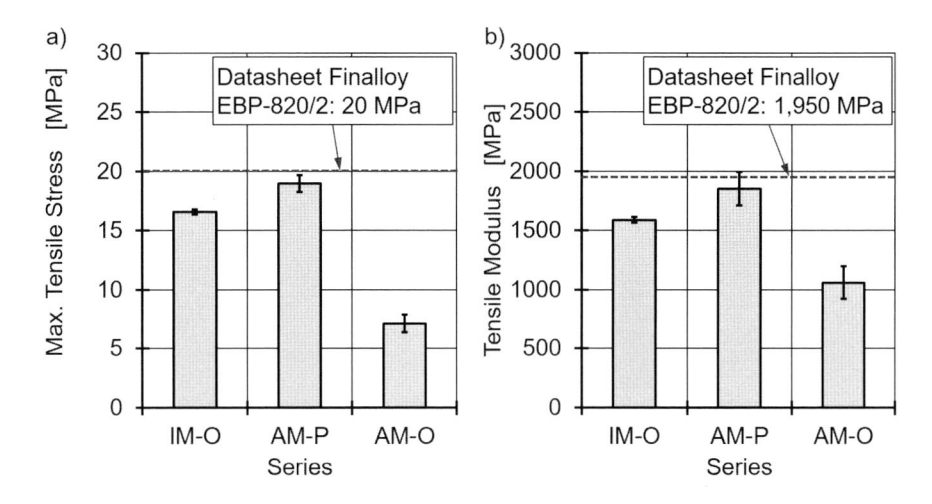

Figure 2.5: Determined maximum tensile stress a) and tensile modulus b)

2.3.2 3-Point bending tests

Stress-strain curves of the carried out 3-point bending tests of exemplary injection moulded and additively manufactured specimens are shown in Figure 2.6. Overall the flexural stress-strain curves show similar behaviour to those of the tensile tests depicted in Figure 2.3. Both IM-O and AM-P show ductile behaviour, reaching a peak in flexural stress before a range of plastic deformation. It has to be noted that neither IM-O nor IM-P specimens break during testing, but slide of the supports due to plastic deformation. This results in notable chatter in the stress strain curve of IM-O specimens. The flexural behaviour of AM-O specimen is characterised by failure before an onset of plastic deformation can occur. Low flexural stresses and strains are reached, compared to IM-O and AM-P. As shown in the image of the fracture zone in Figure 2.6, breaking again occurs between layers. The fracture is observed to not fully separate the specimens, but to be a hinge break with both parts of the specimen being held together by a thin peripheral layer. Similar to the tensile test results, the maximum flexural stress will be used in the following to compare the results of the 3-point bending, taking into account that it coincides with the yield stress for IM-O and AM-P, while accounting for the stress at failure for IM-O.

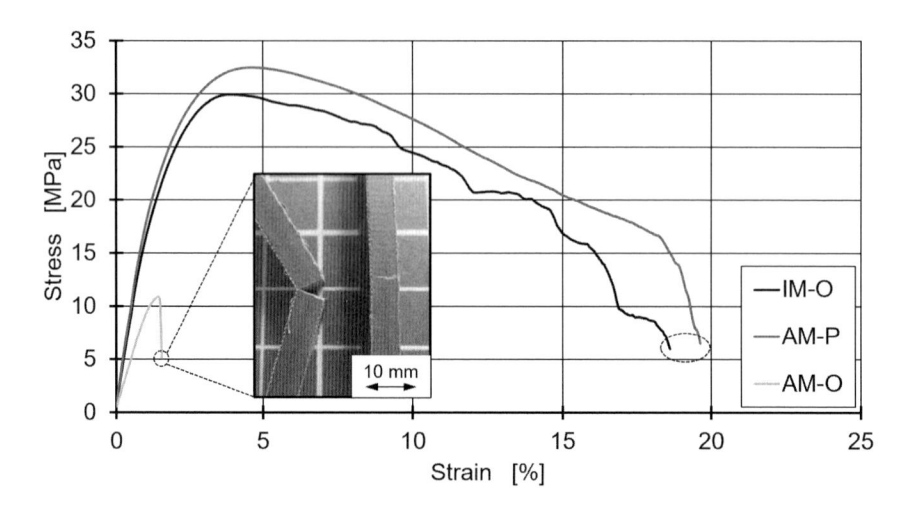

Figure 2.6: Exemplary flexural stress strain curves for the investigated series

The averaged results of the 3-point bending investigations are shown in Figure 2.7, with a) displaying the maximum flexural stress of the series and b) showing the determined flexural modulus. Additionally, manufacturer information is provided as given in the material datasheet [NN19d]. However, no value of the maximum flexural stress is available.

The diagrams show that the highest maximum flexural stress of 33 MPa ± 1.3 MPa at 4.7 % ± 0.1 % flexural strain and the highest flexural modulus at 2,005 MPa ± 102 MPa is reached by AM-P specimen, which are loaded strand-parallel. It hereby surpasses the injection moulded specimens, for which a maximum flexural stress of 30 MPa ± 0.6 MPa at 4.1 % ± 0.1 % flexural strain is determined. Congruent to the observations of the tensile tests, the lowest stresses are reached when additively manufactured specimens are loaded orthogonal to the direction of strand deposition. For such specimens a maximum flexural stress of 11 MPa ± 0.5 MPa at a strain of 1.5 % ± 0.2 % is determined, which corresponds to a third of the stress found for AM-P specimen. The flexural modulus of AM-O specimens is found to be 906 MP ± 75 MPa. The flexural modulus of 2,100 MPa, as given by the datasheet, is not reached by the investigated specimen.

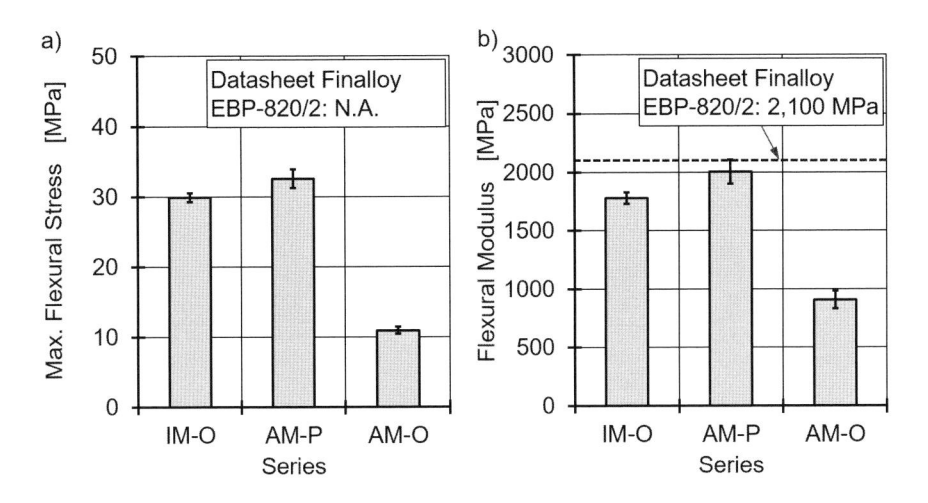

Figure 2.7: Determined maximum flexural stress a) and flexural modulus b)

2.3.3 Charpy impact tests

The results of the notched Charpy impact testing and the value provided by the material datasheet as well as exemplary fractured surfaces are shown in Figure 2.8 a) and Figure 2.8 b). The comparison of the series mirrors those of the tensile and flexural tests in that for a strand parallel load the highest impact strength is determined, followed by the injection moulded and finally orthogonally loaded AM-specimen. AM-P specimen reach an impact strength of 58.8 kJ/m^2 ± 6.6 kJ/m^2, significantly surpassing the impact strength given in the datasheet of 40 kJ/m^2. As depicted in Figure 2.8 b), AM-P specimens do not break completely, but only show an incomplete, partial break. This is based on the sequential failure of individual strands resulting in a successive load-redistribution to still intact strands. For IM-O specimen a notched impact strength of 41.4 kJ/m^2 ± 3.7 kJ/m^2 is determined, confirming to manufacturers specification. The fracture after testing is held together by a thin material layer so that a hinge break is identified. Contrary to IM-O and AM-P, AM-Ospecimen display a complete break at the interlayer boundary and a very low notched impact strength of 3.3 kJ/m^2 ± 0.7 kJ/m^2. It has to be noted that the notched impact strength of IM-O, AM-P and AM-O specimen are determined using different impact pendulums. Due to different dimensions, comparison of the determined notched impact strength is only possible to a limited extend. However, the overall difference in impact behaviour between IM-O, AM-P and AM-O, as characterised by the fracture surfaces, is well represented.

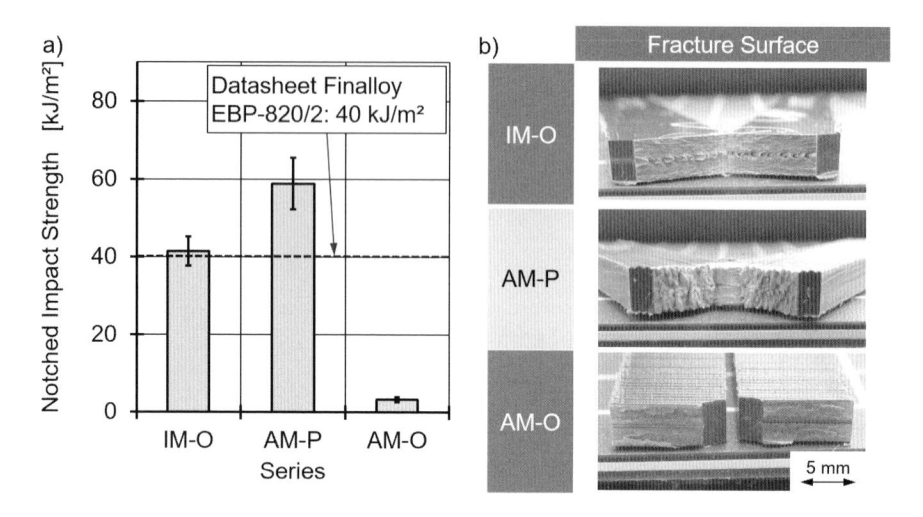

Figure 2.8: Determined notched impact strength a) and respective fracture surfaces b)

2.3.4 Microscopy investigations

In order to further assess the differences in failure behaviour of injection moulded and additively manufactured tensile specimen, microscopy investigations are performed. This includes transmitted light microscopy and FESEM-imaging. Pictures taken by transmitted light microscopy of thin sections are shown in the lower part of Figure 2.9, with the images above showing the specimen's respective failure mode. The image-plane and angle of view of the thin sections are indicated by the dashed line and arrows. The thin sections are prepared so that the center of the specimen gauge area is investigated. Additionally, detailed reflected light microscopy as well as FESEM-images of the tensile specimen fracture surfaces are shown in Figure 2.10. The dashed circles in the fracture surface images in the upper part of Figure 2.10 denote the position ad which the FESEM-images are taken.

When comparing the failure appearance and the thin section of the IM-O specimen in Figure 2.9, the area of stress-whitening is clearly identifiable. This area is marked by a successive darkening of the thin section towards the fracture surface, due to microcracks. More notably however, a heterogeneous specimen structure over the thickness is observed. This structure is divided into three regions, corresponding to those typically found in injection moulded parts manufactured from filled polymers [MG82]. First, on the outside of the specimen a low filler content skin layer. The thickness and thus impact of the skin layer is however neglectable and will therefore no longer be considered. Second, an outer shear layer can be identified. Finally, a spine like core layer can be observed. The development of these layers is based on the shear flow developing during the filling of the mould. These layers are accompanied by different filler orientations, most commonly reported for fibre fillers [MG82, Oss17]. However, such orientations are also applicable to talc fillers [ZM05]. While the talc particles in the outer shear layer are predominantly oriented in the direction of flow, the talc particles in the core layer show no predominant alignment [ZM05, Kun17]. This assessment is reinforced by the FESEM-images of the IM-O specimen's fracture surface depicted in the lower right of Figure 2.10. Shown in the FESEM.image section is the core layer of the specimen. Other than in the thin sections, the melt flow direction of the specimen during manufacturing corresponds to a movement from the left to the right in the

FESEM-image. In this direction a parabolic distribution of the talc particles is identifiable. furthermore, the talc plates shown no predominant orientation.

Talc particles oriented in the direction of load, i.e. in the outer shear layer, create a strengthening effect, which results in higher yield stresses and elongation at break [ZM05]. However, fillers that are aligned orthogonal to the load direction, as in the core layer, do not result in such positive effects and may even lead to a weakening of the material, based on filler-matrix separation [Kun17]. It has to be noted that the flow direction of the thin section extends into the image plane, as the IM-O specimens are taken from the injection moulded plates orthogonal to the direction of flow. Due to these filler orientation, the layers show differing mechanical properties. The core layer exhibits a higher degree of parabolic deformation, as indicated by the protruding center area of the fracture surface. Additionally, the deformation extends through the core layer beyond the stress-whitening zone. Within the stress-whitening zone, cracks in the core layer are noticeable, suggesting damaging of the core layer. In contrast, the outer shear layer shows less deformation and a homogeneous distribution of micro cracks up to the failure surface. Furthermore, right at the failure surface a lighter, frayed area is noticeable, indicating an overload fracture. Thus, it can be assumed that failure is initiated in the core layer, weakening the entire specimen.

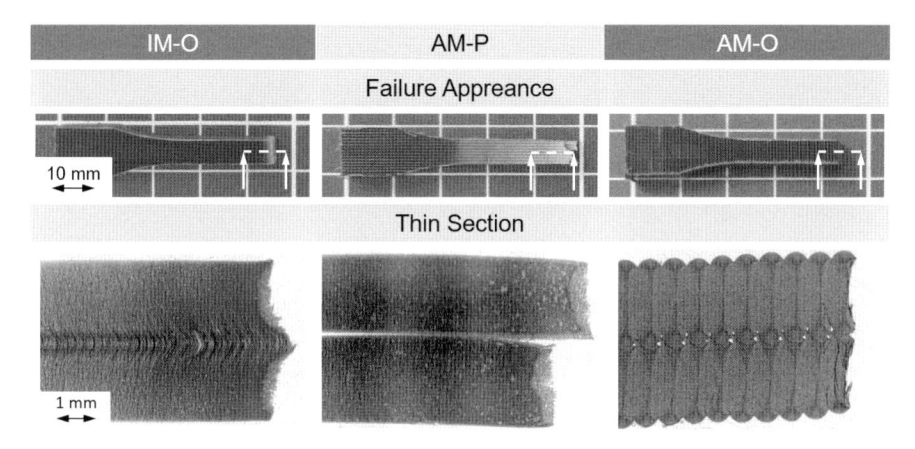

Figure 2.9: Thin sections of the failure zones of tensile specimens

In the center of Figure 2.9 the thin section of of a AM-P specimen is shown. The two strands making up the thickness of the specimen can be seen. Similar to the IM-O specimen, the AM-P specimen exhibits stress-whitening, albeit in a significantly broader range of the gauge length. The stress-whitening is reflected in the thin section by a dark colouring. Contrasting the IM-O thin section, the respective strands of the AM-P specimen shows a homogeneous structure, with similar darkening throughout the strands. The different length of the strands as well was the lighter, frayed area at the fracture indicate a sequential overload failure of the strands. This failure behaviour can be attributed to a homogeneous predominant orientation of the talc particles in the deposition direction of the strand introduced by the strand extrusion [ST14]. This assumption can be substantiated by the FESEM-images of the AM-P specimen fracture surface depicted in Figure 2.10. The image section shows the central area of a failed strand a as denoted by the dashed circle in the fracture surface image. In the FESEM-capture the fracture surface appears frayed, with tendril-like polymer threads extending from the strand, as expected in a ductile failure mode. On the fracture surface, only a few talc particles are identifiably oriented face-sided to the direction of load. Consequently, a predominantly load parallel orientation of the fillers is found. This orientation of the filler leads to an increased yield strength and elongation at break of the AM-P specimens compared to the IM-O specimens as determined by the tensile testing (see Figure 2.3 and Figure 2.4) [ZM05, Kun17].

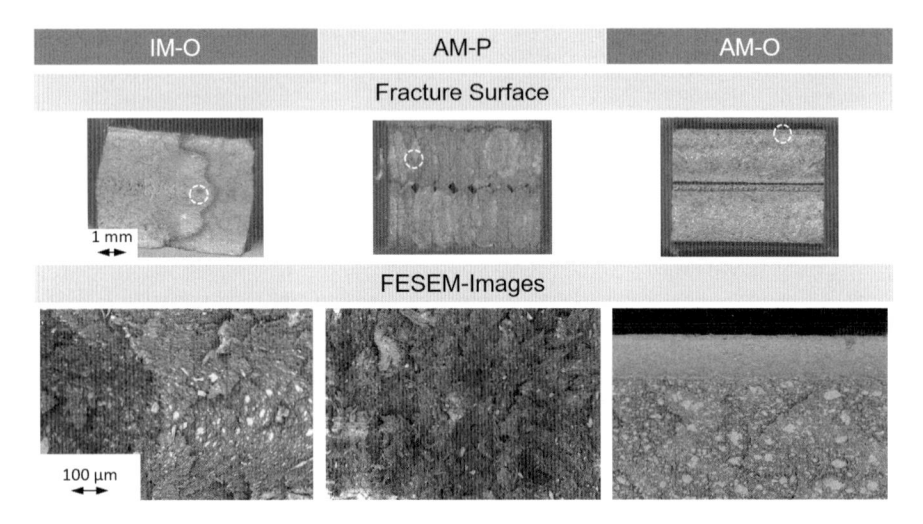

Figure 2.10: Microscopy- and FESEM-images of tensile specimen fracture surfaces

The thin section of a AM-O specimen is shown on the right of Figure 2.9. Several layers of two strand can be identified. Other than the in the thin sections of IM-O and AM-P, no stress-whitening occurred during specimen testing. This is in line with the brittle failure-mode observed during tensile loading, as shown in Figure 2.3. The thin section shows pores located between the two strands and a uneven outer wall, based on the strand deposition, as it is typical for additively manufactured parts. The structure of the individual strands appears homogeneous. Failure occurs at the interface of two layers, running along the interlayer boundary, as it is commonly observed when loading AM-parts parallel to the build-up direction. The fracture starts at valley like point at the outside of a strand, indicating a notch effect facilitating failure. Notably, several dark cracks are visible parallel to the fracture surface, not only in the strand directly at the failure surface, but also several layers away. This could be based on local weak spots in the material, promoted by talc particles oriented orthogonal to the direction of load. That this is the predominant orientation of the fillers is clearly discernible from the AM-O FESEM-image shown in the lower right image of Figure 2.10. The image shows a section of the fracture surface right at the edge of the strand, with the upper, smooth part of the image depicting the strand below the fracture. In the fractured area itself, the talc particles are identifiable by the fact that they stand out brightly from the dark matrix. The orientation of the talc lead to a further weakening of the interlayer boundary based on two reasons. First, filler-matrix separation can easily occur in this orientation, leading to crack initiation [Kun17]. Second, the talc particles inhibit the healing of the polymer layers in areas where they align with the interlayer boundary during deposition, thus reducing the interlayer strength. Consequently, the talc filler is detrimental to the mechanical properties of the AM-specimens tested orthogonal to the direction of strand deposition.

2.3.5 Comparison of test results

The presented results of the mechanical investigations can be used to characterise the general differences and similarities in the mechanical component behaviour of AM- and IM-parts. Additionally, based on the determined values for maximum stresses, strengths and moduli the degree of difference can be assessed. For this purpose Figure 2.11 a) shows the ratio of maximum tensile and flexural stresses as well as impact strengths determined for AM-P and AM-O specimens and those determined for IM-O specimens. Similarly, Figure 2.11 b) shows the tensile and flexural modulus of AM-specimen normalised by the moduli found for IM-specimen. Based on this, a ratio of 1 would describe perfect similarity of AM-and IM-specimens.

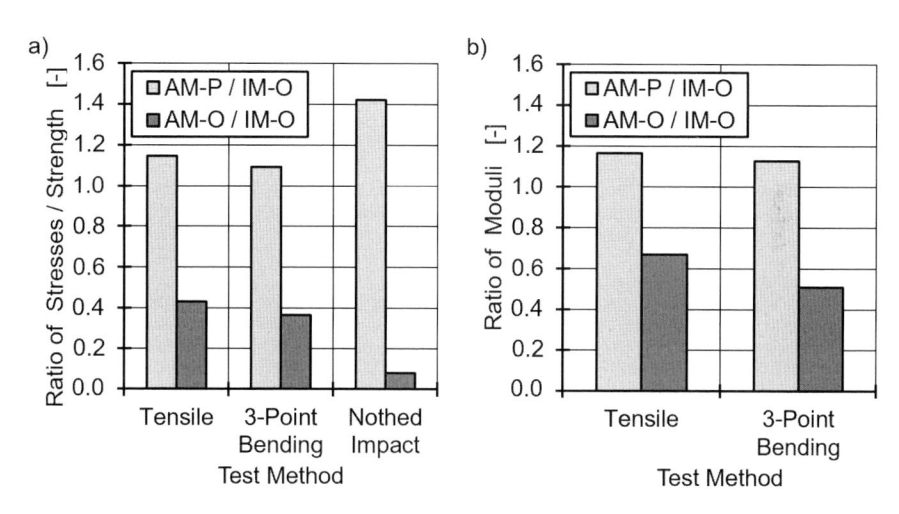

Figure 2.11: Ratio of determined stresses and strength a) and determined moduli b) normalized to injection moulding

A notable observation is that the determined tensile and flexural yield stresses as well as moduli of strand parallel AM-specimen are higher than those of the investigated IM-specimen as denoted by a ratio of 1.15 and 1.1 in Figure 2.11 a). Similarly, the determined ratios of the tensile and flexural moduli are 1.16 and 1.3, as Figure 2.11 b) shows. This can be attributed to the talc filler distribution, as presented in the description of the result of the microscopy investigations in subsection 2.3.4. IM-specimen show a heterogeneous filler distribution, with a core layer, weakening the specimens. Additionally, the specimens are taken orthogonal to the direction of flow, which is a direction exhibiting lower mechanical properties compared to the flow-parallel direction[ZM05]. Contrary to this, in AM-specimen a predominant filler orientation in the direction of strand-deposition is observed, due to the flow conditions during deposition [MGB+18]. These differences in test specimen found by microscope investigations explain the discrepancies in mechanical properties, as higher filler orientations in load direction result in higher yield strength and stiffness. Further investigations of IM-specimen, which are taken parallel to the direction of flow can be carried out to further asses this effect.

Generally, the ratios of yield stresses and moduli shown in Figure 2.11 a) an b) as well as the tensile and flexural stress-strain curves presented in Figure 2.3 and Figure 2.6 show that the properties of AM-parts in direction of strand deposition reflect those of IM-parts. For the investigated PP-material both IM-O and AM-P specimen exhibit a similar ductile tensile and flexural behaviour. Especially, in the elastic region, until the yield strength and yield strain is reached, comparable part behaviour of IM and strand parallel AM is found. From the point of yield onwards, the significantly higher cold-fold flow and elongation at break of AM-parts has to be considered. Based on these findings, a transferability of the mechanical behaviour between AM- and IM-parts in strand direction, in terms of material behaviour, strength and stiffness can be realised, given the prerequisite that the main load direction of the part is parallel to the strand direction of the AM-prototype part. A possible transfer function between AM- and IM- parts can be realized, given further investigations. This function has to consist of reduction factors that transfer properties between manufacturing methods.

In case of a large and complex prototype-parts it is unlikely that mechanical loads are solely introduced parallel to the strands. This is based on complex load cases or restrictions in the orientation of prototype-part during manufacturing. In this case the discrepancy of strand-parallel and interlayer based, strand-orthogonal behaviour has to be considered as the lower limit of part strength and stiffness. First, the behaviour orthogonal to the strand direction is marked by brittle behaviour depicted in Figure 2.3 and Figure 2.6. As such, no yield point is reached, but failure occurs at much lower stresses, as a ratio of 0.43 and 0.37 between AM-O and IM-O maximum tensile and maximum flexural stresses show. Consequently, a transferability of mechanical properties is limited to small deformations before interlayer-failure occurs. The question of the influence of fillers on these mechanical properties remains a part of future investigations, as different matrix-filler combinations could exhibit properties closer to those of IM-parts. Nevertheless, AM-prototypes are suitable to be used in cases in which such low strains suffice. As the ratios of moduli of 0.67 and 0.51 for tensile and flexural modulus show, IM-parts show a stiffer behaviour compared to AM-parts orthogonal to the strand direction. This discrepancy can be addressed by a transfer function, mapping stresses and strains reached by AM-prototypes to those of a IM-part.

The highest discrepancy in mechanical behaviour between IM- and AM-specimen is found for notched impact testing. While the strand parallel impact strengths of AM-specimens significantly surpass those of IM-specimens, given a ratio of 1.42, strand orthogonal impact only yields a ratio of 0.08. As expected, the mechanical behaviour of AM-parts at high strain rates is thus found to be highly dependent on the direction of load. In case of strand-parallel impact load, the specimens only show a partial break, indicating that load are transferred between strands during failure, resulting in high impact strength. Contrary, strand-orthogonal impact yields very low impact strength compared to IM-specimens, due to failure at the layer interface. Based on these discrepancies, currently no transferal of impact properties between AM- and IM-parts seems feasible. However, further investigation into the behaviour of AM-parts subjected to high strain rates have to show, if cases can be identified in which a transferability to the properties of IM-parts is possible, for example by high speed tensile or drop tower testing. Another approach could be an optimisation of the process parameters in order to yield better results for strand-orthogonal impact.

2.4 Conclusions and outlook

In the scope of this investigation three mechanical test methods, tensile testing, 3-point bending and notched Charpy impact testing are used to investigate the mechanical properties of injection moulded specimens as well as from additively manufactured specimen parallel and orthogonal to the direction of strand deposition. The results of the tensile, flexural and impact testing performed show differences regarding the mechanical behaviour, yield and failure stresses as well as modulus and impact strength of injection moulded and additively manufactured parts. As expected, the degree to which the mechanical properties differ is substantially dependent on the direction of load application relative to the direction of strand deposition. In addition, the particular influence of filler materials such as talc on the mechanical properties of additively manufactured parts is exhibited by microscopic investigations of the fracture surfaces of tensile specimens. It is shown that the process induced filler orientation has to be assessed in the comparison of IM- and AM-parts.

Based on the determined tensile and flexural properties, the utilisation of AM-prototypes in cases in which only low strains at low strain rates are required is most promising, as the highest resemblance of part properties can be found in this region. More challenging is the use of AM-parts in destructive testing. Strand parallel load cases show similarities with regard to the mechanical characteristics of injection moulded parts and can thus be compared. However, strand orthogonal load cases exhibit substantially differing mechanical properties to those of injection moulded parts. Therefore, further investigations will focus on the estimation of suitable destructive test cases taking into account these boundary conditions. In addition, a way to transfer the properties of an AM-part to an IM-part is required. A possible approach to be investigated in further research is the development of a transfer function based model, by which the mechanical behaviour ascertained by testing of an AM-part can be transformed to those of a geometrically similar IM-part. As AM-parts are highly anisotropic, a suitable transfer function has to account for the direction of the load introduction. Consequently, in further investigations, the mechanical response of AM-part at loading angles other that parallel and orthogonal to the strand direction will be investigated to derive a relationship between mechanical response and direction of load introduction. Furthermore, not only test specimens are to be investigated, but also sections of geometrically complex parts, to quantify the difference in properties between AM- and IM-structures.

Acknowledgements

The investigations set out in this report are conducted in cooperation with Volkswagen AG, Wolfsburg, Germany, to whom we extend our thanks.

References

[AMO+02] AHN, S.-H.; MONTERO, M.; ODELL, D.; ROUNDY, S.; WRIGHT, P. K.: Anisotropic material properties of fused deposition modeling ABS. *Rapid Prototyping Journal* 8 (2002) 4, p. 248–257

[BLSG04] BELLEHUMEUR, C.; LI, L.; SUN, Q.; GU, P.: Modeling of Bond Formation Between Polymer Filaments in the Fused Deposition Modeling Process. *Journal of Manufacturing Processes* 6 (2004) 2, p. 170–178

[CCMP16] CASAVOLA, C.; CAZZATO, A.; MORAMARCO, V.; PAPPALETTERE, C.: Orthotropic mechanical properties of fused deposition modelling parts described by classical laminate theory. *Materials & Design* 90 (2016), p. 453–458

[CK17] COOGAN, T. J.; KAZMER, D. O.: Bond and part strength in fused deposition modeling. *Rapid Prototyping Journal* 23 (2017) 2, p. 414–422

[FP16] FELDMANN, C.; PUMPE, A.: *3D-Druck – Verfahrensauswahl und Wirtschaftlichkeit // 3D-Druck - Verfahrensauswahl und Wirtschaftlichkeit: Entscheidungsunterstützung für Unternehmen*. Wiesbaden: Springer Fachmedien Wiesbaden, 2016

[GKT19] GEBHARDT, A.; KESSLER, J.; THURN, L.: *3D printing: Understanding additive manufacturing*. Munich and Cincinnati: Carl Hanser Verlag GmbH & Co. KG, 2019

[GRS15] GIBSON, I.; ROSEN, D.; STUCKER, B.: *Additive Manufacturing Technologies*. New York, NY: Springer New York, 2015

[HD03] HOPKINSON, N.; DICKNES, P.: Analysis of rapid manufacturing—using layer manufacturing processes for production. *Proceedings of the Institution of Mechanical Engineers, Part C: Journal of Mechanical Engineering Science* 217 (2003) 1, p. 31–39

[HL16] HOPMANN, C.; LAMMERT, N.: Hybride Fertigungsstrategie für funktionsintegrierte Einzelteile: IKV kombiniert vollständig automatisierte Anlage zur additiven Fertigung mit konventionellen Fertigungstechnike. *Kunststoffe* 2016 (2016) 11, p. 1–5

[HLD+20] HOPMANN, C.; LAMMERT, N.; DAHLMANN, R.; PELZER, L.; HELLMICH, C.: Process and design optimisation in additive manufacturing. *Proceedings of the 30th International Colloquium Plastics Technology*. 2020

[Kun17] KUNKEL, F.: *Zum Deformationsverhalten von spritzgegossenen Bauteilen aus talkumgefüllten Thermoplasten unter dynamischer Beanspruchung*. Otto-von-Guericke-Universität Magdeburg, Dissertation, 2017

[LPN+21] LOVE, L.; POST, B.; NOAKES, M.; NYCZ, A.; KUNC, V.: There's plenty of room at the top. *Additive Manufacturing* 39 (2021), p. 101727

[LSBG02] LI, L.; SUN, Q.; BELLEHUMEUR, C.; GU, P.: Investigation of Bond Formation in FDM Process. *International Solid Freeform Fabrication Symposium* (2002),

[MG82] MENGES, G.; GEISBÜSCH, P.: Die Glasfaserorientierung und ihr Einfluß auf die mechanischen Eigenschaften thermoplastischer Spritzgießteile — Eine Abschätzmethode. *Colloid and Polymer Science* 260 (1982) 1, p. 73–81

[MGB+18] MULHOLLAND, T.; GORIS, S.; BOXLEITNER, J.; OSSWALD, T.; RUDOLPH, N.: Process-Induced Fiber Orientation in Fused Filament Fabrication. *Journal of Composites Science* 2 (2018) 3, p. 45

[NN10]   N.N.: *DIN EN ISO 179-1: Bestimmung der Charpy-Schlageigenschaften – Teil 1: Nicht instrumentierte Schlagzähigkeitsprüfung.* Berlin: Beuth Verlag GmbH, 2010

[NN18]   N.N.: *DIN EN ISO / ASTM 52900: Additive Fertigung - Grundlagen - Terminologie.* Berlin: Beuth Verlag GmbH, 2018

[NN19a]   N.N.: *DIN EN ISO 178: Bestimmung der Biegeeigenschaften.* Berlin: Beuth Verlag GmbH, 2019

[NN19b]   N.N.: *DIN EN ISO 179: Bestimmung der Zugeigenschaften – Teil 1: Allgemeine Grundsätze.* Berlin, 2019

[NN19c]   N.N.: *Processing Data Sheet EBP 820/2 C16.* Processing Data Sheet, Feluy, 2019

[NN19d]   N.N.: *Technical Data Sheet EBP 820/2 C16.* Technical Data Sheet, Feluy, 2019

[Oss17]   OSSWALD, T. A.: *Understanding polymer processing: Processes and governing equations.* Cincinnati and Munich: Hanser Publications, 2017

[Scu20]   SCULPTEO: *The State of 3D Printing 2020 Edition: The data you need to understand the 3D printing world and build your 3D printing strategy.* Report, Villejiuf, 2020

[ST14]   SHAKOOR, A.; THOMAS, N. L.: Talc as a nucleating agent and reinforcing filler in poly(lactic acid) composites. *Polymer Engineering & Science* 54 (2014) 1, p. 64–70

[WRK+08] WENDEL, B.; RIETZEL, D.; KÜHNLEIN, F.; FEULNER, R.; HÜLDER, G.; SCHMACHTENBERG, E.: Additive Processing of Polymers. *Macromolecular Materials and Engineering* 293 (2008) 10, p. 799–809

[ZM05]   ZHOU, Y.; MALLICK, P. K.: Effects of melt temperature and hold pressure on the tensile and fatigue properties of an injection molded talc-filled polypropylene. *Polymer Engineering & Science* 45 (2005) 6, p. 755–763

Abbreviations

Notation	Description
AM	Additive Manufacturing
FESEM	Field Emission Scanning Electron Microscopy
IM	Injection Moulding
-O	Orthogonal
-P	Parallel
PP	Polypropylene

3 Expanding non-planar additive manufacturing with variable layer height to be used on a six-axis robot arm

Ch. Hopmann[1], R. Dahlmann[1], L. Pelzer[1]
[1] Institute for Plastics Processing (IKV) at RWTH Aachen University

Abstract

By producing parts in layers, additive manufacturing (AM) technologies exhibit a high degree of geometric freedom. However, compared to other manufacturing technologies, layer-based processes show disadvantages in surface quality based on the staircase effect. With reduced load capacity in build direction, their mechanical properties are also anisotropic. To address these challenges, a method for non-planar additive manufacturing with variable layer height was developed. Using this method for path planning, it was shown that surface roughness could be reduced to 24 % as compared to conventional AM.

To increase part size, material diversity and manufacturing speed, the implementation of non-planar path planning on a large-scale, screw extrusion AM machine is investigated in this research project. The machine's robot arm enables the use of three additional degrees of freedom, which are used to rotate the extruder according to the slope being manufactured. This way, part quality can be improved further. Additionally, load case tailoring is made available to a large-scale AM technology.

3.1 Introduction

The use of additive manufacturing (AM) technologies has increased noticeably in recent years. With a growth of 21 % in 2020, the industry's yearly revenue is currently 12.6 billion US$ [Rob21]. This is, in part, due to the benefits that AM offers as compared to traditional manufacturing technologies. Because of the layer-based process, complex geometries with internal structures can be produced. Since the process is also tool-less, various geometries can be manufactured on the same machine without the need for part-specific process setups. This is why AM is typically regarded as a very agile manufacturing technology, therefore being oftentimes used in prototyping applications. However, for a few years now, most of the revenue is not done in prototyping anymore [Woh19]. One of the catalysts of the ever-increasing growth of AM technologies is the production of final parts, ready to be used. This, however, raises new challenges for the technology, both in terms of qualitative appearance and mechanical load capacity of the manufactured parts.

Plasticising AM technologies, characterised by an extrusion unit which melts a thermoplastic polymer and deposits strands of material, play an important role in the transition towards end-use parts. With the best-known variant being fused deposition modelling (FDM), plasticising technologies represent the majority of AM technologies being used [Woh19]. This can be attributed to the high material diversity, easy scalability, large potential for automation and low entry cost of the technology [GRS15, RW16]. To take advantage of those benefits, IKV developed the hybrid manufacturing cell [HL16]. This AM machine combines a screw-based extruder and an industrial six-axis robot arm to enable the production of large parts in a short time and with standard pellet material [HP19].

Regardless of price, size or underlying technology, all AM processes exhibit certain disadvantages because of the layer-based process. Most notably, building parts in layers creates a distinct resolution, which results in the approximation of the part's exact surface [DM94, BDH00]. The resulting staircase effect can be seen in Figure 3.1. While being present across all AM technologies and machines, the staircase effect becomes increasingly prominent with the use of larger machines which are geared towards producing larger parts, therefore using larger layer heights. Figure 3.1 shows the staircase effect for a part manufactured using planar slicing (left). The dashed line indicates the target geometry.

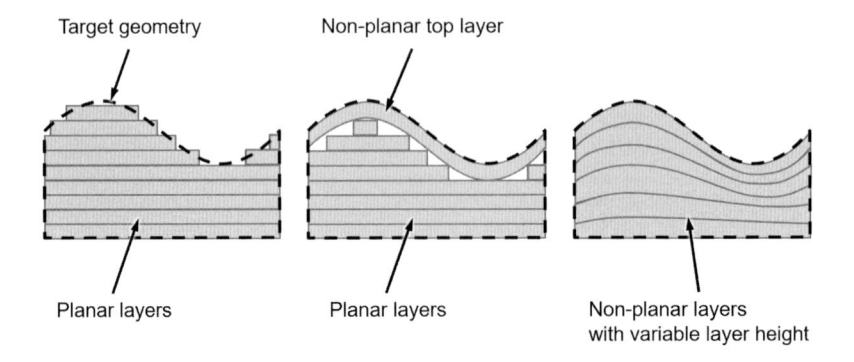

Figure 3.1: Representation of target geometry using planar layers (left), planar layers with non-planar top layers (middle) and non-planar layers with variable layer height (right)

Another disadvantage of the layer-based process, especially in plasticising AM, are the reduced mechanical properties in build direction based on layer bonding. Not only does this reduce the load capacity in this direction, but it also creates anisotropic part properties [CLH07, GWKA15].

To not only address those disadvantages but also turn them into advantages based on the freedom that only plasticising AM technologies provide, non-planar additive manufacturing with variable layer height was developed [PH21]. This method of creating tool paths for AM machines uses the z-axis during the deposition of each layer to create truly curved freeform surfaces. Combined with exact control of the material flow, the variation of layer height inside each layer is possible, enabling non-planar AM without creating weaknesses because of air gaps. Besides creating smooth freeform surface on AM parts, non-planar path planning can also be used to tailor the deposited strands to the part's load case. This way, the inherent anisotropy of AM parts can be actively used to improve mechanical properties.

Previously, the method has been implemented and tested for three-axis FDM machines [PH21]. Here, limitations were found based on the three-axis kinematics, namely a maximum slope which can be manufactured and the nozzle penetration of the part for steeper slopes. Therefore, this article shall demonstrate the expansion of the previous method to the hybrid manufacturing cell, using its six-axis robot arm to increase the technology's degree of freedom. Furthermore, because of the increased size of the machine, the produced parts and the pronounced staircase effect based on larger layer heights, non-planar AM can be even more beneficial here.

3.2 Addressing disadvantages through optimisation in path planning

Conventional AM can actually be described as a 2.5D process. Material is added in a two-dimensional plane. Once the planar layer is completed, the unfinished part and the material applicator are moved relative to each other in the z-direction. This two-dimensional process is repeated until the part is finished. Non-planar AM expands on this by using the z-axis not only during layer changes, but also during the application of a layer, enabling the accurate representation of freeform surfaces through spatially curved layers.

Previous investigations use one or multiple non-planar top layers on top of a planar base part to create an accurate representation of the part's top surface [CAR08, Hua14, LJAT16, Ahl18]. This, however, only addresses surface quality and cannot improve mechanical part properties or layer bonding. Furthermore, air gaps are created in between the non-planar top layers and the planar base layers, which further reduce layer bonding and introduce weak points and stress risers (see Figure 3.1, middle).

To address those disadvantages, the presented method introduces variable layer height to non-planar additive manufacturing (see Figure 3.1, right). Furthermore, every path can be planned in a non-planar fashion, as opposed to only creating a non-planar top layer over planar base layers. Combining those two methods allows for smooth transitions (transition layers) between different slopes and planes in the same part without introducing weak spots through air gaps while still enabling smooth outer freeform surfaces (surface layers). Additionally, the presented method enables tailoring of the part to the given use-case by aligning the deposited strands with the part's load paths (function layers). Figure 3.2 shows a schematic view of the different kinds of layers. If the expected load is not known or a part with reduced anisotropy is favoured, function layers can be placed in a sinusoidal shape, enabling the division of the applied force vectors.

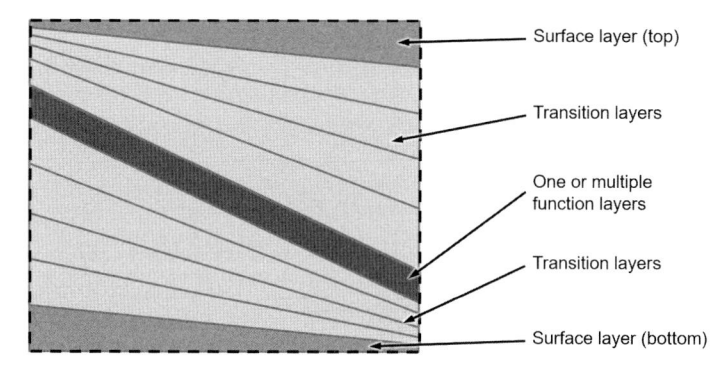

Figure 3.2: Schematic view of a non-planar part. Surfaces are represented accurately through surface layers. A function layer is used to tailor the part to the expected load case.

To achieve those advantages, paths are planned according to six constraints:

1. The layer height can never be smaller than a specified minimum layer height.
2. The layer height can never be larger than a specified maximum layer height.
3. The bottom surface of the part shall be replicated exactly.
4. The top surface of the part shall be replicated exactly.
5. There can be an arbitrary number of mathematical functions describing freeform surfaces inside the part which shall be replicated as closely as possible given the constraints of minimum and maximum layer height.
6. The maximum slope of a non-planar layer can never exceed a specified maximum angle.

Constraints one and two set the range of possible layer heights of each volumetric element to be deposited. A volumetric element is defined as a section of one deposited strand. In planar

AM, this is calculated by multiplying the constant layer height, the strand width, and the feed rate. Here, strand width and feed rate are typically constant within the same feature type of the part (e.g., perimeters, infill). Therefore, the material flow within one feature is constant. For non-planar AM with varying layer height, the size of a volumetric element varies depending on the current slope, the previous freeform surface, and the freeform surface to be approached (Figure 3.3).

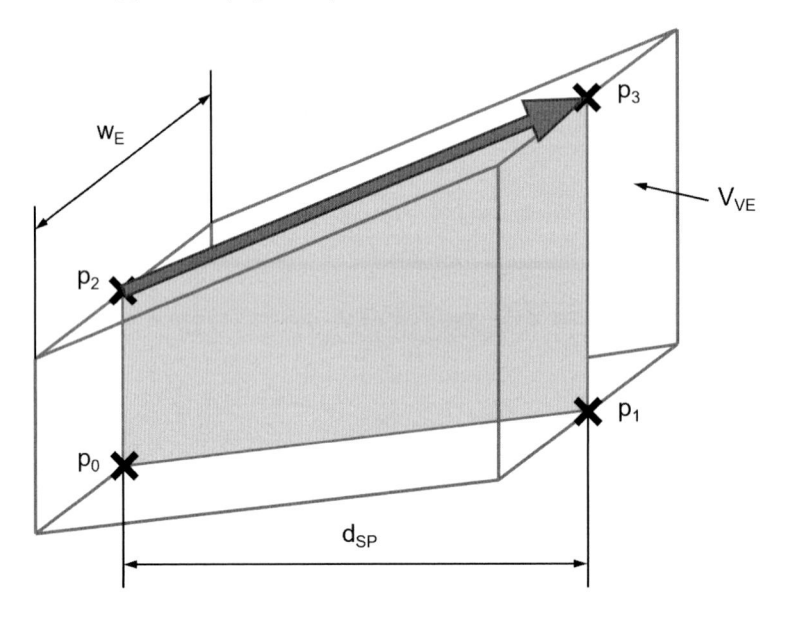

Figure 3.3: Volumetric element for calculating the material to be extruded [PH21]

Therefore, the material to be extruded must be calculated for each section of a move. This is done by calculating the volume of each volumetric element according to Equation 3.1.

$$V_{VE} = \frac{1}{2} \cdot d_{SP} \cdot w_E \cdot (-z(p_0) - z(p_1) + z(p_2) + z(p_3)) \quad (3.1)$$

V_{VE}: Volume of volumetric element
d_{SP}: Euclidian distance in the x-y-plane between support points p_2 and p_3
w_E: Extrusion width
$z(p_{0-3})$: Z coordinate of trapezoid corners

By applying the conservation of volume between the inlet of the extruder and the outlet of the nozzle (see Equation 3.2), the amount of material to be extruded can be calculated according to Equation 3.3 [PH21].

$$V_{VE} = V_F = \pi \cdot \frac{D_F{}^2}{4} \cdot E \qquad (3.2)$$

$$E = \frac{2 \cdot d_{SP} \cdot w_E}{\pi \cdot D_F{}^2} \cdot (-z(p_0) - z(p_1) + z(p_2) + z(p_3)) \qquad (3.3)$$

V_F: Volume of filament entering the extruder
D_F: Filament diameter
w_E: Length of extruded material

Minimum layer height should preferably always be set to 0 mm, regardless of nozzle size. This allows for an arbitrary number of layers to come together in one point, each having a local height of 0 mm. Therefore, the staircase effect can be avoided completely. The maximum layer height is dependent on the nozzle. Just as in planar AM, it should not exceed 80 % of the nozzle's orifice diameter [URL18]. This is because the deposited strand should ideally be pressed against the previous layer, creating a flat contact patch. If the layer height exceeds 80 % of the orifice diameter, this contact surface becomes too small to create reliable adhesion between layers. For any given point inside the part, the largest possible layer height which doesn't violate any other constraint is preferred to reduce manufacturing time.

Constraints three and four ensure that the part's outer surfaces are reproduced exactly. For the bottom of the part, this mostly means creating a layer with a planar underside since the part is manufactured on a planar build platform. Just as the top of the part however, the bottom can be a freeform surface as well. This enables mass customisation through AM by adding features to parts produced using other manufacturing methods. Previously, this has been shown by adding an anti-slip ring to a flat surface of a single board computer case [WSP+19]. Using the presented method, the complexity of the surface to be customised can be increased from planar to freeform surfaces.

To enable load path tailoring of the deposited strands, mathematical functions in the form of Equation 3.4 can be defined, where x, y and z indicate the respective coordinate in three-dimensional space with z_0 describing a constant offset from z=0.

$$z = f(x, y) + z_0 \qquad (3.4)$$

The number of functions and their initial height z_0 is not limited. This way, the part can be tailored to multiple applied forces in different areas. Orienting the deposited strands in the direction of the applied loads is based on the anisotropic properties additively manufactured parts typically exhibit. Depending on the material and the process parameters used, tensile strength is doubled when measured in strand direction as compared to build direction [GWKA15]. Furthermore, studies investigating raster angle of deposited strands in planar parts found that placing strands in the direction of the applied load more than triples the part's tensile strength [Hua14]. Using the presented method, those benefits can be harnessed in all three spatial dimensions. Using this technique, two cases are conceivable. If the part's load case is known, the strands can be placed accordingly using mathematical equations (see Equation 3.4). If the load case is unknown and the goal is to improve mechanical properties in build direction, a sine function can be used to shape the non-planar layers inside the part. This way, the contact surface between the individual layers is increased based on wavelength and amplitude and the applied force vector is split according to the local angle of the sine function, as seen in Figure 3.4. Therefore, part of the applied forces in build direction are absorbed by the strands themselves, and not by the inter-layer bonding.

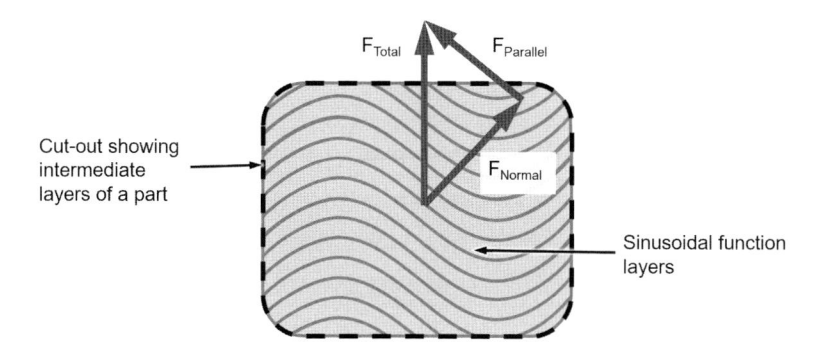

Figure 3.4: Using sinusoidal function layers to increase inter-layer surface and to split the applied force vector, creating a more beneficial load case [PH21]

The last constraint specifies the maximum slope that can be manufactured. This is necessary during calculation of the non-planar layers to determine the number of transition layers between two specified functions or surfaces. The maximum slope depends on the geometry of the nozzle being used as well as the surrounding assembly (e.g., part cooling fans). For the nozzle itself, two geometric features must be considered. A nozzle with a smaller cone angle α is capable of producing steeper slopes. For three-axis systems in particular, the diameter of the nozzle's flat bottom D also must be considered. It directly influences the penetration depth p, which describes how deep the nozzle's cone scratches a non-planar surface. It is calculated using the diameter D and the slope's angle θ according to Equation 3.5 [Ahl18].

$$p = \frac{D}{2} \cdot \sin\theta \qquad (3.5)$$

Next to other collision avoidance measurements, like keeping the local height of each area in memory to be able to move above existing segments of the part being manufactured during travel moves [PH21], the maximum slope constraint is also used to calculate collision free tool paths.

Using all six constraints, the tool path is planned, starting at the bottom of the part. Using non-planar layers and a variable layer height, the first specified load case function is approached. Inside the part, layer height is constantly interpolated to achieve the best orientation of the strands according to the various specified functions. Finally, the top layer representing the part's surface is approached while not violating constraints one, two and six. If at any point not all six constraints can be obeyed, i.e., because an internal function is too steep, therefore violating maximum slope or maximum layer height, the specified functions are approached as closely as possible. Top and bottom surfaces are always replicated exactly.

Using this method of calculating paths for non-planar AM, it has been shown that the quality of top surfaces can be improved significantly. Trials were carried out using a three-axis FDM system equipped with a 0.5 mm airbrush nozzle. This type of nozzle was chosen because of its small cone angle α, allowing for a steep slope angle θ of the manufactured part. For comparison, the same geometry was manufactured using planar slicing with a layer height of 0.25 mm (Figure 3.5, left) and non-planar slicing with variable layer height (Figure 3.5, right). The part's top surface was created using one full wavelength of the sine function according to Equation 3.6 (wavelength: 20 mm, amplitude: 1.2 mm).

$$S_1(x) = 1.2 \cdot \sin(\frac{\pi}{10} \cdot x) \qquad (3.6)$$

Using the laser scanning microscope VK-X210 by manufacturer Keyence Corporation, Osaka, Japan, to measure the surface roughness R_a of both parts, the benefit of using non-planar layers regarding surface quality can be shown. Firstly, both parts are scanned, creating a digital representation of their topology (Figure 3.5, 1). Then, the target function (see Equation 3.6) is subtracted, reducing the freeform surface to a plane (Figure 3.5, 2).

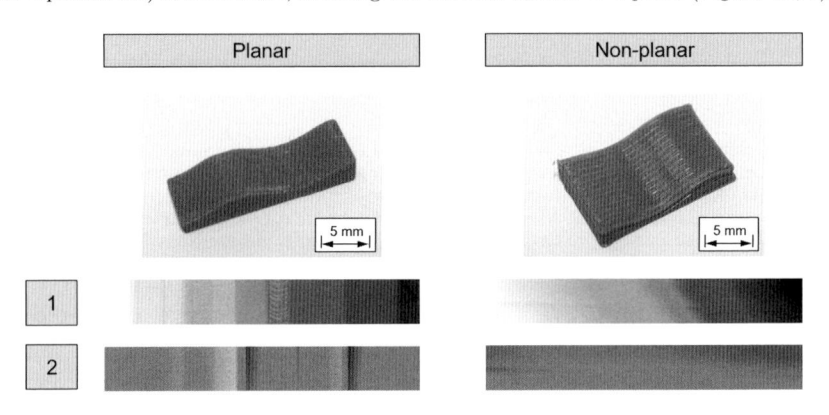

Figure 3.5: Comparison of surface roughness between a planar and a non-planar AM part. Heatmaps indicate the height of the measured topology (1) and the reduced topology, where the target function is subtracted to enable roughness measurements (2) [PH21]

For this plane, the surface roughness R_a can be calculated. While the planar part's surface roughness is 160 µm, the non-planar part shows a surface roughness of only 38.8 µm, or 24 % of the planar part's roughness [PH21].

Current limitations of non-planar AM with variable layer height using three-axis systems are mainly related to the degrees of freedom the kinematic system allows. Firstly, the maximum slope which can be manufactured in a non-planar fashion is limited by the nozzle geometry as well as the surrounding assembly. Secondly, nozzle penetration of the manufactured slope occurs for every non-planar slope, regardless of its angle. This effect becomes more prominent with increasing surface angles [PH21]. It can be mitigated slightly by choosing a suitable nozzle but will always be present as long as the nozzle is not oriented orthogonal to the strand being deposited. Lastly, contact pressure between the extruded material and the previous layer is not homogenously distributed over the nozzle orifice for non-planar layers using a three-axis system [PH21]. This might lead to variations in layer bonding depending on the manufactured slope and therefore influence mechanical properties.

To address those limitations, the degrees of freedom in the kinematic system need to be increased to allow for rotational movement of the extrusion unit around the tool centre point (TCP).

3.3 Non-planar AM using a six-axis robot arm

Previous approaches using more than three degrees of freedom for additive manufacturing were mostly targeting the reduction of support structure [LJ15, RJK16, DDK17, LHY18, XJ20]. By reorienting the nozzle during production, it has been shown that overhangs can be manufactured which typically would require support structure, therefore saving material and manufacturing time. However, those approaches still use planar layers with a fixed layer height. The reorientation is achieved by rotating the coordinate system the planar layers are manufactured in. A different approach is to use an additional rotational axis for changing tools, e.g., to a subtractive tool [LWC14]. This way, additively manufactured parts can be milled to tight tolerances on the same machine. The approach of combining additive and subtractive manufacturing has also been investigated by Li et al. Here, a six-axis robot arm has been used with an FDM toolhead and a milling spindle to achieve an improved surface finish of AM parts [LHY18]. However, just as other methods presented, the advancements in AM are limited to reorienting the coordinate system in order to limit the use of support material and to creating a planar part with a non-planar top surface, similar to the approaches with three-axis systems. Furthermore, the benefit in terms of surface quality has only been judged qualitatively and not quantitatively. One reason for the limited use may have been the challenges in synchronising the robot's movements with the AM toolhead. The only application of non-planar AM using more than three axes where mechanical properties were considered was presented by Fang et al. Here, the concept of orienting the extruded strands along the load path to use anisotropic properties of AM parts is being investigated. However, this approach heavily relies in the orientation of the model to achieve the optimal strand layout, therefore increasing the use of support material to achieve the necessary part orientation on the build platform. Furthermore, the authors note that the staircase effect is increased in comparison to planar AM, since the surface quality is being disregarded in favour of optimising strand orientation regarding the load case [FZZ+20].

To combine the advantages of non-planar AM regarding mechanical properties and accurate reproduction of freeform surfaces in addition to increasing maximum part size, manufacturing speed and material diversity, the approach to non-planar AM presented in chapter 3.2 is expanded to the IKV-developed hybrid manufacturing cell, which is shown in Figure 3.6. This large-volume AM machine consists of a KR 16 six-axis robot arm by Kuka AG, Augsburg, Germany, a custom screw-based extruder, and a custom electrical cabinet for controlling the combined system.

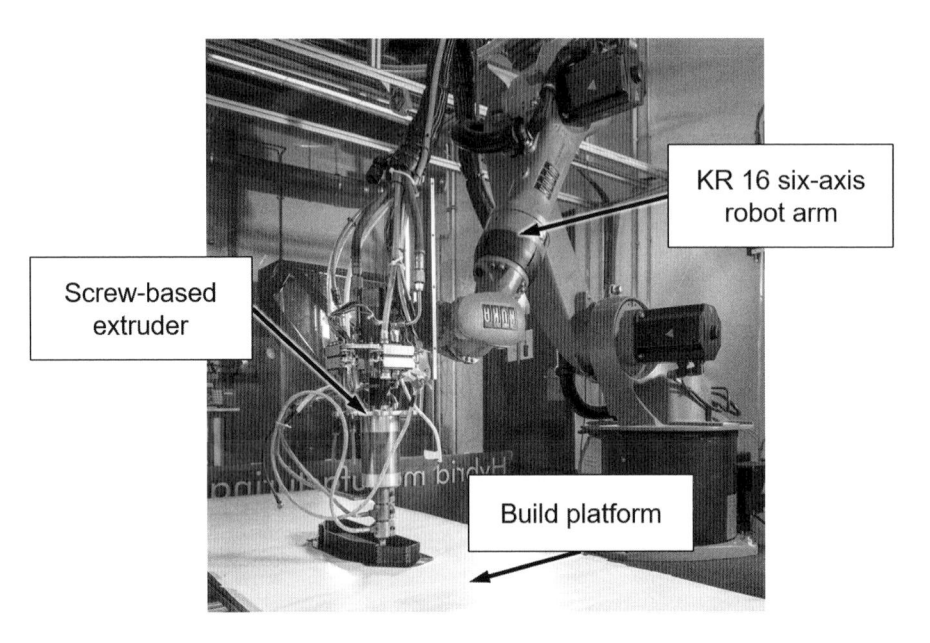

Figure 3.6: The hybrid manufacturing cell combines the build volume and agility of a six-axis robot arm with the material diversity and throughput of a screw-based extruder.

The six-axis robot arm is addressed using Kuka robot language (KRL). Using this language, the TCP target position can be specified in all three spatial axes (x, y, z) as well as the associated rotations around those axes (c, b, a, respectively). The inverse kinematics for setting the actuators in each joint of the robot are solved by the controller. KRL offers two commands for moving the TCP: point-to-point movements (PTP) and linear movements (LIN). PTP moves only account for the accuracy of the movement's start and end point. The move itself is optimised for the six rotational actuators in the robot joints. Therefore, PTP moves are typically carried out in form of an arc. Since in AM, not only the start and end point of a move are relevant but all points along one movement have to be exact to deposit material at the intended location, only LIN moves are used. LIN commands move the TCP from start to end using the shortest route between those points, resulting in a straight, linear movement. LIN moves are specified in KRL as:

```
LIN X [target_X],Y [target_Y],Z [target_Z], A [target_A], B [target_B],
C [target_C] C_VEL
```

with

`target_X,Y,Z`: TCP end position of the move in the specified axis in [mm]

`target_A,B,C`: TCP end orientation of the move around the specified axis in [°].

`C_VEL` specifies the movement speed. It has to be set prior to the LIN command using `$VEL.CP=[movement speed in m/s]`.

The values for `target_X`, `target_Y` and `target_Z` are calculated exactly like the values for x, y and z for non-planar AM using a three-axis system [PH21]. To calculate the orientation of the TCP such that the nozzle is always positioned orthogonal to the strand being deposited, the surface normal must be calculated. To achieve a continuous transition between surface and function layers, transition layers are interpolated between two specified layers (see Figure 3.2). Since those transition layers are not given as parameterised functions, the surface normal cannot be calculated using partial derivation. Instead, the current (x_P, y_P, z_P) and next point to be approached, separated by the slicing resolution r, are used to calculate the direction vectors $\vec{x}$ and $\vec{y}$ (see Equation 3.7 and Equation 3.8). By dividing their cross product by its length, the surface normal $\vec{n}$ is calculated according to Equation 3.9.

$$\vec{x} = \begin{pmatrix} x_P + r \\ y_P \\ z(x_P + r, y_P) \end{pmatrix} - \begin{pmatrix} x_P \\ y_P \\ z(x_P, y_P) \end{pmatrix} \tag{3.7}$$

$$\vec{y} = \begin{pmatrix} x_P \\ y_P + r \\ z(x_P, y_P + r) \end{pmatrix} - \begin{pmatrix} x_P \\ y_P \\ z(x_P, y_P) \end{pmatrix} \tag{3.8}$$

$$\vec{n} = \frac{\vec{x} \times \vec{y}}{|\vec{x} \times \vec{y}|} \tag{3.9}$$

Since the nozzle and the extruded material is rotationally symmetrical around the system's z-axis, only the values for b (rotation around y-axis) and c (rotation around x-axis) have to be considered. They are calculated by projecting the normal vector $\vec{n}$ into the xz- and yz-plane, respectively. This way, every point along the non-planar AM part can be described, including the TCPs orientation and speed.

Initially, the hybrid manufacturing cell was only used for planar AM with constant manufacturing speed. This way, the extruder could be operated with a constant screw speed since the extruded volume had to be constant. For non-planar AM with variable layer height, however, the extruded volume has to be adapted constantly (see chapter 3.2). In this regard, filament-based systems have the benefit of using a single G-Code file processed by one controller to address all three motion axes as well as the extruder. For the hybrid manufacturing cell, the two separate systems for motion and extrusion need to be unified first. Typically, this would be done by introducing a superordinated system which processes the machine code and addresses the subsystems for motion and extrusion accordingly. However, since for safety reasons it is not intended to continuously stream new commands to a Kuka robot arm which get executed immediately, this approach cannot be followed. Instead, the Kuka control unit was defined as the master in this system architecture, with the Kuka machine code containing all necessary commands for motion and extrusion. To achieve synchronicity, the extrusion commands are passed to the extruders control unit via addressable outputs of the Kuka control unit. This way, the extruder's screw speed can be varied during manufacturing, depending on the local layer height needed. Figure 3.7 shows a non-planar part with variable layer height being manufactured on the hybrid manufacturing cell using the additional rotation of the extruder.

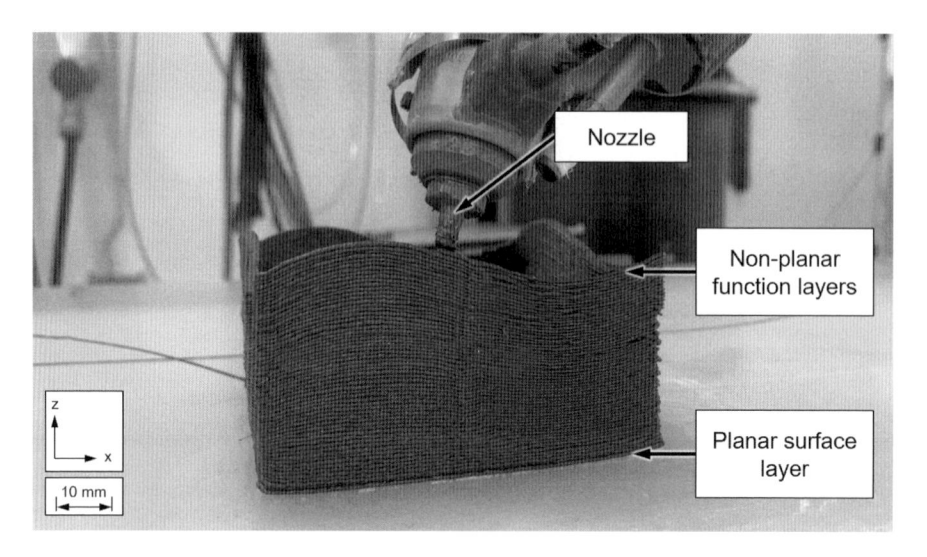

Figure 3.7: A non-planar part with sinusoidal function layers being manufactured on the hybrid manufacturing cell. The extruder is always oriented orthogonally to the deposited strand.

3.4 Trials

To verify the expansion of the non-planar slicer to be able to generate machine code for six-axis kinematic systems and the accompanying control algorithms for managing and synchronizing a screw-based extruder, two test parts are manufactured. For both parts, AKROMID B3 ICF 30 9 AM by manufacturer AKRO-PLASTIC GmbH, Niederzissen, Germany, is used. This material is a 30 % carbon fibre reinforced Polyamide 6.

Firstly, a hollow cube (edge length: 100 mm) with a flat bottom and no top layers is produced. For the whole height of the cube, a sine function with a wavelength of 100 mm and an amplitude of 6 mm is specified. This way, the first layer is deposited with a planar underside, allowing for good adhesion to the build platform. The next layers are then manufactured as transition layers, until the complete sine function can be produced without violating the constraints described in chapter 3.2. Then, as many complete sine functions as possible are manufactured, until the transition towards the flat top layer has to be produced. The test part proves that all calculations are done correctly and that the extruder is always oriented orthogonally towards the deposited strand. However, it also shows that accurate control of material flow using a screw-based extruder is more difficult than compared to a filament-based system. Since a change in screw rotational speed is not immediately followed by a change in extruded material but rather afflicted with a delay, material is not always deposited correctly. This is especially noticeable during transition layers, which typically show the largest variation in layer height (see Figure 3.8). Regarding the walls of the manufactured cube, excess material is visible. Based on the delay between increasing / decreasing screw speed and increasing / decreasing material output, this excess material is missing in other parts of the cube.

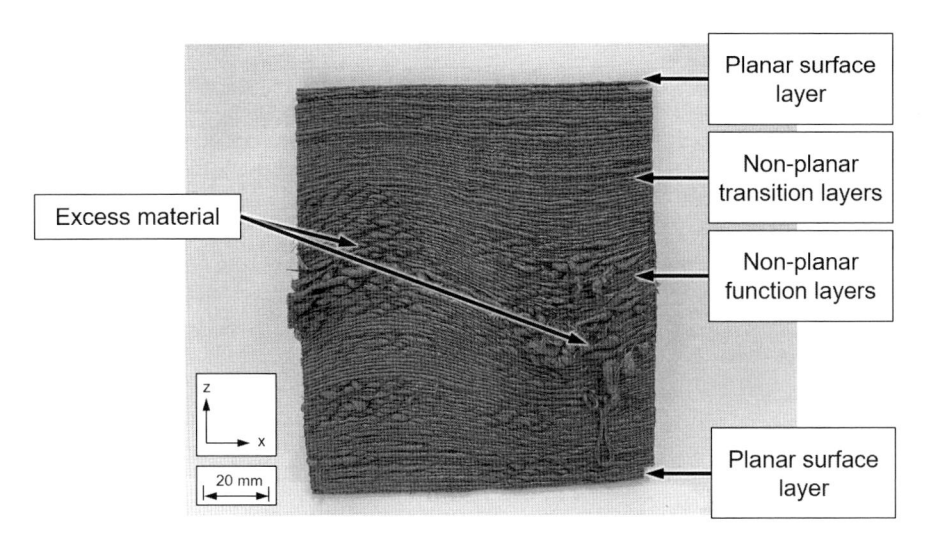

Figure 3.8: Cube with sinusoidal function layers. Because of the output delay of the extruder, material is not always placed in the correct location.

To evaluate the reproduction of freeform surfaces, a second test part is manufactured. This part has a rectangular base of 150 mm by 100 mm and is limited in height by the freeform surface $S_2(x,y)$ described by Equation 3.10. It is shown in Figure 3.9.

$$S_2(x,y) = 20 + 5 \cdot \sin(\frac{x}{20}) + 5 \cdot \sin(\frac{y}{20}) \tag{3.10}$$

During manufacturing, another drawback of screw-based extrusion systems is noticed. FDM systems typically don't fill a part completely with material but rather use a sparse infill. This saves time and material during manufacturing and limits warping of parts due to residual stresses. To create a solid top, the strands of the sparse infill are then bridged by multiple top layers. Since saving time and material are key requirements for the hybrid manufacturing cell and the warpage of parts is even more pronounced with large scale systems, the option of using sparse infill has also been included in the presented slicer. However, because of the pressure at which the material is extruded, bridging over small air gaps is more difficult. Instead of building a bridge, the material is pressed into the gap. Therefore, the second test part was manufactured using a 70 % linear infill, both for internal structures and top / bottom layers. The part itself not only shows the accurate representation of the freeform surface, but because of the strand width of 3 mm, the accurate orientation of the extruder orthogonally to the strands is visible.

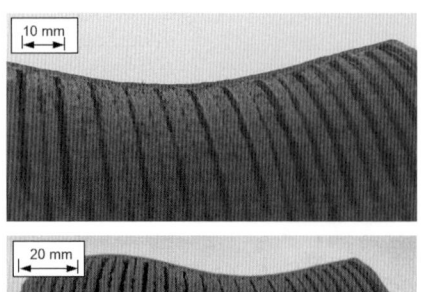

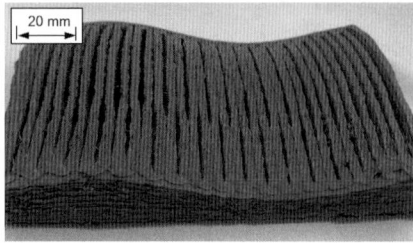

- Material: PA6CF30

- Rectangular base: 150 mm by 100 mm

- Freeform surface:

$$S_2(x,y) = 20 + 5 \cdot \sin\left(\frac{x}{20}\right) + 5 \cdot \sin\left(\frac{y}{20}\right)$$

- Strand width: 3 mm

- Infill: 70 %, linear

Figure 3.9: Part manufactured with a freeform surface. The deposited strands show accurate positioning of the extruder.

3.5 Conclusions and outlook

This article presents the extension of non-planar additive manufacturing with variable layer height to make use of a six-axis kinematic system. This method was already proven to be beneficial in three-axis systems to improve surface quality significantly. By using the additional three rotational degrees of freedom, the method of non-planar AM is improved by increasing maximum possible slope of non-planar layers, eliminating penetration of the nozzle during manufacturing of slopes, and improving contact pressure between the currently deposited strand and the previous layer. Furthermore, by changing the system architecture of the hybrid manufacturing cell, this method enables non-planar AM using screw-based extruders, therefore opening up new possibilities in terms of part size, manufacturing speed and material diversity.

Current limitations of this method are mostly attributable to screw-based extrusion. Because of the delay between setting a different screw rotational speed and the resulting change in material output, material is typically not deposited correctly in applications with varying layer height. Additionally, because of the pressure at which the material is extruded, bridging is not possible. Therefore, sparse infill layers cannot easily be covered by solid top layers. Regarding the used six-axis robot arm, a limitation regarding file size can be noticed. The maximum file size for a contiguous program is around 4 MB, which is sufficient for handling operations where only a few locations have to be approached and the path in between is typically not of concern. For AM applications and especially for non-planar AM, however, several thousand points have to be approached in succession, describing the exact path in order to deposit material in the correct location. Here, a file size of 4 MB only fits very small parts with a low level of complexity.

In future work, the challenge of correct material deposition using a screw-based extruder regardless of delay shall be approached. To address the delay, the non-planar slicer will be expanded by a signal shifting algorithm, such that extrusion commands can be executed earlier. By determining the exact delay based on material used and change in rotational screw speed, the delay in extrusion can be accounted for.

Acknowledgements

Funded by the Deutsche Forschungsgemeinschaft (DFG, German Research Foundation) under Germany's Excellence Strategy – EXC-2023 Internet of Production – 390621612.

References

[Ahl18] AHLERS, D.: *3D Printing of Nonplanar Layers for Smooth Surface Generation.* Universität Hamburg, Master thesis, 2018

[BDH00] BHARATH, V.; DHARAM, P. N.; HENDERSON, M.: Sensitivity Of Rp Surface Finish To Process Parameter Variation. *Solid Free-form Fabrication Proceedings* (2000), p. 251–258

[CAR08] CHAKRABORTY, D.; ANEESH REDDY, B.; ROY CHOUDHURY, A.: Extruder path generation for Curved Layer Fused Deposition Modeling. *Computer-Aided Design* 40 (2008) 2, p. 235–243

[CLH07] CHUNG WANG, C.; LIN, T.-W.; HU, S.-S.: Optimizing the rapid prototyping process by integrating the Taguchi method with the Gray relational analysis. *Rapid Prototyping Journal* 13 (2007) 5, p. 304–315

[DDK17] DING, Y.; DWIVEDI, R.; KOVACEVIC, R.: Process planning for 8-axis robotized laser-based direct metal deposition system: A case on building revolved part. *Robotics and Computer-Integrated Manufacturing* 44 (2017), p. 67–76

[DM94] DOLENC, A.; MÄKELÄ, I.: Slicing procedures for layered manufacturing techniques. *Computer-Aided Design* 26 (1994) 2, p. 119–126

[FZZ+20] FANG, G.; ZHANG, T.; ZHONG, S.; CHEN, X.; ZHONG, Z.; WANG, C. C. L.: Reinforced FDM. *ACM Transactions on Graphics* 39 (2020) 6, p. 1–15

[GRS15] GIBSON, I.; ROSEN, D.; STUCKER, B.: *Additive manufacturing technologies: 3D printing, rapid prototyping and direct digital manufacturing.* New York and Heidelberg and Dodrecht and London: Springer, 2015

[GWKA15] GORSKI, F.; WICHNIAREK, R.; KUCZKO, W.; ANDRZEJEWSKI, J.: Experimental Determination of Critical Orientation of ABS Parts Manufactured using Fused Deposition Modelling Technology. *Journal of Mechanical Engineering* (2015) 15, p. 121–132

[HL16] HOPMANN, C.; LAMMERT, N.: Hybrid Manufacturing Strategy for Functionally Integrated Single Parts. *Kunststoffe international* (2016) 11, p. 39–42

[HP19] HOPMANN, C.; PELZER, L.: Additive Manufacturing of Large Parts by an Extrusion based Hybrid Approach. *Proceedings of the 35th International Conference of the Polymer Processing Society.* Izmir, Turkey, 2019

[Hua14] HUANG, B.: *Alternate Slicing and Deposition Strategies for Fused Deposition Modelling.* Auckland University of Technology, Dissertation, 2014

[LHY18] LI, L.; HAGHIGHI, A.; YANG, Y.: A novel 6-axis hybrid additive-subtractive manufacturing process: Design and case studies. *Journal of Manufacturing Processes* 33 (2018), p. 150–160

[LJ15] LEE, K.; JEE, H.: Slicing algorithms for multi-axis 3-D metal printing of overhangs. *Journal of Mechanical Science and Technology* 29 (2015) 12, p. 5139–5144

[LJAT16] LLEWELLYN-JONES, T.; ALLEN, R.; TRASK, R.: Curved Layer Fused Filament Fabrication Using Automated Toolpath Generation. *3D Printing and Additive Manufacturing* 3 (2016) 4, p. 236–243

[LWC14] LEE, W.-C.; WEI, C.-C.; CHUNG, S.-C.: Development of a hybrid rapid prototyping system using low-cost fused deposition modeling and five-axis machining. *Journal of Materials Processing Technology* 214 (2014) 11, p. 2366–2374

[PH21] PELZER, L.; HOPMANN, C.: Additive manufacturing of non-planar layers with variable layer height. *Additive Manufacturing* 37 (2021), p. 101697

[RJK16] RIEGER, M.; JOHNEN, B.; KUHLENKOETTER, B.: Analysis and Development of the Fused Layer Manufacturing Process using Industrial Robots. *Proceedings of ISR 2016: 47st International Symposium on Robotics.* Munich, 2016

[Rob21] ROBERTS, T.: *Additive manufacturing trend report 2021.* Hubs, 2021

[RW16] RICHTER, S.; WISCHMANN, S.: *Additive Fertigungsmethoden - Entwicklungsstand, Marktperspektiven für den industriellen Einsatz und IKT-spezifische Herausforderungen bei Forschung und Entwicklung.* Begleitforschung AUTONOMIK für Industrie 4.0, Institut für Innovation und Technik (iit), 2016

[URL18] ZUZA, M.: *Everything about nozzles with a different diameter.* URL: https://blog.prusaprinters.org/everything-about-nozzles-with-a-different-diameter_8344/, 13.09.2021

[Woh19] WOHLERS, T.: *Wohlers Report 2019: 3D Printing and Additive Manufacturing State of the Industry.* Wohlers Associates, 2019

[WSP+19] WURZBACHER, S.; SCHMITZ, M.; PELZER, L.; HOHLWECK, T.; TOPMÖLLER, B.; HOPMANN, C.; RÖBIG, M.: Jenseits menschlicher Fähigkeiten: Modellgestützte Prozesseinrichtung durch vollvernetzte Produktion im Spritzgießen. *Kunststoffe* 2019 (2019), p. 142–145

[XJ20] XIAO, X.; JOSHI, S.: Process planning for five-axis support free additive manufacturing. *Additive Manufacturing* 36 (2020), p. 101569

Symbols

Symbol	Unit	Description
a	°	a coordinate (rotary)
b	°	b coordinate (rotary)
c	°	c coordinate (rotary)
D	mm	Diameter of nozzle's flat bottom
D_F	mm	Filament diameter
d_{SP}	mm	Euclidian distance in the x-y-plane between support points
E	mm	Length of extruded material
F_{Normal}	N	Force normal to strand
$F_{Parallel}$	N	Force parallel to strand
F_{Total}	N	Total force being applied to part
$\vec{n}$	–	Surface normal
p	mm	Penetration depth of nozzle
r	mm	Slicing resolution
R_a	μm	Surface roughness
S	–	Surface function
V_F	mm^3	Volume of filament entering the extruder
V_{VE}	mm^3	Volume of volumetric element
w_E	mm	Extrusion width
x	mm	x coordinate (translatory)
$\vec{x}$	–	Direction vector x
y	mm	y coordinate (translatory)
$\vec{y}$	–	Direction vector y
z	mm	z coordinate (translatory)
z_0	mm	Constant offset from z = 0
$z(p_{0-3})$	mm	Z coordinate of trapezoid corners
α	°	Cone angle of nozzle
θ	°	Angle of manufactured slope

Abbreviations

Notation	Description
AM	Additive Manufacturing
FDM	Fused Deposition Modeling
KRL	Kuka robot language
MB	Megabytes
TCP	Tool centre point

Modelling of time-scaled properties of FRP

Moderator: Dr.-Ing. Marc Linus Fecher, Mubea Carbo Tech GmbH

--- **Content** ---

1 Modelling of impact damage and fatigue in composites　　597

M. L. Fecher [1]
[1] Mubea Carbo Tech GmbH

2 Analysis of strain rate dependent material properties using a drop tower for its application in the research of damage prediction on fibre reinforced plastics　　613

Ch. Hopmann[1], J. M. Müller[1]
[1] Institute for Plastics Processing (IKV) at RWTH Aachen University

3 Investigation of the correlation creep/fatigue in Unidirectional fibre reinforced plastics　　629

Ch. Hopmann[1], N. Rozo Lopez[1]
[1] Institute for Plastics Processing (IKV) at RWTH Aachen University

Dr.-Ing. Marc Linus Fecher

Dr.-Ing. Marc Linus Fecher studied mechanical engineering with specialisation in aerospace technology at RWTH Aachen University. After his studies, he worked as Research Associate in the department „Faserverstärkte Kunststoffe und Polyurethane" at IKV as leader of the working group „Flüssigimprägnierverfahren / RTM". In 2016, he moved to industry after successfully completing his PhD at IKV, titled „Analyse des 3D-Faserspritzverfahrens zur Fertigung komplexer und funktionsintegrierter RTM-Strukturbauteile". Since then, Dr. Fecher has been working at Mubea Carbo Tech GmbH, where he has progressed from Lead Engineer Process Development to Head of Development Process Technology, and currently he is associated as a member of the extended executive Board of this company.

1 Modelling of impact damage and fatigue in composites

M. L. Fecher [1]

[1] Mubea Carbo Tech GmbH

Abstract

Fibre reinforced plastics (FRP) also well known as composites are used in a wide range of applications and their market is continuously growing over the last years before the corona crisis [WM20, Gar20].

Beside others, two main fields of application for FRP are the automotive and aerospace industry. In the automotive and aerospace industry FRPs are mainly used due to their high weight specific stiffness and strength and the belonging potential for weight reduction by lightweight composite components. [Nju16, BYG+18]

The weight reduction of aerospace and automotive composite components leads to a reduced fuel consumption, reduced CO2-emissions and higher payloads [Nju16, BYG+18]. Moreover, composites are gaining more and more importance in the automotive sector due to the power train transformation from combustion engines to Battery Electric Vehicles (BEV). Due to the high weight of the batteries and their casing system, composites enable a weight reduction to compensate partially the increased weight of the electric power train and the belonging components [HWS12]. The increasing number of applications for composites in the mentioned sectors lead to a wide range of mechanical challenges. Furthermore, composites are increasingly being used for impact and fatigue dominated load cases.

Applications, which belong to the mentioned category of automotive applications, are amongst others composite wheels, B-pillars, battery skid plates and stabiliser bars (see Figure 1.1).

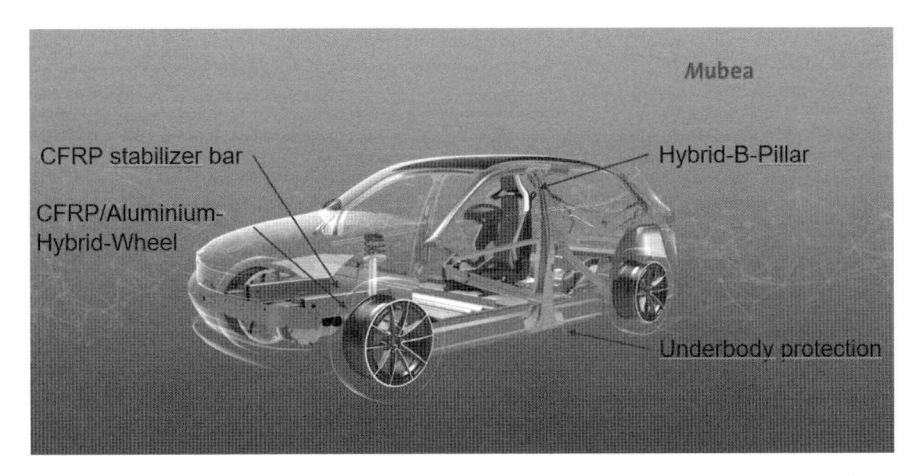

Figure 1.1: Automotive composite applications

The increasing demand to use composite components for such applications requires sufficient methods to design the components by enabling Finite Element Analysis (FEA) and Computer Aided Engineering (CAE) methods. However, the mentioned applications and the dynamic and fatigue load cases are challenging for composites due to their anisotropic and load transferring characteristics to neighboured fibre layers during failure. This material specific behaviour of composites under dynamic and cyclic loads leads to the requirement that FEA and CAE tools are able to predict this behaviour. Moreover, fundamental and high quality material data are necessary to enable the needed calculations.

The Mubea Carbo Tech GmbH (MCT), Salzburg, Austria, is market leader in the manufacturing of complex shaped and functional integrated composite components for automotive applications and therefore a comprehensive knowledge in the field of FEA-calculations for composites was developed over the last years. To transfer this knowledge to use it for aerospace applications currently extensive investigations are carried out in the research project MILAN (Mubea Integrated Lightweight Aviation Composite Components for Next Generation Aircrafts) to generate a material data base to enable the use of the MCT tools for FEA-calculations for aerospace applications. In the following an insight in our development and research activities on analysis methods for impact damage, fatigue damage and degradation will be given.

1.1 Impact damage

As described above the increasing use of composites in automotive and aerospace applications leads to an increased use for applications which are characterised by load cases that are impact dominated or fatigue dominated. Moreover, a combination of both is possible. Good examples for impact damage dominated load cases are the side pole crash while using composite (hybrid) B-pillars or bird strikes on primary structures of air planes like composite landing gears [IOS17, LR13].

The accurate prediction of the mechanical impact performance of a component is driven by various influences like the material model, the modelling approach (e.g. shell, continuum shell or solid elements) and the available database for the chosen material combination.

For components like a hybrid B-pillar the component only has to withstand a single impact load in case of a crash event. For other components a specific fatigue performance after an impact damage is required, caused by the field of application. An example for such a regulation is the automotive CFRP/Aluminium-Hybrid-Wheel. Requirements or regulations on those impact resistances are usually given by the original equipment manufacturers (OEM) or by law e.g. UN/ECE R124 [NN06].

The impact energy is calculated with given equations in consideration of the maximum static wheel load. This impact test must not lead to a sudden air pressure loss or even to a loss of structural integrity of the wheel. This is intended to ensure that the vehicle can be stopped safely, e.g. after driving through a massive pothole with high velocity. Beside stiffness and weight targets, impact damage is often a main design target in the development of composite parts, especially for automotive applications. Nowadays the simulation of impact damage can be done by using sophisticated material models, already available in commercial simulation software packages.

The prediction of the remaining service life after an event of damaging impact, is a big challenge. In contrast to a single crash event (e.g. severe car crash) the composite structure can also be damaged by a barely visible impact damage (BVID). BVIDs are damages of a laminate with little or no surface indication of the damage [NN12]. This kind of damage leads to a local degradation of the material properties and in further service it can result in a significant fatigue damage within the laminate, such as delamination, matrix cracking, fibre pullout and finally fibre fracture.

Although a structure is already damaged, the structural integrity of the structure must be ensured for the remaining use or at least to the next inspection. The aircraft industry calls this attribute damage tolerance as defined in FAA Advisory Circular 25.571-1C for transport aircraft:

"Damage tolerance - The attribute of the structure that permits it to retain its required residual strength for a period of use after the structure has sustained a given level of fatigue, corrosion, accidental or discrete source damage."

For the development based on FEA of such damage tolerant structures, a method has been developed at MCT to assess cumulative damage caused by fatigue. This also enables a considerable reduction of physical tests. In the following chapter this method and the used workflow will be explained more in detail.

1.2 Fatigue damage and degradation

Nowadays, components made of FRP are still developed with static failure criteria assuming pristine material and ideal material properties coming from coupon testing under laboratory conditions. As a result, the consequences of areas that are already damaged in the initial state or at least in the first cycles of loading will not be correctly considered in the structural analysis. This stress-peaks or stress singularities are often wrong interpreted. Granted, these singularities can come from simplifications in the finite element model (boundary conditions, coarse mesh, etc.), but in many cases it can be a local overload of the material. Overloading leads to damage, mainly as micro cracks in the matrix material and, as a result, to a local degradation of the material properties. Since these areas can no longer take the full load, stress redistribution are occurring around the damaged areas. These redistribution starting with the first cycle of loading and unloading until a kind of stress equilibrium around the damage and the external loads is established. Consequently this local events cumulated will lead to a global stiffness reduction and can usually be observed in physical testing of parts made of composite on a testbench.

Static loaded components made of FRP can rightly be viewed as an exception. Furthermore, the stress components itself will be highly non-proportional to external loads, caused by superimposing of external loads (e.g. static weight and dynamic loads), assemblies with contacts (open, close) or other non-linearities.

Fatigue stresses will have in general the following characteristics:

- Non-proportional to external loads
- Multi-axial loading
- 3 dimensional stress state (in-plane and out-of-plane)
- Phase shift of the maxima of the stress components within a cycle
- Not cyclic (application of rainflow-counting algorithm)

Considering this conditions, an accurate prediction of the fatigue life or mechanical performance (degradation) for composite components is very challenging and cannot be done with simple equations. Therefore, the following concept has been developed.

1.2.1 Concept of fatigue and degradation analysis

Comparable to static failure criteria the elaborated fatigue analysis method assesses a composite layer wise in each integration point in a lamina. The expected fatigue life until the beginning of a damage can be calculated, by setting up the method on the result-files of an existing FEA. For a complete fatigue damage and degradation analysis for the intended cycles of use, subsequent calculation loops including FEA must be done. See Figure 1.2. The number of cycles in a loop are dependent on the occurring damage.

The basic idea behind this concept is the modification of well-established static failure criteria for composites, like max. Stress, max Strain, Puck [Puc96] or Cuntze [Cun12]. Currently the max. Stress Criterion and the Puck (2D and 3D) Criterion is implemented. Further criteria will be implemented and compared together when more material data and considerable data of real component testing is available. For the fatigue analysis of metals and adhesives a "critical plane - critical component" criterion is implemented [GPS02].

The user via a simple graphical user interface (GUI) can select the criterion for the method. Additionally it is even possible to use more than one criterion for different materials in parallel. This enables the calculation of fatigue damage of the whole component or assembly in one single fatigue analysis. The framework of the method (Figure 1.2) has a modular structure, so different failure criteria can be relatively simple added or existing modules can be modified when it´s necessary.

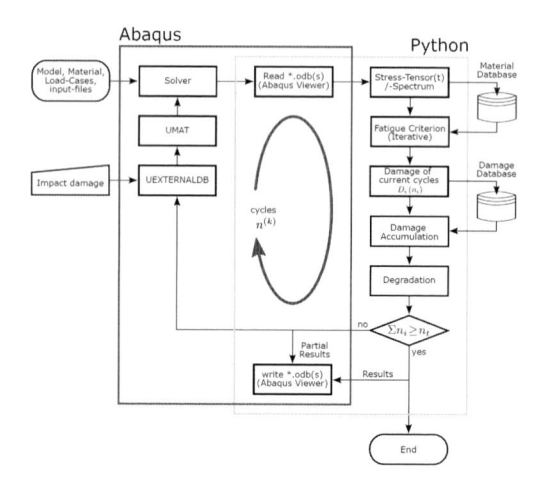

Figure 1.2: Workflow and interfaces of fatigue and degradation analysis

1.2.2 Fatigue stress

Every engineering structure sustains numerous different loading patterns during it´s daily operation. Loading patterns of constant amplitude represent the minority, while the majority of the applied loads are of an irregular nature. A characteristic of the realistic loading is the stress multiaxiality, since, even when the external loads are of a uni axial nature, the developed stress and strain fields are normally more complicated due to the material anisotropy. [Vas11]

Cycle counting is used to summarise irregular load vs. time histories by providing the number of occurrences of cycles of various sizes. The definition of a cycle varies with the method of cycle counting. A significant number of cycle-counting techniques have been proposed over the last 30 years, e.g. ASTM E1049. [Vas11]

As described above, such stress patterns can only be collected from real loading in daily operations with already built parts, e.g. using strain gauges at specific locations of interest. That requires that a prototype part has to be built and then applied in the intended usage to collect the stress data. This procedure is important for ensuring safe life of a component, but can result in expensive design loops.

For MCT as a supplier company, the part development is mainly based on reproducible cyclic loading or at least on load spectra, e.g. external force or moment over a certain number of cycles. This leads almost to a pure virtual development of the composite parts and virtual design loops. The target is to reduce practical tests to a minimum in the design phase and use selective testing for the validation of the analysis and the used development methods. The benefit is a relatively fast development while maintaining low development costs.

Usually such cyclic loads are included in the specifications provided by the OEMs. A good example for a typical cyclic loading is the cornering fatigue or rotating bending test. This test is simulating the lateral forces acting on a wheel in driving around a curve. The wheel rim is fixed rigidly to the test bench and a circulating constant bending moment is applied to the hub mounting area (according [NN06]).

Although the resulting internal stresses are cyclic, the stresses are not proportional (to the cyclic bending moment), out-of-phase and multi axial. As example, Figure 1.3 is showing a typical stress cycle at a specific element of a full CFRP-Wheel in a rotating bending test.

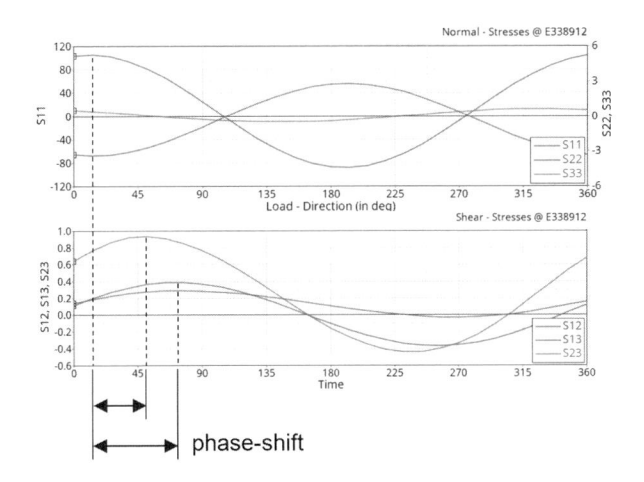

Figure 1.3: Cyclic stresses of a CFRP-Wheel in a rotating bending test

It can be seen in Figure 1.3 that each stress component seems to have its own stress ratio. This is caused by the anisotropy of the laminate, (statically) preloaded structures, complex assemblies with interacting components (contacts) and the external loading. The in-situ stress ratios can be very heterogeneous and can even vary within the integration points (IP) of a lamina. The levels for the definition of the stress ratio are shown in Figure 1.4.

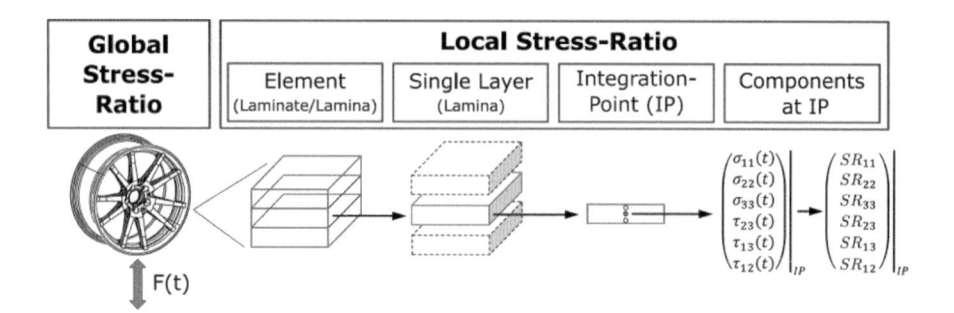

Figure 1.4: Global and local stress ratio

The assumption that the whole part is having the same stress ratio (global stress ratio) can lead to overestimation or underestimation of the local fatigue strength in a fatigue analysis. Consequently, for each material point (integration point) the stress ratio must be determined to use the appropriate local fatigue strength.

1.2.3 Fatigue strength

A fatigue analysis cannot be done by using static strength properties. In cooperation with academic partners, a comprehensive program for material testing of fatigue strength was accomplished. This program for material-characterisation was done for a composite made of a high strength carbon-fibre and a high performance epoxy resin. The specimens were based on standardised static coupon specimens for in-plane material testing. The testing for fatigue strength was done for reversed loads (SR = 1), tensile (SR = 0.1) and compressive (SR = 10) pulsating loads. Based on the experimental data S-N curves have been determined using Basquin´s equation. The fatigue analysis method itself uses piece wise linear constant life diagrams (CLD). A graphical representation is shown in Figure 1.5. Depending on the stress ratio a linear interpolation between the experimentally determined values is used.

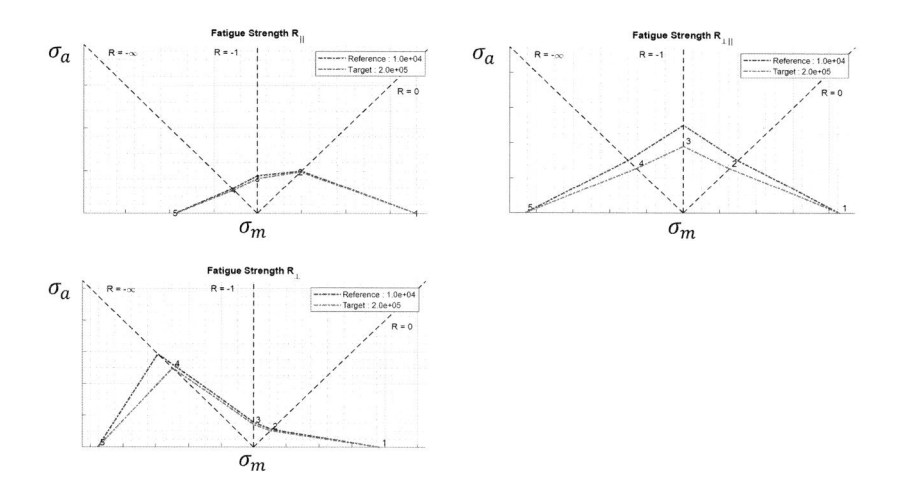

Figure 1.5: Piecewise linear constant life diagrams

1.2.4 Fatigue criterion

The fatigue life prediction of composites and the choice of an appropriate fatigue criterion is still a field of extensive research work. Although a considerable number of fatigue theories and methodologies already have been developed in the last years, extensive test series will be necessary for validation and further development of appropriate and accurate criteria. [Vas11]

The assessment of fatigue damage in the presented method is based on static failure criteria. Figure 1.6 is showing some static failure criteria in comparison to experimental test data. Noticeable are the relatively large deviations of the prediction compared to the experiment, if normal and shear stresses are present at the same time.

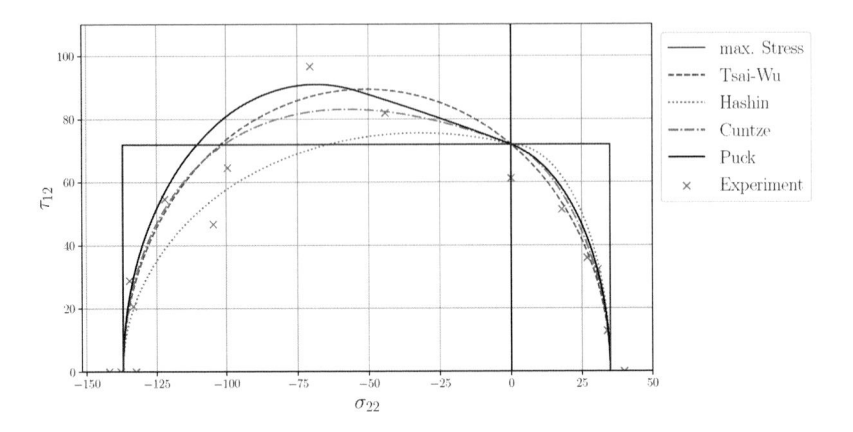

Figure 1.6: Failure envelopes and exemplary test data

In the field of static failure criteria, the World Wide Failure Exercise [KH12] certainly made a substantial contribution. The common failure criteria were examined and compared in order to show their strengths or possible needs for further development.

Because of the general multi axial loading and the three-dimensional stress-state, the implemented criteria in the method are mainly based on the critical plane approach. Puck´s failure criterion for the three-dimensional stress state is using a similar approach. In this criterion, the planes are called action planes and are rotated around the longitudinal axis. On such an action plane three stress components are acting [Puc96]. Puck´s criterion can be used for unidirectional laminae for the three-dimensional stress states (3D) or in the original formulation for shell-like structures with in-plane stress states (2D).

A general explanation of the application of Puck´s criterion on the presented fatigue analysis method will be done on the two-dimensional stress state. The blue line in Figure 1.7 is showing the static failure envelope with static strength. The black dashed line demonstrates the reduced failure envelope depending on the number of cycles, stress-ratio and action plane. With increasing number of cycles, the fatigue strength is decreasing and resulting in a smaller failure envelope. The interaction terms between normal and shear stresses are retained from the static criterion.

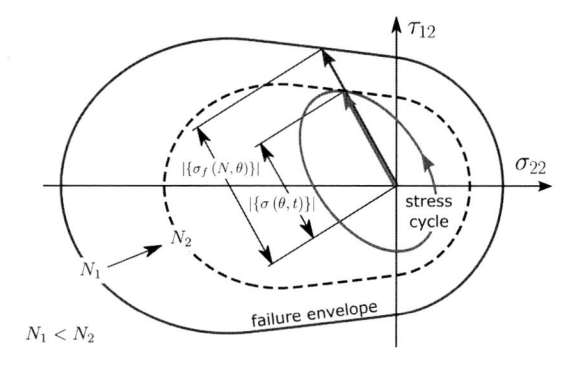

Figure 1.7: Fatigue Failure Envelope

The number of cycles until the beginning of a fatigue damage is determined iteratively. With a higher number of cycles, the area is successively reduced. The algorithm is stopping when a condition for fibre-failure (FF) or inter fibre failure (IFF) according to Puck´s criterion is fulfilled.

A geometrical representation of this criterion for inter fibre failure is given by the stress exposure $f_{E\ IFF}$. If Equation 1.1 is true and the stress cycle touches the failure envelope at least in one point, the tolerable number of cycles is determined. This iterative process must be repeated for each integration point in the model.

$$f_{E\ IFF} = \max N, \theta, t \left(\frac{|\{\sigma(\theta,t)\}|}{|\{\sigma_f(N,\theta)\}|} \right) = 1 \quad (1.1)$$

Analogous procedure is done for the three-dimensional stress state, but additionally with the search for the most critical action plane. This is done in a similar way to the static criterion. The algorithm uses a step size (angular division) of $1°$, which leads to 181 planes in the range of $-90° \leq \theta \leq +90°$. The plane for $+90°$ and $-90°$ can be set as equal, leading to 180 action planes. The transformed stresses of the action planes are shown in Figure 1.8.

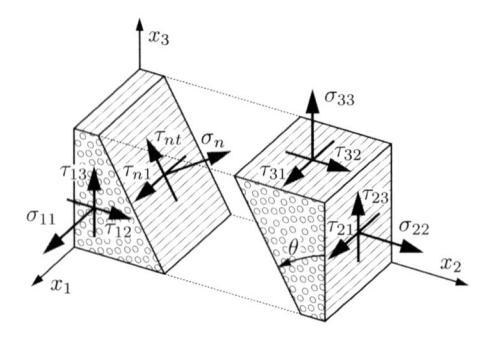

Figure 1.8: Stresses of the action plane

For each of the 180 planes the fatigue strength in dependence of the number of cycles and the stress ratios (SR) has to be determined. Figure 1.8 illustrates the master failure envelope according the theory of Puck.

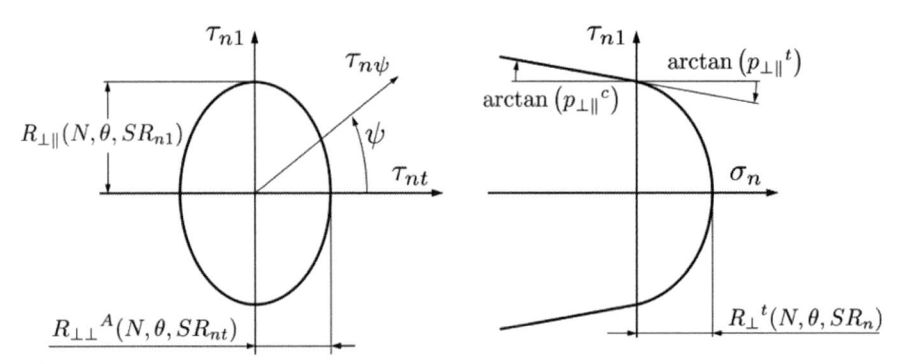

Figure 1.9: Fatigue Strength and Fatigue Failure Envelope for each action plane) $\theta = 0°$, b) $-90° \leq \theta \leq +90°$

Such a failure envelope must be generated for all 180 action planes. To calculate the stress exposure during a cycle, the six time-dependable stress components must be transformed into the stress components of these action planes: $\sigma_n(\theta, t), \tau_{n1}(\theta, t), \tau_{nt}(\theta, t)$.

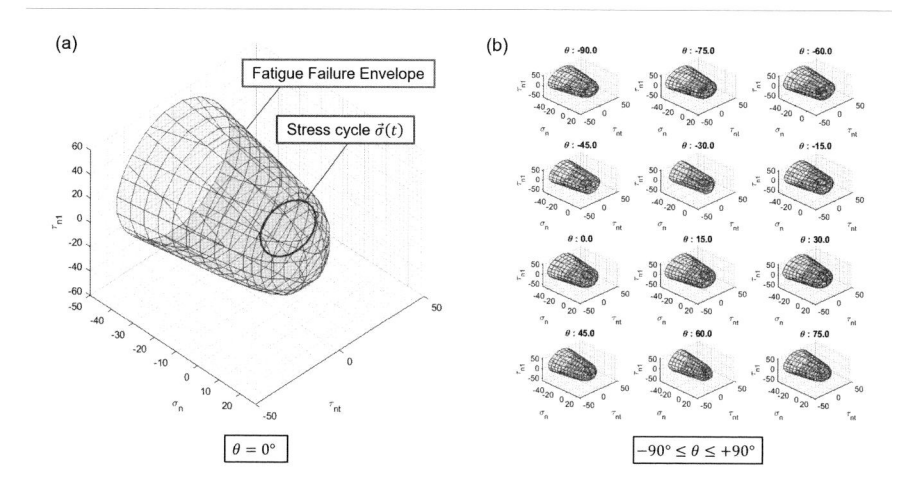

Figure 1.10: Fatigue Failure Envelope and stress cycle: a) $\theta = 0°$, b) $-90° \leq \theta \leq +90°$

The plane with the highest value of stress exposure is considered as the plane with the highest risk for fatigue damage. As already described and illustrated in Figure 1.7, an algorithm is repeating this procedure with an iterative number of load cycles for each integration point and each action plane until the stress exposure $f_{E\ IFF}$ is equal one (see Equation 1.1).

Although a very high number of computations has to be done in this iterative method, thanks to efficient written codes it was possible to reach a considerable reduction of computational time. The runtime is about a few minutes for a typical composite component.

1.2.5 Fatigue damage and accumulation

The damage is calculated using Miner´s rule, also known as Palmgren-Miner linear damage hypothesis. This hypothesis is one of the most widely used and probably the simplest cumulative damage model for damaged caused by fatigue:

$$\sum_{k=1}^{n} \left(\frac{n^{(k)}}{N^{(k)}} \right)^a = D^{(k)} \tag{1.2}$$

The accumulation of damage with different values for the variable a is shown in Figure 1.11. In the presented method, a linear accumulation is used. The use of other values for a must be determined by extensive test series.

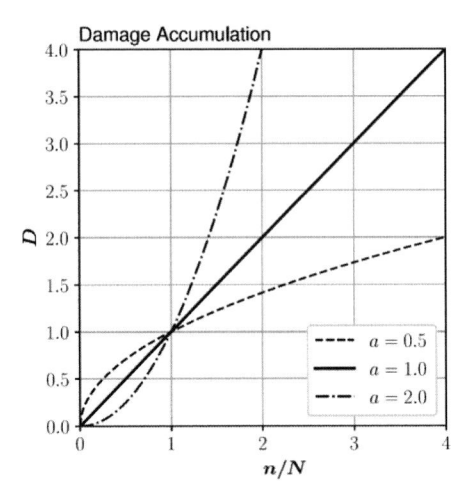

Figure 1.11: Damage accumulation with linear, degressive and progressive accumulation

1.2.6 Degradation of mechanical properties caused by fatigue damage

The local stiffness reduction of a lamina is dependent from the accumulated damage. Therefore, a degradation parameter η is used (see [Puc96, Deu10]) to consider the reduced stiffness according accumulated damage $D^{(k)}$ of the current step k:

$$\eta^{(k)} = \frac{1 - \eta_R}{1 + c \cdot \left(D^{(k)} - 1\right)^\xi} + \eta_R \qquad (1.3)$$

At the onset of damage ($D^{(k)} \geq 1$) the result of this equation starts at $\eta^{(k)} = 1$ and drops with increasing damage until it converges against the residual stiffness η_R. The parameters c and ξ are influencing the shape of the curve and must be determined experimentally.

In application of this method and varying these parameters, it was noticed that the parameters η_R, c and ξ are having a relatively low importance on the result. It was observed that it is less important how, but rather that a degradation begins. Shortly explained, the reason behind is, that degraded material will result in more strain and in higher effective stresses. If this areas cannot supported by stress redistribution to neighbouring material, this circumstance is leading to a fast increase of the damage in this areas and consequently to further degradation.

The degradation parameters are calculated in dependence of the orientation of the most critical plane, which means these parameters are directional. The parameters $\eta_{ij}^{(+)}$ and $\eta_{ij}^{(-)}$ are written to a text file and are therefore provided for the next FEA for the next number of cycles. The lower index denotes which stiffness component will be degraded (11, 22, 33, 12, 13, 23) and the upper sign denotes if it is for positive or negative stress direction. For normal stresses it´s differentiated in tension and compression and for shear stresses in the mathematical positive or negative shear direction. This gives in total a sum of 12 degradation parameters, which are provided for each integration point of the damaged material.

The degradation of the stiffness is done via a user subroutine in the software package Abaqus using the following Equations in a simplified representation:

$$E_{ij}^{(k)} = \eta_{ij}^{(k)(+,-)} \cdot E_{ij}^{(0)} \quad with\ i = j \tag{1.4}$$

$$G_{ij}^{(k)} = \eta_{ij}^{(k)(+,-)} \cdot G_{ij}^{(0)} \quad with\ i \neq j \tag{1.5}$$

$$\nu_{ij}^{(k)} = \eta_{ij}^{(k)(+,-)} \cdot \nu_{ij}^{(0)} \quad with\ i \neq j \tag{1.6}$$

1.2.7 Example CFRP stabiliser bar

The method of the fatigue damage and degradation analysis, as described in the previous sections, will be shown in the following on the example of a stabiliser bar for an automotive application. A typical specification for the development of such an important component for the handling and driving safety of a car is to ensure the stiffness of the stabiliser within a certain value over the service life of the component. The allowable loss of stiffness depends on the application of the composite structure and is usually agreed in advance as design target.

Figure 1.12 is showing the results of a fatigue and degradation analysis of a stabiliser in a test bench condition. The left linkage point is fixed in its movement. On the right linkage point a constant alternating displacement amplitude is applied. The stabiliser is mounted in rubber bearings and rotate relatively freely in these bearings.

The relative simple geometry of this component enables a mesh of high quality. Continuum solid elements with good aspect-ratios are used for the layer-wise representation of the laminate. One solid element for each lamina.

Based on this model the fatigue performance was calculated sequential in more than 20 FEA-steps. See calculation points in Figure 1.12. It also can be seen that the fatigue damage is starting in a very early stage and then turns into a moderate growth.

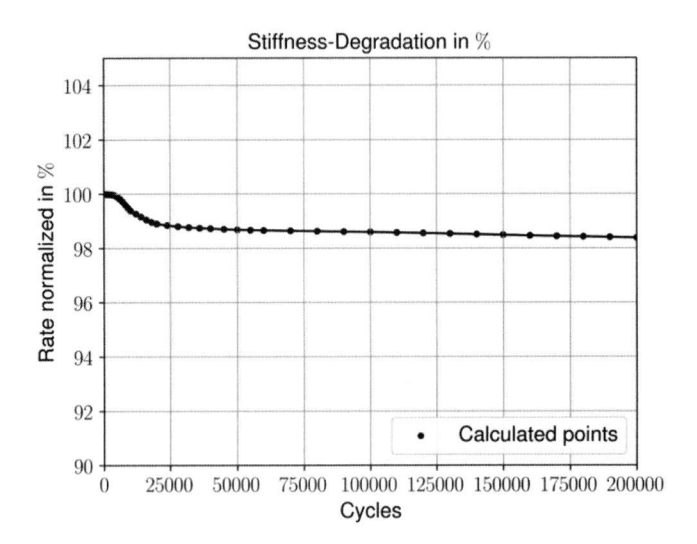

Figure 1.12: Fatigue damage and degradation analysis

The fatigue analysis has been validated on tests with built CFRP-stabiliser bars with same geometry, layup, boundary constraints and loading. The test results have shown a good match of the fatigue damage and comparable stiffness degradation curves.

1.3 Conclusions and outlook

A modern purposeful development of damage and fatigue tolerant composite structures for automotive and aerospace applications have shown the need on comprehensive simulation tools. Reducing the number of physical testing and relocating to virtual testing has a high potential for cost saving and often even more important a significant time saving.

Especially the prediction of fatigue performance after impact damage and ensuring structural integrity is very challenging. Therefore, Mubea Carbo Tech has developed a strategy and method for the prediction of fatigue performance and structural integrity of composite components. The presented method gives the possibility to look beyond the first ply failure and provides a deeper understanding of damage propagation. The implementation of this method for the development of aerospace components is in preparation and therefore an extensive series of material tests is planned. Finally a demonstrator will be built and the method will be validated on this structure.

References

[BYG+18] BACHMANN, J.; YI, X.; GONG, H.; MARTINEZ, X.; BUGEDA, G.; OLLER, S.; TSERPES, K.; RAMON, E.; PARIS, C.; MOREIRA, P.; FANG, Z.; LI, Y.; LIU, Y.; LIU, X.; XIAN, G.; TONG, J.; WEI, J.; ZHANG, X.; ZHU, J.; MA, S.; YU, T.: Outlook on ecologically improved composites for aviation interior and secondary structures. *CEAS Aeronautical Journal* 9 (2018) 3, p. 533–543

[Cun12] CUNTZE, R.: The predictive capability of failure mode concept-based strength conditions for laminates composed of unidirectional laminae under static triaxial stress states. *Journal of Composite Materials* 46 (2012) 19-20, p. 2563–2594

[Deu10] DEUSCHLE, H. M.: *3D failure analysis of UD fibre reinforced composites: Puck's theory within FEA*. Stuttgart University, Dissertation,, 2010

[Gar20] GARDINER, G.: The markets: Automotive (2021). *CompositesWorld* 2020 (2020),

[GPS02] GAIER, C.; PRAMHAS, G.; STEINER, W.: An extended critical plane criterion for general load situations. *Proc. Eighth International Fatigue Congress*volume 2002. Stockholm, Sweden, 2002

[HWS12] HOFER, J.; WILHELM, E.; SCHENLER, W.: Optimal Lightweighting in Battery Electric Vehicles. *World Electric Vehicle Journal* 5 (2012) 3, p. 751–762

[IOS17] IKPE, A. E.; OWUNNA, I. B.; SATOPE, P.: Design optimization of a B-pillar for crashworthiness of vehicle side impact. *JOURNAL OF MECHANICAL ENGINEERING AND SCIENCES* 11 (2017) 2, p. 2693–2710

[KH12] KADDOUR, A. S.; HINTON, M. J.: Benchmarking of triaxial failure criteria for composite laminates: Comparison between models of 'Part (A)' of 'WWFE-II'. *Journal of Composite Materials* 46 (2012) 19-20, p. 2595–2634

[LR13] LEUNG, E.; ROCHETTE, R.: The Development and Certification of Composite Landing Gear Components. *SAMPE UK and Ireland 2013*. 2013

[Nju16] NJUGUNA, J. (EDITOR): *Lightweight composite structures in transport: Design, manufacturing, analysis and performance*. Heidelberg and Oxford: Elsevier Woodhead Publishing, 2016

[NN06] N.N.: *Uniform Provisions Concerning the Approval of: Wheels for Passenger Cars and their Trailers.* , 2006

[NN12] N.N.: *Composite materials handbook*. Warrendale, Pa.: SAE International on behalf of CMH-17 a division of Wichita State University, 2012

[Puc96] PUCK, A.: *Festigkeitsanalyse von Faser-Matrix-Laminaten: Modelle für die Praxis*. München and Wien: Hanser, 1996

[Vas11] VASSILOPOULOS, A. P.: *Fatigue of Fiber-Reinforced Composites*. London: Springer London Limited, 2011

[WM20] WITTEN, E.; MATHES, V.: *Der Markt für Glasfaserverstärkte Kunststoffe (GFK) 2020*: AVK-Martkbericht 2020, 2020

Symbols

Symbol	Unit	Description
$\sigma_{11}, \sigma_{22}, \sigma_{33}$	N/mm^2	Normal stress components
$\tau_{12}, \tau_{22}, \tau_{33}$	N/mm^2	Shear stress components
SR	–	Stress ratio
σ_a, σ_m	N/mm^2	Stress amplitude and mean stress
$\sigma_n, \tau_{n1}, \tau_{nt}$	N/mm^2	Stress components of an action plane
$R_{\parallel}$	N/mm^2	Fatigue strength in longitudinal direction (fibre-parallel)
$R_{\perp}$	N/mm^2	Fatigue strength in transversal direction
$R_{\parallel \perp}$	N/mm^2	Fatigue shear strength
N	–	Number of cycles
$f_{E\ IFF}$	–	Stress exposure for inter fibre failure
$\{\sigma(\theta, t)\}$	N/mm^2	Stress vector for specific plane (θ) and time (t)
$\{\sigma_f(N, \theta)\}$	N/mm^2	Strength vector for specific number of cycles (N) and plane (θ)
Θ	°	Angle of action plane
t	$s, -$	Specific time of stress cycle
$D^{(k)}$	–	Fraction of life consumed by current and previous steps
$n^{(k)}$	–	Number of applied cycles at current step k of the fatigue analysis
$N^{(k)}$	–	Number of cycles to failure at current step k
a	–	Variable for linear, degressive or progressive accumulation
k	–	Index for current step
η	–	Degradation parameter
η_R	–	Parameter for residual stiffness
c, ξ	–	Shape parameter of degradation curve; experimentally determined
E_{ij}	N/mm^2	Elastic modulus
G_{ij}	N/mm^2	Shear modulus
ν_{ij}	–	Poisson ratio

Abbreviations

Notation	Description
MCT	Mubea Carbo Tech GmbH
FRP	Fibre reinforced plastics
CFRP	Carbon fibre reinforced plastics
FEA	Finite Element Analysis
CAE	Computer Aided Engineering
BVID	Barely visible impact damage
GUI	Graphical user interface
OEM	Original Equipment Manufacturer
IP	Integration point
CLD	Constant life diagrams
WWFE	World Wide Failure Exercise
SR	Stress Ratio

2 Analysis of strain rate dependent material properties using a drop tower for its application in the research of damage prediction on fibre reinforced plastics

Ch. Hopmann[1], J. M. Müller[1]
[1]Institute for Plastics Processing (IKV) at RWTH Aachen University

Abstract

Mechanical properties of Polymeric materials exhibit a strong time dependency. This time dependency also accounts for unidirectional endless fibre reinforced plastics (FRP). Characterisation in the high strain rate regime (up to 1000 $1/s$) for endless unidirectional FRP is still a challenging task, since using the conventional split Hopkins pressure bar (SHPB) does not enable the usage of complex specimen fixtures. Hence, it is questionable whether material failure is induced by specimen clamping. Using the drop tower unveil new potential for strain rate dependent material characterisation. In the present research a novel test method for high strain rate material characterisation subjected to shear using a drop tower is presented. The obtained results are discussed with respect to existing experimental data and to local damage of the specimen.

2.1 Introduction

The mechanical properties of polymeric materials are characterised by a strong time dependence. The investigation of viscous properties in the range of durations from seconds to days is carried out with universal testing machines. In addition to the generally well known long-term behaviour of engineering plastics, the ultra-short-term range shows an equally strong dependence on time. In this case, the material usually exhibits a stiffening behaviour with a concomitant increase in strength when the loading rate is increased. The determination of characteristic values at load durations in the range of a few milliseconds (strain rates of up to 1000 $1/s$) means that universal testing techniques cannot be used. The required travel speeds cannot be realised by universal testing machines. Considering the use of a specimen designed for high strain rates, travel speeds of up to 15 m/s are required to achieve strain rates of up to 1000 $1/s$. This concludes the use of special testing technology. Particularly in the context of determining the characteristic values of FRP, it can be seen that the specimen restraint and thus the testing technique has a major influence on the properties measured in the test [Pre11]. It can be deduced from this that the test technique and the clamping systems are essential for the strain-rate-dependent determination of mechanical properties of FRP. Examination of viscous properties enables a better understanding of the damage development under increased strain rates and consequently facilitates improvements in part design.

In the literature, SHPB is widely used to determine mechanical properties at high to very high strain rates (up to 5000$1/s$) [GGR02, TNK07, BDB+14, CTP+16, HD98]. In compression testing using the SHPB, two symmetrical bars are placed in series with the specimen between them. The incident bar is accelerated by a striker during the test. The striker is accelerated by a gas gun. Both bars are instrumented by means of strain gauges (cf. Figure 2.1). With the aid of the SHPB, tests under compressive loading can be realised particularly well. Other load cases are possible, but the clamping of the specimen, e.g. in tensile tests, turns out to be non-trivial. Furthermore, tests under shear loading using Iosipesco or V-notched rail-shear specimens are not possible. The mechanical properties under shear loading are carried out on the SHPB mostly with cross-ply specimens. The measured system response is then back-calculated to a pure shear load. Existing work on mechanical properties determination of FRP under compressive and shear loading raises the question whether damage initiation is induced by the test set up and thus the measured properties should not be understood as intrinsic material properties.

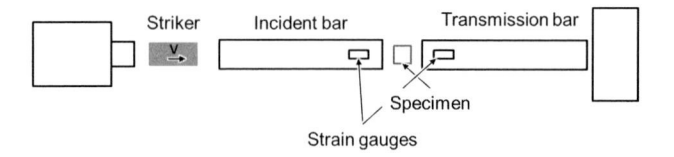

Figure 2.1: Split Hopkins Pressure Bar (SHPB)

In addition, servo-hydraulic machines are used in the determination of properties at high strain rates. However, these machines are primarily suitable for the determination of tensile properties and hardly allow the testing of complex load cases. The use of servo-hydraulic testing technology is not found in the literature for the determination of strain rate-dependent properties of FRP.

An alternative is the drop tower. In the drop tower, a drop frame including drop weight and impactor is dropped from a defined height. In the process, the potential energy of the drop frame is converted into kinetic energy. In order to extend the fall height of the fall carriage, it is possible to use spring packs. With the help of these spring packs, additional potential energy can be introduced into the system, which is used to increase the kinetic energy of the drop frame (cf. Figure 2.2). The energy of the drop carriage is transferred to the specimen by means of an impactor.

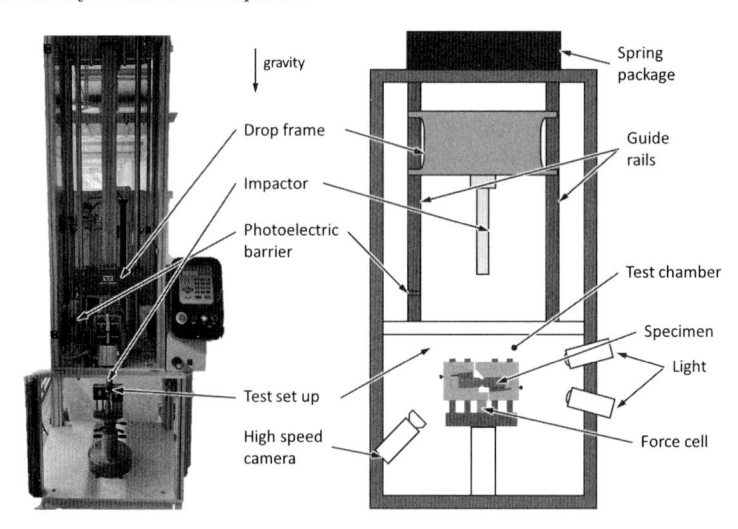

Figure 2.2: Schematic representation of drop tower CAEST 9350

Drop impactors are used primarily for direct testing of materials. This means that the drop frame directly impacts the specimen, thus transferring the energy to the specimen.

However, the testing of FRP requires the use of a complex clamping, especially for the uni-axial compression and shear load cases. This means that complex test setups have to be developed and used. The use of the drop tower allows the use of these test setups and concludes the desired indirect impact of the specimen.

2.2 Strain rate dependence of FRP under shear

Various methods can be found in the literature for determining the strain rate-dependent composite properties under in-plane shear loading. The Iosipescu specimen geometry used in this work has not been previously documented for its behaviour under increased strain rates, thus comparison with other experimental results reported in literature is limited.

Previous tests have been conducted using carbon/epoxy specimens with a fibre orientation of ±45, loaded in both tension and compression. The results of these tests are often described in terms of axial strain or strain rate, which must be taken into account when comparing with shear strain or shear strain rate. As another methodology, off-axis tests were performed in which the fibre orientation was varied between 10° and 60° to the loading direction. Comparisons of these experimental results can only be made to a limited extent, but the strain rate dependence of the composites can be qualitatively evaluated and compared.

In [GGR02], Gilat et al. perform tensile tests with fibre orientations of 45° and 10° at low ($10^{-5}\ s^{-1}$), medium ($1\ s^{-1}$), and high ($400\ s^{-1}$) axial strain rates, and an increase in stiffness and strength can be observed in both sets of tests. For both orientations, the same initial stiffness can be seen for the loading rates of $10^{-5}\ s^{-1}$ and $1\ s^{-1}$, with higher maximum stress occurring at $1\ s^{-1}$ than at $10^{-5}\ s^{-1}$. The highest loading rate studied, $400\ s^{-1}$, is characterised by higher stiffness and higher strength. A nonlinear shape of the stress-strain curve can be seen in all tests. In addition, a composite with orientation of [±45] is investigated at axial strain rates of $9x10^{-5}\ s^{-1}$, $2\ s^{-1}$, and $600\ s^{-1}$, where a similar strain rate dependence is observed, but both higher failure stress and strain at failure are measured. Taniguchi et al. [TNK07] also perform SHPB tensile tests on a [±45] carbon/epoxy composite, comparing quasi-static tests ($1.04x10^{-4}\ s^{-1}$) with dynamic tests ($50\ s^{-1}$, $100\ s^{-1}$). An increase in strength is also observed (77.5% increase from quasi-static to $100\ s^{-1}$), but a linear increase in stiffness is described, with a 77.1% increase.

Berthe et al. [BDB+14] also describe an increase in stiffness and strength of a [±45] composite with increasing strain rate. A range of strain rates is investigated ($10^{-3}\ s^{-1}$, $7x10^{-3}\ s^{-1}$, $0.1\ s^{-1}$, $16\ s^{-1}$, $25\ s^{-1}$, $50\ s^{-1}$) where a step wise increase in stiffness as well as strength is observed, as well as a step wise decrease in elongation at break. The total increase in stiffness between the lowest and highest strain rates is 45%. It should be emphasised that the increase in stiffness between $16\ s^{-1}$ and $25\ s^{-1}$, and between $25\ s^{-1}$ and $50\ s^{-1}$, is significantly greater than between the remaining strain rates.

The behaviour of a [±45] composite under both tensile and compressive loading is studied by Cui et al. [CTP+16]. For this purpose, a method for correcting the fibre alignment during the test is additionally developed. Here, a very similar strain rate dependent behaviour is shown for both tensile and compressive loading. The quasi-static strain rate of $5x10^{-4}\ s^{-1}$ is compared with a dynamic strain rate of $1300\ s^{-1}$ and for both loading types there is a significant increase in strength and stiffness and a decrease in elongation at break.

The strain rate dependent shear properties under compressive loading are also investigated by Hsiao and Daniel [HD98], where 30° and 45° off-axis tests are performed under four different loading rates. For this purpose, off-axis tests of 15°, 30°, 45° and 60° specimens are first performed at the same velocity, from which it can be seen that the deviations of the compressive shear stress-shear strain curves differ only slightly from each other in both the linear and nonlinear regions. As the strain rate is increased, there is an increase in stiffness and strength, with the properties at 30 increasing to the same extent as at 45°. Overall, there is an 80% increase in strength and an 18% increase in stiffness between the quasi-static tests and those at $300\ s^{-1}$. [HD98]

Koerber et al [KXC10] perform both quasi-static and dynamic off-axis compression tests for 15°, 30°, 45° and 60° and calculate the shear properties for these. Again, the qualitative variation of shear stress versus shear strain for the quasi-static and dynamic tests shows little variation among the different angles, but different elongations at break occur depending on the angle. Between the quasi-static tests at $4x10^{-4}\ s^{-1}$ and the dynamic tests between $90\ s^{-1}$ and $350\ s^{-1}$, there was a 25% increase in shear stiffness and a 42% increase in shear strength. All the shear behaviour studies described describe an increase in strength and stiffness for the increase in strain rate, although the factor by which the properties increase varies depending on the source and type of test.

2.3 Digital image correlation

Optical strain measurement via the digital image correlation (DIC) works by comparing image captures of an object during the deformation process, where the strain can be calculated by looking at pixel blocks [ML10]. The procedure is shown schematically in Figure 2.3 [URL20].

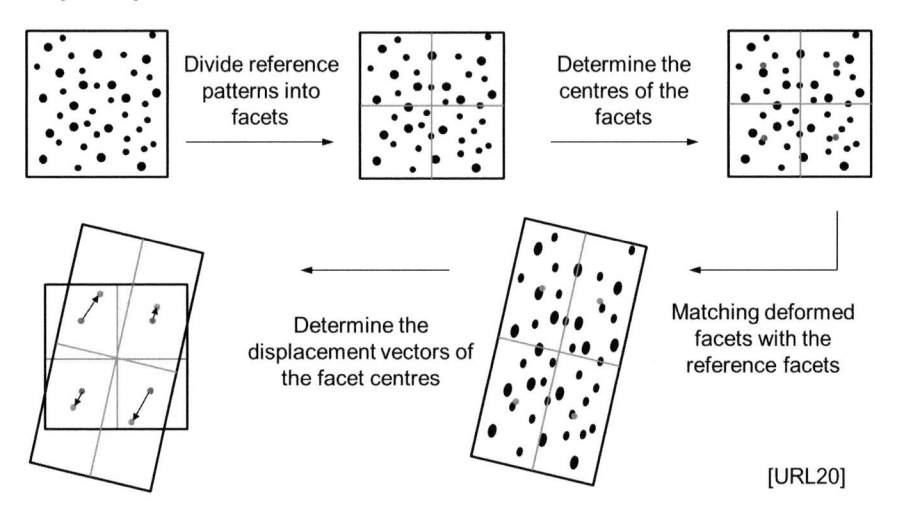

Figure 2.3: Strain calculation with the help of DIC

The required dot pattern is sprayed with the help of an airbrush onto the specimen. While evaluating the spray pattern, one must distinguish between measurement resolution and reliability. A finer dot pattern allows local resolution of the strain measurement. Whereas it also comes along with challenges as the illumination needs to be improved to gather reliable data. A dot size between 3x3 and 7x7 pixels is best suited in Order to satisfy the requirements of resolution and reliability [Reu14a]. It is reacquired that the points have a high grey scale contrast, as this is how the software recognises the points [SOS09]. Finally, the points should cover about 50% of the surface to achieve optimal results [Reu14b]. The strain measurement works as follows: During the deformation process, the specimen with the applied pattern of points is recorded with the help of a camera. Afterwards, the local strain field is determined by GOM correlate (GOM GmbH · a ZEISS company, Braunschweig). Firstly the point pattern is set at the beginning of the experiment as a reference and then compared to the dot patterns of the consecutive pictures. [JIB+18]

In order to calculate the strain as accurately as possible, the reference image is divided into several equally sized sub-areas, so-called facets, as shown in the upper part of Figure 2.3 [URL20]. These areas are recognised by the GOM correlate in the images of the deformed specimen. The bottom left of Figure 2.3, depicts how the vector is then formed from the centre of the reference facet to the centre of the deformed facet. Once this process has been carried out for all facets, the average strain of the specimen can be calculated from the individual displacements of the facets. Since the displacement is calculated as a vector, the strain can be decomposed into its various directional components or the direction of the principal strain can be determined. [JIB+18]

2.4 Material used

The required unidirectional specimens are manufactured using the vacuum assisted resin transfer moulding (VARTM) process. The non-crimped fabric (NCF) used with an area weight $270g/m^2$ consists of fibres of type HTS45 E23 12K. For fibre fixation hot melt adhesive threads (area weight 4 g/m^2) are used. The used epoxy resin is of type Epikote TRAC 06150 by Hexion. During the preform production, 10 layers are stacked and cut to size. The thermal binder in the form of hot melt adhesive threads is then activated by heating and subsequent cooling. This ensures that the individual layers of the fabric are bonded and fixed to each other. After preforming, the preform is placed into the mould which is treated with Loctite Frekote 770 NC by Henkel. The closed mould is vacuumed and the pre-heated at 50°C resin is injected into the mould at a relative pressure of 5 bar. The part is cured at 110°C for 30 min and afterwards demoulded. The unidirectional parts are cut on a diamond saw into rectangular specimens. The notches are grind manually with the help of template into the specimen. The resulting specimen have a fibre volume content of approx. 50%.

2.5 Test setup

Since drop towers have a simple design, a drop tower of the type CEAST 9350 (Instron GmbH, Pfungstadt) available at the IKV is used for the experimental, highly dynamic material characterisation. With the help of the drop tower test velocities in the range from 0.5 m/s to 24 m/s are possible at kinetic energies ranging from 1 J to 1800 J. The drop tower is equipped with a piezo-electric load cell PCB 226c measuring forces of up to 180 kN. For data acquisition the drop tower is equipped with different measuring cards. It allows signal sampling with a frequency of up to 4 MHz. The used impact velocities are shown in Table 2.1.

#	Impact velocity [m/s]	Energy [J]	Shear strain rate [1/s]
1	2.0	20	175
2	4.5	60	394
3	9.9	145	842

Table 2.1: Used test parameters

In Order to measure local shear strains, digital image correlation (DIC) is used. For capturing images, a Photron SA 1.1 is used. With the camera a frame rate of up to 675 000 fps is possible. Hence DIC requires a relatively high pixel density and exposure of the picture taken, the frame rate is limited to around 75 000 fps. At this frame rate a resolution of

256 x 256 pixels is obtained. The start of data and image acquisition is triggered by a light barrier which is actuated by the drop frame.

Since using a drop tower which enables using complex specimen clamping, a self-designed clamping fixture is used. The fixture enables to test specimens according to ASTM D5379 [NN19] (cf. Figure 2.4. The set up consists of two solid aluminium grips. Two adjustable jaws allow to tighten the specimen to realise a defined load introduction into the specimen. The specimens geometry is depicted in Figure 2.5. Load introduction into the test setup is realised by impacting the upper grip via the impactor at given speed and kinetic energy. The region of interest (ROI) depicts the area for strain measurement and hence represents the area for capturing images. During post processing the images are evaluated and the average shear strain within the region of interest is calculated. The materials chosen are selected with the background of reducing the moment of inertia of the moving parts.

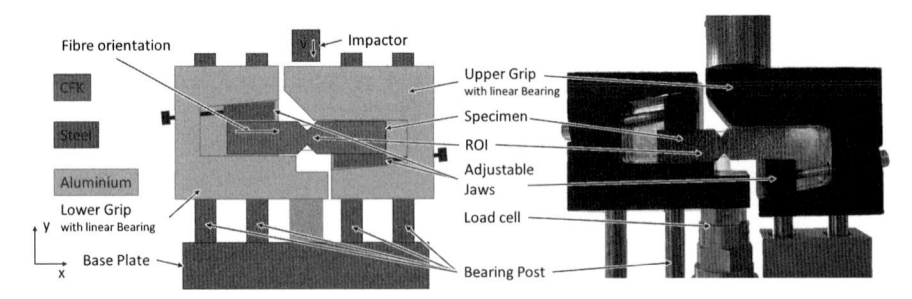

Figure 2.4: Used test setup based on ASTM D5379

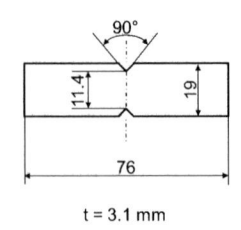

Figure 2.5: Specimen geometry according to ASTM D5379

2.6 Experimental results

In the following the results are presented. Firstly force-time curves are shown and discussed. Secondly the strain measurement is considered, and stiffness and strength of the specimen tested are derived.

Figure 2.6 depicts the force-time behaviour of the tested material. The graph shows the mean value (bold line) for each series and is derived from Figure 2.10 to 2.12. The transparent area marks the range of maximum and minimum measured values within a series. For each series five experiments are considered. Figure 2.6 reveals, that the slope for experiments at 175 1/s and 394 1/s is qualitative equal. This means that the force increases steadily and reaches a first local maximum. The force then drops at the local minimum and rises again. In the case of medium impact velocities, it can be seen, that the second local maximum approximately equal to the first. After reaching the global maximum (strength), the force remains at a constant level in the time range shown. Furthermore, the oscillation of the signal plays a subordinate role. In contrast to the experiments at slow and medium impact velocities are the experiments at high impact velocities. After reaching the first local maxima, which is also the global maximum, the force drops and the curve is superimposed by an oscillation.

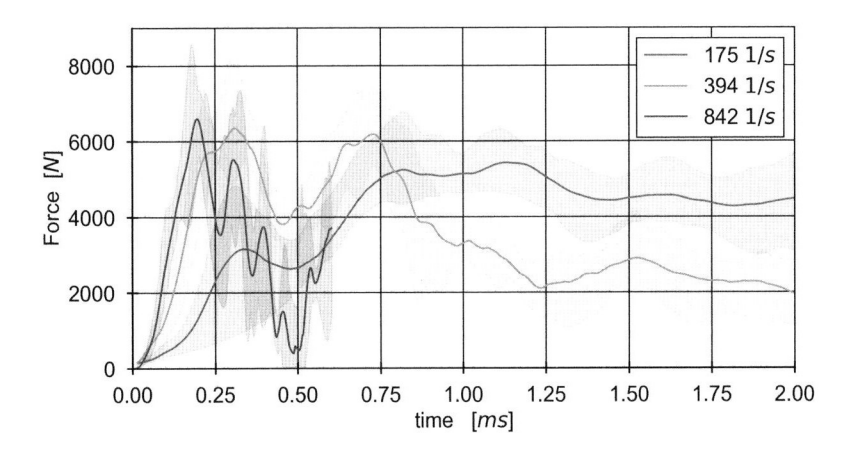

Figure 2.6: Force-time curves of shear experiments at different strain rates

The slope of force-time diagram can be explained by observing the fracture behaviour of the specimen which is depicted in Figure 2.7. Before reaching the local maximum the specimen deforms while the area between the notched experiences an accentuated shear stress. Reaching the first local maximum leads to crack initiation in the notches. According to ASTM D5379 [NN19] the appearance of notch-cracks is acceptable in the quasi-static case. For low and medium impact velocities the force drop after the global maximum is characterised by a pronounced local deformation between the notches and includes the destruction of matrix material due to excessive loading. This deformation is defined as tolerable.

In comparison the high impact velocity experiments exhibit local deformation close by the grips which is induced by the clamping and hence is not a tolerable failure mode of the specimen. This resolves the difference after the first local maximum between high and medium/low impact velocities. This strong material destruction is moreover responsible for the superimposed oscillation. Furthermore, it is obvious that a local stiffness degradation in the area of the clamping/grips also occurs at lower impact velocities. Besides intolerable deformation of the specimen a relative displacement in x-direction of the two grips was observed.

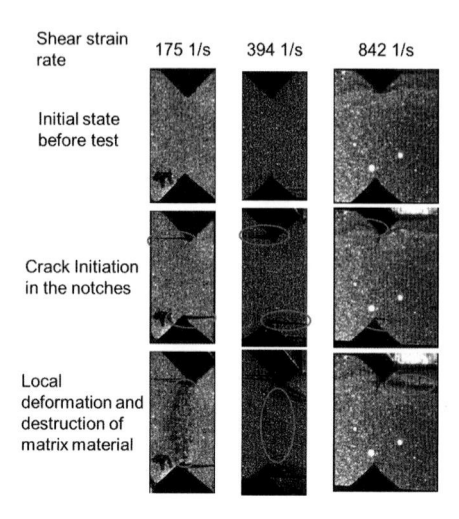

Figure 2.7: Damage development of shear specimen subjected to load

2.6.1 Stiffness and strength

In Figure 2.8 the stiffness and in Figure 2.9 the strength are plotted against the respective strain rate. Stiffness and strength are derived from force-strain curves according to ASTM D5379 (cf. Figure 2.13 to 2.15). For a strain rate of 0.0029 quasi-static experiments were performed on a universal testing machine. These experiments serve only for comparison and hence detailed information about the testing procedure is skipped. For the strength, a significant dependence on the applied strain rate can be seen. Increasing the rate from quasi-static to dynamic loading with a strain rate of $175\ s^{-1}$ results in an increase of the average strength by 25.4%. However, the higher variations in quasi-static loading must be taken into account here, so the real change in strength may be greater or less. The largest increase is observed when the strain rate is further increased to $394\ s^{-1}$. There, there is a 75.1% increase in mean strength from 111 MPa to 194 MPa. No further significant increase in strength can be observed even when the strain rate is further increased to $842\ s^{-1}$.

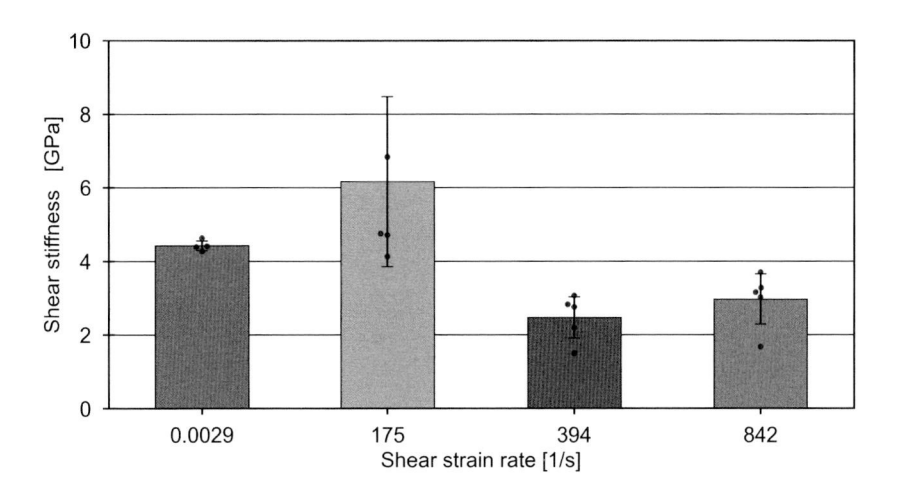

Figure 2.8: Comparison of shear stiffness at different strain rates

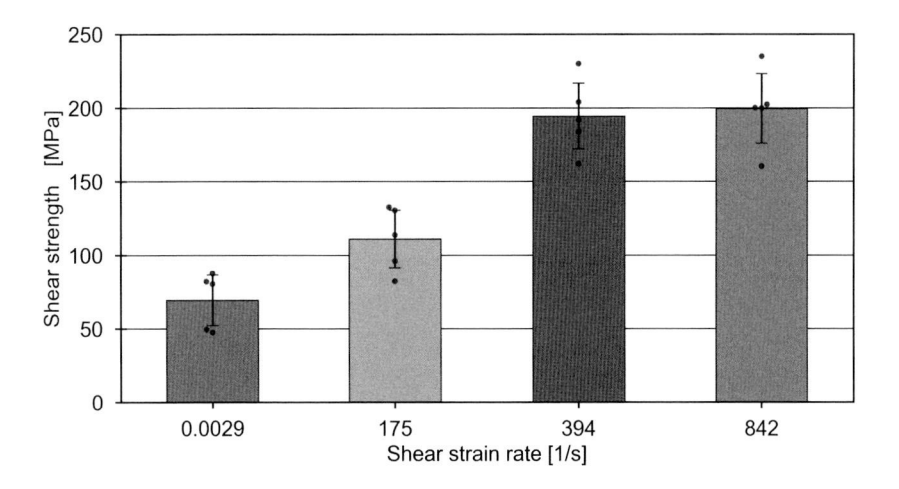

Figure 2.9: Comparison of shear strength at different strain rates

The stiffness also shows strain rate dependence. Increasing the strain rate from the quasi-static range to the dynamic range to $175\ s^{-1}$ results in a 16.9% increase in stiffness, but also an increase in deviations within the test series. Further increase to $394\ s^{-1}$ decreases the stiffness by 40.1% to 2.47 GPa. For the highest strain rate used, $842\ s^{-1}$, a stiffness of 2.97 GPa is measured, which corresponds to an increase of 19.9% compared to the previous strain rate. Thus, except for tests at $175\ s^{-1}$, the dynamic stiffness values are below those of the quasi-static tests.

Thus, for the shear strength, results are shown which are comparable to those in the literature. The gradual increase within the first three strain rates tested, is also seen in the tests performed by other research institutions, only the retention of the strength when increasing the strain rate from $394\ s^{-1}$ to $842\ s^{-1}$ contradicts them. The stiffness increases only for $175\ s^{-1}$ compared to the quasi-static tests, as also reported in the literature. For the higher strain rates, the values decrease below those for quasi-static loading. This does not agree with the observations in literature - a further increase in stiffness is measured.

2.7 Discussion

Shear tests of unidirectional NCF composite laminates loaded at different impact velocities have been performed to investigate the material behaviour and proof the suitability of a drop tower for material characterisation at indirect impacting. In general, the method of indirect impact is suitable to obtain reasonable results. The obtained force-time curves show a low scatter and superimposed oscillations can be neglected. The results are comparable to the literature and show good agreement although the specimen geometry has not been used for mechanical testing at high strain rates.

However, there is still room for improvement. The results reveal that at high strain rates the specimen undergoes local deformation close by the grip. This leads to a reduced strength of the specimen. To overcome this issue an alternative geometry can be used. One possible solution is using the v-notched shear rail set up. This method is also recommended for FRP testing. The main difference is that load introduction into the specimen is performed via the top and bottom surface and not via the front faces of the specimen. This enables an enlarged area for load introduction and hence offers potential to ensure material damage within the ROI.

Besides local damage close by the grips, the stiffness measurement is not in good agreement with the literature. Comparing the standard deviation of strength and stiffness, shows that the standard deviation of the calculated is at least in the same magnitude as the strength and for 175 1/s test even higher. Considering material mechanics and the local morphology of the composite, the strength as a calculated unit is more sensitive against local defects than the stiffness is. Hence, the standard deviation of the stiffness should be smaller than the deviation of the strength. The obtained results at high strain rates do not support this hypothesis. Having a look on the deviations at 0.0029 this trend present. In conclusion, the strain measurement must be critically questioned. To increase the measurement quality, the pattern quality and illumination are parameters which should be adapted. Especially at higher strain rates a frame rate of 75 000 fps leads to a coarse strain curve. Hence increasing the frame rate should be considered. Increasing the frame rate conflicts with the pixel density and hence with the recorded pattern quality. Therefore, adapting these parameters has be performed carefully to increase the strain measurement quality.

2.8 Conclusions and outlook

In this paper, a novel test method for FRP testing subjected to shear at high strain rates is presented. The test method enables the usage of complex specimen geometries with the help of a drop tower since specimen clamping is possible due to indirect impacting. Generally, the results are in good agreement with experimental results in the literature. For the highest strain rate tested, the specimen fails intolerable hence failure occurs out of the gauge section. This issue can be overcome by adopting the clamping geometry.

Based on the findings in this research, current test set up has to be improved. Following improvements ca be performed: (1) increase the stiffness of the test setup, (2) improve bearing to reduce superimposed oscillations, (3) improve specimen clamping. Furthermore, the findings of this research can be extended to further load cases. Especially in area of FRP testing load introduction into the specimen plays a major role.

Acknowledgements

The authors disclosed receipt of the following financial support for the research and authorship of this article: This work was supported by the Deutsche Forschungsgemeinschaft (DFG) [40450244]. We would like to extend our thanks to the DFG. Further thanks are given to Jaan Simon and Hagen Holthusen from the Institute of Applied Mechanics (IFAM) at RWTH Aachen University for the discussions.

Appendix

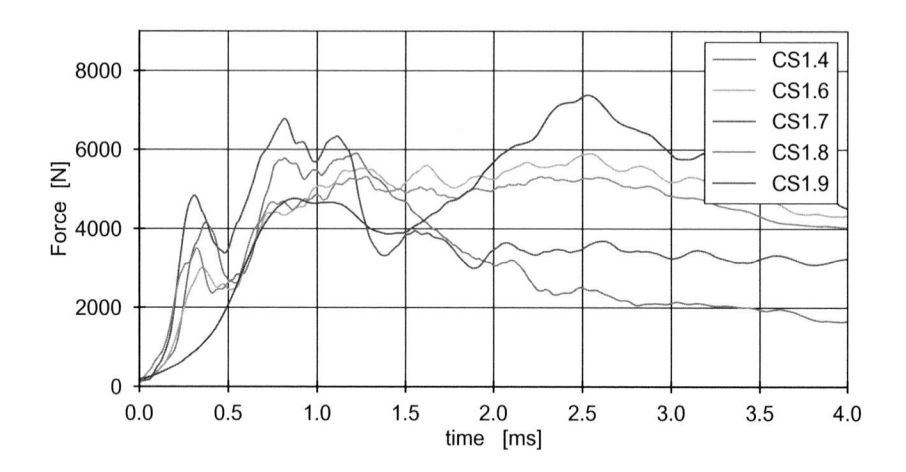

Figure 2.10: Force-time diagram at 175 1/s

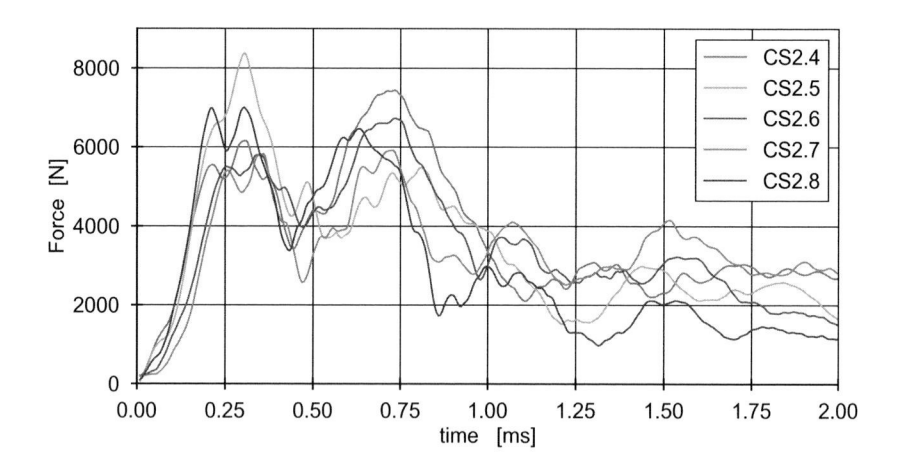

Figure 2.11: Force-time diagram at 394 1/s

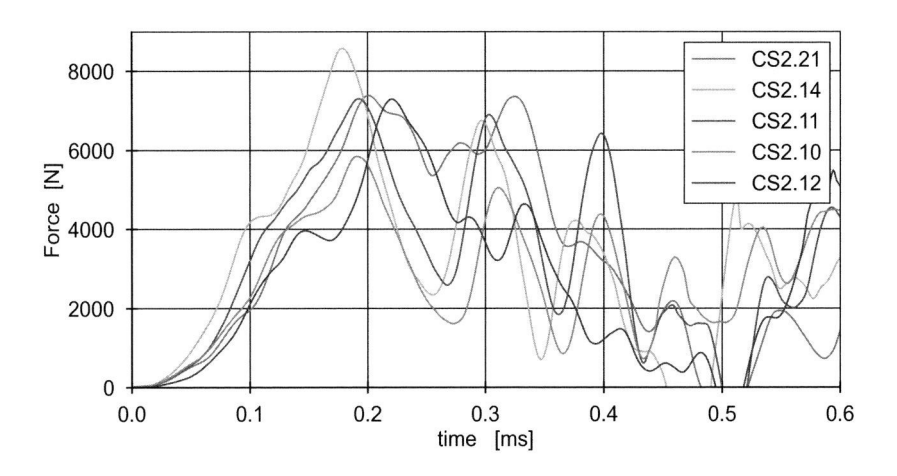

Figure 2.12: Force-time diagram at 842 1/s

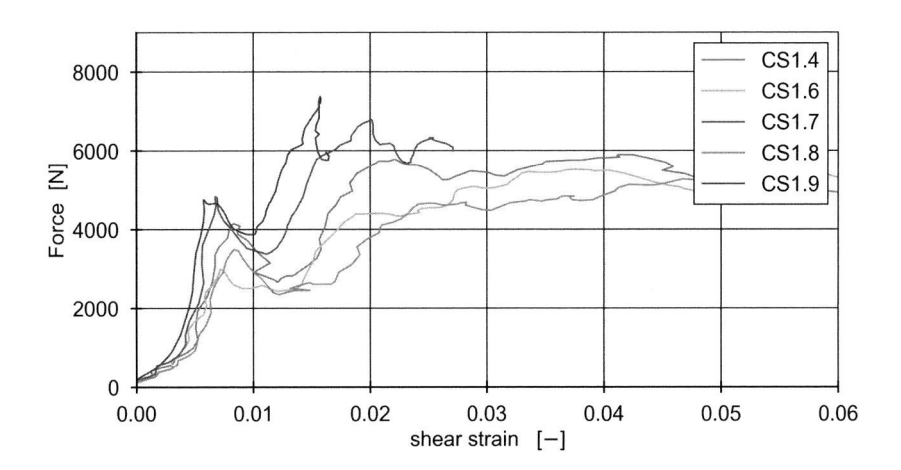

Figure 2.13: Force-shear strain diagram at 175 1/s

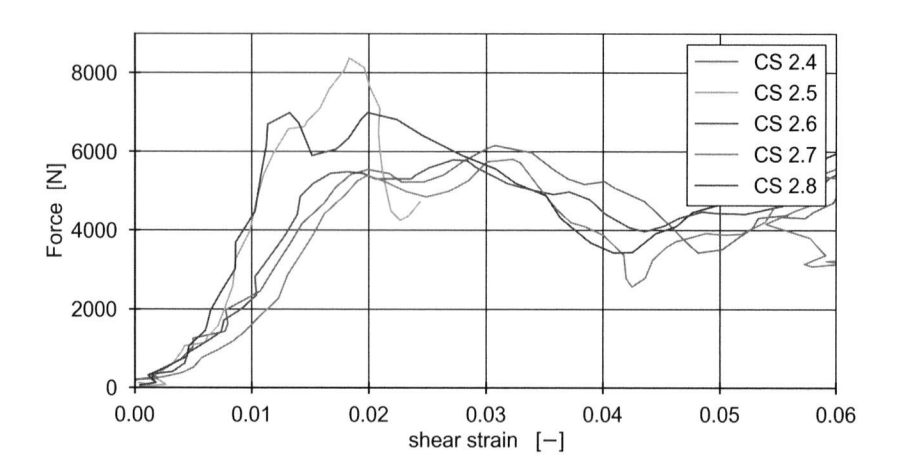

Figure 2.14: Force-shear strain diagram at 394 1/s

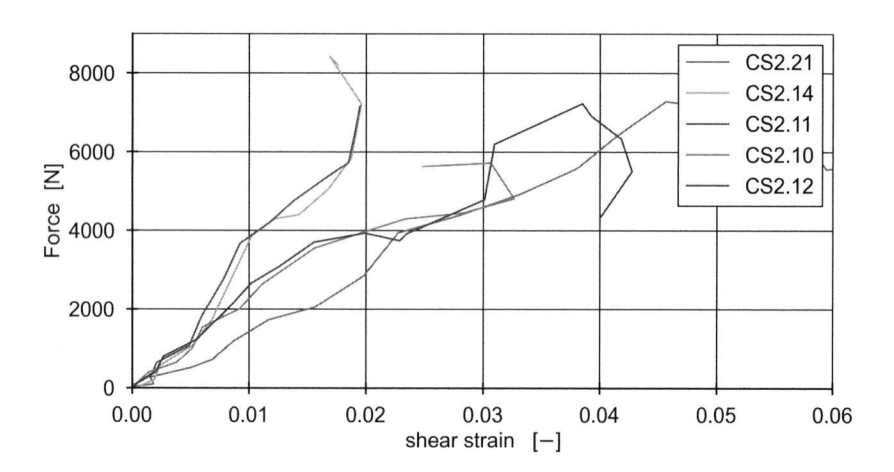

Figure 2.15: Force-shear strain diagram at 842 1/s

References

[BDB+14] BERTHE, J.; DELETOMBE, E.; BRIEU, M.; PORTEMONT, G.; PAULMIER, P.: Dynamic Characterization of CFRP Composite Materials – Toward a Prenormative Testing Protocol – Application to T700GC/M21 Material. *Procedia Engineering* 80 (2014), p. 165–182

[CTP+16] CUI, H.; THOMSON, D.; PELLEGRINO, A.; WIEGAND, J.; PETRINIC, N.: Effect of strain rate and fibre rotation on the in-plane shear response of ±45° laminates in tension and compression tests. *Composites Science and Technology* 135 (2016), p. 106–115

[GGR02] GILAT, A.; GOLDBERG, R. K.; ROBERTS, G. D.: Experimental study of strain-rate-dependent behavior of carbon/epoxy composite. *Composites Science and Technology* 62 (2002) 10-11, p. 1469–1476

[HD98] HSIAO, H. M.; DANIEL, I. M.: Strain rate behavior of composite materials. *Composites Part B: Engineering* 29 (1998) 5, p. 521–533

[JIB+18] JONES, E.; IADICOLA, M.; BIGGER, R.; BLAYSAT, B.; BOO, C.; GREWER, M.; HU, J.; JONES, A.; KLEIN, M.; RAGHAVAN, K.; REU, P.; SCHMIDT, T.; SIEBERT, T.; SIMENSON, M.; TURNER, D.; VIEIRA, A.; WEIKERT, T.: *A Good Practices Guide for Digital Image Correlation.* , 2018

[KXC10] KOERBER, H.; XAVIER, J.; CAMANHO, P. P.: High strain rate characterisation of unidirectional carbon-epoxy IM7-8552 in transverse compression and in-plane shear using digital image correlation. *Mechanics of Materials* 42 (2010) 11, p. 1004–1019

[ML10] MCCORMICK, N.; LORD, J.: Digital Image Correlation. *Materials Today* 13 (2010) 12, p. 52–54

[NN19] N.N.: *Standard Test Method for Shear Properties of Composite Materials by the V-Notched Beam Method.* ASTM International, 2019

[Pre11] PRELLER, F.: *Zum Verhalten von kohlenstofffaserverstärktem Kunststoff bei faserparalleler Druckbelastung.* RWTH Aachen University, Dissertation, 2011

[Reu14a] REU, P.: All about Speckles: Aliasing. *Experimental Techniques* 38 (2014) 5, p. 1–3

[Reu14b] REU, P.: All about speckles: Speckle Size Measurement. *Experimental Techniques* 38 (2014) 6, p. 1–2

[SOS09] SCHREIER, H.; ORTEU, J.-J.; SUTTON, M. A.: *Image Correlation for Shape, Motion and Deformation Measurements: Basic Concepts, Theory and Applications.* Boston, MA: Springer-Verlag US, 2009

[TNK07] TANIGUCHI, N.; NISHIWAKI, T.; KAWADA, H.: Tensile strength of unidirectional CFRP laminate under high strain rate. *Advanced Composite Materials: The Official Journal of the Japan Society of Composite Materials* 16 (2007) 2, p. 167–180

[URL20] LEPAGE, W.: *https://digitalimagecorrelation.org.* URL: https://digitalimagecorrelation.org/, 25.12.2020

Abbreviations

Notation	Description
DIC	Digital image correlation
FRP	Fibre reinforced plastics
NCF	Non-crimped fabric
ROI	Region of interest
VARTM	Vacuum assisted resin transfer moulding

3 Investigation of the correlation creep/fatigue in Unidirectional fibre reinforced plastics

Ch. Hopmann[1], N. Rozo Lopez[1]
[1]Institute for Plastics Processing (IKV) at RWTH Aachen University

Abstract

In this work, the correlation between creep and fatigue in fibre reinforced plastics (FRP) was numerically investigated. For this aim, simulations using a nonlinear viscoelastic material model and without considering damage in the material were performed. The so-called Creep Cycle Jump was used to separate the creep effect from the cycle loading response. To investigate the direct correlation between creep and fatigue, cyclic loading conditions, where no apparent or sudden damage occurs, have been considered. The results of the investigation show that the nonlinear viscoelastic material model resembles the strain accumulation in the matrix material as well as in the FRP. The RVE-model together with the Creep Cycle Jump was able to extract the mutually influencing effects of creep and fatigue in FRP. Thus, the Creep Cycle Jump presents itself as a valuable tool for the investigation of the direct correlation between creep and fatigue.

3.1 Introduction

Fibre reinforced plastics (FRP) have caught the attention of the automotive, aerospace, and energy industries due to their formidable structural performance (e.g. high mechanical strength compared to its low weight). With the introduction of FRP as structural material back in the 1950s, the fatigue response of such kinds of materials has been extensively investigated, gaining special attention from the scientific community. Since this early stage, it was revealed that the understanding of its fatigue behaviour will challenge the field of mechanical analysis due to the complex mechanism involved. Thus, a massive experimental campaign was initiated, and since then a large amount of data has been generated [Vas20]. Despite the extended experimentation, a full understanding of the fatigue across the composite remains unclear. For instance, the applied load conditions influence the way how the damage initiate and propagate through the composite [Kno08, HR73]. For most loading conditions (except for the near-perfect axial loading), the damage in unidirectional-FRP (UD-FRP) is mainly initiated and propagated by the matrix [HR73]. Thus, the matrix cracking and detachment of the fibre/matrix interface takes place within a UD-layer before a fibre fracture forms [PS02]. The literature on damage in FRP is extensive [Gam00, PGBV13, CQ14, KJB+19, MRKH20]. However, no model with physical meaningful parameters has been reported [Vas20].

From a micromechanical point of view, the complex stress/strain distribution within the FRPs, inherently of its inhomogeneous microstructure, remains in an early-stage investigation [Vas20]. Direct measurement of the internal stresses has proved to be limited or not feasible at all [Vas20] and then numerical approaches have been portrayed as a suitable tool for its investigation. One drawback of micromechanical analysis is the high computational effort required to simulate a very large number of loading cycles [RCH20]. However, the so-called "cycle-jump" approaches proved to reduce the computational effort needed for the mechanical analyses of FRP [TCCD07]. Furthermore, this kind of approach showed to be suitable for the analysis of damage accumulation in the FRP during the fatigue cycle [KS01] and proved to resemble the cyclic strain accumulation experienced by the material during cyclic deformation [RCH20].

On the other hand, the current investigations of fatigue modelling in FRP mainly focus on the application of higher-order models with multiple parameters, where most of the parameters have no physical meaning [Vas20]. In addition, most of the fatigue models were developed to describe macroscopic behaviour rather than a micromechanical description of the phenomenon. Since the micromechanical damage of FRP is a successive process, different mechanisms influence each other at the same time. These influences must therefore not be considered individually since their synergistic interaction determines the macromechanical material behaviour. Although it is well-known that creep and fatigue are mutually influencing phenomena [VAT14, MKV19] and it has been extensively studied [PS02, Gue08, MKV19], there is still an incomplete understanding of the separate influence of both mechanisms [Vas20]. Additionally, experimental approaches showed to be insufficient to separate the influences of both mechanisms [MKV19].

In this work, the mutually influencing effects of creep and fatigue in FRP are numerically investigated. Micromechanical simulations, based on a Representative Volume Element (RVE) model were performed. To capture the strain accumulation in the FRP, a nonlinear viscoelastic material model for the matrix in the RVE is implemented. To avoid the influence of damage effects, no damage model was considered and therefore loading conditions with no damage accumulation and lower thermal dissipation were simulated. For the investigation of both separate mechanisms, (e.g. to isolate the strain accumulation effect), a creep cycle jump approach was implemented. For evaluation, the hysteresis shape, and the internal stress distribution in the RVE were considered.

3.2 Description of the nonlinear viscoelastic response in FRP

Fatigue in UD-FRP is influenced by the complex interaction between the fibre strength and fibre distribution as well as by the viscoelastic behaviour of the matrix [Kno08]. It has been shown that the viscoelastic behaviour of the FRP comes from the matrix properties, while the fibres can be assumed as linear elastic [KJB+19, HMKF13, RCH20]. This viscoelastic behaviour is observed in the time-dependent properties e.g. deformation rate-dependent stiffness [Roy01], and in the accumulation of deformations within the molecular structure [Küs12]. To accurately describe the FRP behaviour, it is necessary then a suitable viscoelastic material model for the matrix.

For simplicity, the spring/damper formulation has been widely used for modelling viscoelastic materials [FLO76]. This approach provides an important physical meaning since the stiffness of the elastic response (spring element) corresponds to the elastic modulus of the material. Furthermore, the viscous dissipation is then modelled by a damper element [FLO76]. Thus, a variety of deformation responses can be then obtained by rearranging configurations of the viscoelastic elements (spring and damper). Although the spring/damper model was proposed a long time ago, few applications have been adopted for fatigue analysis [SBVL09]. Nevertheless, this formulation showed to be a powerful tool to describe the material behaviour for both short and long-time responses simultaneously [Küs12, vHaa19].

It was shown that a Generalized Maxwell Model (GMM) (Figure 3.1) can be adopted for the modelling of all kinds of polymeric materials [Bra06, Küs12, vH19, RCH20]. Thus, the time-dependent elastic modulus of the GMM can be represented by a Prony series (see Equation 3.1), where (N) is the number of Maxwell elements. Here, the elastic modulus of each element is given by E_i, while τ_i correspond to its relaxation time. Moreover, E_0 indicates the elastic modulus of the material at the beginning of the deformation. Additionally, a stress dependency can be introduced by considering both $E_i(\sigma)$ and $\tau_i(\sigma)$ as functions of stress levels in the system [Küs12].

$$E(t,\sigma) = E_0 - \sum_{i=1}^{N} E_i(\sigma)(1 - exp(\frac{-t}{\tau_i(\sigma)})) \quad (3.1)$$

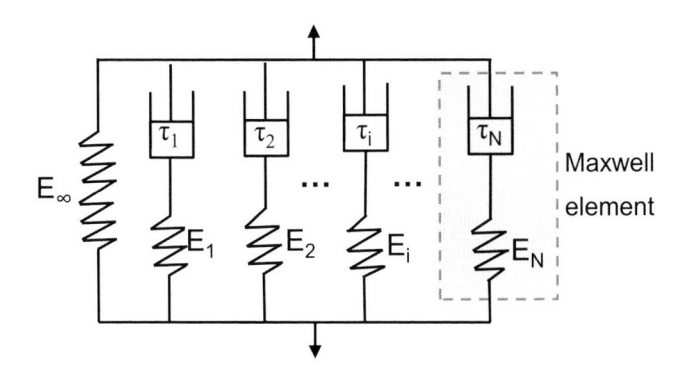

Figure 3.1: Representation of a viscoelastic model based on the Generalized Maxwell Modell

Thus, the matrix and its viscoelastic influence in the FRP can be modelled from a phenomenological basis. To account for a more accurate description, the deformation of the FRP should then be considered from a micromechanical point of view. Thus, the influence of the internal stress distribution, damage propagation, and constituents i.e. fibres and matrix are reflected on the over mechanical response of the FRP.

In this study, we modelled the matrix system within an FRP by a nonlinear viscoelastic constitutive model based on the GMM. This is implemented as a UMAT for ABAQUS [Bra06, Küs12, vH19], a detailed description of the model, its parameters, and calibration can be found in Koch et al. [KJB+19]. Its validation for shear cyclic loading conditions and the adoption of a "switching rule", based on the work of Xia et.al [XSE04c, XSE04a, XSE04b], for the correct representation of the unloading stage are detailed in Rozo Lopez et al. [RCH20].

3.3 Micromechanical simulations

In this work, the viscoelastic effects without the influence of the fibres were investigates using a Single Volume Element (SVE) model. Thus, FE simulations are then performed using the previously described nonlinear viscoelastic constitutive model. Additionally, for the investigation of the viscoelastic effects within an FRP, an RVE is implemented. Thus, the FRP inhomogeneous micromechanical conditions are approximated by the model and FE simulations. Furthermore, during cyclic loading, a certain amount of strain energy is then dissipated due to the viscoelastic response of the matrix [XSE04a, XSE04b]. This led to variations in the hysteresis shape that can be capture by tracking it's the strain energy density components (Figure 3.2). Here (W^d) corresponds to the dissipative strain energy density i.e. viscous dissipation, while (W^{e+}) indicates the positive elastic strain energy density.

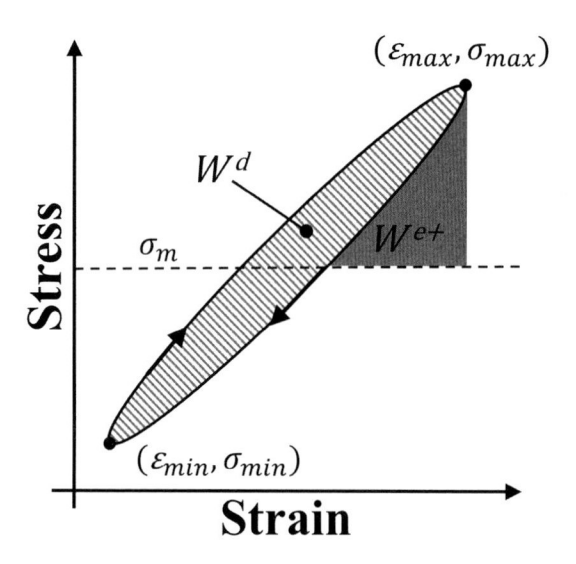

Figure 3.2: Hysteresis shape with its corresponded strain energy components

For the SVE the strain energy density components can be directly calculated from the resultant strain-stress curve. For the RVE, the strain energy density components are then calculated using Hill's principle (see Equation 3.2) and the homogenized stress and strain fields (see Equations 3.3 and 3.4) [RCH20]. For a FE approximation, N_{int} is the number of integration points in the RVE and k corresponds to the index of the kth integration point. Hence, V^k is the volume at the integration point while σ_{ij}^k and ϵ_{ij}^k are the local stress and strain, respectively. For the homogenization, a C++ routine was implemented to read the stress and strain components at each integration point from the ABAQUS result file. A previously developed python script was used for the computation of the dissipative and elastic strain energies densities.

$$W = \frac{1}{2}\sigma_{ij}\epsilon_{ij} = <\sigma_{ij}><\epsilon_{ij}> = <W> \tag{3.2}$$

$$<\sigma_{ij}> \approx \frac{1}{V}\sum_{k=1}^{N_{int}} \sigma_{ij}^k V^k \tag{3.3}$$

$$<\epsilon_{ij}> \approx \frac{1}{V}\sum_{k=1}^{N_{int}} \epsilon_{ij}^k V^k \tag{3.4}$$

3.4 SVE and RVE models

The SVE model is implemented in ABAQUS/standard. A single unit element is modelled by an 8-node linear brick (C3D8). Unidirectional shear boundary conditions are implemented (Figure 3.3). The model account for an epoxy material LY556/H917 that is modelled using the nonlinear viscoelastic model described before.

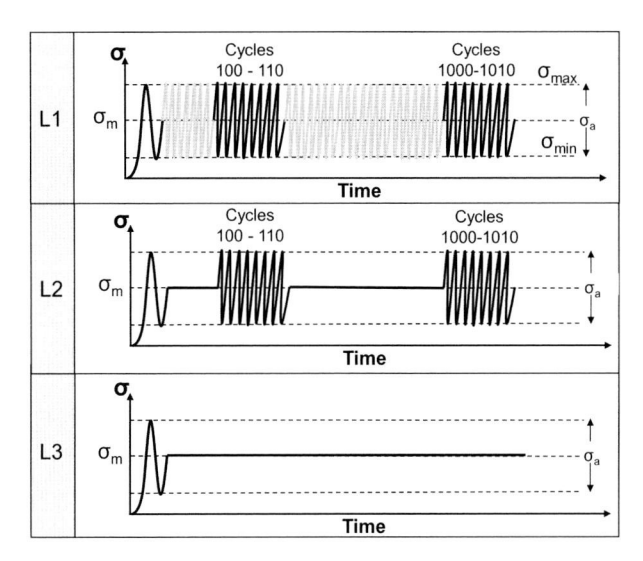

Figure 3.3: SVE-model with unidirectional shear boundary conditions

On the other hand, an RVE model was implemented in ABAQUS/standard. The matrix was modelled by C3D8 elements while the fibres used the version with reduced integration and hourglass control (C3D8R). To avoid the influence of damage effects, no damage model was considered and the fibre/matrix-interphase has been considered as perfectly bonded, i.e. no special interphase conditions were included. A total of 30 fibers were modeled in the RVE for an approximate fibre volume content of 60 % . Furthermore, constraining equations in the boundary nodes of the RVE have been included to reproduce the periodic boundary conditions. To guaranty, the correctness in the shear stress, shear coupling definitions as constraining equations are included [MTL13]. Additionally, due to the unidirectional alignment of the fibres, only two layers of elements along the fibre direction were used. Figure 3.4 presents a coarse-schematic representation of the associated boundary nodes where a force vector is applied (black grid). To simulate the cyclic shear loading, a force in the 1-3 direction is defined.

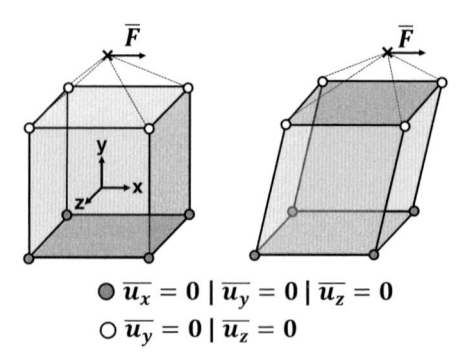

Figure 3.4: RVE-model with coarse-schematic boundary nodes (black grid) and shear load condition in 1-3 direction

● $\overline{u_x} = 0 \mid \overline{u_y} = 0 \mid \overline{u_z} = 0$
○ $\overline{u_y} = 0 \mid \overline{u_z} = 0$

In this study, T700 carbon fibres were modelled as transverse isotropic and linear elastic. The longitudinal shear modulus was modified based on the results obtained in the investigation of the influence of the fibre's longitudinal shear modulus [RCH20]. The values for the material properties are summarized in Table 3.1. The epoxy matrix LY556/H917 has been modelled by the previously described nonlinear viscoelastic model implemented into a user-defined material subroutine (UMAT).

E-Module [MPa]		ν [-]		Shear Module [MPa]	
$E_\parallel$	$E_\perp$	$\nu_\parallel$	$\nu_\perp$	$G_\parallel$	$G_\perp$
230000	15000	0.2	0.5	12000	7000

Table 3.1: Modified material properties Torey T700 carbon fibres, ∥-parallel to the fibre direction, ⊥-transverse to the fibre direction

3.5 Load scenarios

For the investigation of the mutually influencing effects of creep and fatigue, three load conditions were simulated (Figure 3.5). For all the series a non-reversed load ratio of 0.1 ($R = \sigma_{min}/\sigma_{max}$) was used. Furthermore, the loading oscillation frequency f was set at 2 Hz. A maximum stress of 50 MPa for all the cycles is then simulated. For the creep steps, a stress corresponded to the mean stress ($\sigma_m = 25$ MPa) was used. For the L1 simulations, 1010 continuous cyclic loads are simulated.

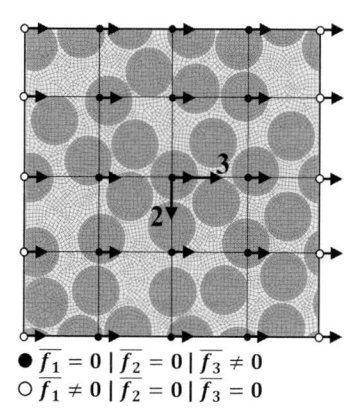

$\bullet \overline{f_1} = 0 \mid \overline{f_2} = 0 \mid \overline{f_3} \neq 0$
$\circ \overline{f_1} \neq 0 \mid \overline{f_2} = 0 \mid \overline{f_3} = 0$

Figure 3.5: Load cases. L1 Continues cycles, L2 Cycle Jump, L3 Creep Simulation

For L2-simulations, a Creep Cycle Jump was applied. This cycle jump approached proved to drastically reduce the computational time for the simulation of cyclic deformations in RVE-models. Also showed to suitable resembled of the cyclic strain accumulation experienced by the material during cyclic deformation [RCH20]. Thus, a first cycle is simulated with a small-time increment. During this step, the unloading viscoelastic properties for the switching rule are determined. Then, a creep step is implemented by a constant shear load where the load magnitude is determined by the mean stress. Furthermore, the creep step acts over a time range t_{creep} determined by the frequency of cyclic deformation f and the chosen number of cycles N_c to be simplified (see Equation 3.5). Thus, after the creep step, the simulation has equivalently undergone a certain number of cycles. As a third step, ten consecutive load cycles are simulated. Two Creep Jump were then performed to obtain the 100th and 1000th loading cycles.

$$t_{creep} = N_c f \tag{3.5}$$

For the L3-simulations, the first cycle is simulated with a small-time increment. Then, a creep step is implemented by a constant shear load which acts over a time corresponded to 1010 cycles.

3.6 Results on the influence of the creep behaviour in the cyclic loading deformation

The cyclic load conditions L1, L2, and L3 were simulated using the SVE-model (Figure 3.6). The results of the L1 simulation show that the nonlinear viscoelastic material model can resemble the strain accumulation typical from this kind on loading conditions [XSE04a, XSE04b]. Furthermore, the model shows a reduction in the strain accumulation rate for long-time cyclic deformation. This is consistent with experimental results that show that this kind of material exhibits a saturation behaviour in terms of strain [TX07a, TX07b, TX08]. On the other hand, the L2 simulation (with the Cycle Jump Approach) resembles correctly the L1 strain curve with the advantage that not all cycles need to be simulated and therefore larger time-steps can be used to speed up the simulation. For the L3 simulation, the creep step using the mean stress can correctly reproduce the path of the mean stress in the L1 and L2 simulations. These results show that the viscoelastic response of the epoxy material can be then compared between cyclic loading simulations (L1 and L2) and creep simulations (L3) in the RVE-model to extract the mutually influencing effects of creep and fatigue in FRP.

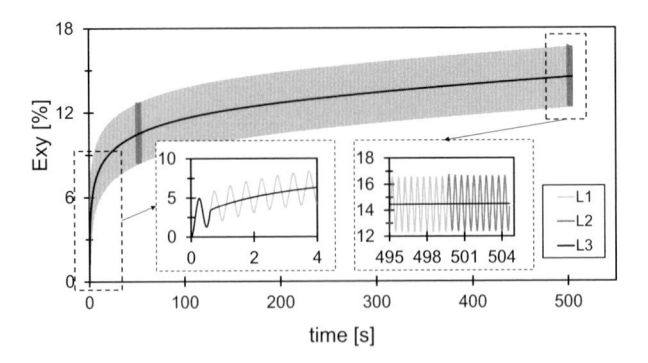

Figure 3.6: Shear strain evolution of the SVE-simulations

Using the RVE-model, the cyclic load conditions L1, L2, and L3 were also simulated (Figure 3.7). The results show that the fibres in the RVE influence the strain accumulation rate. Thus, the fibre stiffness reduces the rate at which the homogenized strain in the RVE is cumulated during cyclic deformations. Nevertheless, some degree of strain accumulation can be observed. This observation supported the assumption for a viscoelastic matrix constitutive model in which the correct replication of deformation depends on the correct selection of the viscoelastic parameters.

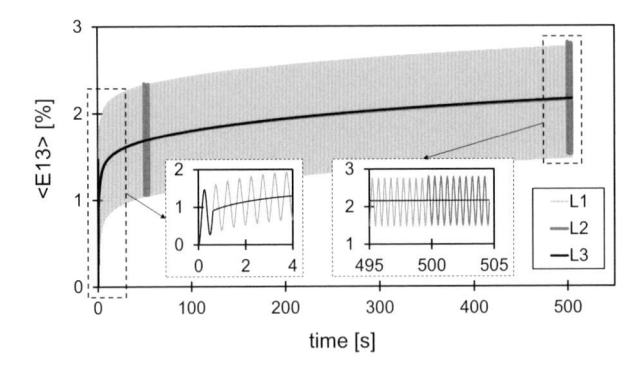

Figure 3.7: Shear strain evolution of the RVE-simulations

Similar to in the SVE-model, the L2 simulation (with the Cycle Jump Approach) resembles correctly the L1 strain curve. Nevertheless, small differences in the strain curve after the creep-step can be observed. Since these strain variations are difficult to analyse directly from the hysteresis curve, the strain energy density components are used (Figure 3.2). For the L3 simulation, the creep step using the mean stress is able to correctly reproduce the path of the mean stress in the L1 and L2 simulation.

Analysing the dissipative strain energy density (W^d) and positive elastic strain energy density (W^{e+}) (Figure 3.8). It is observed that the Cycle Jump Approach (L2) slightly misestimates both energy components. This indicates that the fibres in the RVE influence the stress configuration within the matrix during the creep-step. This leads to the activations of different sets of relaxation times in the viscoelastic model which produce a slightly different strain response. However, this can be corrected with a more meticulous setup of the stress applied during the creep-step. Additionally, this strain development indicates that the internal stress within the RVE's matrix constantly redistributes during the deformation. On the other hand, the viscoelastic model resembles the linearization of the hysteresis shape (Figure 3.8-right). that has been experimentally reported for other authors [TX07a, TX07b, TX08].

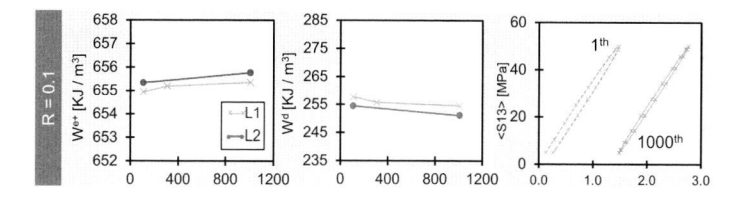

Figure 3.8: Variations in the dissipative strain energy density (W^d), positive elastic strain energy density (W^{e+}) , and hysteresis shape

Thus, for a micromechanical simulation of the stress field, the complete cyclic loading history must be considered to simulate the exact strain energy density. However, a very accurate strain energy density can be obtained by setting up correctly the stress used in the Creep Cycle Jump. On the other hand, this observation opens the question about the internal stress field variations after using the Creep Cycle Jump. Thus, a closer analysis of the internal stress distribution in the matrix of the RVE is carried out after each Creep Cycle Jump (Figure 3.9 and Figure 3.10).

To observe the overall influence of viscoelasticity, the octahedral shear stress (τ_h) was analysed (see Equation 3.6). On the other hand, the hydrostatic contribution (σ_h) (linear elastic effect) is considered by the octahedral normal stress (see Equation 3.7). Here, σ_1, σ_2, and σ_3 correspond to the principal stress components. For better readability, all values in the diagrams have been normalised to the corresponding maximum. Furthermore, the diagrams show the stress state at each integration point in the matrix of the RVE.

$$\tau_h = \frac{1}{3}\sqrt{(\sigma_1 - \sigma_2)^2 + (\sigma_2 - \sigma_3)^2 + (\sigma_3 - \sigma_1)^2} \qquad (3.6)$$

$$\sigma_h = \frac{\sigma_1 + \sigma_2 + \sigma_3}{3} \qquad (3.7)$$

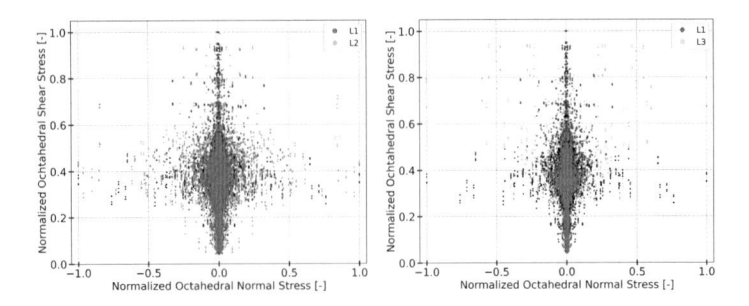

Figure 3.9: Internal stress distribution in the RVE matrix after the first Creep Cycle Jump

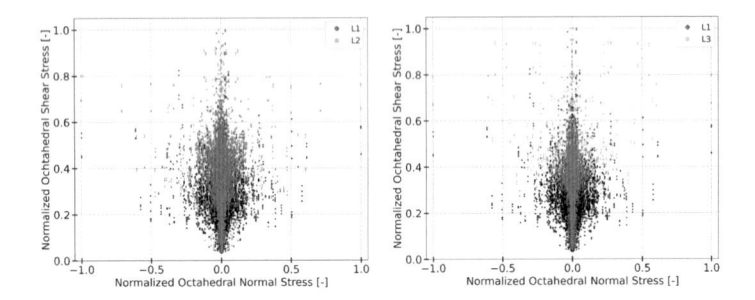

Figure 3.10: Internal stress distribution in the RVE matrix after the second Creep Cycle Jump

The black points are the reference simulation (L1). It can be observed that after the first Creep Cycle Jump (Figure 3.9) the obtained internal stress distribution is similar for both L1 and L2. This indicates that for a short jump (100 cycles) the internal stress distribution is almost unaffected by the Creep Cycle Jump. On the other hand, it can be observed that the Creep Simulation (L3) produces a more compact internal stress distribution. This indicates that the cyclic loading induces a stress reallocation mainly in the normal octahedral component.

After the first Creep Cycle Jump (Figure 3.10), the internal stress distribution for L1 and L2 differs mainly in the octahedral shear stress (viscoelastic component). This variation may come from the lack of viscoelastic dissipation due to the non-calculated cycles. This indicates that for a long jump the internal stress distribution is affected by the Creep Cycle Jump. Although, the homogenized response of the RVE (Hysteresis shape) remains almost unaffected. For the creep simulation (L3) similar behaviour can be observed. This resembles better the internal stresses from L2. This corroborates that the creep response of the epoxy matrix has a clear influence on the internal stress distribution.

3.7 Conclusions and outlook

The results obtained in this investigation support the assumption for a viscoelastic matrix constitutive model for the simulation of cyclic load-deformation of FRP. In addition, the strain accumulation indicates that the stress field within the FRP's matrix redistributes during the cyclic deformation. Thus, for a micromechanical simulation of the stress field, the complete cyclic loading history has to be considered. The model also showed to accurately estimate the strain energy density of the system. Furthermore, the Creep Cycle Jump presents itself as a valuable tool for the investigation of the direct effect/correlation between creep and fatigue.

Although the Creep Cycle Jump is not able to reproduce the stress distribution exactly, it showed that a similar distribution can be obtained with small creep steps. Nevertheless, further validation is needed to define the correct duration of the creep-step.

Acknowledgements

The authors disclosed receipt of the following financial support for the research and authorship of this article: This work was supported by the Deutsche Forschungsgemeinschaft (DFG) [281870175]. We would like to extend our thanks to the DFG.

References

[Bra06] BRANDT, M. A. G.: *CAE-Methoden für die verbesserte Auslegung thermoplastischer Spritzgussbauteile.* RWTH Aachen University, Dissertation, 2006

[CQ14] CARRARO, P.; QUARESIMIN, M.: A damage-based model for crack initiation in unidirectional composites under multiaxial cyclic loading. *Composites Science and Technology* 99 (2014), p. 154–163

[FLO76] FINDLEY, W. N.; LAI, J.; ONARAN, K.: *Creep and relaxation of nonlinear viscoelastic materials - With an introduction to linear viscoelasticity.* Amsterdam and New York and Oxford: North-Holland Publishing Company, 1976

[Gam00] GAMSTEDT, E. K.: Effects of debonding and fiber strength distribution on fatigue-damage propagation in carbon fiber-reinforced epoxy. *Journal of applied polymer science* 76 (2000), p. 457–474

[Gue08] GUEDES, R. M.: Creep and fatigue lifetime prediction of polymer matrix composites based on simple cumulative damage laws. *Composites Part A* 39 (2008), p. 1716–1725

[HMKF13] HOPMANN, C.; MARDER, J.; KÜSTERS, K.; FISCHER, K.: Effect of Cyclical Loading on the Macroscopic Failure Behaviour of Fibre Reinforced Plastics. *29th International Conference of the Polymer Processing Society.* Nuremberg, 2013

[HR73] HASHIN, Z.; ROTEM, A.: A Fatigue Failure Criterion for Fiber Reinforced Materials. *Journal of Composite Materials* 7 (1973), p. 448–464

[KJB+19] KOCH, I.; JUST, G.; BROD, M.; CHEN, J.; DOBLIES, A.; DEAN, A.; GUDE, M.; ROLFES, R.; HOPMANN, C.; FIEDLER, B.: Evaluation and Modeling of the Fatigue Damage Behavior of Polymer Composites at Reversed Cyclic Loading. *Materials* 12 (2019), p. 1727

[Kno08] KNOPS, M.: *Analysis of Failure in Fiber Polymer Laminates - The Theory of Alfred Puck.* Berlin Heidelberg New York: Springer Verlag, 2008

[KS01] KHARRAZI, M. R.; SARKANI, S.: Frequency-dependent fatigue damage accumulation in fiber-reinforced plastics. *Journal of Composite Materials* 35 (2001), p. 1924–1953

[Küs12] KÜSTERS, K.: *Modellierung des thermo-mechanischen Langzeitverhaltens von Thermoplasten.* RWTH Aachen University, Dissertation, 2012

[MKV19] MOVAHEDIRAD, A. V.; KELLER, T.; VASSILOPOULOS, A. P.: Creep effects on tension-tension fatigue behavior of angle-ply GFRP composite laminates. *International Journal of Fatigue* 123 (2019), p. 144–156

[MRKH20] MÜLLER, J.; ROZOLOPEZ, N.; KLEIN, E. A.; HOPMANN, C.: Predicting the damage development in epoxy resins using an anisotropic damage model. *Polymer Engineering and Science* 60 (2020), p. 1324–1332

[MTL13] MOUSSADDY, H.; THERRIAULT, D.; LEVESQUE, M.: Assessment of existing and introduction of a new and robust efficient definition of the representative volume element. *International Journal of Solids and Structures* 50 (2013), p. 3817–3828

[PGBV13] PUPURS, A.; GOUTIANOS, S.; BRONDSTED, P.; VARNA, J.: Interface debond crack growth in tension–tension cyclic loading of single fiber polymer composites. *Composites Part A* 44 (2013), p. 86–94

[PS02] PUCK, A.; SCHÜRMANN, H.: Failure analysis of FRP laminates by means of physically based phenomenological models. In: Hinton, M. J.; Kaddour, A. S.; Soden, P. D. (Editor): *Failure Criteria in Fibre Reinforced Polymer Composites.* Amsterdam: Elsevier, 2002

[RCH20]   ROZOLOPEZ, N.; CHEN, J.; HOPMANN, C.: A micromechanical model for loading and unloading behavior of fiber reinforced plastics under cyclic loading. *Polymer Composites* 41 (2020), p. 3892–390

[Roy01]   ROYLANCE, D.: *Engineering Viscoelasticity*. Cambridge: Massachusetts Institute of Technology, 2001

[SBVL09] SALAS, P. A.; BENSON, D. J.; VENKATARAMAN, S.; LOIKKANEN, M. J.: Numerical Implementation of Polymer Viscoplastic Equations for High Strain-Rate Composite Models. *Journal of Aerospace Engineering* 22 (2009), p. 304–309

[TCCD07] TURON, A.; COSTA, J.; CAMAMHO, P. P.; DAVILA, C. G.: Simulation of delamination in composites under high-cycle fatigue. *Composite Part A* 38 (2007), p. 2270–2282

[TX07a]   TAO, G.; XIA, Z.: An Experimental Study of Uniaxial Fatigue Behavior of an Epoxy Resin by a New Noncontact Real-Time Strain Measurement and Control System. *Polymer Engineering and Science* 47 (2007), p. 780–788

[TX07b]   TAO, G.; XIA, Z.: Ratcheting behavior of an epoxy polymer and its effect on fatigue life. *Polymer Testing* 26 (2007), p. 451–460

[TX08]   TAO, G.; XIA, Z.: Fatigue behavior of an epoxy polymer subjected to cyclic shear loading. *Materials Science and Engineering* 486 (2008), p. 38–44

[Vas20]   VASSILOPOULOS, A. P.: Introduction to the fatigue life prediction of composite materials and structures: past, present, and future prospects. *International Journal of Fatigue* 40 (2020), p. 105512

[VAT14]   VIEILLE, B.; ALBOUY, W.; TALEB, L.: About the creep-fatigue interaction on the fatigue behaviour of off-axis woven-ply thermoplastic laminates at temperatures higher than Tg. *Composite Part B* 58 (2014), p. 478–486

[vH19]   VAN HAAG, J. H. J.: *Modellierung des zeitabhängigen thermomechanischen Steifigkeitsverhaltens spritzgegossener kurz- und langfaserverstärkter Thermoplaste*. RWTH Aachen University, Dissertation, 2019

[XSE04a] XIA, Z.; SHEN, X.; ELLYIN, F.: An Assessment of Nonlinearly Viscoelastic Constitutive Models for Cyclic Loading: The Effect of a General Loading/Unloading Rule. *Mechanics of Time-Dependent Materials* 9 (2004), p. 79–98

[XSE04b] XIA, Z.; SHEN, X.; ELLYIN, F.: Biaxial cyclic deformation of an epoxy resin: Experiments and constitutive modeling. *Journal of Materials Science* 40 (2004), p. 643—654

[XSE04c] XIA, Z.; SHEN, X.; ELLYIN, F.: Cyclic deformation behavior of an epoxy polymer Part II: Predictions of viscoelastic constitutive models. *Polymer Engineering and Science* 45 (2004), p. 103–113

Symbols

Symbol	Unit	Description
E_0	N/mm^2	Initial elastic modulus
E_i	N/mm^2	Elastic modulus Maxwell element
τ_i	s	Relaxation time Maxwell element
W	kJ/m^3	Strain energy density
W^d	kJ/m^3	Dissipative strain energy density
W^{e+}	kJ/m^3	Positive elastic strain energy density
σ_{ij}	N/mm^2	Local stress components
ϵ_{ij}	–	Local strain components
$<W>$	kJ/m^3	Homogenised strain energy density
$<\sigma_{ij}>$	N/mm^2	Homogenised stress components
$<\epsilon_{ij}>$	–	Homogenised strain components
V	m^3	Volume RVE
V^k	m^3	Volume k^{th} integration point
σ_{ij}^k	N/mm^2	Stress component at the kth integration point
ϵ_{ij}^k	–	Strain component at the kth integration point
N_{int}	–	Number of integration points
$E_\parallel$	N/mm^2	Elastic modulus parallel to the fibre direction
$E_\perp$	N/mm^2	Elastic modulus transverse to the fibre direction
$\nu_\parallel$	–	Poisson ratio parallel to the fibre direction
$\nu_\perp$	–	Poisson ratio transverse to the fibre direction
$G_\parallel$	N/mm^2	Shear modulus parallel to the fibre direction
$G_\perp$	N/mm^2	Shear modulus transverse to the fibre direction
t_{creep}	s	Creep step time
f	Hz	Frequency cyclic load
N_c	–	Number of cycles
$\sigma_1, \sigma_2, \sigma_3$	N/mm^2	Principal stress components
σ_h	N/mm^2	Hydrostatic stress
τ_h	N/mm^2	Octahedral shear stress

Abbreviations

Notation	Description
C3D8	8-node liner brick element
C3D8R	8-node liner brick element with reduced integration
FRP	Fibre reinforced plastics
GMM	Generalized Maxwell model
FE	Finite Element
RVE	Representative volume element
SVE	Single volume element
UMAT	User-defined material subroutine

Keywords

A
additive manufacturing	Session 14.1
	Session 14.2
	Session 14.3
administration shell	Session 2.3
analysis	Session 7.3
anisotropy	Session 12.3

B
blends	Session 5.2
blown film	Session 3.3

C
capillary rheometer	Session 6.2
circular economy	Session 2.2
	Session 2.3
	Session 5.2
	Session 5.3
component testing	Session 14.1
	Session 14.3
creep	Session 15.3

D
digital shadow	Session 1.3
digital twin	Session 1.3
	Session 7.2

E
enterprise resource planning	Session 5.1
extensional rheometer	Session 3.2

F
fibre length	Session 11.2

I
influence of humidity	Session 5.2
injection moulding	Session 1.1
	Session 1.2
internal mixer	Session 6.1

L
load, dynamic	Session 15.2

M

material characterisation	Session 3.2
	Session 5.3
material compound	Session 14.3
material data	Session 4.2
material models	Session 9.3
	Session 11.2
material properties	Plenary Session
	Session 8.1
	Session 14.2
	Session 15.2
mechanical characterisation	Session 9.2
mechanical properties	Session 9.3
	Session 11.3
	Session 12.2
mechanical testing	Session 14.2
media influence	Plenary Session
medical technology	Session 13.1
	Session 13.2
mould design	Session 8.2
mould filling process	Session 10.2
	Session 10.3

N

NC programming	Session 14.3

O

optical properties	Session 8.1
	Session 8.3
optimisation	Session 8.2

P

PECVD (Plasma Enhanced Chemical Vapour Deposition)	Session 2.2
plasma	Session 5.2
plasma technology	Session 2.1
plastic waste	Session 5.3
process analysis	Session 6.1
	Session 6.3
process automation	Session 1.1
	Session 10.1
process control	Session 10.1
process model	Session 1.2
	Session 11.3
process optimisation	Session 1.2
	Session 6.1
	Session 6.3
	Session 10.1
	Session 10.2

process optimisation	Session 10.3
process simulation	Session 8.2
	Session 8.3
	Session 11.3
product properties	Plenary Session
production control	Session 4.1
prototypes	Session 14.1

R

recyclate	Session 2.3
	Session 4.2
	Session 10.3
	Session 12.2
recycling	Session 5.2
	Session 5.3
relaxation	Session 6.2
residual stress	Session 8.3
rubber	Session 6.1
	Session 6.3
rubber compounds	Session 6.1
	Session 6.3
rubber rheology	Session 6.2

S

set-up	Session 1.1
	Session 13.1
	Session 13.2
set-up time reduction	Session 1.2
	Session 1.3
simulation	Session 15.3
smart factory	Session 1.3
SMC-processing	Session 11.3
standards	Session 13.2
start-Up	Session 2.1
start-up concept	Session 1.1
stretching (biaxial)	Session 3.2
sustainability	Plenary Session
system planning	Session 2.2

T

temperature distribution	Session 7.3
temperature profile	Session 7.3
tensile shear test	Session 13.3
test methods	Session 15.2
thermoplastic compression moulds	Session 11.2
thermoplastic elastomers	Session 12.3
thermoset	Session 7.2
thixotropy	Session 6.2

tooling design Session 3.2
transfer learning Session 1.2
twin screw extruder Session 5.2

W

weld seam Session 12.2
welding Session 12.2
  Session 12.3

Authors

A
Alizadeh, P.   Institute for Plastics Processing (IKV)
Austermann, J.   Institute for Plastics Processing (IKV)

B
Beaumont, M.   Siemens AG
Beckmann, C.   Fraunhofer Institute for Mechanics of Materials (IWM)
Berg, E.   Institute for Plastics Processing (IKV)
Bergmann, N.   Kautex Textron GmbH & Co. KG
Bibow, P.   Institute for Plastics Processing (IKV)
Block, F.   Institute for Plastics Processing (IKV)
Bölle, S.   Institute for Plastics Processing (IKV)
Bouffier, R.   Kautex Textron GmbH & Co. KG

C
Çelik, H.   Institute for Plastics Processing (IKV)

D
Dahlmann, R.   Institute for Plastics Processing (IKV)
Di Battista, F.   Institute for Plastics Processing (IKV)
Dietrich, M.   WEGENER International GmbH

F
Fatherazi, P.   Institute for Plastics Processing (IKV)
Fecher, M. L.   Mubea Carbo Tech GmbH
Fey, F.   Institute for Plastics Processing (IKV)
Fink, B.   Roche Diagnostics GmbH
Fischer, K.   Institute for Plastics Processing (IKV)
Fritsche, D.   Institute for Plastics Processing (IKV)
Fuchs, J.   Institute for Plastics Processing (IKV)

G
Gebhart, T.   Institute for Plastics Processing (IKV)
Gerads, J.   Institute for Plastics Processing (IKV)
Grüber, D.   Institute for Plastics Processing (IKV)

H
Hesse, M.   HF Mixing Group
Hofmann, J.   Chair for Technology of Optical Systems (TOS)
Hohlweck, T.   Institute for Plastics Processing (IKV)
Holly, C.   Chair for Technology of Optical Systems (TOS)
Hopmann, Ch.   Institute for Plastics Processing (IKV)
Horn, T.   Institute for Plastics Processing (IKV)
Hornberg, K.   Institute for Plastics Processing (IKV)

J

Jaritz, M.	IonKraft GmbH
Jatzlau, P.	RayScan Technologies GmbH
Judt, P.	Hexagon Purus GmbH

K

Kaufholf, C.	Hexagon Purus GmbH
Keuters, M.	tetys GmbH & Co. KG
Kleines, L.	Institute for Plastics Processing (IKV)
Kleinschmidt, D.	Kunststofftechnik Paderborn (KTP)
Kley, Ch.	Institute for Plastics Processing (IKV)
Knupe-Wolfgang, P.	Institute for Plastics Processing (IKV)
Köbel, T.	Institute for Plastics Processing (IKV)
Kölle, O.	Volkswagen Aktiengesellschaft
Kostka, M.	Institute for Plastics Processing (IKV)
Kückelmann, U.	Alba Recycling GmbH

L

Limper, A.	Institute for Plastics Processing (IKV)
Liu, B.	Institute for Plastics Processing (IKV)
Lockner, Y.	Institute for Plastics Processing (IKV)

M

Meyer, W.	Volkswagen Aktiengesellschaft
Müller, J. M.	Institute for Plastics Processing (IKV)
Müller, R.	Institute for Plastics Processing (IKV)

N

Neuhaus, J.	Institute for Plastics Processing (IKV)

P

Peine, A.	W. Köpp GmbH & Co. KG
Pelzer, L.	Institute for Plastics Processing (IKV)

R

Reinhardt, N.	Institute for Plastics Processing (IKV)
Reßmann, A.	tetys GmbH & Co. KG
Richter, F.	Hexagon Purus GmbH
Rozo Lopez, N.	Institute for Plastics Processing (IKV)

S

Sapel, P.	Institute for Plastics Processing (IKV)
Scharf, M.	Phoenix Contact GmbH & Co. KG
Schäuble, R.	Fraunhofer Institute for Microstructure of Materials and Systems (IMWS)
Schmidt, A.	Röhm GmbH
Schmitz, M.	Institute for Plastics Processing (IKV)

S

Schneider, D.	Institute for Plastics Processing (IKV)
Schöll, M.	Institute for Plastics Processing (IKV)
Schön, M.	Institute for Plastics Processing (IKV)
Schöppner, V.	Kunststofftechnik Paderborn (KTP)
Stieglitz, M.	Institute for Plastics Processing (IKV)
Stollenwerk, J.	Chair for Technology of Optical Systems (TOS), Fraunhofer Institute for Laser Technology (ILT)
Strohhäcker, J.	Hexagon Purus GmbH
Strubel, V.	InnovationGreen

W

Wagner, P.	Institute for Plastics Processing (IKV)
Walther, T.	Arburg GmbH + Co. KG
Wang, C.	Institute for Plastics Processing (IKV)
Wang, H.	Aachen Center for Integrative Lightweight Production (AZL)
Weber, J.	Institute for Plastics Processing (IKV)
Weihermüller, M.	Institute for Plastics Processing (IKV)
Wiesel, C.	Institute for Plastics Processing (IKV)
Wipperfürth, J.	Institute for Plastics Processing (IKV)